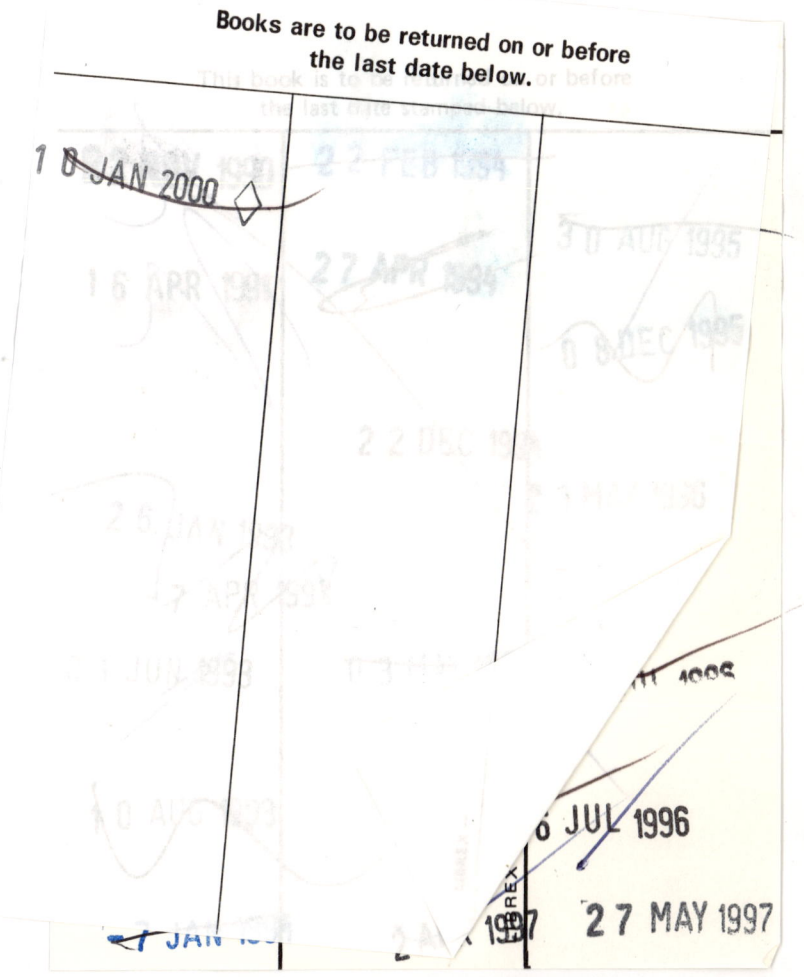

**Books are to be returned on or before
the last date below.**

1 0 JAN 2000

27 MAY 1997

THE
HEAT TREATING
SOURCE BOOK

*A collection of outstanding articles
from the technical literature*

Other Source Books from ASM:

Applications of the Laser in Metalworking
Corrosion
Engineering Applications of Ceramic Materials
Fabrication of Composite Materials
Foundry Technology
Production to Near Net Shape
Quality Control
Rapid Solidification Technology
Reinforced Plastics for Commercial Composites
Selection of Materials for Component Design
Superalloys
Titanium and Titanium Alloys
Tool and Die Failures

THE HEAT TREATING SOURCE BOOK

A collection of outstanding articles from the technical literature

Compiled by
Consulting Editor

Paul S. Gupton
Consultant
Bell and Associates

American Society for Metals
Metals Park, Ohio 44073

Library of Congress Catalog Card No.: 86-71115
ISBN: 0-87170-225-8
SAN 204-7586

PRINTED IN THE UNITED STATES OF AMERICA

Contributors to This Source Book

A. M. AITCHISON
Caterpillar Tractor Co. Ltd.

WALLY L. BAMFORD

FRED J. BARTKOWSKI
Marshall W. Nelson
& Associates, Inc.

J. HOWARD BECK
BTU Engineering Corp.

T. BELL
University of Birmingham
(England)

P. K. BHARGAVA
EMA India Ltd.

ROGER G. BLOCKS
Chem-Al, Inc.

JEFFREY W. BOSWELL
Sunbeam Equipment Corp.

R. C. BRAUN
Manville Products Corp.

CHARLIE R. BROOKS
University of Tennessee

JOHN A. BURGER
Allison Div.
General Motors Corp.

DOMENIC A. CANONICO
Combustion Engineering Inc.

ROGER V. CARTER
Boeing Commercial Airplane Co.

HARRY E. CHANDLER
Metal Progress

DOUGLAS J. CLEARY
Stanwood Corp.

P. COLLIGNON
Vide et Traitement S.A.

A. J. CRAIG, JR.
Homelite

RICHARD CREAL
Heat Treating

THOMAS R. CROUCHER
Progressive Metallurgical
Industries

WILLIAM C. DIMAN
C. I. Hayes Inc.

ROBERT E. DROEGKAMP
Fansteel Metals, Inc.

J. N. DUTTA
New Allenberry Works

DONALD DYKE
Sintered Specialties

DON G. ENSWEILER
Heat Process Associates, Inc.

FRANCIS FAHRENWALD
Fahrenwald Consulting

HOWARD FERGUSON
Lindberg Heat Treating Co.

CARL FIORLETTA
Sciaky Bros., Inc.

STEPHEN FLOREEN
International Nickel Co., Inc.

RICK FREY
M. G. Industries

STEVEN R. FRIED
Bethlehem Steel Corp.

NOBUO FURUKAWA
Sumitomo Electric Industries Ltd.

ARTHUR L. GEARY
Nuclear Metals, Inc.

CAROL A. GIRRELL
Procedyne Corp.

KENNETH D. GLADDEN
Caterpillar Tractor Co.

L. JOSEPH GRAFE
Bloom Engineering Co., Inc.

DONALD GRENDON
Drever Co.

DONALD N. GUY
Lindberg Heat Treating Co.

WILLIAM B. HAMPSHIRE
Lead Industries Association, Inc.

DEAN K. HANINK
Allison Div.
General Motors Corp.

JACK HASSON
E. F. Houghton & Co.

WILLIAM H. HEIL
Timet

WALTER HERMAN
Viking Metallurgical Corp.

MAX HOETZL
Midland-Ross Corp.

LOUIS (NED) E. HUBER, JR.
Kawecki-Berylco Industries, Inc.
Div. of Cabot Corp.

PAUL L. HUBER
Sunbeam Equipment Corp.

NOTE: Affiliations given were applicable at date of contribution.

WILLIAM L. JAMES
Fennell Corp.

JOSEPH E. JAPKA
Procedyne Corp.

NICHOLAS C. JESSEN, JR.
Union Carbide Corp.

PHILIP JOHNSON
Airco Industrial Gases
Div. of BOC Group

K. M. JOSEPH
EMA India Ltd.

W. Q. JUDGE
Engineered Sinterings and
    Plastics

NORMAN O. KATES
Lindberg Corp.

JAMES KELLY
Rolled Alloys

ROY F. KERN
Caterpillar Tractor Co.

FRED W. KLAG
Alloy Engineering Co.

ERHARD KLAR
Glidden Metals
Div. of SCM Corp.

KENT H. KOHNKEN
GTE Sylvania

RALPH J. KOTFILA
McDonnell Aircraft Co.

GEORGE KRAUSS
Colorado School of Mines

A. S. W. KURNEY
Bangla Desh University of
    Engineering and Technology

W. JAMES LAIRD, JR.
Upton Industries, Inc.

ARTHUR L. LaMASTERS
Cleveland Alloy Casting Co.

W. LUTY
Institute of Precision Mechanics

NORMAN C. McCLURE
Commercial Steel Treating Corp.

GERALD J. MacDONALD
Chesmont Engineering Co., Inc.

ROBERT D. McGOWAN
American Metal Climax Inc.

R. M. MALLYA
Indian Institute of Science

CHRISTOPHER F. MASTERS
Kinetic Co.

QUENTIN D. MEHRKAM
Ajax Electric Co.

WILLIAM H. NAYLOR
Sun Oil Co.

H. S. NAYAR
Airco Industrial Gases
Div. of BOC Group

RAYMOND OSTROWSKI
Protection Controls, Inc.

GEORGE OTTO
Maytag Co.

CARL J. OXFORD, JR.
Lear-Siegler, Inc.

B. C. PAI
Regional Research Laboratory
    CSIR

W. H. PARKER
Manville Products Corp.

T. V. PHILIP
Carpenter Technology Corp.

R. M. PILLAI
Regional Research Laboratory
    CSIR

M. MOHAN RAO
Indian Institute of Science

GLENN RATLIFF
Shore Metal Treating

PERCY RAWCLIFFE
Morse Cutting Tools Div.
Gulf & Western Manufacturing

CARL REICHEL
TRW, Inc.

WAYNE SAMUELSON
Shore Metal Treating

OLE SANDVEN
Avco Everett Metalworking
    Lasers

K. G. SATYANARAYANA
Regional Research Laboratory
    CSIR

CHARLES J. SCHOLL
Wyman-Gordon Co.

ROSS B. SHINGLEDECKER
Ladish Co.

THOMAS SIBLEY
Air Products & Chemicals, Inc.

OGLE R. SINGLETON
Reynolds Metals Co.

GRANT E. SPANGLER
Reynolds Metals Co.

RONALD SPITZER
TRW, Inc.

JOHN E. STEIN
Resisto-Loy Co., Inc.

C. A. STICKELS
Ford Motor Co.

SANG-KEE SUH
Ford Motor Co.

S. M. TAPASWI
Elecon Engineering Co. Ltd.

THEODORE K. THOMAS
Honeywell Inc.

DAVID S. THOMPSON
Reynolds Metals Co.

DONALD J. TILLACK
Huntington Alloys, Inc.

RAJ TIWARI
Simplicity Engineers P. Ltd.

ALLEN B. TOWNSEND
Union Carbide Corp.

H. N. UDALL
Thermatool Corp.

DENNIS M. WAGEN
Stanwood Corp.

JAY T. WARE
C-E Air Preheater

RICHARD G. WEBER
Pitney Bowes

R. TERRENCE WEBSTER
Teledyne Wah Chang

JOHN WEST
Hauck Manufacturing Co.

SAMUEL L. WILLIAMS
Rock Island Arsenal

DANIEL S. ZAMBORSKY
Warner & Swasey Co.

G. B. ZUBER
Mine Safety Appliance Co.

PREFACE

Heat treatment is defined as "the controlled heating and cooling of a solid metal or alloy in a way to obtain specific conditions and/or properties." Some treatments harden and strengthen metals; others soften metals and affect other mechanical and physical properties, such as impact strength, ductility, magnetic susceptibility, toughness, machinability, fabricability, and in some instances corrosion resistance.

Heat treatment was first practiced by ancient metalsmiths, whose art ranged from the softening (annealing) of cold worked tools in the Bronze Age to the forging of iron-base swords and daggers in the Egyptian period. It was not until centuries later, however, that the beneficial effects of carbon additions to iron were recognized; current evidence suggests that the iron-carbon relationship was discovered during the Christian era.

The ancient metalsmith later came to be known as the blacksmith, and in many instances he was also the heat treater. Metals are still heat treated today, for the same reasons as in ancient days — namely, to improve or maximize their mechanical properties of strength, hardness, ductility, and toughness.

Most commercially produced metals and alloys respond metallurgically and mechanically to heat treatment. The response varies from the softening of all pure metals and alloys to hardening or strengthening of specific materials. Because of the overwhelming quantity of steels produced and used, we tend to think primarily of iron-base materials when considering heat treating cycles. Iron has an unusual characteristic that it shares with very few metals — namely allotropy, the ability to exist in different crystalline forms. At certain temperatures, it completely recrystallizes and changes from one type of space lattice to another. These changes are entirely spontaneous and reversible and are almost entirely dependent on changing the temperature of the material.

In addition to ferrous alloys, however, nonferrous alloys of aluminum, copper, magnesium, nickel, titanium, and zirconium can be and are frequently strengthened by specific heat treating cycles. Some of these heat treatments are the same as those used for iron-base materials and will be discussed below under specific cycles.

Optimizing hardness, strength, and other mechanical properties by the various heat treating cycles continues to be of primary interest to equipment designers. In addition to optimization, the designer must also consider the thermomechanical effects from temperature gradients imposed on the part during the various heat treating cycles. The fabricability of conventional and modern highly alloyed materials is also of utmost interest to equipment builders. In most cases, prior heat processing greatly affects these fabrications.

The articles in this Source Book were selected and grouped according to the most currently recognized heat treating processes. For those who are not fully familiar with these cycles, a brief description of each follows.

Normalizing is the heating of a ferrous alloy above the upper critical temperature followed by air cooling to ambient temperature. The primary purpose of normalizing is grain size refinement in steels previously exposed to high temperature for forging or other hot working operations.

Annealing is a thermal cycle used primarily for softening metallic materials. It has also been expanded to include heat treating processes whose purposes include homogenization of a composition, elimination of entrapped gas, refinement of grain size, solution treatment for age hardening, and solution treatment for restoration of corrosion resistance. For nonferrous materials, the annealing process is used to eliminate or to reduce the effects of cold working.

Austenitizing is the process of heating ferrous alloys above the transformation range to form FCC austenite. Generally, FCC austenite is a phase that exists only at elevated temperature, since it converts to one of several forms when cooled at various rates.

Quench hardening is the rapid cooling of a steel or an alloy from the austenitizing temperature by immersing it in liquid or gas. The cooling rates required to obtain full hardness are dependent on the alloy composition of a particular steel or iron.

Tempering is a process of heat treating below the transformation temperature in order to restore some degree of ductility to a quench-hardened steel or iron. It permits obtainment of various combinations of mechanical strengths and ductilities. For nonferrous materials, tempering refers to a degree of hardness achieved either by heat treatment or by mechanical working.

Austempering involves the heating of cast irons or steels above the upper critical temperature, followed by quenching them in a molten salt bath held at temperatures above the start of the martensite range. The transformation product is the strong, tough, ductile constituent bainite.

Stress relieving is the reduction of internal stresses in a metal or alloy by heating and holding at a temperature at which the yield strength of the material is greatly reduced. This results in the relaxation of internal stresses to the level of the yield strength, at the temperature to which the component is heated by the creep mechanism.

Carburizing is the absorption and diffusion of carbon into solid ferrous alloys by heating and holding at a temperature above the upper transformation temperature (1650 to 1900 °F) for the specific alloy. Heating is done in a carbonaceous (liquid, solid, or gas) environment. After carburizing, the parts are usually hardened by conventional water quench, followed by tempering.

Carbonitriding, or cyaniding, is a surface treatment carried out above the transformation temperature in either a gaseous atmosphere or a molten salt containing cyanide. Both carbon and nitrogen are simultaneously absorbed and diffused into the material. Quenching causes the formation of a hard wear-resistant case and a soft core.

Nitriding is the introduction of nascent nitrogen into the surface of specific ferrous alloys by holding the alloys at a relatively low temperature (975 to 1050 °F) for an extended period (24 to 72 h). This is a direct conversion process requiring no quenching to produce a hard wear-resistant case.

Nitrocarburizing is a surface-hardening process for ferrous materials in which both carbon and nitrogen in a gaseous atmosphere are simultaneously absorbed and diffused into the surface at an elevated temperature, generally below the lower critical temperature. It is followed by reaustenitizing and quenching.

Precipitation hardening, also called age hardening, involves the elevated-temperature precipitation of a second phase in a specific material that has first been solution treated to dissolve the phase. It is a commonly used process for heat treating aluminum, copper, and nickel alloys.

Sintering is the process by which loose or compressed powders for powder metallurgy parts are bonded by heating at temperatures (930 to 1470 °F) below the melting points of the major constituents. The atmosphere may be vacuum, hydrogen dissociated ammonia, endothermic gas, or several forms of exothermic gases.

The articles and technical papers contained in this Source Book have been selected from the most recent publications involving heat treatment of metals and alloys. The first section discusses current trends in heat treatment; it is followed by a section on part design. The heat treatment of ferrous materials, nonferrous materials, and powder metal parts is covered in separate sections. Production processes such as surface hardening, vacuum methods, and salt bath processing are represented by several articles. Other production considerations, such as furnaces and furnace equipment, production systems, atmospheres and their generation, and quenchants, are fully discussed and explained. Energy conservation, safety and ecology, and testing and quality control round out the material contained in this extensive review. A section containing useful diagrams and tables is also included.

The American Society for Metals extends most grateful acknowledgment to the many authors whose work is presented in this Source Book, and to their publishers.

PAUL S. GUPTON
Consultant
Bell and Associates

CONTENTS

SECTION I: CURRENT TRENDS IN HEAT TREATMENT

SECTION II: FERROUS MATERIALS AND THEIR PROCESSING

SECTION III: SURFACE HARDENING

SECTION IV: POWDER METAL PARTS

SECTION V: HEAT TREATING NONFERROUS METALS

SECTION VI: VACUUM METHODS

SECTION VII: SALT BATH PROCESSING

SECTION VIII: FURNACES AND FURNACE EQUIPMENT

SECTION IX: PRODUCTION SYSTEMS

SECTION X: FLUIDIZED BED PROCESS

SECTION XI: ATMOSPHERES AND THEIR GENERATION

SECTION XII: QUENCHANTS

SECTION XIII: ENERGY CONSERVATION, SAFETY, AND ECOLOGY

SECTION XIV: TESTING AND QUALITY CONTROL

SECTION XV: USEFUL DIAGRAMS AND TABLES

SECTION I
Current Trends in Heat Treatment

Heat Treating

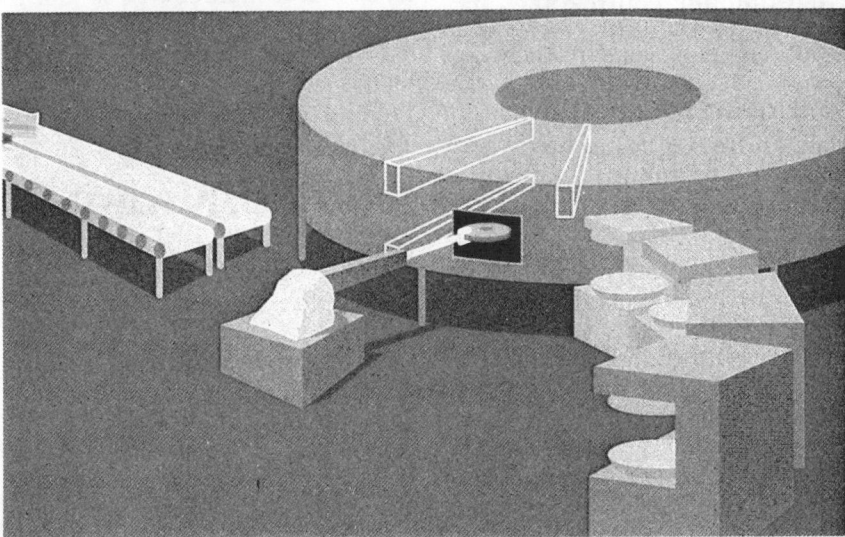

Schematic of a flexible carburizing line conceived by Surface Combustion Div., Midland-Ross Corp. An indexing drive, state of the art microprocessor controls, rotary design, and multiple internal chambers that eliminate the need for trays combine to produce a system that can process parts on a variety of cycles.

Computers and microprocessor based controls continue to play important roles in advancing the state of the art in heat treating/heat processing technology. Surface Combustion Div., Midland-Ross Corp., for example, is looking at the application of flexible manufacturing system (FMS) concepts to hardening and carburizing operations.

Robots are key elements in these proposed systems.

Also spotlighted is Timken Co.'s computer controlled continuous thermal treatment facility (CTTF). Two workers monitor the entire operation.

Other developments in controls promise to save energy in fluidized bed heat treating, narrow temperature variations in car bottom furnaces, ease stress relief of pressure vessels, and automate carbon potential control.

Developments in heat treating processes include a new way to enhance the properties of investment cast and HIP'ed titanium, a new nontoxic salt bath hardening method, expanded applications for ion nitriding/carburizing and ion implantation, and vacuum processing of titanium alloys, magnetic materials, and coated refractory metals at temperatures up to 2800+ F (1540+ C).

Polymer quenchants and nitrogen-methanol and other nitrogen based atmospheres also figure significantly in this forecast.

A Vast Technology

You get a wide screen perspective on the depth and breadth of available heat treating/processing technology from the following sampling of developments in computers/sensors/controls, ion implantation, austempering, vacuum furnaces, fluidized bed furnaces, induction heating systems, salt bath technology, quenchants, atmospheres, and energy conservation.

• This assessment of instrumentation and control technology from Barber-Colman Co. sets the stage: "Digital electronics and microprocessors have made possible the advances in process controllers, programmable controllers, computers, and robotics . . . There is more than enough new technology presently available to keep most companies busy implementing it for the next ten years . . . In the immediate future, there will be increasing interest in developing the communications link between processes and supervisory levels. This technology, known as local area networks, will allow better control of plant operations and lead to paperless operations. Also on the horizon is increased ability to provide service diagnosis from remote locations. In the future, downtime will be minimized through self-diagnostic features, incorporated within the process equipment, and having the ability to transmit this information over the local area network or by telephone modes directly to remote plant or vendor locations."

• Zymet Inc., the company that introduced commercial ion implantation equipment in January 1984, ... is now at the prototype stage with new equipment that combines implantation and coating. "Several new surface treatment and modification capabilities become possible," says Zymet. For example, it will be possible to implant a coating, creating a superior bond with the substrate, and opening up new coating possibilities between insoluble materials. Alloying will also be possible. For example, chromium can be put down in a thin layer on a copper surface before being driven into the copper with ion implantation. The result is a chromium-copper alloy, which is not possible according to the laws of chemistry. Other capabilities include ion beam sputtering and etching, electron beam evaporation deposition (without ion implantation), and ion beam enhanced deposition.

• Sections up to 8 in. (205 mm) thick are being austempered and isothermally transformed, reveals Atmosphere Furnace Co. Only five years ago, section thickness was limited to 0.5 in. (13 mm). The improvement, it is explained, is the result of a better understanding of the mechanism of heat transfer in quenching.

• In Germany, Ipsen Industries International GmbH reports, "Vacuum furnaces with high pressure, inert gas quenching offer extreme uniformity of quenching throughout the load to minimize distortion." Since 1977, Ipsen adds, more than 100 furnaces of this type have been sold in Europe. Latest designs, it is stated, provide broad beam, high velocity gas streams over the length of the furnace chamber. They oscillate crosswise during quenching to avoid stagnant shadow situations, and operate intermittently — as programmed — either from the bottom upward or from the top downward through the load.

• In Canada, Can-Eng Sales Ltd. says its ferritic nitrocarburizing process, "a thermal chemical surface treatment of ferrous materials in a fluidized bed," improves wear and corrosion resistance of parts such as cutting tools, gears, extrusion dies, and injection molding pins. The treatment, it is explained,

involves the diffusional addition of nitrogen and carbon to the part at temperatures within the ferritic range of 1050 to 1150 F (565 to 620 C). A thin, dense, and hard layer is produced over an underlying diffusion zone.

• Induction heat treating robotic cells, predicts Ajax Magnethermic Corp., will become more prominent because they provide flexibility in scheduling parts with minimal tooling change. Programmable control functions and programming capabilities will assure precision handling and heat treating, and better monitoring of critical steps.

• Kolene Corp. was a new system that permits the automatic and monitored addition of chemicals in filtered liquid form to operating molten salt baths on a continuous basis. Benefits other than control of bath chemistry include a reduction in amount of material consumed due to dragout.

• Growing acceptance of polyvinylpyrrolidone (PVP) type polymer quenchants for all types of steels, including oil hardening types, is indicated by Park Chemical Co., adding, "Statistical quench data enable the proper selection of quench conditions. Coupled with improved control and maintenance procedures, greater usage of polymer quenchants, particularly in place of oil and water quenching is forecast." PVP quenchants are also said to "offer some unique and useful quench characteristics for aluminum. Commercial application should come within two to three years."

• Rising emphasis on statistical process control techniques is spurring "demand for reliable atmosphere alternatives," observes the Industrial Gas Div. of Air Products & Chemicals Inc. "Nitrogen based atmospheres with oxygen probe systems combine superior carbon control with operation convenience."

• The changing nature of the energy situation is commented

upon by Atmosphere Furnace Co. Ten years ago, it is pointed out, oil embargoes and curtailments promoted conservation. Future supply was uncertain at best. Today, by comparison, proven reserves of oil and natural gas appear to be adequate, but the cost of energy has shot up dramatically. In many industries, the cost of energy has become a major factor in process and material selection. "Process selection, equipment design, and materials of construction which further reduce energy consumption will continue to be dominant considerations in the heat treating industry," it is predicted.

The Quest for More Control

Nowhere in this technology is innovation more rampant than it is in the application of computers, microprocessors, programmable controllers, and sensors to both discrete functions and to entire systems. In fact, the principles of the machining industry's flexible manufacturing system (FMS) concept and the cellular concept have invaded heat treating. Midland-Ross Corp.'s Surface Combustion Div. for one is an advocate.

"Several systems have already been proposed," says Surface. One, for example, extends a robotic system now in operation to a total product transfer system. Parts are loaded in a preset pattern, transferred by robot to discrete units in the system, delivered to adjacent cells such as an automatic inspection cell, and sent on to final assembly.

In the system now in place, robotics provide flexibility while efficiency is provided by a continuous furnace. The heart of this concept is a robotic load/unload center. Small parts for hardening are conveyed to a scale where they are unloaded automatically into a weighing station. The computer, knowing the exact loading pattern for each part, commands the robot to load and stack baskets of parts to an exact weight for each part, then deliver the parts to a pusher tray hardening furnace. Any combination of 3000 part numbers, according to factory production demands, will be loaded and traced by the production control computer. When parts return to the central load station, the robot unloads stacked baskets of parts into the material transport system according to part number.

The flexible carburizing line is another example cited by Surface. It looks like a conventional rotary hearth. Trays are eliminated via the use of multiple internal chambers made of lightweight structural ceramics. Parts can be processed with a variety of cycles by using the unit's indexing drive system and the memory capacity of state of the art microprocessors. Example: a part halfway around the circle would be indexed to the door by a rapid rotation of the hearth, removed to a press, and the furnace returned to its start position on command from a single process controller.

More Technology — Examples of individual sensors/controls include a microprocessor based gas saver for fluidized bed furnaces, a PC control for heating/cooling, a console for temperature control, a computerized carbon potential controller, and computer supervised digital controllers and appropriate software.

Procedyne Corp. claims its microprocessor based gas saver option for fluidized bed furnaces "has the ability to maintain uniform temperature and surface activity at flow rates as low as one-tenth of current standards." In lengthy case hardening cycles, savings are said to be as high as 50% on a per pound basis.

Selas Corp. of America reports it is getting precise and uniform control of temperature during heating and cooling in large car bottom and hood type furnaces using special pulse fired burners controlled by programmable controllers. Temperature tolerances

of ±5 F (±3 C) are being maintained on large diameter, high chromium rolls. Time controlled and sequential firing cause high velocity products of combustion to be circulated in a selected and controlled manner around the work. During cooling, the same control system passes cooling air only through the burners in addition to cooling air from supplementary jets.

Cooperheat's heat treatment control console acquires accurate thermocouple data during heat treating applications ranging from nuclear steam generator repair to postweld stress relief and the monitoring of large components during furnace cycles.

"Advances in fluidized bed heat treating," reports Procedyne Corp., *"are being made in atmosphere case hardening."* Gas usage, for example, is being slashed.

The console includes a microprocessor computer, a data acquisition system, CRT display, and a printer. Over 200 thermocouples recording temperatures ranging from 32 to 2000 F (0 to 1095 C) can be monitored. Analog and digital recorders provide statistical control, supplying on-line temperature information on all points.

Potential — New versions of computerized carbon potential controllers offered by Ipsen Industries International of West Germany now permit automatic cycle and furnace control after the cycle has been computed by the controller. Development of case depth and profile are displayed on a screen during heat

treatment. Carbon potential control is based on simultaneous measurement of oxygen potential and the CO content.

Use of computer supervised digital controllers via bidirectional digital communications is a major trend in the automation of heat treating, advises Leeds & Northrup Co. "The cost of computers and the software required for applications involving individual digital controllers makes these systems practical in relatively small installations," adds L&N. As of late last year the company started to offer DEC versions of application software for its general purpose single loop controller, which handles process variables like temperature, atmosphere, furnace pressure, and fuel/air ratios. Single loop control means a controller failure affects only one loop, and it can be quickly replaced while work is in process.

Some New Technology

What is described as a new, nontraditional heat treating process for investment cast and HIP'ed titanium alloys that provides strength and fracture toughness capabilities "at least equivalent to those of annealed wrought material" has been introduced by Howmet Turbine Components Corp. A production unit in operation at Howmet's Whitehall, Mich., facility has processed complex net shapes with section thicknesses up to 1.5 in. (38 mm). Dimensional control is said to be "excellent."

The CST process, Howmet explains, eliminates the coarse colony alpha platelet microstructure characteristic of cast titanium which may cause early fatigue cracking. A fatigue resistant, fine Widmanstatten microstructure is produced. It is substantially free of prior beta grain boundary alpha precipitation.

Some comparative properties for 70 F (20 C), 10 million cycle fatigue strength, and for 70 F (20 C) fracture toughness are reported: CST treated castings have a fatigue strength of 100 000 psi (690 MPa) and a fracture toughness of 66 ksi√in. (73 MPa√m). Wrought annealed material has a high cycle fatigue strength of 90 000 psi (620 MPa) and a fracture toughness of 47 ksi√in. (52 MPa√m). And conventionally annealed HIP'ed castings have a high cycle fatigue strength of 62 000 psi (425 MPa) and a fracture toughness of 94 ksi√in. (103 MPa√m).

A new nontoxic salt bath has been developed for case hardening by Heatbath Corp. Case depths up to 0.060 in. (1.5 mm), it is claimed, can be obtained in processing times comparable to those for cyanide salts. Hardness is equal. Other features: chemical analysis is not required for maintenance of the system; a standard salt bath furnace is used; and isothermal quenching can be done directly into a nitrate-nitrite type salt bath without danger.

Future radiant tubes extruded from silicon carbide, predicts East Ohio Gas Co., may be capable of temperatures well in excess of 2000 F (1095 C). "Development work is being done concurrently by at least five companies, each with a proprietary material and production method. This breakthrough would open the door for high temperature carburizing without concern for degradation of alloy tubes."

Ion Processes — Abar plans "continued introduction" of horizontal and vertical furnaces for plasma nitriding, featuring simplified computer control centers for the management of furnace and process functions. Patented, high frequency power supplies will overcome arcing and hollow cathode discharges. Other benefits include better hole penetration and reduced depassivation time for removing oxides on stainless steels.

Seco/Warwick Corp. proposes the use of pulsed plasma as an alternative to conventional dc power as a way to overcome such problems as localized overheating in glow discharge (ion)

nitriding. The company explains, "Pulsed plasma permits operation in the abnormal glow region with a high nitriding potential without heating. Heating can then be done by independently controlled convection and radiation to produce a uniform part temperature."

Adamas Carbide Corp. looks for "extensive research on ion implantation of metals, particularly with nitrogen." The method is deemed suitable for parts made from steel, and high speed steel, and for cemented carbides in wear part applications, but "... does not seem to be suitable for machining applications where cutting tools reach high temperatures in operation."

Mill Systems — A new steel bar quench and tempering operation being built at Quanex Corp.'s LaSalle Steel Co. is expected to be in full operation by May of this year. Direct electric resistance heating is used for both austenitizing and tempering — "in a matter of minutes instead of hours for more conventional methods." It's reported that an ultrafine grain size is produced, providing improved ductility, toughness, and fatigue resistance. Normal heat treated carbon and alloy grades can be water quenched, developing full material hardness in a wide range of sizes and shapes. It is said there is no decarburization or quench cracking during heat treatment, and straightening is not required. In addition, LaSalle will be able to quench and temper steels containing machining additions, including sulfur, lead, selenium, tellurium, and bismuth. Both hot rolled and cold finished bars will be processed.

Timken Co.'s continuous thermal treatment facility at its new Faircrest (Ohio) steel plant boasts a Timken-patented quench and temper operation.

Alloy seamless tubing and bars ranging from 2⅜ to 11 in. (60 to 280 mm) in OD are sorted automatically on an unscrambling/

loading table. Product is then heated in a single strand, roller hearth austenitizing furnace with 22 temperature control zones which are computer controlled to operate within ±10 F (± 6 C). Bars and thin walled tubes are then quenched by an OD progressive spray; thick wall tubing is quenched using Timken's ID/OD process. Product then advances to a double strand, roller hearth tempering furnace having a temperature uniformity of ± 5 F (± 3 C). After tempering, a five roll, computer controlled rotary straightener heat processes product within 200 F (100 C) of the tempering temperature to relieve residual stresses. All computerized controls are housed in a single room high over the floor-level equipment.

Vacuum — C. I. Hayes Inc. plans to debut (in 1985) modular preheat chambers as optional items for its line of vacuum carburizing furnaces with the self-cleaning feature. In self-cleaning, the heating chamber is exposed directly to air for carbon burnout. The added ability to preheat means that productivity is improved because a workload is in the main heat treating zone only during the carburizing-diffusion portion of the cycle.

Lindberg, Unit of General Signal, in anticipation of "the advent of composite materials as a viable substitute for metals" is improving its designs of heat processing equipment for these new materials. "The equipment itself," explains Lindberg, "is essentially a vacuum purge, pressurized cool autoclave.

Processing temperatures in vacuum furnaces are going higher, reports Vacuum Furnace Systems Corp. Many furnaces it is delivering in 1985 are designed for continuous operation at 2650 F (1455 C), and some go as high as 3000 F (1650 C). Applications include the firing of ceramic coatings on refractory metal tubing at temperatures up to 2600 F (1425 C) for 2 h. Some titanium alloys are being fired in excess of 2800 F (1540 C).

In England, Lucas Electrical

Ltd. is developing a "C" version of its Nitrotec process, "which will seriously challenge conventional case hardening where high stress indentation resistance is required for such components as gears."

Atmospheres And Quenchants

In Sweden, studies at the Transmission Div. of Volvo Components Corp. indicate that control of quenching speed is only one source of distortion in the heat treatment of gears. The size of the influence of three variables on distortion is estimated to be: gear design, 50 to 60%; case hardening steel (hardenability), 20 to 30%; and process parameters in case hardening, 5 to 15%. Objectives for future work include: narrower hardenability bands for the steel and new ways of determining hardenability, or closer requirements for the Jominy test method; development of a new quench tank where the spread of distortion in the vertical position is decreased; and development of methods and instruments for measuring and controlling quench speed.

Park Chemical Co. reports that salt bath quenching is preferred "in particular" in the processing of unalloyed austempered ductile irons, and more efficient salt bath quenchants for these materials are being developed. For example, nitrate salts supersaturated with water are in commercial use.

Tenaxol Inc. expects to produce a new polymer quenchant during 1985. Among its features: ability to replace oil quenchants, to be used in "many older operations with minimal modification," and to cover a wide range of solution hardening treatments for nonferrous alloys with a single concentration.

E. F. Houghton & Co. foresees the advance of oil-like polymers "into the heat treating of more metals and into additional applications." An example: quenching SAE 4150 steel after austenitizing in a fluidized bed. In addition, "truly oil-like polymers have continued to prove themselves in quenching crack-prone steels, especially 4150 and 4350."

Atmospheres — A bell furnace flow control system developed by Union Carbide's Linde Div. is designed to improve the quality of annealed nonferrous materials and to reduce the amount and cost of atmosphere. Energy consumption can also be trimmed.

Airco Industrial Gases has developed standard control systems that make it possible to deliver atmospheres of different compositions to the preheat, hot, and cooling zones of open ended furnaces. Zones are separated by slitted, flexible curtains. Zone systems are used in sintering, annealing, neutral hardening, and brazing applications. Efficiency is improved and expensive gases are conserved because the composition of the gas going to each zone can be controlled for optimum results.

East Ohio Gas Co. reports a new application for oxygen probes. They are inserted into atmosphere generator manifolds after the endo gas is cooled. The probe works in a closed loop air/fuel control system, "assuring consistent quality endothermic gas."

Energy Management

East Ohio Gas Co. also reports success in field trials of ceramic recuperators. In an installation at Euclid Heat Treating Co., for example, fuel savings of 45% and a 15% gain in production were realized. The furnace: a Surface rotary retort type rated at 1.8 million Btu/h (530 kW), with three zones. A 1 million Btu/h (295 kW) recuperator was installed in each zone.

Finally, Air Products & Chemical Inc. is of the opinion that "methanol will be readily available in 1985 to heat treaters, and is likely to remain in excess of demand and at competitive prices well into the 90s."

Source: Metal Progress, January 1985, 49-50, 53, 54, 56

Trends in
Heat Processing
Technology

What's notable about this year's forecast is the continued surge toward new combinations of the computer and microprocessor with heat treating processes ranging from induction heat treating to control of an endothermic atmosphere generator.

Other forecast topics include continuous annealing, lead patenting, advances in power supplies for induction heat treating, statistical quality control, polymer quenchants, coal gasification as an energy source, stainless steel annealing, nitrogen base atmospheres, fluidized bed heat treating, use of propane in carburizing and neutral hardening, boronizing, pulse inductive hardening, induction hardening, high frequency surface hardening, gas quenching, direct atmosphere generation for carburizing and hardening, a round furnace design concept, salt baths, ceramic thermal insulation, ion implantation, and electron beam processing.

Main Trend — "With respect to both labor and energy productivity, the largest single influence will be the microelectronic chip," declares Lindberg Corp. "Uses of microprocessors will grow in the heat treat shop, but the revolution will be quiet. Microprocessors are available for almost any use. They can manage complete processes like vacuum furnaces, carburizing, or nitriding systems. More recently, induction hardening and plasma processes. We are expanding the use of computers for order entry, tracking, and certifications."

Comment — Tocco Div., Park-Ohio Industries Inc., adds, "In the present state of the art, the microprocessor and computer have performed primarily a monitoring function in heat treat operations — a passive integrated system. What is now required, and which is the next step, is an active mode of operation or truly interactive mode of control. We believe that induction heating technology, when integrated via the computer technology in a fully inter-reactive operating mode, is the wave of the future ..."

Application — Illinois Gear, a Household International Company, has been operating and controlling batch and pit type carburizing furnaces from a multipoint microprocessor system for the past several months. Features include: data display on a CRT screen; direct couple with an oxygen probe or infrared analyzer; direct reading of carbon; and control of furnace functions with a program with a memory. Results to date: case depth control to within 0.005 in. (0.13 mm); temperature control within 1 to 2 F (0.5 to 1 C); carbon control to within less than 5 points — still being verified with hard data. Close control allows use of boost-diffuse carburizing cycles which save 20 to 30% carburizing time. Carbon potential is controlled while dropping from carburizing to quench temperature.

Application — Sciaky Bros. Inc. reports it is combining the electron beam and computer control "to provide efficient, economical on-line hardening of common carbon bearing steel." Computerized modeling of heat flow is expected to expand usage of the process.

Application — Ipsen Industries points out that the economical manufacture of ball screws dictates the precutting of thread grooves prior to induction hardening. Even the most exact controls in heating power, scanning feed rate, quenching temperature, and concentration of the quenching medium cannot avoid distortion. For relatively long ball screws, ample stock is required to allow for finish grinding of the part.

Ipsen has developed an induction hardening method for ball screws that maintains distortion within ±0.006 in. (±0.15 mm) independent of ball screw length. Called computerized distortion control, the unit will sense, via an onboard computer control system, a change in length of the ball screw during the hardening process. At present length intervals, actual length is compared with required length. In case of variance, hardening depth is proportionately altered via scanning feed rate and/or heating power. The unit is suitable for ball screws up to 6 in. (150 mm) in diameter and a length up to 24 ft (7.3 m).

Comment — "With constant pressure being put on profits, heat treaters will face decisions involving the upgrading of equipment that enhances human and energy productivity. Of interest are: computerization, robotics, and recuperation," observes Hinderliter Heat Treating Inc.

Comment — "The trend to low cost retrofit of heat treating equipment with microprocessor control systems will become more widespread," states Heat Treat Corp. of America. "Units such as the HTC 8200 and HTC 1550 integrated atmosphere and batch furnace control systems, which use any oxygen probe to provide atmosphere carbon control, will be used to provide older equipment with a state of the art update."

Application — Heat-Vac has developed a computer controller exclusively for vacuum furnaces. Some features: "It will not only complete any program a microprocessor can, but can carry out sophisticated programs, such as arc suppression units in ion nitriding furnaces. In event of a

Reprinted from Metal Progress, January 1983, 40, 44, 46, 48, 53, 55, © 1983 American Society for Metals

malfunction the screen will detail the cause and give the remedy. If a power failure occurs it will hold until power is restored, bring the furnace back to the condition it was in before the failure and continue the cycle."

Application — Surface Div. of Midland-Ross has a microprocessor based endothermic atmosphere generator controller that provides an economical method of producing protective atmosphere for heat treating. Automatic CO_2 infrared analysis and a microprocessor controller are teamed up. Endothermic atmosphere composition is maintained throughout the range of generator output. A standard feature: automatic infinite turndown of generator production to match furnace requirements without fluctuations in gas quality. Turndown capability reduces energy consumption and the necessity to flare excess gas production. A microprocessor idling mode program enables the generator to maintain temperature without producing endo gas, saving energy when product gas is required. The microprocessor is also programmed to semiautomatically burn out excess carbon in the generator during offduty periods.

Quenchants — "The heat treating community is becoming more sophisticated in its approach to the use of polymer quenchants," says Tenaxol Inc., adding, "Improved control measures for monitoring polyglycol quenchants (by user or supplier) provide the heat treater with assurance that day to day operations will not be upset by nonuniform results. Also more heat treaters have come to recognize cooling curves as the preferred method to measure initial quenchant performance and changes that may be due to contaminants. This practice has been encouraged in particular by the ASM Heat Treating Division's Quenching and Cooling Committee."

Energy — Wellman Thermal Systems Corp. reports its

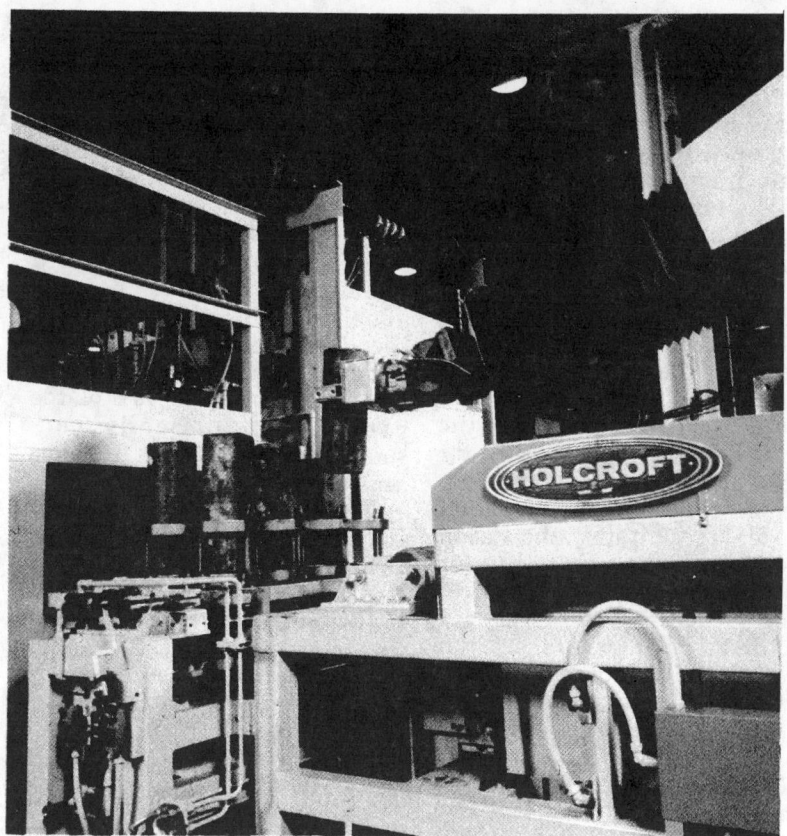

Material handling system for processing 2 ft (610 mm) long ordnance projectiles. The robot receives the projectiles at 1800 F (980 C) from an elevating-type accumulator, then grasps them for transfer to the tray fixture. Before automation, the forging would have first been allowed to cool and then reheated. The robot can handle 240 hot parts per hour. (Holcroft Div., Thermo Electron Corp.)

gasification plants are in use in both the United States and overseas "to provide a reliable and low cost supply of fuel for industrial furnaces and other applications." Producer gas is replacing fuel oil and natural gas. Latest conversions include pit carburizing furnaces; hardening, tempering, and annealing furnaces; and nitrogen generators. Other applications cited include ball forge furnaces, ladle heaters, tundish heating, boilers, and core and mold drying.

Salt Bath — In the patenting of steel rod and wire, salt baths are replacing lead, reports Heatbath Corp. Advantages are said to include more consistent metallurgical properties, "cost per given volume of 10% of that of lead," operating temperatures 100 F (55 C) lower.

Power Supply — Industrial Electric Heating Inc. has a solid state 50 kHz generator it claims is more reliable and more efficient than RF tube type

generators. It is said to be 35% more efficient and operates with 30 to 40 times less voltage. The company says nearly all RF applications can be handled with 50 kHz.

Recycling — Renault reports, "For selected hot forged parts, we use heat contained in the workpieces, immediately after stamping, to obtain the required structure and properties through controlled cooling. These direct treatments are being studied for isothermal annealing of gears, direct quenching, and normalizing of connecting rods. Tight control of the cooling curve is necessary."

EB Hardening — SKF Industries reports it is developing a production version of high energy, electron beam surface hardening of bearing races for Wright-Patterson Air Force Base. The project is in response to a problem with aircraft turbine mainshaft bearing rings. Those conventionally through hardened (the rings are made of M50 steel) are prone to sudden through-section fracture. The new approach being studied begins with bulk ring heat treatment to provide a high fracture toughness core, followed by EB surface hardening of the rolling contact surface to an effective case depth of 0.030 in. (0.76 mm).

Annealing — Air Products & Chemicals Inc. has developed a nitrogen based process for annealing stainless steel which keeps hydrogen levels in the nitrogen down to 20-30% for most parts. Hydrogen can be a major problem. If chromium nitride precipitates, there is a loss in corrosion protection. Air Products says its patented process uses low levels of oxidant additions which preferentially adsorb on the surface of stainless steels and inhibit nitrogen from absorbing into the surface layer. "Reducing hydrogen levels," comments Air Products, "can lead to substantial cost savings over both dissociated ammonia and higher concentration hydrogen-nitrogen atmospheres."

Fluidized Bed — This process is gaining acceptance where batch loadings are the norm, versatility of equipment is needed, and rapid changeover is desirable, reports Can-Eng Ltd. A new application, high temperature carburizing, is being evaluated by the firm. Results are said to be encouraging. The speed of the process seems to be an inherent advantage in this instance. High temperatures do not seem to have deleterious effects on furnace components or the metallurgical properties of parts.

Atmospheres — Research at Ford Motor Co. shows that furnace atmospheres for carburizing and neutral hardening can be produced by introducing propane and air directly into the heat treat furnace at reduced rates. A constant air flow is used while the propane flow is automatically regulated, using the output of a zirconia oxygen sensor as a control signal. Trials have been run in continuous pusher and batch furnaces. Methane (natural gas) or butane may be used instead of propane. With butane, satisfactory atmospheres can be produced at temperatures as low as 800 C (1470 F). Methane is best used at 900 C (1650 F) and above. Ford comments, "Because no gas generator is required and because suitable flow rates are typically one-third to one-half typical flows, 75 to 90% of the hydrocarbon normally consumed in producing furnace atmospheres can be saved."

Gases — Research at Airco Industrial Gases has been centering on the development of new directional atmosphere flow technologies and refinements in the blending and selecting of atmosphere compositions. By optimizing the amount of each constituent in the nitrogen based mixtures, the maximum benefits of the protective atmosphere can be utilized. Altering compositions and directing the correct atmosphere mixture to each furnace zone results in a more efficient use of nitrogen based atmosphere, advises Airco.

Quenchants — "The near and long term future of polymer quenchants is definitely one of growth," remarks E.F. Houghton & Co. A new polymer it has introduced, Aqua-Quench 364, is said to be more like oil than other polymers in its ability to reduce cracking and minimize distortion.

Forecast — Hughes Tool Co. sees "significant improvement in the performance of highly loaded friction bearings from three sources: new and better lubricants, better seals, and new surface treatments for journal surfaces." Hughes, for example, is boronizing surfaces of inner bearing members for a new rock bit design.

Process — A fine grain structure of less than ASTM 12 in hardened areas of martensitic type steels is said to be obtained with an induction hardening process patented by Impulsphysiks GmbH of Germany. The process is based on high peak capacitor discharge of rectangular shaped pulses. Millisecond pulses of up to 30 kW, 3 MVA, at 27.12 MHz are applied to a small inductive loop. The strength of the magnetic field is so high the steel exceeds the Curie temperature of about 730 C (1345 F). There is no smoke or gaseous emissions, and a cooling medium for the part is not required.

Quenching — Ipsen Industries International GmbH of Germany states that "vacuum furnaces with high pressure (up to 5 bar [500 kPa]) gas quenching are increasingly replacing salt bath equipment for hardening high speed steel cutting tools in tool manufacturing plants. Quenching speeds equal or exceed those in salt bath quenching."

Ipsen also reports that classic carburizing cycle times in endogas can be reduced from 25 to 50% via direct atmosphere generation by controlled supply of fuel gas and air into the furnace chamber. Example: a case depth of 0.9 mm (550 HV) is obtained on large gears of low alloy chromium-manganese steel after 3 h at 930 C (1705 F). Similar time savings can be obtained, says the firm, in gas atmospheres of nitrogen and cracked methanol with the aid of an automated computer program (TPS-system) developed in France.

Atmospheres — Liquid Air Corp. reports, "Both furnace manufacturers and the industrial gas companies have collaborated in the rewriting of the NFPA 86C standard on furnace atmospheres. With reissuance of this standard good practices for the use of nitrogen based special atmospheres will be published for the use of the entire heat treating industry."

Furnace Design — Surface Div. of Midland-Ross is developing a round furnace technology that combines conventional pusher tray furnace features with advanced thermal and energy conservation ideas. A round design has less surface area than a conventional rectangular design. Empirical data show that temperature uniformity should be ±7 F (±4 C), or twice the capability of comparable rectangular designs. "Internal geometry," says Surface, "greatly improves roof fan effectiveness because of the unique wind flow across the furnace top, reducing floor space requirements."

Power — New solid state power supplies for heating, forging, and melting have been introduced by Pillar Corp. Its MK-6 line, for example, is dedicated to 10 kHz power supplies that range from 50 to 500 kW and offer a conversion efficiency of 92%. The MK-7 line covers the frequency range of 200 Hz to 10 kHz in kilowatt sizes of 300 to 2500. "This product," says Pillar, "with its unique capacitor coupling, eliminates the need for output transformers and offers conversion efficiencies of 97%."

Inverter — Heating applications like hardening, brazing, and soldering require limited induced penetration, which can be handled with a conventional tube type radio frequency generator. Similar results, it is reported, are being obtained with a 25 kHz solid state inverter developed by Radyne/AKO. Standard alternating current is converted to a controlled direct current voltage by a thyristor controlled rectifier. Operating cost is said to be "at least 30% less than that of the tube type generator."

Fluidized Beds — Procedyne Corp. reports that with refined nitrogen based atmosphere formulations, especially with methanol, one of its fluidized bed furnaces "can perform all of the conventional case hardening treatments of nitriding, nitrocarburizing, carbonitriding, carburizing, neutral hardening, tempering, steam blueing, and stress relieving." Procedyne adds, "With 2000 F (1095 C) units that are 5 ft (1.5 m) across and equally as deep, fluidized beds are capable of handling many of the larger jobs usually reserved for molten salt."

Recycling — Henry E. Sanson & Sons Inc. predicts that "more users of nitrite-nitrate salts will install equipment for salt recovery and recycling. The low melting point salts — usually equal to or close to the MIL-S-10699 B, Class 1 type, are widely used for martempering and austempering of steel or ferrous castings, for tempering or blueing of steel in strip or wire form, and for curing of rubber or plastic extrusions. Treated salts have to be washed with water which dissolves the salts easily and in high concentrations. With at least two wash tanks and counterflow of water and work to be washed, salt concentration in the first wash tank can be high enough (at least 20%) to warrant recovery of salt.

Ceramics — Sauder Energy Systems Inc. believes that high density ceramic fiber modules will be used for both thermal insulation and for structural purposes. Door lintels, frames, and interior furnace areas can be made of high density ceramic fiber fabricated into special shapes. Sauder makes modules with an uncompressed density of 15 lb/ft³ (240 kg/m³). The material can be formed into shapes to fit the application.

Process — New applications for its high frequency method of selective surface hardening are being developed by Thermatool Corp. They include: constant velocity universal joints, a crank for a lawn mower engine, an automotive camshaft, large piston ring grooves, transmission components, and drill bits. The process has been extended to selective surface alloying or inlaying of metallic particles via high frequency melting. Example: selective alloying of turbine blades.

Atmospheres — Union Carbide's Linde Div. has developed technology for atmosphere flow reduction of up to 50%, while maintaining control of the furnace atmosphere, in open end furnaces, such as belts, shaker hearths, and roller hearths. A flow control system for annealing in bell furnaces provides significant reductions when nitrogen based atmo-

spheres are used. The system automatically adjusts flow rate during the cycle to have high purge flows at the start, followed by reduced flow for the balance of the cycle.

Salt Baths — Methyl chloride has long been preferred as a means of maintaining salt baths in a neutral condition for austenitizing low to high alloy steels, points out Park Chemical Co. Unlike some other rectifiers, methyl chloride does not form sludge. However, as a gas, it has a drawback: it must be handled in pressurized containers and introduced slowly to the molten salt bath. Park now has a solid rectifier containing ammonium chloride. The pellet rectifiers are easily stored and handled and can be more accurately measured for use, says the company.

Linings — Insulating Products Div., Babcock & Wilcox reports work is continuing to extend the range of ceramic fiber linings to applications where they aren't being used currently because of severe operating conditions. These efforts include improving fibers, anchoring, and construction techniques. The technology is well advanced. For example, prefabricated sections of ceramic fiber linings are mounted on metal plates and welded directly to a furnace. Fiber modules, cements, and coatings have been developed to allow ceramic fiber refractory lining to be veneered directly to the hot face of a dense refractory lining.

Electron Beam — Using electron beams for surface modification — which includes transformation hardening, glazing, and surface injection — is getting a lot of attention at Leybold Heraeus. In transformation hardening, a steel or alloy is heated to a temperature high enough to cause a crystallographic change, usually followed by an accelerated cool. In glazing, a selected area on a surface is fused. It's reported, for example, that iron carbide oriented toward the surface will enhance

Source: Metal Progress, January 1983, 40, 44, 46, 48, 53, 55

the wear resistance of cast iron. Injection of hard particles holds promise as a way of producing an abrasion resistant surface. For example, titanium carbide can be injected onto the surface of an aluminum alloy at a rate of 100 in.^{2}/min (645 cm^2/min) to a depth of 0.1 in. (2.5 mm) in an out-of-vacuum EB machine.

Ion Implantation — Westinghouse Electric Corp. is of the opinion that ion implantation technology provides a means of engineering the near-surface properties of materials in ways never before achievable by any other techniques. Potentials extend to a number of characteristics related to surface properties, such as friction, wear, corrosion resistance, hardening, bonding, lubrication, and adhesion. Westinghouse, for example, has implanted a number of metalworking and forming tools with nitrogen ions with a dosage of 2×10^{17} ions/cm^2. Two to twelvefold increases in service life were obtained. Examples: a 1Cr-1C tool steel paper slitter, a twofold increase; a WC-6Co synthetic rubber slitter, a twelvefold increase.

Handling — Holcroft reports, "More and more heat treating plants are incorporating sophisticated handling systems in their furnace installations for handling material both inside and outside furnaces. Furnace manufacturers are being asked increasingly to design total systems. Mechanical handling systems and robots will play a greater role in the future. Devices for part location and identification utilizing cameras and possibly lasers will be more widely adopted."

Recuperation — Renewed interest in recuperation for fuel fired heat processing furnaces is predicted by Morgan Construction Co. "Recuperation," the company explains, "will become an integral part of the modernization of the industrial fuel burning/heat process." Reasons: we now know the benefits of recuperation, and we know that energy will never again be cheap.

Bang — The Soviets have what is termed explosive strengthening at the pilot plant level. Claims indicate such applications as railroad frogs, excavator bucket teeth, and wear-prone parts on ore and stone crushers. Uniform strengthening to depths of up to 50 mm (2 in.) is reported. Another technique: heat treating after explosive treatment. A notable increase in ductility is indicated.

Survey Report on Heat Treating/Processing

By Harry E. Chandler

Two leading technologies: ceramic fiber insulation and electric heating. This bell furnace will operate at 2200 F (1205 C) in a hydrogen atmosphere, reports Sauder Energy Systems. Heat storage is said to be one-third that of a furnace lined with insulating firebrick.

Suppliers and users of heat treating/processing equipment (they are listed at the end of this article) taking part in a *Metal Progress* technical-economic survey believe:

1. That the leading trends in this technology include the great emphasis being placed on better process control via such means as the oxygen probe and microprocessor and on energy conservation via such means as improved furnace insulation and waste heat utilization.

2. The total technology of interest at this time includes surface treatments (flame, induction, electron beam, laser, carburizing, boronizing, nitriding, etc.), the fluidized bed process, the vacuum process, electric heating, nitrogen atmospheres, polymer quenchants, multipurpose equipment, steels with narrower-than-standard hardenability bands, higher processing temperatures, more automation in parts handling — particularly robotics, and more sophisticated controls.

3. That incentives for adopting the best of affordable technology usually relate to reducing costs, improving quality, and increasing productivity.

4. That "no unique upturns are expected in heat treating/processing markets when the U.S. economy comes back." Any improvement in the economy, it is believed, will be gradual rather than steep. Specific

Reprinted from Metal Progress, April 1983, 13-16, 18, © 1983 American Society for Metals

markets cited as benefiting from an upturn include gears used in mining and by minimills, oil field tubing, powder metallurgy, specialty alloys, special purpose parts, standard components that are in short supply, and some forms of heat treating/processing equipment.

5. That R&D will stay on its present course when the U.S. economy is restored because, by and large, this activity is not turned down or redirected when business is on a downcycle.

6. That modest improvement in the U.S. economy has started or will take place in 1983, but substantial improvement will not come until sometime in 1984 or in early 1985 at the latest.

7. That for a variety of reasons the U.S. economy, markets served by the heat treating/processing industry, and the industry itself probably will not make it all the way back to prerecession levels.

A closer look at each topic follows.

Assortment of Current And Emerging Trends

Technology in the spotlight could be regarded as a listing of needs for survival in the present and anticipated economic climate. Survey respondents talked in terms of both concepts and specifics.

For example, one supplier believes that most incentives for adopting technology can be related to productivity. He includes more efficient use of energy, reduction in manpower, lower cost of materials, improved part performance, reduction in labor, reduction of in-process inventory, reduction in scrap, improved efficiency in manufacturing processes, and improved quality.

He comments, "There are those who insist that quality is a part of the productivity equation, and frankly I believe rightly so. In the final analysis, the two are closely linked.

Three leading technologies. Top: Boronizing, a surface treatment, was used on the bearing pin of the oil drilling bit head section at left. Head section and bit cone (right) are shown after 111 h service in sand and shale. (Hughes Tool Div., Hughes Tool Co.) Left: Bearing groove being high intensity induction hardened, a selective heat treating process. (Thermatool Corp.) Right: Large hook used in oil drilling equipment is being lowered into a 5000 gal (19 m³) tank containing polymer quenchant. (E. F. Houghton & Co.)

However, past practices and manufacturing attitudes have claimed that you obtain quality at the expense of productivity."

Including quality under the productivity umbrella is important because "this means that effective processing controls must be integrated into the manufacturing process, and be responsive to acknowledge material process variables that are part of the system. Manufacturers are beginning to realize that effective quality control directly and indirectly supports productivity improvement."

Another supplier comments more specifically on the quality matter. "This increased emphasis on quality so apparent in the past five years," he declares, "has extended far beyond final inspection to all of the operations involved in primary metalworking. Heat treatment is no exception, and recent trends have focused on processes with the capability of achieving repetitive mechanical properties with minimum dimensional change ... This battle is now being won with improved atmosphere and temperature sensing equipment in conjunction with computer controlled systems."

Another statement by another commentator: "The common thread running through industrial marketing strategies is an effort to decrease the cost of producing goods by reducing energy costs while improving quality to maintain or gain a selling advantage over competitors."

A supplier states, "Reliable, low cost automatic programmers and process controllers are being used on just about every furnace we build. Even on furnaces with very simple time-temperature programs, the automatic programmers are being applied to assure that the correct cycle is being used. The digital set point and furnace temperature display are also quite popular.

"In addition, ceramic fiber insulation is becoming the rule for furnaces. This is particularly true of batch furnaces as an energy conservation measure since it makes possible total furnace shutdown when required ... because of the material's low heat storage and resistance to thermal shock ..."

A vendor of controls reports, "Among established trends is the replacement of analog instrumentation with digital instrumentation for heat treat process controls for variables such as temperature, metallurgical atmospheres, pressure, and vacuum ... Analog records are still preferred for operations with data logging and quality control requirements."

Handling of parts gets many mentions. For example, "A fair amount of hard automation appears to be a well established trend ... U.S. heat treaters are becoming more competitive in the worldwide market, in part, because of the adoption of more complex automation systems ... Microprocessor-automation systems will be particularly important in continuous processing for volume production operations ..."

A sampling of other technology of current interest:

"For applications where temperature requirements are in a moderate range, electric heating is beginning to emerge as perhaps the preferred method ..."

"Fluidized bed heat treatment has proven its versatility. As capital becomes available, use of this technology will increase."

"To help control distortion in heat treated gears, steels with narrow hardenability bands, one-third standard, are being used."

"The trend toward multipurpose application of a furnace or other processing equipment ... enhances the feasibility of capital investment."

"Look for greater use of such surface treatments as steam oxide, chemical vapor deposition, plasma vapor deposition, and ion nitriding."

"Induction heating is being combined with conventional fuel fired heating for the treatment of tubing."

"More sophisticated material handling devices based on intelligent robots and vision systems will make their appearance. To an increasing degree, such devices will provide total material handling, extending to locations upstream and downstream of the heat treating equipment."

"Interest continues in nitrogen based atmospheres, both for nitrogen-methanol and nitrogen-natural gas systems."

"Industry is demanding more sophisticated technologies for insuring quality control. To be cost effective, the technique must recognize inherent variables in the processing system and the incoming material. Increasing material costs require a system that must handle or be tolerant of lower cost material. The processing technology must have an interreactive mode to automatically accommodate material variation."

"The switch from oil quenching to water based polymer quenchants will continue as knowhow for the latter improves. Another oil shortage will spur this greatly."

Markets and Economy: Where They Fit In

Will there be marked upturns in any markets when the economy bounces back?

Will there be any major changes in heat treating/processing R&D when the economy bounces back?

When will the economy start to improve? Make it all the way back?

Briefly, few if any "marked upturns" are expected. "Bounces back" is also questioned with "crawl back" offered as an alternative in one instance.

Regarding R&D, the major position seems to be, "We don't look for major changes because we have maintained our

Two leading technologies. Top: Automated fluidized bed heat treating furnace. (Procedyne Corp.) Bottom: Automatic continuous vacuum furnace. (C. I. Hayes Inc.) Hayes will introduce a new generation of continuous vacuum carburizers at the 1983 ASM Heat Treating Conference/Workshop. The VBQ series reportedly combines a new level of productivity with reduced maintenance costs.

development commitment throughout the downturn."

The consensus on the last two questions runs like this: "We see modest improvement in the second half of 1983 and continued modest improvement in late 1984 or early 1985."

A mix of views on all three points:

"We definitely expect an upturn in furnace equipment orders for the foundry industry. Many foundries have shut down. Those still in business will be called upon to provide the required capacity."

"A recent market study indicates that about half of the companies with heat treating equipment expect to purchase another furnace in the next few years. Half again expect to replace existing equipment, while the remainder were interested in expanding production capability and in improving operating efficiency."

"One of the first results will be a dramatic increase in demand for raw materials, including metals. In addition, many mining companies will need replacement and more efficient equipment."

"When the economy bounces back I believe there are going to be marked upturns in demand in the consumer product and automotive markets. There is much pentup demand."

"Tool steel heat treating has been particularly depressed and should bounce back strongly when the economy improves."

"We can comment only on the induction heating equipment market. The need for upgrading or replacement has been and remains great ... Replenishing of depleted inventories of standard components may cause a shortlived surge, but we do not expect marked upturns in the general economy or our market. We see a gradual upward trend that will take 18 to 24 months to equal previous sales levels."

R&D is no longer as cyclical as it once was. A comment: "Very few companies can efficiently turn R&D on and off. The activity exists today; it is just not as vigorous as it once was and is probably slanted more toward the 'D,' where incremental improvements are made with less risk and a bit more certain return."

Predictions regarding the direction of the economy are in essential agreement. Differences are in timing. A comment: "We believe the improvement is underway. Custom equipment manufacturers are among the first to feel the effects of an economic recovery because relatively long leadtimes are needed to design and build equipment. With the lowering of interest rates and the general atmosphere more optimistic than it was, those in a position to do so are now starting to make capital equipment purchases. Others in less healthy positions will buy as the economy improves. We see this as a gradual process lasting into 1985 before we reach 'all the way back' status."

"With the automotive upturn already taking place, I look for steady but gradual improvement throughout 1983 with a strong economy in 1984."

"We do not expect any real improvement before late 1983 or early 1984. We do not see a return to previous levels before late 1984 or early 1985."

"We expect slow improvement in the spring and summer of this year; the 1979-80 level should not be reached until the winter of 1984. Consumer related products will be the first to feel the pickup, with durable goods lagging. Capital equipment will be last, except where equipment of advanced design is needed to meet new requirements."

"I do not see the industrial sector improving until 1985. I consider capacity utilization of more than 70% an improvement."

Some Say, 'Never Again No Matter What Happens'

The observation, "Production levels of customers are significantly below those in 1979 and will probably never return to where they once were," keynotes a unique happening which helps to explain the underlying pessimism evident in this survey. Perhaps it is an overreaction to the depressed economy. Perhaps it is realistic. Whatever, prevailing wisdom dictates the following scenario.

"The downturn has taken its toll in total business capacity, and even though the remaining business will be more efficient and cost effective, we doubt that a quick recovery is possible."

"Manufacturing of goods will probably never again reach peak production levels of 1979."

"Some areas will remain at reduced levels and not gain back market share, including automotive, construction equipment, and steel production. Overseas suppliers will be responsible for this."

"... The change in world conditions have made it impossible to return to the way of life we experienced from World War II to 1974."

"... It may never again be the economy of the sixties and seventies, particularly in terms of manufacturing."

"As regards the heat treating/processing market and the economy, I question whether it will ever make it all the way back. I say this because tremendous change is taking place. Foreign competition, changes in production strategy, control of inventories, and new requirements for materials all may erode some of the market."

"Many labor intensive facilities are being moved out of the country. There is also a geographic shift within the United States. Large, vertically organized, concentrated manufacturing facilities are being redistributed to smaller units to provide more effective control, better flexibility, and efficient operation along with improved labor costs. We see many major OEM's re-evaluating in-house products on a make or buy economic analysis, with resultant external sourcing to optimize reductions in components costs."

"The streamlined, efficient, and profitable industries that survive and serve specialty markets are the ones to look to for a market upturn and growth. Firms recognizing foreign sources of raw materials as well as technology should show good growth."

Finally, a forecast garnished with a dash of optimism: "We look for a slow, gradual improvement, with an increase of perhaps 10% over present levels by late 1983. But 'all the way back' must be defined. 1981? 1980? 1979? The answer, I think, without being pessimistic, is 'never again as we knew it.' However, acceptable levels of employment, inflation, productivity, and balance of payments should be attainable, with solid growth by 1985." ⊕

Mr. Chandler is editor of *Metal Progress*. Information for this article was provided by the following:

American Induction Heating Corp.; Air Products & Chemicals Inc.; AGA Gas Inc.; Airco Industrial Gases, Div. Airco Inc.; Benedict-Miller Inc.; Can-Eng Mfg. Ltd.; Drever Co.; East Ohio Gas Co.; Ford Motor Co.; Gas Atmosphere Div., Modern Equipment Co.; C.I. Hayes Inc.; Heatbath Corp.; Heat Treat Corp. of America; Hinderliter Heat Treating Inc.; Holcroft, Div. Thermo Electron Corp.; A.F. Holden Co.; E.F. Houghton & Co.; Hughes Tool Div., Hughes Tool Co.; Illinois Gear, A Household International Co.; Industrial Equipment Div., Westinghouse Electric Corp.; Ipsen Industries, an Alco Standard Co.; Kolene Corp.; Leeds & Northrup Co., Unit of General Signal; Leybold-Heraeus Vacuum Systems Inc.; Linde Div., Union Carbide Corp.; Liquid Air Corp., Industrial Gases Div.; Park Chemical Co.; Pillar Corp.; Procedyne Corp.; Radyne/AKO Corp.; Sauder Energy Systems Inc.; Surface Div., Midland-Ross Corp.; Thermatool Corp., An Inductotherm Co.; Tocco Div., Park-Ohio Industries; TSP Mill Products Div., Xtek Inc.; Vacuum Furnace Systems Corp.

Trends in Heat Processing Technology

Widespread adoption of microelectronic technology coupled with a scientific approach to heating system design is ushering in an age of efficiency.

HEAT TREATING is taking on an aura of sophistication as both suppliers and users acclaim the benefits of microprocessor-based control systems, of maximized efficiency furnace systems, of carefully blended alternative atmospheres, and of selective hardening options such as induction, electron beam, and laser processing.

Microprocessors Are Being Rapidly Adopted

"The most pronounced trend in heat processing technology is the rapid adoption and expanding role of microprocessor-based programmable logic controllers (PLC's) and minicomputers," states Ipsen Industries, an Alco Standard Company. "They are being readily accepted for controlling single batch furnaces, batteries of batch units, and lines with continuous units of varying complexity."

Ipsen's sentiments are echoed by other furnace builders.

Midland-Ross Corp.: "A great deal of automated process control and monitoring is forecast. Our vacuum furnaces, for example, already incorporate a simple-to-operate microprocessor-based PLC system that gives instantaneous process and diagnostic feedback to operators and management. It's anticipated that these devices will soon be used to control entire lines, and may incorporate energy optimizing software."

AGF Inc.: "The trend is toward furnaces equipped with minicomputers and PLC systems. Extensive applications will be found in deep case carburizing and large capacity pit carburizers, batch rotary carburizers, and two-zone basket or tray-form furnaces."

Vacuum Industries Div., GCA Corp.: "The advent of the microprocessor-based programmer has made 'hands off' operation of vacuum furnaces a reality."

Lindberg Corp.: "Advanced sensing devices and microprocessors are being combined to enhance process control and productivity in heat treating. Our Lindcase setup, for example, incorporates a programmable analog processor which times preheat, carburize, diffusion, and quenching cycles in addition to controlling furnace atmosphere and temperature. Control of case depth is maintained within ±0.001 in. (0.03 mm) even when working to specifications as shallow as 0.005 in. (0.13 mm)."

This trend is not simply evidence of technological evolution in action. "Application of this new technology is being forced," stipulates Lindberg Heat

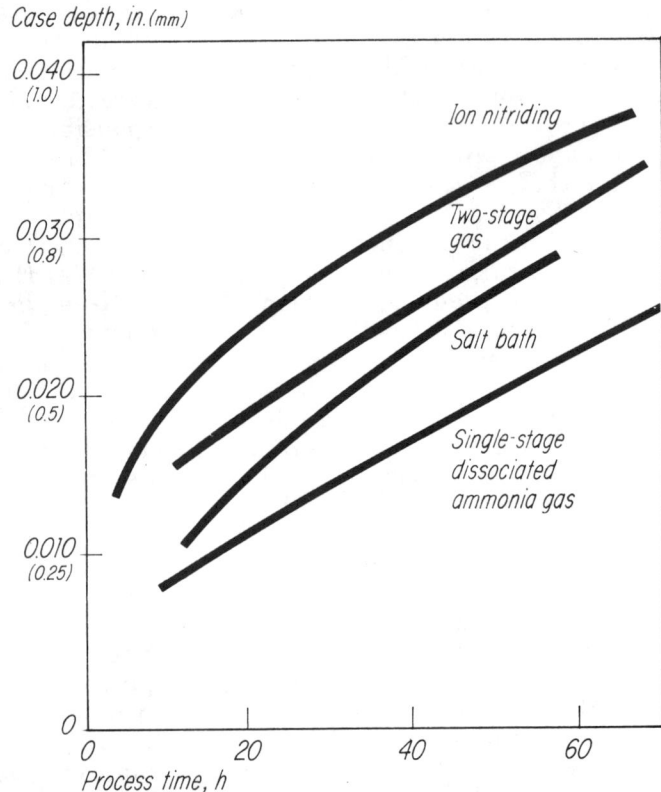

Ion nitriding's higher speed of case development, particularly in the early stages, makes it an alternative to both gas nitriding and salt bath nitriding. (*Advanced Vacuum Systems*)

Reprinted from Metal Progress, January 1980, 77-82, © 1980 American Society for Metals

Treating Co., "by increasing customer and/or consumer demands for quality processing coupled with an increasing risk of product liability."

AGF thinks the trend will help sway user preference toward batch furnaces instead of continuous units. "Continuous installations are very complex and costly. Also consider the shortage of trained technical personnel and the high cost of lost production when a large continuous furnace is shut down. These considerations should make the idea of a number of small, automatically controlled batch furnaces very attractive. For example, if one unit is shut down, production schedules won't be seriously interrupted."

GCA/Vacuum thinks the trend will foster sales of new furnaces. "The relative ease of programming coupled with the consistent operation of these solid state instruments have placed renewed emphasis on the importance of equipment reliability. Manual control by a skilled operator has in the past been a convenient rationale for babying along a tired or obsolete furnace. The high cost of tolerating a rickety furnace has now been put into clear perspective. Heat treaters will find that retrofit of existing equipment with microprocessor-based controls is not cost effective unless the furnace system is in top condition and of modern design."

Case History— International Harvester's system for monitoring and controlling equipment as complex as a two-row continuous carburizing furnace is an example of the state of the art. A microprocessor, interface, and printer team up to monitor parameters such as push time, weight of each basket, temperature in four zones, carbon potential (via oxygen probe signals), and the status of door closure.

"The system will also serve as a maintenance monitor and volume (weight) and energy monitor," says IH. "It will print out the average cost per pound for carburizing per shift, plus energy utilized per pound and the operating cost per hour for a given furnace."

Properties— Users are also taking advantage of microelectronic technology to assess the properties of heat treated parts. Timken Co., for instance, has developed a finite element program for predicting stresses in depth in heat treated bars. "The program can handle dual or multiple heat treating procedures such as carburizing followed by rehardening and tempering with the same ease as solo heat treat operations."

New Heating Systems Are More Energy Efficient

"Energy awareness, coupled with the development of energy conservation and heat recovery equipment, has dramatized the need for using financial indicators such as payback and return on investment in the selection of heat processing equipment," says Holcroft, Div. Thermo Electron Corp.

"In the past, minimizing initial investment was the goal. Now, heat treaters are willing to up initial investment provided the incremental increase will lead to reduced operating expenses associated with energy, operation, and maintenance.

"Life cycle cost," sums up Holcroft, "is slowly becoming a factor in equipment selection."

User demands for efficient heat processing systems have provided the impetus for suppliers to expand their product development efforts along this line.

Midland-Ross Corp., for example, states that energy saving devices have reached a mature stage. "Recuperators are almost universally applied to new gas fired equipment, and multifuel recuperators will become common. More importantly, however, total energy consumption analysis is being performed for heat treat equipment, resulting in numerous novel energy saving innovations."

As an example, the company's Surface Div. recently sold a continuous carburizing line in which the endothermic atmosphere will be burned down within a radiant tube before it leaves the furnace. The device will recover most of the available heat in the atmosphere gas — 17% of the gas required to operate the furnace. The line also includes a device which uses the hot flue gas from the carburizing furnace to heat the associated tempering furnace. The device will save 90% of the fuel needed for heating the tempering unit.

Holcroft cites its Maximum Process Efficiency heat treating systems, which "... recover energy in the stock that is now lost in quenching processes and in part cooling."

And Lindberg Corp. points to the use of programmable demand controllers for reducing electricity charges. "In a small, 1000 ton/yr (900 t/yr) plant, a 15% saving in total electrical power cost per unit weight processed has been obtained, even though demand control is removed after loads reach temperature to permit close process control."

Process Modification— Although the new, more efficient equipment will go a long way toward minimizing energy consumption, innovation on the part of the heat treater-metallurgist should still be encouraged. A modification to processing parameters can often result in an added saving.

Missouri Steel Castings Co., for example, saved fuel and essentially increased heat treating capacity by simply raising the normalizing temperature for carbon and low alloy steel castings from 1650 F to 1700 F (from 900 C to 925 C). Needed time at temperature dropped dramatically from 2 h to ½ h.

Heat Treating Atmospheres For the 80's

"The use of alternative heat treating atmospheres

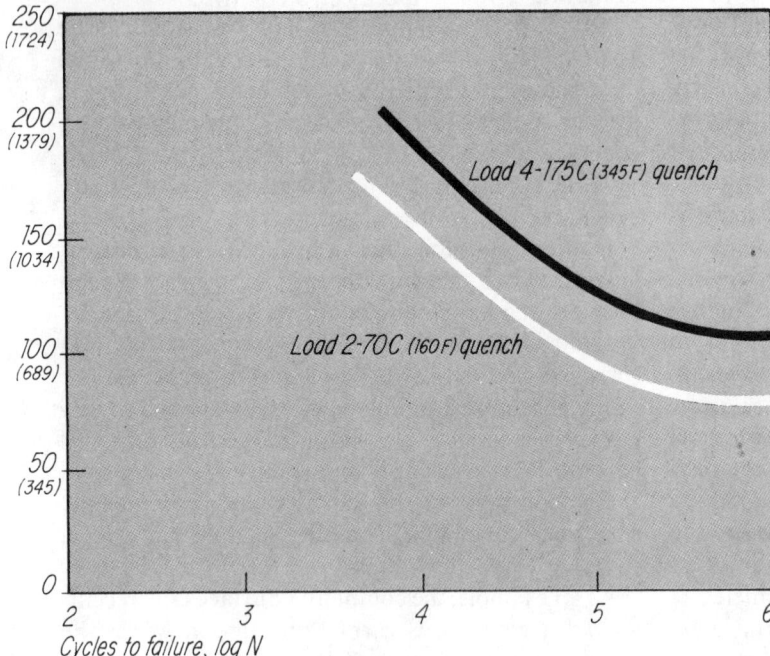

Bending stress, 10³psi (MPa)

Load 4-175C (345F) quench

Load 2-70C (160F) quench

Cycles to failure, log N

Hot oil quenching after carburizing or carbonitriding is said to improve gear fatigue life and uniformity over gears that are press quenched. Distortion is also minimized. Data are for SAE 15B21 gears carburized at 930 C (1700 F).

will significantly expand as natural gas supplies continue to tighten," predicts Lindberg Heat Treating Co. "Key to this expansion will be the increase in the technology base needed to assure process reliability and controllability."

In the U.S., nitrogen-based atmospheres are still receiving the most attention.

Air Products & Chemicals Inc., for example, is developing improved atmospheres for brazing and annealing that consist of "... controlled amounts of 'impurities' added to high purity, hydrogen-nitrogen blends."

The company is also working on a carburizing process that's a variation of an older technique based on the use of modified nitrogen-methanol blends. It's an example of the "expanded technology base" forecast by Lindberg. "By employing recently available analysis and control instrumentation," says Air Products, "we've been able to demonstrate carburization rates faster than those possible with conventional endothermic carriers."

Airco Industrial Gases, Div. Airco Inc., is also developing new techniques that "... will significantly increase the frequency of substitution of nitrogen-based atmospheres for endothermic, exothermic, and dissociated ammonia atmospheres."

Among them: a process that ups productivity by 5 to 15% in ferrous wire, rod, and bar annealing; and an approach to brazing that improves uniformity, eliminates the need for flux and postbraze cleaning operations, and increases belt life.

Benedict-Miller Inc. and Chemetron Gas Div. (now NCG Industrial Gases, Div. Liquid Air Corp. of North America) have jointly developed an automatically controlled process for hardening, medium case carburizing, and carbonitriding that uses nitrogen, methanol, and small amounts of methane to replace generated gases. Users of endothermic atmospheres could reportedly save 15 to 25% in natural gas by switching.

Benedict-Miller explains that a carbon sensor/oxygen probe is the control element. All test runs have been successful with effective case depths meeting requirements for the temperatures and times involved. Results were monitored and checked with a Leco carbon analyzer and a microhardness tester.

Incidentally, several forecasters cite the carbon sensor/oxygen probe as a significant factor in the growth predictions for both nitrogen-based atmospheres and microprocessor-based heat treating control systems.

Drip Feed Method— From Switzerland and Maag Gear-Wheel Co. Ltd. comes a prediction that "the drip feed method of carbon controlled carburizing with methanol and ethylacetate or methanol and acetone will be extensively used in the 80's, particularly with a combined and appropriate application of nitrogen."

The technique can also be used for high temperature carburizing.

Untreated Fuel Gas— Ipsen Industries International GmbH reports it has developed a process for "the generation of a controlled protective or carburizing atmosphere by direct supply of untreated fuel gas into heat treating furnaces."

Benefits are said to include savings in operating

cost and energy consumption combined with faster and often more uniform carburizing than possible with generated endothermic gas.

The process is based on "... the experience that the oxygen potential, as measured with a carbon sensor, is a reliable parameter for carbon potential control under nonequilibrium as well as equilibrium atmosphere conditions."

The West German company says dilution of the atmosphere with nitrogen is possible but generally not required, and may even be undersirable.

Hydrogen— New hydrogen generators that produce ultrapure (99.9999%) gas have been introduced by Matthey Bishop Inc. These units use an economical methanol-water fuel mixture that reportedly brings the cost of the gas down to a level comparable to or even lower than that of commercial grade hydrogen.

MBI thinks this development will help expand the use of ultrapure hydrogen in heat treating and brazing, areas where commercial purity gas has a strong foothold.

Vacuum— The solution heat treating of exhaust valve steels in vacuum is being studied by TRW Inc.'s Valve Div. "Vacuum treating would offer many advantages over conventional solution heat treating; among them, more efficient energy usage and the elimination of the need for gas atmospheres."

High temperature vacuum carburizing has been found to increase fatigue life reports a major truck manufacturer. The property improvement results from the ability to better control surface carbon content. Other benefits: decreases in carburizing time and energy usage and an increase in "alloy" life. This forecaster would like to see the vacuum furnace industry come up with "cost efficient furnaces capable of continuously carburizing large volumes of parts."

Developments in Quenching Technology

"Use of synthetic polymer quenchants is likely to increase in 1980 as heat treaters attempt to reduce oil consumption," predicts E. F. Houghton & Co. Reported advantages of polymers over oil include elimination of smoke, fumes, and fire hazard, less pollution, reduced dragout and evaporation, and lower cost.

Copperweld Steel Co. is one user actively investigating polymer quenchants. The steelmaker adds one advantage to the above list: more uniform properties. "Results of tests on alloy steel bars in diameters of 1 to 6 in. (25 to 150 mm) show that the properties of polymer quenched material are indentical to, and often more uniform than, those of oil quenched bars."

Copperweld also cites a disadvantage: the close

A 13 in. (330 mm) in diameter alloy steel preform in one station of a four-station induction heater. The reciprocating system is also used for reheating superalloy ingots. *(IPE Cheston)*

control of quenchant concentration and temperature needed to assure proper heat treating response.

Inorganics— Heat treaters looking for a quenchant in between water and oil may opt for the water soluble inorganics, says Henry E. Sanson & Sons Inc. "These quenchants give a relatively fast quench with no rancidity problem. They can be used on parts from both induction heating and salt bath operations."

Quench Oils— Sanson reports that use of vacuum quenching is steadily increasing, creating a demand for better low-vapor-pressure oils. The company also predicts that use of special marquenching oils will increase. Furnace efficiency will benefit because these oils can be used over a wide range of temperatures.

The safety problem with quench oils (mentioned earlier) may be reduced with a "new fail safe system for detection of water in quench oil" introduced by Ipsen Industries. The Aquaguard reportedly prohibits quenching of a hot steel charge into oil when it detects a water content exceeding 0.1%, or 30% of that specified by NFPA as the level at which an uncontrollable steam expansion poses a distinct hazard upon quenching.

Tips From Users— A truck builder notes a trend to hot oil quenching after carburizing or carbonitriding. "Hot oil quenching improves fatigue life and uniformity compared with press quenching. Also, distortion is minimized because of the less drastic quench."

Hughes Tool Co. says more attention should be paid to agitation during quenching to better utilize lower alloy steels. "Many facilities offset poor equipment by specifying medium to high alloy steels when a properly quenched low alloy would be satisfactory."

Automakers Keen On Induction Heat Treating

Fiat Auto SPA states that applications for induction hardening are likely to increase at the expense of carburizing and carbonitriding. The reasons: "We can save energy and substitute carbon for alloy steels."

PSA Peugeot-Citroen agrees, but thinks that electron beam and laser hardening should also be considered when switches from atmosphere processes are contemplated.

In the U.S., Tocco Div., Park-Ohio Industries Inc. notes an increasing demand for induction treated parts for diesel engines and new scaled-down gasoline engines. Tocco also points to potential new applications for induction hardening: camshafts made of pearlitic gray or nodular iron or carbon steel (0.50 to 0.60% C); and selectively hardened low carbon steel structural members for automobiles. The latter move would be made for weight reduction reasons.

IPE Cheston cites a trend to adopt simultaneous multifrequency induction heating to attain precise hardness patterns. An example: a hollow flanged drive shaft would be given a thin case ID hardening using 9600 Hz and a deep case pattern on the OD using 3000 Hz.

IPE credits solid state power supplies with making the technique economical.

Heating for Forming— Induction heating of steel billets for forging has a number of advantages over use of fuel-fired furnaces, states Hughes Tool Co. These include reduced scale, no heat wastage when processing is delayed, and lower over-all operating cost. Hughes predicts that this cost benefit will be even greater in the future as oil and gas prices escalate.

New Equipment— Ipsen Industries reports the development of an induction heated rotary retort furnace. "Because the heat is generated within the retort," explains Ipsen, "you can use a heavier wall for longer retort life."

Other benefits: "Because induction can heat the incoming work to process temperature in only a fraction of the retort length needed in conventional designs, productivity is enhanced. Quality is also improved because of induction's inherently more defined control of zone temperatures."

Trefimetaux, a member of the Pechiney Ugine Kuhlmann Group, has developed induction heating technology for annealing wire at speeds ranging from 800 to 2500 m/min (2600 to 8200 ft/min). The wire may be ferrous or nonferrous metal, bare or clad, and of essentially any cross section.

The machine incorporates two unique features.

1. The wire's heated by passing over insulating, deeply scored pulleys and through a torus-shaped inductor supplied with medium frequency power. Efficiency is reportedly 85%.

2. The process is monitored and controlled by an "annealmeter," a device that regulates the amount of power supplied to the inductor based on a nondestructive assessment of wire properties.

Cost Factors Make Salt Baths Attractive

"Capital investment, energy, and versatility advantages will favor the substitution of salt bath heat treating for competitive processes," predicts Henry E. Sanson & Sons.

Kolene Corp. sees the growth being largest in the automotive industry, where U.S. autos are being scaled down and their components are being designed "... to the loading and performance parameters of their European and Japanese counterparts." In these countries, salt bath heat treating, and liquid nitriding in particular, has made significant inroads.

Degussa, a West German manufacturer, points out that liquid nitriding is so versatile that it can make the performance grade in some applications that carburizing can't, and be a lower cost alternative in some applications where carburizing constitutes an overspecification.

A new salt bath nitriding process incorporating salt bath cooling opens up new application areas, says

Degussa. Examples: replacement of hard chromium plating by nitriding, and high temperature nitriding where processing times plummet from 90-120 min to 20-30 min.

Ion Nitriding Gains Commercial Acceptance

Ion nitriding (or plasma or glow discharge nitriding) has already been specified as a substitute for chromium plating. Peugeot-Citroen now ion nitrides AISI 1050 shock absorber rods at a rate of 1500 every 3.5 h. The rods, which measure 12 mm (0.5 in.) in diameter and 300 mm (12 in.) long, are nitrided to a case depth of 10 to 15 μm (400 to 600 micro-in.) and a hardness of HV 1000 in an automatically contolled, batch-type elevator furnace.

The French automaker reports that energy consumption is only 110 Wh/part, compared with 300 Wh/part for chromium plating — an annual saving of 100 MWh based on an output of 3000 parts per day.

Other candidates include tappets and rocker arms.

Peugeot-Citroen concludes by citing current R&D on glow discharge carburizing and carbonitriding, "... areas that show great promise."

Similar efforts are underway in the United States. Advanced Vacuum Systems reports increasing interest in both commercial and laboratory and pilot plant size glow discharge furnaces. "The smaller units have resistance and glow discharge heating capability, and come complete with gas and oil quench systems. They will be capable of ion nitriding, carburizing, and boronizing, as well as vacuum carburizing and straight vacuum quench hardening."

Advantages— Ion nitriding is seen as an attractive alternative to gas and salt bath nitriding for several reasons. AVS mentions its higher speed of case development, particularly in early process stages; and the ease of control of case depth and white layer.

Lindberg Corp. adds, "Treatment temperatures as low as 600 F (315 C) virtually eliminate distortion and alteration of core properties. The process is also nonpolluting."

The Exotic and the Unusual: A Potpourri

Electron beam heat treating, states Sciaky Bros. Inc., has been solving problems that other processes simply can't tackle. Examples are "... difficult to reach surfaces on small, high production parts."

EB systems boast selective hardening capability, excellent energy efficiency, the ability to double as an EB welder, and easy to program computer controls. "This flexibility," says Sciaky, "means that low part production requirements are no longer a deterrent to machine cost justification."

The Laser— A forecaster-researcher predicts that laser beam surface alloying, hardening, and glazing applications will increase thanks to the introduction of "ruggedized" laser systems.

Fiat agrees, citing a trend to replace nodular iron parts with laser surface alloyed, gray iron parts.

And the Indian Institute of Metals reports that laser melt quenching, a process similar to splat cooling, has industrial potential because of its ability to produce corrosion and fatigue resistant, metastable surface layers. "Results of research at Bombay's Bhabha Atomic Research Centre show that the metastable phase can be produced to micrometre depths."

The technique has been used, for example, to produce stress corrosion resistant, martensitic surfaces on titanium alloys. "Any residual stresses arising from incomplete coverage must, however, be removed," says IIM.

Ion Implantation— IIM also sees more applications for ion implantation now that competitively priced equipment is available. "This is not a coating process — there's no interface which could form a plane of weakness, and there's no dimensional change. Implanted nitrogen or carbon will harden the surface and enhance wear resistance."

Boron— "The development of paste boriding compounds means that the process can now be done on automated, continuous lines," states Degussa. The automotive industry in Europe is taking advantage of this development in applications where very hard surfaces are required.

Hughes Tool Co. has developed a boronizing plus carburizing process for bearings that eliminates the need for cobalt base alloys. Success of the process is based on two "discoveries":

1. A carburized and boronized smooth journal surface performs better than a smooth, hard metal alloy pad.

2. A carburized steel surface can be compound boronized in a manner similar to compound carburizing.

Rejuvenation— "Parts which have been subjected to creep, thermal fatigue, or mechanical fatigue of the high stress-low cycle type at elevated temperatures should be amenable to rejuvenation via heat treatment provided that cracking has not started," states Israel Aircraft Industries Ltd.

Damage must be in the form of concentrations of atomic defects. They would be dissipated by a heat treatment tailored to the specific part, alloy, prior heat treatment, and maximum permitted service temperature.

The company reports that some used parts, such as turbine blades and discs, have had their design lifespan increased four-fold by rejuvenation heat treatment. ⊕

NEW TRENDS IN HEAT TREATING

by

R. M. Pillai, B. C. Pai and K. G. Satyanarayana*

ABSTRACT

A closer quality control in the properties of engineering components demands exacting of the specifications for the heat treatment operations. Development in the microelectronics not only gave closer control in operation but helped in revolutionizing the existing heat treating processes. In addition this has helped in the development of newer heat treating processes. This paper briefly outlines the role of microelectronics in quality control of heat treating processes and reviews the recent trends in the heat treatment process including developments in energy saving by modifying the conventionall heat treatment atmospheres and conditions of salt baths, use of new quenchents, newer heat treatment processes like fluidized bed, electron beam treatment, ion implantation, and vacuum and induction heat treatment processes.

Introduction

Effective heat treatment of metals and alloys is an essential phase in the production of dependable components. The steady rising cost of energy coupled with more exacting requirementc for quality control are to be borne in mind by heat treaters. More exacting metallurgical specifications of components call for greater precision in heat treatment operations. Besides necessary equipment to perform the heat treating function, it is equally important to use correct temperatures and holding time so that the resulting properties of the components are consistent with satisfactory performance. Modern heat treating processes require closer control of atmosphere, temperature level and uniformity while complying with the need for reducing operating costs and acceptable environmental conditions for the operator. In view of the changing pattern in heat treatment shops, it is intended in this paper to bring out the recent trends in heat treatment to the attention of Indian heat treating metallurgical industries.

Microelectronics Revolution

Being the hottest subject in today's heat treatment, micro-electronics in general and integrated circuit microprocessor in particular has made a considerable impact. Heat treaters are beginning to realize potential of this emerging technololgy in electronics. Microprocessor oriented control in heat treating processes is able to reduce costs and improve reliability and quality. The new microprocessor based controllers replacing discret analog temperature controllers can perform the functions with electronic speed, high reliability and push button ease. The digital display capability presents set point and process variables readings in a format which can hardly be misinterpreted and can be presented with readabilities to one or two decimal places. Another advantage of digital based instruments is their relatively small size which saves panel space and cuts installation cost. The microprocessor based integrated system concept(1) it a significant step forward in the control of batch process heat treating operations. Hence, heat treating equipment with microprocessor control system will receive greater attention in the future.

A trend towards bidirectional digital communication between instrumentation and a computer is reported and the computer supervises and monitors process parameters. Economics of the Computerization is the main reason why computerized control systems have not been applied to heat treating furnaces until recently. Today, as the rising energy cost curve is nearing a crossover of the dropping computer cost curve it is now economically feasible to consider the advantages of digital computers over conventional instrumentation in the control of furnace process parameters. Computerization in heat treatment results in better temperature control, better records of lot history of each load for future reference, increased production and energy savings.(1-3). For example. it has been reported(2) that accuracy and reproductibility of surface carbon content and case depth are obtained with computerized carbon potential control of carburizing atmospheres.

* The authors are with Materials Division. Regional Research Laboratory (CSIR), Trivandrum 695 019.

More and more heat treating plants are incorporating sophisticated handling systems in their furnace installations for handling both inside and outside furnaces Mechanical handling systems and robots will play a greater role in the future. Devices for part location and identification utilizing cameras and possibly lasers will be more widely adopted.

Energy Saving Developments

The steady rising cost of energy, coupled with more exacting requirements for quality control are parallel trend in the heat treating industry. As energy will never again be cheap, recuperation will become an integral part of the modernization of the industrial fuel burning heat treating furnaces. Producer gas is replacing fuel gas and natural gas. Better burner design and waste heat recovery can lead to reduced gas consumption (58 to 75%) (2). A family of ceramic recuperators for energy conservation in high temperature furnaces have been developed and are reported to save 30-60% fuel, able to handle 1370°C exhaust gas temperatures without dilution and have high corrosion resistance.(2)

A round furnace technology that combines conventional pusher tray furnace features with advanced thermal and energy conservation ideas has been developed and has less surface area than a conventional rectangular design(4). Ceramic fiber is being used as insulation in place of brick or castables in furnace hearths and as replacement of heavy refractory piers resulting in superior furnace performance(2). New solid state power suppliers are reported to eliminate the need for output transformers and offers conversion efficiencies of 97%(4).

Heat contained in the work pieces immediately after stamping of selected hot forged parts is utilised to obtain the required structure and properties through tight controller cooling(4). A new time temperature proportional controller(2) is said to produce staight line temperature control of less than ± 0.5°C Further, it makes the control of carbon concentration easier and reduces the temperature of the wall of the radiant tube heating element itself and stabilizes it to the same degree.

Polymer Quenchant

Polymer based quenchents are getting recognised by heat treaters. Improved control measures for monitoring polyglycol quenchants provide the heat treaters with assurance that day to day operations will not be upset by nonuniform results. A new polymer quenchant(2) is said to have quenching curve close to that of oil and is applicable to crack-prone steels, that could only be quenched in oil upto this time. Quenching of large castings of 4130, 4140 and 4340 alloy steels after austenitizing has given good results. As more heat treaters using polymer quenchant have come to realise cooling curves as the preferred method to measure/assess initial quenchant performance and changes that may be due to contaminants, it is essential to standardize cooling curve data for polymer quenchants(2,4).

Atmospheres

After the first fuel crisis of the early seventies and the natural gas crunch in 1976-77, there was an increase in the production cost of endothermic gas but a much slower rise in the cost of nitrogen. As a consequence, major industrial gas companies quickly recognised the opportunity to use nitrogen based systems as a substitute to generated atmospheres leading to the rapid development of technology for protective atmosphere applications. The best known earlier technique of carburizing(5) uses a mixture of nitrogen-hydrocarbon-oxidant which when passed into the hot zone of a furnace generates a mixture of N_2-CO-H_2. In this technique the carburizing rate is slower and CO level is also too low for a carburizing rate and profile quality than endothermic gas. Endothermic furnace atmosphere for carburizing and neutral hardening can be produced by introducing propane gas and air directly into the heatreating furnace at reduced rates. A constant air flow is used while the propane flow is automatically regulated. Either methane (natural gas) or butane gas may also be used instead of propane. Satisfactory atmosphere can be produced at 800°C and at 900°C and above with butane and methane respectively(4). About 75-90% of the hydrocarbon normally consumed in producing furnace atmosphere can be saved(4). This is because no gas generator is required and suitable flow rates are typically one third to one half of typical flow rates.

Development work carried out to produce nitrogen based gas carburizing mixtures with higher CO levels closer to those endothermic gas has resulted in a CO forming constituent i.e. methanol, which is inexpensive, readily available and easy and safe to handle.

Hence, it is predicted(2) that nitrogen based atmosphere will continue to replace endothermic atmospheres in the hardening and carburizing of steel. Methanol-nitrogen blends duplicating the composition of endothermic gas will lead the way(2,5). The Endomix process(5) using methanol-nitrogen combination employs gaseous nitrogen, derived from a pure liquid source, in conjunction with liquid methanol. Advantages and features of the process include

(1) The atmosphere produced from nitrogen and methanol has flammability limits similar to those of endothermic gas, but greater safety is inherent in the process by virtue of an automatic safety purge.

(2) Uniformity and reproducibility of carbon potential and case depth are found to be atleast as good as those obtained with endothermic gas on full production loads.

(3) Carburizing rate has been shown to be atleast equal to that with propane or natural gas enriched endothermic gas.

(4) In conjunction with infrared and oxygen problem systems, this offers state of the art flexibility and carbon control.

(5) Increased productivity because of the associated greater carburizing rates and shorter cycle times.

(6) Apart from carburizing applications, this process has been successfully utilized for high temperature neutral hardening of AISI M2 tool steel, high temperature annealing and carbonitriding.

However, it is forecast that the cost of methanol is to skyrocket in the next few years and it might not even be readily available to American heat treaters by 1985(2)

A nitrogen based process for annealing stainless steel which keeps hydrogen levels in the nitrogen down to 20-30% for most parts has been developed(4). Hydrocgn can be a major problem. Reducing hydrogen levels can lead to substantial cost savings over both dissociated ammonia and higher concentration H_2-N_2 atmospheres. The patented process uses low level of oxident additions which preferentially absorb on the surface of stainless steels and inhibit nitrogen from absorbing into the surface layer. If chromium nitride precipitates, there is a loss in corrosion protection.

Salt Bath Treatment

A new carburizing process(2) developed by Degussa AG of West Germany is based on a regenerable carburizing bath operated with noncyanide chemicals, a microprocessor controlled salt bath furnace that can handle several programs simultaneously and a salt quench to improve dimensional stability. This process which is expected to create a lot of interest among heat treaeers has been given impetus by the development of an electronic apparatus continuously measuring and controlling the water content in the melt and guaranteeing regulation of quenching rate within a wide range. Thus, the relatively great influence of a water content between 0 and 1% can be used to adjust bath cooling characteristics. Another develolpment work(2) on boriding salt baths is in progress and it is expected to be an economical process against the existing costly ones finding only a modest position among other processes.

Liquid nitrided surface continues to be preferred alternative to high alloy materials and protective plating for wear and corrosion resistant applications. A major US automobile company(2) will bring liquid nitriding on stream for treating carbon steel ball studs that will replace existing stainless steel studs and for replacing induction hardening and Teflon coating of carbon steel shafts. Attempts are also being made to find other applications in which chromium plating would be replaced.

A new salt bath nitriding process, called Melonite which is virtually free from cyanide has been developed. Wear resistance and fatigue properties of ferrous metals treated by this process compare favourably to those achieved with the standard cyanide rich salt bath process (Tuffriding process). Although Tuffriding process was used so far for many automotive and engine components that require high levels of wear and fatigue protection viz., diesel engine cylinder liners, crank shafts, rocker arms, connecting rods and engine valves, the benefits of the process came at a price. The high cyanide content (48%) of the salt bath generated many Kg per day of hazardous waste material. Hence, environmental pressures and Government regulations are responsible for the development of Melonite process and forced industries to switch over to a cyanide free process. This new process uses an oxidizing salt bath of 34-35% cyanate. < 3% cyanide. balance carbonate against a reducing salt bath of 48% cyanate, 40% cyanide used in Tuffriding process.

Salt baths are replacing lead in the patenting of steel rod and wire. Advantages are said to include more consistent metallurgical properties, lower cost per given volume (i.e. 10% of that of lead) and lower operating temperatures (100°F) (4)

Methyl chloride(4) has long been preferred as a means of maintaining salt baths in a neutral condition for austenitizing low to high alloy steels. Unlike some other rectifiers, methyl cholride does not form sludge. However, as a gas, it has a drawback, it must be handled in pressurized containers and introduced slowly to the molten salt bath. The pellet rectifiers (ammonium chloride) are easily stored and handled and can be accurately measured for use.

It is also predicted(4) that more users of nitrite-nitrate salts will install equipment for salt recovery and recycling.

Fluidized Bed Heat Treating

Fluidized bed heat treating (2,7) continues to gain acceptance because of its major advantages such as

1. All types of ferrous and non-ferrous alloys may be treated.

2. High heat transfer rates 120-1200 W/m(2)°C (20-210 Btu/ft(2) hF) can be achieved and thus enabling products to be treated or cooled at speeds very close to those obtained in conventional salt or lead bath equipment.

3. The atmosphere in the heating zone can be immediately adjusted to suit the requirements of the treatment.

4. There is no problem of disposal of fume or effluent

5. The units have high thermal efficiency and low fuel consumption and thus result in low operating costs under service conditions.

6. As the furnace heating up time is relatively short, the bed can be shut down overnight without loosing production time the following day.

7. Fludized solids are non abrasive, noncorrosive and do not wet immersed components while treatment.

8. Expensive gas need not be consumed while there is no work in the bed.

9. High temperature does not seem to have deleterious effects on furnace components or the metallurgical properties of the parts.

10. There are no risks of explosion when loading parts carrying surface oil or moisture as the contaminant simply vaporizes and is removed with the waste gas as in conventional atmosphere furnace.

This process is gaining acceptance where batch loading are the norm, versatality of equipment is needed and rapid changeover is desirable. The relatively high rate of atmosphere circulation required for proper fluidization would render most large atmosphere controlled systems economically unfeasible if recirculation and recuperation were not employed. It is also reported(2.7) that fluidized bed furnace with refined nitrogen based atmosphere formulations, especially with methanol can perform all of the conventional case hardening treatments of nitriding, metrocarburizing, carbonitriding, neutral hardening, tempering steam blueing and stress relieving. Other applications are continuous wire annealing, treatment of tool steels, marquenching of hot work tool steels and heat processing of extrusion dies. Units measuring 1.5m, across and equally as deep are capable of handling many of the larger jobs hitherto reserved for moetan salt(4).

Table I(7) gives the comparison of furnace outputs during carburizing and hardening operations performed by fluidized bed, vacuum furnace, salt bath furnace and protective gas atmosphere furnace. As seen from this table, fludized beds heat treating has advantages and disadvantages as opposed to its other operations counterparts. The relative heating rates of 16 mm steel bar in a salt, lead, fluidized bed and conventional furnaces are shown in Fig. 1A(7) whereas Fig 1B gives the cooling rates for air, oil, water and a fluidized bed(7). The good heating and cooling performance/ characteristics of the fluidized bed furnace is attributed to the turbulent motion and rapid circulation rate of the particles as well as the extremely high solid gas interfacial area.

The temperature uniformity within a fluidized bed has been reported to be within $+2$°C in the active part of 3m(3) volume(7). With careful design and the use of low cost carrier gases as nitrogen, even low temperature surface treatments can be effectively and economically performed. Deep case methanol carbu-

rizing and carbonitriding carried out in fluidized bed furnaces are said to provide outstanding case depth and form low temperature nitriding and nitrocarburizing to have parts with high corrosion resistance and a cosmetic surface appearance viz., cutting tools, gears, guns parts, dies, stamping and autmative components. Fluidized system for annealing copper strip in neutral atmosphere about 9 m long and rated at 1.8 tonnes/hr at a maximum temperature of 540°C is in operation and it is an example of large atmosphere controlled system(2). Clean and efficient ferrous wire processing, high carbon wire bed. The natural gas or propane fired system consists of sections for austenitizing, quenching and transformation which operate at temperature of 900, 315 and 510°C respectively. Wire is heated and cooled by direct immersion in the fluidized particles, which provide heat similar to those achieved in the conventional lead transfer rates, tensile strengths and microstructures patenting system(8).

Electron Beam Treating

Electron Beam (EB) Surface modification which includes transformation hardening, glazing and surface injection is getting a lot of attention. In transformation hardening, a steel or alloy is heated to a temperature high enough to cause a crystallographic change usually followed by an accelerated cool. In glazing, a selected area on a surface is fused. Injection of hard particles holds a promise as a means of producing an abrasion resistant surface. Electron beam heat treating machines accommodate computer control and robotic systems and are environmentally clean and energy efficient (utilizing about 75% of the off-the-wall power in actual working energy(2).

High energy EB surface hardening is being applied to aircraft turbine mainshaft bearing rings. Conventionally hardened rings are reported to have failed suddenly in tests due to through section fracture. Whereas the problem is solved with EB hardened part in which the rolling contact surface is hardened to a case depth of 0.76 mm(2). Titanium carbide can be injected onto the surface of an A1 alloy at a rate of 645 cm(2)/min to a depth of 2.5 mm in an out-of-vacuum machine(4).

Ion Implantation

Although ion implantation is probably five to ten years away for commercial exploitation/acceptance, it provides a means of engineering the near — surface properties of engineering materials in ways never before achievable by any other technique. Its potentials extend to a number of characteristics related surface properties, such as friction, wear, corrosion resistance, hardening, binding, lubrication and adhesion.

In ion implantation, elemental ions are injected into the surface of a material by accelerating an ion beam of a selected species. The process must be carried out

Table – I Comparison of Furnace outputs During carburizing and Hardening Operations, Kg/h (lb/h)

	Fluidized Bed	Vacuum	Salt Bath	Atmosphere
Hardening	150 (330)	144 (320)	70 (155)	137 (300)
Carburizing to care depths:				
0.25 mm (0.01 in)	100 (220)	105 (230)	50 (110)	100 (220)
0.5 mm (0.02 in)	75 (165)	80 (180)	38 (85)	75 (165)
0.75 mm (0.03 in)	50 (110)[1]	60 (130)[2]	24 (50)[3]	55 (120)[4]

1. With 50 Kg (110 lb) load and 10. Kg (22 lb) basket, effective care depth requires 1 h total time in bed

2. With maximum load 200 Kg (440 lb) and 20 Kg (44 lb) basket, requires 3 h cycle (1.25 h recovery + 1.75 h treatment)

3. With net load of 95 Kg (210 lb) in 115 kg (250 lb) charge, requires 4 h total time in salt

4. With maximum load 350 kg (770 lb) and 75 Kg (165 lb) baskets, requires 5 h total time in furnace (1.75 h recovery + 3.25 h treatment)

in a chamber under low pressures. Only the near surface properties of the material being treated are affected. Further, the surface layer is integral with the substrate and there is no coating.

Westinghouse Electric Corporation(2,4) has implanted a number of metalworking and forming tools with nitrogen ions with a dosage of 2 x 10/(17 ions/cm(2) and observed two to twelve fold increase in service life. For example, a 1cr-1C tool steel paper slitter and a WC-6 Co synthetic rubber slitter have experienced 2 and 12 fold increase respectively. In the case of corrosion resistance, the location composition is controlling. It is also reported(9) that the chemical composition of N + C + and B + with Fe deep penetration tion of N, S and B + with Fe deep penetration and the hardening of near-surface layer increase the fatigue strength of steel 30 Kh GSNA more than that obtained by implantation of He + N + and Mn + and maximum increase in fatigue strength and fatigue life at high stress (750-850 MPa) was obtained after implantation by N 1 + and C + respectively.

Thus, for precision parts, altering the surface composition through ion implantation without changing bulk properties or dimensions is a decided advantage.

Vacuum Heat Processing

Hot vacuum furnace(2) are now available for the first time whose interior can be exposed to air, allowing direct loading as in conventional, non-vacuum equipment, saving floor-to-floor time. All the advantages of vacuum remain, including the ability to carburize and carbonitride. The modular concept allows matching to pressure quench and/or oil quench modules, as well as automatic, continuous processing.

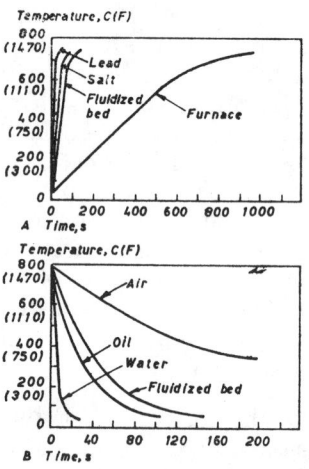

Fig. 1. Relative heating rates (A) and quenching rates (B) are shown for different types of furnaces and different types of quench media. Steel bars with a diameter of 16 mm (0.6 in) were treated.

Vacuum furnaces with high pressure (upto 5 bar (500 KPa)) gas quenching are increasingly replacing salt baths equipment for hardening high speed cutting tools in tool manufacturing plants(2,4). Quenching speed equals or exceeds those in salt bath quenching. High pressure gas quenching in single and/or multiple chamber vacuum furnaces is of interest to auto industry and aircraft industry, particularly for Titanium alloy castings which are normally quenched in water or brine to obtain cooling rates needed for proper aging response. Use of this gas quench can eliminate surface contamination and the need for subsequent cleaning normally encountered in liquid bath quenching, while developing the necessary metallurgical structure.

Induction Heat Treatment

Industry is demanding more system designs to insure quality control. To be cost effective, the systems must certainly recognize the inherent variables of the processing system and the variation in incoming material. Increasing material costs require a system that must handle or be tolerant of lower cost materials which usually have wider variations in specification. The processing technology must have an interreactive mode to automatically accommodate material variation. It is reported(2) that induction heating is being able to meet this requirement because it produces one part at a time and with the development of the proper smart sensors along with interractive non-destructive testing it could accommodate lower cost materials and maintain quality. Recent developments in solid state technology have permitted the use of smaller, compact units and, in many instances, completely, self-contained units with all load matching components supplied in one cabinet. Hogher power ratings permit the reduction of space requirements by a factor of 10 to 1. Induction heat treatment is being applied in-line and teamed with robots for automation and thus slashing processing time. There is also a continuing move toward selective surface hardening as an alternative to thorough hardening or total surface hardening. Use of high intensity induction and high frequency line hardening methods continue to offer a cost effective alternative to EB or laser. Quenching, straightening and machining are eliminated and the processes are extremely fast (< 0.5 sec is sypical). New applications (2,4) include garden shear blade, lawn mower blades and foot plates, constant velocity universal joints, hammer claws, razor knife blades, chisel blades, an automotive crankshaft, large piston ring grooves, transmission components and drill bits.

The process has been extended to selective surface alloying or inlaying of metallic particles via high frequency melting eg. selective alloying of turbine blades. A fine grained structure of less than ASTM 12 in hardened areas of martensitic type steels is said to be obtained with an induction hardening process patented by a German firm(4). The process is based on application of high peak capacitor discharge (millisecond pulses of upto 30 KW, 3MVA at 27.12 MHZ) of rectangular shaped pulses to a small inductive loop. The strength of the magnetic field is so high the steel exceeds the cure temperature of about 730°C. There is no smoke or gaseous emissions and a cooling mediuum for the part is not required.

Other Miscellaneous Developments

Plasma carburizing units with integral oil quenches and capabilities for working volumes 0.6 m wide by 0.9 m long by 0.6 m high will be instroduced in USA(2). This unit will be capable of confering enhanced fatigue and wear properties on parts and providing fast processing times. Software will provide desired surface hardness and case depth for complex shapes such as gears. Instant on/off and neglibible distortion are the other reported features.

The Soviet have developed what is termed explosive strengthening(4) at the pilot plant level. Claims indicate such applications as railroad frogs, excavator buck teeth and wear-prone parts on ore and stone crushers. Uniform strengthening to depths of upto 50 mm is reported.

Conclusion

Shortage of materials and energy, strict environmental regulations, quality products than quantity products and competition are the constant pressures/constraints to which today's metallurgical industries are subjected to. Keeping this and the developments that have taken place elsewhere in the heat treating area, it is hightime for our heat treaters to make use of some of these modern techniques to have efficient utilisation of their equipment, materials, energy and personnel.

References

1. J. R. Swanson, Metal Progress, 120 Nov 1981, P 40-43

2. Metal Progress, 125, No. 1, Jan. 1984, P 31-32 34, 36, 39, 40

3. T. K. Thomas & R. I Grubber, Metal Progress, 120, No 6, Nov 1981, P 32-38

4. Metal Progress, 123 No 1, Jan 1983, P 40, 44, 46; 48; 53, 55

5. Brian J. Sheehy, ibid, 120, Sept 1981, P 120-127

6. Lennard G. Kruger, Metal Progress, 120, Sept 1981, P 45-49

7. Metal Progress, 120, No. 4, Sept 1981, P 132

8. Metal Progress, 120, Sept, 1981, P 138

9. B. G. Vladmirov et al, Poverkhnost 1982 (7), P139-47 OR Chemical Abstracts 98, No 10, 1983 164847

How to Design for Lower Cost, Higher Quality HEAT TREATMENT

By ROY F. KERN

Don't overlook the opportunity to cut heat treating costs through refinements in part design. The objective is to use the most economical steels with minimum loss from distortion and cracking. Good design also promotes metallurgical quality — the key factor in the performance of a part.

A PART THAT IS DIFFICULT to heat treat is not only expensive; it also performs poorly in most instances. This means that emphasis on designing parts for optimum efficiency in heat treatment is almost sure to improve profits through lower direct factory costs and the sales appeal of quality products.

Direct factory costs for scrap and rework on heat treated parts are commonly 2 to 5% of heat treat direct labor. When a part is scrapped because it cracked in heat treating, the loss goes beyond the total investment in it; you must include the profit that would have been made on a good part.

In the early stages of designing a part that is to be heat treated, the engineer should ask himself:

- Has an attempt been made in design to avoid the need for heat treating long bars of small cross section and large thin plates?
- Are section thicknesses as uniform as possible?
- Can the part be made symmetrical about all axes?
- Are holes, grooves, deep splines and keyways held to an absolute minimum?
- Are there suitable locating surfaces and prop-

Mr. Kern is project engineer, Technical Services Div., Caterpillar Tractor Co., East Peoria, Ill.

Reprinted from Metal Progress, October 1967, 143-151, © 1967 American Society for Metals

"When a part is scrapped because it was cracked in heat treating, the loss goes beyond the total investment in it; you must include the profit that would have been made on a good part," Mr. Kern (left) declares. He is conferring with a designer, a practice he recommends as a way to hold down costs of heat treatment.

erly toleranced relationships between planes and diameters where fixture quenching is necessary?

- Have sharp corners been avoided?
- Has provision been made for the removal, prior to heat treatment, of residual stresses due to rolling, forging, casting, welding and straightening?
- Have radical part combinations been avoided?

Is Heat Treating Really Necessary?

Sometimes a part that is particularly troublesome to heat treat may be produced without taking this processing step. Take the 48 in. pinion puller stud in Fig. 1. The part was released as heat treated SAE 4140 steel — the designer stated that it was subject to high tensile loads and axial shock loading. Heat treatment and straightening in the conventional manner were not satisfactory because even a slight waviness in the part caused trouble with the finish and dimensions in machining the 18 in. threads.

Finally, some pieces made of SAE 1144 cold drawn and tempered to 100,000 psi minimum yield point were evaluated with satisfactory results. After a proving ground test confirmed this, thousands were made without a single failure. The change saved 33¢ in material, 95¢ in straightening and 20¢ in heat treating.

Also, parts such as wear plates or strips can often be made from "as-rolled" high carbon or even austenitic manganese steel with a substantial saving over heat treated material.

In one instance, a wear strip was designed into a tractor to prevent a bracket from wearing into the frame as it moved. It was first produced from SAE 8620 carburized and hardened. Field tests revealed that this hardness level was far in excess of what was needed. A change was made to cold finished SAE 1018 with no heat treating. The saving: $2 per part.

The purchase of mill heat treated bars for long shafts is a low cost answer to a difficult heat treating problem. The extra charge for heat treating and machine straightening is only about 4¢ per pound. For best dimensional stability, stress relief after machine straightening is suggested.

Quenching and tempering of large, thin, flat parts should be avoided where possible. One approach is to utilize plate purchased to the required strength level. In the instance of the part shown in Fig. 2, treated plate could not be used because of hardening of burned edges which would seriously impair tool life in milling the entire surface in the welded assembly. Even though straightening costs are more than $2.50 per part, the number required is not large enough to justify the purchase of a quench press.

Uniform Section Thickness

If it is definitely established that heat treating is required, you may find that the one most important factor in design is that sections of the part be as uniform as possible. Ideally all areas should have equal ability to absorb or give up heat. The reason is that a high percentage of distortion problems are due to the impossibility of heating or cooling thin and heavy sections at the same rate.

In the gear design in Fig. 3, the tooth area is much thinner than the hub. Since the heating rate in a furnace on the gear portion was much higher than that of the hub, it tried to "grow" faster, but its attachment to the hub restricted it. The hot metal had to go somewhere, and local hot upsetting took place in the gear. Upon quenching, more metal was hot upset in the attachment zone because the gear cooled so much faster than the hub. As the hub cooled to quenchant temperature, it returned to almost its original size, but the upsetting action that had occurred caused the gear to be smaller. As a result, the gear teeth were not only tapered but also in a high state of residual stress. Lightening the hub, as shown in phantom (Fig. 3), solved the taper problem and reduced scrap from 28% to less than 1%.

Sometimes it is desirable to add stock to a part to make a more uniform section for heat treating. The

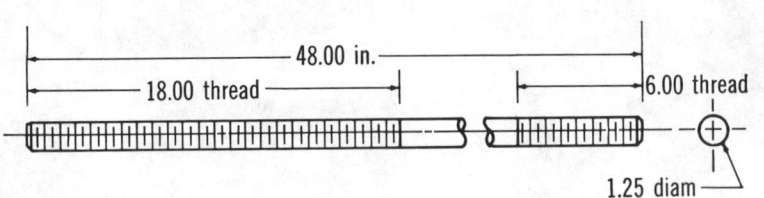

Fig. 1—This part was redesigned from SAE 4140 to cold drawn SAE 1144 to avoid an expensive straightening operation. Total saving was $1.48 per piece.

Fig. 2—Although it costs $2.50 to straighten each of these large flat parts (30 by 48 by 1 in.), the numbers required are not sufficient to justify the purchase of a quenching press. Mill heat treated plate cannot be used because hardening of burned edges would seriously impair tool life in milling the surface in the welded assembly.

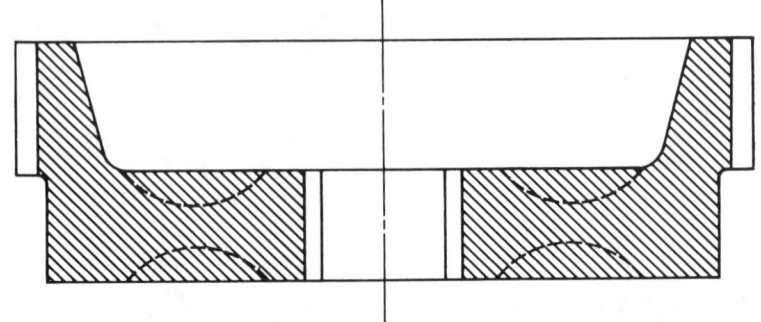

Fig. 3—In this gear design, the tooth area is much thinner than the hub. Lightening the hub, as shown in phantom, solved heat treating problems and reduced scrap from 28% to less than 1%.

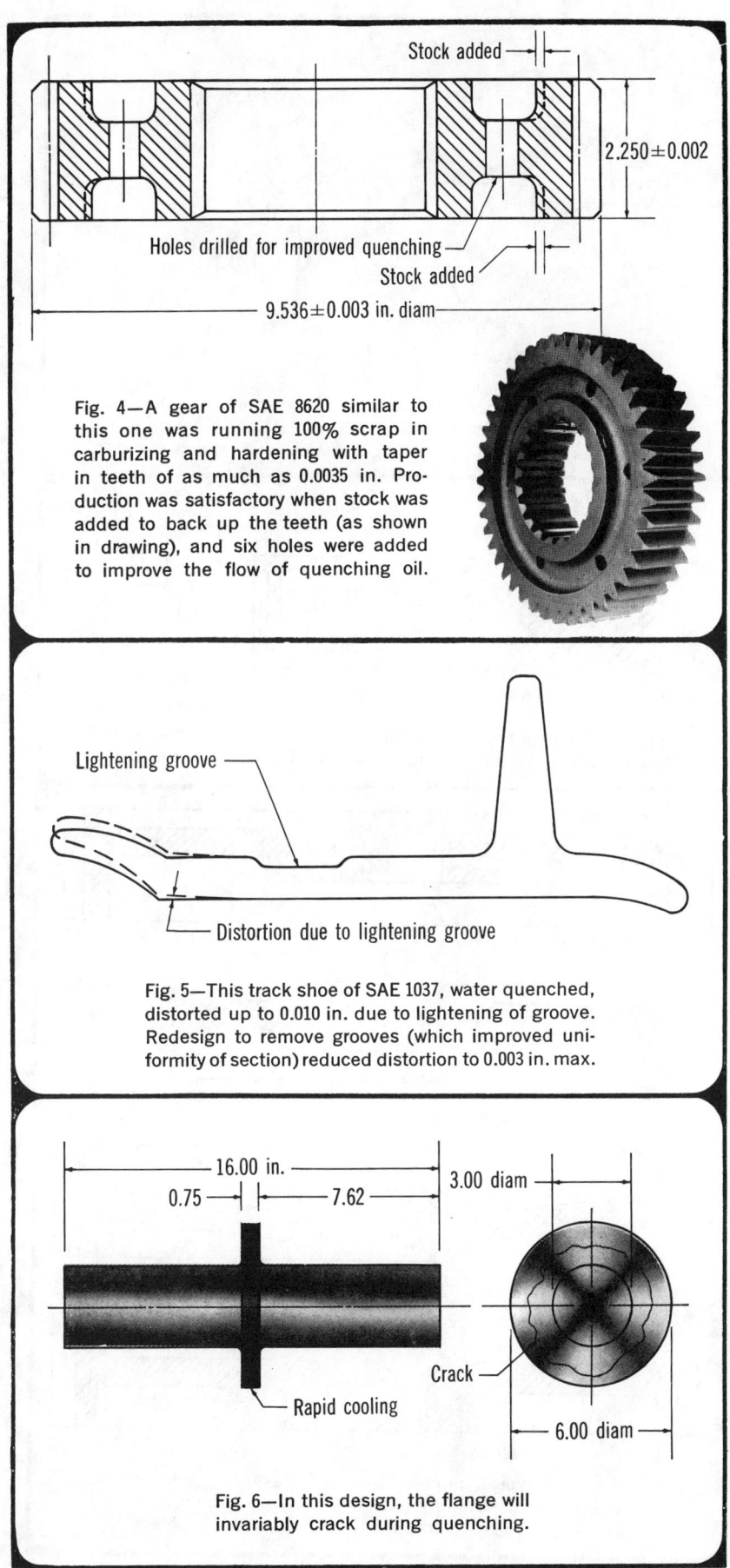

Fig. 4—A gear of SAE 8620 similar to this one was running 100% scrap in carburizing and hardening with taper in teeth of as much as 0.0035 in. Production was satisfactory when stock was added to back up the teeth (as shown in drawing), and six holes were added to improve the flow of quenching oil.

Fig. 5—This track shoe of SAE 1037, water quenched, distorted up to 0.010 in. due to lightening of groove. Redesign to remove grooves (which improved uniformity of section) reduced distortion to 0.003 in. max.

Fig. 6—In this design, the flange will invariably crack during quenching.

gear shown in Fig. 4, made of SAE 8620 steel, carburized and hardened, was running 100% scrap with taper in the teeth of as much as 0.0035 in. By adding stock to back up the teeth (as shown in drawing) and by specifying six holes through the web to provide better flow of quenching oil, the taper was reduced to 0.002 maximum with no rejects.

Another example of where a design was modified for improved uniformity of section to avoid distortion is shown in Fig. 5. The crawler tractor track shoe — made of SAE 1037 steel, water quenched — distorted excessively due to the lightening groove, as shown on the drawing. The groove was removed from the part, and distortion was reduced from 0.010 in. to 0.003 in. max.

A troublesome part is a shaft with a tall, thin flange (Fig. 6). On water quenching, a circumferential rim of the flange chills almost immediately and shrinks, but the flange section adjacent to the shaft remains hot, and is upset by the shrinkage of the rim. Upon further cooling, the shaft and the portion of the flange immediately adjacent to it (which has been upset) cools, and by thermal contraction tries to pull away from the cold, hard rim.

Nearly 100% of shafts with this design will crack in heat treating when a water quench is used, as shown in the drawing. There are several solutions to this problem: make the flange thicker, shield it from the quench, make the shaft portion tubular to even out sections or select a steel that will harden properly in oil.

Sometimes, an unusual problem in uniformity of section crops up in induction hardening of steel. It may be necessary to design for a non-uniform section to even out the heating rate. An example (Fig. 7) is the hardening of large diameter, coarse pitch sprockets. Another induction hardening problem is illustrated by the part in Fig. 8. The thin section around the drilled area severely overheated and cracked. The problem was solved by drilling and tapping the part after induction hardening.

Dimensional changes in heat treating occur because the microstructure of the material is changed. This causes shrinkage or growth. Thus, symmetry of design means symmetry of growth or shrinkage which usually can be allowed for in machining before heat treatment.

Symmetry also affects distortion due to thermal cycling. A classic example is illustrated in Fig. 9. After being heat treated, the gear comes out in tapered or warped condition. The reason: the bottom of the gear cools faster than the top. As it does, it locally upsets hot steel in the web-to-rim radius. When the gear reaches room temperature, this area will be short and will pull a taper in the gear as shown. The modified designs improve symmetry of the part and eliminate warping.

Problems: Holes, Keyways, Grooves

Probably the most damaging thing a designer can do to a part as far as heat treating is concerned is to put a hole, a deep keyway or groove into it. The reason: these configurations are hard to heat uniformly, and they are very difficult to quench.

The problem on heating is illustrated in Fig. 10. The rapidly heated corner of the hole tries to expand faster than the metal around it. Since it is restrained, it cannot grow, so it upsets itself. Cooling to room temperature leaves the metal in a high state of tension, and it may crack even without heating. This difficulty arises in furnace heating, but the problem is not as severe as it is in induction heating. Chamfering or putting a radius around the top of the hole will tend to reduce its heating rate and also distribute shrinkage stress after cooling.

Even if a part is heated uniformly, quenching of a hole, keyway or groove is difficult. The problem is to get quenchant into the hole to prevent a vapor pocket from forming. When a steam (or oil) vapor pocket forms, the sides and the bottom of the hole cool at a slower rate than the surrounding metal. This means the metal in the hole is upset by shrinkage of the surrounding metal and is short when it cools. This results in high tension stresses, possible cracking.

It is difficult to quench even a

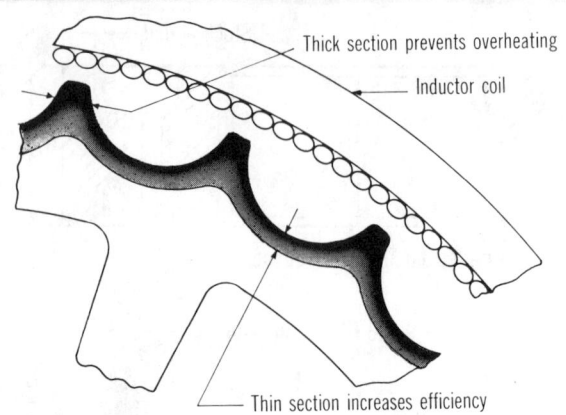

Fig. 7—In induction hardening, a design may have to provide a nonuniform section to even out the heating rate, as illustrated in this large, coarse pitch sprocket.

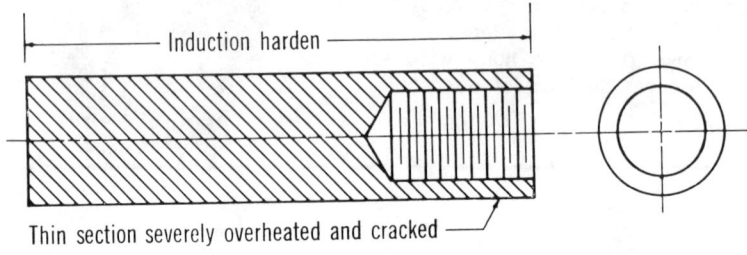

Fig. 8—The problem of cracking in this part during induction hardening was solved by changing to drill and tap after hardening.

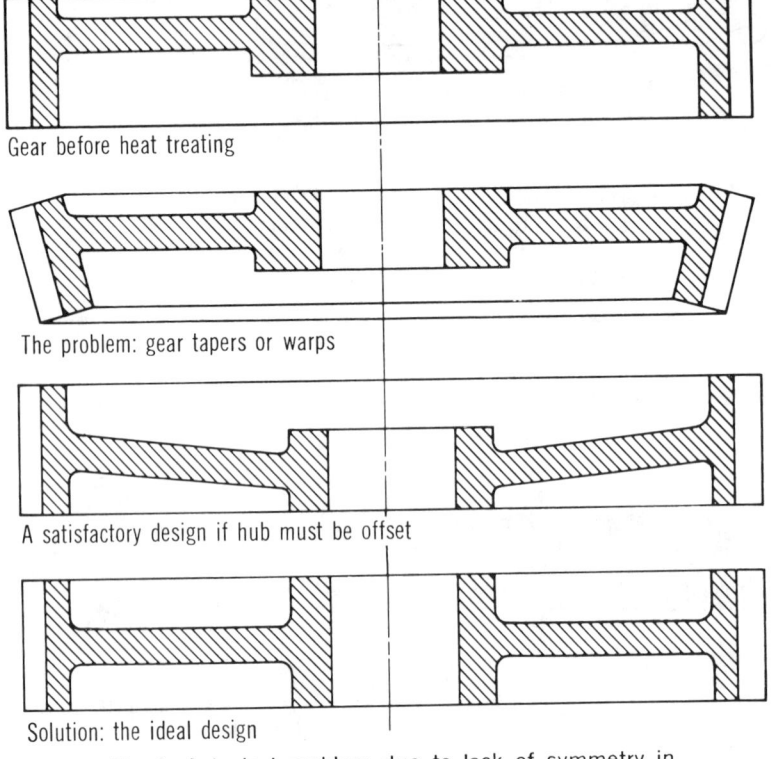

Gear before heat treating

The problem: gear tapers or warps

A satisfactory design if hub must be offset

Solution: the ideal design

Fig. 9—A typical problem due to lack of symmetry in design is illustrated by this gear. It warped during heat treating. Design modifications will solve the problem.

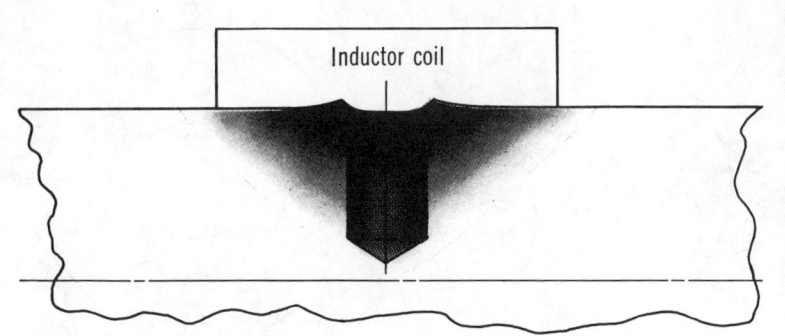

Fig. 10—Uniform heating over a hole is difficult to obtain even with furnace heating. Induction heating is especially troublesome. Metal around hole heats faster than surrounding metal, thus upsetting itself.

Fig. 11—Quenching inside holes is difficult due to formation of large steam pockets resulting in soft spots. Bushings, such as that shown above, usually require an inserted spray or a violent flush-type quench from one end.

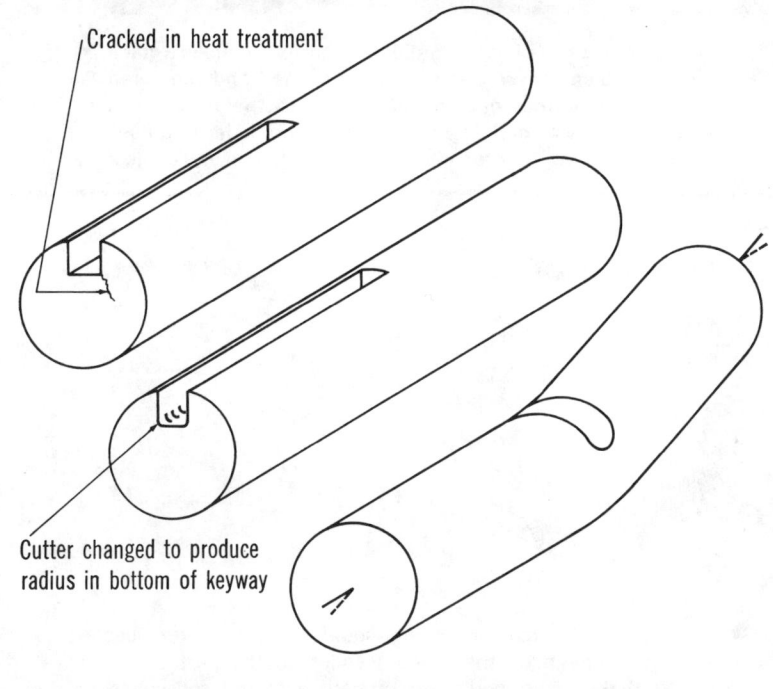

Cracked in heat treatment

Cutter changed to produce radius in bottom of keyway

Fig. 12—Grooves will cause a shaft to warp in heat treating. A keyway with sharp corners often initiates cracks in quenching. The problem is avoided with a radius, as shown.

large hole (Fig. 11). An 8 in. long bushing of SAE 1018 with a hole diameter of 1.75 in. and an outside diameter of 2.75 in. was carburized to a depth of about 0.080 in. and brine quenched. The hardness specified was Rockwell C 60 min on all surfaces, including the bore. After a few pieces were run, it was found that the case hardness was low in the bore — some readings were down to Rockwell C 45. It was suspected that as quenchant tried to enter the bore, it formed steam which kept additional quenchant from entering even with violent agitation.

A test was run: the bushing was welded shut on the ends after carburizing. It was then quenched, and measurements revealed that hardness in the bore was the same as that when the ends were open. This confirmed our belief. Steam was keeping out nearly all quenchant, and the only hardening was being done by the quenchant on the outside. The solution to the soft bore was to quench the bushing with an inserted spray mechanism.

Distortion Becomes a Problem

A groove will usually end up by distorting a shaft in the manner illustrated in Fig. 12. The most dangerous thing about grooves, keyways and holes is that their geometry opens the way for stress concentration. Heat treatment may add a residual tensile stress. It's no wonder that parts often fail in these locations.

The solution is to try to avoid grooves and keyways in highly stressed areas. When they are used, they should be as wide and shallow as possible. A keyway should be produced with a radius rather than a sharp corner (Fig. 12). One possible solution is to use fixtures which make it possible for the hole or the inside of a groove to be quenched first or more drastically than the rest of the part.

Holes create problems, particularly in the webs of gears (Fig. 13). They usually represent an attempt by a designer to reduce weight, but he often ends up with a flat spot for each hole. A good rule to follow: keep the diameter of the hole one-

Source: Metal Progress, October 1967, 143-151

third that of the web width.

Cracks may form around holes during quenching. This was the difficulty with the part containing three ⅝ in. tapped holes shown in Fig. 14. Cracks came from the outside surface into the holes. A new design eliminated the ⅝ in. holes and provisions were made to hold the washer on the end with a bolt in the single large hole.

Provision for Fixture Quenching

A close examination of hand straightening costs in a heat treating department will reveal that this operation is very expensive. Unfortunately, not much can be done about it, at least using time and motion study, because the process is largely an art. Straightening is not only expensive, but it also damages the properties of the part.

On the basis of factory costs alone, the expense of straightening usually justifies the purchase of a quenching press or roller die machine. Such machines are particularly useful because they remove distortion from a part that occurred in heating, and with proper tooling, they can provide uniform and reproducible quenching.

Fixture quenching, however, requires that parts be held accurately in the green (before heat treating) so conformance to the quench tooling is uniform from piece to piece. For press work, these tolerances must ordinarily be no more than ±0.001 in. This includes dimensions on steps between planes, or on flat areas within the same plane.

The part shown in Fig. 15 could not be machined to sufficiently close tolerances in the green to permit use of a plug for accurate fixture quenching. It was completely redesigned as shown and its function was modified somewhat. The heat treating problem was solved.

Parts run in roller dies should ordinarily have at least three diameters that are held to 0.001 in. in the green and concentric within 0.002 in. Such tolerances on a part to be heat treated are difficult for machining people to understand. However, minimized distortion ordinarily more than justifies the cost. Redesign of a part to permit more accurate ma-

Fig. 13—If the designer specifies holes in the web of a gear to reduce weight, heat treatment may produce a flat spot for each hole, as shown on left. A good rule to follow: keep the hole diameter one-third of the web width as illustrated.

Fig. 14—Design of part of SAE 10B40 (bottom) caused a cracking problem when it was water quenched and tempered. Cracks ran from outside diameter into the three ⅝ in. tapped holes. Solution: eliminate the ⅝ in. holes and use a bolt in the large center hole to fasten washer on end of piece.

Fig. 15—The machine shop could not hold tolerances close enough on the cutout section of the part on left so that a plug could be properly inserted for subsequent quenching. The result was a complete redesign of this part (right), its function being somewhat modified.

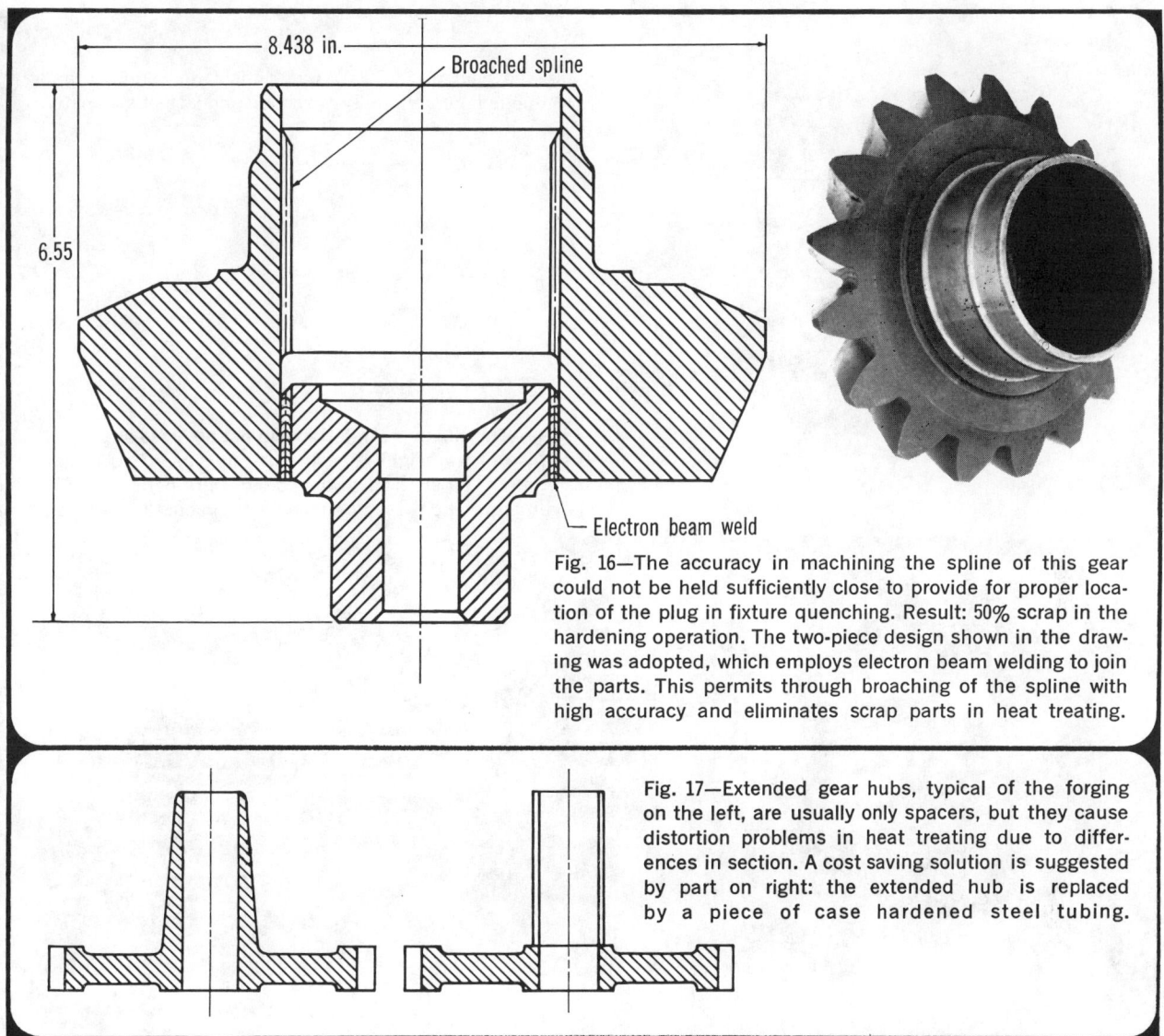

Fig. 16—The accuracy in machining the spline of this gear could not be held sufficiently close to provide for proper location of the plug in fixture quenching. Result: 50% scrap in the hardening operation. The two-piece design shown in the drawing was adopted, which employs electron beam welding to join the parts. This permits through broaching of the spline with high accuracy and eliminates scrap parts in heat treating.

Fig. 17—Extended gear hubs, typical of the forging on the left, are usually only spacers, but they cause distortion problems in heat treating due to differences in section. A cost saving solution is suggested by part on right: the extended hub is replaced by a piece of case hardened steel tubing.

chining to accommodate fixture quenching may be needed (Fig. 16).

Finally, the designer should recognize that, generally speaking, a part cannot be held as close dimensionally in heat treatment as it can, for example, on a hobbing machine. A good rule to follow: where no machining can be done after heat treating, the tolerance on the machining operation should be held to one-third of the tolerance required after heat treating. The two-thirds is available for heat treat distortion.

Avoid Sharp Corners

Sharp corners should be avoided in parts to be heat treated because of the possibility of cracking and spalling. As a minimum, a chamfer — but preferably a radius on a corner — reduces the quenching rate in this area. This diminishes the tendency to upset more hot metal, thereby reducing the tendency to crack. Also, the finish on the face should be maintained as smooth as possible to inhibit crack formation.

While this type of problem is most often met in water quenching, it occasionally turns up in oil quenching grades such as SAE 4145, 4150, 4340.

Residual Stresses Give Trouble

In precision heat treating, the relief of stresses from casting, forging, machining, forming or welding may cause distortion. This problem is particularly troublesome in parts such as spiral bevel or hypoid gears where the tooth configuration tends to distort, resulting in excessively long lapping times.

An anneal or stress relief from a temperature equal to the maximum used in heat treating when the part is as close to being finish machined as possible is the surest cure for the problem. A closely controlled cool from the forging temperature or annealing or normalizing of the rough forging from above the heat treating temperature is good practice.

Too Much in One Piece

Designers often try to get too much of a mechanism into one piece.

Source: Metal Progress, October 1967, 143-151

Fig. 18—The gear of SAE 8617 was difficult to carburize and harden due to poor uniformity of section and lack of symmetry.

Fig. 19—The cluster gear (left) could not be press quenched because of its height. Result: the large gear had excessive tooth taper. The two-piece design permitted the larger gear to be press quenched with practically no taper.

A typical problem is illustrated by the long extended hub on the gear in Fig. 17. Due to the draft required in forging such a part, the hub has a considerable variation in section. In heat treating it, a barrel-shaped bore develops, and there is little that can be done about it. The extended hub is usually just a spacer which could be replaced by a piece of case hardened steel tubing as shown in the suggested modification.

Cluster gears make up another typical class of parts where designers often go too far in trying to get too much into one piece. The result is that the gears are almost impossible to harden without excessive distortion. Likewise, the radical part combination in the gear in Fig. 18 gave heat treating trouble.

The gear and pinion shown in Fig. 19 could not be properly heat treated. The integral part could not be quenched in a press due to its height. Distortion on the big gear was a major problem. With the two-piece design, dimensions of both the gear and pinion could be held accurately in heat treating.

Most engineers who design parts for heat treating — even those with the best of intentions — run into problems. Remember: the defective part is trying to tell a story, and we should try to understand it.

Practically all heat treating problems involve four simple but important laws of metal physics:

1. Steel expands when it is heated and contracts when it cools.

2. Steel will deform permanently when stresses upon it exceed its yield point. This can happen in compression as well as tension.

3. Steel will deform elastically within its elastic limit proportional to the imposed stress, and will break under loads exceeding its strength.

4. Steel occupies a greater volume after hardening than it does before.

There is a real opportunity for cost cutting through design of parts for better heat treating. It often carries with it improved serviceability of products, which means customer satisfaction.

Ferrous Materials and Their Processing

Physical Metallurgy and the Heat Treatment of Steel

By George Krauss, Colorado School of Mines

THE HEAT TREATMENT of steel is based on the physical metallurgical principles which relate processing, properties and structure. In heat treatment, the processing is most often entirely thermal and modifies only structure. Thermomechanical treatments, which modify component shape and structure, and thermochemical treatments, which modify surface chemistry and structure, are also important processing approaches which fall into the domain of heat treatment. Scientific principles link the processing parameters to structure and properties, and are increasingly necessary for proper application of the equipment and instrumentation now available for control of heat treatment processes. Examples of scientific efforts which directly support the technology of heat treatment include characterization of mechanisms of phase transformations which produce desired structures and properties of heat treated parts; determination of phase transformation and annealing kinetics which establish processing times, temperatures and cooling rates for heat treatments; and evaluation of mechanisms of deformation and fracture of the structures produced by heat treatment.

In view of the importance of structure and its formation to heat treatment, the purpose of this article is to describe the various microstructures which form in steels, the various factors which determine the formation of microstructures during heat treatment processing of steel, and some of the characteristic properties of each of the microstructures. Structure-sensitive properties such as strength, ductility and toughness establish the ease of manufacturing, service performance, and limitations to service conditions of heat treated steels.

The descriptions of the microstructures and principles presented here should be considered only introductory, and the references listed at the end of this article should be consulted for more information. The details of the various heat treatments for many grades of steel are presented in the subsequent articles of this section.

THE IRON-CARBON PHASE DIAGRAM

The microstructures which result from heat treatment of steel are composed of one or more phases in which the atoms of iron, carbon and other elements in steel are associated. Figure 1 shows a portion of the iron-carbon phase diagram from pure iron through the carbon concentration of cementite, 6.67 wt %. The temperature and composition ranges in which the various phases exist are shown on the diagram. Alloys containing up to 2 wt % carbon are classified as steels; alloys containing more than 2 wt % carbon are classified as cast irons. The solid lines represent conditions where carbon, when it exceeds its solubility in ferrite and austenite, is present in the form of cementite (Fig. 1). This is invariably the case in steels. The dashed lines represent the conditions where carbon is present as graphite rather than as cementite, a situation much more common in cast irons than in steels.

In steels, the temperatures which are the boundaries of the various phase fields are frequently referred to as critical temperatures. Since the critical temperatures are often identified by changes in slope or thermal *arrests* in heating and cooling curves, they are given the designation "A." If equilibrium conditions are applicable, the designations Ae_1, Ae_3 and Ae_{cm}, or simply A_1, A_3 and A_{cm}, are used as shown in Fig. 1. If heating conditions (which raise the critical temperatures relative to equilibrium) apply, Ac_1, Ac_2 and Ac_{cm} are used, the subscript "c" being derived from the French word *chauffant*. If cooling conditions (which lower the critical temperature relative to equilibrium) apply, the designations Ar_1, Ar_3 and Ar_{cm} are used, the subscript "r" being derived from the French word *refroidissant*. There is hysteresis in the transformation temperatures because continuous heating and cooling leave insufficient time to accomplish the diffusion-controlled phase transformations at the true equilibrium temperatures.

Steels and cast irons contain, in addition to iron and carbon, many other elements which shift the boundaries of the phase fields in the Fe-C diagram. Some alloying elements such as Mn and Ni are austenite stabilizers and extend the temperature range over which austenite is stable. Elements such as Cr and Mo are ferrite stabilizers and restrict the ranges of austenite stability. Therefore, care must be taken in the direct use of the Fe-C diagram to predict phase relationships in commercial alloys which contain elements in addition to Fe and C. Nevertheless, the iron-carbon diagram is the most important reference for understanding the relationships between structure and heat treatment of steels, and, subject to the above limitations, will be used in this article to illustrate the basis for microstructural formation in steels as well as Fe-C alloys.

The phase diagram shown in Fig. 1 assumes equilibrium—i.e., that the carbon and iron have had sufficient time to distribute themselves in the various phases as shown. Sometimes, equilibrium is difficult to achieve, especially in steels which contain elements which diffuse only sluggishly, and, in fact, certain heat treatments such as hardening are designed to prevent formation of equilibrium structures. Thus the fact that equilibrium may not be achieved, together with the shift of the phase-field boundaries by alloying elements, place limitations on the direct use of the Fe-C phase diagram.

Austenite, also referred to as γ-iron, is the face-centered cubic crystal form or phase of iron which is stable at high temperatures. Figure 1 shows that carbon in Fe-C alloys is soluble in austenite up to just over 2 wt %, and that the single-phase austenite field dominates the Fe-C diagram at high temperatures. In all low-alloy steels, therefore, it is possible to produce a single-phase austenite microstructure. This characteristic is perhaps the most important feature of steels in that it enables steels to be hot worked or wrought. Also, cooling from the single-phase austenite field makes possible a wide variety of heat treatments based on transformation of the austenite.

The single-phase austenite, without the obstacles which second phases present to dislocation motion and without the sites for fracture initiation which second-phase particles offer, deforms and recrystallizes readily so that substantial reductions in section size by hot rolling or forging may be accomplished. Traditionally, hot deformation is performed in the upper temperature range of the austenite field. Hot deformation of austenite at lower temperatures or even in the two-phase ferrite-austenite field (controlled rolling), and the addition of small amounts of alloying elements (microalloying) such as niobium and vanadium, which precipitate as fine alloy carbonitrides at low temperatures, are new approaches to processing of steels (Ref 2 and 3). The low-temperature deformation and/or precipitation retard or prevent austenite recrystallization and grain growth, and therefore produce finer

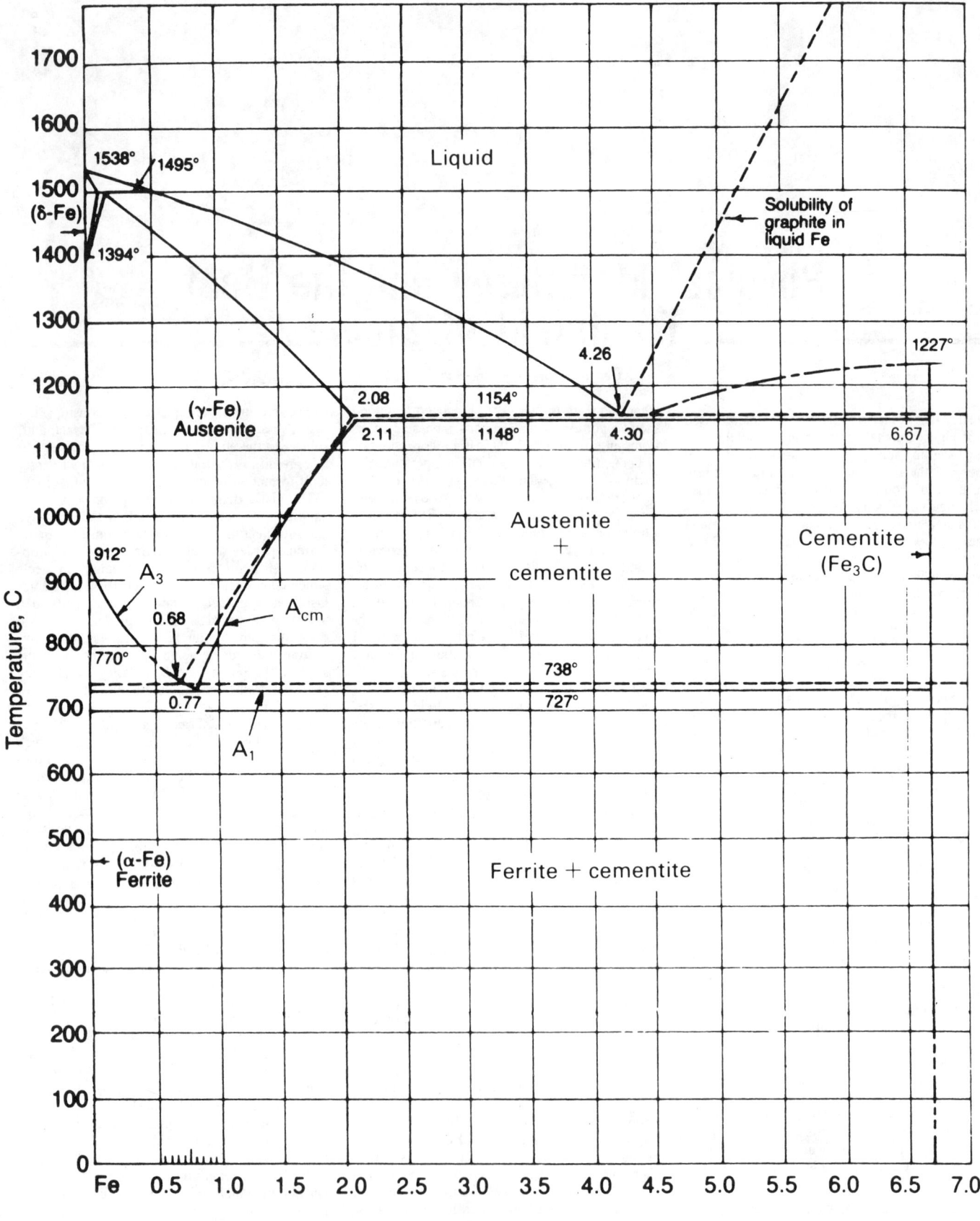

Fig. 1. The Fe-C equilibrium diagram up to 6.67% C. Solid lines indicate Fe-Fe₃C diagram; dashed lines indicate Fe-graphite diagram. (Ref 1)

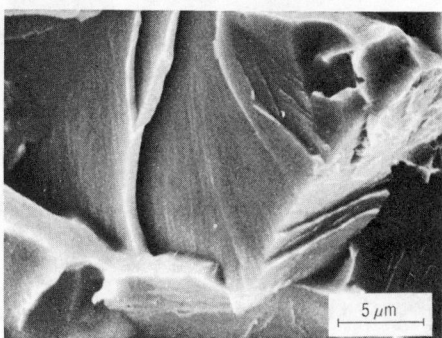

Fig. 2. Scanning electron micrograph showing cleavage of a body-centered cubic microstructure. Courtesy of F. Zia-Ebrahimi.

austenite grains and subsequently fine austenite transformation products during cooling after hot deformation.

Ferrite, also referred to as α-iron, is the body-centered cubic form or phase of iron which is stable at low temperatures. Microstructures in low-carbon steels which consist largely of polycrystalline ferrite are highly formable at room temperature; dislocations move readily on the many slip systems of the body-centered cubic structure (Ref 4). However, at low temperatures, dislocation motion in the body-centered cubic structure is severely restricted (Ref 5 and 6). As a result, ferrite grains fracture in a brittle manner with little plastic deformation at low temperatures. Figure 2 shows an example of the brittle fracture which develops in ferrite stressed at low temperatures and/or high strain rates. The fracture is termed cleavage because it occurs by cleaving or separation across {100} planes of the body-centered cubic structure. Thus cleavage is a direct reflection of the crystal structure of ferrite and presents a major limitation to the use of steels under certain service conditions.

Carbon, because of its small atomic size, is dissolved in the octahedral interstitial sites between iron atoms in ferrite and austenite (Ref 7). When the solubility of carbon in either austenite or ferrite is exceeded, the phase cementite, also referred to as Fe_3C or θ-carbide, forms. The compound cementite has higher strength and lower ductility than either ferrite or austenite and, depending on its morphology and distribution, contributes in a variety of ways to the strengthening, deformation and fracture of steels.

The interstitial sites for carbon in ferrite are much smaller than those in austenite, and therefore the solubility of carbon in ferrite is significantly lower than in austenite. Figure 3 shows an expanded portion of the iron-rich side of the Fe-C diagram. The maximum solubility of carbon in ferrite is only about 0.02 wt % and with decreasing temperature becomes almost negligible. As a result of the decreasing solid solubility with decreasing temperature, on slow cooling, cementite forms on ferrite grain boundaries. If, for some reason, cooling is too rapid for cementite formation, the carbon is trapped in the interstitial sites and contributes to various aging phenomena unique to ferrite steels (Ref 5 and 8). The one process is associated with segregation of carbon atoms to dislocations and grain boundaries, and is referred to as strain aging. The other process is associated with precipitation of fine carbide particles either on dislocations or in the ferrite

matrix and is referred to as quench aging. Figure 4 shows an example of fine dendritic cementite particles which have formed by quench aging on dislocations in the ferrite of a low-carbon steel. Both strain aging and quench aging effectively pin dislocations, and are responsible for the discontinuous yielding of low-carbon steels with largely ferritic microstructures.

PEARLITE AND BAINITE

Figure 1 shows that the austenite in an iron-carbon alloy containing 0.77 wt % carbon must transform to ferrite and cementite at 727 °C. A solid-state reaction in which one phase transforms to two other phases is referred to as a eutectoid reaction. In Fe-C alloys and steels, a unique parallel array of ferrite and cementite lamellae termed pearlite develops as a result of the eutectoid reaction. Figure 5 shows pearlite which has formed in a eutectoid steel; here, the cementite appears white, and the ferrite gray.

Pearlite in a eutectoid steel is nucleated at austenite grain boundaries, and grows as spherical-shaped colonies or nodules into the austenite. Carbon must diffuse to the growing cementite lamellae of the pearlite. Also, iron atoms must rearrange themselves by short-range diffusion from the face-centered cubic structure of austenite to their arrangements in the crystal structures of ferrite and cementite at the interface of the growing pearlite colonies. The rate of transport of carbon and iron atoms is temperature dependent and in-

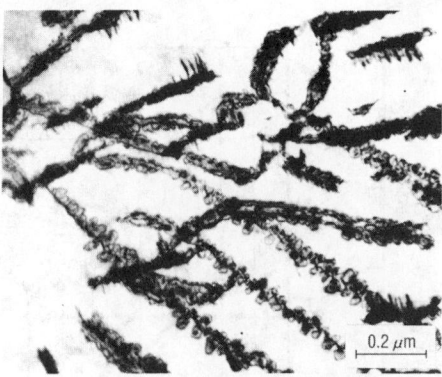

Fig. 4. Transmission electron micrograph showing cementite precipitated on dislocations in an 0.08C-0.63Mn steel aged 115 h at 97 °C. Courtesy of J. E. Indacochea (Ref 9).

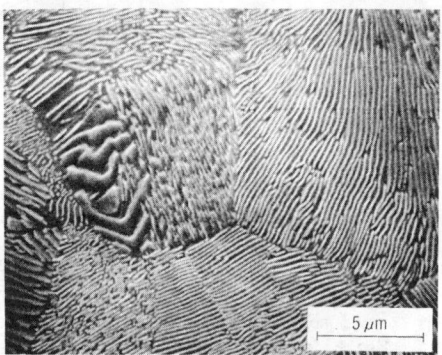

Fig. 5. Scanning electron micrograph showing pearlite in a eutectoid rail steel. Courtesy of F. Zia-Ebrahimi.

creases exponentially with increasing temperature.

At temperatures just below the eutectoid temperature, 727 °C in the Fe-C system, the thermodynamic driving force for the eutectoid reaction (the decrease in free energy per unit volume when austenite is replaced by pearlite) available to offset the increase in energy associated with pearlite colony–austenite interfaces and the ferrite-cementite interfaces within the pearlite colonies is low. As a result, the nucleation rate of colonies is low and the spacing of cementite lamellae within the colonies is large. The coarse interlamellar spacing increases the diffusion distance for carbon, and causes a low rate of growth for those colonies which manage to nucleate. Thus, pearlite transformation at temperatures close to the eutectoid temperatures is sluggish and the pearlite microstructure which forms is relatively coarse. With increased undercooling, the thermodynamic driving force increases, the nucleation rate of pearlite colonies increases, interlamellar spacings decrease, and the growth rate of colonies increases. As a result of the latter changes, the transformation of austenite to pearlite accelerates with decreasing temperature.

Figure 6 shows an isothermal transformation diagram for a eutectoid steel. The diagram shows the beginning and end of the eutectoid transformation of austenite to pearlite for specimens cooled from the single-phase austenite field and held isothermally at temperatures between A_1 and

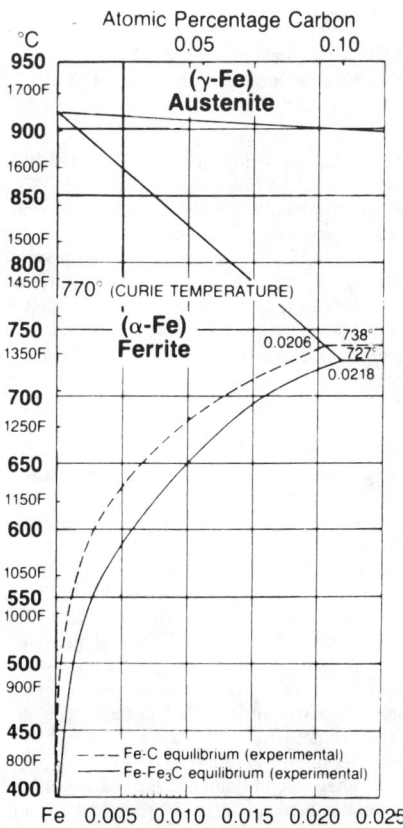

Fig. 3. Fe-rich side of Fe-C diagram, showing extent of ferrite phase field and decrease of carbon solubility with decreasing temperature (Ref 1)

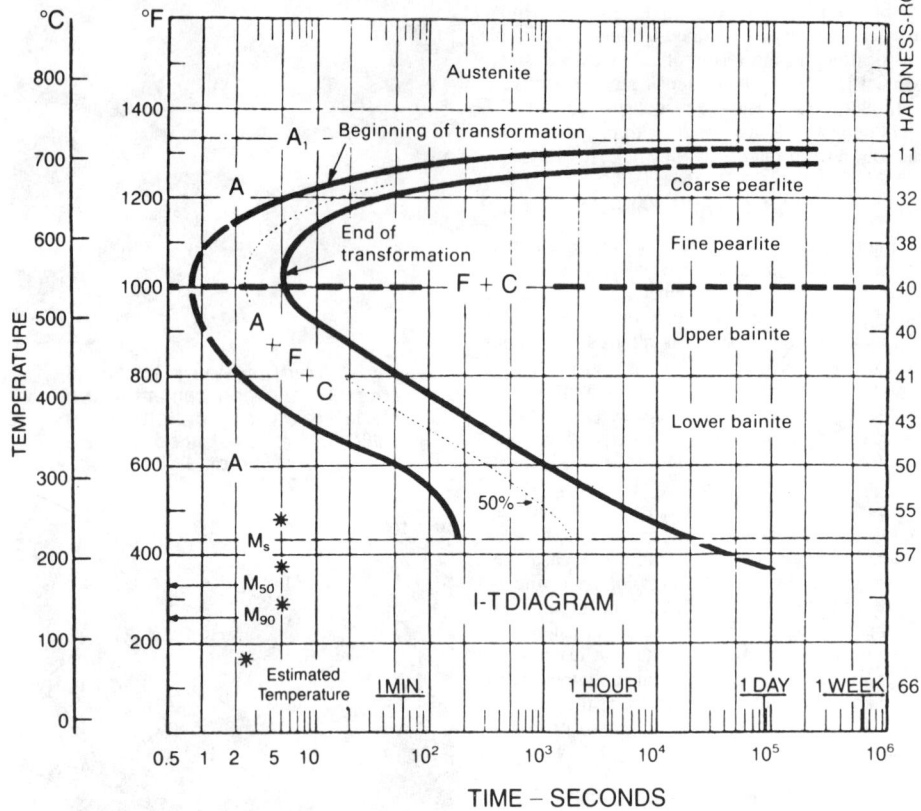

Specimens were austenitized at 900 °C and had an austenite grain size of ASTM No. 6.

Fig. 6. Isothermal transformation diagram for 1080 steel containing 0.79% C and 0.76% Mn (Ref 10)

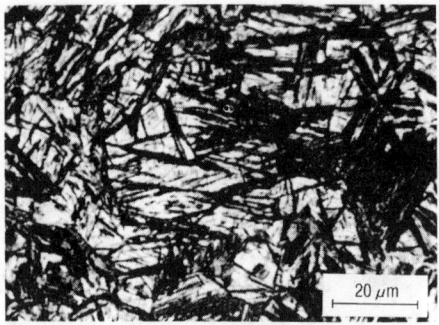

Fig. 8. Light micrograph (nital etch) showing lower bainite (dark plates) formed in 4150 steel. Courtesy of F. A. Jacobs (Ref 12).

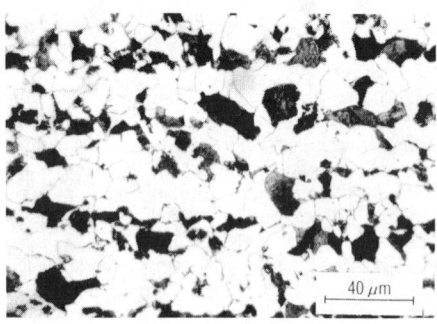

Fig. 9. Light micrograph (nital etch) showing microstructure of proeutectoid ferrite (white) and pearlite (dark) in a 0.17C-1.20Mn-0.19Si steel

Etched in saturated picric acid plus a wetting agent to reveal austenite grain boundaries, then etched in nital to reveal bainite patches.

Fig. 7. Light micrograph showing patches of upper bainite formed in 4150 steel partially transformed at 460 °C. Courtesy of F. A. Jacobs (Ref 12).

540 °C. The acceleration of the transformation with decreasing temperature is apparent.

At temperatures below 540 °C, the diffusion of iron atoms is reduced to the extent that they can no longer be readily transferred even the very short distance across the pearlite/austenite interface. Therefore, the mechanism for the change in crystal structure from austenite to ferrite changes from diffusion to shear. Instead of an atom-by-atom transfer across an interface, large numbers of iron atoms shear or move cooperatively to form lath- or plate-shaped crystals of ferrite. Carbon

diffusion and cementite formation must still occur because of the low solubility of carbon in the body-centered cubic ferrite, but the cementite forms as separate particles rather than as continuous lamellae as in pearlite. The microstructure produced by both shear and diffusion is termed bainite, after Edgar C. Bain, who did much pioneering work in the characterization of austenite transformation and hardenability of steels (Ref 11).

Two forms of bainite develop in steels. One is termed upper bainite because it forms at relatively high temperatures, just below the range of pearlite formation. Upper bainite forms in patches containing many parallel laths of ferrite. Carbon is rejected from the ferrite and concentrates to form relatively coarse cementite particles between the ferrite laths. Figure 7 shows patches of upper bainite formed by partial transformation of the austenite at 460 °C. The austenite which did not transform at 460 °C formed martensite (light background phase) on quenching to room temperature. The general morphology of upper bainite is shown in Fig. 7, but the ferrite laths and cementite particles are too fine to be resolvable in the light micrograph.

The other type of bainite is termed lower bainite because it forms at lower temperatures than does upper bainite. The ferrite takes a plate morphology and the cementite is present as very fine particles within the ferrite plates. Figure 8 shows lower bainite which has formed in a 4150 steel. The bainite plates are at angles with respect to each other, giving an acicular or needle-like appearance to the microstructure rather than the

blocky or feathery appearance of upper bainite. Again the very fine carbide particles in the bainite plates are not resolvable in the light micrograph.

PROEUTECTOID FERRITE AND CEMENTITE

Figure 1 shows that alloys which contain either less carbon (hypoeutectoid steels) or more carbon (hypereutectoid steels) than the eutectoid composition must first form either ferrite or cementite when slowly cooled from the single-phase austenite field. The ferrite or cementite formed before the eutectoid reaction are termed proeutectoid ferrite or proeutectoid cementite.

Figure 9 shows the microstructure of a low-carbon steel after air cooling from austenite. The white grains are proeutectoid ferrite which have nucleated on and grown from the austenite grain boundaries. As the ferrite grains grew, carbon was rejected into the austenite grain boundaries. Eventually, the carbon concentration was sufficient for pearlite formation, and the balance of the microstructure transformed to pearlite. Most of the pearlite colonies appear uniformly black because the light is scattered by the lamellar structures which are too closely spaced to be resolvable in the light micrograph.

The growth of proeutectoid ferrite is dependent on the rejection of carbon atoms into the austenite and the transfer of iron atoms across the ferrite/austenite interface from the face-centered cubic to the body-centered cubic structure. The latter process is dependent on the degree of coherency or disorder in atom arrangement at the interface. Also under some conditions, substitu-

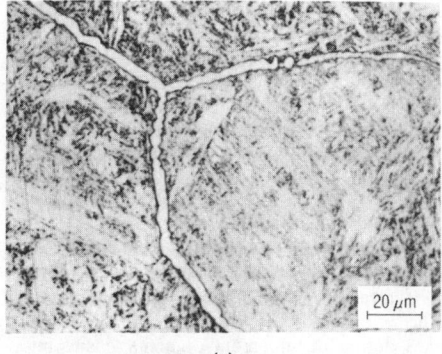

(a)

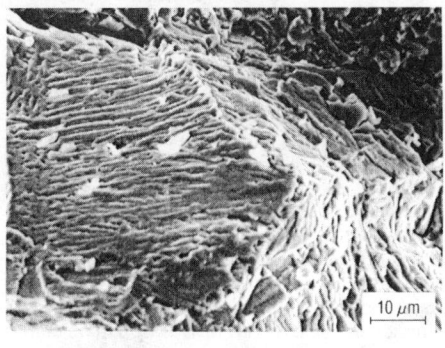

(b)

Fig. 10. (a) Light micrograph (nital etch) showing cementite network on prior austenite grain boundaries in an Fe-1.12C-1.5Cr alloy. (b) Scanning electron micrograph showing cementite interface fracture morphology in same alloy. Courtesy of T. Ando.

tional alloying elements must be incorporated into the ferrite structure if they are ferrite stabilizers or rejected from the ferrite if they are austenite stabilizers. Recent experimental and theoretical work on the effects of alloy element partitioning and interface structure on the formation of proeutectoid ferrite is reviewed in Ref 13.

Generally under conditions of slow cooling, the proeutectoid ferrite grows uniformly into austenite and an equiaxed ferrite grain structure develops as shown in Fig. 9. However, if the austenite in hypereutectoid steels is rapidly cooled, the transfer of iron atoms across ferrite/austenite interfaces is restricted, and the diffusion-controlled growth of ferrite is replaced by a shear mechanism. As a result a plate-shaped morphology of ferrite, frequently referred to as acicular or Widmanstätten ferrite, develops in rapidly cooled low-carbon steels. Substitutional alloying elements such as manganese tend to retard the formation of equiaxed ferrite grains and promote acicular ferrite formation.

In hypereutectoid steels, proeutectoid cementite nucleates and grows on austenite grain boundaries during cooling from the austenite phase field. Figure 10(a) shows a continuous network of proeutectoid cementite which has formed on austenite grain boundaries of an Fe-Cr-C alloy. The balance of the microstructure is martensite which formed on quenching after the cementite network had developed. The interface between

the proeutectoid cementite and martensite is quite brittle. Figure 10(b) shows fracture that has followed proeutectoid cementite interfaces which are characterized by many flat facets with intervening ledges.

Initial proeutectoid cementite growth appears to depend only on diffusion of carbon and therefore proceeds very rapidly. Later stages of cementite growth require partitioning of substitutional alloying elements such as chromium and therefore are very sluggish (Ref 14). The very rapid initial growth of proeutectoid cementite may occur even during oil quenching for hardening and is associated with the intergranular fracture often observed in high-carbon steel quenched from temperatures above A_{cm}. Figure 11 shows an example of 52100 steel quenched from 1000 °C (Ref 15). The intergranular fracture facets are quite smooth, in contrast to Fig. 10(b), and no cementite is visible in this scanning electron micrograph. The presence of small amounts of cementite is, however, established by Auger electron spectroscopy, an analytical technique capable of determining chemical compositions of very thin layers (Ref 16).

In view of the brittleness which continuous networks of proeutectoid cementite impart, hypereutectoid steels are reheated intercritically into the austenite/cementite two-phase field for annealing (if maximum ductility and machinability are desired) or for hardening (if wear and fatigue

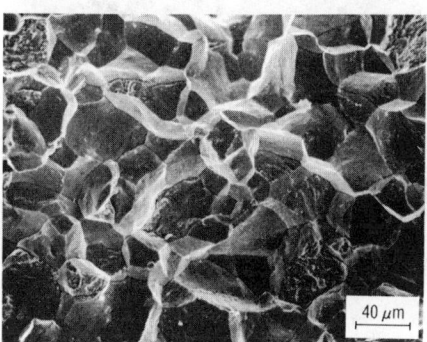

Fig. 11. Scanning electron micrograph showing intergranular fracture in 52100 steel oil quenched from 1000 °C (Ref 15)

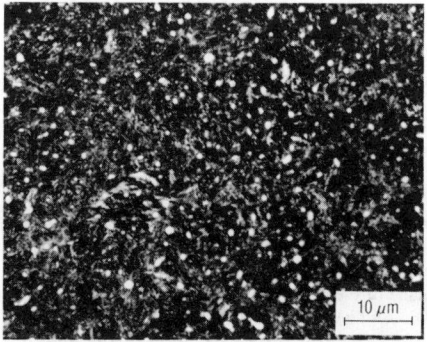

Fig. 12. Light micrograph (nital etch) of a high-carbon bearing steel, showing spheroidized cementite particles in a matrix of martensite produced by intercritical austenitizing and quenching. Courtesy of J. Bruce Kelley.

Fig. 13. Scanning electron micrograph of fracture surface of 52100 steel intercritically austenitized at 800 °C and oil quenched. Arrows point to fine spherical carbide particles. (Ref 15)

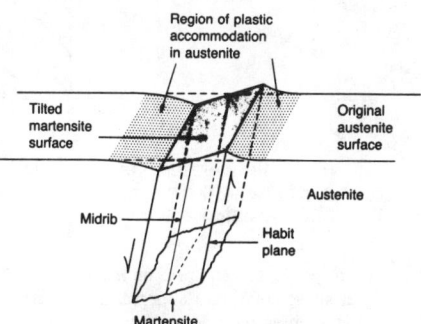

Fig. 14. Schematic illustration of shear and surface tilt associated with formation of a martensite plate. Courtesy of M. D. Geib (adapted from Ref 19).

resistance are required). During the intercritical heating, proeutectoid cementite networks as well as the lamellae of cementite in pearlite partially dissolve and spheroidize. Figure 12 shows the microstructure of an intercritically austenitized and hardened bearing steel. The spheroidized cementite particles (white) are dispersed in a matrix of martensite (dark). When a hardened steel with a microstructure similar to that shown in Fig. 12 is fractured, a transgranular fracture morphology (see Fig. 13) develops rather than intergranular fracture (Fig. 11). The fracture is initiated at the fine spherical carbide particles and the toughness is related to spacing of the particles (Ref 17).

MARTENSITE

Martensite is the phase formed in steels by a diffusionless, shear transformation of austenite, and is the base structure for hardened steels. Martensite is not shown on the Fe-C diagram because it does not form under equilibrium conditions; generally rapid cooling to temperatures well below A_1 is required to form martensite. As expected from the Fe-C diagram, martensite eventually decomposes to a mixture of ferrite and cementite if heated below A_1.

Shear or the displacive, cooperative movement of many atoms has already been mentioned as a mechanism by which bainite and acicular proeutectoid ferrite form. The formation of the latter structures, however, occurs under conditions such that carbon diffusion accompanies the formation of body-centered cubic ferrite. When martensite forms, even the carbon atoms cannot diffuse. Thus

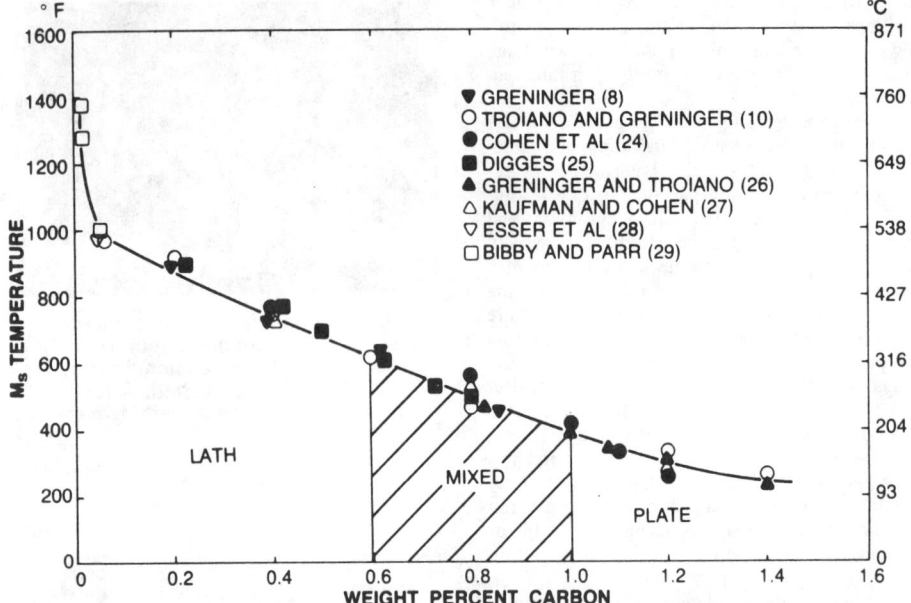

Fig. 15. M_s temperature as a function of carbon content in steels. Composition ranges of lath and plate martensite in Fe-C alloys are also shown. (Ref 2; investigations indicated are identified by their numbers in that reference.)

the carbon atoms are trapped in the octahedral interstitial sites, creating a supersaturated ferrite with a body-centered tetragonal crystal structure. The higher the concentration of carbon atoms, the greater the tetragonality (Ref 18).

Figure 14 shows schematically a martensite plate which has formed in austenite adjacent to a free surface. The martensite surface is tilted by the shear transformation, and the austenite plane along which the martensite forms is termed the habit plane. In order to accomplish the shape deformation shown, not only must the face-centered cubic austenite lattice transform to the body-centered tetragonal lattice of martensite, but the martensite crystal once formed must accommodate itself to the constraints of the surrounding bulk austenite and the restrictions imposed by the plane-strain deformation parallel to the habit plane (Ref 19). This accommodation is accomplished by slip or twinning of the martensite plate, and as a result martensite in steels contains a high residual density of dislocations and/or fine twins.

The martensitic transformation is characterized by athermal kinetics—i.e., the amount of martensite formed is independent of time and is a function only of the amount of undercooling below the M_s temperature, the temperature at which martensite starts to form on cooling in a given steel. The following equation has been developed (Ref 20) for estimating the volume fraction of martensite, f, formed by quenching to any temperature, T_q:

$$f = 1 - \exp - [0.011 (M_s - T_q)]$$

Thus, if the M_s of a given steel is known, the amount of martensite formed on quenching to any temperature below M_s can be established.

The M_s temperature is a function of the carbon and alloying-element content of a steel, and a number of relationships have been developed to relate M_s to composition (Ref 7). Figure 15 shows M_s as a function of carbon content. The decrease in M_s with increasing carbon content is related to the increased shear resistance produced by in-

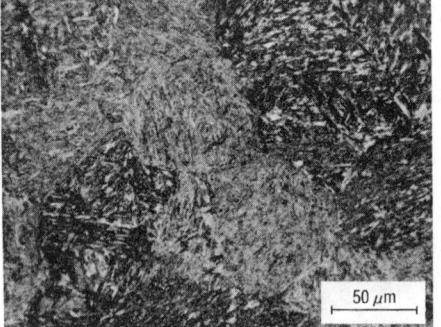

Fig. 16. Light micrograph (nital etch) showing lath martensite in 4340 steel oil quenched from 940 °C and tempered at 350 °C (Ref 23)

creasing amounts of carbon in solid solution in the austenite. An important consequence of low M_s temperature, according to the above equation, is the reduced amount of martensite which forms on cooling to room temperature. Therefore, large volume fractions of austenite may be retained in high-carbon steels.

Figure 15 indicates that two types of martensite form in carbon steels. The two categories are based on morphology and microstructural characteristics of the martensite (Ref 7 and 21). The lath morphology forms in low- and medium-carbon steels, and consists of regions or packets where many fine laths or board-shaped crystals are arranged parallel to one another. The habit plane of the laths is close to but not exactly {111}. The width of most of the laths is less than 0.5 μm—i.e., below the resolution of the light microscope, and therefore the microstructure appears very uniform, with only the largest laths resolvable. Figure 16 demonstrates the above characteristics of lath martensite in a 4340 steel. Electron microscopy is required to show that the fine structure of lath martensite consists of a high density of tangled dislocations and that retained

austenite is present as thin films between the martensite laths (Ref 22).

The plate morphology of martensite forms in high-carbon steels and consists of martensite plates which form at angles with respect to each other on either {225}$_\gamma$ or {259}$_\gamma$ habit planes. Figure 17 shows a plate martensite microstructure in an Fe-1.39C alloy cooled to room temperature. Consistent with the low M_s of this alloy, a large amount of retained austenite is present.

The fine structure of plate martensite consists of thin twins, about 10 nm thick, and/or dislocation arrays typical of low-temperature plastic deformation. The impingement of nonparallel plates during development of a martensite microstructure sometimes causes microcracks to form in the martensite (Ref 25). Examples of microcracks are shown in the large plate of Fig. 17. The density of microcracks in plate martensite is reduced by formation of martensite in fine-grained austenite, by lowering the carbon concentration of the austenite by intercritical austenitizing (thereby developing a more parallel martensite morphology and less impingement), and by tempering.

The carbon range in which a mixed morphology of lath and plate martensite forms is sensitive to alloy content and is not well known. Even in the range of carbon contents where lath martensite forms, there is a gradual decrease in the definition of packets with increasing carbon content (Ref 26).

TEMPERED MARTENSITE

As-quenched martensite is supersaturated with carbon, has a very high interfacial energy per unit volume associated with the fine laths or plates of the martensitic microstructure, contains a high density of dislocations which store considerable strain energy, and may coexist with retained austenite. As a result of these characteristics, martensitic microstructures are quite unstable, and decompose when heated. A practical benefit of the decomposition is increased toughness, and for this reason almost all hardened steels are heated to some temperature below Ac_1, a heat treatment process which is referred to as tempering.

A wide range of microstructures may be produced by tempering of martensite. Carbon atoms rearrange themselves into various configurations

Fig. 17. Light micrograph (aqueous 10% sodium bisulfide etch) showing plate martensite and retained austenite in an Fe-1.39C alloy (Ref 24)

Table 1. Tempering reactions in steel

Temperature range, °C	Reaction and symbol (if designated)	Comments
−40 to 100	Clustering of 2 to 4 carbon atoms on octahedral sites of martensite (A1); segregation of carbon atoms to dislocations and boundaries	Clustering is associated with diffuse spikes around fundamental electron diffraction spots of martensite
20 to 100	Modulated clusters of carbon atoms on (102) martensite planes (A2)	Identified by satellite spots around electron diffraction spots of martensite
60 to 80	Long period ordered phase with ordered carbon atoms (A3)	Identified by superstructure spots in electron diffraction patterns
100 to 200	Precipitation of transition carbide as aligned 2-nm-diam particles (T1)	Recent work identifies carbides as eta (orthorhombic, Fe_2C); earlier studies identified the carbides as epsilon (hexagonal, $Fe_{2.4}C$)
200 to 350	Transformation of retained austenite to ferrite and cementite (T2)	Associated with tempered-martensite embrittlement in low- and medium-carbon steels
250 to 700	Formation of ferrite and cementite; eventual development of well-spheroidized carbides in a matrix of equiaxed ferrite grains (T3)	This stage now appears to be initiated by chi-carbide formation in high-carbon Fe-C alloys
500 to 700	Formation of alloy carbides in Cr-, Mo-, V- and W-containing steels. The mix and composition of the carbides may change significantly with time (T4)	The alloy carbides produce secondary hardening and pronounced retardation of softening during tempering or long-time service exposure around 500 °C
350 to 550	Segregation and cosegregation of impurity and substitutional alloying elements	Responsible for temper embrittlement

and structures within the martensite crystals even at temperatures well below 100 °C (Ref 27). Tempering between 100 °C and Ac₁ produces various types of carbide-particle dispersions as well as major changes in the matrix martensite. The reactions which produce the carbides have long been recognized and are classified as stages of tempering: T_1, T_2, etc. The reactions which depend on very short-range rearrangement of carbon atoms in the as-quenched martensite prior to carbide formation have only recently been studied, and to distinguish those reactions from the carbide-forming reactions it has been suggested that they be classified as aging reactions: A_1, A_2, etc. (Ref 28 and 29).

Table 1 lists the various reactions and microstructural changes which may be developed by tempering steel (Ref 29). The aging and tempering classifications serve primarily to mark microstructures which form on the way to equilibrium, ultimately a microstructure which consists of spheroidized carbide particles dispersed in a matrix of equiaxed ferrite grains. Many of the reactions or microstructural states require further characterization, some occur concurrently, and others may yet be discovered. The reactions are controlled by diffusion of carbon, iron, and/or alloying elements, and therefore steel composition, time and temperature determine where a given tempering treatment stops in the sequence of structural changes indicated in Table 1.

Significant increases in toughness are achieved by tempering at temperatures above 150 °C. In general, subject to the development of various embrittlement phenomena, as tempering temperature increases, toughness increases and hardness decreases. Therefore, in applications where high hardness must be retained, tempering is performed at relatively low temperatures, usually between 150 and 200 °C. Figure 18 shows the fine structure which develops in martensite as a result of low-temperature tempering. A portion of a single plate of martensite in an Fe-1.22C alloy tempered at 150 °C is shown. The fine dark streaks mark positions of rows of very fine carbide particles, each about 2 nm in diameter, which have precipitated from the supersaturated martensite (Ref 30). The particles themselves are masked by the strain which accompanies the precipitation and which causes the dark contrast in the transmission electron micrograph. The carbide is not the equilibrium Fe₃C, but a transition carbide, first designated as epsilon-carbide with a hexagonal structure as identified by x-ray dif-

Fig. 18. Transmission electron micrograph showing fine structure in a plate of martensite in an Fe-1.22C alloy (Ref 30)

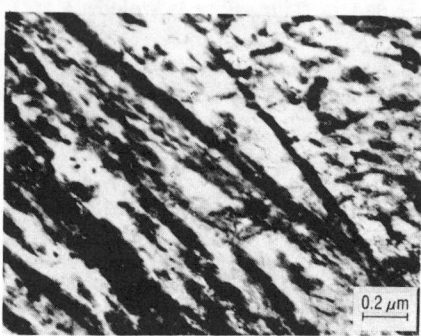

Fig. 19. Transmission electron micrograph showing chi-carbide in the martensitic microstructure of an Fe-1.22C alloy tempered at 350 °C (Ref 33)

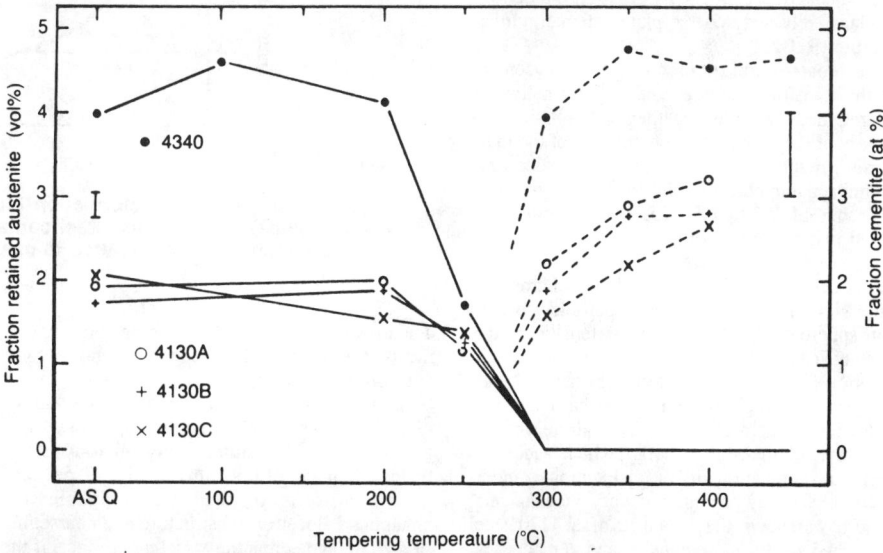

Fig. 20. Retained austenite and cementite as a function of tempering temperature in several medium-carbon steels (Ref 34)

fraction (Ref 31), but more recently designated eta-carbide with an orthorhombic structure as identified by electron diffraction (Ref 32). Both the epsilon-carbide and eta-carbide have carbon contents substantially higher than that of cementite.

Steels tempered to develop the fine transition carbides show a modest but significant increase

in toughness. The hardness, however, remains high because of the extremely fine carbide dispersion and the retention of much of the dislocation substructure introduced by the martensitic transformation.

In steels tempered between 200 and 350 °C, the transition carbide is replaced by cementite or χ-carbide, and retained austenite transforms to

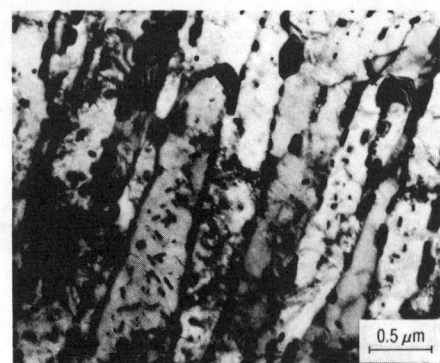

Fig. 21. Transmission electron micrograph showing microstructure of 4130 steel water quenched from 900 °C and tempered 500 min at 650 °C. Courtesy of F. Woldow.

ferrite and cementite. The χ-carbide is a complex carbide with a monoclinic structure which forms in tempered high-carbon martensites and is eventually replaced by cementite. Figure 19 shows a dense distribution of carbide particles, identified as χ-carbide, in an Fe-1.22C alloy tempered at 350 °C (Ref 33). The carbide particles are considerably coarser than the transition carbides in Fig. 19, and are present at the interfaces of the martensite plates as well as within the plates.

Figure 20 shows that small amounts of austenite are present even in medium-carbon steels, that the austenite is stable throughout the tempering-temperature range in which the transition carbide forms, and that the austenite begins to transform at temperatures above 200 °C. Austenite in medium-carbon steels is retained between martensite laths and, when it transforms on tempering, produces relatively coarse plates of interlath cementite (Ref 22).

The coarse carbides produced by replacement of the transition carbides and transformation of the retained austenite, together with a limited recovery of the dislocation substructure of the martensite, reduce impact toughness. This decrease in impact toughness produced by tempering in the range of 250 to 400 °C is referred to as tempered martensite embrittlement.

Tempering at temperatures above 400 °C produces substantial coarsening of the microstructure. Not only do the cementite particles coarsen and spheroidize, but also the martensitic matrix is significantly altered. The laths are almost dislocation-free and are now ferrite because all carbon has completely precipitated as carbides. The reduction in dislocation density is driven by the reduction of the strain energy which accompanies the elimination of the dislocations, and is accomplished by various recovery mechanisms. Figure 21 shows the structure of a 4130 steel quenched to form martensite and tempered at 650 °C. Carbide particles are present within and between the remanent martensite laths. The lath morphology, although coarsened, persists because of the pinning of the lath boundaries by carbides. In alloy steels such as 4130, which contains nominally 1% Cr and 0.2% Mo, various alloy carbides, in addition to cementite, form during high-temperature tempering. The intralath carbides in Fig. 21 have a specific habit plane and have been identified as Mo_2C (Ref 35). A number of other carbides of chromium, molyb-

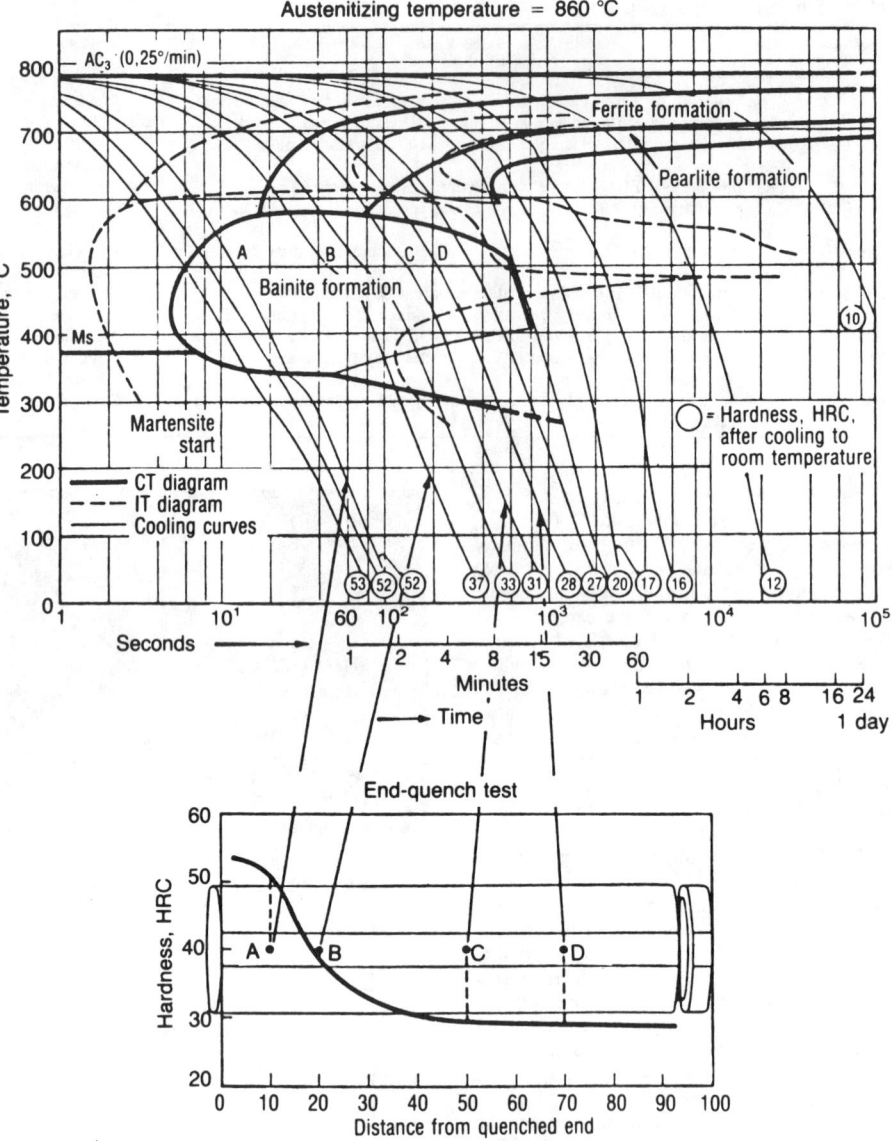

Fig. 22. Continuous transformation (solid lines) and isothermal transformation (dashed lines) diagrams for steel containing nominally 0.4 C, 1.0 Cr and 0.2 Mo. Several cooling rates are related to positions on a Jominy end-quench specimen. (Ref 38)

denum, vanadium and other carbide-forming elements may also develop depending on the composition of the steel.

As tempering temperature increases above 400 °C, hardness and strength drop rapidly and toughness improves significantly. In alloy steels, the development of fine alloy carbide dispersions offsets the softening which accompanies the changing dislocation substructure and coarsening of the lath and cementite structure. In fact, if the alloy carbide dispersions are sufficiently fine and dense, an increase in hardness may develop. This increase in hardness due to alloy carbide precipitation high in the tempering-temperature range is referred to as secondary hardening.

As noted, toughness increases significantly with increasing tempering temperature. However, if impurities such as phosphorus, antimony and tin are present in a steel, these elements may segregate to grain boundaries and/or carbide-matrix interfaces and cause large reductions in impact

toughness (Ref 36). This phenomenon develops during tempering in, or slow cooling through, the temperature range 350 to 550 °C, and is referred to as temper embrittlement. The impurity atom segregation may be accompanied by the co-segregation of the substitutional alloying elements present in steels (Ref 37).

TRANSFORMATION DIAGRAMS

The previous sections have shown that the transformation of austenite produces a wide variety of microstructures in response to such factors as steel composition, temperature of transformation, and cooling rate. In order to characterize the conditions which produce the various microstructures, two types of transformation diagrams have been developed. Isothermal transformation (IT) diagrams are based on the austenite decomposition at constant temperatures, while continuous transformation (CT)

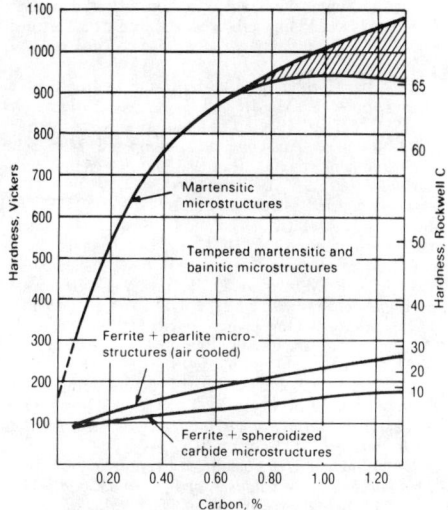

Fig. 23. Hardness as a function of carbon content for various microstructures in steels. Cross-hatched area shows effect of retained austenite. (After Ref 39)

diagrams follow microstructural development as a function of cooling rate. Most heat treatments are performed by continuous cooling, and therefore CT diagrams more accurately than IT diagrams represent the conditions encountered in commercial practice.

An example of an IT diagram for a eutectoid steel has already been shown in Fig. 6. Figure 22 shows CT (solid lines) and IT (dashed lines) diagrams for a medium-carbon alloy steel, 4140, containing nominally 0.4% C, 1% Cr and 0.2% Mo. Superimposed on the diagram are various cooling rates, some of which are related to positions on a Jominy end-quench specimen. The more rapid cooling rates produce microstructures of higher hardness as indicated.

The microstructures in the medium-carbon steel (Fig. 22) are more varied than in the eutectoid steel (Fig. 6) in that proeutectoid ferrite forms. Also the alloying elements in the 4140 steel significantly retard formation of ferrite and pearlite and thereby increase the range of cooling rates which form martensite and bainite. Comparison of the IT and CT diagrams in Fig. 22 shows that all of the phase transformations are shifted to lower temperatures and longer times by continuous cooling.

The CT diagrams provide the basis of hardenability, the technology which is concerned with estimating the depth and distribution of martensite in hardened components as a function of cooling rate and composition. Figure 22 shows that an important effect of alloying is to retard the formation of microstructures of low hardness. Therefore, martensite formation may be accomplished by less-severe quenching with the advantages of lower residual surface tensile stresses, reduced distortion and/or prevention of quench cracking. For a given quench, alloying increases depth of martensite formation in a part.

SUMMARY: CARBON CONTENT, MICROSTRUCTURE AND PROPERTIES

Figure 23 shows hardness as a function of carbon content for the various types of microstruc-

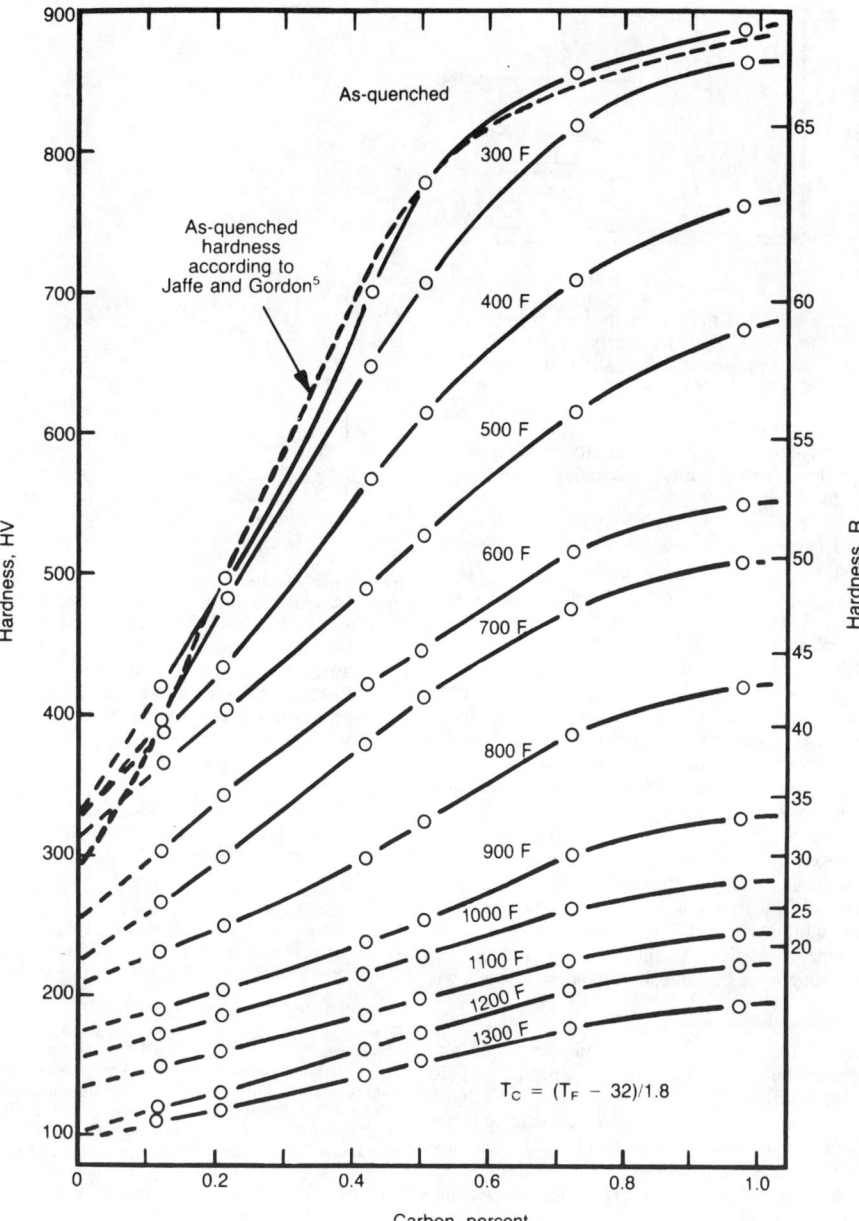

$$T_C = (T_F - 32)/1.8$$

Fig. 24. Hardness as a function of carbon content of martensite in Fe-C alloys tempered at various temperatures (Ref 40)

tures which may be formed by heat treatment of steel. More detail regarding the effect of tempering is shown in Fig. 24. Hardness is a readily measured property which in general is directly proportional to strength and inversely proportional to ductility.

Figures 23 and 24 demonstrate the great versatility of steels and the major effect of carbon content on establishing mechanical properties. Not all of the microstructures may be readily formed in all steels. For example, in low-carbon steels which have very low hardenability, it may be impossible to produce fully martensitic structures in all but the very thinnest sections. Low-carbon steels are therefore invariably used with ferrite-pearlite microstructures where the high ductility of ferrite is beneficial for cold working and formability. At the other extreme, medium- and high-carbon steels alloyed with chromium, nickel and/

or molybdenum may have such high hardenability for martensite or bainite formation that it is very difficult to form ferrite-pearlite microstructures except under conditions of very slow cooling or in very heavy sections.

Figures 23 and 24 show that the higher the carbon content the higher the hardness of a given microstructure. The microstructures with the highest hardness are formed by transformations which involve shear—i.e., the martensite or bainite transformations or the tempering of the shear-produced martensite microstructures. The shear transformations produce fine crystals (laths or plates), supersaturate the structure with carbon or create very fine carbide dispersions, and introduce a high dislocation density into the product phases. The lower-strength microstructures are produced by diffusion-controlled transformations or microstructural changes which pro-

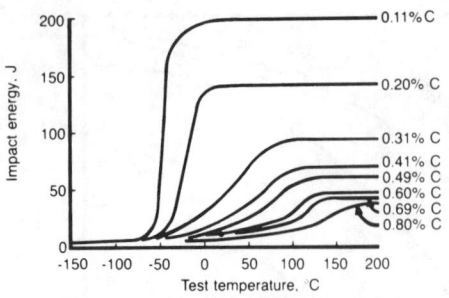

Fig. 25. Change in impact transition curves with increasing pearlite content in normalized carbon steels (Ref 41)

duce coarse ferrite-carbide microstructures without developing extensive dislocation substructures in the ferrite.

In addition to hardness and strength, fracture resistance or toughness is a major consideration in the selection of steels and heat treatments for severe applications. Figures 25 and 26 show CVN impact toughness for ferrite-pearlite and tempered martensitic microstructures, respectively. For the ferrite-pearlite steels, increasing carbon content reduces both the energy absorbed during ductile fracture and raises the transition temperature at which brittle, cleavage fracture occurs. Thus very-low-carbon steels with largely ferrite microstructures are best suited for applications which require high toughness.

Figure 26 shows that increasing tempering temperature increases impact toughness for a given hardened steel, but that increasing carbon content drastically reduces toughness after all tempering treatments. The effects of phosphorus and tempered martensite embrittlement in lowering the toughness of hardened steels are also shown in Fig. 26.

The very low toughness of high-carbon steels with either pearlite or tempered martensitic microstructures limits their use to applications where high hardness is of benefit to wear and fatigue resistance but where impact, tensile loading is not a major service condition. Examples of such applications are railroad rails produced from eutectoid steels with fully pearlitic microstructures and bearings produced from 52100 steel that has been heat treated by intercritical austenitizing, oil quenching and low-temperature tempering to produce tempered martensitic microstructures of high hardness.

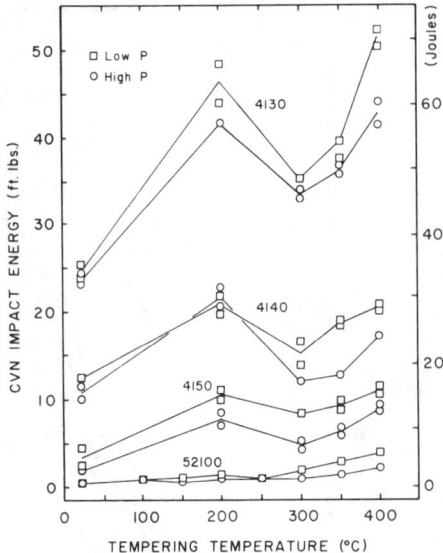

High phosphorus levels are approximately 0.02%. Low phosphorus levels are approximately 0.002% for the 41xx steels and 0.009% for the 52100 steel.

Fig. 26. CVN impact energy absorbed at room temperature as a function of tempering for medium- and high-carbon steels (Ref 42 and 43)

REFERENCES

1. *Metals Handbook*, 8th Ed., Vol 8, ASM, Metals Park, OH, 1973
2. *Thermomechanical Processing of Microalloyed Austenite*, edited by A. J. DeArdo, G. A. Ratz and P. J. Wray: TMS-AIME, Warrendale, PA, 1982
3. *Deformation, Processing, and Structure*, edited by G. Krauss: ASM, Metals Park, OH, 1984
4. *Theory of Dislocations*, by J. P. Hirth and J. Loth: McGraw-Hill, New York, 1968
5. *The Physical Metallurgy of Steels*, by W. C. Leslie: McGraw-Hill, New York, 1981.
6. *Mechanical Properties of BCC Metals*, edited by M. Meshii: TMS-AIME, Warrendale, PA, 1982
7. *Principles of Heat Treatment of Steel*, by G. Krauss: ASM, Metals Park, OH, 1980

8. The Quench-Aging of Low-Carbon Iron and Iron-Manganese Alloys: An Electron Transmission Study, by W. C. Leslie: *Acta Metallurgica*, Vol 9, 1961, p 1004-1022
9. "Dual Phase Behavior and Aging of a Renitrogenized Steel," by J. E. Indacochea: M.S. Thesis, Colorado School of Mines, Golden, CO, 1978
10. *Atlas of Isothermal Transformation and Cooling Transformation Diagrams*: ASM, Metals Park, OH, 1977, p 28
11. Historical Account of the Contribution of E. C. Bain, by H. W. Paxton and J. B. Austin: *Metallurgical Transactions*, Vol 13, 1972, p 1035-1042
12. "The Combined Effects of Phosphorus and Carbon on Hardenability and Phase Transformation Kinetics in 41XX Steels," by F. A. Jacobs: M.S. Thesis, Colorado School of Mines, Golden, CO, 1982
13. *Solid-Solid Phase Transformations*, edited by H. I. Aaronson, D. E. Laughlin, R. F. Sekerko and C. M. Wayman: TMS-AIME, Warrendale, PA, 1982
14. Development and Application of Growth Models for Grain Boundary Allotriomorphs of a Stoichiometric Compound in Ternary Systems, by T. Ando and G. Krauss: *Metallurgical Transactions*, Vol 14A, 1983, p 1261-1269
15. Microstructure and Fracture of 52100 Steel, by K. Nakazawa and G. Krauss: *Metallurgical Transactions A*, Vol 9A, 1978, p 681-689
16. The Effect of Phosphorus Content on Grain Boundary Cementite Formation in AISI 52100 Steel, by T. Ando and G. Krauss: *Metallurgical Transactions*, Vol 12A, 1981, p 1283-1290
17. The Relationship of Microstructure to Fracture Morphology and Toughness of Hardened Hypereutectoid Steels, by G. Krauss: in *Microstructure and Residual Stress Effects on the Properties of Case Hardened Steels*, TMS-AIME, Warrendale, PA, 1984
18. Effect of Carbon on the Volume Fractions and Lattice Parameters of Retained Austenite and Martensite, by C. S. Roberts: *Transactions of AIME*, Vol 197, 1953, p 203-204
19. The Crystallography of Martensite Transformations, by B. A. Bibby and J. W. Christian: *JISI*, Vol 197, 1961, p 122-131
20. A General Equation Prescribing the Extent of the

Austenite-Martensite Transformation in Pure Iron-Carbon Alloys and Plain Carbon Steels, by D. P. Koistinen and R. E. Marburger: *Acta Metallurgica*, Vol 7, 1959, p 59-60
21. The Morphology of Martensite in Iron-Carbon Alloys, by A. R. Marder and G. Krauss: *Transactions of ASM*, Vol 60, 1967, p 651-660
22. Retained Austenite and Tempered Martensite Embrittlement, by G. Thomas: *Metallurgical Transactions A*, Vol 9A, 1978, p 439-450
23. Tempered Martensite Embrittlement in SAE 4340 Steel, by J. P. Materkowski and G. Krauss: *Metallurgical Transactions A*, Vol 10A, 1979, p 1643-1651
24. Microcracking Sensitivity in Fe-C Plate Martensite, by A. R. Marder, A. O. Benscoter and G. Krauss: *Metallurgical Transactions*, Vol 1, 1970, p 1545-1549
25. Microcracking in Fe-C Acicular Martensite, by A. R. Marder and A. O. Benscoter: *Transactions of ASM*, Vol 61, 1968, p 293-299
26. The Morphology of Microstructure Composed of Lath Martensite in Steels, by T. Maki, K. Tsuzaki and I. Tamura: *Transactions of the Iron and Steel Institute of Japan*, Vol 20, 1980, p 207-214
27. Winchell Symposium on Tempering of Steel: *Metallurgical Transactions*, Vol 14A, 1983, p 985-1146
28. Early Stages of Aging and Tempering of Ferrous Martensites, by G. B. Olson and M. Cohen: *Metallurgical Transactions*, Vol 14A, 1983, p 1057-1065
29. Tempering and Structural Change in Ferrous Martensites, by G. Krauss: in *Phase Transformations in Ferrous Alloys*, TMS-AIME, Warrendale, PA, 1984
30. A Study of the Early Stages of Tempering in an Fe-1.22%C Alloy, by D. L. Williamson, K. Nakazawa and G. Krauss: *Metallurgical Transactions A*, Vol 10A, 1979, p 1351-1363
31. Structural Transformations in the Tempering of High Carbon Martensitic Steel, by K. H. Jack: *JISI*, Vol 169, 1951, p 26-36
32. Crystal Structure and Morphology of the Carbide Precipitated for Martensitic High Carbon Steel During the First Stage of Tempering, by Y. Hirotsu and S. Nagakura: *Acta Metallurgica*, Vol 20, 1972, p 645-655
33. Chi-Carbide in Tempered High Carbon Martensite, by C.-B. Ma, T. Ando, D. L. Williamson and G. Krauss: *Metallurgical Transactions A*, Vol 14A, 1983, p 1033-1045
34. Determination of Small Amounts of Austenite and Carbide in a Hardened Medium Carbon Steel by Mössbauer Spectroscopy, by D. L. Williamson, R. G. Schupmann, J. P. Materkowski and G. Krauss: *Metallurgical Transactions A*, Vol 10A, 1979, p 379-382
35. *Steels Microstructure and Properties*, by R. W. K. Honeycombe: Edward Arnold Publishers, Ltd., London, and ASM, Metals Park, OH, 1982
36. Temper Brittleness—An Interpretive Review, by C. J. McMahon, Jr.: in *Temper Embrittlement in Steel*, STP 407, ASTM, 1968, p 127-167
37. The Thermodynamics of Interactive Co-Segregation of Phosphorus and Alloying Elements in Iron and Temper-Brittle Steels, by M. Guttman, Ph. Dumonlin and M. Wayman: *Metallurgical Transactions A*, Vol 13A, 1982, p 1693-1711
38. *Atlas zur Wärmebehandlung der Stähle*, Vol 1-4: Max-Planck-Institut für Eisenforschung, in cooperation with the Verein Dentscher Eisenhüttenlente, Verlag Stahleisen, M. B. H., Düsseldorf, 1954-1976
39. *Alloying Elements in Steel*, 2nd Ed., by E. C. Bain and H. W. Paxton: ASM, Metals Park, OH, 1961
40. Hardness of Tempered Martensite in Carbide and Low Alloy Steels, by R. A. Grange, C. R. Hribal and L. F. Porter: *Metallurgical Transactions A*, Vol 8A, 1977, p 1775-1785
41. The Optimization of Microstructures in Steel and Their Relationship to Mechanical Properties, by F. B. Pickering: in *Hardenability Concepts With Applications to Steel*, edited by D. V. Doane and J. S. Kirkaldy, TMS-AIME, Warrendale, PA, 1978, p 179-228
42. "A Study of Mechanisms of Tempered Martensite Embrittlement in Low-Alloy Medium-Carbon Steels," by F. Zia-Ebrahimi: Ph.D. Thesis, Colorado School of Mines, Golden, CO, 1982
43. "The Effects of Phosphorus and Tempering on the Fracture of AISI 52100 Steel," by D. L. Yaney: M.S. Thesis, Colorado School of Mines, Golden, CO, 1981

Stress-Relief Heat Treating of Steel

By Domenic A. Canonico
Director of Metallurgical and
Materials Laboratory
C-E Power Systems
Combustion Engineering Inc.

STRESS-RELIEF HEAT TREATING is used to relieve stresses that remain locked in a structure as a consequence of a manufacturing sequence. This definition separates stress relief heat treating from postweld heat treating in that the goal of postweld heat treating is to provide, in addition to the relief of residual stresses, some preferred metallurgical structure or properties (Ref 1 and 2). For example, most ferritic weldments are given postweld heat treated to improve fracture toughness of heat-affected zones. Moreover, austenitic and nonferrous alloys are frequently postweld heat treated to improve resistance to environmental damage.

Stress-relief heat treating is the uniform heating of a structure or portion thereof to a suitable temperature below the transformation range (Ac_1 for ferritic steels), holding at this temperature for a predetermined period of time, followed by uniform cooling (Ref 2 and 3). Care must be taken to ensure uniform cooling, particularly when a component is composed of variable section sizes. If the rate of cooling is not constant, new residual stresses can result that are equal to or greater than those

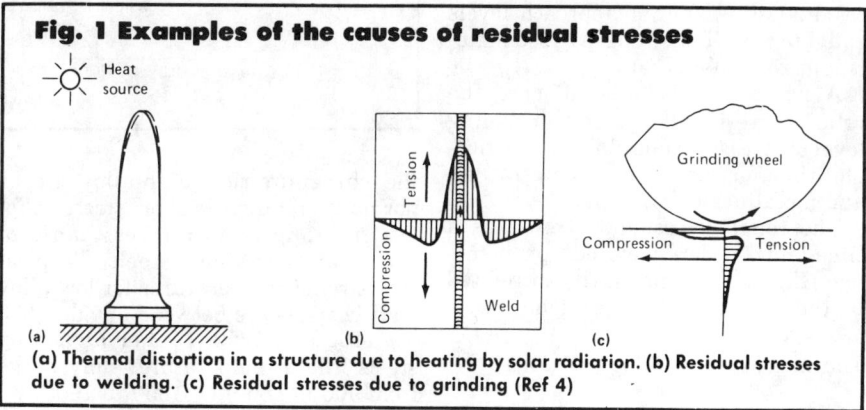

Fig. 1 Examples of the causes of residual stresses

(a) Thermal distortion in a structure due to heating by solar radiation. (b) Residual stresses due to welding. (c) Residual stresses due to grinding (Ref 4)

for which the heat treating was intended to relieve.

Stress-relief heat treating can reduce distortion and high stresses from welding that affect service performance. The presence of residual stresses can lead to stress corrosion cracking near welds and in regions of a component that has been cold strained during processing.

Residual stresses in a ferritic steel cause significant reduction in resistance to brittle fracture. In a material that is not prone to brittle fracture,

such as an austenitic stainless steel, residual stresses can be sufficient to provide the stress necessary to promote stress corrosion cracking even in environments that appear to be benign (Ref 4).

There are many sources of residual stress; they can occur during processing of the material from ingot to final product form (Ref 4 and 9). Residual stresses can be generated during rolling, casting or forging; during forming operations such as shearing, bending and machining; and during fabrication, in particular, welding. Residual stresses are present whenever a component is stressed beyond its elastic limit and plastic flow occurs. Bending a bar during fabrication at a temperature where recovery cannot occur (cold forming, for example), will result in one surface location containing residual tensile stresses, whereas a location 180° away will contain residual compressive stresses (Ref 5). Quenching of thick sections results in high residual compressive stresses on the surface of the material. These high compressive stresses are balanced by residual tensile stresses in the internal areas of the section (Ref 6).

Grinding is another source of residual stresses; these can be compressive or tensile in nature, depending on the grinding operation. Although these stresses tend to be shallow in depth, they can cause warping of thin parts (Ref 7).

The cause of residual stresses that has received the most attention in the open literature is welding. The residual stresses associated with the steep thermal gradient of welding can occur on a macroscale over relatively long distances (reaction stresses) or can be highly localized (microscale) (Fig. 1). Welding usually results in localized residual stresses that approach levels equal to the yield strength of the material at room temperature.

A number of factors influence the relief of residual stresses, including level of stress, permissible (or practicable) time for their relief, temperature, and metallurgical stability.

The relief of residual stresses is a time-temperature related phenomenon (Fig. 2), parametrically correlated by the Larson-Miller equation:

Thermal effect $= T(\log t + 20)(10^{-3})$

where T is temperature (Rankin) and t is hours. It is evident in Fig. 2 that similar relief of residual stresses can be achieved by holding a component for longer periods of time at a lower temperature. For example, holding a piece at 595 °C (1100 °F) for 6 h provides the same relief of residual stress as heating at 650 °C (1200 °F) for 1 h.

Relief of residual stresses represents typical stress-relaxation behavior, in which the material undergoes microscopic (sometimes even macroscopic) creep at the stress-relief temperature. Creep-resistant materials, such as the chromium-bearing low-alloy steels and

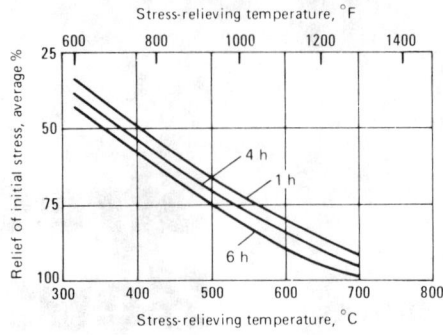

Fig. 2 Illustration of the relationship between time and temperature in the relief of residual stresses in steel (Ref 3)

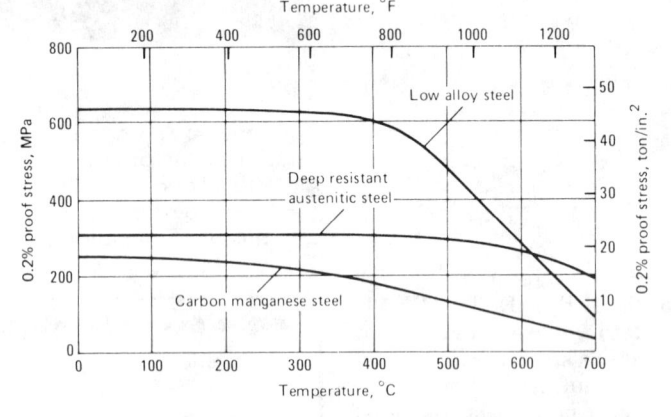

Fig. 3 Variation of the yield strength with temperature for three generic classes of steel (Ref 10)

the chromium-rich high-alloy steels, normally require higher stress-relief heat treating temperatures than conventional low-alloy steels. Typical stress-relief temperatures for low-alloy ferritic steels are between 595 and 675 °C (1100 and 1250 °F). For high-alloy steels, these temperatures may range from 900 to 1065 °C (1650 to 1950 °F).

For high-alloy steels, such as the austenitic stainless steels, stress relieving is sometimes done at temperatures as low as 400 °C (750 °F). However, at these temperatures, only modest decreases in residual stress are achieved. Residual stresses can be significantly reduced by stress-relief heat treating those austenitic materials in the temperature range from 480 to 925 °C (900 to 1700 °F). At the higher end of this range, nearly 85% of the residual stresses may be relieved. Stress relief heat treating in this range, however, may result in sensitizing susceptible material. This metallurgical effect can

lead to stress-corrosion cracking in service (Ref 8). Frequently, solution-annealing temperatures at around 1065 °C (1950 °F) are used to achieve a reduction of residual stresses to acceptably low values.

Some copper alloys may fail by stress-corrosion cracking due to the presence of residual stresses. These stresses are usually relieved by mechanical or thermal stress-relief treatments. Stress-relief heat treating tends to be favored because it is more controllable, less costly, and also provides a degree of dimensional stability. Stress-relief heat treating of copper alloys is usually carried out at relatively low temperatures, in the range from 200 to 400 °C (390 to 750 °F) (Ref 9).

Resistance of a material to the reduction of its residual stresses by thermal treatment can be estimated with a knowledge of the influence of temperature on its yield strength. Figure 3 provides a summary of the yield strength to temperature relationship for three generic classes of steels. The room temperature yield strength of these materials provides an excellent estimate of the level of localized residual stress that can be present in a structure. To relieve the residual stress requires that the component be heated to a temperature where its yield strength approaches a value that corresponds to an acceptable level of residual stress. Holding at this temperature can, through the reduction of strain due to creep, further reduce the residual stress. Uniform cooling after residual-stress heat treating is mandatory if these levels of residual stress are to be maintained.

REFERENCES

1. The Metallurgical Effects of Residual Stresses, by N. Bailey: in *Residual Stresses,* The Welding Institute, 1981, p 28-33
2. *Metallurgy and Weldability of Steels,* by C. E. Jackson *et al.:* Welding Research Council, 1978
3. Fundamentals of Welding: *Welding Handbook,* 7th Ed., Vol 1, American Welding Society, 1976
4. *Defects and Failures in Pressure Vessels and Piping,* by Helmut Thielsch: Reinhold Publishing Corporation, NY, 1965, p 311
5. *Mechanical Metallurgy,* 2nd Ed., by G. E. Dieter: McGraw-Hill, Inc., 1976
6. *Residual Stresses and Fatigue in Metals,* by J. O. Almen and P. H. Black: McGraw-Hill Book Company, 1963
7. Machining: *Metals Handbook,* 8th Ed., Vol 3, American Society for Metals, 1967, p 260
8. Properties and Selection: Stainless Steels, Tool Materials and Special Purpose Metals: *Metals Handbook,* 9th Ed., Vol 3, ASM, 1980, p 47-48
9. Properties and Selection: Nonferrous Alloys and Pure Metals: *Metals Handbook,* 9th Ed., Vol 2, ASM, 1979, p 255-256
10. Thermal Stress Relief and Associated Metallurgical Phenomena, by C. G. Saunders: in *The Welding Institute Research Bulletin,* Vol 9, No. 7, Part 3, 1968

SELECTED REFERENCES

1. Welding and Brazing: *Metals Handbook,* 8th Ed., Vol 6, American Society for Metals, Metals Park, OH, 1971, p 213
2. Classification and Nomenclature of Internal Stresses, by E. Orowan: in *Symposium on Internal Stresses in Metals and Alloys,* The Institute of Metals, London, 1948, p 47-59
3. Stress Relieving of Weldments, by E. R. Parker: *Welding Journal,* Vol 36, No. 10, Oct 1957, p 433-S
4. The Effect of Residual Stresses on Fracture, by J. D. Harrison and R. H. Leggatt: in *Residual Stresses,* The Welding Institute, 1981, p 17-20
5. Residual Stresses in Welded Plates, by N. R. Nagaraja and L. Tall: *Welding Journal,* Vol 40, No 10, 1961, p 468-S

Normalizing of Steel

By Samuel L. Williams
General Engineer
Rock Island Arsenal

NORMALIZING OF STEEL is a heat treating process that is often considered from both thermal and microstructural standpoints. In the thermal sense, normalizing is an austenitizing heating cycle followed by cooling in still or slightly agitated air. Typically, the work is heated to a temperature about 55 °C (100 °F) above the upper critical line of the iron – iron carbide phase diagram, as shown in Fig. 1; that is, above Ac_3 for hypoeutectoid steels and above A_{cm} for hypereutectoid steels. To be properly classed as a normalizing treatment, the heating portion of the process must produce an austenitic phase (face-centered cubic crystal structure) prior to cooling. Typical normalizing temperatures for many standard steels are given in Table 1.

Normalizing is also frequently thought of in terms of microstructure. The areas of the microstructure that contain about 0.8% carbon are pearlitic (lamellae of ferrite and iron carbide). The areas that are low in carbon are ferritic (body-centered cubic atomic structure). In hypereutectoid steels, proeutectoid iron carbide (iron carbide in excess of that within the pearlite structure) can be present in the microstructure. Air-hardening steels are excluded from the class of normalized steels because they do not exhibit the "normal" pearlitic microstructure that characterizes normalized steels.

Uses. A broad range of ferrous products can be normalized. All of the standard low-carbon, medium-carbon, and high-carbon wrought steels can be normalized, as well as many castings. Austenitic steels, stainless steels and maraging steels either cannot be normalized, or usually are not normalized.

The purpose of normalizing varies considerably. Normalization may increase or decrease the strength and hardness of a given steel in a given product form depending on the thermal and mechanical history of the product. Actually, the functions of normalizing may overlap with or be confused with those of annealing, hardening and stress relieving. Improved machinability, grain-structure refinement, homogenization, and modification of residual stresses are among the reasons for which normalizing is done. Homogenization of castings by normalizing may be done in order to break up or refine the dendritic structure and facilitate a more even response to subsequent hardening. Similarly, for wrought products, normalization can obliterate banded grain structure due to hot rolling, as well as large grain size or mixed large and small grain size due to forging practice. The details of normalizing treatments applied to three

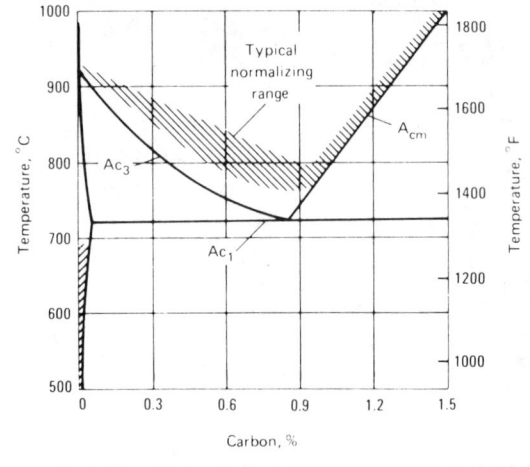

Fig. 1 Partial iron - iron carbide phase diagram, showing typical normalizing range for plain carbon steels

Table 1 Typical normalizing temperatures for standard carbon and alloy steels

Based on production experience, normalizing temperature may vary from as much as 27 °C (50 °F) below to as much as 55 °C (100 °F) above indicated temperature. The steel should be cooled in still air from indicated temperature.

Grade	Temperature °C	°F	Grade	Temperature °C	°F
Plain carbon steels			**Standard alloy steels (continued)**		
1015	915	1675	4817	925	1700
1020	915	1675	4820	925	1700
1022	915	1675	5046	870	1600
1025	900	1650	5120	925	1700
1030	900	1650	5130	900	1650
1035	885	1625	5132	900	1650
1040	860	1575	5135	870	1600
1045	860	1575	5140	870	1600
1050	860	1575	5145	870	1600
1060	830	1525	5147	870	1600
1080	830	1525	5150	870	1600
1090	830	1525	5155	870	1600
1095	845	1550	5160	870	1600
1117	900	1650	6118	925	1700
1137	885	1625	6120	925	1700
1141	860	1575	6150	900	1650
1144	860	1575	8617	925	1700
			8620	925	1700
			8622	925	1700
Standard alloy steels			8625	900	1650
1330	900	1650	8627	900	1650
1335	870	1600	8630	900	1650
1340	870	1600	8637	870	1600
3135	870	1600	8640	870	1600
3140	870	1600	8642	870	1600
3310	925	1700	8645	870	1600
4027	900	1650	8650	870	1600
4028	900	1650	8655	870	1600
4032	900	1650	8660	870	1600
4037	870	1600	8720	925	1700
4042	870	1600	8740	925	1700
4047	870	1600	8742	870	1600
4063	870	1600	8822	925	1700
4118	925	1700	9255	900	1650
4130	900	1650	9260	900	1650
4135	870	1600	9262	900	1650
4137	870	1600	9310	925	1700
4140	870	1600	9840	870	1600
4142	870	1600	9850	870	1600
4145	870	1600	50B40	870	1600
4147	870	1600	50B44	870	1600
4150	870	1600	50B46	870	1600
4320	925	1700	50B50	870	1600
4337	870	1600	60B60	870	1600
4340	870	1600	81B45	870	1600
4520	925	1700	86B45	870	1600
4620	925	1700	94B15	925	1700
4621	925	1700	94B17	925	1700
4718	925	1700	94B30	900	1650
4720	925	1700	94B40	900	1650
4815	925	1700			

typical production parts are given in Table 2, which also lists the reasons for normalizing and gives some of the mechanical properties obtained in the normalized and tempered condition. Comparisons of typical hot rolled or annealed mechanical properties versus typical normalized properties are presented in Table 3. Depending on the mechanical properties required, normalizing may be substituted for conventional hardening when the size or shape of the part is such that liquid quenching might result in cracking, distortion or excessive dimensional changes. Thus, parts that are of complex shape or that incorporate sharp changes in section may be normalized and tempered, provided that the properties obtained are acceptable.

The rate of heating generally is not critical for normalizing; on an atomic scale, it is immaterial. In parts having great variations in section size, however, thermal stress can cause distortion.

Time at temperature is critical only in that it must be sufficient to cause homogenization. Sufficient time must be allowed for solution of carbides, if present, and/or for movement of alloy atoms to obtain a desired final structure. Generally, time sufficient for complete austenitization is all that is required. One hour at temperature, after the furnace recovers, per inch of part thickness, is considered to be very liberal. Parts often can be austenitized adequately in much less time (with a saving in energy).

The rate of cooling significantly influences both the amount of pearlite and the size and spacing of the pearlite lamellae. At higher cooling rates, more pearlite forms, and the lamellae are finer and more closely spaced. Both the increased amount of pearlite and the greater fineness of the pearlite result in higher strength and higher hardness. Conversely, lower cooling rates result in softer parts. The effect of mass on hardness (via its effect on cooling rate) is illustrated by the data in Table 4. In any part having both thick and thin sections, the potential exists for variations in cooling rate, and thus for variations in strength and hardness as well as an increase in the probability of distortion or even cracking. Cooling rate sometimes is enhanced purposely with fans, to increase strength and hardness of parts or to decrease the time required, following the furnace operation, for sufficient cooling of parts to permit convenient handling.

After parts have cooled evenly to "black heat" below Ar_1 (the parts are no longer red, as when they were removed from the furnace), they may be water or

Table 2 Typical applications of normalizing and tempering

Part	Steel	Heat treatment	Properties after treatment	Reason for normalizing
Cast 50-mm (2-in.) valve body, 19 to 25 mm (³/₄ to 1 in.) in section thickness	Ni-Cr-Mo	Full annealed at 955 °C (1750 °F), normalized at 870 °C (1600 °F), tempered at 665 °C (1225 °F)	Tensile strength, 620 MPa (90 ksi); 0.2% yield strength, 415 MPa (60 ksi); elongation in 50 mm or 2 in., 20%; reduction in area, 40%	To meet mechanical-property requirements
Forged flange	4137	Normalized at 870 °C (1600 °F), tempered at 570 °C (1060 °F)	Hardness, 200 to 225 HB	To refine grain size and obtain required hardness
Valve-bonnet forging	4140	Normalized at 870 °C (1600 °F) and tempered	Hardness, 220 to 240 HB	To obtain uniform structure, improved machinability and required hardness

oil quenched to decrease the total cooling time. In heavy sections, cooling of the center material to "black heat" may require considerable time. Thermal shock, residual thermally induced stress, and resultant distortions are factors to be considered. The microstructure remains essentially unaffected by the increased cooling rate, provided that the entire mass is below the lower critical temperature, Ar_1, although changes involving precipitates may occur.

Carbon Steels. Table 1 lists typical normalizing temperatures for some standard grades of carbon steel. These temperatures can be interpolated to obtain values for carbon contents not listed.

Steels containing 0.20% C or less usually receive no treatment subsequent to normalizing. However, medium-carbon or high-carbon steels are often tempered after normalizing to obtain specific properties, such as a lower hardness for straightening, cold working or machining. Whether or not tempering is desirable depends primarily on carbon content and section size. Table 3 presents typical mechanical properties of selected carbon and alloy steels in the hot rolled, normalized and annealed conditions. A low-carbon or medium-carbon steel of thin section may be harder after normalizing than a high-carbon steel of large section size subjected to the same treatment.

Alloy Steels. For alloy steel forgings, rolled products and castings, normalizing is commonly used as a conditioning treatment before final heat treatment. Normalizing also refines the structures of forgings, rolled products and castings that have cooled nonuniformly from high temperatures.

Table 1 lists typical normalizing temperatures for some standard alloy steels.

Alloy carburizing steels, such as 3310 and 4320, usually are normalized at temperatures higher than the carburizing temperature, to minimize distortion in carburizing and to improve machining characteristics. Carburizing steels of the 3300 series sometimes are double normalized with the expectation of minimizing distortion; these steels are tempered at about 650 °C (1200 °F) for intervals of up to 15 h to reduce hardness to below 223 HB for machinability. Carburizing steels of the 4300 and 4600 series usually can be normalized to a hardness not exceeding 207 HB, and therefore, need not be tempered for machinability.

Hypereutectoid alloy steels, such as 52100, are normalized for partial or complete elimination of carbide networks, thus producing a structure that is more susceptible to 100% spheroidization in the subsequent spheroidize annealing treatment. The spheroidized structure provides improved machinability and a more uniform response to hardening.

Forgings

When forgings are normalized before carburizing or before hardening and tempering, the upper range of normalizing temperatures is used. However, when normalizing is the final heat treatment, use is made of the lower range of temperatures.

Furnaces. Either batch-type or continuous furnaces may be used for normalizing steel forgings. In a continuous furnace, forgings to be normalized are usually placed in shallow pans, and a pusher mechanism at the loading end

of the furnace transports the pans through the furnace. Furnace burners, located on both sides of the furnace, fire below the hearth, and combustion products rise along the walls of the work-zone muffle and exhaust into the roof of the furnace. No atmosphere control is used. Combustion products enter the work zone through ports lining both sides of the entire hearth. A typical furnace is 9 m (30 ft) long and has 18 gas burners (or 9 oil burners) on each side. For purposes of temperature control, such a furnace is divided into three 3-m (10-ft) zones, each having a vertical thermocouple extending into it through the roof of the furnace.

Processing. Small forgings are usually normalized as received from the forge shop. They are placed or piled loosely on the pans to a maximum depth of 75 mm (3 in.). A typical furnace has five pans in each of the three furnace zones. Heating is adjusted so that the work reaches normalizing temperature in the last zone. After passing through the last zone, the pans are discharged onto a cooling conveyor. The work, while still in the pans, is cooled in still air to below 480 °C (900 °F); it is then discharged into tote boxes, in which it cools to room temperature. Total furnace time is approximately 3½ h, but during this period the work is held at the normalizing temperature for only 1 h.

Normalizing of large open-die forgings usually is performed in batch-type furnaces pyrometrically controlled to narrow temperature ranges. Forgings are held at the normalizing temperature long enough to allow complete austenitizing and carbide solution to occur (usually one hour per inch of section thickness), and then are cooled in still air.

Table 3 Properties of selected carbon and alloy steels in the hot rolled, normalized and annealed conditions

AISI grade(a)	Condition or treatment	Tensile strength MPa	ksi	Yield strength MPa	ksi	Elongation(b), %	Reduction in area, %	Hardness, HB	Izod impact strength J	ft·lb
1015	As rolled	420	61	315	46	39.0	61	126	111	82
	Normalized at 925 °C (1700 °F)	425	62	325	47	37.0	70	121	116	85
	Annealed at 870 °C (1600 °F)	385	56	285	41	37.0	70	111	115	85
1020	As rolled	450	65	330	48	36.0	59	143	87	64
	Normalized at 870 °C (1600 °F)	440	64	345	50	35.8	68	131	118	87
	Annealed at 870 °C (1600 °F)	395	57	295	43	36.5	66	111	123	91
1022	As rolled	505	73	360	52	35.0	67	149	81	60
	Normalized at 925 °C (1700 °F)	485	70	360	52	34.0	68	143	117	87
	Annealed at 870 °C (1600 °F)	450	65	315	46	35.0	64	137	121	89
1030	As rolled	550	80	345	50	32.0	57	179	75	55
	Normalized at 925 °C (1700 °F)	520	76	345	50	32.0	61	149	94	69
	Annealed at 845 °C (1550 °F)	465	67	340	50	31.2	58	126	69	51
1040	As rolled	620	90	415	60	25.0	50	201	49	36
	Normalized at 900 °C (1650 °F)	590	86	375	54	28.0	55	170	65	48
	Annealed at 790 °C (1450 °F)	520	75	355	51	30.2	57	149	44	33
1050	As rolled	725	105	415	60	20.0	40	229	31	23
	Normalized at 900 °C (1650 °F)	750	109	425	62	20.0	39	217	27	20
	Annealed at 790 °C (1450 °F)	635	92	365	53	23.7	40	187	17	13
1060	As rolled	815	118	485	70	17.0	34	241	18	13
	Normalized at 900 °C (1650 °F)	775	113	420	61	18.0	37	229	13	10
	Annealed at 790 °C (1450 °F)	625	91	370	54	22.5	38	179	11	8
1080	As rolled	965	140	585	85	12.0	17	293	7	5
	Normalized at 900 °C (1650 °F)	1010	147	525	76	11.0	21	293	7	5
	Annealed at 790 °C (1450 °F)	615	89	375	55	24.7	45	174	6	5
1095	As rolled	965	140	570	83	9.0	18	293	4	3
	Normalized at 900 °C (1650 °F)	1015	147	500	73	9.5	14	293	5	4
	Annealed at 790 °C (1450 °F)	655	95	380	55	13.0	21	192	3	2
1117	As rolled	485	71	305	44	33.0	63	143	81	60
	Normalized at 900 °C (1650 °F)	465	68	305	44	33.5	54	137	85	63
	Annealed at 860 °C (1575 °F)	430	62	280	41	32.8	58	121	94	69
1118	As rolled	520	76	315	46	32.0	70	149	109	80
	Normalized at 925 °C (1700 °F)	480	69	320	46	33.5	66	143	104	76
	Annealed at 790 °C (1450 °F)	450	65	285	41	34.5	67	131	106	79
1137	As rolled	625	91	380	55	28.0	61	192	83	61
	Normalized at 900 °C (1650 °F)	670	97	395	58	22.5	49	197	64	47
	Annealed at 790 °C (1450 °F)	585	85	345	50	26.8	54	174	50	37
1141	As rolled	675	98	360	52	22.0	38	192	11	8
	Normalized at 900 °C (1650 °F)	705	103	405	59	22.7	56	201	53	39
	Annealed at 815 °C (1500 °F)	600	87	355	51	25.5	49	163	34	25
1144	As rolled	705	102	420	61	21.0	41	212	53	39
	Normalized at 900 °C (1650 °F)	665	97	400	58	21.0	40	197	43	32
	Annealed at 790 °C (1450 °F)	585	85	345	50	24.8	41	167	65	48
1340	Normalized at 870 °C (1600 °F)	835	121	560	81	22.0	63	248	93	68
	Annealed at 800 °C (1475 °F)	705	102	435	63	25.5	57	207	71	52
3140	Normalized at 870 °C (1600 °F)	890	129	600	87	19.7	57	262	54	40
	Annealed at 815 °C (1500 °F)	690	100	425	61	24.5	51	197	46	34
4130	Normalized at 870 °C (1600 °F)	670	97	435	63	25.5	60	197	86	64
	Annealed at 865 °C (1585 °F)	560	81	360	52	28.2	56	156	62	46
4140	Normalized at 870 °C (1600 °F)	1020	148	655	95	17.7	47	302	23	17
	Annealed at 815 °C (1500 °F)	655	95	415	61	25.7	57	197	55	40
4150	Normalized at 870 °C (1600 °F)	1155	168	735	107	11.7	31	321	12	9
	Annealed at 815 °C (1500 °F)	730	106	380	55	20.2	40	197	25	18
4320	Normalized at 895 °C (1640 °F)	795	115	465	67	20.8	51	235	73	54
	Annealed at 850 °C (1560 °F)	580	84	425	62	29.0	58	163	110	81
4340	Normalized at 870 °C (1600 °F)	1280	186	860	125	12.2	36	363	16	12
	Annealed at 810 °C (1490 °F)	745	108	470	69	22.0	50	217	51	38

(a) All grades are fine grained except for those in the 1100 series, which are coarse grained. (b) In 50 mm or 2 in.

Table 3 (continued)

AISI grade(a)	Condition or treatment	Tensile strength MPa	ksi	Yield strength MPa	ksi	Elongation(b), %	Reduction in area, %	Hardness, HB	Izod impact strength J	ft·lb
4620	Normalized at 900 °C (1650 °F)	575	83	365	53	29.0	67	174	133	98
	Annealed at 860 °C (1575 °F)	510	74	370	54	31.3	60	149	94	69
4820	Normalized at 860 °C (1580 °F)	755	110	485	70	24.0	59	229	110	81
	Annealed at 815 °C (1500 °F)	680	99	465	67	22.3	59	197	93	69
5140	Normalized at 870 °C (1600 °F)	795	115	470	69	22.7	59	229	38	28
	Annealed at 830 °C (1525 °F)	570	83	295	43	28.6	57	167	41	30
5150	Normalized at 870 °C (1600 °F)	870	126	530	77	20.7	59	255	32	23
	Annealed at 825 °C (1520 °F)	675	98	355	52	22.0	44	197	25	19
5160	Normalized at 860 °C (1575 °F)	955	139	530	77	17.5	45	269	11	8
	Annealed at 810 °C (1495 °F)	725	105	275	40	17.2	31	197	10	7
6150	Normalized at 870 °C (1600 °F)	940	136	615	89	21.8	61	269	36	26
	Annealed at 815 °C (1500 °F)	665	97	410	60	23.0	48	197	27	20
8620	Normalized at 910 °C (1675 °F)	635	92	355	52	26.3	60	183	100	74
	Annealed at 870 °C (1600 °F)	535	78	385	56	31.3	62	149	112	83
8630	Normalized at 870 °C (1600 °F)	650	94	430	62	23.5	54	187	95	70
	Annealed at 845 °C (1550 °F)	565	82	370	54	29.0	59	156	95	70
8650	Normalized at 870 °C (1600 °F)	1025	149	690	100	14.0	45	302	14	10
	Annealed at 795 °C (1465 °F)	715	104	385	56	22.5	46	212	29	22
8740	Normalized at 870 °C (1600 °F)	930	135	605	88	16.0	48	269	18	13
	Annealed at 815 °C (1500 °F)	695	101	415	60	22.2	46	201	40	30
9255	Normalized at 900 °C (1650 °F)	935	135	580	84	19.7	43	269	14	10
	Annealed at 845 °C (1550 °F)	775	112	485	71	21.7	41	229	9	7
9310	Normalized at 890 °C (1630 °F)	905	132	570	83	18.8	58	269	119	88
	Annealed at 845 °C (1550 °F)	820	119	440	64	17.3	42	241	79	58

(a) All grades are fine grained except for those in the 1100 series, which are coarse grained. (b) In 50 mm or 2 in.

Axle-Shaft Forging. In forging an axle shaft made of fine-grain 1049 steel, only one end of the forging bar was heated to upset the wheel-flange section. When the part was examined in cross section from the flanged end to the cold end, the following metallurgical conditions were revealed:

The hot worked flanged area of the axle exhibited a fine-grain structure as a result of the hot working at the forging temperature (approximately 1100 °C or 2000 °F). However, a section adjacent to the flange, which also had been heated to the forging temperature but which had not been hot worked, exhibited a coarse-grain structure. Nearer the cool end of the shaft, a zone that reached a temperature of about 700 °C (1300 °F) exhibited a spheroidized structure. The cold end of the shaft retained its initial fine grain size throughout the forging operation.

In subsequent operations, this shaft was to be mechanically straightened, machined, and induction hardened. Because of the mixed grain structure, these operations posed several problems. The coarse-grain area adjacent to the flange was extremely weak in the transverse direction, and there was a possibility that fracture would occur if this section were subjected to a severe straightening operation. The spheroidized area would not respond adequately to induction hardening, because the solution rate of this type of carbide formation was too sluggish for the relatively rapid rate of induction heating. Furthermore, the mixed metallurgical structure would present difficulties in machining. Consequently, normalizing was required in order to produce a uniformly fine-grain structure throughout the axle shaft prior to straightening, machining and induction hardening.

Low-Carbon Steel Forgings. In contrast to the medium-carbon axle shaft discussed in the preceding paragraphs, forgings made of carbon steels containing 0.25% C or less are seldom normalized. Only severe quenching from above the austenitizing temperature will have any significant effect on their structure or hardness.

Structural Stability. Normalizing and tempering is also a preferred treatment for promoting the structural stability of low-alloy heat-resistant alloys, such as AMS 6304 (0.45 C, 1 Cr, 0.5 Mo, 0.3 V), at temperatures up to 540 °C (1000 °F). Wheels and spacer rings used in the "cold" ends of aircraft gas turbine engine compressors are typical of parts subjected to such treatment to promote structural stability.

Effects on Mechanical Properties. Differences in mechanical properties obtained by normalizing and tempering and by quenching and tempering result from differences in the rate of cooling from the austenitizing temperature, and hence are functions of the hardenability of the steel and the section size of the part. For air cooling, just as for liquid quenching, a larger section size requires a higher alloy content if a given hardness is to be maintained. Attainment of the same hardness in a normalized part as in a liquid-quenched part also requires a higher alloy content, to compensate for the smaller hardening response elicited

Table 4 Effect of mass on hardness of normalized carbon and alloy steels

All data are based on single heats. Sources: data for 3310, 3140 and 4063 are from *Modern Steels and Their Properties*, 6th Ed., Bethlehem Steel Corp., 1966; all other data are from *Modern Steels and Their Properties* (Handbook 3310), Bethlehem Steel Corp., Sept 1978.

Grade	Normalizing temperature °C	°F	Hardness, HB, for bar with diameter, mm (in.), of: 13(1/2)	25(1)	50(2)	100(4)
Carbon steels, carburizing grades						
1015	925	1700	126	121	116	116
1020	925	1700	131	131	126	121
1022	925	1700	143	143	137	131
1117	900	1650	143	137	137	126
1118	925	1700	156	143	137	131
Carbon steels, direct-hardening grades						
1030	925	1700	156	149	137	137
1040	900	1650	183	170	167	167
1050	900	1650	223	217	212	201
1060	900	1650	229	229	223	223
1080	900	1650	293	293	285	269
1095	900	1650	302	293	269	255
1137	900	1650	201	197	197	192
1141	900	1650	207	201	201	201
1144	900	1650	201	197	192	192
Alloy steels, carburizing grades						
3310	890	1630	269	262	262	248
4118	910	1670	170	156	143	137
4320	895	1640	248	235	212	201
4419	955	1750	149	143	143	143
4620	900	1650	192	174	167	163
4820	860	1580	235	229	223	212
8620	915	1675	197	183	179	163
9310	890	1630	285	269	262	255
Alloy steels, direct-hardening grades						
1340	870	1600	269	248	235	235
3140	870	1600	302	262	248	241
4027	905	1660	179	179	163	156
4063	870	1600	285	285	285	277
4130	870	1600	217	197	167	163
4140	870	1600	302	302	285	241
4150	870	1600	375	321	311	293
4340	870	1600	388	363	341	321
5140	870	1600	235	229	223	217
5150	870	1600	262	255	248	241
5160	860	1575	285	269	262	255
6150	870	1600	285	269	262	255
8630	870	1600	201	187	187	187
8650	870	1600	363	302	293	285
8740	870	1600	269	269	262	255
9255	900	1650	277	269	269	269

Table 5 Typical mechanical properties of normalized alloy steel sheet

Grade	Thickness mm	in.	Tensile strength MPa	ksi	Yield strength(a) MPa	ksi	Elongation(b), %	Hardness, HRC
4130	4.9	0.193	835	121	585	85	14	25
4335(c)	4.6	0.180	1725	250	1240	180	8	48
4340(c)	2.0	0.080	1860	270	1345	195	7	50

(a) At 0.2% offset. (b) In 50 mm or 2 in. (c) Modified: 0.40% Mo, 0.20% V.

by air cooling compared with that brought about by liquid quenching.

Hardness is not the only property that is affected by differences in cooling rate. Other mechanical properties may be expected to differ from the normalized and tempered condition to the quenched and tempered condition, even when surface hardness is about the same for both treatments. For example, although tensile strength will be about the same, yield strength, elongation, and reduction in area will be lower in the normalized and tempered condition than in the quenched and tempered condition.

Multiple normalizing treatments are employed (*a*) to obtain complete solution of all lower-temperature constituents in austenite by the use of high initial normalizing temperatures (for example, 925 °C or 1700 °F), and (*b*) to refine final pearlite grain size by the use of a second normalizing treatment at a temperature closer to the Ac_3 temperature (for example, 815 °C or 1500 °F) without destroying the beneficial effects of the initial normalizing treatment.

Locomotive-axle forgings made of carbon steel to AAR Specification M-126, Class F (ASTM A236, Class F), containing 0.45 to 0.59% C and 0.60 to 0.90% Mn, are double normalized to obtain a uniformly fine grain structure along with other exacting mechanical-property requirements. Forgings made of a low-carbon steel (0.18% C) with 1% Mn intended for low-temperature service are double normalized to meet sub-zero impact requirements.

Bar and Tubular Products

Frequently, the finishing stages of hot-mill operations employed in making steel bar and tube produce properties that closely approximate those obtained by normalizing. When this occurs, normalizing is unnecessary and may even be inadvisable. Nevertheless, the reasons for normalizing bar and tube products are generally the same as those applicable to other forms of steel.

There is an additional reason, however, for normalizing bar and tube products. When these products are cold finished by a sequence of cold reductions with high subcritical anneals between passes, some spheroidization occurs. In such instances, the product is sometimes normalized before the last cold reduction pass. Normalizing eliminates the spheroidization that earlier passes and anneals may have generated and restores the pearlitic structure beneficial to machinability in low-carbon and medium-carbon grades of carbon or alloy steel.

Tubes are easier to normalize than bars of equivalent diameter, because the lighter section thickness of tubes permits more rapid heating and cooling. These advantages help minimize decarburization and promote more nearly uniform microstructures in tube products.

Furnaces. Continuous furnaces of the roller-hearth type are widely used for normalizing tube and bar products, especially in long lengths. Batch-type furnaces or other types of continuous furnaces are satisfactory if they provide some means for rapid discharge and separation of the load to permit free circulation of air around each tube as it cools. Continuous furnaces should have at least two zones, one for heating and one for soaking. Cooling facilities should be ample, so that uniform cooling can proceed until complete transformation has occurred. If tubes are packed or bundled during cooling from a high temperature, the purpose of normalizing is defeated and a semiannealed or a tempered product results.

Generally, protective atmospheres are not used in roller-hearth continuous furnaces for normalizing bar or tube products. The scale that forms during normalizing is removed by acid pickling or abrasive blast cleaning.

Alloy Steels. Although the principles involved in normalizing alloy steel bar, tube and pipe products are the same as those for normalizing carbon steels, application of these principles is sometimes more complex for alloy steels. Some alloy grades require more care in heating, to prevent cracking from thermal shock. They also require longer soaking times because of lower austenitizing and solution rates. For many alloy steels, rates of cooling in air

to room temperature must be carefully controlled. Certain alloy steels are forced-air cooled from the normalizing temperature in order to develop specific mechanical properties. This is a normalizing treatment only in the microstructural sense discussed near the beginning of this article.

Castings

In industrial practice, steel castings may be normalized in car-bottom, box, pit or continuous furnaces; the same heat treating principles apply to each type of furnace.

Loading. Furnaces are loaded with castings in such a manner that each casting will receive an adequate and uniform heat supply. This may be accomplished by stacking castings in regular order or by interspersing large and small castings so that load concentration in any one area is not excessive. At normalizing temperatures, the tensile strength of steel is greatly reduced, and heavy unequal sections may become distorted unless bracing and support are provided. Accordingly, small and large castings may be arranged so that they support each other.

Loading Temperature. When castings are charged, the temperature of the furnace should be such that the thermal shock will not cause metal failure. For the higher-alloy grades of steel castings, such as C5, C12 and WC9, a safe furnace temperature for charging is 315 to 425 °C (600 to 800 °F). For lower-alloy grades, furnace temperatures may be as high as 650 °C (1200 °F). For cast carbon steels and low-alloy steels with low carbon contents (low hardenability), castings may be charged into a furnace operating at the normalizing temperature.

Heating. After the furnace has been charged, the temperature is increased at a rate of approximately 220 °C/h (400 °F/h) until the normalizing temperature is reached. Depending on steel composition and casting configuration, a reduction in the rate of heating to approximately 28 to 55 °C/h (50 to 100 °F/h) may be necessary to avoid cracking. Extremely large castings should be heated more slowly, to prevent development of extreme temperature gradients.

Soaking. After the normalizing temperature has been reached, castings are "soaked" at this temperature for a period of time that will ensure complete austenitization and carbide solution. The duration of the soaking

period may be predetermined by microscopic examination of specimens held for various times at the normalizing temperature.

Cooling. After the soaking period, the castings are unloaded and allowed to cool in still air. Use of fans, air blasts or other means of accelerating the cooling process should be avoided.

Sheet and Strip

Hot rolled steel sheet and strip (about 0.10% C) are normalized primarily to refine grain size, minimize directional properties and develop desirable mechanical properties. Uniformly fine equiaxed ferrite grains normally are obtained in hot rolled sheet and strip by finishing the final hot rolling operation above the upper transformation temperature. However, if part of the hot rolling operation is performed on steel that has transformed partially to ferrite, the deformed ferrite grains usually will recrystallize and form abnormally coarse-grain patches during the self-anneal induced by coiling or piling at temperatures of 650 to 730 °C (1200 to 1350 °F). Also, relatively thin hot rolled material, if it is inadvertently finished well below the upper transformation temperature and coiled or piled while it is too cold to self-anneal, may possess directional properties. These conditions are unsuitable for some types of severe press drawing applications and may be corrected by normalizing.

Normalizing also may be used to develop high strength in alloy steel sheet and strip if the products are sufficiently high in carbon and alloy contents to enable them to transform to fine pearlite or martensite when cooled in air from the normalizing temperature. In general, the hardened material is tempered to attain an optimum combination of strength and ductility. Typical mechanical properties of normalized 4130, modified 4335 and modified 4340 steel sheet are given in Table 5.

Processing. The normalizing operation consists of passing the sheet or strip through an open, continuous furnace in which the material is heated to a temperature approximately 55 to 85 °C (100 to 150 °F) above its upper transformation temperature, 845 or 900 °C (1550 to 1650 °F), thus obtaining complete solution of the original structure with the formation of austenite, and then air cooling the material to room temperature.

Furnace Equipment. Normalizing furnaces are designed to heat and cool sheets singly or two in a pile. They are built in the form of long, low chambers, and usually comprise three sections: a preheating zone (12 to 20% of the total length); a heating, or soaking, zone (about 40% of the total length); and a cooling zone, which occupies the remaining 40 to 50% of the length.

Heating Arrangements. Normalizing furnaces usually are heated with gas or oil and do not employ protective atmospheres. Therefore, sheets are scaled during heat treatment. Burners are arranged along each side of the heating zone; they usually are above the conveyor, but occasionally are both above and below it. The furnace roof, which is higher in the preheating and soaking zones than in the cooling zone, is usually built in sections. In most furnaces, both the preheating zone and the cooling zone are heated by the hot gases from the heating zone. However, both of these zones may be equipped with burners for more accurate temperature control. Air is excluded by regulating the draft to maintain a slight pressure within all zones.

Conveyor-Type Furnaces. In modern furnaces of the conveyor type (the only type suitable for treating short lengths), sheets are carried through each of the three zones on rotating disks made of heat-resistant alloys. These disks have polished surfaces, which prevent them from scratching the sheets, and are staggered to ensure uniform heating. The disks are mounted on water-cooled shafts, which are driven by variable-speed motors through chains and sprockets or shafts and gears. These furnaces may be up to 2.5 m (100 in.) wide and from 27 to 61 m (90 to 200 ft) long. Fuel consumption is 2.3 to 5.2 million kJ per tonne (2.0 to 4.5 million Btu per ton) of steel treated, and production rates vary from 2.7 to 10.9 tonnes (3 to 12 tons) per hour.

Normalizing in a three-zone conveyor-type furnace equipped with pyrometric controls is a relatively simple operation. If scratching of sheets is to be avoided, the sheets are brought to the charging table and hand laid, one or more at a time, on a rider or conveyor sheet. Heavy sheets are normalized singly, but lighter sheets may be stacked two in a pile. To control heating and retard scaling, single sheets may be laid on a rider sheet and covered with a cover sheet. Sheets are carried by disk-rollers into the preheating zone, where they absorb heat rapidly because of the large temperature differential between the sheets and the interior of the furnace and because of the large surface-to-volume ratio. As the sheets become heated and the temperature differential is reduced, the rate of heat absorption slackens. After traveling 4½ to 6 m (15 to 20 ft), the sheets enter the soaking zone at a temperature several degrees below the normalizing temperature. Heating is completed in the soaking zone, which is maintained at a constant temperature, and sheets are held at the required temperature for a time sufficient to convert the microstructure to austenite before they are passed into the cooling zone. The sheets emerge from the cooling zone at a temperature that can be varied between 150 and 540 °C (300 and 1000 °F), and are conveyed for a short distance on the runout table, where, after being cooled rapidly in air, they are carefully removed from the rider sheet. The trip through such a furnace is carried out at a uniform speed of 0.03 to 0.10 m/s (5 to 20 ft/min) and requires 5 to 20 min to complete.

Catenary Furnaces. The catenary, or free-loop, type of furnace is designed for continuous normalizing of cold reduced steel unwound from coils; it does not have rolls or any other type of conveyor for supporting the material passing through the heating zone. The heating zones of catenary furnaces range in length from 6 to 15 m (20 to 50 ft). The preheating and cooling zones usually are shorter than those in conveyor-type furnaces, and for some kinds of work may be omitted entirely. At their exit ends, catenary furnaces may incorporate pickling or other descaling equipment for removing surface oxides formed on the steel during normalizing.

Heat Treating of Ultrahigh-Strength Steels

By T. V. Philip
Specialist, Tool and Alloy Metallurgy
Carpenter Technology Corporation

ULTRAHIGH-STRENGTH STEELS are heat treated by use of equipment and techniques similar to those employed for heat treating constructional alloy steels. The ultrahigh-strength steels ordinarily are quenched and tempered to specified hardnesses, but for critical applications it may be necessary to pull tensile specimens to ensure that a required combination of strength and ductility has been achieved. In still other instances, it may be necessary to conduct impact or fracture-toughness tests to ensure that a required level of resistance to brittle fracture has been attained.

Ultrahigh-strength steels are described in Volume 1 (Ninth Edition) of this Handbook. In that article, only those commercial structural steels capable of a minimum yield strength of 1380 MPa (200 ksi) are considered; and general characteristics, heat treatment and properties are presented for three types of alloys from among the broad class of ultrahigh-strength constructional steels. These three alloy types are (a) medium-carbon low-alloy steels, (b) medium-alloy air-hardening steels and (c) high-alloy hardenable steels.

The present article gives a brief description of the heat treatment and associated properties of the same types of steels, basically reproduced from the earlier article. Chemical compositions are given in Table 1. For a more detailed discussion of these ultrahigh-strength steels, the reader is referred to the original article.

Medium-Carbon Low-Alloy Steels

The medium-carbon low-alloy steels considered in this article are types 4130, 4140, 4340, 6150, 8640 and two modifications of 4340—namely, 300M and D-6a. These steels are readily hot forged. To avoid stress cracks resulting from air hardening, the forged part should be slowly cooled in a furnace or in an insulating medium. Prior to machining, usual practice is to normalize at 870 to 925 °C (1600 to 1700 °F) and temper at 650 to 675 °C (1200 to 1250 °F) or, if the steel is a deep air-hardening grade, to anneal by furnace cooling from 815 to 845 °C (1500 to 1550 °F) to about 540 °C (1000 °F). These treatments impart moderately hard microstructures suitable for machining. A very soft spheroidized structure can be obtained by full annealing. Such a structure is less well suited for machining than the normalized-and-tempered structure. However, for severe cold forming operations such as spinning, deep drawing and wiredrawing, the soft and ductile spheroidized structure is preferred. If blanks for parts are produced by flame cutting, they are annealed before being formed or machined. Welded parts, especially if complex, are stress relieved, or hardened and tempered, immediately after welding.

Reprinted from Metals Handbook, 9th Edition, Vol. 4, 119-129, © 1981 American Society for Metals

Table 1 Compositions of ultrahigh-strength steels described in this article

Designation or trade name	C	Mn	Si	Cr	Ni	Mo	V	Co
Medium-carbon low-alloy steels								
4130	0.28-0.33	0.40-0.60	0.20-0.35	0.80-1.10	· · ·	0.15-0.25	· · ·	· · ·
4140	0.38-0.43	0.75-1.00	0.20-0.35	0.80-1.10	· · ·	0.15-0.25	· · ·	· · ·
4340	0.38-0.43	0.60-0.80	0.20-0.35	0.70-0.90	1.65-2.00	0.20-0.30	· · ·	· · ·
300M	0.40-0.46	0.65-0.90	1.45-1.80	0.70-0.95	1.65-2.00	0.30-0.45	0.05 min	· · ·
D-6a	0.42-0.48	0.60-0.90	0.15-0.30	0.90-1.20	0.40-0.70	0.90-1.10	0.05-0.10	· · ·
6150	0.48-0.53	0.70-0.90	0.20-0.35	0.80-1.10	· · ·	· · ·	0.15-0.25	· · ·
8640	0.38-0.43	0.75-1.00	0.20-0.35	0.40-0.60	0.40-0.70	0.15-0.25	· · ·	· · ·
Medium-alloy air-hardening steels(b)								
H11 Mod	0.37-0.43	0.20-0.40	0.80-1.00	4.75-5.25	· · ·	1.20-1.40	0.40-0.60	· · ·
H13	0.32-0.45	0.20-0.50	0.80-1.20	4.75-5.50	· · ·	1.10-1.75	0.80-1.20	· · ·
9Ni-4Co steels								
HP 9-4-20	0.16-0.23	0.20-0.40	0.20 max	0.65-0.85	8.50-9.50	0.90-1.10	0.06-0.12	4.25-4.75
HP 9-4-30	0.29-0.34	0.10-0.35	0.20 max	0.90-1.10	7.0-8.0	0.90-1.10	0.06-0.12	4.25-4.75

(a) Phosphorus and sulfur contents may vary with steelmaking practice. Usually, these steels contain no more than 0.035 P and 0.040 S; 9Ni-4Co steels are specified to have 0.10 max P and 0.10 max S. (b) ASTM A681; composition ranges utilized by some producers are narrower.

4130

Type 4130 is a water-hardening alloy steel of low-to-intermediate hardenability.

Heat Treatments. The following standard heat treatments apply to type 4130 steel:

- *Normalize:* Heat to 870 to 925 °C (1600 to 1700 °F) and hold for a period that depends on section thickness; air cool. Tempering at 480 °C (900 °F) or above is often done after normalizing to increase yield strength.
- *Anneal:* Heat to 830 to 860 °C (1525 to 1575 °F) and hold for a period that depends on section thickness or furnace load; furnace cool.
- *Harden:* Heat to 845 to 870 °C (1550 to 1600 °F) and hold, then water quench; or heat to 860 to 885 °C (1575 to 1625 °F) and hold, then oil quench. Holding time depends on section thickness.
- *Temper:* At least ½ h at 200 to 700 °C (400 to 1300 °F); air cool or water quench. Tempering temperature and time at temperature depend mainly on desired hardness or strength.
- *Spheroidize:* Heat to 760 to 775 °C (1400 to 1425 °F) and hold 4 to 12 h; cool slowly.

Properties. Table 2 summarizes the typical properties obtained by tempering water-quenched and oil-quenched 4130 steel bars at various temperatures. Because 4130 steel has low hardenability, section thickness must be considered when heat treating to high

Table 2 Typical mechanical properties of heat treated 4130 steel

Tempering temperature		Tensile strength		Yield strength		Elongation in 50 mm or 2 in., %	Reduction in area, %	Hardness, HB	Izod impact energy	
°C	°F	MPa	ksi	MPa	ksi				J	ft·lb
Water quenched and tempered(a)										
205	400	1765	256	1520	220	10.0	33.0	475	18	13
260	500	1670	242	1430	208	11.5	37.0	455	14	10
315	600	1570	228	1340	195	13.0	41.0	425	14	10
370	700	1475	214	1250	182	15.0	45.0	400	20	15
425	800	1380	200	1170	170	16.5	49.0	375	34	25
540	1000	1170	170	1000	145	20.0	56.0	325	81	60
650	1200	965	140	830	120	22.0	63.0	270	135	100
Oil quenched and tempered(b)										
205	400	1550	225	1340	195	11.0	38.0	450	· · ·	· · ·
260	500	1500	218	1275	185	11.5	40.0	440	· · ·	· · ·
315	600	1420	206	1210	175	12.5	43.0	418	· · ·	· · ·
370	700	1320	192	1120	162	14.5	48.0	385	· · ·	· · ·
425	800	1230	178	1030	150	16.5	54.0	360	· · ·	· · ·
540	1000	1030	150	840	122	20.0	60.0	305	· · ·	· · ·
650	1200	830	120	670	97	24.0	67.0	250	· · ·	· · ·

(a) 25-mm (1-in.) diam round bars quenched from 845 to 870 °C (1550 to 1600 °F). (b) 25-mm (1-in.) diam round bars quenched from 860 °C (1575 °F)

hardness or strength. Effects of mass on typical properties of heat treated 4130 steel are indicated in Table 3.

4140

Type 4140 steel is similar in composition to 4130 except for a higher carbon content, which imparts greater hardenability and strength. When any ultrahigh-strength steel is heat treated to high strength levels, it is subject to hydrogen embrittlement during operations such as acid pickling and cadmium or chromium electroplating.

Ductility can be restored in thin sections by baking for 2 to 4 h at 190 °C (375 °F), and in sections more than 38 mm (1½ in.) thick by baking for 23 h at 190 °C.

Heat Treatments. The following standard heat treatments apply to 4140 steel:

- *Normalize:* Heat to 845 to 900 °C (1550 to 1650 °F) and hold for a period that depends on section thickness; air cool.
- *Anneal:* Heat to 845 to 870 °C (1550

Table 3 Effects of mass on typical properties of heat treated 4130 steel

Round bars oil quenched from 845 °C (1550 °F) and tempered at 540 °C (1000 °F); 12.83-mm (0.505-in.) diam tensile specimens cut from center of 25-mm diam bar, and from mid-radius of 50- and 75-mm diam bars

Bar size		Tensile strength		Yield strength		Elongation in 50 mm or 2 in., %	Reduction in area, %	Surface hardness, HB
mm	in.	MPa	ksi	MPa	ksi			
25	1	1040	151	880	128	18.0	55.0	307
50	2	740	107	570	83	20.0	58.0	223
75	3	710	103	540	78	22.0	60.0	217

Table 4 Typical mechanical properties of heat treated 4140 steel

12.7-mm (¹/₂-in.) diam round bars, oil quenched from 845 °C (1550 °F)

Tempering temperature		Tensile strength		Yield strength		Elongation in 50 mm or 2 in., %	Reduction in area, %	Hardness, HB	Izod impact energy	
°C	°F	MPa	ksi	MPa	ksi				J	ft·lb
205	400	1965	285	1740	252	11.0	42	578	15	11
260	500	1860	270	1650	240	11.0	44	534	11	8
315	600	1720	250	1570	228	11.5	46	495	9	7
370	700	1590	231	1460	212	12.5	48	461	15	11
425	800	1450	210	1340	195	15.0	50	429	28	21
480	900	1300	188	1210	175	16.0	52	388	46	34
540	1000	1150	167	1050	152	17.5	55	341	65	48
595	1100	1020	148	910	132	19.0	58	311	93	69
650	1200	900	130	790	114	21.0	61	277	112	83
705	1300	810	117	690	100	23.0	65	235	136	100

Table 5 Effects of mass on typical properties of heat treated 4140 steel

Round bars oil quenched from 845 °C (1550 °F) and tempered at 540 °C (1000 °F); 12.83-mm (0.505-in.) diam tensile specimens cut from center of 25-mm diam bars, and from mid-radius of 50- and 75-mm diam bars

Diameter of bar		Tensile strength		Yield strength		Elongation in 50 mm or 2 in., %	Reduction in area, %	Surface hardness, HB
mm	in.	MPa	ksi	MPa	ksi			
25	1	1140	165	985	143	15	50	335
50	2	920	133	750	109	18	55	202
75	3	860	125	655	95	19	55	293

to 1600 °F) and hold for a period that depends on section thickness or furnace load; furnace cool.
- *Harden:* Heat to 830 to 870 °C (1525 to 1600 °F) and hold; oil quench. (For water quenching, which is rarely used, hardening temperatures are 815 to 845 °C, or 1500 to 1550 °F.) Holding time depends on section thickness.
- *Temper:* At least ¹/₂ h at 175 to 230 °C (350 to 450 °F) or 370 to 675 °C (700 to 1250 °F); air cool or water quench. Tempering temperature and time at temperature depend mainly on desired hardness. To avoid blue brittleness, 4140 usually is not tempered

between 230 and 370 °C (450 and 700 °F).
- *Spheroidize:* Heat to 760 to 775 °C (1400 to 1425 °F) and hold 4 to 12 h; cool slowly.

Properties. Table 4 summarizes the mechanical properties obtained by tempering oil-quenched 4140 steel at various temperatures. Because 4140 is not a deep-hardening steel, section size should be considered, especially when specifying heat treatment for high strength levels. The effects of mass on hardness and tensile properties of 4140 steel are shown in Table 5.

4340

Type 4340, the most popular steel in this class, is a deep-hardening steel. In thin sections, the steel is air hardening, although in practice it is usually oil quenched. When 4340 is heat treated to tensile strengths greater than about 1400 MPa (about 200 ksi), it is subject to hydrogen embrittlement. Parts exposed to hydrogen, such as in pickling or electroplating, should be baked at 185 to 195 °C (365 to 385 °F) for at least 8 h, and for 23 h if thicker than 38 mm (1¹/₂ in.), as soon as possible after the pickling or plating operation.

Heat Treatments. The following standard heat treatments apply to 4340 steel:
- *Normalize:* Heat to 845 to 900 °C (1550 to 1650 °F) and hold for a period that depends on section thickness; air cool.
- *Anneal:* Heat to 830 to 860 °C (1525 to 1575 °F) and hold for a period that depends on section thickness or furnace load; furnace cool.
- *Harden:* Heat to 800 to 845 °C (1475 to 1550 °F) and hold 15 min for each 25 mm (1 in.) of thickness (minimum, 15 min); oil quench to below 65 °C (150 °F); or quench into fused salt at 200 to 210 °C (390 to 410 °F), hold 10 min, then air cool to below 65 °C (150 °F).
- *Temper:* At least ¹/₂ h at 200 to 650 °C (400 to 1200 °F); air cool. Temperature and time at temperature depend mainly on desired strength or hardness.
- *Spheroidize:* Preheat to 690 °C (1275 °F) and hold 2 h, increase temperature to 750 °C (1375 °F) and hold 2 h, cool to 650 °C (1200 °F) and hold 6 h, furnace cool to about 600 °C (1100 °F), and finally air cool to room temperature. An alternative schedule is to heat to 730 to 750 °C (1350 to 1375 °F) and hold several hours, then furnace cool to room temperature.
- *Stress relieve:* After straightening, forming or machining, parts may be stress relieved at 650 to 675 °C (1200 to 1250 °F).

Properties. Through hardening of 4340 steel can be achieved by oil quenching round sections up to 75 mm (3 in.) in diameter, and by water quenching larger sections (up to the limit of hardenability). The influence of section size on tensile properties of oil-quenched and water-quenched 4340 is indicated by the data in Table 6.

Hardness of type 4340 as a function

Table 6 Effects of mass on mechanical properties of 4340 steel

Data from *Alloy Digest* and from A. M. Hall, Sr., "Introduction to Today's Ultrahigh-Strength Structural Steels", STP 498, American Society for Testing and Materials, Philadelphia, 1971

Section diameter		Tensile strength		Yield strength		Elongation in 50 mm or 2 in., %	Reduction in area, %	Hardness, HB
mm	in.	MPa	ksi	MPa	ksi			
Oil quenched and tempered(a)								
13	½	1460	212	1380	200	13	51	...
38	1½	1450	210	1365	198	11	45	...
75	3	1420	206	1325	192	10	38	...
Water quenched and tempered(b)								
75	3	1055	153	930	135	18	52	340
100	4	1035	150	895	130	17	50	330
150	6	1000	145	850	123	16	44	322

(a) Austenitized at 845 °C (1550 °F); tempered at 425 °C (800 °F). (b) 75-mm (3-in.) diam bar austenitized at 800 °C; 100- and 150-mm (4- 6-in.) diam bars austenitized at 815 °C (1500 °F). All sizes tempered at 650 °C (1200 °F). Test specimens taken at mid-radius

Fig. 1 Variation for hardness with tempering temperature of 4340 steel

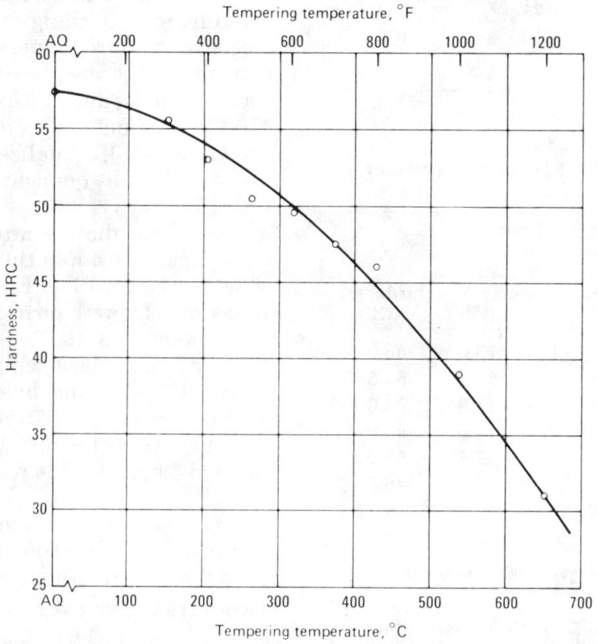

All specimens oil quenched from 845 °C (1550 °F) and tempered 2 h at temperature

of tempering temperature is plotted in Fig. 1. Typical mechanical properties of oil-quenched 4340 are given in Table 7. Additional data on mechanical properties (notch toughness and fracture toughness) of this steel tempered to different hardnesses are given in Table 8.

300M

Alloy 300M is basically a silicon-modified (1.6% Si) 4340 steel, but is slightly higher in carbon and molybdenum and also contains vanadium. The steel exhibits deep hardenability. Because of its high silicon and molybdenum contents, 300M is more prone to decarburization than the steels so far described; and during heat treating, care should be exercised to avoid decarburization. When heat treated to strength levels higher than 1380 MPa (200 ksi), 300M is susceptible to hydrogen embrittlement, and thus parts should be properly baked after plating (same baking procedure as that used for type 4340).

Heat Treatments. The following standard heat treatments apply to 300M steel:

- *Normalize:* Heat to 915 to 940 °C (1675 to 1725 °F) and hold for a period that depends on section thickness; air cool. If normalizing is to enhance machinability, recharge into a tempering furnace at 650 to 675 °C (1200 to 1250 °F) before the steel reaches room temperature.
- *Harden:* Austenitize at 855 to 885 °C (1575 to 1625 °F). Oil quench to below 70 °C (160 °F); or quench in salt at 200 to 210 °C (390 to 410 °F) and hold 10 min, then air cool to 70 °C or below.
- *Temper:* Two to four hours at 260 to 315 °C (500 to 600 °F); double tempering recommended. This tempering procedure produces the best combination of high yield strength and high impact properties. Tempering outside the range given above results in severe deterioration of properties.
- *Spheroidize:* Heat to about 775 °C (1430 °F) and hold for a period that depends on section thickness or furnace load. Cool to 650 °C (1200 °F) at a rate no faster than 5.5 °C/h (10 °F/h), then cool to 480 °C (900 °F) no faster than 11 °C/h (20 °F/h), and finally air cool to room temperature. The same schedule is recommended for annealing.

Properties. Variations in hardness and mechanical properties of 300M with tempering temperature are presented in Table 9. This steel has deep hardenability, so that heat treated bars 75 mm (3 in.) in diameter have essentially the same tensile properties as bars 25 mm (1 in.) in diameter. Reductions in tensile strength, ductility and impact strength, however, are observed in heat treated bars 145 mm (5¾ in.) in diameter. Variations in properties of 300M with section size are presented in Table 10.

D-6a and D-6ac

Ladish D-6a was designed primarily for use at room-temperature tensile strengths of 1800 to 2000 MPa (260 to 290 ksi). It is deeper hardening than 4340. The alloy is called D-6a when produced by air melting in an electric furnace, and D-6ac when produced by air melting followed by vacuum arc remelting. Except for some improvements in mechanical properties of D-6ac due to melting practice, the characteristics of the two steels are identical.

Table 7 Typical mechanical properties of 4340 steel

Oil quenched from 845 °C (1550 °F) and tempered at various temperatures

Tempering temperature °C	°F	Tensile strength MPa	ksi	Yield strength MPa	ksi	Elongation in 50 mm or 2 in., %	Reduction in area, %	Hardness HB	HRC	Izod impact energy J	ft·lb
205	400	1980	287	1860	270	11	39	520	53	20	15
315	600	1760	255	1620	235	12	44	490	49.5	14	10
425	800	1500	217	1365	198	14	48	440	46	16	12
540	1000	1240	180	1160	168	17	53	360	39	47	35
650	1200	1020	148	860	125	20	60	290	31	100	74
705	1300	860	125	740	108	23	63	250	24	102	75

Table 8 Notch toughness and fracture toughness of 4340 steel tempered to different hardnesses

Hardness, HB	Equivalent tensile strength(a) MPa	ksi	Charpy V-notch impact energy J	ft·lb	Plane-strain fracture toughness MPa√m	ksi√in.
550	2040	296	19	14	53	48
430	1520	220	30	22	75	68
380	1290	187	42	31	110	100

(a) Estimated from hardness

Table 9 Typical mechanical properties of 300M steel

Round bars, 25 mm (1 in.) in diameter, oil quenched from 860 °C (1575 °F) and tempered at various temperatures

Tempering temperature °C	°F	Tensile strength MPa	ksi	Yield strength MPa	ksi	Elongation in 50 mm or 2 in., %	Reduction in area, %	Charpy V-notch impact energy J	ft·lb	Hardness, HRC
90	200	2340	340	1240	180	6.0	10.0	17.6	13.0	56.0
205	400	2140	310	1650	240	7.0	27.0	21.7	16.0	54.5
260	500	2050	297	1670	242	8.0	32.0	24.4	18.0	54.0
315	600	1990	289	1690	245	9.5	34.0	29.8	22.0	53.0
370	700	1930	280	1620	235	9.0	32.0	23.7	17.5	51.0
425	800	1790	260	1480	215	8.5	23.0	13.6	10.0	45.5

Table 10 Effects of mass on tensile and impact properties of 300M steel

Round bars, normalized at 900 °C (1650 °F), oil quenched from 860 °C (1575 °F), and tempered at 315 °C (600 °F)

Bar diameter mm	in.	Tensile strength MPa	ksi	Yield strength MPa	ksi	Elongation in 50 mm or 2 in., %	Reduction in area, %	Charpy V-notch impact energy when tested at +21 °C (+70 °F) J	ft·lb	−46 °C (−50 °F) J	ft·lb	−73 °C (−100 °F) J	ft·lb
25	1	1990	289	1690	245	9.5	34.1	30	22	26	19	24	18
75	3	1940	281	1630	236	9.5	35.0	26	19	19	14	12	9
150	5¾ ..	2120	308	1800	261	7.3	22.3	12	9	9	7	7	5

Heat Treatments. The following standard heat treatments apply to D-6a and D-6ac steels:

- *Normalize:* Heat to 870 to 955 °C (1600 to 1750 °F) and hold for a period that depends on section thickness; air cool.

- *Anneal:* Heat to 815 to 845 °C (1500 to 1550 °F) and hold for a period that depends on section thickness or furnace load; furnace cool to 540 °C (1000 °F) at a rate no faster than 28 °C/h (50 °F/h), then air cool to room temperature. Hot forgings may be annealed as follows for maximum machinability: Immediately after forging, charge the parts into a 650 °C (1200 °F) furnace and hold 12 h, increase temperature to 900 °C (1650 °F) and hold for a period that depends on section size, cool to 650 °C (1200 °F) and hold 10 h, and finally air cool to room temperature.

- *Harden:* Austenitize at 845 to 900 °C (1550 to 1650 °F) for ½ to 2 h. Sections no larger than 25 mm (1 in.) in thickness or diameter may be air cooled. Larger sections may be oil quenched to 65 °C (150 °F); or salt quenched to 205 °C (400 °F), and then air cooled. For optimum dimensional stability, "aus-bay" quench into a furnace or salt bath at 525 °C (975 °F), equalize the temperature, then quench in an oil bath held at 60 °C (140 °F), or quench in 205 °C (400 °F) salt and air cool. The cooling rate during quenching significantly affects fracture toughness. For high fracture toughness, especially in heavy sections, austenitize at 925 °C (1700 °F), "aus-bay" quench to 525 °C (975 °F), equalize temperature, and oil quench to 60 °C (140 °F).

- *Temper:* Immediately after hardening, temper 2 to 4 h in the range 200 to 700 °C (400 to 1300 °F), depending on desired strength or hardness.

- *Spheroidize:* Heat to 730 °C (1350 °F) and hold 5 h, increase temperature to 760 °C (1400 °F) and hold 1 h; furnace cool to 690 °C (1275 °F) and hold 10 h; furnace cool to 650 °C (1200 °F) and hold 8 h; air cool to room temperature.

- *Stress relieve:* Heat to a temperature from 540 to 675 °C (1000 to 1250 °F) and hold for 1 to 2 h; air cool.

Properties. Typical room temperature hardness of D-6a steel bar as a function of tempering temperature is plotted in Fig. 2; other typical mechanical properties of D-6a bar are given in Table 11. Tensile properties of heat treated D-6ac billet material are given in Table 12.

6150

Type 6150 can be considered an ultrahigh-strength steel, although as a constructional steel it is not as popular as the other steels in this class. It is a shallow-hardening steel. Parts made of 6150 can be readily welded; after welding, parts should be normalized, then hardened and tempered to the desired hardness.

Fig. 2 Variation for hardness with tempering temperature for D-6a steel

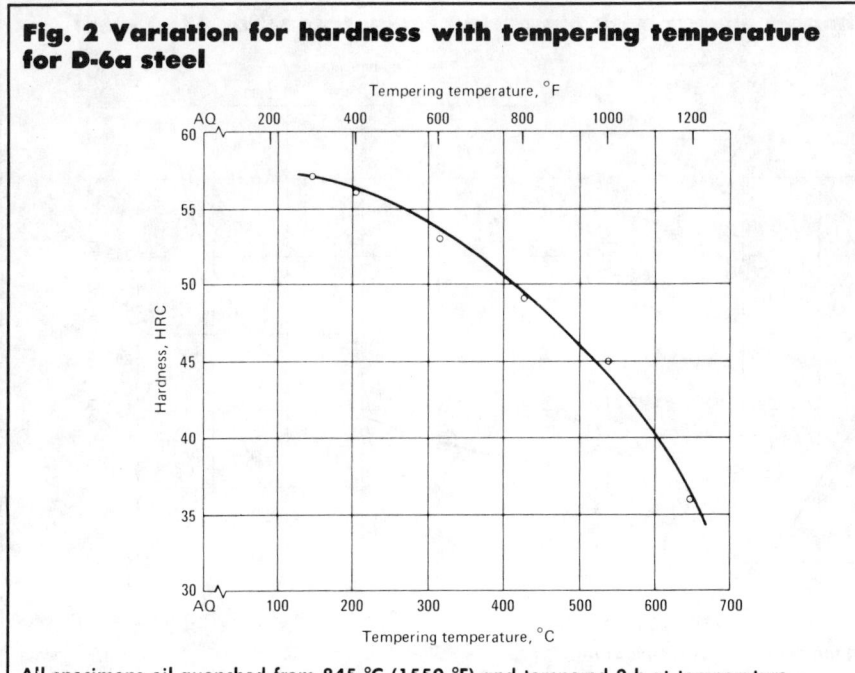

All specimens oil quenched from 845 °C (1550 °F) and tempered 2 h at temperature

Table 11 Typical mechanical properties of D-6a steel bar

Normalized at 900 °C (1650 °F), oil quenched from 845 °C (1550 °F) and tempered at various temperatures

Tempering temperature		Tensile strength		Yield strength		Elongation in 50 mm or 2 in., %	Reduction in area, %	Charpy V-notch impact energy	
°C	°F	MPa	ksi	MPa	ksi			J	ft·lb
150	300	2060	299	1450	211	8.5	19.0	14	10
205	400	2000	290	1620	235	8.9	25.7	15	11
315	600	1840	267	1700	247	8.1	30.0	16	12
425	800	1630	236	1570	228	9.6	36.8	16	12
540	1000	1450	210	1410	204	13.0	45.5	26	19
650	1200	1030	150	970	141	18.4	60.8	41	30

Table 12 Typical tensile properties of double tempered D-6ac billet

Austenitized 1 h at 900 °C (1650 °F), quenched in fused salt at 205 °C (400 °F) and held 5 min, then air cooled to room temperature. Tempered 1 h at 205 °C; second temper, 4 h at indicated temperature

Second tempering temperature		Tensile strength		Yield strength		Elongation in 50 mm or 2 in., %	Reduction in area, %
°C	°F	MPa	ksi	MPa	ksi		
480	900	1686.5	244.6	1540.3	223.4	11.1	40.0
510	950	1652.7	239.7	1519.7	220.4	13.2	44.1
540	1000	1613.4	234.0	1483.8	215.2	13.7	47.2

Heat Treatments. The following heat treatments apply to 6150 steel:

- *Normalize:* Heat to 870 to 955 °C (1600 to 1750 °F) and hold for a period that depends on section thickness; air cool.

- *Anneal:* Heat to 845 to 900 °C (1550 to 1650 °F) and hold for a period that depends on section thickness or furnace load; furnace cool.

- *Harden:* Austenitize at 845 to 900 °C (1550 to 1650 °F); oil quench.

- *Temper:* At least ½ h at 200 to 650 °C (400 to 1200 °F). Tempering temperature and time at temperature depend on desired final hardness or strength.

- *Austemper:* Austenitize in a salt bath at 845 to 900 °C (1550 to 1650 °F); quench in a salt bath at 230 to 315 °C (450 to 600 °F), hold 20 to 30 min, then oil quench or air cool to room temperature.

- *Martemper:* Austenitize in a salt bath at 845 to 870 °C (1550 to 1600 °F); quench in a salt bath at 230 to 260 °C (450 to 500 °F), equalize, then air cool or quench to room temperature. Temper to desired hardness.

- *Spheroidize:* Heat to 800 to 830 °C (1475 to 1525 °F), hold until heated through, furnace cool to 650 °C (1200 °F) and hold several hours, then cool slowly to room temperature.

Properties. Typical mechanical properties of small-diameter round sections of 6150 tempered at various temperatures are given in Table 13. Hardness and Izod impact energy as functions of tempering temperature are plotted in Fig. 3. The effects of section size on tensile properties and hardness are given in Table 14.

8640

Type 8640 steel was especially designed for maximum hardenability and best combination of properties with minimum alloying additions. It is an oil-hardening steel, but may be water hardened if precautions are taken to prevent cracking.

Heat Treatments. The following standard heat treatments apply to 8640 steel:

- *Normalize:* Heat to 870 to 925 °C (1600 to 1700 °F) and hold for a period that depends on section thickness; air cool.

- *Anneal:* Heat to 845 to 870 °C (1550 to 1600 °F) and hold for a period that depends on section thickness or furnace load; furnace cool.

- *Harden:* Austenitize at 815 to 845 °C (1500 to 1550 °F); quench in oil or water.

- *Temper:* At least ½ h at 200 to 650 °C (400 to 1200 °F). Temperature and time at temperature depend on desired hardness.

- *Spheroidize:* Heat to 705 to 720 °C (1300 to 1325 °F) and hold several hours; furnace cool.

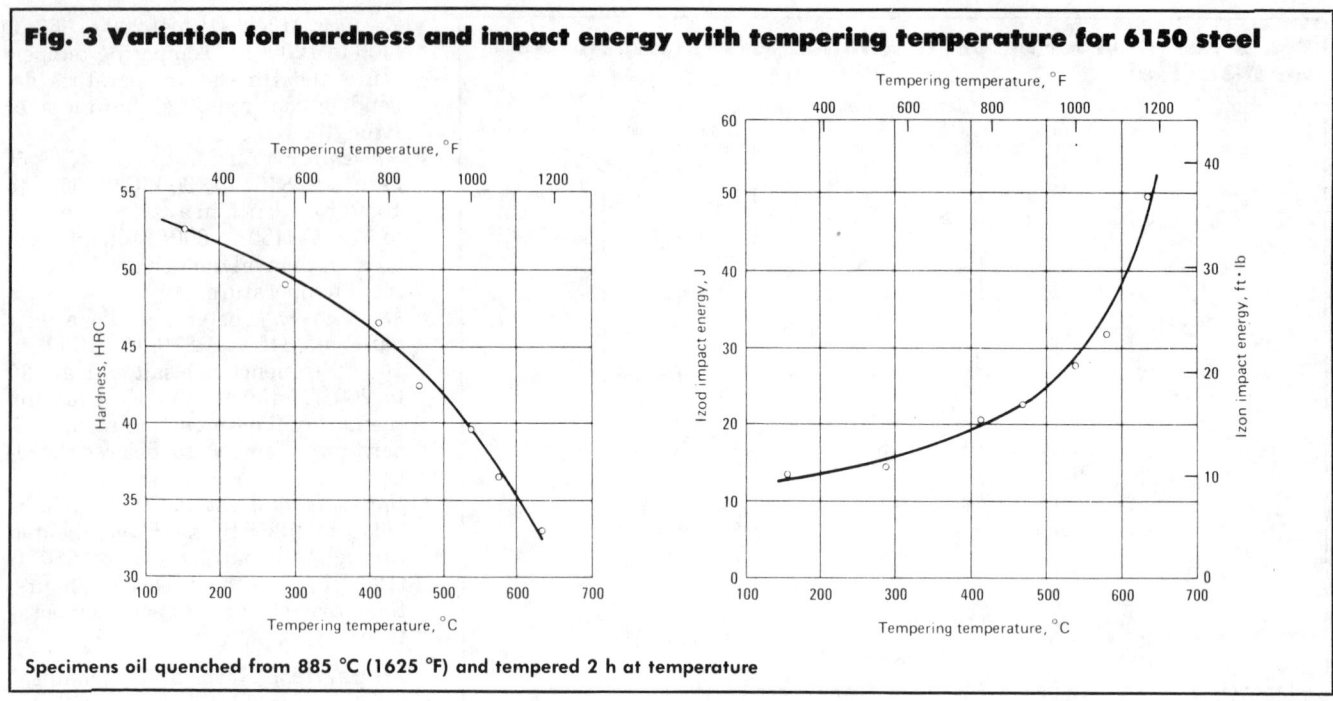

Fig. 3 Variation for hardness and impact energy with tempering temperature for 6150 steel

Specimens oil quenched from 885 °C (1625 °F) and tempered 2 h at temperature

Table 13 Typical room-temperature tensile properties of heat treated 6150 steel

Tempering temperature °C	°F	Tensile strength MPa	ksi	Yield strength MPa	ksi	Elongation in 50 mm or 2 in., %	Reduction in area, %	Hardness, HB	Izod impact energy J	ft·lb
Round bars, 14 mm (0.55 in.) in diameter(a)										
205	400	2050	298	1810	263	1	5	610	···	···
260	500	2070	300	1810	263	4	12	570	···	···
315	600	1950	283	1720	250	7	27	540	···	···
370	700	1770	257	1620	235	10	37	505	9	7
425	800	1585	230	1490	216	11	42	470	14	10
480	900	1410	204	1340	195	12	44	420	16	12
540	1000	1250	182	1210	175	13	46	380	20	15
595	1100	1150	167	1080	157	16	47	350	28	21
Round bars, 25 mm (1 in.) in diameter(b)										
425	800	1570	228	1450	210	10	37	461	···	···
480	900	1360	197	1210	175	11	41	401	···	···
540	1000	1180	171	1030	150	12	45	341	···	···
595	1100	1030	150	875	127	15	50	302	···	···
650	1200	920	133	760	110	19	55	262	···	···
705	1300	810	118	660	96	23	61	235	···	···

(a) Normalized at 870 °C (1600 °F), oil quenched from 860 °C (1575 °F) and tempered at various temperatures. (b) Oil quenched from 860 °C and tempered at various temperatures

Properties. Variations in typical properties of heat treated round sections of 8640 with tempering temperature are given in Table 15. Variations in properties with section size (mass effect) are given in Table 16.

Medium-Alloy Air-Hardening Steels

Heat treatments for the ultrahigh-strength steels H11 Modified (H11 Mod) and H13, which are also known as 5% Cr hot work die steels, are discussed in this section. These steels are similar in composition, heat treatment and many properties. They have deep hardenability and can be hardened through in large sections by air cooling. Air hardening results in minimal residual stresses after hardening. Both H11 Mod and H13 are secondary-hardening steels, and thus develop optimum properties when tempered at temperatures above the secondary-hardening peaks at about 510 °C (950 °F). These high tempering temperatures provide substantial stress relief and stabilization of properties so that the steels can be used advantageously at elevated temperatures. This also enables heat treated parts to be warm worked, or preheated for welding, at temperatures as high as 55 °C (100 °F) below the prior tempering temperature. Both H11 Mod and H13 steels are subject to hydrogen embrittlement, and parts should be baked after any exposure to environments where hydrogen may be absorbed. Baking for 24 h or longer at 190 °C (375 °F) or higher is recommended.

Because H11 Mod and H13 are air-hardening steels, forged parts must be cooled slowly after forging to prevent stress cracking. After forging, parts should be charged into a furnace at about 790 °C (1450 °F), soaked until

Table 14 Effects of mass on typical properties of heat treated 6150 steel

Round bars, oil quenched from 830 °C (1525 °F) and tempered at 540 °C (1000 °F); 12.83-mm (0.505-in.) diam tensile specimens taken from center of 25-mm bars and from mid-radius of 50- and 75-mm diam bars

Bar size		Tensile strength		Yield strength		Elongation in 50 mm or 2 in., %	Reduction in area, %	Hardness, HB
mm	in.	MPa	ksi	MPa	ksi			
25	1	1185	172	1040	151	14	45	341
50	2	1170	170	1030	149	13	48	341
75	3	1090	158	950	138	13	47	331

Table 15 Typical room-temperature mechanical properties of 8640 steel

Tempering temperature		Tensile strength		Yield strength		Elongation in 50 mm or 2 in., %	Reduction in area, %	Impact energy		Hardness	
°C	°F	MPa	ksi	MPa	ksi			J	ft·lb	HB	HRC
Round bars, 13.5 mm (0.53 in.) in diameter(a)											
205	400	1810	263	1670	242	8.0	25.8	11.5(b)	8.5(b)	555	55
315	600	1585	230	1430	208	9.0	37.3	15.6(b)	11.5(b)	461	48
425	800	1380	200	1230	179	10.5	46.3	27.8(b)	20.5(b)	415	44
540	1000	1170	170	1050	152	14.0	53.3	56.3(b)	41.5(b)	341	37
650	1200	870	126	760	110	20.5	61.0	96.9(b)	71.5(b)	269	28
Round bars, 25 mm (1 in.) in diameter(a)											
425	800	1382	200.5	1230	179	10	46	27(c)	20(c)	415	44
480	900	1250	181	1120	162	13	51	41(c)	30(c)	388	42
540	1000	1070	155	940	137	17	56	54(c)	40(c)	331	36
595	1100	1020	148	910	132	16	57	73(c)	54(c)	302	32
650	1200	865	125.5	760	110.5	20	61	83(c)	61(c)	269	28

(a) Oil quenched from 830 °C (1525 °F) and tempered at indicated temperature. (b) Izod. (c) Charpy V-notch

Table 16 Effects of mass on typical properties of heat treated 8640 steel

Oil quenched from 830 °C (1525 °F) and tempered at 540 °C (1000 °F)

Bar size		Tensile strength		Yield strength		Elongation in 50 mm or 2 in., %	Reduction in area, %	Surface hardness, HB
mm	in.	MPa	ksi	MPa	ksi			
25	1	1070	155	940	137	17	56	331
50	2	910	132	770	112	18	57	293
75	3	860	125	710	103	19	58	277

Table 17 Typical longitudinal mechanical properties of H11 Mod steel

Air cooled from 1010 °C (1850 °F); double tempered, 2 + 2 h at indicated temperature

Tempering temperature		Tensile strength		Yield strength		Elongation in 50 mm or 2 in., %	Reduction in area, %	Charpy V-notch impact energy		Hardness, HRC
°C	°F	MPa	ksi	MPa	ksi			J	ft·lb	
510	950	2120	308	1710	248	5.9	29.5	13.6	10.0	56.5
540	1000	2010	291	1675	243	9.6	30.6	21.0	15.5	56.0
565	1050	1850	269	1565	227	11.0	34.5	26.4	19.5	52.0
595	1100	1540	223	1320	192	13.1	39.3	31.2	23.0	45.0
650	1200	1060	154	850	124	14.1	41.2	40.0	29.5	33.0
705	1300	940	136	700	101	16.4	42.2	90.6	66.8	29.0

the temperature is uniform, and then slowly cooled, either in the furnace or in an insulating medium such as ashes, lime, mica or silocel. When the forgings have cooled, they should be spheroidize annealed. Weldments, especially heavy-section weldments, should be cooled slowly in a furnace heated to the preheating temperature, or in an insulating medium, immediately after welding. After being cooled, weldments should be given a full spheroidizing anneal.

H11 Mod

This steel is a modification of the martensitic hot work die steel H11, the significant difference being that H11 Mod has a slightly higher carbon content.

Heat Treatments. The following standard heat treatments apply to H11 Mod steel:

- *Normalize:* Generally not recommended. For effective homogenization, heat to about 1065 °C (1950 °F), soak 1 h for each 25 mm (1 in.) of thickness, air cool. Anneal immediately after the steel reaches room temperature. NOTE: There is a possibility that H11 Mod may crack during this treatment, especially if the surface is significantly decarburized.

- *Anneal:* Heat to 845 to 885 °C (1550 to 1625 °F) in a furnace, preferably one with controlled atmosphere, and hold to equalize temperature; cool very slowly in the furnace to about 480 °C (900 °F), then more rapidly to room temperature. This treatment should produce a fully spheroidized microstructure free of grain-boundary carbide networks.

- *Harden:* Preheat to 760 to 815 °C (1400 to 1500 °F), then raise the temperature to 995 to 1025 °C (1825 to 1875 °F) and hold 20 min plus 5 min for each 25 mm (1 in.) of thickness (minimum, 25 min); air cool in still air. This can be conveniently done in a neutral salt bath or a controlled-atmosphere furnace. For a few applications, oil quenching from the low end of the hardening temperature range may be done. Air cooling, which produces less distortion than oil quenching, is more commonly employed.

- *Temper:* At the secondary hardening peak temperature of about 510 °C (950 °F) for maximum hardness and strength, or preferably above the secondary-hardening peak to temper

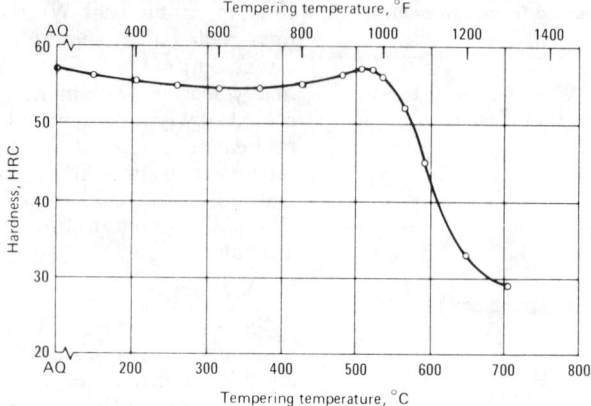

All specimens air cooled from 1010 °C (1850 °F) and double tempered, 2 + 2 h at temperature

Table 18 Effect of billet size and melting method on typical transverse strength and ductility of H11 Mod steel

Air cooled from 1010 °C (1850 °F); triple tempered, 2 + 2 + 2 h at 540 °C (1000 °F)

Billet size	Melting method	Tensile strength		Reduction in area, %
		MPa	ksi	
150 by 150 mm (6 by 6 in.)......	Air	1965	285	16.1
	VAR	1985	288	25.7
300 by 300 mm (12 by 12 in.)....	Air	1972	286	7.2
	VAR	2013	292	19.7

Table 19 Typical room-temperature longitudinal mechanical properties of H13 steel

Round bars, oil quenched from 1010 °C (1850 °F) and double tempered, 2 + 2 h at indicated temperature

Tempering temperature		Tensile strength		Yield strength		Elongation in 4 D, %	Reduction in area, %	Charpy V-notch impact energy		Hardness, HRC
°C	°F	MPa	ksi	MPa	ksi			J	ft·lb	
527	980	1960	284	1570	228	13.0	46.2	16	12	52
555	1030	1835	266	1530	222	13.1	50.1	24	18	50
575	1065	1730	251	1470	213	13.5	52.4	27	20	48
593	1100	1580	229	1365	198	14.4	53.7	28.5	21	46
605	1120	1495	217	1290	187	15.4	54.0	30	22	44

back to a lower hardness or strength with improved ductility and toughness. A minimum of 1 h at temperature should be allowed, but parts preferably should be double tempered (2 h at temperature, cool to room temperature, then 2 h more at temperature). Triple tempering is more desirable, especially for critical parts. For high-temperature applica-

tions, parts should be tempered at a temperature above the maximum service temperature to guard against unwanted changes in properties during service.

- *Stress relieve:* Heat to 650 to 675 °C (1200 to 1250 °F); cool slowly to room temperature. This treatment is often used to achieve greater dimensional accuracy in heat treated parts by

stress relieving rough-machined parts, then finish machining, and finally heat treating to the desired hardness.

- *Nitride:* For increased wear resistance, finish machined and heat treated parts may be nitrided. The nitriding operation can be considered as the second temper of a double tempering operation. The parts should be gas or liquid nitrided at about 525 °C (980 °F). The nitrided case depth depends on time at temperature. For example, gas nitriding in 20 to 30% dissociated ammonia for 8 to 48 h normally produces a case depth of about 0.2 to 0.35 mm (0.008 to 0.014 in.).

Properties. Variation of hardness with tempering temperature for H11 Mod is plotted in Fig. 4. Variations in typical room-temperature longitudinal mechanical properties with tempering temperature are given in Table 17. As an indication of the deep air hardenability of this steel (to depths greater than 305 mm, or 12 in.), the transverse tensile strength and ductility obtained in large billets of air-melted and vacuum-arc-remelted H11 Mod are given in Table 18, which also shows the improvement in ductility that results from vacuum arc remelting. Each value is the average of four tests, two from the top and two from the bottom of the ingot.

H13

The main difference in composition between H11 Mod and H13 is the higher vanadium content of the latter (see Table 1); this leads to a greater dispersion of hard vanadium carbides, which results in higher wear resistance. H13 parts may be nitrided for additional wear resistance. Also, H13 has a slightly wider range of carbon content than H11 Mod. Depending on the producer, the carbon content of H13 may be near the high or low side of the accepted range, and for a given heat treatment the properties will vary correspondingly.

Heat Treatments. The following standard heat treatments apply to H13 steel:

- *Normalize:* Not recommended for H13. Some improvement in homogeneity can be obtained by preheating to about 790 °C (1450 °F), heating slowly and uniformly to 1040 to 1065 °C (1900 to 1950 °F) and holding 1 h for each 25 mm (1 in.) of

Fig. 5 Variation for hardness with tempering temperature for H13 steel

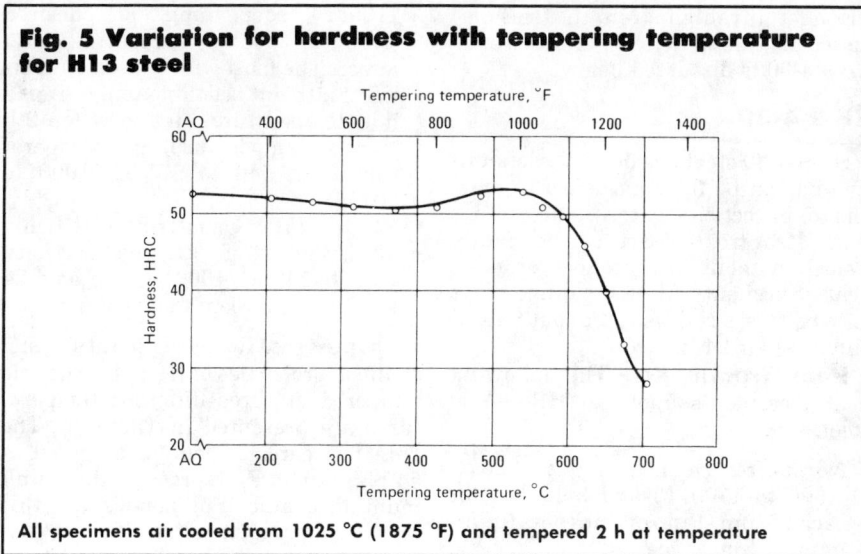

All specimens air cooled from 1025 °C (1875 °F) and tempered 2 h at temperature

Table 20 Room-temperature mechanical properties of HP 9-4-30 steel

Property	Typical value for hardness of 49-53 HRC(a)	Typical value for hardness of 44-48 HRC(b)	Minimum value(c)
Tensile strength	1650-1790 MPa (240-260 ksi)	1520-1650 MPa (220-240 ksi)	1520 MPa (220 ksi)
Yield strength(d)	1380-1450 MPa (200-210 ksi)	1310-1380 MPa (190-200 ksi)	1310 MPa (190 ksi)
Elongation in 4 D	8-12%	12-16%	10%
Reduction in area	25-35%	35-50%	35%
Charpy V-notch impact energy	20-27 J (15-20 ft·lb)	24-34 J (18-25 ft·lb)	24 J (18 ft·lb)
Fracture toughness (K_{Ic})	66-99 MPa$\sqrt{m}$ (60-90 ksi$\sqrt{in.}$)	99-115 MPa$\sqrt{m}$ (90-105 ksi$\sqrt{in.}$)	...

(a) Oil quenched from 845 °C (1550 °F), refrigerated to −73 °C (−100 °F) and double tempered at 205 °C (400 °F). (b) Same heat treatment as (a) except double tempered at 550 °C (1025 °F). (c) For sections forged to 75 mm (3 in.) or less in thickness (or to less than 0.016 m², or 25 in.², in total cross-sectional area), quenched to martensite and double tempered at 540 °C (1000 °F)

thickness, and then air cooling. Just before or just as the steel reaches room temperature, it should be recharged into a furnace and given a full spheroidizing anneal. NOTE: There is a risk of cracking during this treatment, especially if done in a furnace where the atmosphere is not controlled to prevent surface decarburization.

- *Anneal:* Heat uniformly to 860 to 900 °C (1575 to 1650 °F) in a controlled-atmosphere furnace, or with the part in a neutral compound, to prevent decarburization, and hold to equalize temperature; cool very slowly in the furnace to about 480 °C (900 °F), then cool more rapidly to room temperature. This treatment should result in a fully spheroidized microstructure free from grain-boundary carbide networks.
- *Harden:* Preheat to 790 to 815 °C (1450 to 1500 °F), then raise the temperature uniformly to 995 to 1025 °C (1825 to 1875 °F) and soak 20 min plus 5 min for each 25 mm (1 in.) of thickness (minimum, 25 min); air cool in still air. For a few applications, oil quenching from the low side of the hardening temperature may be done, but at the risk of distortion or cracking. Air cooling is preferred, and usually is done from the high side of the hardening temperature range.
- *Temper:* At the secondary-hardening peak of about 510 °C (950 °F) for

maximum hardness and strength, but preferably at a higher temperature to temper back to a lower level of hardness or strength with improved toughness and ductility. Double tempering—2 h at temperature, air cool, then 2 h more at temperature—is recommended; for critical parts, triple tempering may be desirable.

- *Stress relieve:* Heat to 650 to 675 °C (1200 to 1250 °F) and soak 1 h or more; cool slowly to room temperature. This treatment often is used to achieve greater dimensional accuracy in heat treated parts by stress relieving rough-machined parts, then finish machining, and finally heat treating to the desired hardness.

- *Nitride:* Finish-machined and heat treated parts may be nitrided. Because it is carried out at about the normal tempering temperature, nitriding can serve as the second temper in a double tempering treatment. The nitrided case depth depends on time at temperature. For example, gas nitriding at 510 °C (950 °F) for 10 to 12 h produces a case depth of 0.10 to 0.13 mm (0.004 to 0.005 in.), whereas gas nitriding for 40 to 50 h results in a case depth of about 0.3 to 0.4 mm (0.012 to 0.016 in.). For selective nitriding, copper plating is preferred for stopping off areas that are not to be nitrided; stop-offs containing lead should be avoided, because lead has been found to embrittle H13 steel.

Properties. The properties presented in this section are for H13 with a carbon content in the middle of the composition range (for composition range, see Table 1). Somewhat different properties should be expected when the carbon content is near the high end or the low end of the range.

Variation of hardness with tempering temperature for H13 is plotted in Fig. 5. Typical room-temperature longitudinal mechanical properties of bars tempered to different hardness levels are given in Table 19. H13 has deep hardenability, although its hardenability is slightly less than that of H11 Mod. For example, an H13 bar 330 mm (13 in.) in diameter and 2743 mm (108 in.) long, when fast air cooled from 1010 °C (1850 °F), had an as-quenched hardness of 45 HRC.

9Ni-4Co Steels

During the 1960's, Republic Steel Corporation introduced a family of four HP 9-4 (9Ni-4Co) steels having high fracture toughness when heat treated to very high strength levels. Among these, HP 9-4-20 and HP 9-4-30 are commercially available. They nominally contain 0.20 and 0.30% C, respectively (see Table 1 for chemical compositions). With the increase in carbon content, the attainable strength increases but with corresponding decreases in toughness and weldability. The high nickel content of 9% provides deep hardenability and toughness, and the 4% Co prevents retention of excessive austenite in heat treated parts. The carbide-forming elements, chromium and molybdenum, also impart hardenability, but the amounts of these carbide formers are adjusted to provide a fairly flat response to tempering without pronounced secondary hardening and its attendant reduction in toughness.

HP 9-4-20, although it has good weldability and fracture toughness, cannot develop a yield strength of 1380 MPa (200 ksi), which was selected as the criterion for ultrahigh-strength steels discussed in this article. Therefore, only HP 9-4-30 is discussed below.

HP 9-4-30

HP 9-4-30 steel has deep hardenability and can be fully hardened to martensite in sections up to 150 mm (6 in.) thick. Heat treated parts can be readily welded without any preheat or postheat treatment. After welding, parts may be stress relieved at about 540 °C (1000 °F) for 24 h.

Heat Treatments. The following heat treatments apply to HP 9-4-30 steel:

- *Normalize:* Heat to 870 to 925 °C (1600 to 1700 °F) and hold 1 h for each 25 mm (1 in.) of thickness (minimum, 1 h); air cool.
- *Anneal:* Heat to 620 °C (1150 °F) and hold 24 h; air cool.
- *Harden:* Austenitize at 830 to 860 °C (1525 to 1575 °F) and hold 1 h for each 25 mm (1 in.) of thickness (minimum, 1 h); water or oil quench. Complete the martensitic transformation by refrigerating at least 1 h at −87 to −60 °C (−125 to −75 °F); let warm to room temperature.
- *Temper:* At 200 to 600 °C (400 to 1100 °F), depending on desired strength; double tempering preferred. The most widely used tempering treatment is double tempering (2 h at temperature, air cool, then 2 h more at temperature) at a temperature from 540 to 575 °C (1000 to 1075 °F).
- *Stress relieve:* Usually required only after welding of restrained sections. Heat to 540 °C (1000 °F) and hold 24 h; air cool to room temperature.

Properties. Room-temperature mechanical properties of HP 9-4-30 double tempered at three different temperatures are presented in Table 20. The data for material double tempered at 540 °C (1000 °F) represent the minimum mechanical properties for this condition; properties listed for the other conditions may be considered typical.

REFERENCES

The data presented in this article were extracted from numerous sources as referenced in the original article by this author on Ultrahigh-Strength Steels, Metals Handbook, 9th Ed, Vol 1, p 421 to 443.

Control of Distortion in Tool Steels*

Edited by Daniel S. Zamborsky
Corporate Metallurgist
Warner & Swasey Co.

DISTORTION in tool steel parts includes all irreversible changes in size and shape that result from processing, from heat treatment, and from temperature variations and loading in service. A basic understanding of distortion is important for two reasons. First, most tool steel parts must interact with other parts in service, and excessive distortion may prevent them from interacting in the intended manner. Second, finishing operations for correcting distortion not only are expensive but also may destroy some desirable properties and introduce others that are undesirable.

Changes in size or shape of tool steel parts may be either reversible or irreversible. Reversible changes are those caused by stressing in the elastic range or by temperature variations that neither cause changes in the metallurgical structure nor induce stresses that exceed the elastic range. Under such conditions, the initial dimensional values can be restored by a return to the original state of stress or temperature.

The upper limit of reversible dimensional change in a tool steel is determined by the stress required to initiate deformation (that is, the elastic limit corresponding to a preselected value of plastic strain), the elastic deformation per unit stress (modulus of elasticity), the effect of temperature on these properties, the coefficient of thermal expansion and the temperature-time combinations at which stress relief and phase changes occur.

For practical purposes the modulus of elasticity of all tool steels, regardless of composition or heat treatment, is 210 GPa (30×10^6 psi) at room temperature. Therefore, if a tool steel part deforms excessively under service loading but returns to its original dimensions when the load is removed, a change in grade or type of tool steel or in heat treatment will not be useful. To counteract excessive elastic distortion it is necessary to (a) reduce the applied stress by increasing the section size or (b) use a tool material with a higher modulus of elasticity (such as cemented tungsten carbide).

Irreversible changes in size or shape of tool steel parts are those caused by stresses that exceed the elastic limit or by changes in metallurgical structure (most notably, phase changes). Such irreversible changes sometimes can be corrected by thermal processing (annealing, tempering or cold treating) or by mechanical processing to remove excess material or to redistribute residual stresses.

Nature and Causes of Distortion

Distortion is a general term encompassing all irreversible dimensional changes. There are two main types: size distortion, which involves expansion or contraction in volume or linear dimensions without changes in geometrical form; and shape distortion, which entails changes in curvature or angular relations, as in twisting, bending, and/or nonsymmetrical changes in dimensions. Frequently, both size distortion and shape distortion (illustrated schematically in Fig. 1) occur during a heat treating operation.

Size distortion is the result of a change in volume produced by a change in metallurgical structure during heat

*This article has been condensed from the more complete discussion of this complicated subject provided by Bernard S. Lement in his book Distortion in Tool Steels (Ref 1).

treatment. Shape distortion results from either residual or applied stresses. Residual stresses developed during heat treatment are caused by thermal gradients within the metal (producing differing amounts of expansion or contraction) by nonuniform changes in metallurgical structure and by nonuniformity in the composition of the metal itself, such as that due to segregation.

Changes in metallurgical structure during heat treatment of tool steels are produced by the three steps described below.

The first step involves heating an annealed structure (usually consisting of ferrite and spheroidal carbides, commonly called spheroidite) to about 800 °C (1450 °F) or higher to change the ferrite to austenite and to dissolve all or most of the spheroidal carbides to the austenite. For plain carbon or low-alloy tool steels, austenitizing results in a contraction in volume. The extent of volumetric contraction decreases with increasing amounts of carbon present in the composition. This can be approximated as follows:

$$V_{SA} = -4.64 + 2.21 \, (\%C) \qquad \text{(Eq 1)}$$

where V_{SA} is the volume change in percent that occurs when spheroidite transforms to austenite. By use of this equation, it can be estimated that, if heated to a temperature high enough to dissolve all of the carbon in the austenite, a 0.50% carbon tool steel would exhibit a volume change of -3.53%, a common type containing 1% carbon would exhibit a change of -2.43%, and a very-high-carbon type containing 1.5% carbon would exhibit a change of -1.33%. However, tool steels having carbon contents higher than that of the eutectoid composition are normally austenitized at temperatures only high enough to dissolve the eutectoid amount of carbon. Under these circumstances, 1% carbon and 1.5% carbon tool steels would exhibit changes in volume of -2.77 and -2.53%, respectively, after austenitizing. These percentages are less than that calculated directly from Eq 1 because an allowance must be made for the volume occupied by undissolved carbides, which is about 3.5% for the 1.0% carbon steel and about 12% for the 1.5% carbon steel.

The second step involves cooling fast enough to cause the austenite to transform to martensite. The steel expands on transformation, the amount of expansion being in inverse proportion to

Fig. 1 Size and shape distortion in hardening

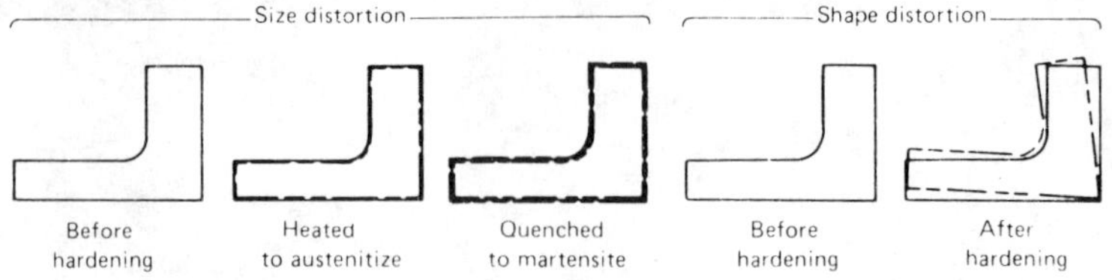

Table 1 Microconstituents in various tool steels after hardening

Steel	Hardening treatment	As-quenched hardness, HRC	Martensite, vol %	Retained austenite, vol %	Undissolved carbides, vol %
W1	790 °C (1450 °F), 30 min; WQ	67.0	88.5	9	2.5
L3	845 °C (1550 °F), 30 min; OQ	66.5	90	7	3.0
M2......	1225 °C (2235 °F), 6 min; OQ	64	71.5	20	8.5
D2	1040 °C (1900 °F), 30 min; AC	62	45	40	15

the amount of carbon in solution in the austenite:

$$V_{AM} = 4.64 - 0.53 \, (\%C) \qquad \text{(Eq 2)}$$

where V_{AM} is the percent volume change that occurs when austenite transforms to martensite. By use of Eq 2, it can be estimated that a 0.5% carbon tool steel would exhibit a volume increase for this transformation of 4.37% and that 1.0 and 1.5% carbon steels would exhibit increases of 4.07% and 3.71%, respectively, if austenitized at the normal austenitizing temperature (only 0.8% carbon, the eutectoid amount of carbon, in solution and again allowing for the volume occupied by undissolved carbides).

Equations 1 and 2 can be used to calculate the net change in dimensions in a tool steel when it is heat treated to transform it from an annealed to a fully hardened (martensitic) state. For the examples referred to above, normal heat treatment would produce net volume increases of $-3.53 + 4.37 = 0.84\%$ in the 0.5% carbon tool steel, $-2.77 + 4.07 = 1.30\%$ in the 1.0% carbon steel and $-2.53 + 3.71 = 1.18\%$ in the 1.5% carbon steel. Net changes in linear dimensions would be about one third the corresponding net changes in volume.

The third step involves reheating the freshly formed martensite to relatively low temperatures (tempering) to increase toughness and reduce lattice stress. Tempering produces various changes in metallurgical structure, depending on temperature and time at temperature.

After very long times at room temperature or shorter times at temperatures up to 200 °C (400 °F), the high carbon martensite in plain carbon and low-alloy tool steels decomposes into low-carbon martensite (about 0.25% carbon) plus epsilon carbide, with an accompanying contraction in volume. At higher tempering temperatures, 200 to 430 °C (400 to 800 °F), the martensite decomposes into ferrite plus cementite.

Transformation of the maximum amount of austenite to martensite on quenching usually requires continuous cooling to below the martensite-finish temperature (M_f), which for a eutectoid tool steel is about -50 °C (-60 °F). To prevent cracking of very large or very intricate pieces, it is common practice to remove the tool from the quenching medium and begin tempering it while it is still slightly too warm to hold comfortably in the bare hands (about 60 °C, or 140 °F). Under these conditions, a substantial proportion of the structure (10% or more) may still be austenite. Most alloying elements lower the M_f temperature. Consequently, more austenite is retained at room temperature in the more highly alloyed tool steels. On tempering at increasing temperatures in the range 120 to 260 °C (250 to

Table 2 Typical dimensional changes in hardening and tempering

Tool steel	Hardening treatment Temperature °C	°F	Quenching medium	Total change in linear dimensions, % after quenching	°C °F	150 300	205 400	260 500	315 600	370 700	425 800	480 900	510 950	540 1000	565 1050	595 1100
O1	815	1500	Oil	0.22		0.17	0.16	0.18	...	...	...	...	...	...	...	...
O1	790	1450	Oil	0.18		0.09	0.12	0.13	...	...	...	...	...	...	...	...
O6	790	1450	Oil	0.12		0.07	0.10	0.14	0.10	0.00	−0.05	−0.06	...	−0.07	...	...
A2	955	1750	Air	0.09		0.06	0.06	0.08	0.07	...	0.05	0.04	...	0.06	...	...
A10	790	1450	Air	0.04		0.00	0.00	0.08	0.08	0.01	0.01	0.02	...	0.01	...	0.02
D2	1010	1850	Air	0.06		0.03	0.03	0.02	0.00	...	−0.01	−0.02	...	0.06	...	...
D3	955	1750	Oil	0.07		0.04	0.02	0.01	−0.02	...	...	...	...	...	...	...
D4	1040	1900	Air	0.07		0.03	0.01	−0.01	−0.03	...	−0.4	−0.03	...	0.05	...	...
D5	1010	1850	Air	0.07		0.03	0.02	0.01	0.00	...	0.3	0.03	...	0.05	...	...
H11	1010	1850	Air	0.11		0.06	0.07	0.08	0.08	...	0.3	0.01	...	0.12	...	...
H13	1010	1850	Air	−0.01		...	...	...	...	...	...	0.00	...	0.06	...	...
M2	1210	2210	Oil	−0.02		...	...	...	...	...	...	...	−0.06	0.10	0.14	0.16
M41	1210	2210	Oil	−0.16		...	...	...	...	...	...	...	−0.17	0.08	0.21	0.23

500 °F), increasing amounts of this retained austenite transform to bainite for some tool steel compositions with an accompanying expansion in volume.

Depending on the alloy content of the tool steel, all, some, or none of the retained austenite will transform during tempering. In some highly alloyed tool steel compositions, cementite redissolves at tempering temperature of 540 to 595 °C (1000 to 1100 °F) to form alloy carbides, which induces an additional expansion in volume. The formation of alloy carbides during tempering is characteristic of tool steels containing large amounts of carbide-forming elements such as chromium, molybdenum and tungsten, which are found in high-speed tool steels.

Size Distortion in Tool Steels

Typical volume percentages of martensite, retained austenite and undissolved carbides are given in Table 1 for four different tool steels quenched from their recommended austenitizing temperatures.

Typical changes in linear dimensions for several tool steels are given in Table 2. As shown in this table, some tool steels such as A10 show very little size change when hardened and tempered over the entire range from 150 to 600 °C (300 to 1100 °F).

Other types, such as the M2 and M41 high speed steels, expand about 0.2% (2 mm/m, or 0.002 in./in.) when hardened and tempered in the temperature range of 540 to 595 °C (1000 to 1100 °F) to develop full secondary hardness. Although the information in Table 2 is useful in comparing size distortion in several tool steels, the factor of shape distortion makes it impossible to use these data alone to predict dimensional changes of a particular tool made from any of these steels. Densities and thermal expansion characteristics for several classes of tool steels are presented in Table 3.

Shape Distortion in Tool Steels

The strength of any tool steel decreases rapidly above about 600 °C (1100 °F). At the austenitizing temperature, the yield strength is so low that plastic deformation often occurs simply from the stresses induced in the part by gravity. Therefore, long parts, large parts and parts of complex shape must be properly supported at critical locations to prevent sagging at the hardening temperature.

Rapid heating increases shape distortion, especially in large tools and in complex tools containing both light and heavy sections. If the rate of heating is high, light sections will increase in temperature much faster than heavy sections. Likewise, the outer surfaces in heavy sections will increase in temperature much faster than the interior. Differences in thermal expansion due to the differences in temperature between light and heavy sections or between surface and interior in heavy sections will be enough to set up large stresses in the material. Under these stresses, the hotter regions will deform plastically, to relieve the thermally induced stress.

Eventually, the hotter portions will reach the furnace temperature, while the cooler portions will continue to increase in temperature. At this point, a decrease in thermal differential begins, which will cause a partial reversal in thermal stress which produced plastic deformation when the temperature differential was high. This may cause the part to undergo further plastic deformation, but to a lesser extent than the deformation caused by the initial high temperature differential. Such deformation will occur in a different direction.

Slow heating minimizes distortion by keeping temperature differentials low and thermal stresses within the elastic range of the material throughout the heating cycle. Ideally, all heat treatment of tool steel parts should start from a cold furnace to provide the greatest freedom from shape distortion during heating. Starting from a cold furnace is neither very practical nor energy efficient unless heat treating is being done in a vacuum furnace. When heat treating in fused salt or an atmosphere furnace, preheating the parts at an intermediate temperature prior to heating them to the austenitizing temperature provides the best compromise.

During quenching, large temperature differences between surface and interior, and between light and heavy sections, can cause severe shape distortion, because of thermal stress and mechanical stress produced by a martensitic transformation. This problem is most severe if the hardenability of the steel is so low that a fast cooling rate is required to obtain full hardness. In such a situation, especially when making a large or complex part, it may be best to substitute a high-hardenability, air-hardening tool steel, which re-

Table 3 Density and thermal expansion of selected tool steels

Type	Density Mg/m³	Density lb/in.³	μm/m·K from 20 °C to: 100 °C	205 °C	425 °C	540 °C	650 °C	μin./in. °F from 68 °F to: 200 °F	400 °F	800 °F	1000 °F	1200 °F
W1	7.84	0.283	10.4	11.0	13.1	13.8(a)	14.2(b)	5.76	6.13	7.28	7.64(a)	7.90(b)
W2	7.85	0.283	···	···	14.4	14.8	14.9	···	···	8.0	8.2	8.3
S1	7.88	0.255	12.4	12.6	13.5	13.9	14.2	6.9	7.0	7.5	7.7	7.9
S2	7.79	0.281	10.9	11.9	13.5	14.0	14.2	6.0	6.6	7.5	7.8	7.9
S5	7.76	0.280	···	···	12.6	13.3	13.7	···	···	7.0	7.4	7.6
S6	7.75	0.280	···	···	12.6	13.3	···	···	···	7.0	7.4	···
S7	7.76	0.280	···	12.6	13.3	13.7(a)	13.3	···	7.0	7.4	7.6(a)	7.4
O1	7.85	0.283	···	10.6(c)	12.8	14.0(d)	14.4(d)	···	5.9(c)	7.1	7.8(d)	8.0(d)
O2	7.66	0.277	11.2	12.6	13.9	14.6	15.1	6.2	7.0	7.7	8.1	8.4
O7	7.8	0.283	···	···	···	···	···	···	···	···	···	···
A2	7.86	0.284	10.7	10.6(c)	12.9	14.0	14.2	5.96	5.91(c)	7.2	7.8	7.9
A6	7.84	0.283	11.5	12.4	13.5	13.9	14.2	6.4	6.9	7.5	7.7	7.9
A7	7.66	0.277	···	···	12.4	12.9	13.5	···	···	6.9	7.2	7.5
A8	7.87	0.284	···	···	12.0	12.4	12.6	···	···	6.7	6.9	7.0
A9	7.78	0.281	···	···	12.0	12.4	12.6	···	···	6.7	6.9	7.0
D2	7.70	0.278	10.4	10.3	11.9	12.2	12.2	5.8	5.7	6.6	6.8	6.8
D3	7.70	0.278	12.0	11.7	12.9	13.1	13.5	6.7	6.5	7.2	7.3	7.5
D4	7.70	0.278	···	···	12.4	···	···	···	···	6.9	···	···
D5	···	···	···	···	···	12.0	···	···	···	···	6.7	···
H10	7.81	0.281	···	···	12.2	13.3	13.7	···	···	6.8	7.4	7.6
H11	7.75	0.280	11.9	12.4	12.8	12.9	13.3	6.6	6.9	7.1	7.2	7.4
H13	7.76	0.280	10.4	11.5	12.2	12.4	13.1	5.8	6.4	6.8	6.9	7.3
H14	7.89	0.285	11.0	···	···	···	···	6.1	···	···	···	···
H19	7.98	0.288	11.0	11.0	12.0	12.4	12.9	6.1	6.1	6.7	6.9	7.2
H21	8.28	0.299	12.4	12.6	12.9	13.5	13.9	6.9	7.0	7.2	7.5	7.7
H22	8.36	0.302	11.0	···	11.5	12.0	12.4	6.1	···	6.4	6.7	6.9
H26	8.67	0.313	···	···	···	12.4	···	···	···	···	6.9	···
H42	8.15	0.295	···	···	···	11.9	···	···	···	···	6.6	···
T1	8.67	0.313	···	9.7	11.2	11.7	11.9	···	5.4	6.2	6.5	6.6
T2	8.67	0.313	···	···	···	···	···	···	···	···	···	···
T4	8.68	0.313	···	···	···	11.9	···	···	···	···	6.6	···
T5	8.75	0.316	11.2	···	···	11.5	···	6.2	···	···	6.4	···
T6	8.89	0.321	···	···	···	···	···	···	···	···	···	···
T8	8.43	0.305	···	···	···	···	···	···	···	···	···	···
T15	8.19	0.296	···	9.9	11.0	11.5	···	···	5.5(c)	6.1	6.4	···
M1	7.89	0.285	···	10.6(c)	11.3	12.0	12.4	···	5.9(c)	6.3	6.7	6.9
M2	8.16	0.295	10.1	9.4(c)	11.2	11.9	12.2	5.6	5.2(c)	6.2	6.6	6.8
M3, class 1	8.15	0.295	···	···	11.5	12.0	12.2	···	···	6.4	6.7	6.8
M3, class 2	8.16	0.295	···	···	11.5	12.0	12.8	···	···	6.4	6.7	7.1
M4	7.97	0.288	···	9.5(c)	11.2	12.0	12.2	···	5.3(c)	6.2	6.7	6.8
M7	7.95	0.287	···	9.5(c)	11.5	12.2	12.4	···	5.3(c)	6.4	6.8	6.9
M10	7.88	0.255	···	···	11.0	11.9	12.4	···	···	6.1	6.6	6.9
M30	8.01	0.289	···	···	11.2	11.7	12.2	···	···	6.2	6.5	6.8
M33	8.03	0.290	···	···	11.0	11.7	12.0	···	···	6.1	6.5	6.7
M36	8.18	0.296	···	···	···	···	···	···	···	···	···	···
M41	8.17	0.295	···	9.7	10.4	11.2	···	···	5.4	5.8	6.2	···
M42	7.98	0.288	···	···	···	···	···	···	···	···	···	···
M46	7.83	0.283	···	···	···	···	···	···	···	···	···	···
M47	7.96	0.288	10.6	11.0	11.9	···	12.6	5.9	6.1	6.6	···	7.0
L2	7.86	0.284	···	···	14.4	14.6	14.8	···	···	8.0	8.1	8.2
L6	7.86	0.284	11.3	12.6	12.6	13.5	13.7	6.3	7.0	7.0	7.5	7.6
P2	7.86	0.284	···	···	13.7	···	···	···	···	7.6	···	···
P5	7.80	0.282	···	···	···	···	···	···	···	···	···	···
P6	7.85	0.283	···	···	···	···	···	···	···	···	···	···
P20	7.85	0.283	···	···	12.8	13.7	14.2	···	···	7.1	7.6	7.9

(a) From 20 to 500 °C (68 to 930 °F). (b) From 20 to 600 °C (68 to 1110 °F). (c) From 20 to 260 °C (68 to 500 °F). (d) From 38 °C (100 °F)

quires only a slow cooling rate to fully harden.

However, if lower-hardenability steels requiring liquid quenching are used, fixturing and pressure die quenching can help minimize distortion. Long symmetrical parts should be fixtured, and should be quenched in the vertical position with vertical agitation of the quench mediums.

Special Techniques for Controlling Shape Distortion

Special quenching procedures such as martempering and austempering may also be useful for controlling distortion in parts which have an appropriate configuration and been made of material having appropriate hardenability. In martempering, parts are quenched in hot molten salt fast enough to avoid transformation to high-temperature transformation products such as ferrite or pearlite. The parts are held at a bath temperature in the range from slightly above to slightly below the M_s just long enough to equalize the interior and surface temperatures. The parts are then removed from the bath and allowed to air cool to room temperature. Slow cooling through the martensitic transformation range reduces distortion as compared with rapid quenching. Martempered tools must be given the usual tempering treatment.

Austempering can be used to reduce distortion if a hardness no higher than 57 HRC is acceptable for the application. In austempering, parts are also quenched in hot molten salt but by temperature selection are forced to transform into bainite rather than martensite. Bainite forms at temperatures above those at which martensite forms. The parts must be held long enough at a temperature above the M_s (usually about 230 °C, or 450 °F) to permit the austenite to transform to lower bainite. When air cooled to room temperature, austempered tools exhibit less shape distortion and generally require no subsequent tempering.

Besides being reduced through control of rates of heating and cooling, shape distortion can be reduced by employing a localized method of heating and quenching such as flame hardening, induction hardening, electron beam or laser hardening to treat only that portion of the tool that must be hardened.

Fig. 2 Typical diameter changes during heat treatment for high speed steel bars

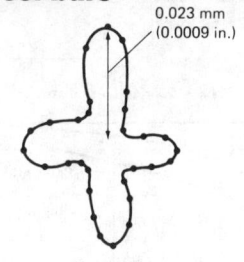

(a)   Conventional process

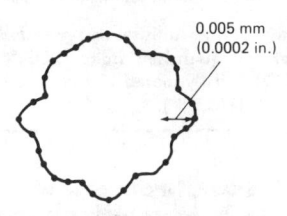

(b)   Special process

Drawings produced by calculation from precision measurements of diameter. Charts are plots on polar coordinates depicting variations in diameter after heat treatment for a bar that was round within ±1.25 μm (±0.00005 in.) before heat treatment. (Courtesy of Latrobe Steel Co., Latrobe, Pennsylvania)

Controlling out-of-roundness is important for certain precision applications, such as class C and D cutting hobs made of high speed steels. Class C and D hobs must be held to close size limits because they are not ground to size after heat treatment, but rather are used in the unground condition.

Normal size distortion in hardening and tempering can be accommodated by making the tool slightly oversize or slightly undersize, as required, before heat treating. High speed steel bars, however, have been observed to go out-of-round as much as 0.05 mm (0.002 in.) during heat treatment. The pattern of size distortion shown in Fig. 2(a) can occur. It appears to be related to the initial shape of the cast ingot and to the specific primary-mill processing used to reduce the ingot into bars. By changing steelmaking, forging and rolling procedures, out-of-roundness has been reduced to the smaller differential pattern shown in Fig. 2(b), where the difference between high and low points is

only 0.005 mm (0.0002 in.). High speed steel bars made this way are marketed by a few tool steel producers as "close tolerance hob stock". An even better method of combating out-of-roundness is to use high speed tool steel bars made from hot isostatically pressed powders, which maintain the best possible symmetry during conventional heat treatment.

Stabilization involves reducing the amount of retained austenite in heat treated material. Retained austenite that can slowly transform and produce distortion if the material is later heated or subjected to stress. Stabilization also reduces internal (residual) stress, which makes distortion in service due to stress relaxation less likely to occur. Stabilization is most important for tools that must retain their exact size and shape over long periods of time (i.e., gauges, blocks).

If the tool steel chosen provides the required hardness after tempering at a relatively high temperature, it is possible to reduce the amount of retained austenite and the internal stress by multiple tempering. Initial tempering reduces internal stress and conditions the retained austenite so that it can transform to martensite on cooling from the tempering temperature. Usually, a second or third retempering is necessary to reduce the internal stress set up by the transformation of retained austenite.

Single or repeated cold treatment to a temperature below M_f will cause most of the retained austenite to transform to martensite in plain carbon or low-alloy tool steels that must be tempered at low temperatures to achieve the hardness required. Cold treatment may be applied either before or after the first temper. If, however, the tools tend to crack because of the additional stress induced by dimensional expansion during cold treatment, it is generally prudent to apply cold treatment after the tools have been tempered the first time. When cold treatment is applied after the first temper, the amount of retained austenite that transforms during the cold treatment may be considerably less than desired because some of the austenite may have been stabilized by tempering prior to cold treating. Cold treatment is usually done in a commercial refrigeration unit capable of attaining −70 to −95 °C (−100 to −140 °F). Tools must be retempered promptly after return to room temperature following cold treatment to reduce internal stress and increase the toughness of the newly formed martensite.

For some tools, a small percentage of retained austenite is desirable to improve toughness and provide a favorable internal stress pattern that will help the tool withstand service stresses. For these tools, a full stabilizing treatment may actually result in tools that are unfit to perform their required functions.

Powder Metallurgy Steels

In recent years, tool steels with improved properties have been produced by the powder metallurgy (P/M) process. In this process, molten metal alloy is solidified as a fine powder by spraying it into a chamber filled with inert gas. Steel containers are filled with these powder particles, evacuated of gases, sealed, heated, and white hot, isostatically pressed to full density. The resulting compact is rolled or forged to size on conventional steel-mill equipment or, in some instances, is used as compacted to make tools.

P/M tool steels have two major advantages: complete freedom from macrosegregation and porosity and uniform distribution of extremely fine carbides. These characteristics provide deeper hardening and faster response to hardening conditions (see Fig. 3). The latter is important, particularly for molybdenum high speed steels, which tend to decarburize rapidly at austenitizing temperatures. P/M products also show less out-of-roundness distortion in large-diameter bars (see Table 4).

When sulfur is added to P/M tool steels, they exhibit a very fine homogeneous distribution of sulfides. This uniform sulfide distribution promotes better machinability. After heat treating, the refined hardened and tempered P/M tool steels grindability and greater toughness than conventionally processed (cast and wrought) tool steels.

As of 1979, the following AISI compositions of high speed steels were available in P/M form: M2, M2 with high sulfur and high carbon, M3 class 2 with sulfur, M4, M35 with sulfur, M42 and T15. P/M steels can be substituted for their conventional counterparts in all applications, and are particularly advantageous when heavy sections are required.

The freedom from gross segregation provided by the P/M process makes it possible to readily fabricate new higher-

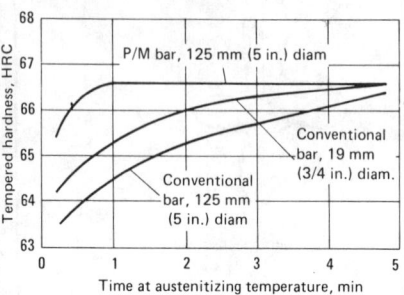

Fig. 3 Comparison of response to hardening for P/M and conventionally produced bars of M2S (HC) tool steel

Hardness at mid-radius was evaluated for bars oil quenched from 1200 °C (2200 °F) and tempered 2 + 2 + 2 h at 550 °C (1025 °F).

Table 4 Out-of-roundness distortion in large-diameter bars of M2S tool steel

Bar diameter mm in.	Production method	Typical out-of-roundness(a) mm in.
75 3	P/M	0.008 0.0003
	Conventional	0.020 0.0008
125 5	P/M	0.013 0.0005
	Conventional	0.033 0.0013
190 7.5	P/M	0.015 0.0006
	Conventional	0.051 0.0020

(a) Maximum diameter minus minimum diameter after normal hardening treatment

alloy tool steels compositions. One type now available, which contains 1.50 C, 3.75 Cr, 3.00 V, 10 W, 5.25 Mo and 9.00 Co, is reported to have the highest hot hardness of any high speed steel.

Surface Treatments

In many applications, service life of tool steels can be increased by surface treatments.

Oxide coatings, provided by treatment of the finish-ground tool in an alkali-nitrate bath or by steam oxidation, prevent or reduce adhesion of the tool to the workpiece. Oxide coatings have doubled tool life—particularly in machining of gummy materials such as soft copper and non-free-cutting low-carbon steels.

Plating of finished high speed steel tools with 0.0025 to 0.0125 mm (0.1 to

0.5 mil) of chromium also prolongs tool life by reducing adhesion of the tool to the workpiece. Chromium plating is relatively expensive, and precautions must be taken to prevent tool failure in service due to hydrogen embrittlement.

Electroless nickel plating has been used successfully as a replacement for chromium plating, both in routine production and for salvage plating operations on tool steel parts. Because plating by this method is accomplished by means of chemical reduction, it does not depend on any galvanic coupling between dissimilar metals, and there is no electrolysis involved. Therefore, there is no danger of hydrogen embrittlement. Plated hardness is in the high Rockwell 50's range, with good, uniform plated thickness on all surfaces, and the plated surfaces have a low coefficient of friction.

Carburizing is not recommended for high speed steel cutting tools because the cases on such tools are extremely brittle. However, carburizing is useful for applications such as cold work dies that require extreme wear resistance and that are not subjected to impact or highly concentrated loading. Carburizing is done at 1040 to 1065 °C (1900 to 1950 °F) for short periods of time (10 to 60 min) to produce a case 0.05 to 0.25 mm (0.002 to 0.010 in.) deep. The carburizing treatment also serves as an austenitizing treatment for the whole tool. A carburized case on high speed steels has a hardness of 65 to 70 HRC, but does not have the high resistance to softening at elevated temperatures exhibited by normally hardened high speed steel.

Nitriding successfully increases the life of all types of high speed steel cutting tools. However, gas nitriding in dissociated ammonia produces a case that is too brittle for most applications. Liquid nitriding for about 1 h at 565 °C (1050 °F) provides a light case, increasing both surface hardness and resistance to adhesion. For nitrided high speed steel taps, drills and reamers used in machining annealed steel, fivefold increases in life have been reported, with average increases of 100 to 200%. Obviously, if this nitrided case is removed when the tool is reground, the tool must then be retreated, which reduces the cost advantage of the process.

In addition, special surface-treatment processes, such as aerated nitriding baths, improve resistance to adhesive wear without producing excessive brittleness. Sulfur-containing nitriding baths provide a high-sulfur surface layer for additional resistance to seizing.

Sulfide Treatment. A low-temperature (190 °C, or 375 °F) electrolytic process using sodium and potassium thiocyanate provides a seizing-resistant iron sulfide layer. This process can be used as a final treatment for all types of hardened tool steels without much danger of overtempering.

Maraging Steels

Certain high-nickel maraging steels are being used for special noncutting tool applications; 18Ni(250) is the type most frequently used. For more demanding applications, the higher-strength 18Ni(300) is often preferred. Where maximum abrasion resistance is required, any of the maraging steels can be nitrided.

Maraging steels achieve full hardness—nominally 50 HRC for 18Ni(250), 54 HRC for 18Ni(300) and 58 HRC for 18Ni(350)—by means of a simple aging treatment, usually 3 h at about 480 °C (900 °F). Because the development of hardness does not depend on cooling rate, full hardness can be developed uniformly in massive sections, with almost no distortion. Decarburization is of no concern in these alloys, because they contain very little carbon and because their aging temperature is relatively low. However, if the long-time service temperature exceeds the aging temperature, maraging steels overage and undergo significant drop in hardness.

The 18Ni(300) grade is used for aluminum die-casting dies and cores, aluminum hot forging dies, dies for molding plastics, and various support tooling used in extrusion of aluminum. In die casting of aluminum, maraging steel dies can be used at higher hardness than is possible with dies made of 1113 tool steel because maraging steel is not as prone to heat checking. Because the aging process results in very little size change, it is possible to machine the intricate impressions for plastic molding dies to final size prior to final hardening.

For molding extremely abrasive types of plastics, the higher surface hardness provided by 18Ni(350) maraging steel is desirable.

REFERENCE

1. *Distortion in Tool Steels,* by B. S. Lement: American Society for Metals, 1959

SELECTED REFERENCES

- "Tool Steels", (a Steel Products Manual): American Iron and Steel Institute, March 1978
- *Source Book on Industrial Alloys and Engineering Data:* American Society for Metals, Metals Park, OH, 1978, p 251–292
- *The Metallurgy of Tool Steels,* by P. Payson: John Wiley & Sons, Inc., 1962
- *Metallurgy and Heat Treatment of Tool Steels,* by R. Wilson: McGraw-Hill, London, 1975
- *Tool Steels,* by G. A. Roberts and R. A. Cary: American Society for Metals, 1980
- *Tool Steel Simplified,* Revised Ed., by F. R. Palmer et al.: Chilton Book Co., Radnor, PA, 1978

Heat Treating of Maraging Steels

By Stephen Floreen
Research Fellow
International Nickel Company, Inc.

MARAGING STEELS are ultrahigh-strength steels that differ from conventional steels in that they are not hardened by carbon content. Instead these steels are strengthened by precipitation of intermetallic compounds produced by age hardening a matrix of very-low-carbon martensite. Carbon, in fact, is an impurity in maraging steels and is kept at the lowest possible concentration.

Compositions of the common grades of maraging steel are given in Table 1. Grades have been developed that provide specific levels of yield strength ranging from 1030 to 3450 MPa (150 to 500 ksi).

The absence of carbon and the use of intermetallic precipitation to achieve hardening produce several unique characteristics of maraging steels that set them apart from conventional steels. Hardenability is of no concern. The low-carbon martensite formed after annealing is relatively soft, about 30 to 35 HRC. During age hardening there is only a very slight dimensional change. Thus, fairly intricate shapes can be machined in the soft condition and then hardened with a minimum of distortion.

Physical Metallurgy

Phase transformations in maraging steels are illustrated in Fig. 1. Plotted on the metastable diagram (Fig. 1a) are the austenite-to-martensite transformation on cooling and the martensite-to-austenite reversion on heating. The equilibrium diagram (Fig. 1b) shows that at higher nickel contents the equilibrium phases at low temperatures are austenite and ferrite.

In the metastable diagram, no phase transformations occur until the M_s temperature is reached and martensite is formed. Even very slow cooling of heavy sections produces only martensite, with no problems of low hardenability due to the large section size. The remaining alloying elements in the steel do, of course, alter the M_s temperatures significantly from those shown in Fig. 1, but the characteristic independence of cooling rate is not altered.

Table 1 Nominal compositions of commercial maraging steels

| Grade | Composition(a), % | | | | |
	Ni	Mo	Co	Ti	Al
18 Ni (200)	18	3.3	8.5	0.2	0.1
18Ni (250)	18	5.0	7.75	0.4	0.1
18Ni (300)	18	5.0	9.0	0.65	0.1
18Ni (350)	18	4.2(b)	12.5	1.6(b)	0.1
18Ni (Cast)	17	4.6	10.0	0.3	0.1

(a) For all grades, carbon content is no more than 0.03%. (b) Some producers use a combination of 4.8% Mo and 1.4% Ti, nominal.

For most grades of maraging steel, M_s temperatures are on the order of 200 to 300 °C (390 to 570 °F) and the steel is fully martensitic at room temperature.

Age hardening is produced by heat treating for several hours at temperatures typically on the order of 480 °C

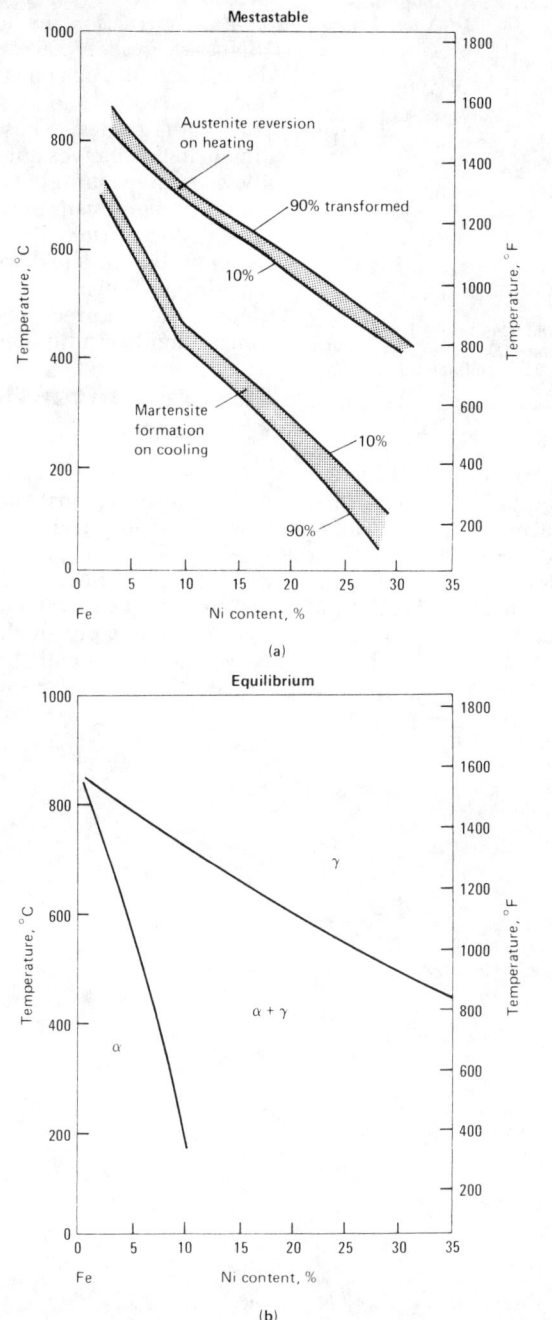

Fig. 1 Metastable and equilibrium phase relationships in the Fe-Ni system

aging steels usually is the result of (*a*) overaging by coarsening of the precipitate particles, and (*b*) reversion to austenite. Very substantial amounts of austenite, on the order of 50%, can eventually be formed.

Maraging steels normally contain little or no austenite after standard heat treatments, but sometimes austenite is deliberately formed. For example, if maraging steel is to be used in an application where overaging in service is expected, as in tooling for die casting of aluminum, the material will be slightly overaged beforehand. This minimizes overaging in service and resultant surface tensile stresses. Extreme overaging to form large amounts of reverted austenite also has been employed as an intermediate treatment to enhance response to cold working or to minimize the effects of thermal gradients during hot working and subsequent storage of extraordinarily heavy sections.

Overaged maraging steels are expected to show good resistance to brittle fracture and stress-corrosion cracking. Unfortunately, there appears to be considerable heat-to-heat variability in these characteristics when overaging heat treatments are used. Therefore, it is difficult to recommend specific overaging treatments that will produce consistent mechanical properties. In general, if a specific yield strength is required, it is better to use a maraging steel designed to produce that yield strength by conventional aging than to use an overaged higher-strength alloy.

Heat Treating

Conventional heat treatments for the standard grades of maraging steel are given in Table 2. Alloys with higher titanium contents are susceptible to formation of TiC films at austenite grain boundaries after holding at temperatures on the order of 900 to 1100 °C (1650 to 2000 °F). These films can severely embrittle the alloy when it is subsequently age hardened, leading to low-energy fractures along prior austenite grain boundaries. Prolonged annealing in this temperature range should be avoided for all compositions.

Solution Treatment. Maraging steels normally are solution annealed (austenitized) one hour for each 25 mm (1 in.) of section size. Atmosphere control may be necessary to minimize surface damage. Ordinarily, dry hydrogen or dissociated ammonia atmospheres

(900 °F). During this stage the equilibrium phase diagram becomes important. That is, with prolonged holding at 480 °C, the structure tends toward the equilibrium phases, primarily ferrite and austenite. Fortunately, the precipitation reactions that cause hardening proceed much more rapidly than the

reversion reactions that produce austenite and ferrite. Thus, very substantial increases in strength are achieved before reversion takes place. With longer aging times or higher temperatures, hardness will reach a maximum and then start to drop.

Overaging. Softening in aged mar-

Table 2 Heat treatments and typical mechanical properties of standard 18Ni maraging steels

Grade	Heat treatment (a)	Tensile strength MPa	ksi	Yield strength MPa	ksi	Elongation in 50 mm or 2 in., %	Reduction in area, %	Fracture toughness MPa√m	ksi√in.
18Ni (200)	A	1500	218	1400	203	10	60	155-200	140-220
18Ni (250)	A	1800	260	1700	247	8	55	120	110
18Ni (300)	A	2050	297	2000	290	7	40	80	73
18Ni (350)	B	2450	355	2400	348	6	25	35-50	32-45
18Ni (Cast)	C	1750	255	1650	240	8	35	105	95

(a) Treatment A: solution treat 1 h at 820 °C (1500 °F), then age 3 h at 480 °C (900 °F). Treatment B: solution treat 1 h at 820 °C (1500 °F), then age 12 h at 480 °C (900 °F). Treatment C: anneal 1 h at 1150 °C (2100 °F), age 1 h at 595 °C (1100 °F), solution treat 1 h at 820 °C (1500 °F) and age 3 h at 480 °C (900 °F)

are used. The cooling rate after annealing is of little consequence because it has no effect on either microstructure or properties. It is essential, however, that the steel be cooled to room temperature before it is age hardened. If this is not done, the steel may contain untransformed (retained) austenite and may be much softer than expected.

Age hardening normally is done at 455 to 510 °C (850 to 950 °F) for 3 to 12 h. In typical treatments at 480 °C (900 °F), grades 18Ni(200), 18Ni(250) and 18Ni(300) are held 3 to 6 h and grade 18Ni(350) is held 6 to 12 h. The 350 grade also is aged for 3 to 6 h at 495 to 510 °C (925 to 950 °F). For applications such as die casting tooling, aging at temperatures on the order of 530 °C (985 °F) is employed.

The standard age-hardening heat treatments listed in Table 2 produce 0.04% contraction in length in the 18Ni(200) grade, 0.06% contraction in the 18Ni(250) grade and 0.08% contraction in both the 18Ni(300) and 18Ni(350) grades. These very small dimensional changes during hardening allow many maraging steel components to be finish machined in the annealed condition. The finished parts then can be hardened without further machining. When greater dimensional accuracy is required, an allowance for contraction is readily made.

Cleaning After Heat Treatment

For removal of oxide films formed by heat treatment, grit blasting is the most efficient technique. Maraging steels can be chemically cleaned by pickling in sulfuric acid or by duplex pickling, first in hydrochloric acid and then in nitric-plus-hydrofluoric acid. As with conventional steels, care must be taken to avoid overpickling.

Cold Treating of Steel

By Norman C. McClure
Chief Metallurgist—Plant 2
Commercial Steel Treating Corp.

COLD TREATING of steel is a supplemental treatment that may be used for such purposes as enhancing transformation of austenite to martensite and improving stress relief of castings and machined parts. Generally, 1 h of cold treatment is adequate for each inch of cross section.

Hardening and Retained Austenite

Whenever hardening is to be done during heat treating, complete transformation from austenite to martensite generally is desired prior to tempering. From a practical standpoint, however, conditions vary widely and 100% transformation rarely, if ever, occurs. Cold treating may be useful, in many instances, for improving the percentage of transformation and thus enhancing properties.

During hardening, martensite develops as a continuous process from start (M_s) to finish (M_f) through the martensite-formation range. Except in a few highly alloyed steels, martensite starts to form at well above room temperature. In many instances, transformation is essentially complete at room temperature. Retained austenite tends to be present in varying amounts, however, and, when considered excessive for a particular application, must be transformed to martensite and then tempered.

Cold Treating vs Tempering. Immediate cold treating without delays at room temperature or at other temperatures during quenching offers the best opportunity for maximum transformation to martensite. In some instances, however, there is a risk that this will cause cracking of parts. Therefore, it is important to ensure that the grade of steel and the product design will tolerate immediate cold treating rather than immediate tempering. Some steels must be transferred to a tempering furnace when still warm to the touch, to minimize the likelihood of cracking. Design features such as sharp corners and abrupt changes in section create stress concentrations and promote cracking.

In most instances, cold treating is not done before tempering. In several types of industrial applications, tempering is followed by deep freezing and retempering without delay. For example, such parts as gages, machineways, arbors, mandrils, cylinders, pistons, and ball and roller bearings are treated in this manner for dimensional stability. Several freeze-draw cycles are used for critical applications.

Cold treating is also used to improve wear resistance in such materials as tool steels, high-carbon martensitic stainless steels and carburized alloy steels for applications where the presence of retained austenite may result in excessive wear. Transformation in service may cause cracking and/or dimensional changes that can promote failure. In some instances, more than 50% retained austenite has been observed. In such cases, no delay in tempering after cold treatment is permitted, or cracking can develop readily.

Process Limitations. In some applications in which explicit amounts of retained austenite are considered beneficial, cold treating might be detrimental. Moreover, multiple tempering, rather than alternate freeze-draw cycling, is generally more practical for transforming retained austenite in high speed and high-carbon/high-chromium steels.

Hardness Testing. Lower than expected HRC readings may indicate excessive retained austenite. Significant increases in these readings as a result of cold treatment indicate conversion of austenite to martensite. Superficial hardness readings, such as HR15N, may show even more significant changes.

Precipitation-Hardening Steels. Specifications for precipitation-hardening steels may include a mandatory deep freeze after solution treatment and prior to aging.

Shrink Fits. Cooling the inner member of a complex part to below ambient temperature may be a useful way of providing an interference fit. Care must be taken, however, to avoid the brittle cracking that may develop when the inner member is made of heat treated steel with high amounts of retained austenite, which converts to martensite on subzero cooling.

Stress Relief

Residual stresses often contribute to part failure, and residual stresses frequently are the result of temperature changes that produce thermal expansion and phase changes—and, consequently, volume changes.

Under normal conditions, temperature gradients produce nonuniform dimensional and volume changes. In castings, for example, compressive stresses develop in lower-volume areas, which cool first, and tensile stresses develop in areas of greater volume, which are last to cool. Shear stresses develop between the two areas. Even in large castings and machined parts of relatively uniform thickness, the surface cools first and the core last. In such cases, stresses develop as a result of the phase (volume) change between those layers that transform first and the center portion, which transforms last.

When both volume and phase changes occur in pieces of uneven cross section, normal contractions due to cooling are opposed by transformation expansion. The resulting residual stresses will remain until a means of relief is applied. This type of stress develops most frequently in steels during quenching. The surface becomes martensitic before the interior does. Although the inner austenite can be strained to match this surface change, subsequent interior expansions place the surface martensite under tension when the inner austenite transforms. Cracks in high-carbon steels arise from such stresses.

The use of cold treating has proved beneficial in stress relief of castings and machined parts of even or nonuniform cross section. The following are features of the treatment:

- Transformation of all layers is accomplished when the material reaches -84 °C (-120 °F).
- The increase in volume of the outer martensite is somewhat counteracted by the initial contraction due to chilling.
- Re-warm time is more easily controlled than cooling time, thus allowing equipment flexibility.
- The expansion of the inner core due to transformation is somewhat balanced by the expansion of the outer shell.
- The chilled parts are more easily handled.
- The surface is unaffected by low temperature.
- Parts that contain various alloying elements and that are of different sizes and weights can be chilled simultaneously.

Advantages of Cold Treating

Unlike heat treating, which requires that temperature be precisely controlled to avoid reversal, successful transformation through cold treating depends only on the attainment of the minimum low temperature and is not affected by lower temperatures. As long as the material is chilled to -84 °C (-120 °F), transformation will occur; additional chilling will not cause reversal.

Time at Temperature. After thorough chilling, additional exposure has no adverse effect. When heat is used, holding time and temperature are critical. In cold treatment, materials of different compositions and of different configurations may be chilled at the same time even though each may have a different high-temperature transformation point. Moreover, the warm-up rate of a chilled material is not critical as long as uniformity is maintained and gross temperature-gradient variations are avoided.

The cooling rate of a heated piece, however, has a definite influence on the end product. Formation of martensite during solution heat treating assumes immediate quenching to ensure that austenitic decomposition will not result in the formation of bainite and cementite. In large pieces comprising both thick and thin sections, not all areas will cool at the same rate. As a result, surface areas and thin sections may be highly martensitic, and the slower-cooling core may contain as much as 30 to 50% retained austenite. In addition to incomplete transformation, subsequent natural aging induces stress and also results in additional growth after machining.

Aside from transformation, no other metallurgical change takes place as a result of chilling. The surface of the material needs no additional treatment. The use of heat frequently causes scale and other surface deformations that must be removed.

Equipment for Cold Treating

A simple home-type deep freezer can be used for transformation of austenite to martensite. Temperature will be approximately -18 °C (0 °F). In some instances, hardness tests can be used to determine if this type of cold treating will be helpful. Dry ice placed on top of the work in a closed, insulated container also is commonly used for cold treating. The dry ice surface temperature is -78 °C (-109 °F), but the chamber temperature normally is about -60 °C (-75 °F).

Mechanical refrigeration units with circulating air at approximately -87 °C (-125 °F) are commercially available. A typical unit will have the following dimensions and operational features: chamber volume, up to 2.7 m^3 (95 ft^3); temperature range, 5 to -95 °C (40 to -140 °F); load capacity, 11.3 to 163 kg/h (25 to 360 lb/h); and thermal capacity, up to 8870 kJ/h (8400 Btu/h).

Although liquid nitrogen at -195 °C (-320 °F) may be employed, it is used less frequently than any of the above methods because of its cost.

Stress Relieving of Austenitic Stainless Steels

AUSTENITIC STAINLESS STEEL has good creep resistance; consequently, it must be heated to about 900 °C (1650 °F) to attain adequate stress relief. In some instances, heating to the annealing temperature may be desirable. Holding at a temperature lower than about 870 °C (1600 °F) results in only partial stress relief. The most effective stress-relieving results are achieved by slow cooling. Quenching or other rapid cooling, as is normal in the annealing of austenitic stainless steel, will usually reintroduce residual stresses.

Selection of Treatment

Selection of an optimum stress-relieving treatment is difficult, because heat treatments that provide adequate stress relief can impair the corrosion resistance of stainless steel, and heat treatments that are not harmful to corrosion resistance may not provide adequate stress relief. To avoid specifying a heat treatment that might prove harmful, ASME Code neither requires nor prohibits stress relief of austenitic stainless steel.

Metallurgical characteristics of austenitic stainless steels that may affect the selection of a stress-relieving treatment are discussed below.

- *Heating in the range from 480 to 815 °C (900 to 1500 °F):* Chromium carbides will precipitate in the grain boundaries of wholly austenitic unstabilized grades. In partially ferritic cast grades, the carbides will precipitate initially in the discontinuous ferrite pools rather than in a continuous grain-boundary network. After prolonged heating such as is necessary for heavy sections, however, grain boundary carbide precipitation will occur. For cold worked stainless, carbide precipitation may occur as low as 425 °C (800 °F); for types 309 and 310, the upper limit for carbide precipitation may be as high as 900 °C (1650 °F). In this condition, the steel is susceptible to intergranular corrosion. By using stabilized or extra-low-carbon grades, these intergranular precipitates of chromium carbide can be avoided.

- *Heating in the range from 540 to 925 °C (1000 to 1700 °F):* The formation of hard, brittle sigma phase may result, which can decrease both corrosion resistance and ductility. During the times necessary for stress relief, sigma will not form in fully austenitic wrought, cast or welded stainless. However, if the stainless is partly ferritic, the ferrite may transform to sigma during stress relief. This is generally not a problem in wrought stainless steels, because they are fully austenitic; however, some wrought grades—particularly types 309, 309Cb, 312 and 329—may contain some ferrite. Furthermore, the composition of most austenitic stainless welds and castings is intentionally adjusted so that ferrite is present as a deterrent to cracking. The niobium (columbium)-containing cast grade CF-8C normally contains 5 to 20% ferrite, which is more likely to transform to sigma than the niobium (columbium)-free ferrite in the unstabilized CF-8 grade.

- *Slow cooling an unstabilized grade (other than an extra-low-carbon grade):* Through either of the above temperature ranges, slow cooling may allow sufficient time for these detrimental effects to take place.

- *Heating at 815 to 925 °C (1500 to 1700 °F):* The coalescence of chromium carbide precipitates or sigma phase will occur, resulting in a form less harmful to corrosion resistance or mechanical properties.

- *Heating at 955 to 1120 °C (1750 to 2050 °F):* This annealing treatment

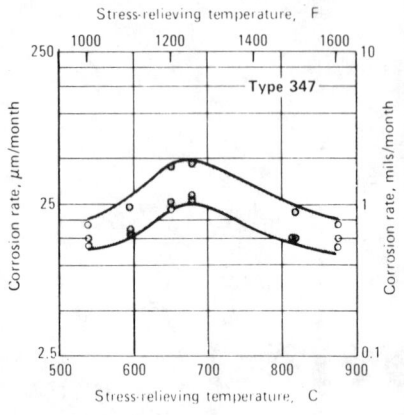
Fig. 1 Effect of stress relieving on corrosion rate of type 347 stainless steel

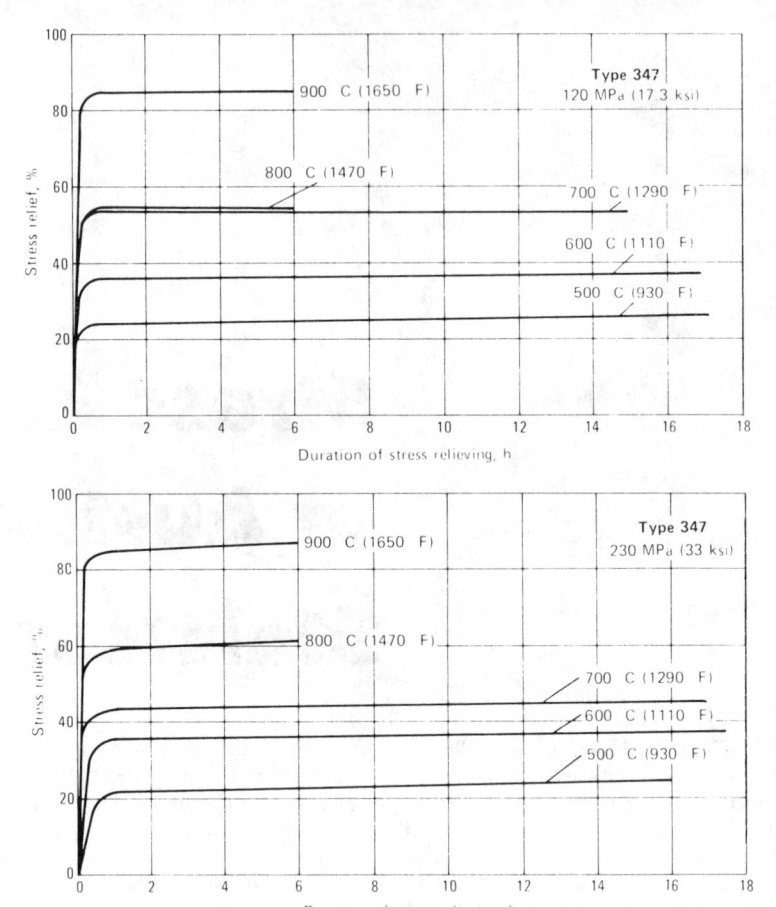

Fig. 2 Stress relief obtained in type 347 stainless steel, as a function of temperature, initial stress, and time at temperature

causes all grain-boundary chromium carbide precipitates to redissolve and transforms sigma back to ferrite, as well as fully softening the steel.

• *Stress relieving to improve the notch toughness:* Unlike carbon and alloy steels, austenitic steels are not notch sensitive. Consequently, stress relieving to improve notch toughness would be of no benefit. Notch-impact strength may actually be decreased if the steel is stress relieved at a temperature at which chromium carbide is precipitated or sigma phase forms.

Although stabilized alloys do not require high-temperature annealing to avoid intergranular corrosion, the stress-relieving temperature exerts an influence on the general corrosion resistance of these alloys. Figure 1 shows the effect of stress relieving for 2 h at various temperatures on the corrosion rate of type 347 stainless steel in boiling 65% nitric acid. The corrosion resistance of type 347 in boiling nitric acid is better when the material is treated at 815 to 870 °C (1500 to 1600 °F) than when treated at 650 to 705 °C (1200 to 1300 °F).

Figure 2 shows how the percentage of stress relief increases with an increase in stress-relieving temperature for type 347 stainless steel. These data also demonstrate the relative unimportance of holding time.

General Recommendations. In the selection of the proper stress-relieving treatment, consideration must be given also to the specific material used, fabrication procedures in-

volved, and to the design and operating conditions of the equipment. Stress relieving generally is not advisable unless the service environment is known or suspected to cause stress corrosion. If stress relieving seems warranted, due regard should be given the metallurgical factors and their effect on the steel in the intended service. The use of stabilized or extra-low-carbon grades is advantageous in view of the greater latitude allowed in stress relieving.

Table 1 gives suggested stress-relieving treatments for service applications and environments. Because of the varying degrees of stress relief that may be required, number of different grades of stainless in use, many fabricating procedures that may be employed, and the multitude of service requirements, many alternative treatments are indicated in Table 1 to allow selection of the stress-relieving treat-

ment best suited to particular circumstances.

Results Obtained by Various Treatments

Inadequate Stress Relief. Austenitic stainless steels have in many instances been stress relieved at temperatures normally used for carbon steels (540 to 650 °C, or 1000 to 1200 °F). Although at these temperatures virtually all residual stress is relieved in carbon steel, only 30 to 40% of the residual stress is relieved in austenitic stainless (Fig. 2). Because the treatment does not provide adequate stress relief, stainless stress relieved in this temperature range is often susceptible to stress corrosion. Table 2 shows the residual stresses remaining in solid austenitic stainless steels after being stress relieved for various times at tem-

Table 1 Stress-relieving treatments for austenitic stainless steels

Application or desired characteristics	Suggested thermal treatment(a) Extra-low-carbon grades, such as 304L and 316L	Stabilized grades, such as 318, 321 and 347	Unstabilized grades, such as 304 and 316
Severe stress corrosion	A, B	B, A	(b)
Moderate stress corrosion	A, B, C	B, A, C	C(b)
Mild stress corrosion	A, B, C, E, F	B, A, C, E, F	C, F
Remove peak stresses only	F	F	F
No stress corrosion	None required	None required	None required
Intergranular corrosion	A, C(c)	A, C, B(c)	C
Stress relief after severe forming	A, C	A, C	C
Relief between forming operations	A, B, C	B, A, C	C(d)
Structural soundness(e)	A, C, B	A, C, B	C
Dimensional stability	G	G	G

(a) Thermal treatments are listed in order of decreasing preference. A: anneal at 1065 to 1120 °C (1950 to 2050 °F), slow cool. B: stress relieve at 900 °C (1650 °F), slow cool. C: anneal at 1065 to 1120 °C (1950 to 2050 °F), quench(f) or cool rapidly. D: stress relieve at 900 °C (1650 °F), quench or cool rapidly. E: stress relieve at 480 to 650 °C (900 to 1200 °F), slow cool. F: stress relieve at below 480 °C (900 °F), slow cool. G: stress relieve at 205 to 480 °C (400 to 900 °F), slow cool (usual time, 4 h per inch of section). (b) To allow the optimum stress-relieving treatment, the use of stabilized or extra-low-carbon grades is recommended. (c) In most instances, no heat treatment is required, but where fabrication procedures may have sensitized the stainless steel the heat treatments noted may be employed. (d) Treatment A, B or D also may be used, if followed by treatment C when forming is completed. (e) Where severe fabricating stresses coupled with high service loading may cause cracking. Also, after welding heavy sections

peratures ranging from 595 to 1010 °C (1100 to 1850 °F).

Annealing and Water Quenching. Numerous instances have been reported in which satisfactory service was obtained for vessels and parts that were stress relieved by being annealed (at 1065 to 1120 °C, or 1950 to 2050 °F) and water quenched. However, it is unlikely that these products were subjected to service environments conducive to severe stress corrosion, because a water quench will almost always reintroduce high residual stresses. One instance in which stress corrosion was caused by annealing and quenching is described below.

Example 1. Thirty type 316 stainless vessels were quench annealed and placed in the same type of service at different locations. All but two of the vessels gave many years of excellent service. These two failed within 2 months because of the presence of chlorides in the environment at their particular locations; chlorides were absent at the locations of the other vessels.

Intergranular Corrosion. In a number of instances, partially stress-relieved stainless steel parts have failed through intergranular corrosion.

Example 2. Type 316 stainless steel hardware used in coastal steam stations was partially stress relieved at 620 to 650 °C (1150 to 1200 °F). Failure by intergranular attack in seawater occurred in less than 6 months.

Example 3. A type 304 stainless steel heat exchanger was partially stress relieved at 650 °C (1200 °F) for 2

Table 2 Stresses in austenitic stainless steel after various treatments

Temperature °C	Treatment Temperature °F	Time, h	Residual stress MPa	ksi
After welding 23.5-cm (9.25-in.-) OD, 16.5-cm (6.5-in.-) ID pipe				
As welded			205 to 175	30.0 to 25.7
595	1100	16	140	20.0
595	1100	48	140	20.0
595	1100	72	160	23.0
650	1200	4	150 to 165	21.5 to 24.0
After welding 12.7-cm-(5-in.-) OD, 10.2-cm-(4-in.-) ID pipe				
As welded			125 to 100	18.5 to 14.7
650	1200	4	95 to 105	13.7 to 15.3
650	1200	12	110	16.0
650	1200	36	108	15.6
900	1650	2	nil	nil
1010	1850	1	nil	nil

h and furnace cooled. Failure by intergranular attack occurred in 7 days.

Prevention of Stress Corrosion by Stress Relieving. A number of instances have been recorded in which beneficial effects were derived from an adequate stress-relief treatment.

Example 4. Heaters made of type 316L failed after a few weeks of service in contact with acid organic chloride and ammonium chloride, whereas heaters that had been stress relieved at 955 °C (1750 °F) were completely free of stress-corrosion cracking after 4 years of service under the same conditions.

Example 5. Two type 316L stainless steel vessels were used in 85% phosphoric acid service. One was not stress relieved and experienced extensive stress corrosion. Stress relieving the

other vessel at 540 °C (1000 °F) completely prevented the stress corrosion. This illustrates that, even though a stainless steel component may not be completely stress relieved, reducing the stress level may prevent stress corrosion.

Stress relief of unstabilized grades of stainless at 900 °C (1650 °F) will result in some intergranular carbide precipitation, and in some instances a small amount of intergranular attack may be encountered. However, failure after a few years by intergranular attack is preferable to failure within a few weeks by stress-corrosion cracking. Moreover, the intergranular attack probably could be avoided by using an extra-low-carbon or stabilized grade of austenitic stainless steel.

Heat Treating of Stainless Steels and Heat-Resisting Alloys

Heat Treating of Stainless Steels

AUSTENITIC STAINLESS STEELS

Austenitic stainless steels may be divided into five groups: (*a*) conventional austenitics, such as types 301, 302, 303, 304, 305, 308, 309, 310, 316 and 317; (*b*) stabilized compositions, primarily types 321, 347 and 348; (*c*) low-carbon grades, such as types 304L, 316L and 317L; (*d*) high-nitrogen grades, such as AISI types 201, 202, 304N and 316N, and the Nitronic series of alloys; and (*e*) highly alloyed austenitics.

Conventional austenitics cannot be hardened by heat treatment but will harden as a result of cold working. These steels are usually purchased in an annealed or cold worked state. Following welding or thermal processing, a subsequent re-anneal may be required for optimum corrosion resistance, softness and ductility. During annealing, chromium carbides, which markedly decrease resistance to intergranular corrosion, are dissolved. Annealing temperatures, which vary somewhat with the composition of the steel, are given in Table 1.

Because carbide precipitation can occur at temperatures between 425 and 900 °C (800 and 1650 °F), it obviously is desirable that the annealing temperature should be safely above this limit. Moreover, because all carbides should be in solution before cooling begins, and because the chromium carbide dissolves slowly, the highest practical temperature consistent with limited grain growth is selected. This temperature is in the vicinity of 1095 °C (2000 °F).

Cooling from the annealing temperature must be rapid, but it must also be consistent with limitations of distortion. Whenever considerations of distortion permit, water quenching is used, thus ensuring that dissolved carbides remain in solution (because it precipitates carbides more rapidly, type 310 invariably requires water quenching).

FERRITIC STAINLESS STEELS

Ferritic stainless steels may be divided into two groups: (*a*) conventional ferritics, such as types 409, 430, 434 and 446; and (*b*) low-interstitial ferritics, such as types 439, 444, E-BRITE, SEA-CURE, AL 29-4C and AL 29-4-2. Ferritic stainless steels are not hardened by quenching but rather develop minimum hardness and maximum ductility, toughness and corrosion resistance in the annealed and quenched condition. Therefore, the only heat treatment applied to ferritics is annealing. This treatment relieves stresses developed during welding or cold working and provides a more homogeneous structure by dissolving transformation products formed during welding. Postweld heat treatment of low-interstitial ferritic stainless steels is generally unnecessary and is frequently undesirable. Table 2 summarizes current annealing practices for ferritic grades.

MARTENSITIC STAINLESS STEELS

Heat treating of martensitic stainless steel is essentially the same as for plain carbon or low-alloy steels, in that maximum strength and hardness depend chiefly on carbon content. The principal metallurgical difference is that the high alloy content of the stainless grades causes the transformation to be so sluggish, and the hardenability to be so high, that maximum hardness is produced by air cooling in the center of sections up to approximately 30.5 cm (12 in.) thick.

Annealing. Temperatures and resulting hardnesses for process (subcritical) annealing, full

Reprinted from *Metals Handbook*, Desk Edition, 28.60-28.63, © 1985 American Society for Metals

Table 1. Recommended annealing temperatures for austenitic stainless steels

UNS No.	Designation	Temperature(a) °C	°F
Conventional grades			
S30100, S30200, S30215	301, 302, 302B	1010 to 1120	1850 to 2050
S30300, S30323	303, 303Se	1010 to 1120	1850 to 2050
S30400, S30500, S30800	304, 305, 308	1010 to 1120	1850 to 2050
S30900, S30908	309, 309S	1040 to 1120	1900 to 2050
S31000, S31008	310, 310S	1040 to 1065	1900 to 1950
S31600	316	1040 to 1120	1900 to 2050
S31700	317	1065 to 1120	1950 to 2050
Stabilized grades			
S32100	321	955 to 1065	1750 to 1950
S34700, S34800	347, 348	980 to 1065	1800 to 1950
N08020	Carpenter 20Cb-3	925 to 955	1700 to 1750
Low-carbon grades			
S30403	304L, 304LN	1010 to 1120	1850 to 2050
S31603, S31703	316L, 316LN, 317L	1040 to 1110	1900 to 2025
High-nitrogen grades			
S20100, S20200	201, 202	1010 to 1120	1850 to 2050
S30451	304N	1010 to 1120	1850 to 2050
S31651	316N	1010 to 1120	1850 to 2050
S24100	Nitronic 32, Carpenter 18Cr-2Ni-12Mn	1010 to 1065	1850 to 1950
S24000	Nitronic 33	1040 to 1095	1900 to 2000
S21904	Nitronic 40, Carpenter 21Cr-6Ni-9Mn	980 to 1175	1800 to 2150
S20910	Nitronic 50, Carpenter 22Cr-13Ni-5Mn	1065 to 1120	1950 to 2050
S21800	Nitronic 60	1040 to 1095	1900 to 2000
S28200	Carpenter 18-18 PLUS	1040 to 1095	1900 to 2000
Highly alloyed grades			
	317LM, 317LX, 317L PLUS, 317LMO, 7L4	1120 to 1150	2050 to 2100
	JS700, JS777	1065 to 1150	1950 to 2100
N08904	904L, AL-4X, 2RK65	1075 to 1125	1965 to 2055
N08028	Sanicro 28	...	...
N08366	AL-6X	1205 to 1230	2200 to 2250
S31254	254 SMO	1150 to 1205	2100 to 2200

(a) Temperatures given are for annealing a composite structure. Time at temperature and method of cooling depend on thickness. Light sections may be held at temperature for 3 to 5 min per 2.5 mm (0.10 in.) of thickness, followed by rapid air cooling. Thicker sections are water quenched. For many of these grades, a postweld heat treatment is not necessary. For proprietary alloys, alloy producers may be consulted for details. Although cooling from the annealing temperature must be rapid, it must also be consistent with limitations of distortion.

Table 2. Recommended annealing treatments for ferritic stainless steels

UNS No.	Designation	Treatment temperature °C	°F
Conventional ferritic grades			
S40500	405	650-815	1200-1500
S40900	409	870-900	1600-1650
S43000	430	705-790	1300-1450
S43020	430F	705-790	1300-1450
S43400	434	705-790	1300-1450
S44600	446	760-830	1400-1525
Low-interstitial ferritic grades			
S43035	439	870-925	1600-1700
S44400	444	955-1010	1750-1850
S44626	E-BRITE	760-955	1400-1750
S44660	SEA-CURE, SC-1	1010-1065	1850-1950
...	AL 29-4C	1010-1065	1850-1950
S44800	Al 29-4-2	1010-1065	1850-1950
S44635	MONIT	1010-1065	1850-1950

Note: Postweld heat treating of low-interstitial ferritic stainless steels is generally unnecessary and frequently undesirable. Any annealing of these grades should be followed by water quenching or very rapid cooling.

be austenitized at the high end of the temperature range. For alloys that are to be tempered above 565 °C (1050 °F), the low side of the austenitizing range is recommended, because it enhances ductility and impact properties.

PRECIPITATION-HARDENING STAINLESS STEELS

Recommended procedures for homogenization, austenite conditioning, transformation cooling and precipitation hardening (age-tempering) of a semiaustenitic precipitation-hardening stainless steel are given in Table 5. For procedures pertaining to other precipitation-hardening grades, see Metals Handbook, 9th Edition, Volume 4.

Heat Treating of Heat-Resisting Alloys

STRESS RELIEVING

Stress relieving of heat-resisting alloys and refractory metals frequently entails a compromise; the desirability of maximum relief of residual stress must be weighed against possible effects deleterious to high-temperature properties and corrosion resistance.

True stress relieving of wrought material usually is confined to alloys that are not age-hardenable. Thus, the time and temperature cycles may vary considerably, depending on the metallurgical characteristics of the alloy and on the type and magnitude of residual stresses developed by previous fabricating processes.

Stress-relieving temperatures are usually below the annealing or recrystallization temperatures. Typical cycles for wrought alloys are listed in Table 6; temperatures at least 25 °C (50 °F) higher or lower than those listed are usually satisfactory.

ANNEALING

When applied to heat-resisting alloys, annealing implies full annealing—that is, complete re-

Table 3. Annealing temperatures and procedures for wrought martensitic stainless steels

Type	Process (subcritical) annealing Temperature(a), °C	Hardness	Full annealing Temperature(b)(c), °C	Hardness	Isothermal annealing(c) Procedure(d)	Hardness
403, 410	650-760	82-92 HRB	830-885	75-85 HRB	Heat to 830 to 885 °C; hold 6 h at 705 °C	85 HRB
414	650-730	99 HRB-24 HRC	Not recommended		Not recommended	
416, 416(Se)	650-760	86-92 HRB	830-885	75-85 HRB	Heat to 830 to 885 °C; hold 2 h at 720 °C	85 HRB
420	675-760	94-97 HRB	830-885	86-95 HRB	Heat to 830 to 885 °C; hold 2 h at 705 °C	95 HRB
431	620-705	99 HRB-30 HRC	Not recommended		Not recommended	
440A	675-760	90 HRB-22 HRC	845-900	94-98 HRB	Heat to 845 to 900 °C; hold 4 h at 690 °C	98 HRB
440B, 440C,	675-760	98 HRB-23 HRC	845-900	95 HRB-20 HRC	Same as 440A	20 HRC
440F	675-760	98 HRB-23 HRC	845-900	98 HRB-25 HRC	Same as 440A	25 HRC

(a) Air cool from temperature; maximum softness is obtained by heating to temperature at high end of range. (b) Soak thoroughly at temperature within range indicated; furnace cool to 790 °C; continue cooling at 15 to 25 °C/h to 595 °C; air cool to room temperature. (c) Recommended for applications in which full advantage may be taken of the rapid cooling to the transformation temperature and from it to room temperature. (d) Preheating to a temperature within the process annealing range is recommended for thin-gage parts, heavy sections, previously hardened parts, parts with extreme variations in section or with sharp re-entrant angles, and parts that have been straightened or heavily ground or machined to avoid cracking and minimize distortion, particularly for types 420 and 431, and 440A, B, C and F.

annealing and isothermal annealing are given in Table 3. Full annealing is an expensive and time-consuming treatment; it should be used only when required for subsequent severe forming. Types 414 and 431 do not respond to full or isothermal annealing procedures within a reasonable soaking period.

Austenitizing temperatures, soaking times, quenching media and tempering temperatures are summarized in Table 4. The 440 grades require all the extra precautions that are taken, to prevent quench cracking, during quenching of high-hardenability steels. When maximum corrosion resistance and strength are desired, the steel should

Table 4. Procedures for hardening and tempering wrought martensitic stainless steels to specific strength and hardness levels

Type	Austenitizing Temperature °C	°F	Quenching medium(c)	Tempering temperature °C min	°C max	°F min	°F max	Tensile strength MPa	ksi	Hardness, HRC
403, 410	925 to 1010	1700 to 1850	Air or oil	565	605	1050	1125	760 to 965	110 to 140	25 to 31
				205	370	400	700	1105 to 1515	160 to 220	38 to 47
414	925 to 1050	1700 to 1925	Air or oil	595	650	1100	1200	760 to 965	110 to 140	25 to 31
				230	370	450	700	1105 to 1515	160 to 220	38 to 49
416, 416(Se)	925 to 1010	1700 to 1850	Oil	565	605	1050	1125	760 to 965	110 to 140	25 to 31
				230	370	450	700	1105 to 1515	160 to 220	35 to 45
420	985 to 1065	1800 to 1950	Air or oil(e)	205	370	400	700	1550 to 1930	225 to 280	48 to 56
431	985 to 1065	1800 to 1950	Air or oil(e)	565	605	1050	1125	860 to 1035	125 to 150	26 to 34
				230	370	450	700	1210 to 1515	175 to 220	40 to 47
440A	1010 to 1065	1850 to 1950	Air or oil(e)	150	370	300	700	...	...	49 to 57
440B	1010 to 1065	1850 to 1950	Air or oil(e)	150	370	300	700	...	...	53 to 59
440C, 440F	1010 to 1065	1850 to 1950	Air or oil(e)	...	160	...	325	...	...	60 min
				...	190	...	375	...	...	58 min
				...	230	...	450	...	...	57 min
				...	355	...	675	...	...	52 to 56

(a) Preheating to a temperature within the process annealing range is recommended for thin-gage parts, heavy sections, previously hardened parts, parts with extreme variations in section or with sharp re-entrant angles, and parts that have been straightened or heavily ground or machined, to avoid cracking and minimize distortion, particularly for types 420, 431, and 440A, B, C and F. (b) Usual time at temperature ranges from 30 to 90 min. The low side of the austenitizing range is recommended for all types subsequently tempered to 25 to 31 HRC; generally, however, corrosion resistance is enhanced by quenching from the upper limit of the austenitizing range. (c) Where air or oil is indicated, oil quenching should be used for parts more than 6.4 mm ($^1/_4$ in.) thick; martempering baths at 150 to 400 °C (300 to 750 °F) may be substituted for an oil quench. (d) Generally, the low end of the tempering range of 150 to 370 °C (300 to 700 °F) is recommended for maximum hardness, the middle for maximum toughness, and the high end for maximum yield strength. Tempering in the range of 370 to 565 °C (700 to 1050 °F) is not recommended, because it results in low and erratic impact properties and poor resistance to corrosion and stress corrosion. (e) For minimum retained austenite and maximum dimensional stability, a subzero treatment −75 °C ± 10 °C (−100 °F ± 20 °F) is recommended; this should incorporate continuous cooling from the austenitizing temperature to the cold transformation temperature.

Table 5. Recommended heat treating procedures for a semiaustenitic precipitation-hardening stainless steel (UNS S17400)

Homogenization. 1175 ± 15 °C (2150 ± 25 °F), 2 h + 30 min per 25 mm (1 in.)(a)

Austenite conditioning (solution treatment). 1040 ± 15 °C (1900 ± 25 °F), 30 min + 30 min per 25 mm (1 in.)(a)

Transformation cooling. To below +30 °C (+90 °F)

Precipitation hardening. To obtain minimum tensile strengths shown, use the following treatments for wrought alloys(b):

Tensile strength MPa	ksi	Treatment
1310	190	.... 1 h at 480 ± 5 °C (900 ± 10 °F)
1170	170	....4 h at 495 ± 5 °C (925 ± 10 °F)
1070	155	....4 h at 550 ± 5 °C (1030 ± 10 °F)
1030	150	....4 h at 565 ± 5 °C (1050 ± 10 °F)
1000	145	....4 h at 580 ± 5 °C (1075 ± 10 °F)
930	135	....4 h at 620 ± 5 °C (1150 ± 10 °F)

(a) To prevent cracking and ensure uniform properties, cool as follows: 75 mm (3 in.) and less, oil quench or air cool; 75 to 150 mm (3 to 6 in.), air cool; 150 mm (6 in.) and over, air cool under cover. *All parts must be cooled to below +30 °C (+90 °F)* prior to the precipitation-hardening cycle. (b) If hardness exceeds maximum specified, reheat treat at a slightly higher temperature for a minimum of 30 min.

Table 6. Typical stress-relieving and annealing cycles for wrought heat-resisting alloys

Alloy	Stress relieving Temperature °C	°F	Holding time per inch of section, h	Annealing(a) Temperature °C	°F	Holding time per inch of section, h
Iron-base and iron-nickel-chromium alloys						
RA-330	900	1650	1(b)	1110(c)	2025(c)	$^1/_4$(d)
19-9 DL	675(e)	1250(e)	4	980	1800	1
A-286	(f)	(f)	...	980	1800	1
Discaloy	(f)	(f)	...	1035	1900	1
Nickel-base alloys						
Astroloy	(f)	(f)	...	1135	2075	4
Hastelloy B	(f)	(f)	...	1175	2150	1
Hastelloy C	(f)	(f)	...	1215	2225	1
Hastelloy W	(f)	(f)	...	1175	2150	1
Hastelloy X	(f)	(f)	...	1175	2150	1
Incoloy 800	870	1600	1$^1/_2$	980	1800	$^1/_4$
Incoloy 800H	...	...	...	1175	2150	...
Incoloy 825	...	...	...	980	1800	...
Incoloy 901	(f)	(f)	...	1095	2000	2
Inconel 600	900	1650	1	1010	1850	$^1/_4$(d)
Inconel 601	...	...	...	980	1800	...
Inconel 625	870	1600	1	980	1800	1
Inconel 690	...	...	...	1040	1900	$^1/_2$
Inconel 718	(f)	(f)	...	955	1750	1
Inconel X-750	880(g)	1625(g)	...	1035	1900	$^1/_2$
Nimonic 80A	(f)	(f)	...	1080	1975	2
Nimonic 90	(f)	(f)	...	1080	1975	2
René 41	(f)	(f)	...	1080	1975	2
Udimet 500	(f)	(f)	...	1080	1975	4
Udimet 700	(f)	(f)	...	1135	2075	4
Waspaloy	(f)	(f)	...	1010	1850	4
Cobalt-chromium-nickel-base alloys						
L-605 (HS-25)	(h)	(h)	...	1230	2250	1
N-155 (HS-95)	(h)	(h)	...	1175	2150	...
S-816	(h)	(h)	...	1205	2200	1
Refractory metals(j)						
Ta-10W	1205(k)	2200(k)	1	1425(k)	2600(k)	1
FS-80	1095(k)	2000(k)	1	1315(k)	2400(k)	1
FS-82	1095(k)	2000(k)	1	1315(k)	2400(k)	1
Mo-0.5 Ti	1095(m)	2000(m)	$^1/_2$	1315(m)(n)	2400(m)(n)	1
TZM	1205(m)	2200(m)	1	1425(m)(n)(p)	2600(m)(n)(p)	1

(a) Minimum hardness is achieved by cooling rapidly from the annealing temperature, to prevent precipitation of hardening phases. Water quenching is preferred, and is usually necessary for heavy sections; air cooling is preferred for heavy sections of Waspaloy, Udimet 500, Udimet 700 and Inconel X-750, because water quenching causes cracking. However, for complex shapes subject to excessive distortion, oil quenching is often adequate and more practical. Rapid air cooling usually is adequate for parts formed from strip or sheet. Rapid cooling from the annealing or solution treating temperature does not suppress the aging reaction of some alloys, such as Astroloy; these alloys become harder and stronger. (b) Time given is minimum; some plants use as long as 3 h per inch. (c) Nominal temperature; 1035 to 1175 °C (1900 to 2150 °F) is commonly used. (d) Short time is required for prevention of grain coarsening. (e) Nominal temperature; 650 to 705 °C (1200 to 1300 °F) is permissible. (f) Full annealing is recommended, because intermediate temperatures cause aging. (g) Used only for stress equalizing of warm worked grades. (h) Full annealing is recommended, if further fabrication is performed; otherwise, material can be stress relieved at approximately 55 °C (100 °F) below annealing temperature. (j) Annealing temperatures depend on prior plastic deformation, degree of cold work, alloy content and interstitial purity. Annealing temperatures given are those most frequently used for cold worked sheet or plate; in many instances, more precise determination of the recrystallization temperature is necessary for a specific application. (k) Heat and cool in vacuum or inert-gas atmosphere. (m) Heat and cool in hydrogen or vacuum. (n) Seldom used as finished product in annealed condition, because recrystallization raises the ductile-brittle transition temperature, resulting in brittleness at low temperatures. (p) For vacuum-arc-cast material with a minimum of 50% cold work.

---

crystallization and the attainment of maximum softness. The practice is usually applied to wrought alloys of the nonhardening type. For a majority of the hardenable alloys, annealing cycles are the same as those used for solution treating. However, the two treatments serve different purposes. Annealing is used mainly to increase ductility (and reduce hardness) to facilitate forming or machining, prepare for welding, relieve stresses after welding, produce specific microstructures, or soften age-hardened structures by re-solution of second phases. Solution treating is intended to dissolve second phases to produce maximum corrosion resistance or to prepare for aging. Additionally, it will homogenize microstructure prior to aging.

Annealing practices vary considerably among different plants. Representative annealing temperatures, holding times, and cooling procedures are given in Table 6.

Table 7. Typical solution-treating and aging cycles for wrought heat-resisting alloys

Alloy	Solution treating Temperature °C	°F	Time, h	Cooling procedure	Aging Temperature °C	°F	Time, h	Cooling procedure
Iron-base alloys								
A-286	980	1800	1	Oil quench	720	1325	16	Air cool
Discaloy	1010	1850	2	Oil quench	730	1350	20	Air cool
					650	1200	20	Air cool
N-155	1175	2150	1	Water quench	815	1500	4	Air cool
Nickel-base alloys								
Astroloy	1175	2150	4	Air cool	845	1550	24	Air cool
	1080	1975	4	Air cool	760	1400	16	Air cool
Hastelloy B	1175	2150	1/2	(a)	(b)	(b)	...	...
Hastelloy B-2	1065	1950	1/2	Rapid quench	...	...	...	...
Hastelloy C-4	1065	1950	1/2	Rapid quench	...	...	...	...
Hastelloy C-276	1120	2050	1/2	Rapid quench	...	...	...	...
Hastelloy N	1175	2150	1/2	Rapid quench	...	...	...	...
Hastelloy S	1065	1950	1/2	Rapid quench	...	...	...	...
Hastelloy C	1220	2225	1	(a)	(b)	(b)	...	...
Hastelloy W	1175	2150	1	(a)	(b)	(b)	...	...
Hastelloy X	1175	2150	1	(a)	...	...	...	...
Inconel 901	1095	2000	2	Water quench	790	1450	2	Air cool
					720	1325	24	Air cool
Inconel 600	1120	2050	2	Air cool	...	...	...	...
Inconel 601	1150	2100	1	Air cool	...	...	...	...
Inconel 617	1175	2150	2	(a)	...	...	...	...
Inconel 625	1150	2100	2	(a)	...	...	...	...
Inconel 706	925-1010	1700-1850	...	...	845	1550	3	Air cool
					720	1325	8	Furnace cool
					620	1150	8	Air cool
	925-1010	1700-1850	...	...	730	1350	8	Furnace cool
					620	1150	8	Air cool
Inconel 718	980	1800	1	Air cool	720	1325	8	Furnace cool
					620	1150	8	Air cool
Inconel X-750 (AMS 5667)	855	1625	24	Air cool	705	1300	20	Air cool
Inconel X-750 (AMS 5668)	1150	2100	2	Air cool	845	1550	24	Air cool
					705	1300	20	Air cool
Nimonic 80A	1080	1975	8	Air cool	705	1300	16	Air cool
Nimonic 90	1080	1975	8	Air cool	705	1300	16	Air cool
René 41	1065	1950	1/2	Air cool	760	1400	16	Air cool
Udimet 500	1080	1975	4	Air cool	845	1550	24	Air cool
					760	1400	16	Air cool
Udimet 700	1175	2150	4	Air cool	845	1550	24	Air cool
	1080	1975	4	Air cool	760	1400	16	Air cool
Waspaloy	1080	1975	4	Air cool	845	1550	24	Air cool
					760	1400	16	Air cool
Cobalt-base alloys								
Haynes 25; L-605	1230	2250	1	Rapid air cool	(b)	(b)	...	...
Haynes 188	1175	2150	1/2	Rapid air cool	...	...	...	...
Haynes 556	1175	2150	1/2	Rapid air cool	...	...	...	...
S-816	1175	2150	1	(a)	760	1400	12	Air cool
Stellite 6B	1230	2250	1	Air cool	...	...	...	...

Note: Alternate treatments may be used to improve specific properties. (a) To provide an adequate quench after solution treating, it is necessary to cool below about 540 °C (1000 °F) rapidly enough to prevent precipitation in the intermediate temperature range. For sheet metal parts of most alloys, rapid air cooling will suffice. Oil or water quenching is frequently required for heavier sections that are not subject to cracking. (b) Aging occurs in service at elevated temperatures.

SOLUTION TREATING AND AGING

Solution treating and aging practices for iron, nickel- and cobalt-base heat-resisting alloys are summarized in Table 7.

PROTECTIVE ATMOSPHERES

Protective atmospheres are used in annealing or solution treating if heavy oxidation cannot be tolerated. If oxidation can be tolerated, because of subsequent stock removal, heat-resisting alloys can be solution treated in air or in the normal mixture of air and combustion products found in gas-fired furnaces. However, refractory metals must always be heat treated in a vacuum or in an inert-gas atmosphere (argon, helium, or an ArHe mixture) or hydrogen. In some cases, ceramic coatings are used to prevent surface attack.

Exothermic Atmosphere. A lean and dilute exothermic atmosphere is relatively safe and economical. The surface scale formed in such an atmosphere can be removed by pickling or by salt bath descaling and pickling. Such an atmosphere, formed by burning fuel gas with air, contains about 85% nitrogen, 10% carbon dioxide, 1.5% carbon monoxide, 1.5% hydrogen, and 2% water vapor. This atmosphere will produce a scale rich in chromium oxides.

Endothermic atmospheres prepared by reacting fuel gas with air in the presence of a catalyst are not recommended, because of their carburizing potential. Similarly, the endothermic mixture of nitrogen and hydrogen formed by dissociating ammonia is not used, because of the probability of nitriding.

Dry hydrogen (dew point, −50 °C [−60 °F] or lower) is used in preference to dissociated ammonia for bright annealing of heat-resisting alloys. Hydrogen is not recommended for bright annealing of alloys containing significant amounts of elements (such as aluminum or titanium) that form stable oxides not reducible at normal heat treating temperatures and dew points. Hydrogen is not recommended for annealing or solution treating alloys that contain boron, because of the danger of deboronization through formation of boron hydrides. Nor can hydrogen be used for heat treating niobium and tantalum, because of its embrittling effect.

Dry argon (dew point, −50 °C [−60 °F] or lower) should be used if no oxidation can be tolerated. It is mandatory that this type of atmosphere be used in a sealed retort or sealed furnace chamber. A purge of at least ten times the volume of the retort is recommended before the retort is placed in the furnace. The argon must be kept flowing continually during and after the treatment until the workpieces have cooled nearly to room temperature, to prevent the formation of an oxide film.

Heat-resisting alloys containing stable-oxide formers such as aluminum and titanium, with or without boron, must be bright annealed in a vacuum or in a chemically inert gas such as argon. If used, argon must be pure and dry—dew point, −50 °C (−60 °F) or lower.

Heat Treating of Cast Irons

Introduction to Heat Treating of Cast Irons

CAST IRONS may be compared with steels in their reactions to hardening. However, because cast irons (except white iron) contain graphite and substantially higher percentages of silicon, they require higher austenitizing temperatures. The graphitizing effect of silicon is so powerful that an unalloyed gray iron may become completely graphitized below the A_1 temperature during heating for austenitizing. Thus, some high-silicon irons require longer intervals at the austenitizing temperature for reabsorption of the desired carbon content in austenite. This interval is extended because silicon retards the absorption of carbon in austenite. The lower-silicon irons respond best to heat treatment.

HARDNESS MEASUREMENTS

Conventional hardness measurements on cast irons always indicate lower values than the true hardness of the metal matrix. This discrepancy, which is more pronounced in gray iron than in ductile and malleable irons, occurs because conventional hardness readings are composite values that reflect the hardnesses of both the matrix material and soft graphite.

Figure 1(a) shows the relation between observed HRC readings and those converted from microhardness values for five gray irons of different carbon equivalents. Hardness measurements were taken at two laboratories after quenching and after tempering of each iron. The data in Fig. 1(a) show why the observed values obtained by conventional hardness testing may be misleading, and help to explain the good wear resistance of gray irons with apparently low hardness. Note that there is a correlation with carbon equivalent for all five irons tested and that the discrepancy between observed and converted hardness values diminishes at the lower hardness level.

Another comparison between observed and converted HRC values for gray and ductile irons

is shown in Fig. 1(b). These irons were quenched in water from 900 °C (1650 °F) and tempered at 425 °C (800 °F) for 2 h.

Heat Treating of Gray Irons

THE HEAT TREATMENT most frequently applied to gray iron, with the possible exception of stress relieving, is annealing. Annealing of gray iron consists of heating the iron to a temperature high enough to soften it, and/or to minimize or eliminate massive eutectic carbides, thus improving its machinability. This heat treatment reduces mechanical properties substantially, however. It will reduce the grade level approximately to the next lower grade; for example, the properties of a class 40 gray iron will be diminished to those of a class 30 gray iron. Figure 2 shows the effect of full annealing on a tensile strength of class 30 gray iron arbitration bars.

ANNEALING

Gray iron commonly is subjected to one of three annealing treatments, each of which involves heating to a different range of temperature. These treatments are ferritizing annealing, medium (or "full") annealing and graphitizing annealing.

Ferritizing Annealing. For most gray irons, a ferritizing annealing temperature between 705 and 760 °C (1300 and 1400 °F) is recommended. The furnace temperature profile must be such that castings are sure to reach the set temperatures. Precise temperatures within this range depend on the exact composition of the iron.

Medium ("full") annealing is usually performed at temperatures between 790 and 900 °C (1450 and 1650 °F). This treatment is used when a ferritizing anneal would be ineffective because of the high alloy content of a particular iron. It is recommended, however, that the efficacy of temperatures at or below 760 °C (1400 °F) be tested before a higher annealing temperature is adopted as part of a standard procedure.

Graphitizing Annealing. If the microstructure of gray iron contains massive carbide particles, higher

annealing temperatures are necessary. Graphitizing annealing may have the purpose simply of converting massive carbide to pearlite and graphite.

To break down massive carbide with reasonable speed, temperatures of at least 870 °C (1600 °F) are required. With each additional 55 °C (100 °F) increment in holding temperature, the rate of carbide decomposition doubles; consequently, it is general practice to employ holding temperatures of 900 to 955 °C (1650 to 1750 °F). However, at 925 °C (1700 °F) and above, the phosphide eutectic present in irons containing 0.10% P or more may melt.

NORMALIZING

Gray iron is normalized by being heated to a temperature above the transformation range, held at this temperature for a period of about 1 h per inch of maximum section thickness, and cooled in still air to room temperature. Normalizing may be used to enhance mechanical properties, such as hardness and tensile strength, or to restore as-cast properties that have been modified by another heating process, such as graphitizing or the preheating and postheating associated with repair welding.

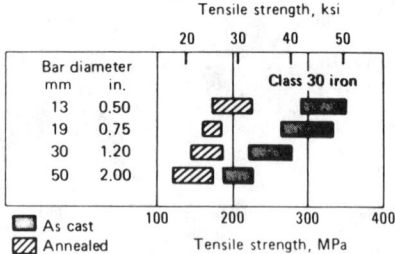

Fig. 2. Effect of annealing on tensile strength of class 30 gray iron

Specimens were arbitration bars from 31 heats. Bars were annealed at 925 °C (1700 °F) for 2 h, plus 1 h per 25 mm (1 in.) of section over 25 mm, and cooled at a maximum rate of 160 °C/h (285 °F/h) from 925 to 565 °C (1700 to 1050 °F). Cooling continued from 565 °C at a maximum rate of 130 °C/h (230 °F/h) to 200 °C (390 °F); bars were then air cooled to room temperature.

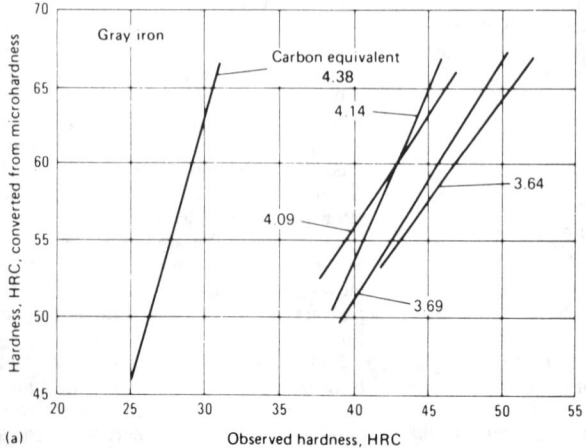

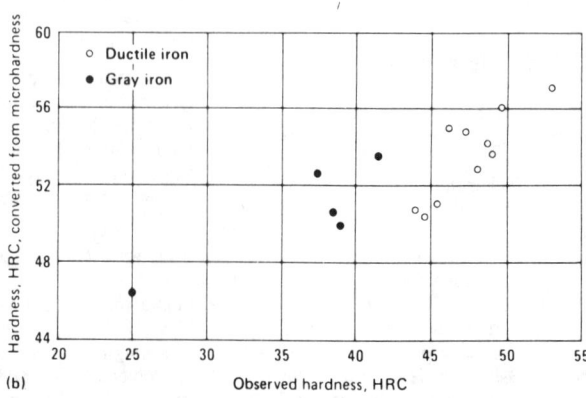

(a) Relation, as influenced by carbon equivalent, for gray iron containing type 3 graphite. (b) Relation for gray and ductile irons quenched in water from 900 °C (1650 °F) and tempered 2 h at 425 °C (800 °F).

Fig. 1. Relations between observed and converted hardness values for gray and ductile irons

Reprinted from Metals Handbook, Desk Edition, 28.52-28.54, © 1985 American Society for Metals

Table 1. Effect of air cooling from various temperatures on typical properties of gray irons

Condition(a)		Unalloyed iron(b)			Alloyed iron(c)			Combined carbon, %
		Hardness, HB	Tensile strength MPa	ksi	Hardness, HB	Tensile strength MPa	ksi	
As cast		207	265	38.1	212	265	38.7	0.84
Air cooled from:								
°C	°F							
540	1000	202	210	30.4	212	275	39.7	0.82
595	1100	190	255	36.8	210	275	40.0	0.86
650	1200	138	195	28.2	202	265	38.6	0.80
705	1300	125	180	26.4	187	265	38.4	0.81
760	1400	131	190	27.4	170	235	34.2	0.60
815	1500	152	205	29.7	212	295	42.6	0.76
870	1600	152	205	29.7	217	305	44.2	0.81
925	1700	152	205	29.6	223	290	42.4	0.82
980	1800	152	210	30.1	255	340	49.0	0.80

(a) Specimens 30 mm in diameter by 180 mm (1.2 in. in diameter by 7 in.) were held at temperature for 1 h before being cooled in still air to room temperature. (b) As-cast composition: 3.15 total C, 0.54 combined C, 2.59 Si, 0.09 P, 0.135 S, 0.88 Mn, 0.01 Cr, 0.10 Ni. (c) As-cast composition: 3.33 total C, 0.84 combined C, 2.27 Si, 0.076 P, 0.122 S, 0.72 Mn, 0.44 Cr, 0.36 Ni, 0.28 Mo.

The temperature range for normalizing gray iron is approximately 885 to 925 °C (1625 to 1700 °F). Heating temperature has a marked effect on microstructure and on mechanical properties such as hardness and tensile strength. This is demonstrated in Table 1.

HARDENING AND TEMPERING

Gray irons are hardened and tempered to improve their mechanical properties, particularly strength and wear resistance.

Austenitizing. In hardening gray iron, the casting is heated to a temperature high enough to promote formation of austenite, held at that temperature until the desired amount of carbon has been dissolved, and then quenched at a suitable rate. Heating for austenitizing may be accomplished in a salt bath or in an electrically heated, gas-fired or oil-fired furnace.

The temperature to which the casting must be heated is determined by the transformation range of the particular gray iron of which it is made. The transformation range can extend more than 55 °C (100 °F) above the A_1 (transformation-start) temperature. A formula for determining the approximate A_1 transformation temperature of unalloyed gray iron is:

$$°C: 730 + 28.0 \, (\% \, Si) - 25.0 \, (\% \, Mn)$$

$$°F: 1345 + 50.4 \, (\% \, Si) - 45.0 \, (\% \, Mn)$$

Chromium raises the transformation range of gray iron. In high-nickel, high-silicon irons, for example, each percent of chromium raises the transformation range by about 40 °C (72 °F). Nickel, on the other hand, lowers the critical range. In a gray iron containing from 4 to 5% Ni, the upper limit of the transformation range is about 710 °C (1310 °F).

Provided that recommended limits are not exceeded, the higher the casting is heated above the transformation range, the greater will be the amount of carbon dissolved in the austenite (Fig. 3) and the higher will be the hardness of the casting after quenching.

Quenching. Oil is the quenching medium most frequently used for gray iron. Water generally is not a satisfactory quenching medium for furnace-heated gray iron; it extracts heat so rapidly that distortion and cracking are likely in all except small parts of simple design. Recently developed water-soluble polymer quenches can provide the convenience of water quenching along with lower cooling rates, which can minimize thermal shock.

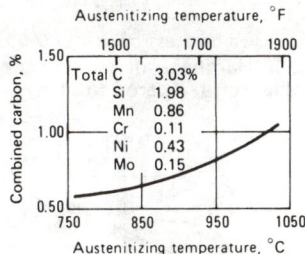

Specimens were furnace heated and water quenched. Combined carbon by difference.

Fig. 3. Increase in combined carbon with increase in austenitizing temperature for gray iron

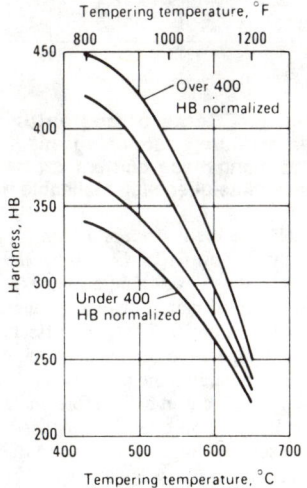

Fig. 4. Hardness of normalized ductile iron tempered at various temperatures

The least severe quenching medium is air. Unalloyed or low-alloy gray iron castings usually cannot be air quenched, because the cooling rate is not high enough to form martensite. However, for irons of high alloy content, forced-air quenching is frequently the most desirable cooling method.

Tempering. After quenching, castings usually are tempered at temperatures well below the transformation range for about 1 h per inch of thickest section. As the quenched iron is tempered, its hardness decreases, whereas it usually gains in strength and toughness.

FLAME HARDENING

Flame hardening is the method of surface hardening most commonly applied to gray iron. Both unalloyed and alloyed gray irons can be successfully flame hardened. However, some compositions yield much better results than others. One of the most important aspects of composition is the combined carbon content, which should be in the range of 0.50 to 0.70%.

INDUCTION HARDENING

Gray iron castings can be surface hardened by the induction method when the number of castings to be processed is large enough to warrant the relatively high equipment cost and the need for special induction coils.

Heat Treating of Ductile Irons

WHEN MAXIMUM DUCTILITY and good machinability are desired and high strength is not required, ductile iron castings are generally given a full ferritizing anneal. The microstructure is thus converted to ferrite and spheroidal graphite.

Two different annealing cycles may be used satisfactorily. Selection of one or the other will depend on the type of heat treating equipment that is available. These two cycles are:

- Hold at 900 to 955 °C (1650 to 1750 °F) for 1 h plus 1 h or more per inch of section thickness. For thin-section castings containing 2.20 to 2.70% Si, holding at 955 °C (1750 °F) for 1 to 3 h is sufficient. In heavy-section castings where chill has formed on corners, holding at 955 °C (1750 °F) for 3 to 8 h may be required. Cool to 690 °C (1275 °F) in any convenient manner (but uniformly, if residual stress is to be avoided), and hold at 690 °C (1275 °F) for 5 h plus 1 h per inch of casting section.
- Hold at 900 to 955 °C (1650 to 1750 °F) as above, but furnace cool to 650 °C (1200 °F) so that the cooling rate between 790 and 650 °C (1450 and 1200 °F) does not exceed 20 °C/h (35 °F/h).

A shorter, subcritical annealing cycle can be used when carbides can be tolerated and maximum impact properties are not required.

NORMALIZING OF DUCTILE IRON

Normalizing can result in a considerable improvement in tensile properties and may be used in production of ductile iron of types 100-70-03 and 120-90-02. The following temperatures and minimum holding times are recommended for normalizing unalloyed ductile iron (see Fig. 4):

Section			Time,
mm	in.	Temperature	h
Under 13	Under 1/2	870 °C (1600 °F) min	1
13 to 25	1/2 to 1	940 °C (1725 °F)	1
Over 25	Over 1	940 °C (1725 °F)	2

HARDENING AND TEMPERING OF DUCTILE IRON

A temperature of 845 to 925 °C (1550 to 1700 °F) is normally used for austenitizing commercial castings and produces the highest as-quenched hardness. Oil is preferred as a quenching medium, to minimize stresses, but water or brine may be used for simple shapes. Complicated castings may have to be quenched in oil at 80 to 100 °C (180 to 210 °F) to avoid cracks.

To relieve quenching stresses, castings should be tempered immediately after quenching. Tempered hardness depends on as-quenched hardness level, alloy content and tempering time, as well as on temperature. Precise data on ductile irons are not available for use in drawing precise tempering curves such as those for steels. Figure 5 can be used as a first approximation by the heat treater. More accurate control of hardness can be attained by close control of material composition and heat treating cycle.

SURFACE HARDENING OF DUCTILE IRON

Ductile iron responds readily to surface hardening by flame or induction processes. Because of the short heating cycle in these processes, the pearlitic types of ductile iron, 80-60-03 and 100-70-03, are preferred. Irons without free ferrite in their microstructures respond almost instantly to flame or induction heating and require very little holding time at the austenitizing temperature in order to be fully hardened.

Heat Treating of Malleable Irons

FERRITIC AND PEARLITIC malleable irons are both produced by annealing white iron of controlled composition. Thus, annealing is an essential part of the manufacturing process for these irons.

The annealing treatment involves three important steps. The first causes nucleation of graphite and is initiated during heating to a high holding temperature and occurs very early during the holding period.

The second step consists of holding at 900 to 970 °C (1650 to 1780 °F); this step is called first-stage graphitization (FSG). During FSG, massive carbides are eliminated from the iron structure. At this point, the iron is rapidly cooled to 725 to 740 °C (1340 to 1360 °F) prior to entering second-stage graphitization.

The third step in annealing consists of slow cooling through the allotropic transformation range of the iron; this step is called second-stage graphitization (SSG). During SSG, a completely ferritic matrix free of pearlite and carbides is obtained when the cooling rate is 2 to 17 °C/h (3 to 30 °F/h). This cooling rate, which depends on the silicon content of the iron and the temper carbon nodule count, may be increased to 85 °C/min (150 °F/min) to form a pearlitic matrix. Oil quenching from the FSG temperature following completion of that step will produce a martensitic matrix.

HARDENING AND TEMPERING OF PEARLITIC MALLEABLE IRON

A typical procedure for producing a hardened pearlitic malleable iron consists of (a) air quenching castings after first-stage annealing, which results in retention of about 0.75% combined carbon in the matrix; (b) reheating and holding for 1 h at 845 to 870 °C (1550 to 1600 °F)

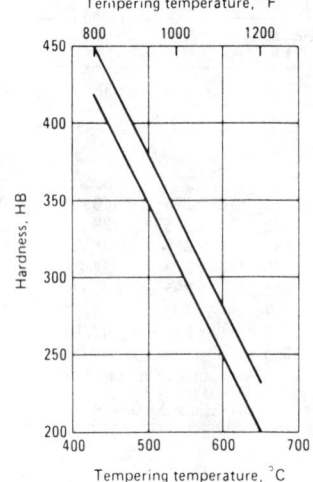

As-quenched hardness more than 500 HB.

Fig. 5. Hardness of oil-quenched ductile iron tempered for 1 h

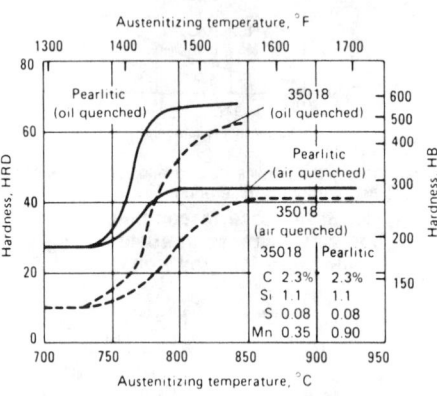

Fig. 6. Effects of austenitizing temperature, quenching medium and manganese content on hardness of as-quenched malleable iron

to reaustenitize the matrix; and then (c) quenching in heated (80 to 105 °C; 180 to 220 °F) and agitated oil, thus developing a matrix consisting of martensite and bainite with a hardness of 555 to 627 HB. Figure 6 shows the effects of austenitizing temperature, quenching medium and manganese content on the hardness of ferritic and pearlitic malleable iron both before and after heat treating. If direct oil quenching is used, caution must be exercised to prevent cracking due to high combined carbon.

Tempering treatments consist of cycles of no less than 2 h at temperature to ensure uniformity

of product. Tempering times must also be adjusted for section thickness and quenched microstructures. Fine pearlite and bainite require longer tempering times than that for martensite. In general, final hardness is controlled with process controls approximately the same as those encountered in heat treatment of medium-carbon and higher-carbon steels.

Heat Treating of Austenitic Irons

FLAKE-GRAPHITE corrosion-resistant austenitic cast irons are susceptible to work hardening during machining and require careful cooling from the casting operation and/or subsequent heat treating operations. Compositions of these irons are given in Table 2.

Stress Relieving. For most applications, it is recommended that austenitic cast irons be stress relieved at 620 to 675 °C (1150 to 1250 °F), for 1 h per inch of section, to remove residual stresses resulting from casting or machining, or both. Stress relieving should follow rough machining, particularly for castings that must conform to close dimensional tolerances, that have been extensively welded, or that are to be exposed to high stresses in service.

Holding of castings at 480 °C (900 °F) for 1 h per inch of thickness will remove about 60% of the stress; stress relieving at 675 °C (1250 °F) will remove almost 95%. It is usually acceptable to cool castings in air at a rate of 1 to 2 h per inch of section thickness, although furnace cooling produces maximum stress relief. Stress relieving does not affect tensile strength, hardness or ductility.

Spheroidize Annealing. Castings with hardnesses above 190 HB may be softened by heating to 980 to 1040 °C (1800 to 1900 °F) for 1/2 to 5 h, except those alloys containing 4% or more chromium. Excessive carbides cause this high hardness and may occur in rapidly cooled castings and thin sections. Annealing dissolves or spheroidizes carbides. Although it lowers hardness, spheroidize annealing does not adversely affect strength.

High-Temperature Stabilization. Except for castings of alloy type 1, which are not recommended for service above 430 °C (800 °F), castings used for either static or cyclic service at 480 °C (900 °F) or above should be given a stabilization heat treatment. This stabilization treatment consists of holding at 760 °C (1400 °F) for 4 h minimum or at 870 °C (1600 °F) for 2 h minimum, furnace cooling to 540 °C (1000 °F), and then cooling in air.

Table 2. Compositions of flake-graphite corrosion-resistant austenitic cast irons

| Type | Composition, % | | | | | |
	TC(a)	Si	Mn	Ni	Cu	Cr
1(b)	3.00 max	1.00-2.80	0.50-1.50	13.50-17.50	5.50-7.50	1.50-2.50
1b	3.00 max	1.00-2.80	0.50-1.50	13.50-17.50	5.50-7.50	2.50-3.50
2(c)	3.00 max	1.00-2.80	0.50-1.50	18.00-22.00	0.50 max	1.50-2.50
2b	3.00 max	1.00-2.80	0.50-1.50	18.00-22.00	0.50 max	3.00-6.00(d)
3	2.60 max	1.00-2.00	0.50-1.50	28.00-32.00	0.50 max	2.50-3.50
4	2.60 max	5.00-6.00	0.50-1.50	29.00-32.00	0.50 max	4.50-5.50
5	2.40 max	1.00-2.00	0.50-1.50	34.00-36.00	0.50 max	0.10 max(e)
6(f)	3.00 max	1.50-2.50	0.50-1.50	18.00-22.00	3.50-5.50	1.00-2.00

(a) Total carbon. (b) Type 1 is recommended for applications in which the presence of copper offers corrosion-resistance advantages. (c) Type 2 is recommended for applications in which copper contamination cannot be tolerated, such as handling of foods or caustics. (d) Where some machining is required, 3.0 to 4.0 Cr is recommended. (e) Where increased hardness, strength and heat resistance are desired, and where increased expansivity can be tolerated, Cr may be increased to 2.5 to 3.0%. (f) Type 6 also contains 1.0% Mo.

SECTION III
Surface Hardening

Vacuum Carburizing and Carbonitriding of Wrought and P/M Ferrous Steels

By DR. RICHARD G. WEBER, Manager
Manufacturing Process Engineering
Pitney Bowes
Stamford, Conn. 06904

Since its inception partial pressure carburizing and carbonitriding in a vacuum furnace has been gaining popularity. Carburizing or carbonitriding in a vacuum furnace does not involve generated endothermic atmospheres with reactive gas additions. Consequently, simultaneous reactions involving CO, CO_2, H_2O, H_2 and CH_4 are eliminated. In a vacuum furnace only one gas reaction dominates. When propane is used the reaction is $C_3H_8 + 3Fe \rightarrow 3Fe(C) + 4H_2$, and if methane is used the reaction is $CH_4 + Fe \rightarrow FeC + 2H_2$. In addition, the carbon potential in a vacuum furnace is not determined by the atmosphere, but is a function of the carbon saturation of the steel and its time at temperature, as shown in Fig. 1. The difference between propane and methane carbon potential is due to the availability of carbon during the cracking reactions previously given. At and below 1600°F (871°C) carburizing or carbonitriding with methane is not recommended since it has a lower cracking efficiency on steel parts than propane.

The data presented in this article are intended to serve several purposes. The case depth data can be used as a guide to heat treaters who are initiating vacuum furnace carburizing and carbonitriding. Furthermore, the case depth data for vacuum processing are compared to case depths achieved by carburizing and carbonitriding in an endothermic atmosphere furnace. The results are given for wrought AISI 1010 steel and a powder metallurgy steel (Fe-0.3%C). The P/M alloy is included because Pitney-Bowes uses many P/M parts as well as wrought parts in their business machines. Shown in Fig. 2 are some of the 350 active P/M parts which range in size from 1/8″ (.32 cm) diameter to over 2½″ (6.4 cm) diameter. These parts, in many cases, are vital components which must be precisely processed to ensure reliability. This need for precise case hardening in the range .003″ to .010″ (.008 cm to .025 cm) at a tolerance of ±.001″ (±.003 cm) precipitated the use of a microprocessor controlled vacuum furnace. The vacuum furnace used to establish the case depth results is a standard production furnace with integral quench manufactured by C.I. Hayes, Fig. 3. The atmosphere case carburizing and carbonitriding was performed in a sealed quench furnace.

The curves shown in Fig. 4 summarize the average carburizing and carbonitriding case depth data at 1600°F (871°C) for AISI 1010 steel and P/M alloy, Fe-0.3%C. Each curve is based on apparent case depths at a Vickers microhardness of 700. For each curve a constant (K) was determined using the equation, Case Depth = (constant) (total time)$^{1/2}$. These curves are only a guide however, and heat treaters should develop similar data based on their own production parts.

The total time for vacuum case hardening is composed of the carburizing or carbonitriding time (CT) and the diffusion time (DT). At 1600°F (871°C) 'DT' is related to 'CT' by the equation DT = 2/3 CT.

Fig. 4 shows that the vacuum processing is faster than

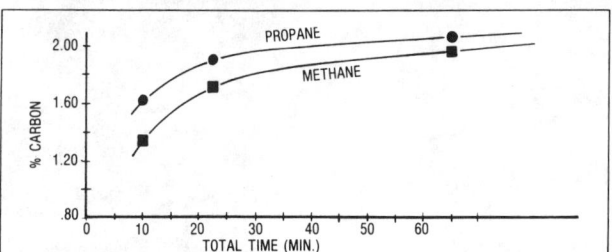

Fig. 1 Vacuum carburizing of .003 shim stock, % carbon vs. total time at 1600°F (871°C) using propane and methane.

Fig. 2 Some examples of P/M parts which require precise case hardening.

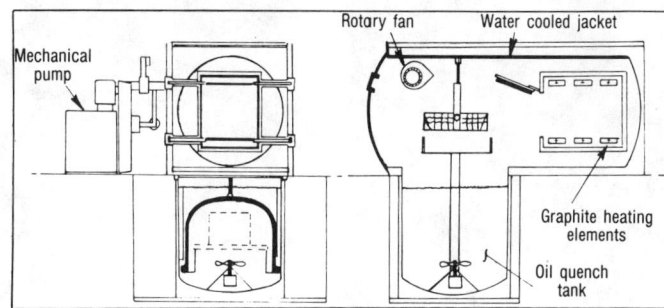

Fig. 3 Diagram of vacuum carburizing furnace designed with integral oil quench.

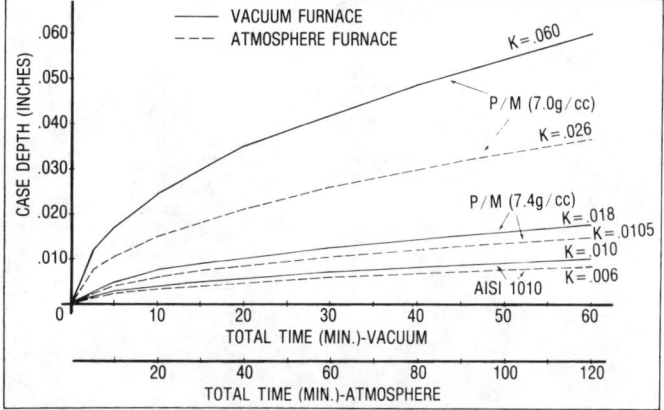

Fig. 4 Vacuum and atmosphere furnace carburizing and carbonitriding of P/M and AISI 1010 materials at 1600°F (871°C) (case depth versus time) in comparison.

processing in the endothermic atmosphere furnace. This occurs because the carbon potential is greater during vacuum furnace carburizing and carbonitriding, Fig. 1, than conventionally achieved in an endothermic atmosphere furnace. The nominal composition of the endothermic atmosphere used for this study was 20% CO + 20% H_2 + 1% CO_2 + N_2 balance. Additions to the endothermic atmospheres were made with methane so that the carbon potential was about 1%.

The curves in Fig. 4 also show that after the same processing time, the case depths of P/M parts are much greater than wrought AISI 1010 parts. The enhanced case depth is the result of the porosity inherent in P/M parts. At a density of 7.0g/cc (.025 oz/16.387 cu in.) porosity is 11% and at 7.4g/cc (.026 oz/16.387 cu in.) it is 6%. Furthermore, the curves show that greater case depths occurred on P/M parts after vacuum furnace processing than after atmosphere furnace processing. However, the case depth differences were found to increase as the P/M porosity increased. This effect cannot be explained entirely by the differences in carbon potential. It is believed that in the vacuum furnace, gas diffusion within pores is augmented by the initial evacuation of the pores and the continued gas partial pressure processing. In an atmosphere furnace, partial pressure processing is not possible.

In summary, it can be concluded that vacuum furnace carburizing or carbonitriding of wrought parts can be equivalent to atmosphere furnace processing. When case carburizing or carbonitriding P/M parts however, the part porosity must be a factor under consideration, especially during vacuum furnace processing.■

Production Experience with High-temperature Carburising

A. M. AITCHISON   **Caterpillar Tractor Co. Ltd.**

Where relatively deep case depths are involved, the use of elevated carburising temperatures enables substantial reductions in furnace cycle times, albeit at the expense of added maintenance costs. In a reappraisal of carburising procedures at Caterpillar Tractor, the adoption of a treatment temperature of 1010°C contributes to an 80% increase in production capability. The author outlines the background to the change, operating experience, benefits and drawbacks.

INTRODUCTION

The Glasgow plant of Caterpillar Tractor Company has a large heat treatment section employing over 100 people. Processes carried out include straight hardening (using oil, synthetic and water quenching), induction hardening, carbonitriding, annealing and stress relieving, as well as carburising.

Carburising is conducted in eight pit-type furnaces, two of 760mm (30in) diameter × 1220mm (48in) deep and six measuring 1070mm (42in) diameter × 1220mm (48in) deep. All are heated by gas-fired radiant tubes, the carburising medium being natural gas with nitrogen as carrier. Oxygen probes are used for furnace atmosphere control and case depth checks are by testpiece fracture, ultrasonic measurement and microstructural examination.

The majority of components carburised are cold-extruded bushings (SAE 10B16) requiring a case depth (to 0.4%C) in the range 2.2-2.8mm (0.088-0.112in). These are furnace-cooled after carburising and reheated for quenching.

In 1971 it was decided that further investment in capital equipment would be required unless a significant improvement could be realised in the output achieved from the existing carburising plant. The problem was divided into three areas of investigation:

1. Furnace uptime and utilisation.
2. Furnace loading.
3. Cycle time.

Furnace uptime and utilisation

It was found that 40% of available burden hours was being lost due to equipment downtime and idling between charges. The solution was to initiate a

Based on the paper presented at the Wolfson Heat Treatment Centre seminar "Innovations in Surface Heat Treatment" at the University of Aston in Birmingham on May 19th 1982.

preventive maintenance programme in which each carburiser was completely serviced once a year, each service taking 3-4 weeks. Additional carburising pots were also purchased in order to eliminate delay between charges. As a result, a subsequent audit showed furnace availability increased to 88.5% and utilisation to 90%.

Furnace loading

Prior to the investigation, a charge comprised two standard pots, both loaded with two layers of bushings with a space above. A simple redesign, with one pot larger than the other, allowed five layers per charge—a 25% increase in production.

These two aspects highlight the scope available for improvement merely by questioning and modifying established shop practice. As described below, examination of ways in which cycle times might be reduced involved a more fundamental approach which illustrated the virtues of high-temperature carburising.

CYCLE TIMES

In this part of the investigation, heating and cooling times to and from the carburising temperature were examined as well as the duration of the actual carburising stage.

Heating and cooling times

For a given charge weight and furnace, heating and cooling times should be reasonably constant. However, wide discrepancies were found due to such variables as burner and tube condition, exhaust fan condition, and inefficiencies in the operation of dampers and motorised valves.

Before the investigation, carburising was normally carried out at 955°C, with carburising gas introduced at 850°C. Since the total cycle time was fixed, variations in time to reach the carburising temperature led to variations in the time the charge spent at that temperature. Thus inconsistent case depths and high rework percentages were experienced. At the higher carburising temperatures contemplated, the effect of such variations would be even greater. In order to overcome the problem:

1. Carburising additions were not introduced until the full process temperature had been attained and the carburising stage was timed from that point.
2. Furnace operators and supervisors were instructed to check heating and cooling rates against an ideal standard.

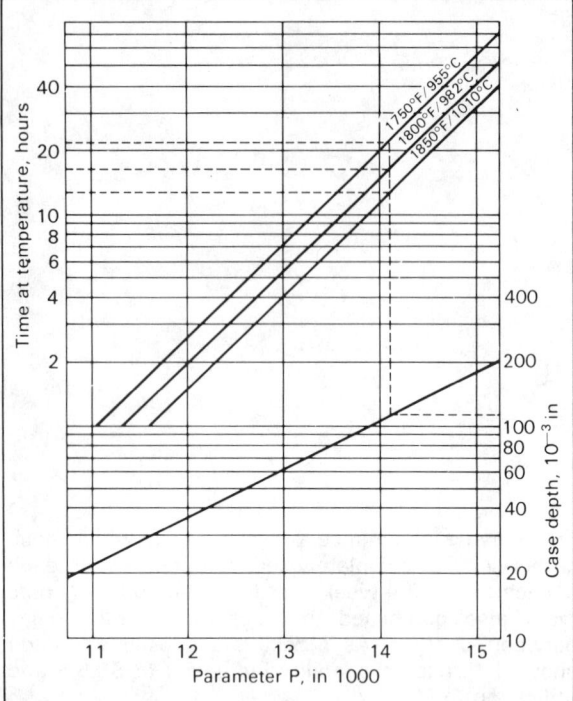

Fig. 1. Time/temperature/case depth relationships (after Lewis[1]).

Time at carburising temperature

The paper "A practical way to determine carburising time and temperature" by Lewis[1] was the basis for our investigations. This provides a mathematical model for predicting the depth of carburised case for any combination of time and temperature if other operating variables are reasonably constant. A graph *(Fig. 1)* relating these parameters can be constructed based on the parametric equations:

$$P = T(5 + \log t)$$

and

$$\log D = A + \frac{P}{4000}$$

where

T = absolute temperature ($°F + 460$)
t = time at temperature (hours)
D = case depth (10^{-3} in)
A = constant dependent on shop practice

The bold line in the lower part of *Fig. 1* reflects shop practice at Caterpillar for the grade of steel involved, showing the relationship between P and case depth measured on a consistent basis. The broken line projected vertically to the 955°C isotherm indicates that the desired case depth can be achieved in 22 hours at that temperature, two hours less than in our original practice. The predicted times for the same case

Table 1. The effect of modifications to carburising procedures on production capability

Carburising conditions	Original procedure at 955°C	Modified procedure at 1010°C
Parts loaded	428	535
Cycle time, h	29.5	20.5
Parts/h	14.5	26.1
Increase in production rate at 1010°C		80%

depth at 980 and 1010°C are 17.5 and 13 hours respectively.

The first step was to increase carburising temperature to 980°C; with an 18-hour cycle (12-hour boost/6-hour diffuse), the results were as predicted. After a six-month period, the process temperature was increased to 1010°C. Following a number of trials, a 9-hour boost/4-hour diffuse cycle was adopted.

OPERATING EXPERIENCE

The total impact of implementing the procedures suggested by the three phases of the investigation was an increase in production capability of 80% without capital investment in new equipment *(Table 1)*. However there are drawbacks:

Deterioration of furnace hardware

At 1010°C, the physical properties of the standard alloys used for furnace hardware are at the operational limits with regard to creep and impact resistance. Accordingly, after service of 12-18 months, furnace furniture and baskets required replacement. The support alloy hardware failed due to sagging and fan blades chipped and fractured. Baskets failed by the breaking off of lifting lugs in a brittle manner (much to the alarm of operators raising the 3400 kg load!). Records of life of these components were not maintained prior to the elevation of carburising temperature. However, there is no doubt that our expenditure on these alloys is now higher.

Premature failure of radiant tubes

These tubes are failing by fusion at hot spots. Apart from the elevated operating temperature, a more significant factor is the deposition of soot on the outside of the tube. At 1010°C, using nitrogen/natural gas additions only, most of the carburising occurs by direct cracking of the hydrocarbon, leading to precipitation of carbon on the alloy.

Atmosphere control by oxygen probe

With a nitrogen/natural gas atmosphere, oxygen potential readings were so low that no natural gas was called for during the boost cycle. This problem was circumvented by making a standard gas addition, using the probe to check furnace condition. Results obtained under these circumstances are satisfactory from a production standpoint (i.e. metallurgical properties comply with engineering requirements). However, currently we are experimenting with additions to increase the CO/CO_2 content of the atmosphere with the aim of *controlling* natural gas addition by oxygen probe.

CONCLUSIONS

Our experience of high-temperature carburising shows that processing at 1010°C results in very dramatic reductions in furnace time but with increased alloy and maintenance costs. It may be that a carburising temperature of 980°C offers the best compromise between productivity and economic operation.

REFERENCE

1. **Lewis C. F.** A practical way to determine carburizing time and temperature. *Metal Progress. Sept. 1969, Vol. 96, No. 3, 90, 93.*

AUTHOR'S ADDRESS

Mr. Aitchison is Heat Treatment Superintendent at Caterpillar Tractor Co. Ltd., P.O. Box 162, Glasgow G2 1JP, Scotland.

RECENT TRENDS IN GAS CARBURIZING METHODS AND CONTROL SYSTEMS

By
S. M. Tapaswi*

Abstract

Gas carburizing has become a most popular method compared to pack carburizing or salt bath carburizing. This is mainly because of the ease it offers and accuracy it gives in controlling the carbon potential in the furnace. Various sources of carbon are used in gas carburizing using different systems. These developments are reviewed keeping in mind Indian conditions and emphasizing the betterments in the control system as well.

Introduction

Gas carburizing enjoys considerable popularity amongst other carburizing methods such as pack carburizing and salt bath (or liquid) carburizing in the heat treatment shops all over. The reason for this popularity could be attributed to the ease it offers in controlling the carbon potential in the furnace and thereby the surface carbon contents and carbon gradient in the case. Over the years there has been tremendous research done on gas carburizing methods resulting in very fast development of this process. It is our intention here to review in brief the progress made in the field of gas carburizing and its controlling systems and consider the suitability of the various methods under Indian conditions.

In the process of gas carburizing carbon from the gaseous medium is diffused on the surface of machined components at high temperature. The components are generally made from low carbon and low alloy steels to yield a tough and ductile core and a high wear resistant case with high hardness.

Gas Carburizing Furnaces

The equipments used in the process of gas carburizing are generally classified as given in fig. No. 1.

Applications such as those where large number of similar components of case depth, say between 0.5 to 2.5 mm are intended to be gas carburized economically for production rates of 100 kg/hr and above, usually prefer continuous type furnace e.g. automobile gears and components. In such furnaces components are charged from one end and they pass through pre-heating, purging, soaking and carburizing, diffusion zones.

Then follows the cooling or transformation zone in case of two stage processes or the oil quenching zone in case of direct hardening steels. Last in the continuous line, comes the tempering zone. The components charged continuously from one end come out one after other from other end in duly heat treated condition.

In the case of batch type furnaces, which are generally preferred for small batches of drastically varying case depths such as large industrial duty gears, components for material handling equipments, components of machine tools etc, small batches are loaded on the heat resisting fixtures in the furnaces, preheated soaked and carburized at required temperature, diffused, furnace cooled or cooled in separate cooling chamber or quenched directly in oil in directly quenched batches. A brief contact of the charge with air is unavoidable in pit type furnace as against horizontal sealed quench furnace where quenching is also done under protective atmosphere.

Gas Carburizing Methods

Though the gas carburizing methods are more or less comparable in all furnaces we shall restrict ourselves mainly for pit type batch furnaces as basis for comparison of gas carburizing methods. The gas carburizing methods in pit type furnaces are classified as below:

i)   Endo-gas generator method
ii) Drip Feed Method
iii) Cylinder gas method
iv) Granulate process.

In the process of gas carburizing the carbon that is to be introduced on the surface of the components is obtained from variety of sources such as natural gas, butane (LPG) or propane, organic liquids of pure hydro-carbon type like terpenes, dipentene or benzene or oxygenated hydrocarbons like alcohols (methanol, iso-propyl alcohols) glycols and ketones (acetone, ethylacetate) etc. and solids carburizers like charcoal and coke etc. So much interest has been raised amongst various research workers, that they have even tried kerosene and diesel also as alternative sources of carbon for gas carburizing operation.

Even though the starting source is different the end

---

* Dy. Chief Metallurgist, Elecon Engineering Company, Limited, Vallabh Vidyanagar, Gujarat 388 120. Presently: Works Manager, Mettaco Engineering Co. P. Ltd., Halol 389 350.

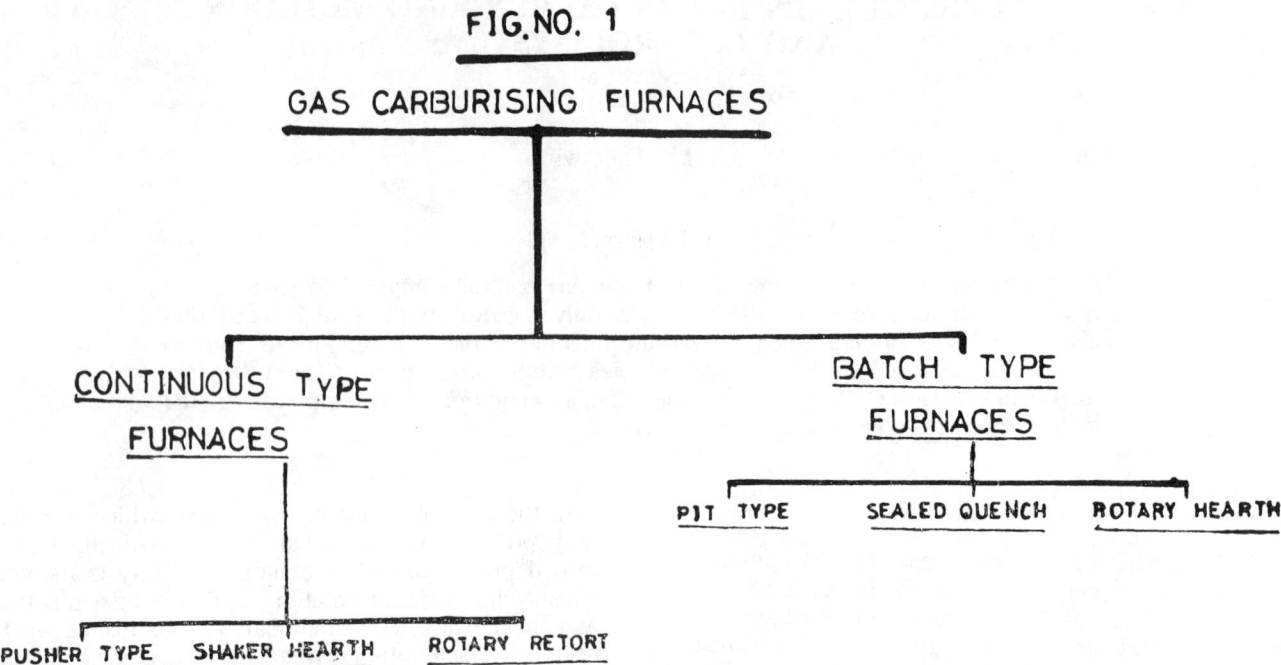

FIG. NO. 1

GAS CARBURISING FURNACES

CONTINUOUS TYPE FURNACES

PUSHER TYPE   SHAKER HEARTH   ROTARY RETORT

BATCH TYPE FURNACES

PIT TYPE   SEALED QUENCH   ROTARY HEARTH

product formed is generally carbon monoxide gas (CO) which decompes as per the reaction.

$$2 CO \rightleftharpoons C + CO_2$$

This nascent carbon is readily absorbed on the surface of the steel. It is very interesting to note that when Haywood[1] tried to carburize steel samples in solid carburizing agents but totally in the absence of oxygen the results were negative, which emphasizes the fact that the above reaction forms the basis of all carburizing processes. The carbon from the furnace atmosphere is used for the following purposes:

a) For conditioning the walls of retort and fixtures and bringing them to equilibrium value.
b) For adjusting the carbon potential of the furnace.
c) For carbon absorption on the components which depends on their surface area.

At the start of the carburizing cycle demand for (a) & (b) is very large but once the equilibrium is reached the demand drops to only (b) & (c). Thus the gas carburizing method must be adoptable to the above dynamic conditions of carbon demand in the furnace.

Endo-gas Generator Method: Though many new and better methods are being developed practically in the European countries majority of the heat treatment shops operate with this method.

The method employs a endo-thermic gas generator which separately generates the atmosphere required in the furnace. In the countries like North America & Italy where piped natural gas is available, it is used as the starting gas for endo-gas generators. Other countries use town gas or bottled hydrocarbons such as propane or butane as a feed-stock for endothermic gas generator.

The generator as shown in Fig. No. 2 consists of a retort of Nickel bearing heat resisting material which acts as catalyst (additionally porous refractories impregnated with Nickel oxide is also used) and is maintained at high temperature (around 1050°C). Air and hydrocarbons are charged in this retort at desired flow rate and proportion, through carburettor and mixing pumps to yield a endo-thermic gas of apporximate composition given in Table No. 1. Generally dew point controllers are provided on the endo-gas generators to regulate the composition of the endo gas obtained from the generators.

For the endo-generators to be really economical considering the high capital investment they should be at least of capacity of 20 Nm³/hr. Hence in most of the heat treatment shops one generator serves three or more furnaces, which means that all carburizing furnaces shall come to a stand still if one generator is out of order. Hence either a spare generator or enough spares almost to build entirely a new generator are needed to be maintained by HT Shops. Also though regulated gas is obtained from the generator a separate carbon potential control instrument is required on individual furnaces for controlling the flow of directly charged hydro-carbons in the furnace.

Drip-feed Method: In this method liquid hydrocarbons are charged directly in the furnace at high temperature on plates of Nickel bearing materials, which then crack and yield a gas of desired composition. There are two processes in this category (i) single liquid method and (ii) two liquids methods.

Single Liquid Method: In the single liquid method various liquids such as isopropylalcohol, methanol, terpentine, benzene, acetone etc. are blended in desired proportions which when charged in furnace will yield

TYPE OF ATMOSPHERE	COMPOSITION (vol.%)					
	H_2	CO	CH_4	CO_2	N_2	Others
ENDO-GAS FROM PROPANE	31.0	23.0	—	—	BAL.	—
ENDO-GAS FROM NATURAL GAS	40.0	20.0	—	—	BAL.	—
ENDO-GAS FROM TOWN GAS	60.0	16.0	—	—	BAL.	—
EXO-GAS	14.0	7.0	—	5.0	BAL.	—
TOWN GAS ONLY	55.0	6.0	25.0	2.0	BAL.	—
NATURAL GAS ONLY	—	—	90.0	1.0	BAL.	—
GAS FROM CARBO DRIP	48.0	20.0	20.0	0.0	BAL.	$C_mH_n = 0.2$
	58.0	25.0	25.0	0.3		0.6
METHANOL ONLY	65.6	32.3	0.6	0.3	BAL.	$H_2O = 0.9$
ETHYL ACETATE ONLY	65.6	32.3	0.6	0.3	BAL.	$H_2O = 0.9$
CHARCOAL BASED	1.2	34.7	—	—	BAL.	—

TABLE. NO. 1

TYPICAL COMPOSITIONS OF CARBURISING ATMOSPHERES

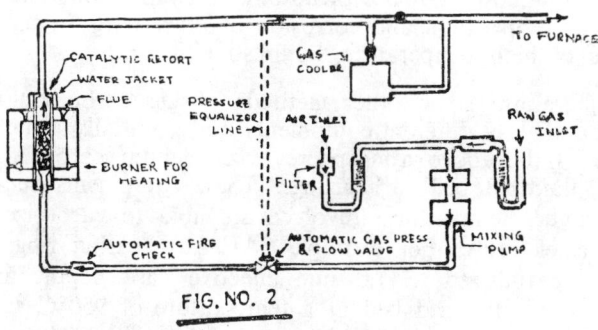

FIG. NO. 2

FLOW DIAGRAM OF ENDO-GAS GENERATOR.

a gas of required carbon potential. Normally varying carbon potential in such furnaces is difficult but a little control is obtained by varying the amount of liquid going into the furnace. However, control becomes very difficult when charges of different surface area are to be carburized and then decarburization or excessive carbon in the case or soot formation results. Automatic control of carbon potential is considered impossible with this method since the composition of the furnace atmosphere frequently changes and the C-CO₂-CO or H₂-CH₄H₂O relations remain no more valid. Earlier hot wire resistance method which gives 'C' potential directly was tried which has inherent problems on higher sides of range[2].

The latest development in the single liquid system has been a new process introduced by Ing. Buro Frey, West Germany using Carboprocessor (type. C) control system. In this method only one liquid hydro-carbon

such as benzene or acetone and air are mixed in the furnace itself to form a gas of desired carbon potential. The process is economical since instead of carrier gas it uses atmospheric air for control of carbon potential. One such gear heat treating unit is working in Friedrikshafen in West Germany using benzene and air. The electronic control system required should be very fast sensing and reacting type and involves central micro processing unit[3].

Two Liquids Method: In two liquids method, under which the patented Carbo-Maag process is covered, two liquid hydro-carbons are directly charged in the gas carburizing furnace where thy crack to give a atmosphere of desired carbon potential. One of the liqiud, which is generally methanol, is continuously charged in the furnace to form a carrier or neutral gas and it maintains positive pressure in furnace. The second liquid, which is rich in its carbon contents i.e. either ethyl acetate or acetone is then added as and when desired to maintain the carburizing potential as required.

A gas sampling device continuously takes out gas sample from the furnace. As per the original method, the dew point of this gas sample is determined by a peltronic unit 'CM3'. The actual value at a particular time will be compared with the set value and a control signal will be given to the solenoid valve carrying carburizer liquid to 'close' or 'open' at a preset flow rate. This desired carbon potential will be maintained in the furnace without problems of sooting or decarburization. These two liquids are so selected that by varying their proportion of mixing there is no effective change in the volume percentages of CO & H₂ formed, which is very important for proper control of C-potential. The cracking and carburizing reactions of acetone & ethyl acetate are given in Fig. No. 3. Due to difference in the kinetics of these reactions non-equilibrium quantities of CO₂ or CH₄ are generated, which give control problems when using CO₂ infrared analyser for adjustment of carbon potential and ethyl acetate as carburizer liquid.

The carbon availability in the carbo-maag method is 2.33 times higher than in the endo-generator method[4]. Also the carbon transfer coefficients are higher in comparison to endothermic gas i.e. 285 x 10⁻⁷ cm/s for methanol, ethyl acetate system against 114 x 10⁻⁷ cm/s for endothermic gas[5]. This gives an added advantage to the carbo-maag process as compared to the endo-generator process.

Cylinder Gas Method: This is comparatively a recent method and as will be seen from the number of papers presented in the 35th (1979) Heat Treatment Conference at Wiesbaden in West Germany, lot of development work is being done on this method in U.S.A. & Europe. There are few installations in HT Shops based on this method but the sucess of this method depends on the economic viability of the process since the bottled or cylinder gases are generally expensive.

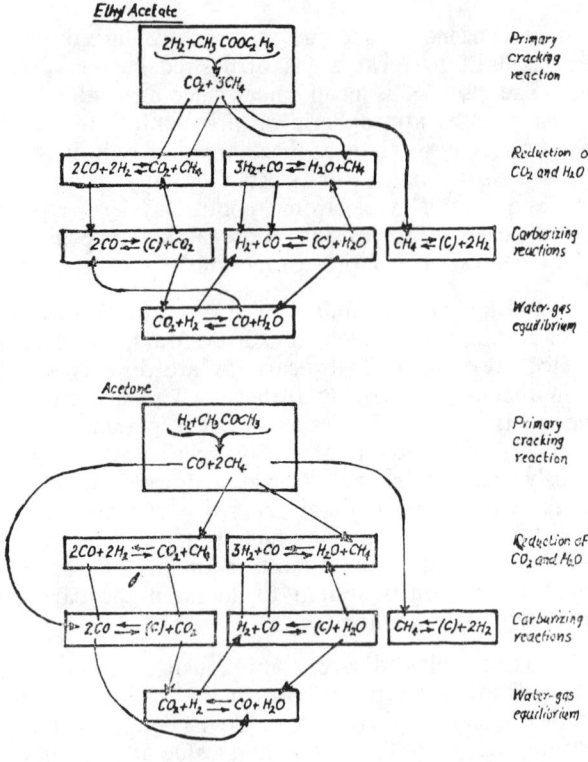

CHEMICAL REACTIONS IN CARBO-MAAG PROCESS

FIG. NO. 3

There are few variations possible in this method. In one of the methods called methane CO_2 CAP process patented by M/s. Air products & Chemicals, U.SA. — a blend of Nitrogen, natural gas and carbon dioxide is used to replace the endothermically generated atmospheres. Nitrogen & carbondioxide are stored in large storage tanks at the place of installations and these are regularly filled by the Company through tankers. These gases are passed through the vapourizers to the control panel where natural gas is also supplied. With the help of flow meters and a separate controlling instruments these gases are suitably mixed and charged in the furnace to give the desired carbon potential. In U.S.A. there are many gas suppliers who evaluate the existing carburizing systems and give technical know-how on the cylinder gas method e.g. Air Products, Airco Industrial Gas Division, Burdett Oxygen Co., Liquid Air Inc. & Union Carbides Linde Divn. These companies install hardware and on site liquid nitrogen tank or generator to supply liquid nitrogen for the process[6].

In another process a carrier gas based on gaseous mixture of Nitrogen and methanol is used[7]. Fig. No. 4 shows the principle of the method called endomix system developed by BOC Furnaces[8]. Liquid Nitrogen and Methanol are stored in large tanks. Nitrogen is mixed with natural gas or propane through a flow control panel and is introduced in the furnace. Similarly liquid methanol is also passed into the fur-

nace under nitrogen pressure. The composition of the resulting gas is analysed and through a automatic carbon control instrument signal is given to the solenoid valve supplying natural gas or propane.

These processes offer following advantages:

1. Elimination of gas generator, its capital expenditure and maintainance problems.
2. Easier and flexible control of the carbon potential and carbon availability.
3. Increase in productivity by eliminating conditioning time during start-ups, following idle periods and is not effected by power failures or mechanical breakdown.

Granulate Process: This process is still under experimental stages only. In this process a solid carburizing medium is kept in containers in the furnace space. Air in the furnace reacts with this solid carburizer to form carbondioxide at lower temperature and carbon monoxide at higher temperature.[2] The CO formed reacts with steel surfaces and gives a mild carburizing action. Introduction of additional hydrocarbon liquid in the furnace helps to build the carbon potential but control of carbon potential is still difficult. The process may become obsolete like the charcoal based atmosphere generators due to principal disadvantages of high operating costs, absence of automatic control, intermittance of operation and corrosion of alloy parts in the area of high temperature combustion[9].

There are some other methods of gas carburizing using slightly different equipments but essentially using one of the above atmospheres viz. vacuum carburizing, fluidized bed carburizing. These equipments due to higher temperature involved are able to carburize at faster rates. For example SAE8620 bearing rings were carburized to 1.0 mm effective case depth in $1\frac{1}{2}$ hrs in fluidized bed at a temperature of 950°C as against 6 hrs required by conventional gas carburizing at 925°C normally[10].

Automatic Control Systems

The success of carburizing treatment depends mainly on the control of carbon potential during the process. Industrial users realised that in order to avoid expensive rejections and rework and to achieve the desired case micro-structure, surface and case carbon contents were extremely important e.g. in the gear components requiring automatic tooth-flank grinding on high speed machines. This realization was the driving force for the developments from uncontrollable pack carburizing process to the present day automatic systems using oxygen probes.

For the basic carburizing reactions F. Neumann[11] has calculated the tolerance for carbon potential by varying the parameters given in table No. 2 & 3, theoretically for single variables.

$$C + CO_2 \rightleftharpoons 2CO \quad K = \frac{p^2CO}{pCO_2.a_c} = P\frac{\%CO^2}{\%CO_2a}$$

Deviations	Reference	Carbon level Tolerance at 0.8%C
$+0.02\%\ CO_2$	0.02/0.20	$+0.0625\%\ C$
$+0.01\%CO_2$	0.01/0.20	$+0.0313\%\ C$
$+0.50\%\ CO$	0.5/23.0	$+0.027\ \%\ C$
$+1.00\%\ CO$	1.0/23.0	$+0.054\ \%\ C$
$+10$ Torr (P)	10/760	$+0.0082\%\ C$
$+10°C$ (T)	10/925	$+0.065\ \%\ C$

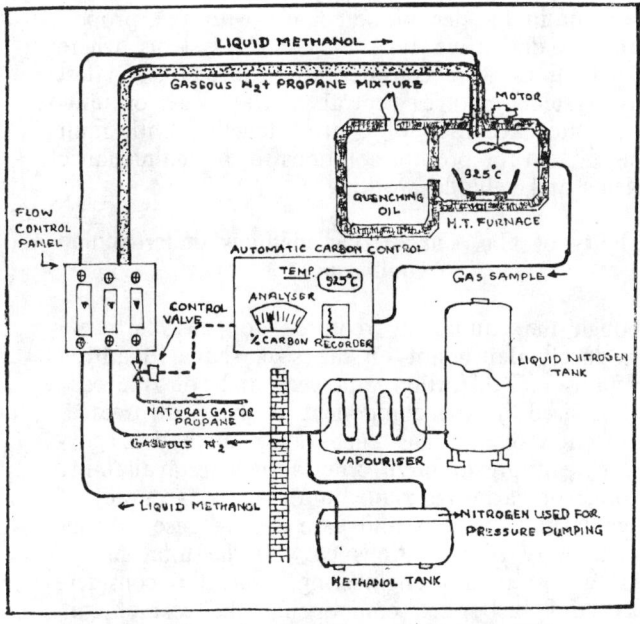

FIG. NO. 4   A TYPICAL ENDO-MIX SYSTEM

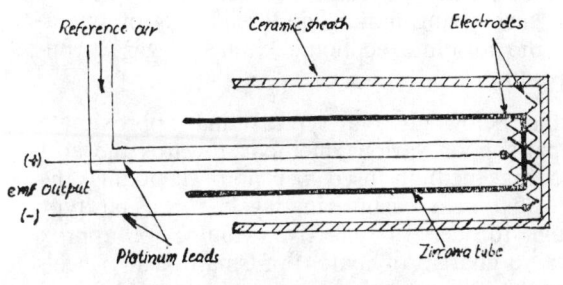

FIG. NO. 5

SCHEMATIC ARRANGEMENT OF OXYGEN PROBE

Thus in practice where two or more such reactions are in equilibrium the carbon potential of the furnace atmosphere is governed by the composition of the furnace gas, temperature and pressure within the furnace. Generally temperature and pressure are held constant and the carbon potential varies directly with $\%CO_2$, $\%CO$, $\%CO_2/\%CO_2$, $\%H_2O$, $\%H_2$, $\%CO_2$, $\%CH_4$ or partial pressure of O_2 in the furnace. And therefore to measure carbon potential, simple gas analysers like Orsat apparatus to the present day gas chromatographs are used. Long analysis time and relatively large errors make Orsat analysers unsuitable for accurate and automatic control. Some heat treatment shops do use gas chromatographs as well today. Though majority of the gases present in the furnace atmosphere can be analysed by gas chromatograph generally only one or two are used for control. Gas chromatograph would be more suitable for development work rather than routine production. Hot wire analysers, which uses the principle of change in the resistance depending on the solubility of carbon, are also found unsuitable for automatic control due to the errors obtained especially at higher carbon potential range.

The principal instruments used to-day for measurements and automatic control are dew point analysers, infrared analysers and oxygen probes.

New Point Analysers: There are many instruments available for checking the dew points of furnace atmospheres such as dew cup instruments, fog chamber instruments, chilled mirror instruments, chilled metal instrument. But in practice peltier cell type and Li-Cl type instruments are used for automatic controls.

In peltronic unit, as used in the original carbo-maag process, sample gas enters the measuring head which houses the dew point mirror. The latter is refrigerated by a peltier cell down to the dew point temperature and a photo-electric cell reacting on the reflective power of the mirror keeps the mirror temperature

Table No. : 3

$$C + H_2O \rightleftharpoons H_2 + CO \quad K = \frac{pH_2.pCO}{pH_2O.a_c} = P\frac{\%H_2.\%CO}{\%H_2O.a_c}$$

Deviations	Reference	Carbon level Tolerance at 0.8%C
$+2°CTp$ (dew point)	2/-10	$+0.1125\%\ C$
$+0.50\%\ CO$	0.5/23	$+0.0135\%\ C$
$+1.00\ CO$	1.0/23	$+0.027\ \%\ C$
$+1.00\ H_2$	1.0/31	$+0.02\ \ \%\ C$
$+10$ Torr (P)	10/760	$+0.016\ \%\ C$
$+10°C$ (T)	10/925	$+0.047\ \%\ C$

Source: Tool & Alloy Steels, May 1981, 161-167

finely adjusted such that dew precipitation remains constant. A resistance wire in the mirror transmits the temperature (actual dew point) to the compensating recorder and control system.

In another dew point measuring instrument hygroscopic nature of lithium chloride salt is used. Dry lithium chloride will absorb water at temperature and dissolve forming a saturated solution. This solution may in turn be heated to the temperature at which the evaporation tendency of the moisture just matches the absorbtion tendancy of the salt. This temperature is directly related to the dew point temperature.

With the recent developments, an accuracy of measurement $+ 0.25°C$ and reproducibility of $+ 0.1°C$ can be obtained with peltronic units as against $+1°C$ for Li-Cl type units. Even then the disadvantages such as frequent cleaning of filters, change in dew point with dirt, condensation of water in pipe lines, long purge time and lack of method for accurate calibration does not make the instruments ideally suitable for automatic control.

Infrared Analysers: These analysers are based on the principle that when infrared radiations are passed through a mixture of mono or di-atomic gases, these radiations are absorbed depending on the volume percentage of gas. Normally $\%CO_2$ in the furnace gas is measured to co-relate the carbon potential. With latest infrared CO_2 gas analysers reproducible and readable measurements within $0.005\% CO_2$ are obtained with a range of 0.0 to 0.5% CO_2. Response time are rapid i.e. approximate 20 seconds in good instruments. The instrument can be easily and frequently calibrated with standardization gases. The disadvantages include higher investment costs, complicated electronic system needing trained electronic specialist and the other draw backs as with all other gas sampling systems such as filteration, water condensate removal, leakages in sample lines etc. Generally single point to three point analysers are used with single or multirange instruments.

Oxygen Probes: The principal carburizing reaction can be re-written as

$$CO_2 \rightleftarrows CO + \tfrac{1}{2} O_2$$

$$\therefore \ K = \frac{pO_2^{\frac{1}{2}} \cdot PCO}{p \ CO_2}$$

Thus it will be seen that the partial pressure of oxygen in the furnace atmosphere can be directly related to carbon potential. The latest development in the atmosphere controlling instruments is oxygen probes or carbon sensors. The principle of the method is (shown in Fig. No. 5) that when a stabilized Zirconia tube (solid electrolyte) separates gases of two different oxygen contents viz. furnace atmosphere and room air it gives rise to an e.m.f. proportional to the difference of partial pressures of oxygen in these two gases.

F. Neumann[12] has given equation

$$E = 0.0496. \ T. \ \log p1/p2$$

Where E = emf across the wall of zirconia tube in mv

   T = Temperature °X

   p1&p2: Partial pressure of oxygen on the side 1 & 2 of the zirconia tube.

Platinum electrodes attached to both sides of the tube measure the emf which is subsequently amplified. This emf is proportional to the partial pressure of oxygen in the furnace which changes according to the carbon potential of the atmosphere. Thus measurement and control of this emf allows control of carbon potential[13]. This new technique gets rid of gas sampling pump, filters, lines etc. and measures carbon potential just as simply as a thermocouple measures temperature. Further work is being done on this technique to reduce the fragility of the oxygen probe and increase its life.

Regarding the control instruments also there is a considerable improvement. Old autocontrol systems were On-Off type controllers giving a fairly large band width for fluctuations of controlling parameters. Betterment in the technique came with the proportional and derivative type of control systems where the fluctuations were considerably reduced. The latest control systems involves central micro-process or units giving better accuracy of control together with fault finding system for prompt solutions to the maintenance problems and reliable working.

Suitability of Gas Carburizing methods under Indian conditions

Though the equipment required for gas carburizing is largly dependant on the size, shape, quantity, restrictions on distortion allowed and on the case depth desired in the component to be heat treated, the selection of the gas carburizing method is governed by entirely different criteria such as availability and price of carburizing medium, desired accuracy of surface carbon contents and case depths, ease of control, flexibility of operations etc. If the total manufacturing cost of any engineering product is considered involving carburized components, the cost of carburizing medium falls below even 1.0 percent and hence majority of manufacturers give more importance to ease and quality of control obtained by using a particular carburizing mdium and reliability of operation than the absolute economic value of gas carburizing medium.

The table below gives a typical and approximate cost comparison of various methods towards the cost of carburizing medium used per hour (including the capital cost/hr) in a carburizing cycle for a pit type carburizing furnace of effective volume of approx $0.5m^3$, for a charge of batch of small gear wheels (Wt. 500 kgs) for achieving a case depth of 1.0 mm at 920°C considering present prices of L.P.G., Methanol, Acetone, Liquid Nitrogen, Capital cost of generators, gas supply cubicle etc. under Indian conditions

Sr. No.	Gas Carb. Method	Cost of carburizing medium/hour	Capital Investment required.
1.	Endo-Gas method with LPG as feed stock	Rs. 20.80	Rs. 1,90,000/- *
2.	Drip-feed method Single Liquid	Rs. 14.90	Rs. 17,000/- **
3	Drip-feed method Carbo-maag	Rs. 13.10	Rs. 26,000/-
4.	Endo-mix cylinder gas method	Rs. 24.50	Rs. 2,00,000/-***

* Exclusive of cost of LPG tank installation.

** Using carburizing fluid of M/s. Killicks Nixon Ltd. Bombay.

***Inclusive of cost of installing liquid N_2 tank, vapourizers, panel etc.

but excluding the price & over-head charges of carbon potential controlling instrument on the furnace which would be almost same whichever method is used.

In the endo-gas and endo-mix gas the cost rates would get substantially reduced if the system is to operate for more than one furnace e.g. for 2 furnaces using one endo-generator the cost of carburizing medium will reduce from Rs. 20.80/hr to Rs. 14.50/hour and it will be still lesser for 3 furnaces and so on.

Thus, we can conclude that for HT Shops having lower requirements of carburizing gas drip-feed process offers most economical operations. Also the availability of LPG in India to small consumers is far from satisfactory and the capital investments required are high. However, for larger H.T. Shops endo-generator method still offers an economical solution. The cylinder gas method especially the one using methanol appears quite promising under Indian conditions and technically also offers many advantages. However, the economic viability of the process must be established by the gas suppliers by offering know-how and simple and better distribution and storage facilities.

References

1. F. W. Haywood : "Lecture on gas carburizing". Harterie-Technische Mitteilungen Band 11 Heft. 2-1956

2. Urs. Wyss : Gear meterials & heat treatment — Maag hand book.

3. Ihg. Peter Frey: Personal correspondance, 1980.

4. Urs. Wyss : 'Drip-feed carburizing' — Metal Progress November/1978.

5. **F. Neumann & U. Wyss : 'Aufkohlungswirkung von gasgemischen im system $H_2/CH_4/H_2O$ — CO/CO_2 — N_2' Internationalizer Gesprach uber warmebehandlungs frage Aug. 70.**

6. Edward A. Huntress : Nitrogen : All purpose atmosphere American Machinist April/80 PP 127

7. S. Janesn : Warmebehandlung Unter Synthetischen Endogas aus stickstoff und methanol — Lecture notes 35th Harterei Kolloquium at Wiesbaden Oct. 79

8. Groynn Bowes: 'Is Nitrogen economically viable as furnace atmosphere' Engineering Aug. 79 Pp 1017.

9. Metals Hand Book : ASM Publications Vol. 2 — 1971 PP 79

10. Peter J. Mullins : 'Heat treating tries new ways' IAMI-4/1978.

11. F. Neumann : 'Metallurgische Gesichtspunkte Zur Prozess Kontrolle bei der Gasaufkohlung von stahl' XXIX Harterei Kolloquim. Oct. 1973

12. F. Neumann : 'Der Potentialbegriff und seine Aussage im Rasmen thermochemischer Prozesss' 33th Harterei-Kolloquium Oct. 77

13. Timothy M. Meyers : 'Experience with carbon sensor in furnace atmosphere control' Metal Progress Apr/78 PP 36.

An overview—

High Temperature Carburizing

by Frederic J. Mahler

Part II

Atmosphere Control

Another area of concern is in process control. One worry has been with the accuracy of CO_2 control for high temperature application. Due to the nature of the relationship between carbon potential, percent CO_2, and temperature, the equilibrium CO_2 level for a given carbon potential is considerably lower for 1,900°F than for 1,700°F.

Frederic J. Mahler is associated with the technical center, capital goods division, of the Midland-Ross Corporation, Toledo, Ohio. Equipment used in the experiments described was manufactured by the Surface Division of Midland-Ross.

The main concern is with the ability to measure this level accurately. Figure 9 should help to shed some light on the situation. Assuming a target of 1.0 percent carbon potential in the atmosphere and an infrared instrument accuracy of ±1 percent of scale, two instruments of different ranges are compared.

For a CO_2 analyzer of 0-0.5 percent range, typical of what might be used in an automatic CO_2 controller designed for 1,700°F operation, the rated accuracy of the instrument would result

in a spread of ±0.04 percent carbon at the 1,700° level. If the same instrument was used to obtain a 1.0 percent carbon potential at 1,900°F, however, the rated accuracy would result in a spread of ±0.10 percent carbon, probably unacceptably wide.

If, on the other hand, a CO_2 analyzer of 0-0.1 percent range were used in the controller, the rated accuracy would result in a spread of ±0.02 percent carbon at 1,700°F and only ±0.03 percent at 1,900°F.

This 1,900° accuracy would be, in

Figure 9

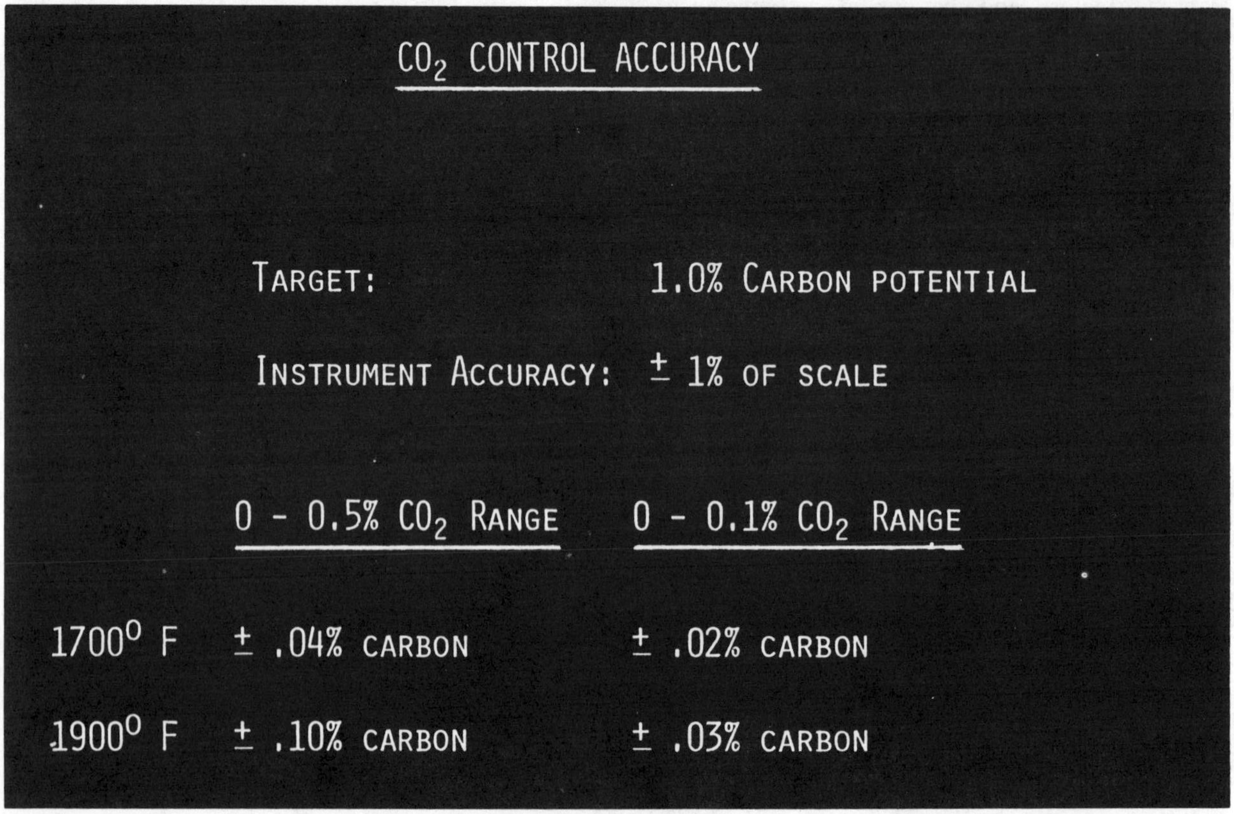

CO₂ CONTROL ACCURACY

TARGET: 1.0% CARBON POTENTIAL

INSTRUMENT ACCURACY: ± 1% OF SCALE

	0 - 0.5% CO₂ RANGE	0 - 0.1% CO₂ RANGE
1700° F	± .04% CARBON	± .02% CARBON
1900° F	± .10% CARBON	± .03% CARBON

108

fact, better than the 1,700° accuracy was with the 0-0.5 percent range instrument. So, CO_2 control problems of accuracy at higher temperatures can be overcome.

Dew point control, unlike CO_2 control, is based upon a linear relationship and high process temperatures would merely lower the equilibrium dew point level for a given carbon potential. The accuracy of control at the new temperature, however, would be unchanged.

Similarly, for oxygen probe control, the millivolt output of the probe would be different for a given carbon potential at higher temperatures, but the accuracy would be unaffected.

It should be pointed out, however, that the life of the probe at higher temperatures may be significantly affected. At least one manufacturer of oxygen probes rates their instrument up to 2,000°F in a carburizing atmosphere.

Thus, we have identified many of the problems associated with high temperature carburizing. There may well be other problems identified as development work continues. These will need to be resolved.

It is useful also to take a look at some of the incentives which are justifying the work toward solving these problems.

In order to examine the incentives, detailed theoretical analyses of a standard batch carburizer and a typical continuous carburizer were performed and compared for 1,700°F and 1,900°F operation. The results of these analyses indicate significant advantages in the following areas:

1. Increased production through decreased cycle time.
2. Reduced furnace size.
3. Decreased energy consumption.
4. Cost reduction.

The key factor in the above areas is the reduction of required cycle times at higher temperatures. It is important to point out, however, that all of these factors are heavily dependent upon the case depth to which carburizing is being done, which will be demonstrated in the following discussion.

Batch Carburizer

For the batch comparison a batch furnace with load size 30″ x 48″ x 30″ was used. For this analysis the maximum currently rated loadings for

1,700°F and 1,900°F operation were used.

It is significant to note that the current 1,900°F loading for the furnace is a little more than half of the maximum rated 1,700°F loading. With design modifications this rating could probably be upgraded, which would further improve the incentives for high temperature carburizing.

Also for this analysis, it was assumed that each furnace was heated by gas fired radiant tubes with recuperation. The carburizing cycles examined used forced air cooling to 1,550°F and holding for one half hour prior to quenching. Insulating material was upgraded for 1,900°F operation and heat transfer factors were adjusted accordingly.

Figure 10 is a plot of the gross energy consumption of the furnace at each temperature, expressed in gross BTU per gross pound processed, versus effective case depth. These curves indicate that above roughly 0.065″ effective case depth 1,900°F carburizing becomes more energy efficient despite the smaller load size. At a case depth of 0.085″ effective the gross energy savings would be about 9 percent.

Figure 10

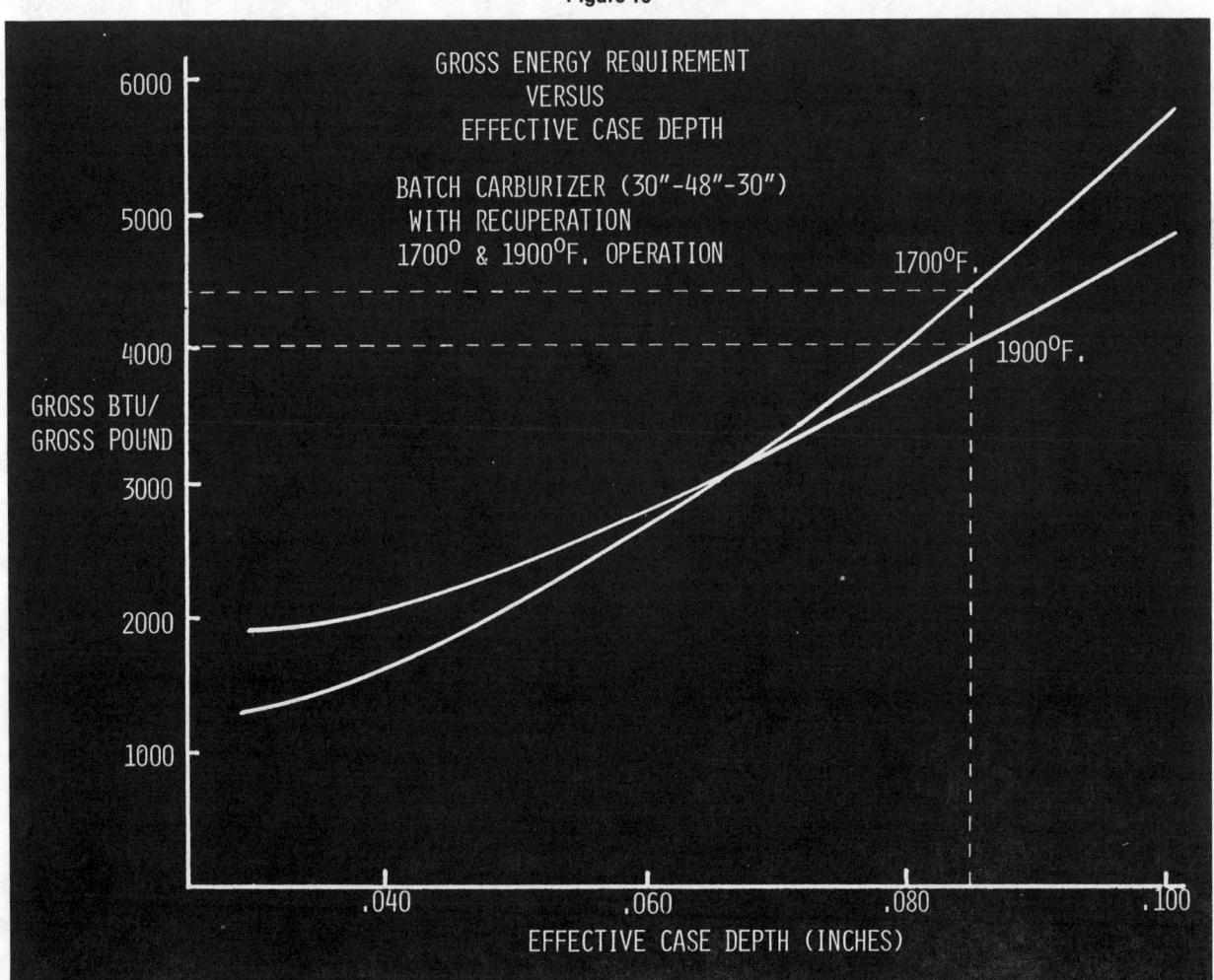

A plot of maximum gross production rate versus effective case depth is presented in Figure 11 for the furnace at each temperature. These data assume the time between discharging one load and charging the next to be negligible. They indicate that above 0.037″ effective case depth 1,900°F carburizing offers increased production rates over 1,700°F. At a case depth of 0.060″ effective the increased production would be about 37 percent.

For the purpose of illustrating the trade-off between increased production rates and energy requirements, a plot of factors of comparison versus effective case depth is presented in Figure 12. These factors are simply the 1,900°F values divided by the 1,700°F values from the two previous figures. This model considers only basic costs and does not include such things as maintenance costs, etc.

This plot helps illustrate the advantages in production capabilities and fuel utilization that 1,900°F carburizing would have over 1,700°F. The 1.0 line on the graph indicates the break-even point.

As indicated before, above about 0.037″ effective case, the 1,900°F carburizer provides increased production by significant factors. For example, at a case depth of 0.085″ effective, the factor would be 1.5, indicating a 50 percent increase in maximum gross production.

In terms of BTU per gross pound processed, above about 0.065″ effective, 1,900°F carburizing offers improved efficiency of fuel utilization. For example, at 0.085″ effective, the factor would be about 0.9, indicating a decrease in required BTU per pound of about 10 percent.

It is important to add that these figures are based on furnaces running at maximum rated capacities. So, the furnace must be kept full in order to realize the maximum benefits.

Continuous Carburizer

For this analysis a 2-row pusher-tray carburizer was used as a model. Typical basic design and sizing procedures were used to determine the furnace configuration needed at both 1,700°F and 1,900°F to produce 1,000 pounds per hour gross at an effective case depth of 0.060″.

Gas fired radiant tube heating was used and recuperation was provided on the heating zones. Also, insulation

Figure 11

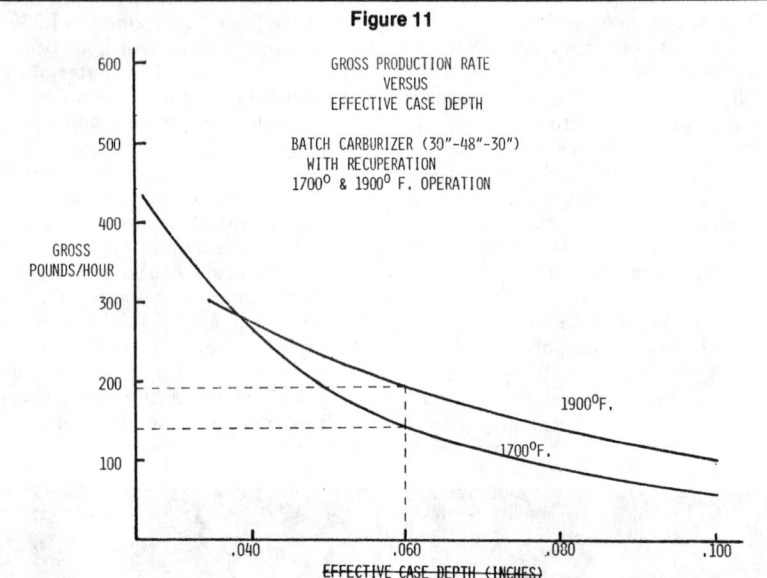

GROSS PRODUCTION RATE
VERSUS
EFFECTIVE CASE DEPTH

BATCH CARBURIZER (30″-48″-30″)
WITH RECUPERATION
1700° & 1900° F. OPERATION

Figure 12

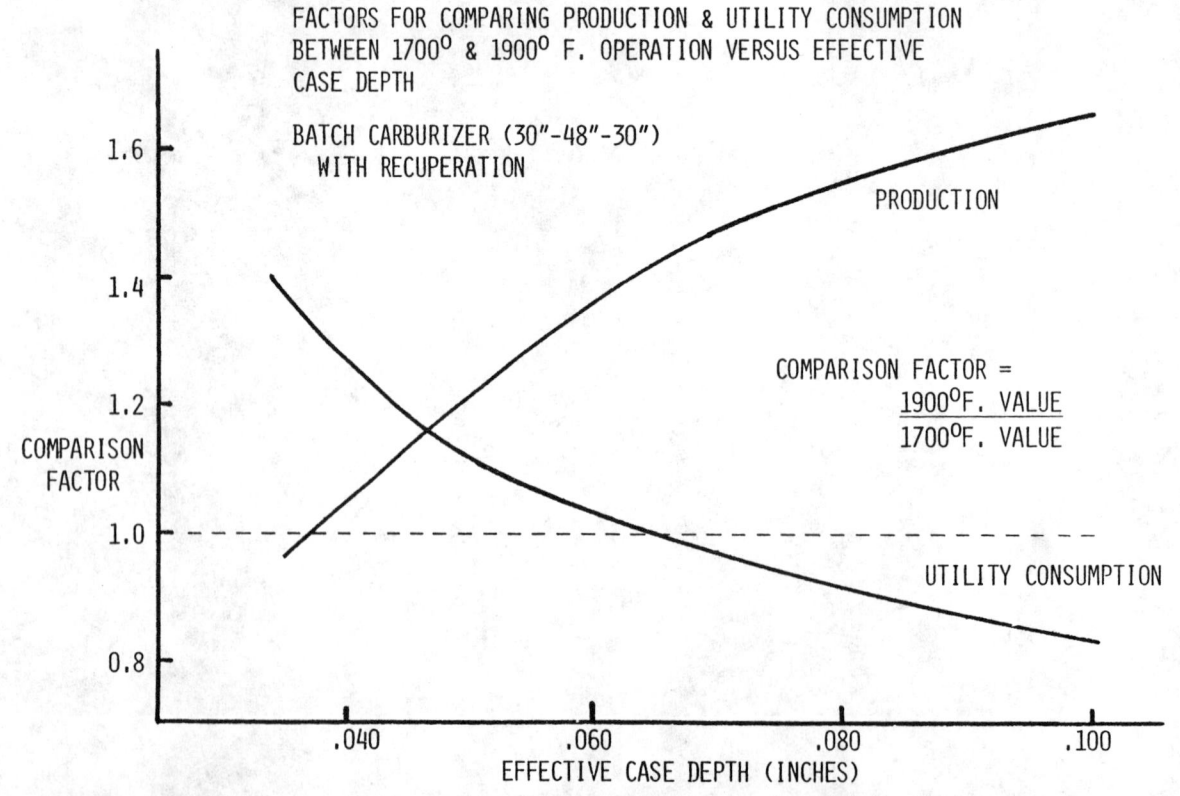

FACTORS FOR COMPARING PRODUCTION & UTILITY CONSUMPTION BETWEEN 1700° & 1900° F. OPERATION VERSUS EFFECTIVE CASE DEPTH

BATCH CARBURIZER (30″-48″-30″) WITH RECUPERATION

PRODUCTION

COMPARISON FACTOR =
$$\frac{1900°F. \text{ VALUE}}{1700°F. \text{ VALUE}}$$

UTILITY CONSUMPTION

was upgraded and heat transfer factors adjusted accordingly.

Figure 13 is a schematic diagram illustrating the differences in required zone lengths and distribution in order to meet these specifications at the two temperatures. The first thing one might notice is that there are only three zones in the 1,900°F furnace.

This is due to the fact that the total carburizing length has been reduced from 10 trays to 3. It would be unnecessary and, in fact, unfeasible to divide three trays into two zones. A normal zone would contain no less than about three trays due to construction and design restraints.

Notice also the increased heating zone length. It obviously takes longer to heat a load to 1,900°F. The increased cooling zone length is required to provide sufficient separation between the 1,900°F carburizing zone and the 1,550° cooling zone.

The large 350°F differential and radiation between zones would require the extra distance to get zone equalization at the lower temperature.

Once these initial calculations were set up for this model, it wasn't difficult to extend the furnace in single tray increments to compare energy requirements for deeper case specifications. The resulting plot of total gross furnace energy consumption versus effective case depth at the two temperatures is shown in Figure 14.

This illustrates that above about 0.060″ effective case depth, a 1,900°F furnace uses less energy per pound to process a given case depth and is considerably shorter than a 1,700°F fur-

Figure 13

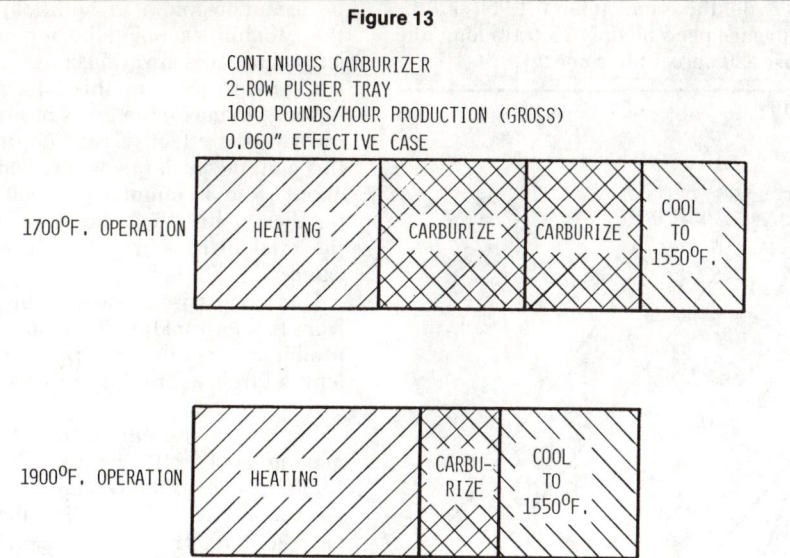

Figure 14

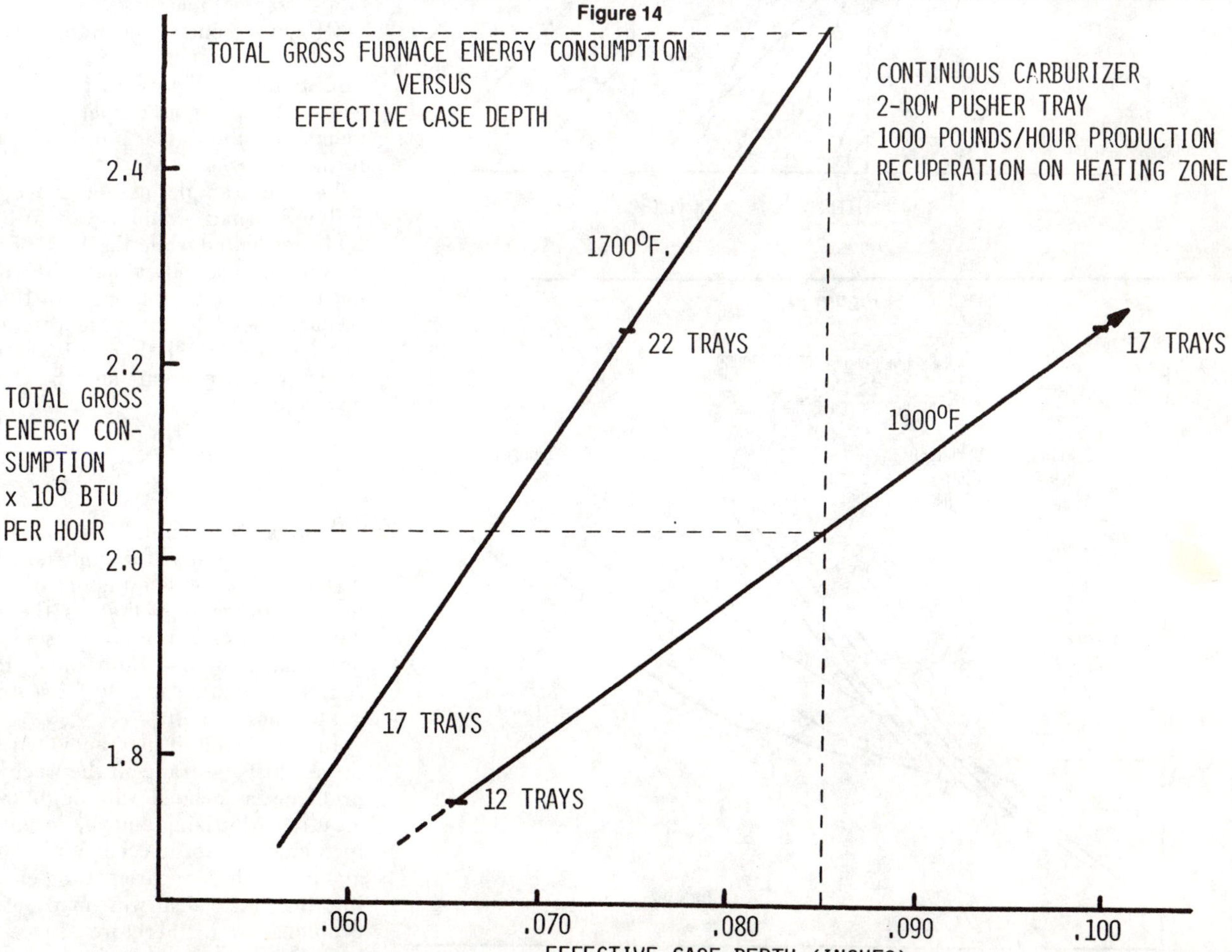

Source: Heat Treating, February 1979, 18, 20, 22, 24, 26

nace. The spread between the two temperatures for energy consumption and furnace size increases as the case depth increases.

For example, the maximum length a furnace of this type can be, roughly 27 trays, results in a maximum of 0.085" effective case depth at 1,000 pounds per hour, running at 1,700°F. To do the same job at 1,900°F, the furnace need be only 15 trays long and uses 20 percent less energy.

This comparison, thus far, has been restricted to a constant production rate of 1,000 pounds per hour. It would be useful to know, in addition, what the capabilities would be for various length furnaces at various push rates.

Figure 15 presents this information in terms of maximum gross production rates versus effective case depth. For this plot, the push rate was varied from about 15 to 30 minutes per push. Two families of lines are shown for which the total furnace tray length is indicated.

One comparison that can be made from this information is the maximum production obtainable from a given length furnace for a given case specification.

For example, an 18-tray furnace making 0.060" effective case depth at 1,700°F can produce up to 1,130 pounds per hour. Converting that furnace to 1,900°F operation would increase the maximum production capability to about 1,680 pounds per hour, or a 48 percent increase.

Of course, one would want to know what the impact on energy consumption would be. Figure 16 presents the gross BTU per pound energy requirements for the same two families of furnaces.

For the example just described, the 1,700°F furnace would use about 1,735 BTU per pound while the 1,900°F furnace would use only about 1,315 BTU per pound. So, even though an 18-tray furnace uses 22 percent more energy per hour at 1,900°F than at 1,700°F, it is using 24 percent less BTU per pound processed.

Figure 15

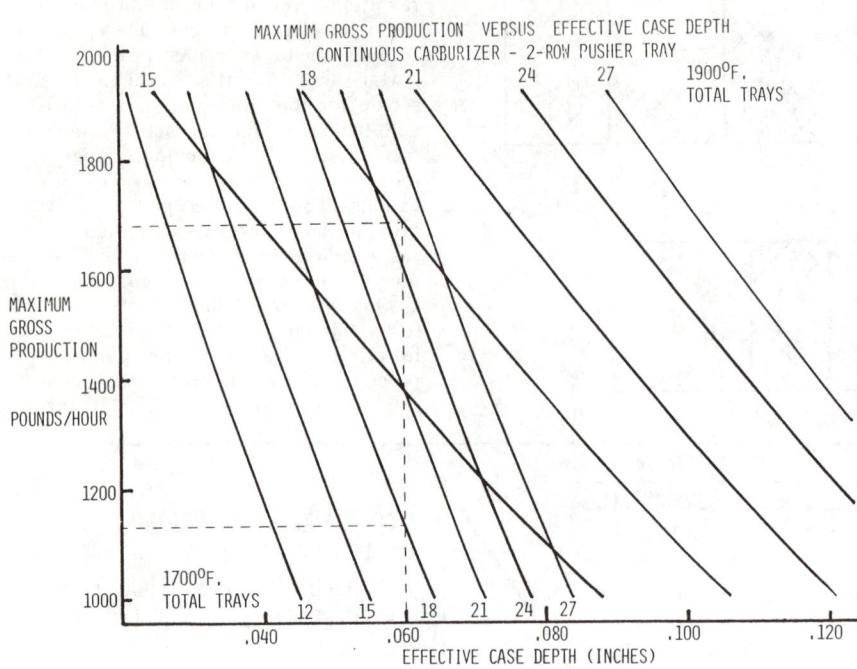

Figure 16

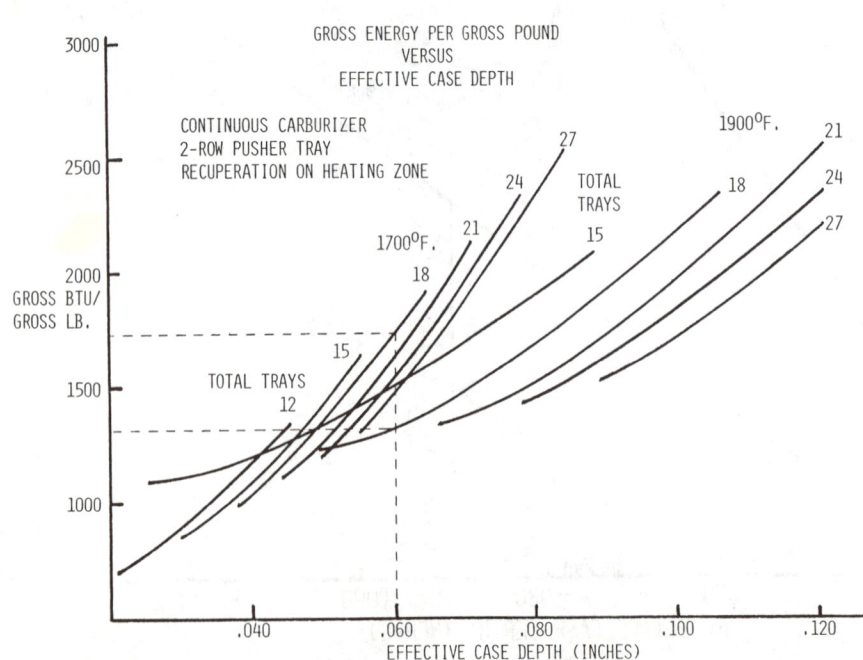

Conclusion

Figures 10 through 16 all illustrate the great potential for high temperature carburizing. The important point to keep in mind is that as the case depth specifications increase, so do the potential benefits. Equipment must run at maximum capacity, however, to get the most benefit.

In conclusion, it may be stated that there exists a trade-off between the problems associated with high temperature carburizing and the incentives for wanting it to succeed. We are now at a point where we are getting enough information to deal with many of the problems and, therefore, successful high temperature carburizing practice on a broad scale may become a reality in the not-too-distant future. ■

High-frequency Selective Surface Hardening

H. N. UDALL   Thermatool Corp.

An alternative to laser and electron beam techniques where a beam is traversed along a line to be hardened, the new high-frequency method of selective surface hardening enables a straight or curved line, or a pattern of lines, to be produced with a single energy pulse of less than one second duration. The heated area is self-quenched rapidly by the adjacent cold metal, thus providing a high surface hardness. This article outlines the principles and typical applications.

INTRODUCTION

Many engineering components in general use today are through-hardened, or case-hardened over the whole surface, even though the hardness may only be required in a specific area.

In recent years there has been substantial interest in selectively surface hardening small areas of a part in order to save energy, improve part performance, shorten production time, etc. Considerable work has been conducted in this field using laser or electron beam equipment, and recently Thermatool Corp. announced a new method of selective surface hardening using high-frequency contact resistance technology. For over twenty years, the basic process has been used to heat the edges of tubes, pipes, and other products for continuous seam welding. Now the technology has been adapted for selective surface hardening.

PRINCIPLE

Fig. 1 illustrates the basic principle. High-frequency current, typically in the 300 to 400 kHz range, is supplied directly to the workpiece through two small contacts, one at each end of the area to be hardened. The current then passes through a "proximity conductor" located close to the surface to be heated. This concentrates the current and confines its heating effect to the surface of the part directly under the conductor, as shown in the cross-section in *Fig. 1.* Because of the current concentration, the heating time is very short — usually less than ½ second. Therefore, heating is highly localised and as soon as the current flow ceases, the heated area is self-quenched very rapidly, to below the martensite transformation temperature, by the adjacent large mass of cold material.

The hardened line need not be straight as in *Fig. 1* since it mirrors the shape of the proximity conductor. Thus if this conductor is curved, the hardened line will be curved. This is illustrated by the part A in *Fig. 2(a)*, which also shows

Based on the paper "Selective surface hardening by high-frequency resistance heating" presented at the Wolfson Heat Treatment Centre conference "Advances in Surface Heat Treatment" at the National Exhibition Centre, Birmingham, on September 30th 1982.

some other typical samples hardened by the technique. Power supplies from 80 to 200 kW are currently available for this process. At the present time, lines up to 915 mm (36 in) long can be produced with a typical power requirement of 80-230 W/mm^2 (50-150 kW/in^2) depending on size and heating time. Typical line widths are from 3 to 16 mm (⅛ to ⅝ in) with a depth of hardening from 0.5-1.0 mm (0.02-0.04 in).

APPLICATIONS

The part B in *Fig. 2(a)*, AISI 01 tool steel, exhibits a hardened line on the surface 14.5 mm (0.57 in) wide by 1.0 mm (0.040 in) deep. *Fig. 2(b)* shows cross-sections of this part; the hardness of the stripe is 62 HRC.

Component C in *Fig. 2(a)* is made from AISI 1075, through-hardened and tempered to 47 HRC for use as a spring. Two lines have been selectively hardened to 60 HRC in order to resist localised wear. *Fig. 2(c)* shows sections of this hardened area. Because of the very rapid heating and quenching inherent in this process, there is no significant

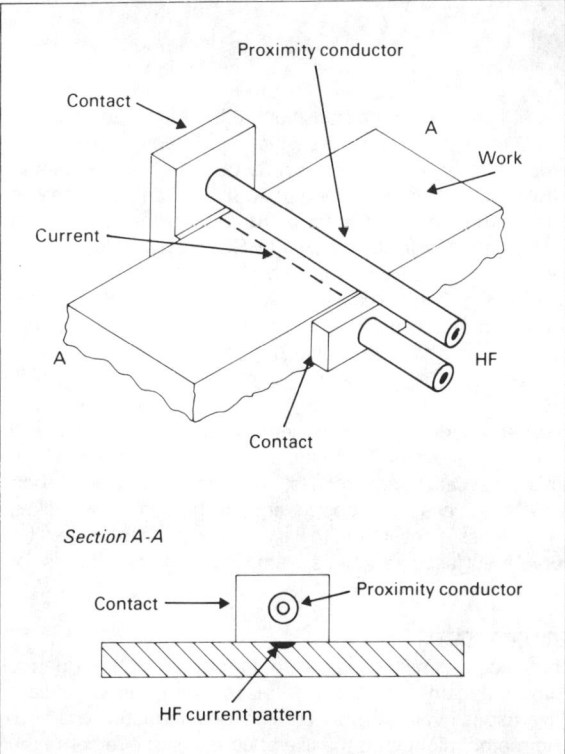

Fig. 1. Basic elements of selective surface hardening by high-frequency resistance heating.

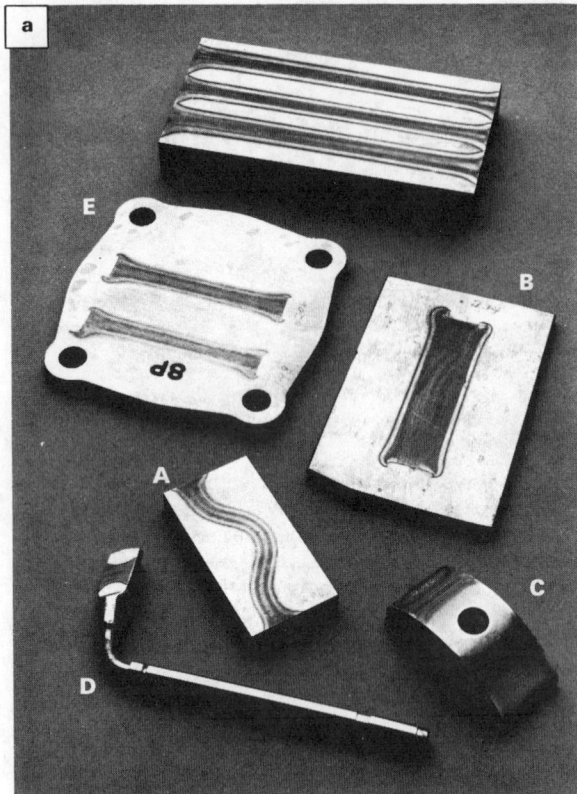

Fig. 2. (a) Some applications of high-frequency selective hardening. (b) Sections through the hardened area of part B, AISI 01 tool steel. (c) Sections through the hardened area of part C, AISI 1075 steel. (d) A section through part E, a grey cast iron, showing part of the hardened line on the left.

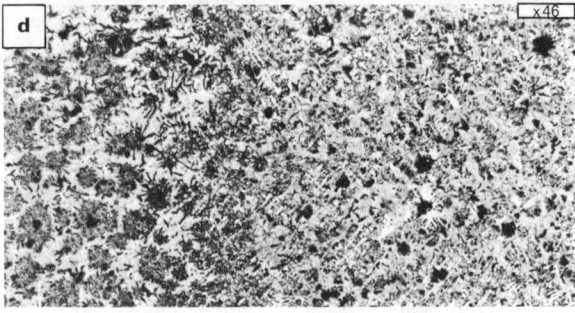

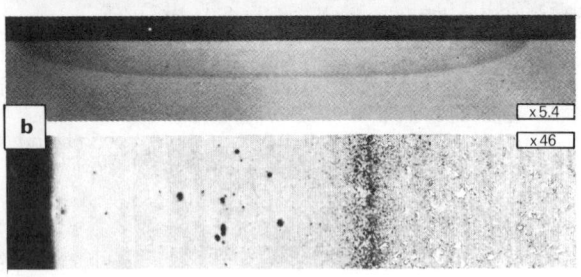

loss of hardness in the transition heat-affected zone.

Part D in *Fig. 2(a)* is a lever which has a wear point on the paddle and which would normally be case carburised over its whole surface. It is made of AISI 1117 and a stripe has been hardened at the wear point. Due to the low carbon content, the hardness is only 43 HRC; however, life testing has shown that this is satisfactory in this particular case.

Part E in *Fig. 2(a)* is a cast iron plate which has had two wear stripes hardened on its surface. The cross-section of the stripe in *Fig. 2(d)* shows that the pearlitic matrix has been replaced by a considerably harder, fine martensitic structure.

Powder metallurgy components can also be hardened successfully by this technique. Two compositions which have given satisfactory results, both contained 0.8% carbon with, in one case, 2% copper and, in the other, 2% nickel. The process is not limited to flat surfaces but can be used for curved surfaces, such as cam lobes or the bores of cylinders.

EQUIPMENT

The use of small contacts to introduce high-frequency current directly into the workpiece often has significant advantages over the more conventional induction method. Thermatool pioneered the use of contacts to overcome the limitations of induction coils in many continuous seam-welding operations, and there are now some 400 such installations in operation throughout the world in sizes from 60 to 600 kW. The same power supply which has been production-proven in steel mills, automotive plants, and other industrial environments is also used for the selective surface hardening process, with the addition of specialised controls and tooling designed for the part to be heat treated. Since the total hardening cycle is completed in less than one second, very high-production rates can be achieved, generally considerably in excess of those attainable by laser or electron beam equipment. In addition, no vacuum chamber is required and no special surface coating is needed.

SUMMARY

This new, patented, selective surface hardening technique is very fast and lends itself to high-production parts such as in the automotive or appliance industries. It employs a rugged, industrially-proven high-frequency power supply and can be used with a variety of interchangeable contact systems to harden a number of different parts. Compared with conventional heat treating, energy consumption is minimal, and distortion is generally very low.

AUTHOR'S ADDRESS

Mr. H. N. Udall, M.A.(Cantab), C.Eng., M.I.Mech.E, is Director of Research at Thermatool Corp., 280 Fairfield Avenue, Stamford, Connecticut 06902, USA.

Advances in Equipment for Plasma Nitriding and Carburising

P. COLLIGNON   Vide et Traitement S.A.

Users and manufacturers of plasma nitriding plant, Vide et Traitement have introduced a number of innovations in equipment design, operation and control to enhance the productivity and repeatability of glow-discharge heat treatment. These advances are combined in new equipment, based on vacuum tempering furnaces, where a dual heating system offers further benefits in nitriding whilst extending the range of application of plasma processing to high-temperature thermochemical treatments such as carburising.

INTRODUCTION
Whilst the principles and metallurgical results of plasma nitriding are well understood, some difficulties have arisen in the operation of these thermochemical treatment plants. Our experience as both users and manufacturers of furnaces has provided an opportunity to examine various aspects of ionic bombardment in order to improve the productivity and reliability of plasma heat treatment equipment.

Our studies have been directed towards solving the problems met in performing commercial heat treatment. Generally speaking, the difficulties are harder to overcome in this industrial sector where the parts to be treated are varied. A contract heat treater has to cope with orders covering hundreds or even thousands of parts, but he also has to face the tricky problem of one or a few parts calling for special treatment.

Let us first analyse the problems to be resolved in order to achieve proper utilisation of the equipment.

PROBLEMS
In the course of operations we have discovered a number of limitations. The main problems are:
1. The difficulty of treating components with small-diameter holes.
2. The necessity of stopping-off certain holes in order to avoid the "hollow-cathode' phenomenon (local over-heating). The drawback here entails the use of costly tools and involves long preparation times before treatment of parts.
3. The lack of automation necessitates qualified staff and constant furnace supervision.

Based on the paper presented at the Wolfson Heat Treatment Centre seminar "Innovations in Surface Heat Treatment" at the University of Aston in Birmingham on May 19th 1982.

4. Extensive precautions need to be taken in order to ensure homogeneous temperature distribution within the workload:
   (a) charging similar parts.
   (b) loading parts symmetrically in the furnace.
   (c) varying the loading density of parts from the periphery to the centre of the charge.

SOLUTIONS PROPOSED
The following sections outline some of the advances in equipment design and control which we have introduced to overcome the problems listed in the foregoing.

Generator design
The plasma generator developed incorporates a transistorised undulator, working at high frequency (1,000 to 5,000 Hz) and producing a cyclic ratio and variable amplitude signal. A characteristic of the generator is its very short break times (of the order of

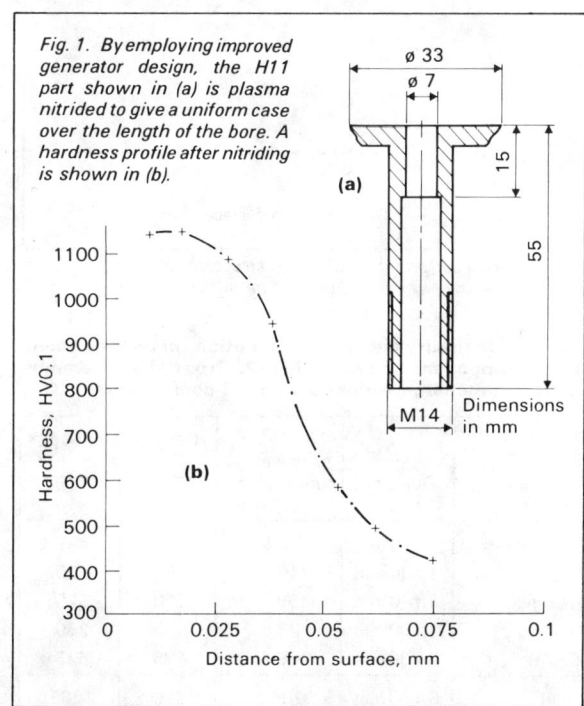

Fig. 1. *By employing improved generator design, the H11 part shown in (a) is plasma nitrided to give a uniform case over the length of the bore. A hardness profile after nitriding is shown in (b).*

Fig. 2. Precision engineering part for a hydraulic engine.

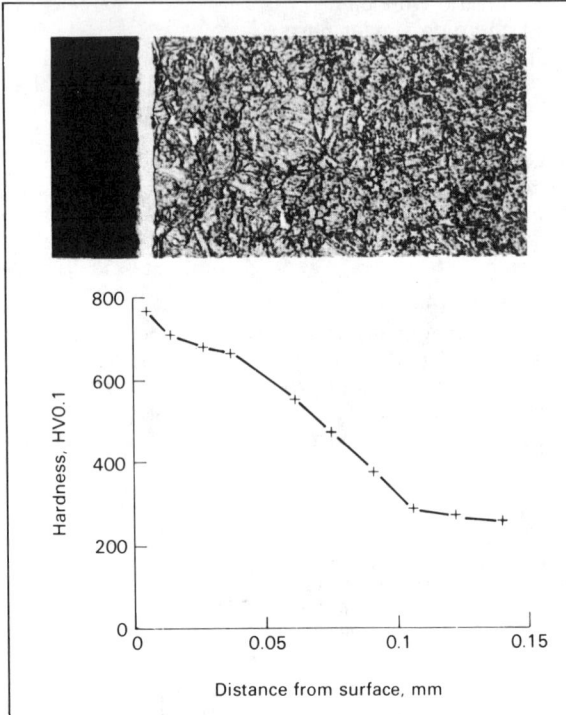

Fig. 3. Hardness profile and case structure obtained on the component shown in Fig. 2 after plasma nitriding.

Table 1. Electricity and gas consumptions in the treatment of the component shown in Fig. 2. Gross charge weight 1384 kg; net charge weight 972 kg (81 components).

Stage	Duration	Power consumption kWh	Gas consumption, litres	
			N₂	H₂
Evacuation	10 min	0.8	—	—
Depassivation	30 min	10	—	80
Heating	3 h 40 min	158	110	320
Holding	2 h 00 min	28.5	95	280
Cooling	2 h 30 min	9.2	175	525
Total	8 h 50 min	206.5	380	1205

one microsecond) which prevent arc formation. These features together produce a very stable plasma and eliminate electrical problems over the whole pressure range.

The use of modular generators with transistorised switching ensures greater flexibility in use. They allow work to be done over a broad pressure range from 0,5 to 8 mbar. In addition, the use of a high-frequency impulse current does away with the normal arc-break system; such a technique has enabled us to solve the difficult problem of "hollow cathodes" and to undertake the treatment of small-diameter bores.

Some examples of industrial applications benefitting from these innovations are illustrated in *Figs. 1 & 2*. The diagram in *Fig. 1(a)* shows a part manufactured from H11 steel for the moulding industry; this requires nitriding of the bore in order to overcome binding problems. The plasma treatment parameters are:

—Number of parts: 200
—Cycle: 90 minutes at 530°C
—Pressure: 7 mbar
—Gaseous addition: 40%N₂/60%H₂

This treatment gives a nitrided layer which is uniform over the whole length of the bore, 60 microns deep and with a surface hardness of 1100 HV *(Fig. 1(b))*.

Fig. 2 shows a precision engineering component made from SAE 4135 steel and pre-treated to 880 N/mm² (90 kg/mm²). Mass nitriding of this part is done without the stopping-off of the bores normally necessary in plasma nitriding to avoid over-heating. By thus eliminating the costly procedures which would otherwise be required, the treatment results in an appreciable gain in productivity. Metallurgical examination shows a nitride compound layer of 5 microns and a diffusion layer of 0.12 mm *(Fig. 3)*. Electricity and gas consumption figures are reported in *Table 1*.

Automation

The use of these furnaces by non-specialised staff and the large number of parameters to be controlled call for complete automation of the plant. The introduction of a microprocessor equipped with industrial peripherals into the operation of a plasma nitriding installation brings improved performance by optimising the various paramaters and running the entire system without human intervention.

Liaison with the operator is achieved via a screen-keyboard terminal (shown on the left in *Fig. 4*). All the information required for the smooth running of the process is set out in clear language on the screen and the choice of responses is explained. This information is recorded on an alpha-numerical printer.

After charging the furnace, the operator, through a conversational system, gives the computer the treatment parameters, namely:

—Duration
—Temperature
—Pressure
—Rate of temperature rise
—Gas mixture composition

Then the various phases of the automatic cycle are:

(a) Establishing the initial vacuum.
(b) Ionic "depassivation" of the parts in hydrogen, with variation of power as a function of the state of the surface.
(c) Controlled temperature rise with choice between different gas mixtures (neutral or reactive).
(d) Introduction of the treatment gas, each of the constituents of which is accurately controlled by mass flowmeters.

(e) Maintenance of temperature level.

(f) Cooling, which may be slow (in a vacuum) or accelerated (with forced circulation of the gases).

Furnace design

In order to reduce losses, which at temperatures of 400 to 600°C and at pressures of 1 to 10 mbar are mainly due to radiation, we have built hot-wall furnaces. This reduces energy consumption by a factor of from 2 to 3 and has enabled us to use generators of half the capacity for the treatment of the same-size load, whilst improving the homogeneity of temperature significantly. In order to improve productivity, we have equipped furnaces with a device for gas cooling the load at the end of the cycle. This comprises a two-speed centrifugal turbine for forced circulation of the gas and a finned-tube-type water heat exchanger.

Apart from this, we have opted for simple and rational solutions in the design of furnaces. For example, instead of manufacturing a bottom-loading furnace, we have gone for a bell-type furnace with built-in handling and two work bases *(Figs. 4 & 5)*. In this way we have been able to reduce investment by more than 30%, simplify handling, and reduce floorspace requirements.

With the aims of improving furnace specifications and solving the problem of very complicated parts (pressure-cast mouldings, dirty or crude foundry parts), we have developed and completed a series of new furnaces for plasma processing. These are based on vacuum tempering furnaces *(Figs. 6 & 7)*, rather than using a traditional vacuum furnace and relying on plasma alone to heat up the parts (transformation of the kinetic energy of the ions into heat).

The new type of equipment combines two distinct heating systems: a patented exchanger block resistor and the plasma. The major advantage of this technique is the ability to heat up parts to be treated uniformally (±2½°C) by convection in a neutral gas, regardless of the shape or dimensions of the parts. Heating by plasma alone cannot achieve the same level of temperature accuracy/uniformity, particularly when parts of very different shapes and dimensions are included in the same load.

The efficiency of this dual method of heating is maximised by utilising convection in a gas as heating medium. This overcomes the problem of achieving temperature uniformity when heating by radiation alone in vacuum at temperatures up to 600°C (i.e., temperatures at which, in most cases, plasma nitriding is carried out). Furthermore, it counteracts the slow heating rates associated with radiation at lower temperatures and the need to equalise temperature for longer periods because of masking effects between parts.

Once the parts have reached treatment temperature, the neutral heating gas is evacuated and replaced by the process gas which is ionised by the high-frequency impulse current generator. Maximum power is necessary for raising the temperature of the parts; thereafter, the generator only has to ensure temperature maintenance (compensation for the thermal losses of the furnace) and ionisation. For a load of 2 to 300 kg, a few kW are sufficient.

Apart from the fact that the quality of the treatment, in which uniformity is an essential factor, is distinctly improved, this system also overcomes the sometimes awkward problem of depassivation of the parts, the plasma only being applied at the treatment temperature. In *Fig. 8,* comparison of the two cycles shows the time

Fig. 4. Bell-type plasma nitriding furnace shown with the microprocessor keyboard and display unit. The furnace is equipped with a built-in handling device, for lifting and sideways movement of the bell, and an accelerated cooling system.

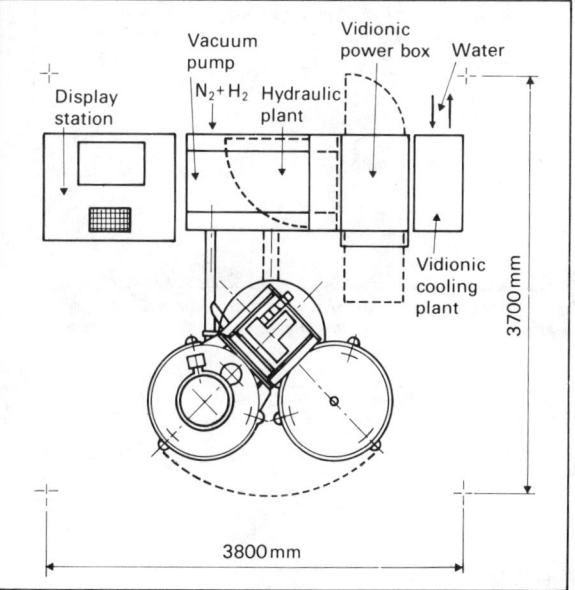

Fig. 5. Plan view showing the floorspace occupied by a bell-type plasma nitriding furnace with a charge volume of 800 mm diameter × 1500 mm high with two work bases.

Fig. 6. A vacuum tempering furnace suitable for plasma nitriding.

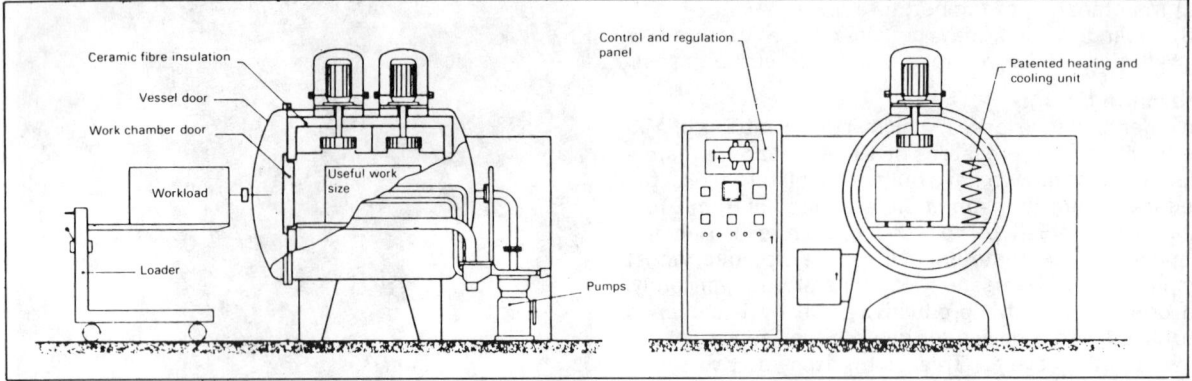

Fig. 7. Schematic diagram of a vacuum tempering furnace equipped for plasma nitriding (workspace dimensions 500 × 600 × 900 mm).

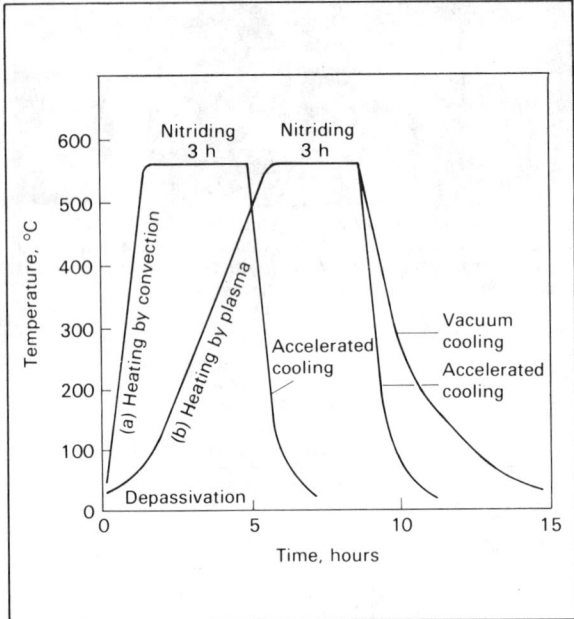

Fig. 8. Comparison of overall cycle times for a 3-hour nitriding treatment when heating by (a) convection and (b) ionic bombardment.

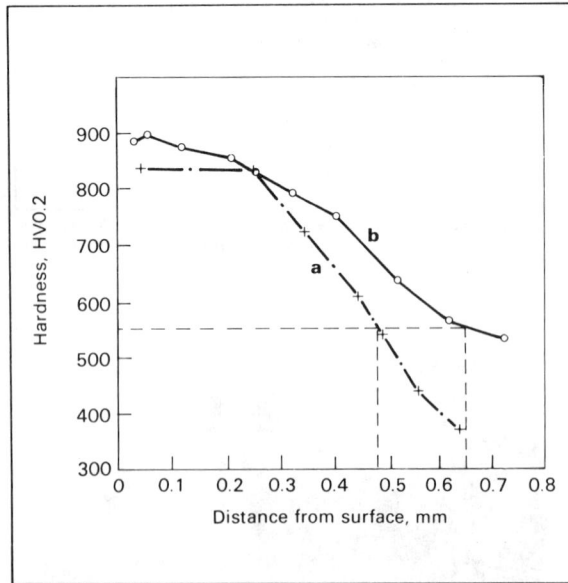

Fig. 9. Hardness profiles through plasma-carburised cases: (a) Steel 20M5 (0.16/0.22% C; 1.1/1.4% Mn) carburised for 90 minutes at 800°C; (b) Steel 16NC6 (0.12/0.17% C; 0.6/0.9% Mn; 0.85/1.15% Cr; 1.2/1.6% Ni) carburised for 120 minutes at 843°C.

savings yielded by this innovation over the the duration of a nitriding treatment for a 430 kg load made up of parts of different mass. It can be seen that in the case of heating by ionic bombardment it is necessary to raise temperature progressively in order to ensure good uniformity of load temperature. It will also be noted that heating by convection eliminates the depassivation phase.

In the case of high-temperature thermochemical treatments (carburising and carbonitriding) there is no reason why such a system should not be used in a vacuum furnace. Benefits then arise from:

(a) additional heating which occurs by radiation at temperatures above 700°C

(b) a generator designed to break down the gases of the furnace atmosphere electrically and to activate the surface reactions.

Carburising trials using such equipment, with both samples and production components, have achieved very successful results in terms of case development and attractive economics. Indeed, the characteristics of plasma processing, particularly the electrical dissociation of the gas and the activation of surface reactions, allow carburising to be conducted at low temperatures *(Fig. 9)*. Thereby, finer case structures are obtained, the risk of distortion of parts is limited, and mechanical properties are improved.

CONCLUSIONS

After years of operation, the first generation of plasma nitriding furnaces has been seen to exhibit a number of limitations. However, recent significant advances in electronics have provided solutions to these problems. Innovations in electrical power supply and the use of microprocessors herald the second generation of equipment in which the greater stability of the plasma and the greater degree of process control are prime improvements.

Now glow discharge technology is a reliable industrial tool used on a large scale and no longer restricted to nitriding treatments and low temperatures. The new patented system described in this article is also capable of operating at high temperatures for thermochemical processes like carburising where rapid case build-up allows short cycle times. This, combined with the flexibility of the system, provides a sound basis for the manufacture of high-production industrial furnaces.

AUTHOR'S ADDRESS

Dr. Collignon is Research Manager at Vide et Traitement S.A., Place Charles Andrieu, 60530 Neuilly-en-Thelle, France.

INDUCTION HARDENING

By P. K. Bhargava* and K. M. Joseph**

Introduction

Since its inception in the 1930's, the induction process of surface hardening has shown a spectacular growth in application. Normally its use is favoured by its quality characteristics in economical mass production. Occasionally, however, induction hardening has been the only practical method which could be used.

Surface hardening by induction method has substantial advantages over other conventional methods of heat treatment. The only limitation is the huge initial capital investment; but where production rate and quality are justified, the investment is repayed soon.

After more than 35 years of practical use, the induction hardening process has achieved a high degree of technical perfection and has made considerable contribution in automation of many operational processes in mass production lines.

The full potential of induction hardening can be realised in the field of fully automated heating process which can directly be fitted into any production line.

Today more and more manufacturing units are realising the significance and economics of induction hardening and thereby increasing their production, quality, saving in energy and reducing costs.

Principle of Induction Hardening

Heat for hardening metals can be generated within the part itself by electromagnetic induction. When alternating current flows through a work coil (also called an Inductor Coil), a highly concentrated alternating magnetic field is produced near and within the work coil. The strength of the magnetic field depends on the current density and the magnetic permeability of the immediate vicinity. This magnetic field induces potential in the workpiece placed in or near the work coil and since the work piece represents a closed circuit, the induced voltage drives a current through the work piece. The resistance of the work piece to the flow of induced current (eddy currents) causes heating by I^2R losses.

Due to high frequency current (Skin heating effect). the heating takes place instantaneously on surface and subsequent quenching of heated work piece either by spray quenching or dip quenching, hardens the surface.

The pattern of hardening by induction is determined by (a) The profile of inductor coil in relation to the profile of component to be hardened (b) Number of turns of the coil (c) Operating frequency and alternating current power input (d) Coupling between inductor coil and the component (e) Heating time and quenching time (f) Pressure, temperature and direction of quenching media etc.

Metallurgical Introduction to Surface Hardening by Induction

The term hardening used in heat treatment refers to the process of heating and subsequent cooling by which the steel is hardened. Steel of pro-eutectoid composition initially consisting of ferrite and pearlite in its structure when heated to a temperature above upper critical temperature, gets transformed to "austenite". From this austenitic temperature, the steel is quenched in some medium whereby heat is removed nite". From this austenitic temperature, the steel is cooled at a rate greater than critical cooling rate, transformation occurs by a shear mechanism that produces Martensite — a body centered tetragonal structure — in which carbon is dissolved, producing a highly stressed state and has a hardness upto Rockwell C 65 depending upon the carbon content of the steel.

In induction hardening, because of the possibility of transmitting high frequency and high power input into the workpieces in a very short time, the surface of steel is heated by induced current (Skin effect phenomena) to above critical temperature very quickly, thereby transforming the surface structure into austenite, without affecting the core of the steel, which on immediate quenching gives martensite structure. In view of the very short heating time, there is least tendency for decarburisation, grain growth, distortion, surface oxidation or the core of the component getting affected. These are some advantages not possible with other conventional methods of heating.

Excellent electrical, electronic and mechanical controls, design of work coil and quench fixture in an induction hardening equipment, makes it possible to achieve uniform surface hardness, uniform case depth and uniform hardened structure on all components. Also unlike other conventional heating equipments absolutely clean working conditions could always be maintained around induction hardening machines.

Possibility of stress relieving/tempering by induction immediately after induction hardening operation in the same working cycle helps homogenising the hardened structure further, thus producing metallurgically sound and quality products. Sometimes stress relieving/tempering operation could even be avoided altogether due to left over heat in the component.

Difference between high frequency (or radio frequency) Induction Hardening and Medium Frequency Induction Hardening

High frequency hardening (i.e. frequencies between 300 K C/s — 500 K C/s is generally used when shallow case depths are required particularly on small cross section components (i.e. case depths between 0.3 mm

* Director of EMA India Ltd., Kanpur.
**Executive, Technical, EMA India Ltd., Kanpur.

Reprinted with permission from Tool & Alloy Steels, June 1979, 195-198, 200-203, © 1979 Alloy Steel Producers Association of India

upto 1.5 mm). Sometimes high frequency hardening becomes necessary due to profile of the component.

Medium frequency hardening (i.e. frequencies between 1 K C/s — 10 K C/s) is generally used when deeper case depths are required (i.e. depths between 1 mm — 10 mm). Selection of frequency depends mainly on case depth requirement and cross section of the component.

High frequency power is delivered mainly by valve type electronic generators while medium frequency power can either be delivered by a Motor-Generator Set or a Static type Electronic Convertor.

Inductor coils when used with high frequency induction hardening are generally made from small cross section copper tubes while inductor coils when used with medium frequency hardening, are generally made with bigger cross section copper tubes. Special iron cores or concentrators are available for use on both H.F. and M.F. inductor coils.

Main advantages of induction hardening over other conventional methods

1. Surface hardening by induction greatly improves torsional, bending, fatigue and wear strength.

2. The process has made it possible to reduce costs considerably by substituting cheaper plain carbon steels for more expensive alloy steels.

3. Heating is extremely rapid enabling substantial economies to be realised. Heat transfer is as high as 5 to 6 times faster as compared to any other conventional heating.

4. Partial or selective hardening of component is possible without affecting other parts of the component, thus increasing production, saving energy and improving quality.

5. The equipment does not require long starting period and since the heat is produced within the component, there are no stand by energy losses. Stand by energy losses in conventional heating systems are in the order of 2 to 5 times the useful heat.

6. Excellent and accurate electrical and electronic controls guarantee uniform quality of hardening on mass production lines. Possibility of transmitting high power inputs into the components in a very short time, thus enabling to process greater number of components in shortest possible time.

7. Induction hardening process affects only the workpiece, thus least possibility of operator suffering from any effect of heat.

8. Induction hardening process is a very neat and clean process. No messing with oil or smoke. Equipments are very compact and thus can directly be fitted into any production line saving space, material handling and other costs substantially.

9. Due to very short heating times and absence of any reactive furnace media there is almost no scale or distortion in induction hardening of majority of components, thus eliminating any further cleaning or machining operations after hardening operation.

10. Possibility of using the same hardening equipment for other induction heating applications e.g. Brazing, Soldering, Billet heating, Melting etc. by incorporating only minor changes and some additional toolings.

11. Because of the inbuilt well protected automatic controls, induction hardening machines once set up by skilled personnels can be operated by semi-skilled or even unskilled operators.

Description of main parts of a complete induction hardening equipment

Induction Hardening equipments can mainly be categorised in two groups :

(A) Medium frequency induction hardening equipment.

(B) High frequency/Radio frequency induction hardening equipment.

(A) A complete medium frequency induction hardening equipment shall consist of following equipment parts :

1. **Main Power Source :** It can be a motor generator set or a static convertor of adequate power and frequency.

The motor generator type power source is a vertical water cooled design.

The Static type Generator is a solid state electronic design with thyristor controls. Main M. F. electrical components are water cooled.

2. **Starting Equipment for Power source :**

(a) For starting a motor-generator, generally an automatic star delta starter is used.

(b) For starting static convertor, no particular starter is needed. It can directly be connected to main switch gear of adequate capacity.

3. **Cooling water Protection Units for Power source :**

These units incorporate special adjustable type flow switches and thermostatic controls and are generally installed at the cooling water outlet of the Motor Generator. If water flow drops below specified limit or max. outlet temperature of water exceeds the max. allowed limit, this unit trips the power source.

4. **Medium Frequency Control Panel:** This panel controls the medium frequency power before it is delivered to the inductor coil and generally incorporates the following parts :

(a) Main medium frequency power contactor.

(b) Excitation Unit for motor generator.

(c) Measuring instruments for M.F. current, Excitation current, M. F. Voltage, M. F. power & power factor.

(d) Parallel M.F. water cooled capacitor bank.

(e) Electronic protection unit for motor generator bearings etc.

(f) Automatic voltage regulator for M.F. voltage.

(g) Bus bar system with cooling water protection devices.

(h) Various electrical control elements like control transformers rectifiers, control contactors, work hour meter, push buttons, indicating lamps, programme switches etc.

5. **M.F. water cooled Workhead Transformer with variable ratios**: This transformer is connected to the M.F. control panel via water cooled flexible leads. The inductor coil is connected to this transformer.

The function of this transformer is maching of impedance or stepping down the generator M.F. voltage to suit the single turn low impedance inductors thereby optimising power transfer from the generator to the inductor coil most efficiently. The stepped down voltage can further be varied by variable ratios available on the transformer to suit a variety of inductor-job combinations.

6. **Coaxial Cable and Flexible Water Cooled Leads**: Coaxial cable is required to connect the generator M.F. output terminals to the M.F. control panel.

Water cooled flexible leads are required to connect the M.F. output terminals of bus bar in M.F. control panel to the M.F. workhead matching transformer which generally is placed on the hardening machine/work station.

7. **Hardening Machines/Work stations**: These are generally designed to suit hardening of a particular component or a group of components. These generally consist of :

(a) The workholding fixtures.

(b) Inductor, matching transformer and adjustable table assembly.

(c) Feed scanning arrangement of various kinds to bring various parts of the component under the influence of the inductor and quench assembly. Variable feed scanning speeds are automatically selectable as per requirement.

(d) Programming of various types of atomatic process cycles via suitable arrangement e.g. micro switches and knockers, multi timers, potentiometers, programme switches etc. Variable power levels are automatically selectable as per the requirement.

(e) Provision for accurate quenching via solenoid valves and pressure switches.

(f) Magnetic clutch & brake units for accurate stopping and starting of machine movements.

(g) Electrical control panel.

8. **Various Toolings:** e.g. Work holding fixture to suit individual components, inductor coil for hardening individual component, quench assemblies to suit individual components etc.

9. **Quench Medium recooling and recirculating equipment:** In cases where instead of plain water quenching, some soluble oil or synthetic quenching medium is used, the quench medium cooling and circulating system has to be separated from the plain cooling water circuit. For this purpose this special self-contained closed circuit type equipment for recooling and recirculating the special quenching media is used. It is available in various capacities to suit individual requirements.

10. **Cooling water recooling and recirculating equipment**: Since the major high frequency components of any induction hardening equipment are water cooled, it is essential to have an effective cooling water recooling and recirculating equipment and facilities. This system can be via a refrigeration unit if required or via cooling towers installed outdoors or closed circuit cooling system to be installed inside the works and near the induction hardening equipment. The system is available in various capacities to suit individual capacities.

(B) A complete High Frequency Induction Hardening Equipment generally consists of following equipment parts :

1. **High Frequency Generator as Main Power Source:** These are generally valve type H. F. generators incorporating all electronic controls of modern design. The main oscillator valve is either water cooled or air-cooled.

2. **H. F. Transformers or H. F. tuning circuit/tank circuit.** This is an external water cooled transformer connected to H.F. generator via a special H.F. coaxial cable. This transformer can be placed on the hardening machine/work station and it carries the inductor coil and the quench assembly.

3. **Hardening Machines/Work Stations**: These are same as for medium frequency equipments specially made to suit the job.

4. **Various Toolings**: Same as for medium frequency equipments and as required for the particular component.

5. **Quench Medium Recooling and Recirculating equipment:** Same as for medium frequency equipment.

6. **Cooling water recooling and recirculating equipment:** Same as for medium frequency equipment.

In addition to above, wherever the cooling water has more than 100 ppm hardness, it becomes essential a watersoftening plant also for medium frequency as well high frequency. induction hardening equipments to avoid any blocking up of cooling coils or tubes with hard water scale.

Considerations in Induction Hardening

1. Metallurgical considerations: The process of induction hardening depends upon the rapid heating of the surface of the steel component followed by quenching. Because of this, direct hardening steels are necessary and there is a wide variety of steels which are suitable for induction hardening. The surface hardness obtained is primirily a function of the carbon content of the steel and is little influenced by the presence of alloying elements. This fact makes the use of plain carbon steels possible for many applications.

When induction hardening is envisaged, it is generally possible to use plain carbon steels instead of expensive alloy steels or carbourising steels, which would have been necessary if other conventional hardening methods were employed. The possibility of a change of material is always worth the consideration because of the following facts:

(a) Carbon steels are cheaper than alloy steels.

(b) Carbon steels are more readily available.

(c) Carbon steels are generally superior from the point of view of machinibility.

(d) Carbon steels are much less prove to quench cracking than alloy steels.

(e) By replacing carburising steel with direct hardening plain carbon steels, the carburising process can be eliminated.

These facts can, in many cases, offer appreciable economies.

Although plain carbon steels can generally be used satisfactorily for induction hardening, the use of alloy steels is sometimes advisable in the interests of obtaining enhanced core properties such as in the case of gears. Generally speaking, however, the case hardening of steels with a core strength exceeding 80 tons per sq. inch is not recommended.

The optimum degree of surface hardness which can be obtained from any steel is a function of the carbon content ad curves have been published which clearly indicate this point. These curves show that the lowest carbon content for a RC 50 surface hardness is 0.35%. These results can be obtained only under the most favourable conditions and for production purposes where a surface hardness of RC 55/60 is required it is more advisable to use a carbon content of 0.40% to 0.45%.

The hardening of steel is brought about by the carbon in the heated layers going into solution and although this is a rapid process at hardening temperatures, it is dependent upon both temperature and time. Induction heating being very rapid, it is sometimes desirable to use somewhat higher temperatures than would be employed when using furnace heating. This is particularly true in the case of alloy steels where the diffusion of the alloying elements takes longer than that of the carbon. When treating alloy steels by induction methods, therefore, it may be necessary to employ temperatures some 50 to 200°C in excess of these used with furnace heating where a "Soak" period is involved. Plain carbon steels, on the other hand, need be heated very little in excess of the conventional hardening temperature.

The three factors influencing the induction hardening of alloy steels may therefore be summarised as follows:

(a) The amount and nature of alloying element present in the steel.

(b) The length of Induction heating cycle.

(c) The previous condition of steel.

The greater the quantity of alloying elements, the longer will be the heating cycle required for complete diffusion. Alternatively, higher temperature can be employed for shorter periods since the rate of diffusion is a function of both time and temperature.

2. Selection of Power and Frequency: Generally selection of power and frequency for surface hardening by induction depends on the following data:

1. Cross section and profile of the job.

2. Depth of hardening required.

3. Production rate.

4. Material specifications.

However the first two of the above points play a major role.

A number of formulae, theories, graphs etc. have been published from time to time for selection of power and frequency and when calculations are required, the same can be referred to.

However, given below are two tables for generals reference:

GOOD: Indicate frequency that will most efficiently heat the material to austenising temperature for specified depth.

FAIR: Indicates a frequency that is lower than the optimum but high enough to heat the material to austenising temperature for the specified depth. Lower efficiency main disadvantage.

POOR: Indicates a frequency that will overheat the surface unless low energy input is used. Efficiency is bad, production is low.

3. Selection of Power Control Equipment: Once the power and frequency requirement has been found out, the next step is to design and select the right power control equipment such as motor generator, starters, control panels, impedence matching transformer etc. of proper specifications and ratings as per the requirement. From the various data available for induction hardening, these specifications and ratings can easily be designed and selected.

4. Selection of hardening Machine/Work Station: After the power, frequency and power control equipment details are completed, the next step is to choose a proper and efficient hardening machine/work station suiting to the components to be hardened. Various

Depth of Hardening mm	Diameter mm	Motor Generator (M.F.)			Valve Type Generator R. F. over 200 KHz
		1 KHz	4 KHz	10 KHz	
0.5 mm to 1.25 mm	6 mm to 12 mm	—	—	—	GOOD
	12 mm to 19 mm	—	—	—	GOOD
1.25 mm to 2.5 mm	19 mm to 25 mm	—	—	FAIR	GOOD
	25 mm to 50 mm	—	FAIR	GOOD	FAIR
	over 50 mm	—	GOOD	GOOD	POOR
2.5 mm to 5 mm	19 mm to 50 mm	FAIR	GOOD	FAIR	POOR
	50 mm to 100 mm	GOOD	GOOD	FAIR	
	over 100 mm	GOOD	GOOD	POOR	

TABLE 2

Power density required KW per sq. inch for Surface hardening

Frequency K c/s	Depth of hardening	Input KW per sq. inch		
		Low	optimum	High
500	0,015 to 0,045	7	10	12
	0,045 to 0,090	3	5	8
10	0,060 to 0,090	8	10	16
	0,090 to 0,120	5	10	15
	0,120 to 0,160	5	10	14
3	0,090 to 0,120	10	15	17
	0,120 to 0,160	5	14	16
	0,160 to 0,280	5	10	14
1	0,200 to 0,280	5	10	12
	0,280 to 0,360	5	10	12

types of such hardening machines are available depending upon the shape and size and production rate of the components. Vertical & horizontal progressive scanning type machines with variable feed speeds and variable power levels with pre set values and timings are available. Special turn table type machines are available, special gear hardening machines with possibility of tooth by tooth hardening, are available. In fact an induction hardening machine can be custom built to suit any individual requirement with variable automatic hardening programmes. If required, an induction hardening machine today can also, be designed and fitted with numerically controlled hardening programmes with digital readouts facility etc.

Process Control in Induction Hardening

Induction Hardening systems feature many process controls depending upon the type of components and their production rate requirements. Possibility of excellent electrical and electronic controls, feedback systems, predetermined time sequences, accurate location of inductor coil and quench and stable power supply guarantee consistent hardening and metallurgical results.

However, described below are some of the process factors which play the major part in the technique of induction hardening:

1. **Design of Inductor Coil:** This plays a very important role in any induction hardening operation as it is through it, that the power is distributed into component. The design of the inductor coil mainly depends on following data:

(a) Shape and size of component.
(b) Depth of hardness and profile of hardened zone required.
(c) Production rate requirement.
(d) Power & frequency rating of the equipment.
(e) Process of induction hardening e.g. Surface hardening, through hardening, hardening in stationary position, hardening by progressive method or hardening by single shot method etc.

Following types of inductor coils are generally designed for induction hardening depending upon the job requirements:

1. Inductor coils having single turn.
2. Inductor coils having multi turns.
3. Inductor coils in series.
4. Tunnel type coils.

Source: Tool & Alloy Steels, June 1979, 195-198, 200-203

5. Pancake coils.
6. Iron cored inductors.
7. Inductors with built in quench etc.

Most commonly used coils are single turn or ring type coils for all plain cylindrical type workpieces using mostly progressive hardening method under rotation. Skid type inductors are used mostly for continuous heating such as parts mounted on a turntable or a conveyer belt. Solid copper coils are used where it generally becomes necessary to have a built in quench. Iron cored inductors are used where power needs to be concentrated into small areas of the work piece with maximum efficiency without affecting nearby areas. Even spot heating is possible with such inductors. The ratio of power density realised with Iron cored inductors compared with work coils of conventional type is in the region of 5 to 1. In single shot hardening technique, mostly cored inductors are used.

2. **Heating Time:** In induction hardening process, heating time is very vital as this time allows the workpiece to attain correct hardening temperature as well as correct metallurgical transformation in its structure. Incorrect heating time can lead to defective hardening. Further, heating time is often governed by production rate and required case depth. Generally heating time of parts heated in a fixed position relating to the inductor for 10 secs or less is controlled to within $\pm$ 0.1 secs. For heating times greater than 10 secs but less than 60 secs, it is controlled to within $\pm$ 0.2 secs and heating time over 60 secs is controlled to within $\pm$ 10 secs. Needless to say special multi timers are used in induction hardening equipments to control heating times.

3. **Power Density:** Power density coupled to heating time must be controlled carefully if heating is to be uniform from part to part. In modern induction hardening equipments power density or control of medium/high frequency voltage is automatically controlled to within $\pm$ 1% independent of mains supply voltage by employing automatic electronic voltage regulators to ensure uniform heating from part to part.

4. **Delay Time or Soaking Time:** Due to the fact that in induction hardening using high frequency currents, the maximum heat concentrates only on the surface of the component and in order for heat to penetrate to the required depth and to allow the structure of steel to transform homogenously, it becomes necessary to allow a slight delay or soak time which is the period the power gets switched off and the quench jets strike the heated area. This delay or soak time can very accurately be controlled by using special timers.

5. **Feed Speeds and Power Levels:** When progressively hardening long cylindrical workpieces like rear axle shafts or camshafts where different diameters and profiles are to be hardened in one and same hardening cycle only, it becomes necessary in an induction machine to make provision for automatic progressive feed speeds changeover as well as automatic power level changeovers. Provision is also required sometimes to have automatic increase or decrease in quench pressure. In all modern induction hardening machines, this auto-

matic programming is provided with cam bars, knockers, micro switches, potentiometers, which are interchangeable and adjustable to suit hardening of various types of parts. If production requirements are high, even electronic digital scanners are available which do away with cam adjustments required from time to time.

6. **Quenching:** The next most important control in an induction hardening process is the method of quenching the heated part. When spray quenching, the quench delay, type of quench pressure, quench direction and velocity are all important factors. The design of quench assembly is generally dependent upon the inductor with which it is used, shape and size of the component, direction and velocity of quench spray required.

The most common methods of quenching in induction hardening process are :

1. Spray Quenching

2. Immersion Quenching

Occasionally air quenching, self quenching and submerged quenching methods are also used depending upon the requirement.

In spray quenching, the quench generally is a hollow ring through which the quench medium is circulated. The ring is drilled with a series of holes on the surface facing the workpiece so that the quench medium emerges in a form of spray which impinges on to the surface of workpiece. Water is the most common medium used for this type of quenching although, when less severe conditions are required, a mixture of soluble oil and water can be used.

In spray quenching, the spray ring is usually arranged to move with the coil and these two parts of the assembly are kept as close as possible to prevent heat loss. Naturally the control of the quench is bound up with the pressure of water and in this connection, pressures between 2 and 5 ATU (kg/cm²) are considered suitable.

Of equal importance is the angle of incidence between the quenching spray and surface of the work. It has been found in practice that the most consistent results for progressive hardening have been obtained if the spray holes are drilled so that the jets strike the surface being hardened at an angle of between 5 to 10 degrees and away from the heating coil otherwise build up of water occurs between the coil and the quench ring causing patchy or uneven surface hardness. In many cases however the quench assemblies are designed for more angled quenching if required.

With immersion quenching the workpiece is heated in the inductor coil and suitable arrangements are made for the part to be jettisoned into a tank containing the appropriate quenching medium which may be any one of the following: (a) Brine solltion (b) cold water or hot water (c) Soluble oil solution (d) pure oil. Further arrangements is made for proper state of agitation in the quench tank. A further point to be borne in mind is, whether using immersion quenching or spray quenching, that when components are hardened on a production basis, precautions must be taken to ensure that the temperature of quenching medium does not rise unduly, thus influencing the degree of hardness obtained. This condition may arise due to continuous spray quenching of heated parts or continuous immersion of heated parts in quench tank. To counteract this, the quench medium collecting tank is made of adequate size and arrangement is provided for proper circulating, cooling and filtering of quenching medium. For such recirculation and cooling of quench medium, self contained fully automatic units are available alongwith induction hardening machines.

7. **Cooling of Equipment :** The volume, pressure, condition and temperature of cooling water supplied for the cooling of motor generators, high frequency valve generators, transformers, inductors, busbars and capacitors etc. used in induction hardening equipments, must be controlled within specified limits so as to avoid serious damage. Meters and gauges are provided to indicate these limits, high and low pressure switches are provided to control pressure limits, special flow control witches are provided to control flow of cooling water, special temperature control units are used to check specified temperature of outgoing cooling water. These protections are integrated electrically into the automatic induction hardening cycle circuit of the equipment and in case of any preset limits are exceeded, these protection units trip the high frequency power immediately.

Some of the components which are commonly induction hardened in various industries

1. **Wheel type Tractor manufacturing industry :** Camshafts, Crankshafts, Rear Axle shafts, Front Wheel Spindle shafts, Gears, Shifter Forks, Splined shafts, Transmission shafts, Rocker pads etc.

2. **Motor car and Truck manufacturing industry :** Camshafts, Crankshafts, Rear Axle Shafts, Front Wheel Spindle Shafts, Various Transmission Gears, End Pieces, Differential Housings, Cylinder Heads, Splined Shafts, Pump Shafts, King Pins, Universal Joints, Shifter Forks, Rocker Arms, Push Rods, Starter Gear Rings, Gudgeon Pins, Valve seats etc.

3. **Machine Tool Industry :** Lathe Beds, Machine columns, Transmission Gears, various types of Shafts.

4. **Earthmoving Equipment Manufacturing Industry :** Track Rollers, Track Roller shafts, Track Links, Track shoes, Track Pins, Sprockets, Idlers, Large gears, various types of shafts, Hydraulic Cylinder Piston Rods, Track Bushes, Trunions etc.

5. **Miscellaneous Industry :** Hand Tools like pliers, screw drivers, chisels, cutters etc., Hacksaw Blades, Hammer Heads, Bolt Heads, Ice skates, Rotor spindles, Motor shafts, Typewriter keys, Pump spindles, etc. etc.

Characteristics of some of the induction hardened parts:

Axle shafts: With very few exceptions, all axle shafts in passenger car, trucks and off-road vehicles are surface hardened by induction. Although some use a portion of the hardened surface as a bearing on shafts, the primary purpose of induction hardening is to put the surface under compressive stress. This results in increasing bending and torsional fatigue life as much as 200%. Induction hardened Axles consist of a hard, high strength 'tube' soft core, in other words a strong axle with better torsional strength and ductile core. In addition to substantially improving strength, induction hardening saves substantial costs per passenger car shaft. Most shafts are of plain medium carbon steel and surface hardened after machining to a deep case depth of 5-8 mm depending on cross section. Hardness is in the Rc 50s after draw. Such deep case depth improves yield strength of axle shafts considerably. In some axle shafts, the driven end has a thrust face which must be hardened. In induction hardening process this face can also be hardened in the same cycle in which the main portion of the shaft is hardened, thus reducing number of operations and costs.

Transmission shafts and spindles : Today's transmission shafts specially in automatic transmissions, are required to have excellent bending and torsional strength besides surface hardening for wear resistance. With induction hardening, through excellent electrical controls, the above requirements are best achieved. Methods adopted generally are progressive hardening or single shot hardening.

The wheel spindles, when induction hardened, give deep and uniform case hardening specially in filler areas. This gives excellent wear properties and longer life. Further due to induction hardening the surface is put under compression, thereby improving fatigue life.

Gears : Reliability and minimum noise are main requirements for gears used in transmissions. Therefore, keeping distortion as low as possible is of utmost importance. Induction hardening is just the process which gives minimum distortion. Following advantages inherent with this process result in low distortion:

1. Only areas requiring hardening are heated.

2. Heating is rapid with minimum effect on adjacent areas.

3. There is minimum material expansion during heating and contraction during quenching.

4. Uniform hardening of all contact areas results in high wear resistance property.

Exhaust Valve Seats: Induction hardened Valve seats withstand valve seat recession, an acute and serious problem which results from the use of non-leaded fuel. Induction hardened valve seats are distortionless, eliminate the need for special machining operations before or after treatment. Induction hardened seats have no effect on the exhaust valve durability and perform satisfactorily when either leaded or non-leaded fuels are used.

Hardening valve seat is essential because valve seat recession can be a very serious problem. Recession occurs when the exhaust valve wears into mating seat and becomes excessive in engines operated at high speed under high loads. Thus valve failure and poor engine performance often results.

Crankshafts : Through the Crankshaft pass all the explosive forces generated in the combustion chambers of internal combustion engines that propel autos, tractors, and other vehicles. Because the horse power of such engines are rising everyday, the crankshafts are required to transmit larger forces at higher speeds. In other words requiring greater strengths and wear resistance.

In earlier days conventional methods were being used e.g. heating in a furnace and quenching in salt baths, nitriding etc. However, with all these methods serious problems like un-uniform and inadequate hardness and distortion were experienced. Through years of research and development induction hardening of crankshafts journals and fillets was developed.

Today by induction hardening of crankshafts, automobile manufacturers have been able to produce crankshafts of much superior quality more economically achieving following major advantages.

1. Fillets and journals are hardened to a higher hardness with uniform hardness and case depth. Induction hardening strengthens fillets and raises wear resistance of journals in crankshafts.

2. Only the portions requiring hardening are heated, thus leaving the rest of the portions of crankshaft soft for easy machining and balancing.

3. Least distortion and absolutely clean and scale free operation.

4. Due to almost distortion free operation, sometimes there is no need for straightening crankshaft after induction hardening.

5. Possibility of induction tempering immediately after induction hardening in the same sequence, thus improving metallurgical properties.

6. Induction hardening process has enabled manufacturers to reduce the size and weight of the crankshafts which are less sensitive to torsional vobrations and stresses and help reduce bending moments.

7. Induction hardened crankshafts have much improved mechanical strength, increased fatigue strength and much reduced metallurgical stresses.

"Induction heat treating always pays a bonus".

NITRIDING OF STEELS — AN OVERVIEW

by

A.S.W. Kurney*, M. Mohan Rao**, and R.M. Mallya**

Abstract

Among the thermochemical processes for the improvement of surface properties of machine components, nitriding holds an important position in Industry. A considerable amount of information generated as a result of application of nitriding over 50 years and related research activities is spread over the general metallurgical literature. This paper presents a review of the conventional practices of gas and liquid nitriding. Also, the more recent technique of Ion Nitriding or Glow Discharge Nitriding is described.

Introduction

A hard, wear resistant surface with a tough core are essential properties for the satisfactory performance of many engineering components. It is not possible to impart such a combination of properties only by thermal and/or mechanical treatments. A component treated for maximum hardness becomes too brittle whereas the same component treated for maximum toughness will not be hard enough. Consequently, several case hardening processes have been developed by which this highly desirable combination of properties can be attained commercially. Nitrided layers are superior to layers obtained by other case hardening techniques. Nitriding can now be performed in various saturating media and numerous modifications of the process have been developed to meet specific requirements. Earlier reviews are devoted to specific nitriding processes. This paper presents a review and assessment of the conventional nitriding processes and the more recent technique of Ion Nitriding.

Advantages of Nitriding

1. All the case hardening processes involve a change in the chemical composition of the surface layers through incorporation of elements carbon and/or nitrogen. The processes of carburising, cyaniding etc., depend upon a heat treatment after the surface composition has been changed to impart the desired hardness to the surface. In the case of nitriding, the hardness of the surface layer is due to the formation of inherently hard compounds. For this reason no further heat treatment is necessary.

2. Nitriding of steels is performed at temperatures much lower than those employed in carburising and so the properties of the core are not altered.

3. The nitrided parts are free from internal stresses and therefore free from ageing effects.

4. The hardness, wear-resistance and corrosion resistance of nitrided components are higher than those of other steel treating processes.

5. Since quenching or further heat treatment are not required there is negligible distortion. Hence it is suited for treatment of machine components.

In general the time required to complete the process is much longer than with carburising, requiring from 50-90 hours to produce a maximum case depth of from .50 to 0.75 mm. Also, nitriding requires close control of processing parameters. The cost of nitriding process is also a limitation to its use. However, nitriding offers an extremely valuable combination of properties which justifies its use for components subjected to severe service conditions.

Steels for Nitriding

The alloying elements commonly used in commercial steels, aluminium, chromium, vanadium and molybdenum, greatly improve the nitriding characteristics. These elements combine with nitrogen to form nitrides which are stable at nitriding temperatures. Molybdenum in addition, reduces the risk of embrittlement at nitriding temperatures. Other alloying elements such as nickel, copper, silicon and manganese, have little if any, effect on nitriding characteristics. Most of the steels that are commonly used for nitriding contain combinations of aluminium, chromium, molybdenum and vanadium. The carbon content is somewhat higher than is employed for carburising grades. The following are some of the steels that have been nitrided for specific applications.

1. Aluminium containing low alloy (nitralloys).

2. Medium carbon, chromium containing low alloy steels of the SAE 4100, 4300, 5700, 6100, 8600, 9300 and 9800 series.

Paper presented at the National Seminar on "Heat Treatment and Furnace Technology" Indian Institute of Science, Bangalore, '79.

* Formerly Research Scholar, Department of Metallurgy, Indian Institute of Science, Bangalore and now with Bangla Desh University of Engineering and Technology, Dacca.

** Department of Metallurgy, Indian Institute of Science, Bangalore, India.

3. Hot work die steels such as H11, H12, and H13,

4. Ferritic and Martensitic stainless steels of the 400 series.

5. Austenitic stainless steels of the 300 series.

6. Precipitation hardening stainless steels such as 17-4 PH and 17-7 PH.

The Process

Atomic nitrogen is necessary for nitriding and nitriding can now be performed in various saturating media, solid, liquid or gaseous. Nitriding in solids, like pack carburisng has been reported[5] recently but this has not yet gained acceptance in industry. Although nitriding at temperatures above 600°C is being studied, gas and liquid nitriding at temperatures below 600°C is widely used in industry.

A fourth state of matter, the plasma state is also being currently used in nitriding of engineering components. The plasma state produced in gaseous NH_3 or N_2-H_2 mixtures either in a glow discharge at 1 to 10 torr pressure created by a d.c. potential or in a r.f. discharge at 5 to 20 torr could be used for nitriding. However, it has been shown that r.f. discharge is not very effective in the nitriding of steels[6]. Ionic glow discharge nitriding is finding increasing applications in the nitriding of ferrous materials.

Gas Nitriding

In this process, nitriding is carried out by placing properly degreased steel parts inside a furnace heated to a temperature of 500-565°C for times ranging from 10-100 hrs. The steel parts are held in containers made of inconel or high nickel alloys. Prior to nitriding, the required core properties must be developed by hardening and tempering. Anhydrous ammonia gas is circulated around the parts to be nitrided, preferably with the aid of a blower. Anhydrous ammonia partially dissociates on the surface of the workpiece releasing nascent nitrogen which combines with the elements in the steel to form nitrides. The process is controlled by metering the flow of ammonia and checking the percentage of dissociation. The percentage dissociation is increased through a decrease in the flow rate or an increase in the temperature. In recent times greater efficiency has been achieved by fluid bed nitriding.

Single stage Nitriding: Nitriding may be carried out using either a single stage or a double stage process. In single stage nitriding, the dissociation of ammonia generally ranges from 15-30 percent and the

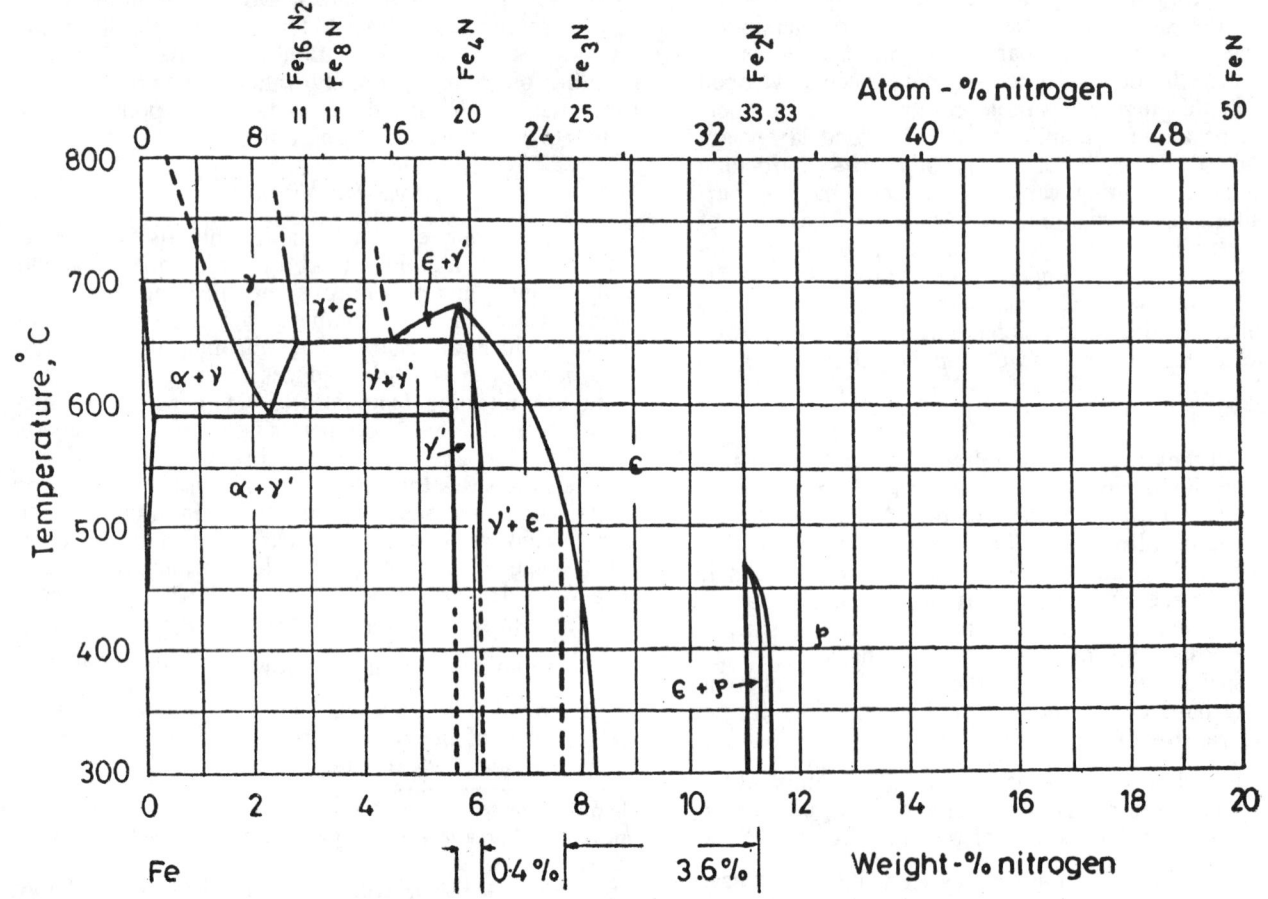

FIG. 1   Fe-N PHASE DIAGRAM

temperature from 500-525°C. This process produces a brittle nitrogen rich layer known as the "white layer" which may extend to a depth of 0.05 mm. This is followed by a zone of diffusion layer. The constitution of the layer is evident from Iron-Nitrogen constitution diagram in Fig. (1). There are a series of phases of progressively decreasing nitrogen content from the outside to the core. viz.,

1. $Fe_2N|Fe_3N$ — ε nitride
2. Fe_4N — γ'— nitride
3. A nitrogen enriched diffusion zone.

In long nitriding cycles, there is a tendency for the formation of an undesirable white layer (comprising of the two nitrides) about 0.05 mm thick, causing brittleness and spalling in service.

Double stage Nitriding: The double stage process also known as the Floe process[7] has the advantage that the depth of the "White Layer" can be minimised to .005 to .015 mm. The first stage of the double stage process consists of treatment at 525°C with 20% dissociation of ammonia for 5-10 hours. In the second stage the temperature is raised to 550°C and the dissociation of ammonia is increased to 80-85 percent for the remainder of the cycle. An auxiliary ammonia dissociator is necessary to obtain the high percentage of dissociation in the second stage. The second stage serves as a diffusion cycle permitting the high nitrogen surface layer to diffuse inward similar to the diffusion in gas carburised cases.

The Floe process has done much to decrease the thickness of the white layer, but unfortunately it has not been possible to eliminate it. Dashfield[8] has discovered a means of completely removing the white layer, by soaking in hot cyanide solution. A typical solution cycle would consist of soaking the nitrided components in a solution of 1 kg. of Sodium cyanide in 1 litre of water heated to 75 to 90°C. Removal could be speeded up using shorter soaks followed by cleaning with 220 mesh aluminium oxide grit at 80 psi pressure. Vapour blasting and blast cleaning using fine glass-beads at various pressures could also be used to satisfy specific finish requirements.

Modified Processes: It has been observed that the use of additions of other components to gas mixtures markedly changes the composition of the surface layer and the kinetics of nitriding. Addition of oxygen, water vapour and air to the furnace atmosphere has been shown to accelerate the nitriding process. It has been shown that with 4 per cent oxygen in the ammonia atmosphere the nitriding is accelerated[9]. Surface passivation during nitriding of steels containing Chromium can be combated by addition of chlorine and its compounds to the furnace atmosphere. The positive action of chlorine in nitriding of these steels is not due to disruption of oxide films but primarily due to an increase in the energy of the interacting components, which leads to intensive electron emission from the treated surface and larger number of negative ions of ammonia in the zone of adsorption[10]. Carbon tetra-

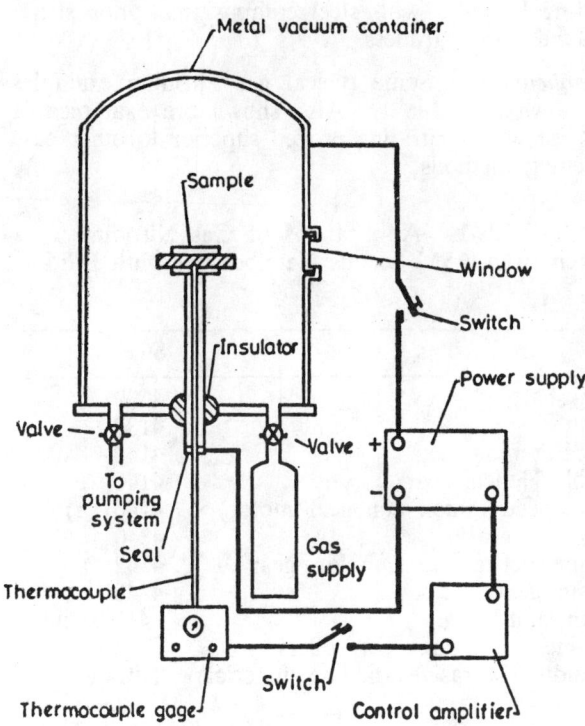

Fig. 2. LABORATORY SET-UP FOR IONITRIDING.

chloride has proved a better depassivator of chromium containing steels than ammonium chloride. The presence of hydrogen in the nitriding medium retards the nitriding process and causes decarburisation and embrittlement.[11]

An examination of the beneficial white layer produced in the cyanide baths revealed that it contained ε —nitride incorporating traces of carbon and/or oxygen in solution. Consequently, various processes have been developed for generating a carbon bearing ε nitride or ε carbonitride in a gaseous atmosphere. The Ipsen Company of U.S.A. has developed in 1970 a nitriding process in 50 per cent ammonia and 50 per cent propane or endothermal gas. This is called "Nitemper" in U.S.A., "Nikotriding" in FRG and "Naitemper" in Japan. In using an endothermal gas-ammonia mixture certain precautions must be taken to ensure explosion free operation. In recent years a West German firm has introduced nitriding with carburising and oxidising gases—the so called "Nitrok" process. In this, the use of exothermal gases eliminates the danger of explosion.

It has more recently been suggested that the special gas additions are superfluous and that suitable control of ammonia flow in a treatment at 575°C produces a "beneficial" white layer[14] presumably consisting of pure ε nitride.

The case depth and case hardness, the two criteria referred to in the control of case properties will vary

Source: Tool & Alloy Steels, September 1983, 333-340

not only with the duration and other conditions of nitriding but also with steel composition, prior structure and core hardness.

Applications: Some typical gas nitriding examples are shown in Table 1. Also shown are examples of parts for which nitriding proved superior to other case hardening methods.

TABLE-1. Applications of Gas Nitriding
(Taken from ASM Metals Handbook, Eighth Edition. Vol. 2 '64)

Part	Steel
Crankshaft	4130
Piston ring	4130
Helical timing gear	4140
Double Helical gear	4140
High Speed Pinion (on gear motor)	4140 (a)
Oil pump gear	4340
Marine helical transmission gear	4142
Torque gear	4340
Loom shuttle	410 stainless
Bushings (For conveyor rollers handling abrasive alkaline material)	Nitralloy 135 Type G (b)
Gear	AMS 6470 (c)

a. Steel 4140 substituted for 8620 and nitriding proved superior to gas carburizing of 8620.
b. Substitution of Nitralloy 135 type G and Nitriding proved superior to carburized bushings.
c. Steel AMS 6470 substituted for 3310 and nitriding proved superior to carburized 3310.

Liquid Nitriding

Nitriding in Fused Salt Baths: This technique was developed in the Federal Republic of Germany and was earlier called "soft nitriding". This process is now called the "Tenifer Process" in FRG and the "Tufftride Process" in U.K. This uses molten cyanide as the case hardening medium at temperatures similar to those employed in gas nitriding. Although the salts are the same as those in liquid carburising. in view of the lower temperatures used, liquid nitriding adds more nitrogen and less carbon to the steel, various compositions of these baths are in use for treatment of a wide variety of carbon, low alloy and tool steels. Liquid nitriding provides the same advantages as gas nitriding. In addition, liquid nitriding is able to provides the same advantages as gas nitriding. In addition, liquid nitriding is able to produce a satisfactory nitrided case on carbon steels. Gas nitriding is to be prefered where heavier case depths are required. Some applications are illustrated in Table 2 where liquid nitriding was preferred to other case hardening processes.

A commercial bath for liquid nitriding is composed of 60 to 70 percent by weight of sodium salts (NaCN, Na_2CO_3 and NaCNO) and 40 to 30 percent by weight of potassium salts (KCN, K_2CO_3, KCNO and KCl) operating at around 570°C. The molten salt bath

should be "aged" by being held at 570-590°C for at least 12 hours.

Table 2. Applications of liquid Nitriding
(Taken from ASM Metals Handbook, Eighth Edition. Vol. 2, '64)

Part	Steel
Thrust Washer	1010 Nitriding preferred to Carbo nitriding
Seat Bracket	1020 Nitrided steel preferred to same steel cyanide treated.

Ageing decreases the cyanide content and increases the cyanate and carbonate contents of the bath. It has been shown that the rate of nitriding and the properties of the nitrided layer depend on the cyanate content of the bath.

For satisfactory performance, the following operating procedures are important in liquid nitriding:

i) The bath should be aged and proper control of the cyanate content of the bath should be maintained.

ii) Appropriately cleaned, degreased and pre-heated articles should be placed in the bath.

iii) The bath should be analysed periodically and necessary additions should be made to restore compositions. Bath must be relieved of oxidation products.

iv) Overheating of the bath above 600°C should be avoided.

v) Salts should be changed after 3-4 months of operation.

vi) Generally, Titanium crucibles or titanium plated crucibles give best results. It is advisable to keep the baths covered when not in use.

Operating temperature of liquid baths is 570°C. At higher temperatures carbides containing nitrogen are formed. Salt consumption increases and properties of core and case deteriorate. At lower temperatures nitrides containing carbon are formed, and the process slows down and surface properties deteriorate due to formation of undesirable nitride phase. Overheating[15] causes what is ksown as "Cracking". Cyanates decompose to cyanide and carbonate and bath performance deteriorates.

The main disadvantage of the above salt baths is the problem of toxicity of the salts and the problem of waste disposal. Consequently attempts have been made to replace these salts by non-toxic chemicals. Only partial success has been achieved with cyanate and urea as starting bath compositions. The cyanate based baths are used in the "New Tenifer Process". The operating bath compositions are identical and the pro blem of waste disposal remains. The nitriding results obtained with this bath are identical.

The bath containing urea as the starting composition, in the process of heating undergoes simmersation, condensation and interacts with other components in the bath forming cyanides. A typical composition of the bath is 54 per cent urea and 46 percent soda to obtain sodium cyanate and 48 percent urea and 52 percent potash to obtain potassium cyanate. Each kilogram of salt mixture produces 0.575 kg. salt and 0.425 kg. gaseous products. Because of the evolution of gases the bath must have a reliable ventilation system.

Special Processes: Several special liquid nitriding processes have been developed using additives. to accelerate the chemical activity of the bath and to improve properties obtained. The chief among these are:

i) Liquid Pressure Nitriding: In this process anhydrous ammonia is forced through a cyanide-cyanate bath and maintained at a pressure of 1 to 30 psi. A new bath does not require ageing. Maintenance of the bath in the required ratio range is greatly simplified. Positive results have also been obtained by blowing ammonia through urea based baths.

ii) Aerated Bath Nitriding: In this process, measured amounts of air are pumped through a molten bath. The introduction of the air provides agitation and enhances the rate of nitriding by forming more cyanate. This is also known as activated nitriding.

iii) Sulfocyaniding: The steel is saturated simultaneously with carbon, nitrogen and sulphur. Sulphur in amounts from 2-25 percent is added to cyanide based baths. The structure of a sulfocyanided layer is the same as that obtained in liquid nitriding except that it has a thin sulphide layer on the surface[16]. This is also known as the "Sursulf" or "Soft Nitriding" process. Sulphocyaniding lowers the co-efficient of friction. improves wear resistance, antiscoring properties and fatigue limit.

Nitriding in Solutions: Substances which release nitrogen during heating have been used. The parts and the inductor for heating are placed in the solution. High frequency current heats the part and its surroundings to the desired temperature. A vapour jacket is created around the part, in which the nitrogen bearing compounds dissociates. The atomic nitrogen diffuses into the steel. Substances used include 10-15 percent solution of NH_3, NH_4Cl, nitrophenol, aniline and acetamide in glycerine. This technique has not been widely used because of the difficulty of:

a. Stabilising temperature

b. Maintaining the necessary concentration of dissolved substance and

c. treating a large number of parts at a time.

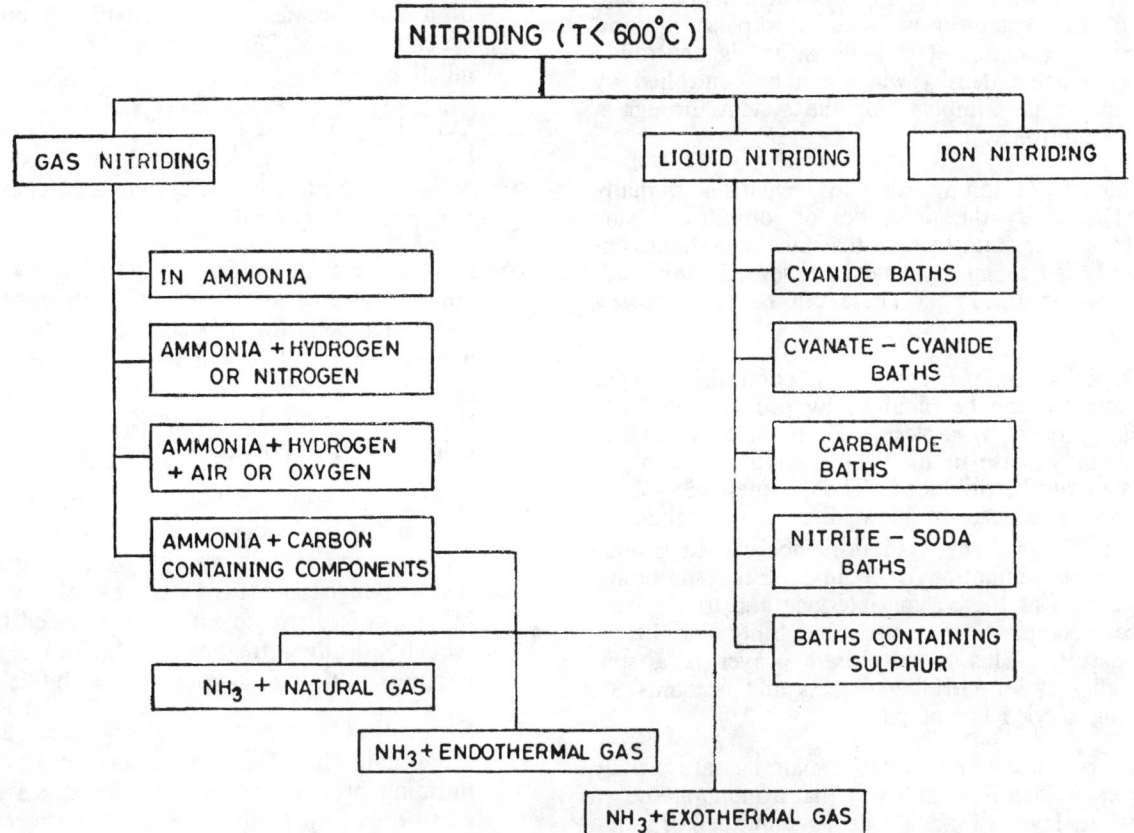

FIG. 3. COMMERCIALLY POPULAR NITRIDING PROCESSES

Source: Tool & Alloy Steels, September 1983, 333-340

Nitriding in Molten Neutral Salts: Ammonia is blown through neutral salt baths and utilises the advantage of heating liquid baths. This has a shorter processing time and is used for small scale production in U.S.A.

Ion-Nitriding

The search for sources of more active nitrogen led to the development of nitriding in ionised gases. This process is known as Ion-nitriding, Glow discharge nitriding or Plasma-nitriding. Although the first patent on the stabilisation of the glow-discharge was obtained by Berghaus[17] in 1932, commercial use of the process in the Western World started only in the 1960's. It is now widely used throughout the world.

The main elements of the equipment, shown in Figure (2), used in this process are:

i) Vacuum container with the pumping system

ii) A power supply with controller

iii) A gas-dispensing system.

In this process, a low pressure gas (nitrogen, nitrogen+ hydrogen or nitrogen+NH_3 at 1 to 10 torr) is ionised by the application of a suitable d.c. voltage between two electrodes. The workpiece is made the cathode and the grounded metal container acts as the anode. A glow completely surrounds the article irrespective of its distance from the anode. Positive nitrogen ions are attracted by the cathodic workpiece and hit its surface with tremendous kinetic energy. The release of this energy heats the workpiece up and nitriding also occurs. The temperature is controlled by the ion-current density which can be controlled by varying the voltage imposed on the system through a feed-back controller.

The success of ion-nitriding led scientists in many countries to study the properties of ion-nitrided surfaces. It is interesting to note that all these studies independently led to an agreement on several major advantages of ion-nitriding. These can be summarised as follows:

(1) Superior, almost completely controllable layer structure can be obtained by proper control of the process variables; a thin tenacious monophase γ-phase or a thick monophase ε-phase layer can be produced. The thickness of γ-phase layer produced in glow discharge depends on rate of sputtering and does not exceed 8 μm. This self-limitation is because of the sputtering reaction at the plasma interface and to the narrow composition range of stability of this γ-phase[18]. This mono-phased γ-layer is a speciality of ion-nitriding process and accounts for long service life of parts.

Carbon-steels and cast-iron materials are usually ion-nitrided in such a way that a monophased white layer of about 0.010 mm to 0.015 mm thickness is produced. The diffusion layer of these steels is quite soft and therefore a slightly thicker white layer with good wear and gliding properties is preferable. ε nitride layer thickness can be pre-set between 6 μm and 50 μm depending upon the intensity of the sputtering[19].

(2) Shorter processing time: For example, with hot-working tool-steels 40-60 hrs of gas nitriding at 530°C gives similar thickness as 20 hrs of ion-nitriding at the same temperature. This accelerated rate of nitriding can be attributed to one of the following:

   i. Appearance in the discharge plasma of the high energy ions.

   ii. More effective nitrogen transfer from the gaseous medium to the metal surface.

   iii. A predominance of volume diffusion.

(3) Uniform thickness: The glow-discharge is uniform over the entire cathodic surface areas independent of their distance to the anode. This together with complete cleaning and depassivation by bombardment guarantees complete and uniform transfer of nitrogen into the workpiece.

(4) Easy-masking: The tin-plating, copper-plating and liquid glass used as a protective coating in case of conventional nitriding cannot be used in ion-nitriding. Areas can be protected from ion-nitriding by sheet-materials (dielectric materials) or by special sleeves slit a few hundredth of a mm, because glow-discharge cannot form in such gaps. This method of protection is much cheaper.

(5) Denitriding Possibilities: This can be achieved by keeping the nitrided sample in the discharge with only hydrogen. Gaseous ammonia forms and denitriding results.

(6) Low-temperature nitriding: Nitriding at a temperature as low as 350°C has been done. The strong temperature dependence of the amount of nitrogen being supplied by the various nitrogen carrying media makes low-temperature nitriding treatments in gas or liquid media difficult or even impossible[20] [21] [22]

Low temperature nitriding is particularly important for cold-working tools for rolling, bending etc., and certain special steels. It is vital that the high core hardness of tools developed by proper heat treatment is not reduced by subsequent nitriding treatment. And so low temperature nitriding offers new possibilities.

(7) Economic Considerations: Energy and gas costs are usually much lower than those of other nitriding processes. Only workpieces are heated to treatment temperature and treatment times are much less than those with other nitriding techniques.

Wages for handling, cleaning and stacking the pieces prior to nitriding are comparable in all nitriding processes. It should also be noted that degreasing should precede ion-nitriding. Easy masking, no post-nitriding cleaning requirements results in considerable economics. Waste disposal and water treatment is not required.

Finally maintenance and repair costs are lower. Except for vacuum pumps there are no moving parts and modern semiconductor techniques guarantee long life for electric units. All these advantages have led to increasing use of ion-nitriding in industries. The heaviest piece ever nitrided is a pressure barpin, which is used in heavy forging machines[23]. This piece weighed 28,000 kgms had a length of 185 cm and 140 cm dia. The smallest pieces to be ion-nitrided are the balls of ball-pens about 0.75 mm dia. The longest piece nitrided is a 1100 cm. long extrusion cylinder. However, majority of pieces that are nitrided in glow-discharge are quite small parts like valves, gears, drills, taps etc.

Summary

The more important modifications of the nitriding processes are shown in Fig. 3. Various modifications of nitriding, particularly ion-nitriding have opened up renewed thinking on the potentialities of nitriding.

References

1. A. A. Yurgenson, Protective Coating of Metals, 1 (1969) P. 40.

2. P. Birk, Traitement Thermique, 93 (1975) P. 37.

3. Ya. D. Kogan, Metal i Term Obrabotka Metal, 3, (1974) P.2.

4. Ya. M. Lakhtin Metal i Term Obrabotka Metal, 3, (1974) P.8.

5. A. Oldewartel, Z. Wirtsch, Fertingung, 70, No. 9 (1975) P.46.

6. M. Hudis, J. Appl. Phys. 44 (1973) 1489.

7. U.S. Patent No. 2, 437, 249.

8. D.A. Dashfield, Metal Progress (1964) P. 82.

9. H. J. Grabke, Arch. Eisenhuttenw, 44, No. 8 (1973) P 603.

10. Ya. M. Lakhtin and Ya. D. Kogan, Met. Sci. Heat treatment 10 (1977) P. 842.

11. A. A. Yurgenson, Mashgiz, Sverdlovsk (1962).

12. T. Bell and S. Y. Lee, Heat treatment, The Metals Society 1975, P. 99.

13. Dawes C. and Tranter D. F., Metallurgia and metal Ferming, 40 (1973) P. 58.

14. Dawes C. and Tranter D. F., Metals Technology, 5, (1978) P. 278.

15. Metals Handbook, ASM 8th edition Vol. 2 (1964) 148.

16. J. C. Gregory, Heat-treatment of metals, 2 No. 2 (1975) P. 55.

17. B. Berghaus, German Patent DRP 668-639 (1932).

18. V. Kiritchenko and T. Bell, Heat Treatment of Metals, 4 (1978) P. 90.

19. B. Edenhofer, Heat Treatment of Metals, No. 2 (1974) P. 59.

20. A. N. Minkewitsch, 'Thermochemical Surface Treatment of Steel', VEB-Verlog Technik, Berlin 1953.

21. B. Finnern, 'Gas and Bath nitriding' Carl Hanser Verlag, Munich, 1965.

22. B. Edenhofer, Fachberichte for Obertlachentechnik 12, No. 4 (1974) P. 97.

23. B. Edenhofer, The Metallurgist and Materials Technologist, (1976) 425.

Gaseous Ferritic Nitrocarburizing

By T. Bell
Hanson Professor of
   Industrial Metallurgy
Department of Metallurgy &
   Materials
University of Birmingham
and the ASM Committee on
   Nitrocarburizing*

FERRITIC NITROCARBURIZING PROCESSES are thermochemical treatments that involve diffusional addition of both nitrogen and carbon to the surface of ferrous materials at temperatures completely within the ferrite-phase field. Cycle times are usually less than 3 h; these processes are termed "short-cycle" nitriding. The primary object of such treatments is usually to improve antiscuffing characteristics of ferrous engineering components by providing the surface with a compound layer—really, a surface zone—exhibiting good wear/friction-resistant properties. In addition, fatigue characteristics can be considerably improved, particularly when nitrogen is retained in solid solution in the "diffusion zone" beneath the compound layer. This retention normally is achieved by quenching in oil or water from the treatment temperature. Corrosion resistance provided by the compound zone is an important secondary benefit. Some distinction should be drawn between two different processes with similar names—ferritic nitrocarburizing and carbonitriding. Carbonitriding is performed with the steel in an austenitic phase, above 760 °C (1400 °F). Ferritic nitrocarburizing is performed in the ferritic range, below 675 °C (1250 °F).

Nitrocarburizing is used on a wide range of engineering components, such as rocker-arm spacers, textile machinery gears, pump cylinder blocks and jet nozzles, which are treated for wear resistance. Components such as crankshafts and drive shafts are treated to improve fatigue properties.

Ferritic nitrocarburizing has been successfully applied to most ferrous materials, including wrought and sintered plain carbon and alloy steels, stainless steels and cast irons. The most marked improvement in both antiscuffing and fatigue properties, relative to untreated material, is found with plain low-carbon steels. Consequently, these materials form the pri-mary basis for discussion throughout this article.

Early nitrocarburizing processes used molten cyanide-base salts to confer property improvements. Concern about over-all environmental aspects of heat treat processing with cyanide-base salts created intense interest in development of cyanide-free nitrocarburizing treatments as technically and economically viable alternatives to cyanide-base processes. Detailed metallurgical studies of materials treated by the cyanide nitrocarburizing processes led to the introduction of a variety of gaseous, vacuum and cyanide-free salt bath nitrocarburizing treatments. Many of these processes are described in other articles. This article concentrates on gaseous ferritic nitrocarburizing.

Preliminary Treatments. The surface to be nitrocarburized must be free of contaminants such as oxides, scales, oil and decarburization, if opti-

*Kenneth D. Gladden, Senior Development Engineer, Caterpillar Tractor Co.; Donald N. Guy, Project Manager, Technology Center, Lindberg Heat Treating Co.

Reprinted from Metals Handbook, 9th Edition, Vol. 4, 264-269, © 1981 American Society for Metals

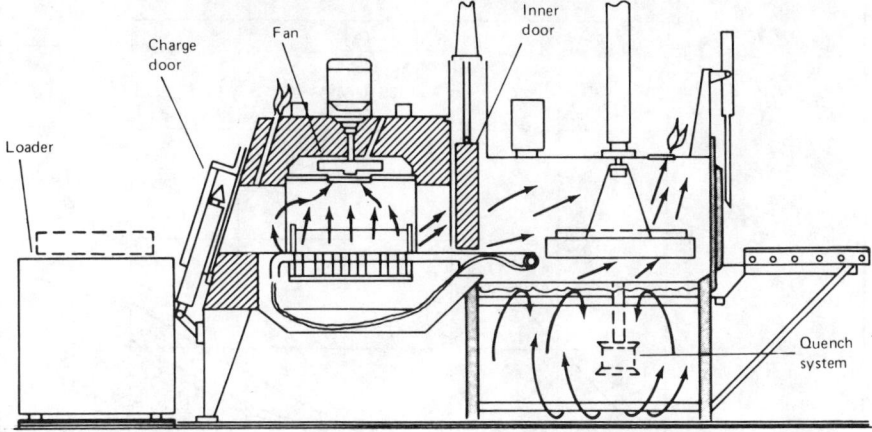

Fig. 1 Typical furnace for gaseous nitrocarburizing

Loader

Charge door

Fan

Inner door

Quench system

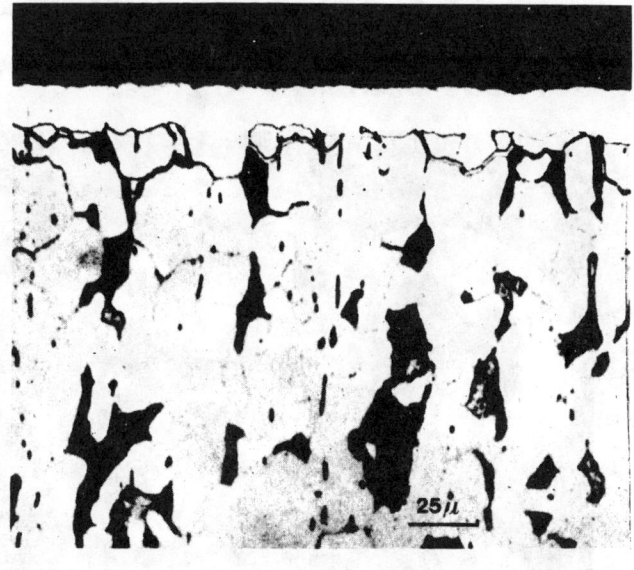

Fig. 2 AISI 1015 material after 3 h of gaseous nitrocarburizing in an ammonia/endothermic gas mixture at 570 °C (1060 °F) followed by oil quenching

25 μ

ation of core properties during the nitrocarburizing process.

Gaseous Nitrocarburizing

Gaseous nitrocarburizing commonly employs sealed-quench batch furnaces of the same design used for carburizing and carbonitriding (Fig. 1). Furnace operating temperatures are low enough to maintain steels in the ferritic condition. The atmosphere employed consists of ammonia diluted with a carrier gas. In one process, the atmosphere is formed from equal amounts of ammonia and endothermic gas, American Gas Association (AGA) type 302. In another process, a typical atmosphere consists of 35% ammonia and 65% refined exothermic gas (AGA type 201, nominally 97% nitrogen), which may be enriched with a hydrocarbon gas. High-purity nitrogen is used as a diluent in a variant of this process. Gaseous nitrocarburizing is performed near 570 °C (1060 °F), a temperature just below the austenite range for the Fe-N system. Treatment times generally range from 1 to 5 h.

The properties produced by gaseous nitrocarburizing are similar to those produced by salt bath nitrocarburizing, and the process can be applied to most ferrous alloys including carbon steels, cast iron, stainless steels and tool steels. An advantage of the gaseous nitrocarburizing processes over salt bath processes is the elimination of environmental problems associated with the handling and disposal of toxic cyanide waste salts. Also, the annoyance of salt entrapment in holes is avoided.

The objective of the process is to produce a thin layer of iron carbonitride and nitrides, the "white layer" or compound zone, with an underlying diffusion zone containing dissolved nitrogen and iron (or alloy) nitrides. The white layer enhances surface resistance to galling, corrosion and wear. The diffusion zone increases the fatigue endurance limit significantly, especially in carbon and low-alloy steel.

The white layer is composed primarily of the epsilon carbonitride phase, with other nitrides and oxides. The exact composition is a function of the nitride-forming elements in the material and the composition of the atmosphere.

The white layer has a reduced tendency to spall compared to the white

mum results are to be obtained. Vapor degreasing is adequate for most applications. It may be necessary, especially on high chromium materials and cast irons that have been burnished or otherwise highly finished, to grit blast with fine abrasive, to apply a light phosphate coating, or both, before nitrocarburizing. Highly finished surfaces are often associated with superficial metal flow at the surface, with the result that the initiation of nitriding is difficult. Surfaces having very low sur-

face roughness can respond well to nitriding, provided that surface burnishing has been avoided. Finishing techniques with good cutting action are required.

Preliminary heat treatments range from simple stress relieving in order to control distortion to hardening and tempering in order to increase the core strength of the material. Stress relief and tempering temperatures should be at least 25 °C (45 °F) above the nitrocarburizing temperature to prevent alter-

Fig. 3 Electron microprobe traces of nitrogen, carbon, and oxygen in compound layer formed by 3 h of gaseous nitrocarburizing in ammonia/endothermic gas mixture

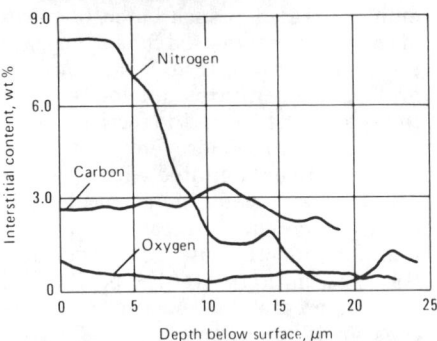

Fig. 4 Modified four-ball wear tests on AISI 1015 material

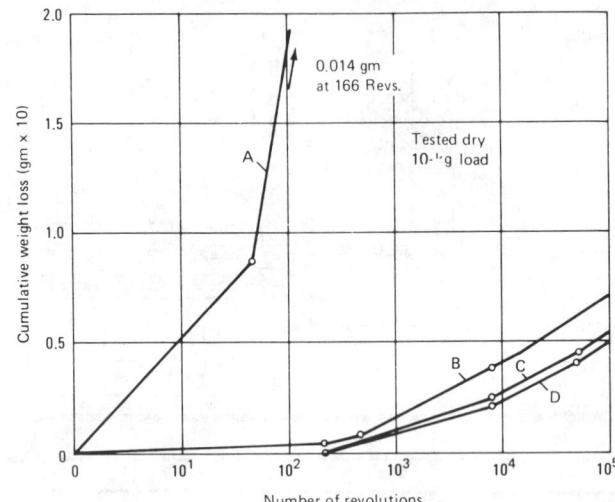

(a) Untreated. (b) Gaseous nitrocarburizing for 3 h, atmosphere cooled. (c) Gaseous nitrocarburizing for 3 h, oil quenched. (d) Gaseous nitrocarburizing for 8 h, oil quenched

layer formed during conventional gas nitriding with ammonia. This has been attributed by most investigators to an essentially single-phase white layer of the less brittle epsilon phase, but is also due in part to the fact that white layers are generally thinner in nitrocarburizing. The atmosphere conditions required to produce the less brittle phases appear to be affected by both carburizing gases such as CO and CH_4 and oxidizing gases such as air, CO_2 and H_2O.

Furnace Condition and Safety Precautions. Batch furnaces with integral oil quenches are ideally suited for performing gas nitrocarburizing; however, the over-all condition of the furnace is somewhat more critical than when the same furnace is used for other heat treating operations such as hardening, carburizing and carbonitriding. The hot chamber temperature should be controllable within ±5 °C (±10 °F) at 570 °C (1060 °F) throughout the entire volume. Thermocouple and instrument systems designed to operate at the higher temperatures of other heat treating processes are not always adequate for lower temperatures. Gas leaks in the furnace and around doors must be minimized. *Safety precautions must be carefully considered and rigorously enforced, because the process involves a combustible atmosphere that is explosive when operated below the self-ignition temperature.* Only minor gas leaks can be allowed, and double pilots should be provided at all doors. An interlock between door operation and pilot function provides added safety. Precautions must be taken to ensure

Fig. 5 Blanked gear shift gates, made from low carbon steel, after gaseous nitrocarburizing

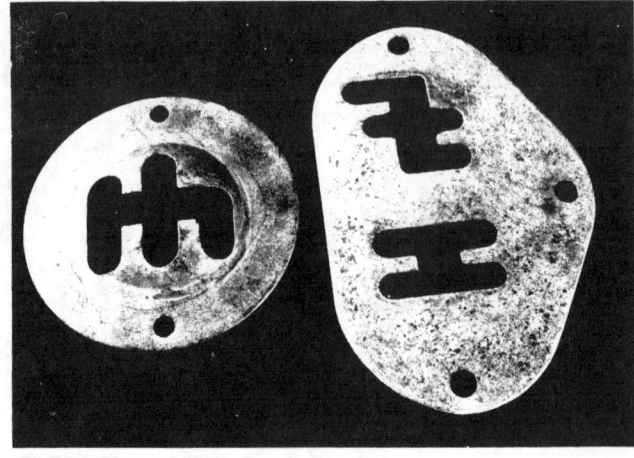

that gas burn-off ports are properly sized, free of clogging, well vented from the building and equipped with dependable pilots. All outside doors must be equipped with flame screens of sufficient capacity to cover the entire door opening.

The usual precautions also must be practiced with the quench oil: (a) adequate extinguisher equipment, (b) assurance that the oil is free from water, and (c) maintenance of adequate temperature controllers and over-tempera-ture devices to ensure that the oil does not become overheated. Steps also must be taken to ensure that atmosphere flow is sufficient to maintain positive pressure in the furnace during quenching, to prevent the egress of air through burn-off or small leaks. At start-up, atmosphere gas may be introduced by heating the furnace above self-ignition temperature—760 °C (1400 °F)—and introducing atmosphere in the usual manner or by purging all the air from the furnace with nitrogen before introducing the reactive atmosphere. On shutdown, the furnace should be heated above self-ignition temperature of 760 °C (1400 °F) before burning the furnace out, or it should be purged with nitrogen.

Process. The basic consideration behind all gaseous nitrocarburizing processes is the type of atmosphere that can be used to cause carbon and nitrogen to be added simultaneously to the surface of ferrous materials and so produce the desired epsilon phase.

Ammonia, the most readily available source of active nitrogen, catalytically dissociates at 570 °C (1060 °F) on ferrous surfaces according to the following reaction:

$$NH_3 \rightarrow (N)_{Fe} + \frac{3}{2} H_2 \qquad \text{(Eq 1)}$$

The nitrogen in the active condition diffuses into the work being treated. After nucleation of the compound layer, saturation of the epsilon (ε) carbonitride can be described by the reaction:

$$NH_3 \rightarrow (N)_{\epsilon} + \frac{3}{2} H_2 \qquad \text{(Eq 2)}$$

The nitriding potential of the atmosphere is given by the expression:

$$\theta_N = \frac{pNH_3}{(pH_2)^{3/2}} \qquad \text{(Eq 3)}$$

where pNH_3 and pH_2 are the partial pressures of ammonia and hydrogen, respectively, in the nitrocarburizing atmosphere. Changes in nitrogen content of the epsilon carbonitride are related not only with changes in the value of θ_N, but also are affected by the activity of nitrogen within the carbonitride phase:

$$\theta_N = \frac{{}^a(N)}{K} \qquad \text{(Eq 4)}$$

where K is the equilibrium constant for the nitriding reaction.

The activity of nitrogen is reduced by incorporating carbon into the lattice. If reduction in activity is minimal, the value of θ_N, the nitriding potential of the atmosphere, can be assumed to dictate the nitrogen content of the epsilon carbonitride phase. However, neither the range of suitable θ_N values nor the optimum value for specific ferrous materials have been established.

A particularly suitable source of carbon for gaseous nitrocarburizing is endothermic gas, the composition of which can be readily adjusted to provide a high degree of flexibility over the control of the carburizing potential. It

also is widely used in heat treating for gas carburizing operations. Endothermic gas contains sufficient free oxygen to assist the rate of conversion of the epsilon nitride phase into an oxygen-bearing carbonitride structure. The reaction:

$$2\,CO \rightarrow (C)_{\epsilon} + CO_2 \qquad \text{(Eq 5)}$$

controls saturation of this epsilon phase. The activity of carbon in the compound layer is given by:

$$^a(C)_{\epsilon} = \frac{(pCO)^2}{pCO_2} \cdot K_1 \qquad \text{(Eq 6)}$$

where $^a(C)_{\epsilon}$ is the activity of carbon in the carbonitride phase, pCO and pCO_2

are the partial pressures of carbon monoxide and carbon dioxide, respectively, and K_1 is the equilibrium constant for the reaction. In the absence of quantitative data on the effect of nitrogen on the activity of carbon, the carburizing potential:

$$\theta_C = \frac{(pCO)^2}{pCO_2} \qquad \text{(Eq 7)}$$

has to be assumed to dictate the carbon content of the epsilon carbonitride phase.

Testing Results. The compound layer formed on AISI 1015 steel by gaseous nitrocarburizing in an ammonia/endothermic gas mixture is shown in

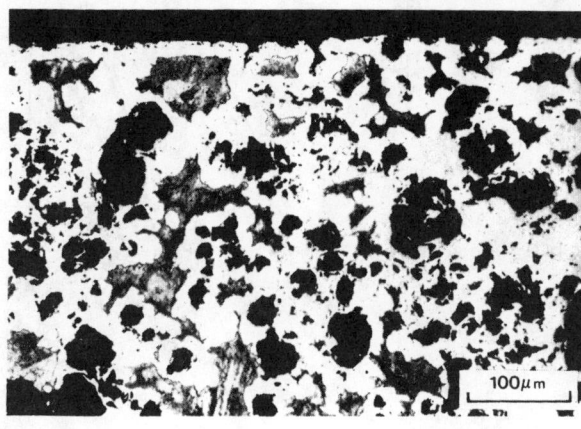

Fig. 6 Metallographic appearance of sintered iron after a 3 h gaseous controlled nitrocarburizing treatment followed by oil quenching

100 μm

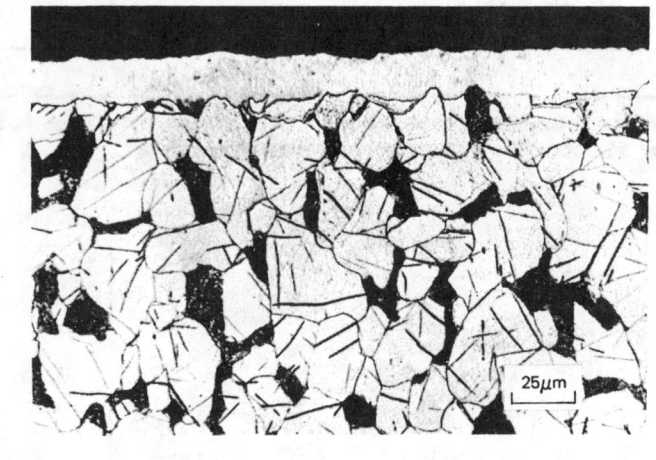

Fig. 7 Structure of mild steel after 2 h of gaseous nitrocarburizing followed by atmosphere cooling

25 μm

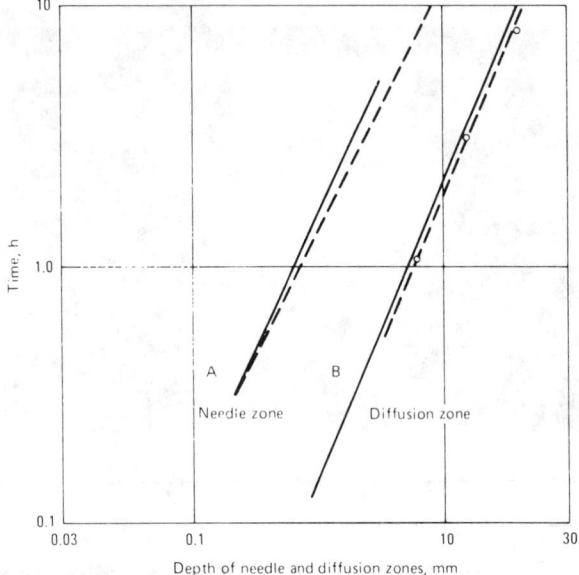

Fig. 8 Needle zone precipitate depth for salt-bath and gaseous nitrocarburized unalloyed low-carbon steel, compared to nitrogen diffusion depth

___treatment 1, cyanide-base salt-bath treatment, low-carbon steel; _ _ _ gaseous nitrocarburized AISI 1015; A, optical metallography; B, nitrogen analysis; and o, from microhardness plot AISI 1015.

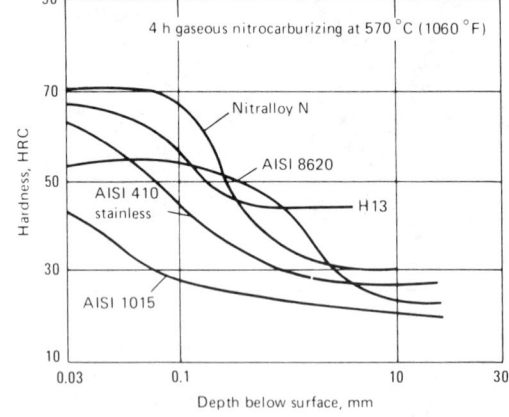

Fig. 9 Microhardness profiles of the diffusion zone for a series of steels after gaseous nitrocarburizing

HRC values were converted from microhardness values.

Fig. 2. The layer is somewhat more dense than that formed by salt bath nitrocarburizing. X-ray diffraction analysis has confirmed the predominance of epsilon carbonitride phase in the compound layer, while microprobe analysis (Fig. 3) confirms that carbon and nitrogen contents of the compound layer are essentially within the epsilon phase limits. See the article on liquid nitriding in this volume.

A number of reports clearly demonstrate the improvement in wear resistance and antiseizure properties resulting from gaseous nitrocarburizing treatments involving the use of ammonia and endothermic gas. This improvement in wear characteristics un-

der dry running conditions relative to untreated material is illustrated in Fig. 4, which shows results of modified four-ball wear tests on gaseous nitrocarburized AISI 1015 steel after various treatments.

Figure 5 illustrates stamped gear shift gates of low carbon steel, which have been nitrocarburized in an atmosphere consisting of 50% ammonia and 50% endothermic gas to prevent wear and fretting of the unfinished blanked edges.

Controlled Nitrocarburizing. A possible limitation of the process employing 50% ammonia and 50% endothermic gas is that optimum processing conditions for all classes of material, including cast irons, tool steels and stainless steels, are not ensured with a single ammonia/endothermic gas input ratio. A further, and perhaps more serious, limitation is that reproducibility may be impaired with variable loads and from furnace to furnace. These difficulties can be overcome largely by use of an infrared monitoring and control system.

With controlled nitrocarburizing, variable loads and materials can be accommodated, as well as the requirement for reproducible component growth characteristics. Controlled treatment involves processing in an atmosphere with a predetermined nitriding and carburizing potential, and controlling the nitrocarburizing potential within the furnace by infrared gas analysis of the ammonia content. Carburizing potential is controlled by a similar analyzer. An example of sintered iron treated by this process is presented in Fig. 6. The epsilon compound zone is present both on the outermost surface and also on the internal surfaces of the pores.

Vacuum Nitrocarburizing. This process is described in the article on gas nitriding.

Diffusion Zone Characteristics and Fatigue Properties

Diffusion zone characteristics are essentially independent of the type of nitrocarburizing media. During any ferritic nitrocarburizing treatment, only nitrogen diffuses inward from the carbonitride compound zone because the ferrite is normally already at its equilibrium concentration with respect to carbon. The diffusion zone beneath the compound layer on oil-quenched

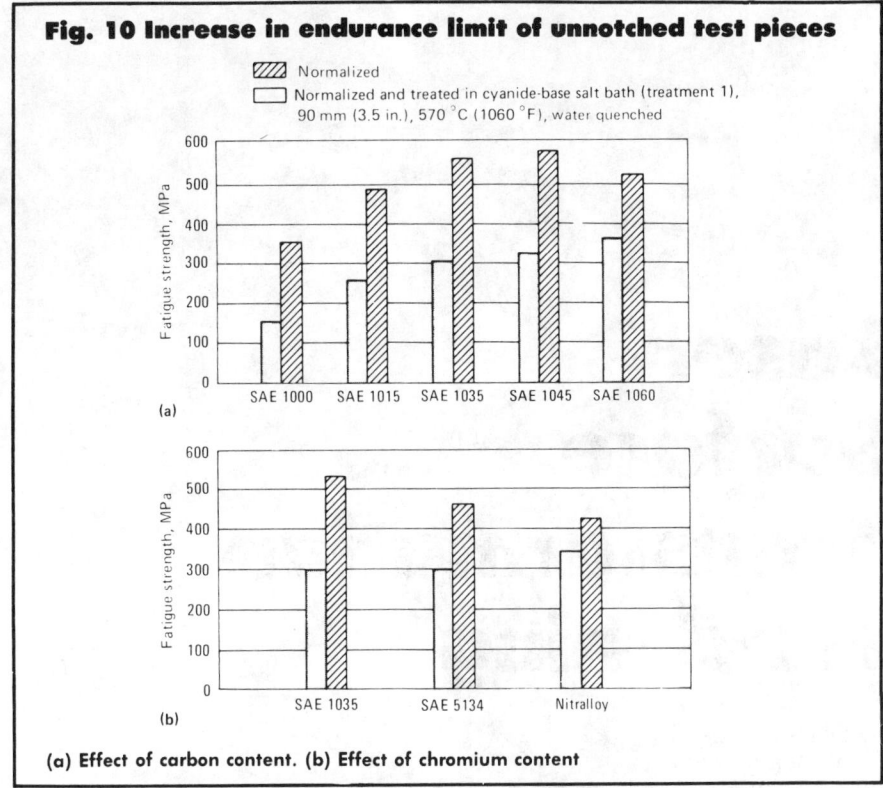

Fig. 10 Increase in endurance limit of unnotched test pieces

Legend:
- ▨ Normalized
- ☐ Normalized and treated in cyanide-base salt bath (treatment 1), 90 mm (3.5 in.), 570 °C (1060 °F), water quenched

(a) Effect of carbon content. (b) Effect of chromium content

Table 1 Endurance limits of nitrocarburized steel

Section size		Endurance limit after:					
		Salt bath treatment(a)		Gaseous treatment(b)		Vacuum treatment(c)	
mm	in.	MPa	ksi	MPa	ksi	MPa	ksi
5	0.2	480	70	490	71	480	70
8	0.3	450	65	435	63	455	66
13	0.5	400	58	340	49	355	51

Note: Tests were run at 10^7 cycles on AISI 1015 steel specimens nitrocarburized at 570 °C (1060 °F) for 2 h. (a) Salt bath treatment, water quench. (b) Gaseous treatment in endothermic ammonia gas mixture, oil quench. (c) Subatmospheric pressure treatment, methane/ammonia gas mixture, oil quench

The depth of the diffusion zone lessens as the level of nitride-forming elements in the material increases, because of a drop in the rate of diffusion of nitrogen in the parent lattice. At the same time, however, the hardness of the diffusion zone rises as the alloy content increases. This rise is due to submicroscopic precipitation of alloy nitrides in a manner identical to conventional gas nitriding. These features are illustrated in Fig. 9, for a series of steels that have received a 4 h gaseous nitrocarburizing treatment at 570 °C (1060 °F) followed by oil quenching.

The improvement in fatigue strength of nitrocarburized materials, as determined with unnotched Wöhler test specimens, depends on the hardness and depth of the diffusion zone. The potential for improvement in fatigue strength lessens with increasing carbon and alloy content (Fig. 10).

As would be expected, salt, gaseous and vacuum treatments all increase fatigue resistance to the same extent, for similar treatment times and section sizes, provided the quench rate is sufficiently rapid to retain the nitrogen in solid solution. Table 1 shows this condition is fulfilled for AISI 1015 in section sizes of up to 8 mm (0.3 in.). Above this size, properties are controlled by the nature of the quenching media, in which instance vacuum nitrocarburizing produces comparable properties to gaseous nitrocarburizing followed by atmosphere cooling. Quench rate has no effect on fatigue properties conferred by alloy nitride precipitates.

To take advantage of the antigalling resistance conferred by the white layer produced by nitrocarburizing, contact stresses cannot be so high as to exceed the yield strength of the metal under the nitride layer. When contact stresses are high, the underlying metal must be strengthened, either by increasing nitrided case depth or by employing another case hardening method, such as carbonitriding. Because wear processes are complex phenomena, tests on actual parts or part assemblies are usually necessary to determine whether case hardening by nitrocarburizing will produce satisfactory results.

samples is indistinguishable from the original matrix material. After any form of nitrocarburizing, however, atmosphere cooling results in formation of needle-like precipitates of Fe_4N nitride within the diffusion zone. Figure 7 illustrates the compound zone and diffusion zone of a gaseous nitrocarburized mild steel sample that has been atmosphere cooled from the treatment temperature. The depth of the needle precipitate zone fails to reflect the true depth of nitrogen penetration. Consequently, a microhardness profile, as a function of depth on a quenched material, gives the best indication of the diffusion zone thickness in the absence of detailed nitrogen layer analysis. In Fig. 8, the needle zone precipitate depth for salt bath and gaseous nitrocarburized unalloyed low-carbon steel is compared with the true nitrogen diffusion distance as measured by detailed chemical analysis and (in gaseous nitrocarburized material) by microhardness. It can be seen that the true depth of the diffusion zone in these low-carbon steels is several times greater than the visible needle zone depth.

Laser Surface Transformation Hardening

By Ole Sandven
Chief Metallurgist
Avco Everett Metalworking Lasers

(handwritten note: replaces flame heat.)

SURFACE TRANSFORMATION HARDENING of ferrous materials is an established process widely used to enhance the mechanical properties of highly stressed machine parts, such as gears and bearings. Surface hardening increases the wear resistance of the material and, under favorable circumstances, increases the fatigue strength due to residual compressive stresses that are induced in the workpiece surface by the transformation hardening process. The surface hardening process is not fundamentally different from conventional through hardening of ferrous materials. In both processes, increased hardness and strength are obtained by quenching the material from the austenite region to form hard martensite. Surface hardening differs from conventional through hardening in that only a thin surface layer is heated to austenitization temperatures prior to quenching, leaving the interior of the workpiece essentially unaffected.

Because ferrous materials are fairly good heat conductors, it is necessary to use very intense heat fluxes to heat the surface layer to austenitization temperatures without unduly affecting the bulk temperature of the workpiece.

This heat input is commonly obtained by the use of very hot flames or by high frequency induction heating. By selectively heating the workpiece surface to austenitization temperatures, desired surface hardening is obtained by application of a quench medium to the hot surface, or by self quenching.

Self quenching occurs when the cold interior of the workpiece constitutes a sufficiently large heat sink to quench the hot surface by heat conduction to the interior at a rate high enough to allow martensite to form at the surface.

In recent years, industrial lasers have become available for metalworking uses, including surface hardening. A laser can generate very intense energy fluxes at the workpiece surface and the resulting temperature profiles in the workpiece usually can be made steep enough to negate the need for external quench media. The laser beam is a beam of light, which is essentially independent of the workpiece, easily controlled, requires no vacuum, and generates no combustion products. It is ideally suited, therefore, for this purpose.

Fundamentals of Laser Surface Hardening

When a laser beam impinges on a surface, part of its energy is absorbed as heat at the surface. If the power density of the laser beam (usually given in watts per square centimetre) is sufficiently high, heat will be generated at the surface at a rate higher than heat conduction to the interior can remove it, and the temperature in the surface layer will increase rapidly. In a very short time, a thin surface layer will have reached austenitizing temperatures, whereas the interior of the workpiece is still cool. Even with a relatively moderate power density of 500 W/cm² (3300 W/in.²), temperature gradients of 500 °C/mm (25 °F/mil) can be obtained. By moving the laser beam over the workpiece surface (see Fig. 1), a point on the surface within the path of the beam is rapidly heated as the beam passes. This area is subsequently cooled rapidly by heat conduction to the interior after the beam has passed. By selecting the correct power density and speed of the laser spot, the material will harden to the desired depth.

A relatively broad area beam, usual-

140

Reprinted from Metals Handbook, 9th Edition, Vol. 4, 507-517, © 1981 American Society for Metals

ly in the shape of a square or a rectangle, is used in the laser hardening process. The power density of a focused laser beam used for hardening is much lower than the power density of the small, intense focused spots used for welding and cutting. The power density is typically in the 1000 to 2000 W/cm² (6400 to 13 000 W/in.²) range, occasionally as high as 5000 or as low as 500 W/cm² (32 000 to 3200 W/in.²).

The resulting depth of case will depend on the hardening response of the material, but it will rarely be more than 2.5 mm (0.1 in.). For steel with low hardenability, such as plain carbon steel, the depth of case obtainable is much smaller, varying from perhaps 0.25 mm (0.01 in.) in mild steels to 1.3 mm (0.05 in.) in a medium carbon steel. Because of the very high heating and cooling rates obtainable, it is possible to harden steels not normally considered hardenable, such as SAE 1018. For the same reason, the hardness obtainable by the laser hardening process can, in some instances, be slightly higher than that considered possible with conventional methods.

Ferrous materials are not good absorbers of infrared and far-infrared electromatic radiation. For instance, polished pure iron has an absorptivity of about 4% at room temperature. Grit-blasted cast iron has an absorptivity of 25% at room temperature, and this value increases to about 40% at 800 °C in an inert atmosphere. The formation of oxides on the ferrous surface can increase absorptivity beyond these values, but efficient use of the laser energy demands the introduction of a controlled high-value absorbing coating on the material surface. Chemical coatings, such as manganese phosphate and paints of graphite, silicon, and carbon, have all been used successfully. Some of these coatings may burn off during the heating process, and some may leave a residue which in itself can be an indicator of the maximum surface temperature reached. In any event, the absorptivity of these coatings at the beginning of the heating cycle is high (90% or better) and continues to be higher than that of the bare material throughout the temperature excursion.

For shorter wavelength (1.06 μm near-infrared) radiation, as would be the case in (yttrium-aluminum-garnet) (YAG) laser transformation hardening, the room temperature absorptivity may be as high as 60%, but an absorb-ing coating is still recommended for efficient use of the laser energy.

The major advantages of laser surface hardening include: close control of the power input with modern metalworking lasers; the laser can provide high power density, which in turn minimizes the total energy input and, thereby, dimensional distortion; and, the ability of the laser to reach normally inaccessible areas on the workpiece surface. Because no vacuum or protective atmosphere enclosure is needed, and the distance from the workpiece to the last optical element of the laser system can be quite long, it is possible to process very large or irregular-shaped workpieces.

On the negative side, the depth of case obtainable is limited to about 2.5 mm (0.1 in.), usually less than half of this, and the capital cost of the equipment may be high. Therefore, careful analysis of a potential application for laser hardening is needed to ascertain the cost-effectiveness of the process.

Metallurgy of Laser Hardening

As stated earlier, the surface hardening process is not fundamentally different from the conventional hardening processes because the same metallurgical reactions occur. In surface hardening, however, the heating portion of the process cycle must, out of necessity, be much shorter than that of conventional hardening. This is particularly true for laser hardening, where heating to the austenitizing temperatures occurs within seconds or even fractions of a second. In fact, in laser surface hardening the heating period is frequently shorter than the cool-down time; hence, the expression "up quench" is often used.

The response of ferrous materials to rapid cool-down from the austenite region has been studied in great detail for many decades and is well understood. The same cannot be said for the heating period of the process. Basically, the problems associated with these high heating rates are that the formation of austenite as well as the redistribution of carbon, necessary to form a homogeneous γFe-C solid solution, are processes that require small but finite time intervals. The kinetics of these processes under very high heating rates are still somewhat uncertain, making the design of a laser surface hardening process more difficult than that of a con-

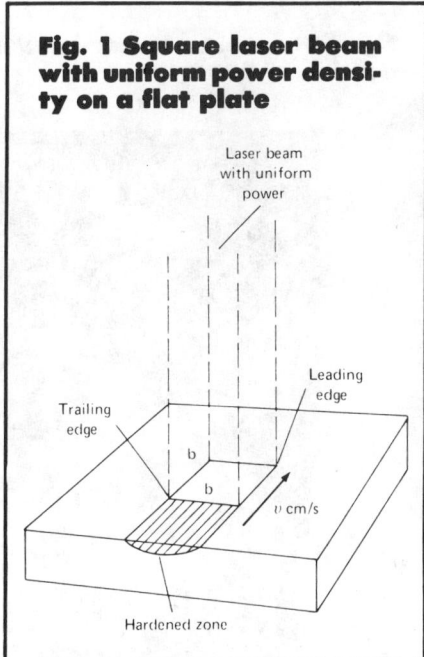

Fig. 1 Square laser beam with uniform power density on a flat plate

ventional through hardening procedure.

The basic reaction taking place during the heating period is the transformation of the bcc (α) iron into fcc (γ) iron. This occurs by nucleation and growth of the new phase in the matrix of the old phase. In slow heating, the process will start at A₁ (723 °C or 1335 °F) in a carbon steel and will be complete at the A₃ line. However, when the heating rate is high, the system is far from equilibrium conditions and the A₃ line will tend to be displaced upward to higher temperatures. Thus, although the temperature may be sufficiently high to form austenite under condition of slow heating, the same temperature level may be insufficient even to initiate austenitization under high heating rates. Laser hardening parameters are, therefore, usually designed to give peak temperatures well above those employed in conventional hardening to ensure austenitization.

The equilibrium room temperature structure of iron and steel will contain carbon in the form of iron carbide or graphite as a separate phase. To bring about hardening upon quenching, this carbon must be uniformly dissolved in the austenite. To do so, the carbon must be redistributed by diffusion into areas that originated from practically carbon-free ferrite.

This is a time-dependent process even at the high temperatures used in laser hardening and, under certain con-

Fig. 2 Structure of laser-hardened ductile cast iron at magnification of 250 ×

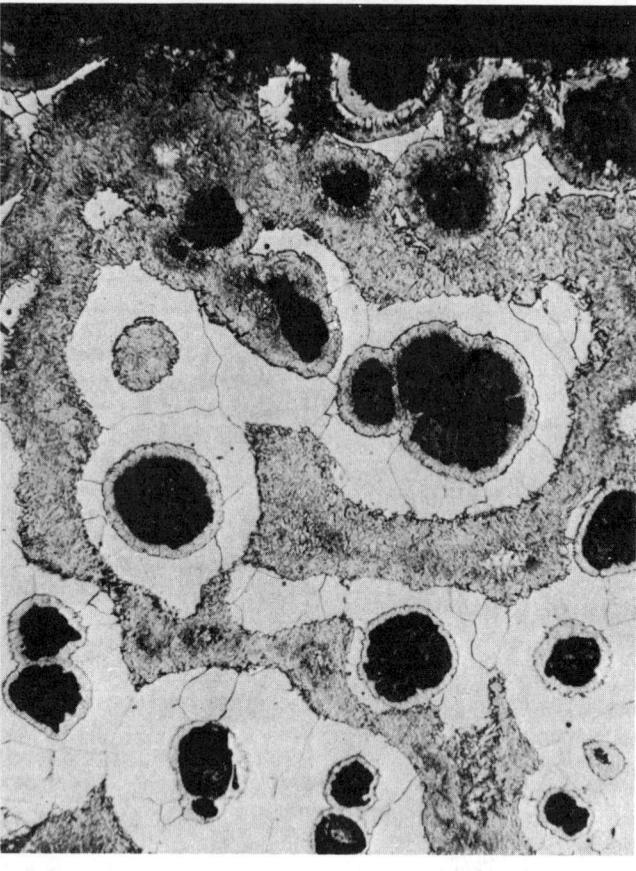

Fig. 3 Influence of processing parameters on heat penetration in laser surface transformation hardening

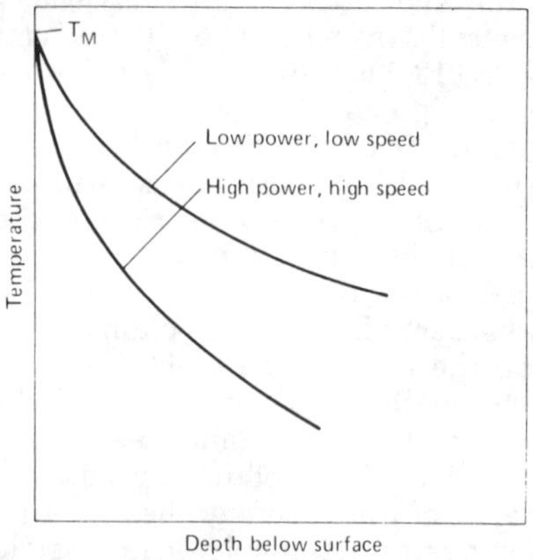

ditions, laser processing may occur too rapidly to allow for complete diffusion. This problem is obviously most prevalent where the carbon distribution in the starting material is nonuniform, such as in coarse pearlitic structures, structures containing proeutectoid cementite, spheroidized materials, and cast irons, particularly cast iron with a high content of free ferrite. Figure 2 shows the structure of a ductile cast iron after laser processing. The extent of carbon diffusion that occurred during the processing is clearly revealed by the region of martensite surrounding the graphite nodules.

The intrinsically high heating rates associated with laser hardening, combined with the need to allow sufficient time above the A_3 point to form homogeneous austenite, result in high peak temperatures. Under such conditions, there may be a tendency to form a thin surface layer, 5 to 20 μm (195 to 785

micro-in.) thick, of a phase that has yet to be categorized accurately. Retained austenite is found in laser hardened materials as revealed by x-ray analysis, but it is not clear how variation in laser processing parameters influence the amount of retained austenite.

The upper limit for the surface temperature in laser processing is set by the melting point of the material, because surface melting is undesirable in most instances. On simple plane surfaces, this is easy to avoid, but on workpieces with more complex design the surface temperature may vary across the beam-impingement area, even if the power density over the beam cross section is uniform. This can occur when the surface is curved, and in particular, when the beam strikes an area with abrupt angles. The projected power density will then be nonuniform, and the heat flow uneven, so that part of the illuminated surface may reach the

melting point before the remaining area has reached sufficient temperatures to harden to the required depth. Sharp edges and corners can be troublesome, because they tend to concentrate the heat flow and lead to blunting of the edges by melting.

Another problem associated with some laser surface hardening applications is the necessity to overlap hardening passes. This may occur at the closure of a path around a cylindrical workpiece, or in the overlap zone between parallel paths being processed sequentially. Because lateral heat flow from the moving laser spot is unavoidable, backtempering will take place by the heat flow into areas already hardened by previous passes or, in the case of closed paths, at the start of the pass. The lower the processing speed, the more pronounced the effect, because relatively more heat will have time to diffuse into the previously hardened area. Even at high speeds and with the application of external heat sinks or water cooling, this effect cannot be entirely eliminated. The use of special optics may eliminate the necessity of forming overlap areas. However, this will always entail illuminating the entire width of the work area with laser radiation, and the available power of the laser limits the size of the workpiece that can be processed by the use of such optics. In many applications, the formation of a backtempered zone is not detrimental, provided that the processing is done in such a way that the backtempered area is located in a position of low service stresses or wear. Thus, for example, the interior walls of internal combustion engines may be hardened by straight or spiral laser passes, leaving unhardened material between the passes without harmful effects.

Heat Flow in Laser Hardening

In laser surface transformation hardening, thermal energy is generated by absorption of the laser radiation at the surface. The increase of temperature in the interior of the workpiece is by way of conduction only; no sources of thermal energy exist below the surface. Thus, if the rate of absorbed power and the thermal properties of the material are known, it is possible, at least in principle, to calculate the temperature distribution in the workpiece. This is of considerable value, because it is then possible to predict the results of laser processing in advance and to calculate the optimum processing parameters, such as power density, processing speed and spot size.

The long wavelength electromagnetic radiation (infrared) from a typical carbon dioxide laser is not efficiently absorbed by ferrous metals at room temperatures. It is, therefore, necessary to coat the workpiece with a substance that will aid in absorbing the laser energy. Commonly used coating materials are manganese phosphate, graphite or carbon-black paint. The paint, in the form of flat, black spray paint, is by far the most convenient coating to use. It is easy to apply and is fairly insensitive to variation in coating thickness.

When laser energy is absorbed at the surface at a rate of 500 W/cm^2 (3200 W/in.^{2}) or more, surface temperature rises very rapidly because the conduction of heat to the interior cannot keep up with the influx of energy to the surface. The higher the input flux, the more rapidly the temperature rises in the surface layer, and as a consequence, the temperature gradient in the workpiece will be steeper. The maximum surface temperature allowable is the melting point of the material, although in practical applications, temperatures should be held well below this value. This clearly acts as a constraint on the depth of austenitization that can be achieved. If lower laser power density is used and the processing speed is correspondingly decreased, the surface temperature will rise more slowly, and the temperature gradient will be less steep. This allows

austenitization to a greater depth, as shown in Fig. 3. However, the rate of cooling by self quenching will be slower, and it may be insufficient to allow the material to harden fully. Thus, the combined effects of melting temperature and hardenability act to impose a limit on the obtainable depth of case regardless of the power available for the processing. To obtain good self quenching, it is generally necessary to use high power density and high processing speed for steels of low hardenability. Steels of high hardenability can be processed to greater depth by relatively low power densities and low processing speed. For many such steels, the slow processing rate will be necessary to give the material time to form homogeneous austenite. Thus, deep case depth in plain carbon steels may be difficult to achieve. It is possible to use external quench procedures, thereby obtaining deeper case depths in steels of low hardenability. This is achieved at the cost of greater dimensional distortion, because the total power input increases with decreasing speed for constant maximum surface temperature. If the absorbed power density, laser spot dimensions, processing speed and thermal properties of the material are known, the temperature distribution in the workpiece can be calculated by means of several expressions found in the literature. The simplest expression is obtained if it is assumed that heat only flows normal to the workpiece surface, that is one-dimensional heat flow. If the workpiece is large in this direction, the temperature is given by:

$$T = T_o + 2Q/K \sqrt{\alpha t_D} \ \text{ierfc} \ \frac{\delta}{\sqrt{\alpha t_D}}$$

(Eq 1)

where T is temperature in centigrade, T_o is room temperature in centigrade, Q is absorbed power in W/cm^2, K is thermal conductivity in W/cm C°, α is thermal diffusivity in cm^2/s, δ is depth below the surface in cm, and t_D is dwell time in s.

Dwell time is the amount of time a given spot on the surface will be exposed to the laser beam. This is, therefore, equal to the length of the laser spot in the direction of travel divided by the speed of travel. The expression ierfcx is the integrated complementary error function defined by:

$$\text{ierfcx} = \int_{x}^{\infty} \text{erf} \times \text{dy} = \int_{x}^{\infty} (1-\text{erfx}) \ \text{dy}$$

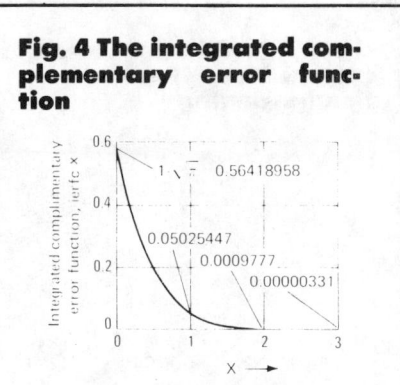

Fig. 4 The integrated complementary error function

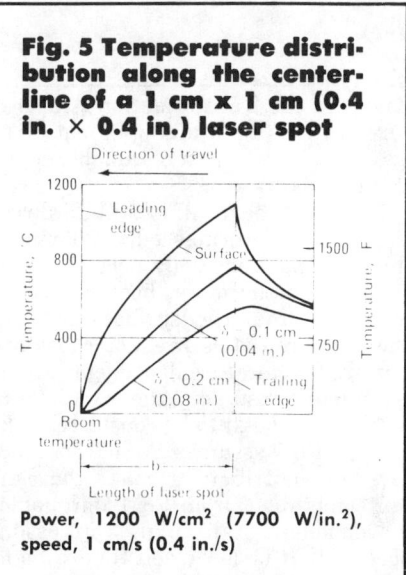

Fig. 5 Temperature distribution along the centerline of a 1 cm x 1 cm (0.4 in. × 0.4 in.) laser spot

Power, 1200 W/cm^2 (7700 W/in.^{2}), speed, 1 cm/s (0.4 in./s)

where erfx is the error function:

$$\text{erfx} = \frac{2}{\sqrt{\pi}} \int_{0}^{x} e^{-Y^2} \ \text{dy}$$

These functions are tabulated in many sources and can easily be evaluated. Figure 4 shows a plot of ierfcx.

If we apply Eq 1 to a moving laser spot on a plane surface, it becomes clear that as the leading edge of the spot reaches a given spot on a plane surface, the temperature will rapidly start to increase, and the maximum temperature will be reached at the trailing edge of the spot, as shown in Fig. 5. A similar, but smaller, temperature increase will be experienced at points below the surface. If Eq 1 is solved for various values of the depth δ under the trailing edge of the spot, a temperature profile can be constructed representing the maximum temperature conditions in the vicinity of the surface (see Fig. 6).

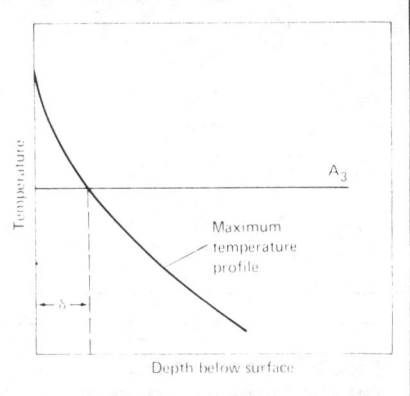

Fig. 6 Predicting depth of case in laser transformation hardening

Table 1 Thermal properties of ferrous materials

Material	Transformation temperature °C	°F	Thermal diffusivity(a) cm²/s	in.²/s	Thermal conductivity(a) W/cm·°C	W/in.²·°F
1025	850	1560	0.073	0.011	0.39	0.085
1045	800	1470	0.074	0.011	0.38	0.083
1078	725	1335	0.065	0.010	0.36	0.079
4140	790	1455	0.070	0.011	0.36	0.079
5130	780	1435	0.066	0.010	0.36	0.079
Gray cast iron	...	...	0.099 to 0.148	0.015 to 0.023	0.46 to 0.57	0.10 to 0.0125

(a) To obtain values in SI units, mm²/s and W/m·K, multiply by 100.

Knowing the transformation temperature A_3, it is then a simple matter to determine the largest obtainable depth of hardened case δ.

In reality, heat diffuses in all directions into the workpiece from the moving laser spot, not only normal to the surface. Furthermore, heat is lost by radiation from the surface, and the thermal properties are not constants but are temperature dependent. Finally, in many instances, the workpiece is not large enough to be considered infinite, as Eq 1 assumes. All of these factors can contribute to make the estimates obtained from Eq 1 unreliable. Nevertheless, at high processing speeds (that is, low dwell time, t_D) and for reasonably thick specimens and large laser spots, Eq 1 gives a good estimate for depth of obtainable case. For a laser spot with dimensions 1.27 by 1.2 cm (0.5 by 0.47 in.), Eq 1 can be safely used down to a processing speed of 1 cm/s (0.4 in./s) provided that the workpiece is at least 0.6 cm (0.2 in.) thick. It is possible to develop expressions that take into account the three-dimensional nature of the heat flow, but such expressions become very tedious and time consuming in use, and it is often not worth the extra work.

By a relatively simple expansion of Eq 1, the cooling rate of any given point in the workpiece can be estimated if only one-dimensional heat flow is considered. This is done by means of the expression:

$$T = T_o + 2Q/K \left\{ \sqrt{\alpha t} \text{ ierfc} \frac{\delta}{2\sqrt{\alpha t}} \right.$$

$$\left. - \sqrt{\alpha (t - t_D)} \text{ ierfc} \frac{\delta}{2\sqrt{\alpha t - t_D}} \right\}$$

(Eq 2)

In this equation, the symbols have the same meaning as in Eq 1 except that t is the time elapsed since the leading edge passed over the spot for which we need to know the temperature; and $t - t_D$ is the time elapsed since the trailing edge passed over the spot, that is, the time elapsed after cooling by self quenching began. When using this equation, it is very important that the workpiece be large enough to provide an adequate heat sink.

Equation 2 can be used in conjunction with C-T diagrams to predict whether or not the cooling rate by self quenching will be adequate for hardening under a given set of circumstances. To aid in the use of Eq 1 and 2, Table 1 gives the value of the conductivity and the diffusivity for commonly used materials. The listed values are the average values from room temperature up to 1000 °C (1830 °F), and are given in units convenient for calculation of temperatures by Eq 1 and 2 when the power input (Q) is in W/cm². It should be noted that Q is not the applied power density but the estimated absorbed power in W/cm².

Processing Parameters

Many factors influence the results of laser surface transformation hardening including size of the laser spot, power density, processing speed, thermal properties of the material and laser hardenability of the material. The last factor encompasses the material's response to rapid heating and quenching and is, in part, dependent on the starting condition of the material, that is, whether the material is normalized, annealed, etc.

The parameters of principal interest are power density and processing speed. It is obvious that processing speed should be as high as possible to attain high production rates. However, the speed at which the laser spot moves over the surface is not a good measure of the production rate by itself, because the dimensions of the spot normal to the direction of travel are equally important in determining the area coverage rate. Furthermore, another important factor in determining the results of processing is the time under the beam that a given spot on the surface experiences, that is, the spot dimension in the traveling direction divided by the speed, commonly referred to as dwell time. The relative dimensions of a rectangular spot do not influence the coverage rate as long as the power density stays constant. Therefore, the area of the laser spot that will give a specific result is limited by the available power. An exception to this is when the spot is very narrow in the direction of travel and/or when the speed is very low (or equivalent, the dwell time, t_D, very long). Under such conditions, lateral heat losses become large and the spot dimensions influence the results. The conditions under which this effect becomes noticeable is:

$$B = vb/4\alpha < 3.5$$

where b is the spot dimension in the direction of travel, v is the speed of the spot and α is the diffusivity, all in cgs units.

The following are general guidelines for choice of processing conditions:

- Usable power densities in laser surface hardening are in the 500 to 5000 W/cm² (3200 to 32 000 W/in.²) range. Corresponding dwell times are in the range 0.1 to 10 s. For carbon steels, the power density is usually from 1000 to 1500 W/cm² (6400 to 9600 W/in.²), and the dwell time 1 to 2 s.
- Materials with high hardenability can be processed at low power density and high dwell time (low speed), whereas materials with low harden-

- ability should be processed at high power density and low dwell times.
- Rectangular or square laser spots with uniform power density are most suitable in obtaining uniform hardened case.
- High power density and low dwell time give shallow case, but high cooling rates. The reverse is true for low power densities.
- Maximum surface temperature is proportional to the square root of the speed. Hence, a doubling of the power density requires a quadrupling of the speed to obtain equivalent maximum surface temperatures.
- Increasing the power density results in lower total energy input for the same maximum surface temperature.
- Steel with normalized, annealed or spheroidized structures, steel with proeutectoid cementite, cast irons and steels with stable alloy carbides require longer dwell times than steels that have been hardened and tempered.
- Small workpieces will require higher power densities and lower dwell times than large pieces, unless external quenching media are used.

Metalworking Lasers

Several models of metalworking lasers of both domestic and foreign manufacture are commercially available. The majority of these are of either the neodymium, YAG solid state type or the carbon dioxide gas type. These lasers may have pulsed or continuous output power. Both types, whether pulsed or continuous wave, can be used for transformation surface hardening.

The power output of metalworking lasers at the present time is in the 50-W to 15-kW range, and more powerful systems will be available in the future. Because the cost of a laser system is approximately proportional to its maximum power output, the minimum size of a laser system for a potential industrial application should be determined on the basis of such factors as required production rate, minimum practical spot size and power density necessary to achieve the desired results. It should be noted, however, that the laser is a versatile tool and can be used for a number of other metalworking processes such as welding, cutting and hardfacing. If a laser system is to be used in a variety of applications, it may be ben-

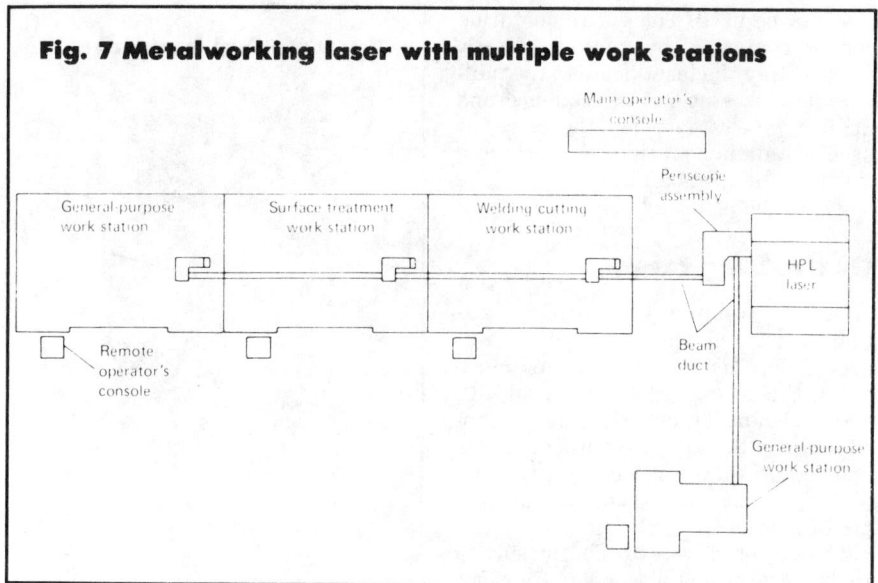

Fig. 7 Metalworking laser with multiple work stations

eficial to have as much power available as possible.

Lasers emit electromagnetic radiation in the infrared portion of the spectrum, and the laser beam from these machines is invisible. The carbon dioxide laser emits radiation with 10.6-μm wavelength, and this radiation is easily absorbed in a variety of nonmetallic substances. A thin sheet of lucite between the operator and the beam/workpiece interaction zone is sufficient to absorb potentially harmful stray radiation. The output of the YAG laser has a much shorter wavelength (1.064 μm), and therefore, the operator is required to wear special colored protective eyeglasses. YAG lasers are limited to relatively low power levels. Therefore, in metalworking processes requiring more than 500 W of power, the carbon dioxide laser is usually employed because it can deliver much higher continuous output.

The primary output beam from the laser rarely is used in metalworking applications. Instead, the output beam is directed and shaped by optical systems to generate a laser spot of the desired size and shape on the workpiece surface. Such an arrangement allows substantial flexibility in the use of a laser system. Because the coherent radiation from a laser has low loss of power with distance, the laser itself can be situated at a considerable distance from the work area. Furthermore, different types of metalworking applications can be performed with the same laser by changing the optical system. Several jobs can be performed simulta-

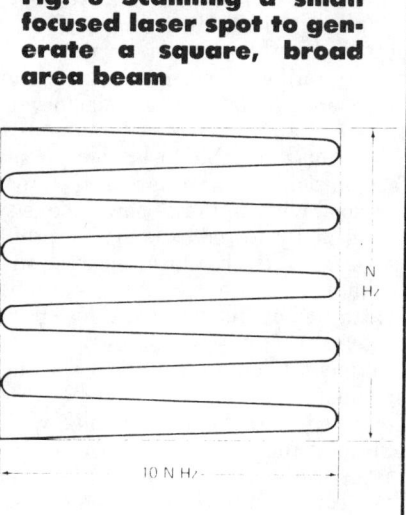

Fig. 8 Scanning a small focused laser spot to generate a square, broad area beam

neously with one laser by using several work stations, each with its own optical system designed for a specific application. Figure 7 shows a typical laser system arrangement. In this instance, each work station has its own control console, allowing manual or automatic control of beam power, duration of power delivery to the workpiece, rate of increase of power at the start of the run, rate of decrease at the end (ramp-up and ramp-down) and manipulation of the workpiece fixture protective gas flow. The laser power is automatically maintained at the desired level by a feedback control device in the laser. Positive feedback from temperature sensors, monitoring the temperature in the beam workpiece interaction zone, also can be used. The entire operation can be controlled by microprocessors. By directing the laser beam to the individual work stations in sequence and utilizing the time when the power is used at another station for workpiece manipulation, maximum usage of the laser can be achieved.

Optical Systems

In laser welding and cutting operations, a focused beam of intense power is used. For laser transformation hardening, this focused spot is replaced with a broad beam of much lower power density, partly because the hardening process requires lower power densities and partly to obtain reasonable rates of area coverage.

The simplest way to obtain such a spot is to position the workpiece surface in such a way that it intercepts the beam some distance from the focal plane of a converging beam. However, the power density distribution of a spot obtained in this manner is rarely sufficiently uniform and at the same time large enough to give satisfactory results.

The most suitable laser spot for surface hardening is a square or rectangular spot with uniform power density, sometimes referred to as top hat profile (see Fig. 1). Such a laser spot requires reshaping of the output laser beam that is attained by the use of various optical systems.

Transmission optical elements, such as lenses and windows, can be used. However, because of the long wavelength of the radiation typical for CO_2 lasers, these elements must be made from special materials such as zinc selenide to avoid excessive absorption.

Reflective optical components are less fragile and better adapted to industrial environment. They consist of flat, spherical or parabolic mirrors made from copper or molybdenum, which have excellent reflective characteristics to laser radiation. By the use of these mirrors, the beam can be reshaped and redirected to suit the particular application requirement.

The simplest way to form a square or rectangular beam is to use mirrors or lenses to form a focused spot and then to scan this spot rapidly back and forth in two perpendicular directions, as shown in Fig. 8. If scanning is done at a sufficient rate, the result will be equivalent to spreading the power of the focused spot over the area being

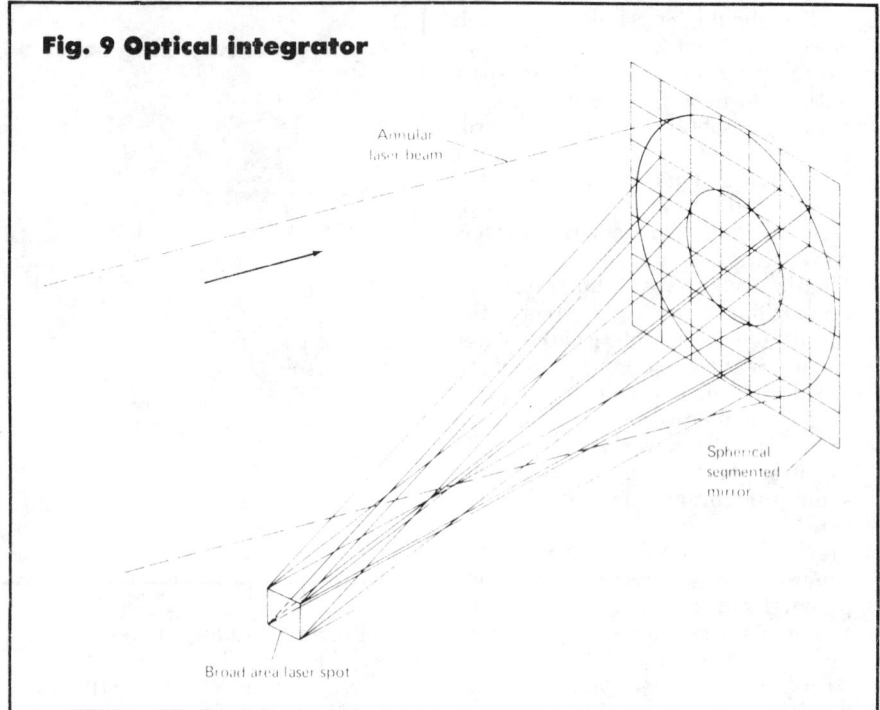

Fig. 9 Optical integrator

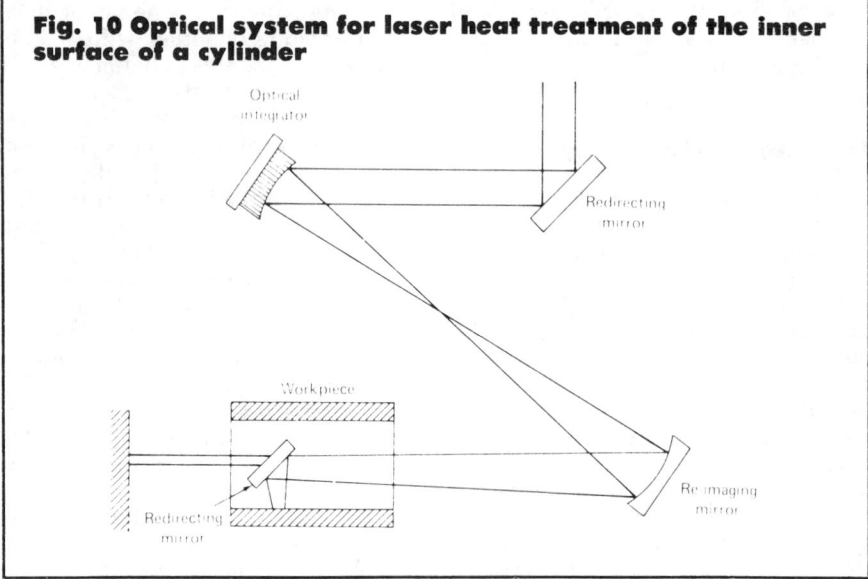

Fig. 10 Optical system for laser heat treatment of the inner surface of a cylinder

Table 2 Laser hardening of 1045 plate

Speed mm/min	in./min	Power, kw	Measured case mm	in.	Calculated case mm	in.
510	20	2500 . . . . 0.52	0.020		0.04	0.002
510	20	3000 . . . . 1.02	0.040		0.82	0.030
510	20	3600 . . . . 1.37	0.055		1.56	0.060
760	30	3000 . . . . 0.24	0.010		0	0
760	30	3600 . . . . 0.66	0.025		0.60	0.024
760	30	4150 . . . . 1.24	0.050		1.08	0.043

scanned, because the thermal lag of the metal will act to even out the power input to a uniform power density. The scanning can be performed conveniently by electromechanical vibration of one or more of the mirrors in the optical

Fig. 11 Toric mirrors for treating cylinders

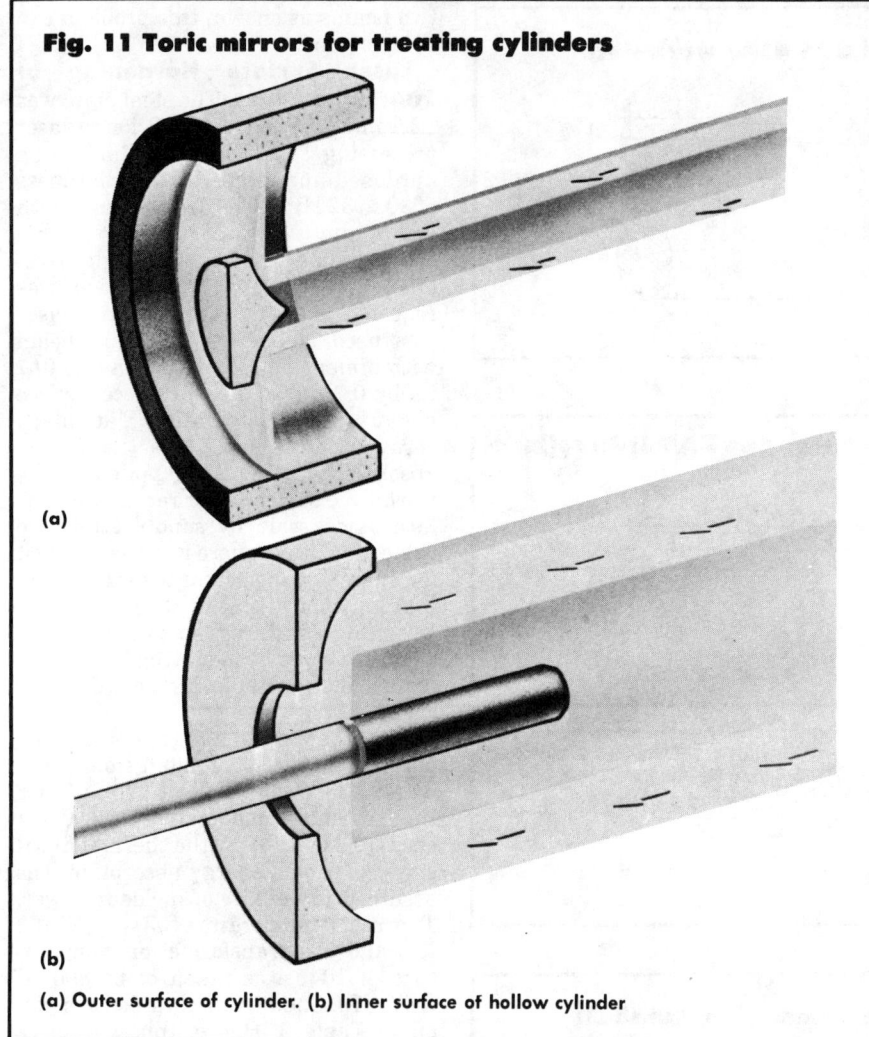

(a)

(b)

(a) Outer surface of cylinder. (b) Inner surface of hollow cylinder

Fig. 12 Using two equal laser beams to surface harden a semicircular groove

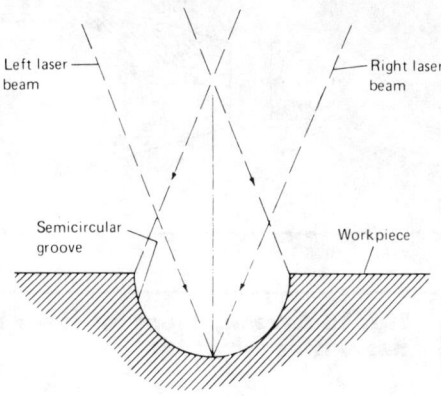

Left laser beam

Right laser beam

Semicircular groove

Workpiece

Fig. 13 Laser heat treating SAE 1045

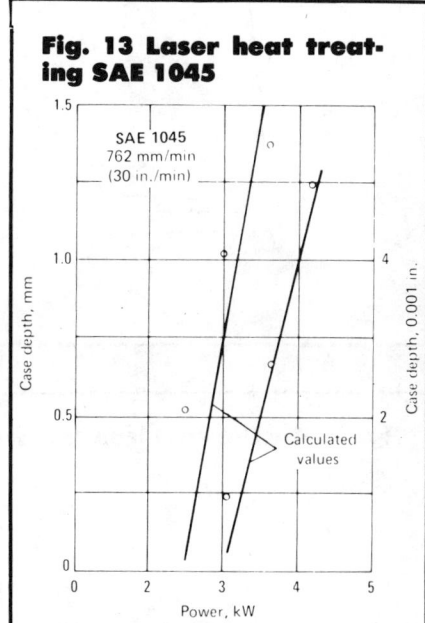

SAE 1045
762 mm/min
(30 in./min)

Calculated values

Case depth, mm

Case depth, 0.001 in.

Power, kW

system. In this way, it is possible to generate spots of desired size and shape on the workpiece surface.

Another method of forming a square or rectangular beam is to use a device known as an optical integrator. This device, shown in Fig. 9, is an array of flat mirrors mounted on a spherical surface placed to intercept the output laser beam. Each mirror forms an image of the part of the output beam that it intercepts, and all of the images from the array will be formed in the same position. In this way, the output beam is reshaped to form the desired square or rectangular spot in the focal plane of the integrator, having uniform power density over the illuminated area on the workpiece surface.

Both scanning optics and integrator optics form beams that can be redirected by flat mirrors or refocused by appropriately curved mirrors. In this way, the laser spot can be generated in

the exact shape at desired location. Figure 10 shows how a spot can be projected on the inner wall of a tube.

For workpieces of cylindrical design, special optical components called toric mirrors can be used. These mirrors take advantage of the fact that the output beam from the laser can be made hollow or annular in cross section. Directing the beam by suitable mirrors onto the workpiece causes a continuous band of laser irradiated surface to form around the periphery of the workpiece, either on the inner surface of a hollow cylinder or on the outer surface of a cylinder, as shown in Fig. 11. The complete surface can be laser treated by moving the workpiece in the axial direction through the ring-shape laser spot. No backtempering is encountered because no start/stop or parallel spiral zones are generated. The relatively large area of the ring-shape laser spot limits the size of workpieces that can be

processed with this technique to about 25 mm (1 in.) diameter for each 8 kW of power available.

Another technique that can be used for nonplanar surfaces is beam splitting. The laser beam is split into two equal parts by a copper prism, and the individual parts of the beam are directed at different angles to the workpiece surface by suitable reflective mirrors. Figure 12 shows the application of this technique to heat treating a wide semicircular groove. If only a single beam, normal to the bottom of the groove, had been used, the angle of incidence of the laser beam at the left and right top of the groove would have been too shallow to generate sufficient heat. By using

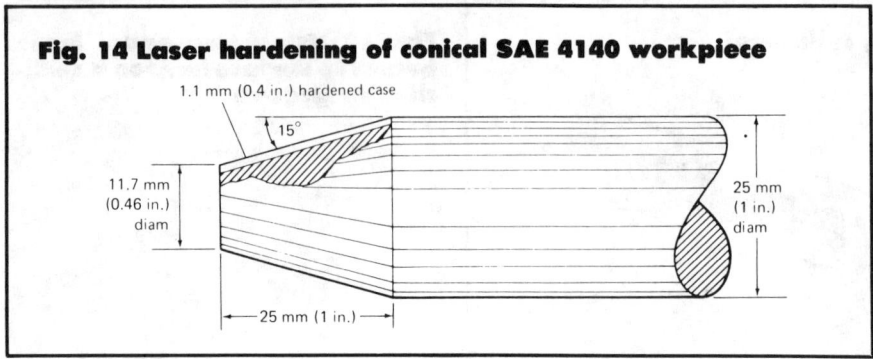

Fig. 14 Laser hardening of conical SAE 4140 workpiece

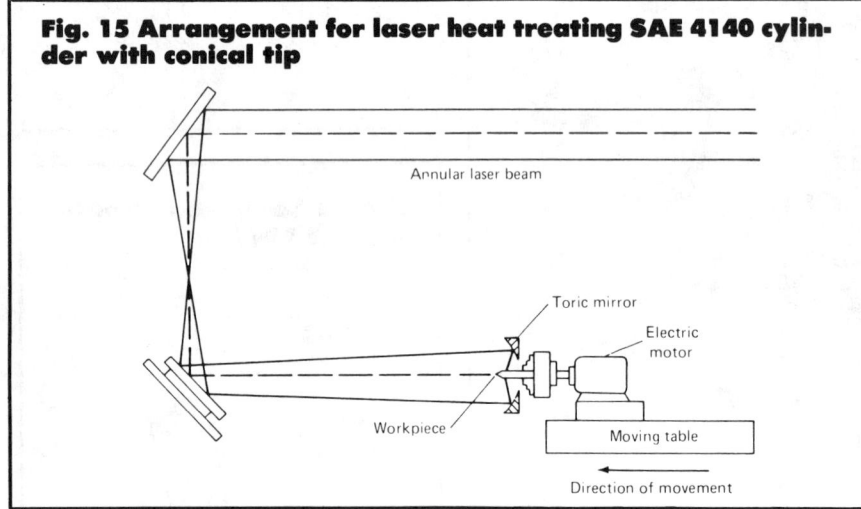

Fig. 15 Arrangement for laser heat treating SAE 4140 cylinder with conical tip

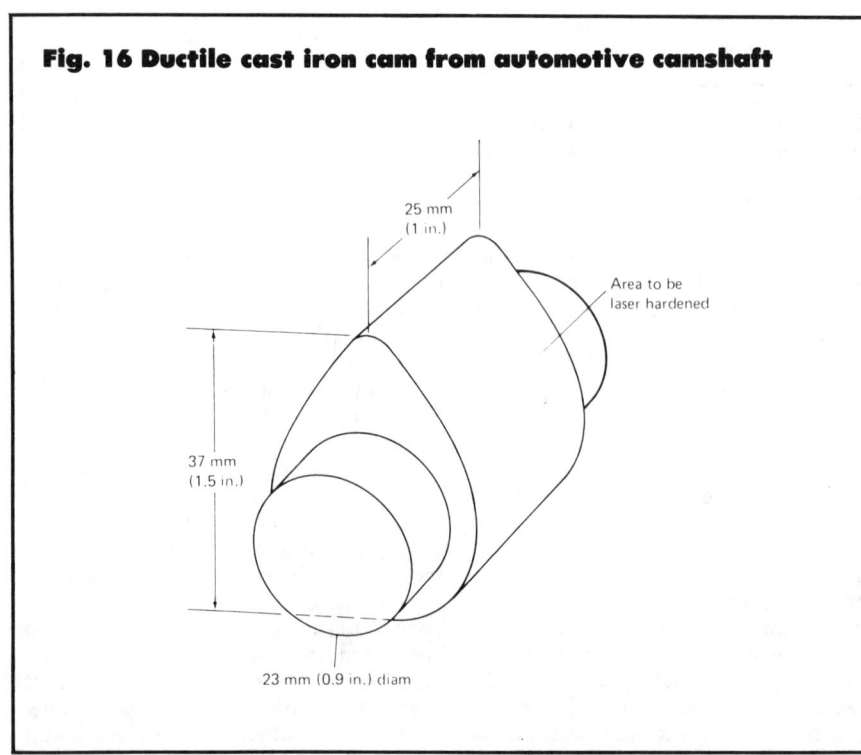

Fig. 16 Ductile cast iron cam from automotive camshaft

two beams as shown, this problem can be overcome.

Laser Surface Hardening of 1045 Steel Plate. The steel plate was 12.5 mm (0.5 in.) thick. Prior to laser processing, the material had been hardened and tempered to a hardness of 30 to 32 HRC. The desired case depth was 1 mm (0.04 in.); hardness at this depth should be 45 HRC. Single hardened band, 18 mm (0.7 in.) wide, was required. A 15-kW carbon dioxide laser was used, delivering a square beam with dimensions 18 mm by 18 mm (0.7 in. by 0.7 in.) to the workpiece surface by an optical integrator. Flat, black spray paint was used as the energy absorbing coating and applied to the workpiece surface after removal of surface oxide scale by sandblasting. No protective atmosphere was used, except for a fan blowing across the interaction zone to remove smoke. The workpiece was moved with respect to the stationary laser beam by mounting it on a controlled motion table in the work station.

The approximate range of processing parameters was calculated from Eq 1, using 0.38 W/cm·°C (0.083 W/in.·°F) for the thermal conductivity and 0.074 cm²/s (0.01 in.²/s) for the thermal diffusivity of 1045. Energy absorption was assumed to be 85% of incident power. The room temperature was 20 °C (68 °F), and the transformation temperature of 1045 was taken to be 800 °C (1470 °F). The results obtained are given in Table 2. Hence, this workpiece could be laser processed to the required case depth at a speed of 762 mm/min (30 in./min) (coverage rate 136 cm²/min or 21 in.²/min) at a power of 4.1 kW.

The relationship between calculated and measured case depth is shown in Fig. 13. As expected, using the simple one-dimensional Eq 1, the correlation between calculated and observed results are better at the higher speed.

Laser Surface Hardening of 4140 Cylinder with Conical Top. The object was to surface harden the conical part of the workpiece shown in Fig. 14 to increase the wear resistance. The minimum required case depth was 1.1 mm (0.044 in.); the hardness at this depth should be 45 HRC. The workpiece had been hardened and tempered to 40 to 42 HRC prior to laser processing to obtain strength and ductility.

Because the workpiece was cylindrical and had a maximum diameter of 25 mm (1 in.), a toric mirror could be used in conjunction with a 15-kW CO_2 laser.

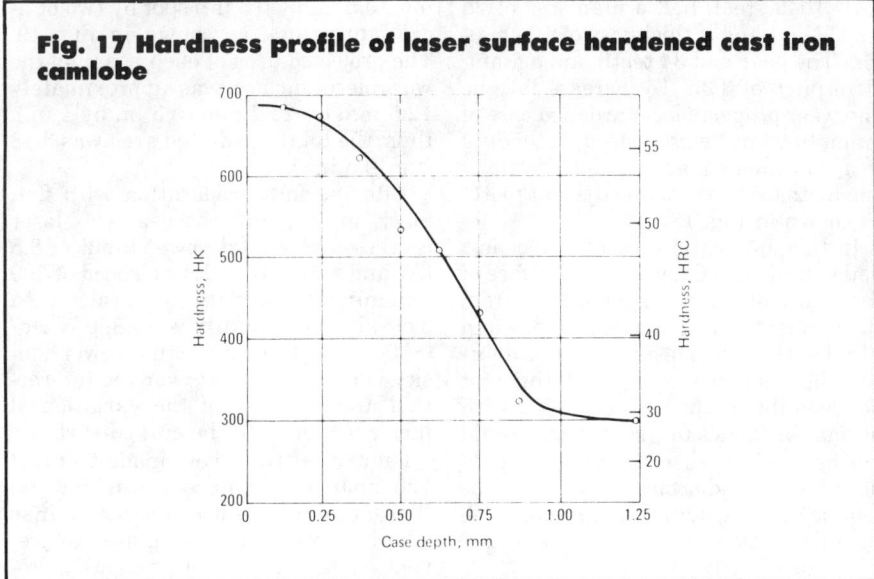

Fig. 17 Hardness profile of laser surface hardened cast iron camlobe

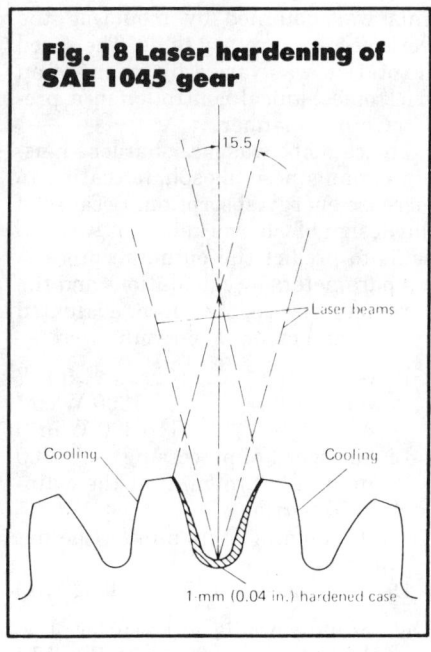

Fig. 18 Laser hardening of SAE 1045 gear

Fig. 19 Dual beam optics for laser gear heat treating

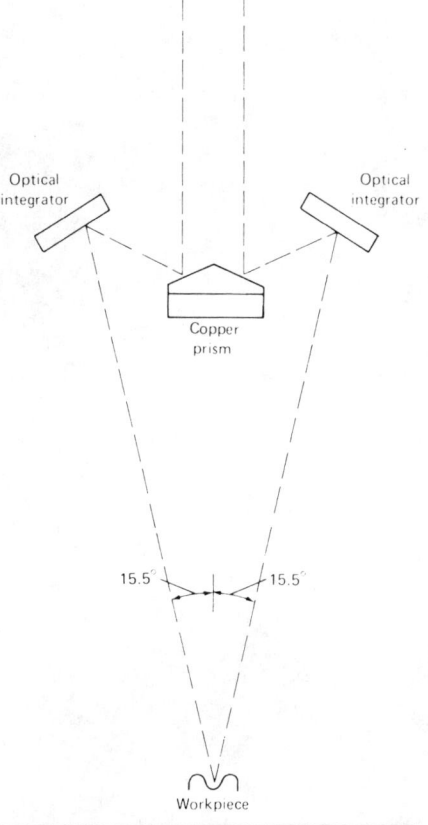

The processing arrangement is shown in Fig. 15. In this fixture, the workpiece was rotated at 1300 rpm under the ring-shape laser spot formed by the toric mirror. This was done to ensure uniform power density around the periphery of the workpiece. The ring-shape spot had a width of 5 mm (0.2 in.), and hardening of the desired area could be obtained by moving the workpiece in the axial direction. In this way, the entire area could be covered without forming any overlap zones.

Because the area under the ring-shape laser spot will increase as the beam sweeps from the tip of the cone to the cylindrical part of the workpiece, the power input had to be increased linearly with the distance in order to maintain constant power density.

Theoretical calculations, using Eq 1 with a thermal conductivity of 0.35 W/cm·°C (0.083 W/in.·°F) and a thermal diffusivity of 0.070 cm^2/s (0.01 in.^{2}/s), showed that at a processing speed of 229 mm/min (9 in./min) and a power density of 1620 W/cm^2 (10 400 W/in.^{2}) would give a surface temperature of 1360 °C (2480 °F). This is only an estimate, as Eq 1 assumes a plane workpiece rather than a cylindrical one. To process at a slower rate and at lower power density would give deeper heat penetration, but the self-quenching rate would be lower. On the other hand, processing at higher speed and power would lead to shallower heat penetration.

The workpiece was coated with flat, black spray paint and processed at the following parameters:

- Axial speed: 221 mm/min (8.7 in./ min); or 3.8 mm/s (0.15 in./s)
- Power: 3500 W at the tip, increasing to 7600 W at the cylindrical portion
- Rate of power increase: 620 W/s

The hardened case obtained ranged from 1.65 mm (0.06 in.) at the tip to 2 mm (0.08 in.) at the cylindrical portion of the workpiece. The surface hardness was 58 to 59 HRC.

Laser Surface Hardening of Cast Iron Camshaft Lobes. The surface of the lobes of an automotive camshaft made from ductile cast iron (see Fig. 16) was to be surface hardened to increase wear resistance. The desired case depth, defined as the depth where the hardness was 50 HRC, was 0.5 to 1.0 mm (0.02 to 0.04 in.).

A 15-kW CO$_2$ laser was used for the processing. The optical system delivered a focused spot, with a diameter of 10 mm (0.4 in.) to the workpiece. This spot was scanned over a distance of 22 mm (0.9 in.) normal to the direction of processing and 25 mm (1 in.) in the direction of processing. The frequency of scanning was 125 Hz in the normal direction and 700 Hz in the processing direction, forming a rectangular spot 22 mm by 25 mm (0.9 in. by 1.0 in.) on the camlobe surface.

To obtain an even hardened case around the periphery of the camlobe, it was necessary to vary the angular speed of rotation of the lobe under the laser beam. The reason is that the angle of incidence of the laser beam to the workpiece changed during rotation, from nearly normal incidence at the

cylindrical portion of the lobe to a grazing incidence of only 20 to 30° at the flat portion. Furthermore, at constant rotational speed, the linear speed of processing would vary as the lobe rotated.

This was obtained by mounting the workpiece on a rotary table. The speed of rotation was varied by means of an electromechanical controller in a predetermined manner.

The camlobe was laser hardened, using a manganese phosphate coating to increase energy absorption. Because of the design of the workpiece, it was difficult to predict the optimum processing parameters by calculations and the parameters were, therefore, evaluated by trial and error. The results were:

- Power input.................9 kW
- Power density .... 1600 W/cm^2 (10 300 W/in.^{2})
- Linear speed of processing .... 760 mm/min (30 in./min) at the cylindrical portion 180 mm/min (7 in./min) at the flat portion
- Depth of case.. 0.55 mm (0.022 in.)

The hardness profile of the surface layer of the camlobe is shown in Fig. 17.

Laser Surface Hardening of a Large Gear. The gear, made from SAE 1045 steel, had a diameter of 28 cm (11 in.) and a thickness of 10 cm (4 in.). The gear had 34 teeth and a diametral pitch of 3.35. To increase fatigue and wear properties, a hardened case of 1 mm (0.04 in.) was desired, extending in a continuous manner from the tip of one tooth to the tip of an adjacent tooth, as shown in Fig. 18.

In this application, two laser beams had to be directed toward the surface of the gear teeth at an angle of 15.5° from the normal to the root area, as shown in Fig. 18. These two beams must be abutting but not overlapping at the root between the teeth. A single beam with normal incidence to the root area could not be used because the angle of incidence at the adjacent fillets would be too shallow to generate a hardened case in this critical area, where fatigue cracks are likely to originate.

The output beam from a 15-kW CO_2 laser was split into two equal parts by a reflective copper wedge. Each beam was then directed to the workpiece surface in the form of a 12.5 mm by 12.5 mm (0.5 in. by 0.5 in.) spot by two optical integrators, as shown in Fig. 19. The projected area of each beam on the workpiece surface was approximately 12.5 mm by 25.4 mm (0.5 in. by 1 in.); thus, the total irradiated area was 3.18 cm^2 (0.5 in.^{2}).

After surface preparation with flat, black spray paint, the gear was laser processed at a total power input of 8.8 kW and with a translation speed of 500 mm/min (20 in./min). The calculated over-all power density was 1380 W/cm^2 (8900 W/in.^{2}), but the actual power density varied somewhat over the interaction area because of the variation of incidence angle to the curved surface.

Sequential runs were made on teeth 120° apart to minimize heat distortion. Water cooling was used on tooth flanks adjacent to the processing area to prevent backtempering of previously processed teeth. The obtained case depth (45 HRC) was 1.2 to 1.3 mm (0.046 to 0.052 in.) on mid flank and at the root, and 1 mm (0.04 in.) at the fillets. Surface hardness was 59 to 60 HRC.

Electron-Beam Heat Treating

By Carl Fiorletta
Supervisor, Electron Beam
  Heat Treating Systems
Sciaky Bros., Inc.

ELECTRON-BEAM HEAT TREAT-ING is a selective hardening process in which the surface of a hardenable ferrous alloy is heated rapidly above the transformation temperature of the alloy by direct bombardment or impingement of an accelerated stream of electrons. At the end of a heating cycle of 0.5 to 2.5 s, the flow of electrons is stopped abruptly to allow the part or workpiece being processed to self-quench and to form a martensitic structure with a compressive stress on the surface of the hardened area. The electron-beam hardening process normally is applied to finish-machined or ground surfaces. Because the buildup of energy is rapid and well controlled, postheat treatment operations such as grinding or straightening are not required in most instances.

Application Criteria

A part or workpiece to be heat treated is considered a suitable candidate for electron-beam heat treating if it meets the following criteria:

- The material must contain adequate carbon to produce satisfactory case hardness. Figure 1 presents relationships between carbon content and case hardness for plain carbon steels.
- The stream of electrons must have line-of-sight access to the area requiring heat treatment and a beam-impingement angle of at least 25°. Guidelines for acceptable part configurations are shown in Fig. 2.
- The component being heat treated may be processed in a vacuum envelope or chamber, or at pressures up to 1 atm in air or inert gas. Vacuum chambers in high-production heat treating systems typically have interior volumes of 0.02 to 0.11 m³ (0.8 to 4.0 ft³).
- The surface to be heat treated should be machined or ground to final dimensions. If grinding is required after heat treating, removal of 0.05 to 0.25 mm (0.002 to 0.010 in.) of stock normally is sufficient.
- To prevent magnetic interaction or unintentional deflection of the electron beam, the component being heat treated must be demagnetized prior to hardening. Demagnetization normally is required if the part has been fixtured with magnetic clamps or chucks in operations prior to heat treating.
- The mass of the part must be suffi-

Fig. 1 Relationship between maximum hardness and carbon content for plain carbon steels

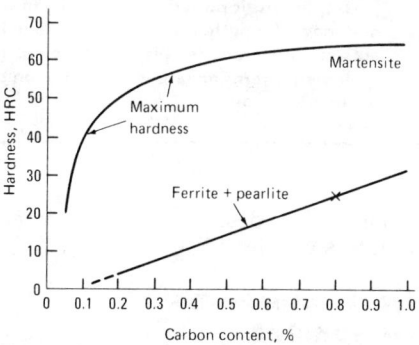

cient to self-quench the heat treated area. The ability of a part to self-quench is determined by its composition as well as its configuration. Steels of high hardenability (such as AISI 4150) can be through hardened in many instances. For plain carbon steels, however, five to eight units of mass beneath the heated area are required for each unit of mass of hardened case. The self-quench phe-

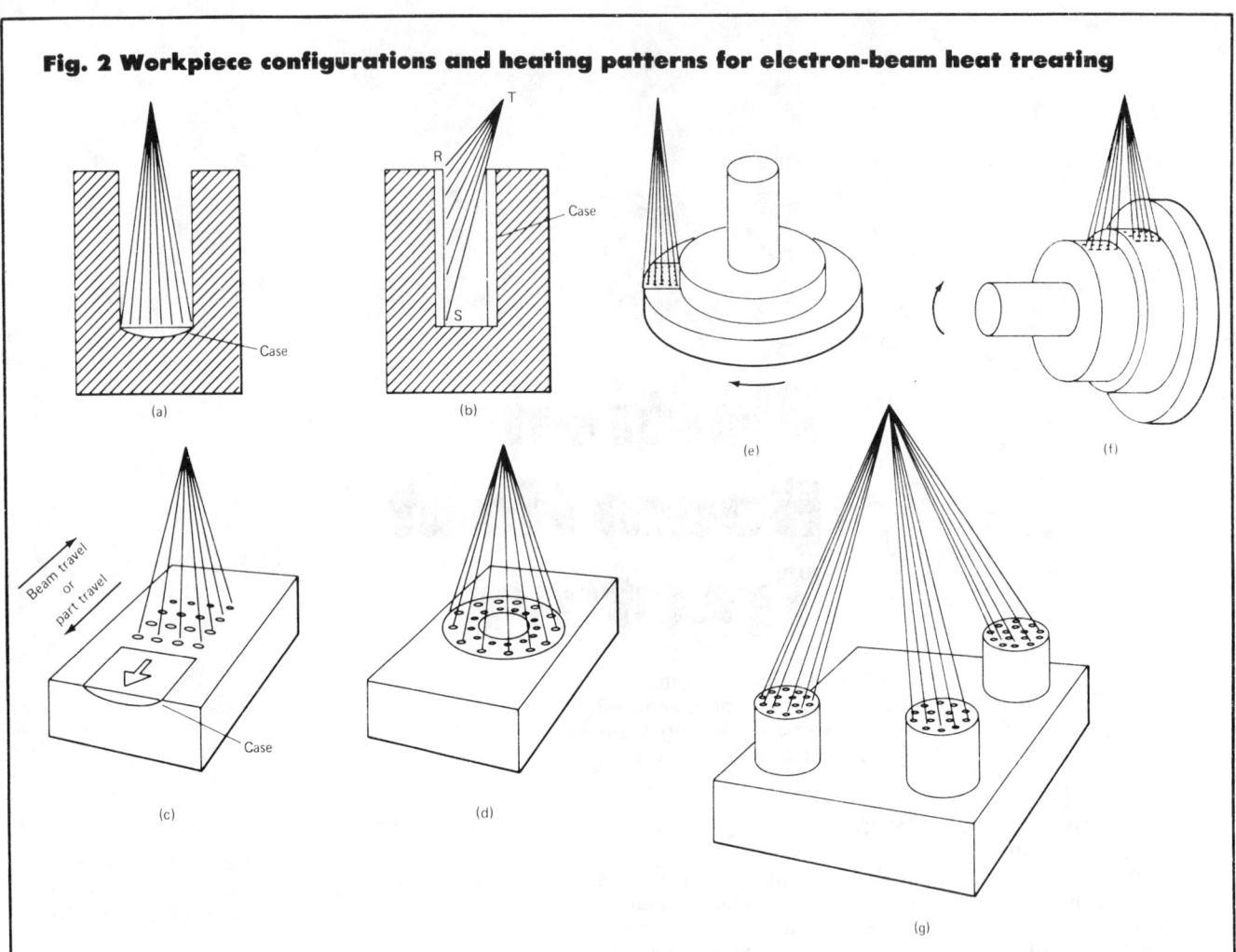

Fig. 2 Workpiece configurations and heating patterns for electron-beam heat treating

(a) Display static pattern within cavity in workpiece. (b) Maintain angle of workpiece rotation, RST, at 25° minimum. (c) Display static pattern and move the pattern or the workpiece to heat treat large areas. (d) Display static pattern; this annular pattern has well-defined inside and outside diameters. (e) Display static pattern and rotate workpiece. (f) Display more than one pattern and rotate workpiece. (g) Display multiple patterns on one workpiece or on a small group of workpieces for simultaneous hardening; patterns may be similar or dissimilar in geometric shape.

nomenon is described in detail later in this article.

Electron-Beam Equipment

In electron-beam heat treating, a highly concentrated beam of high-velocity electrons is used to heat selective surface areas. The electron beam is produced by an electron gun such as the one schematically represented in Fig. 3. Depicted is a 42-kW gun used for welding as well as for heat treating.

Beam Control. Free electrons escape from the filament when it is resistance heated to a very high tempera-ture. These electrons are accelerated and collimated into a dense, extremely energetic beam by the accelerating potential between the cathode and the anode. The high-energy beam thus formed passes through a small-diameter hole in the anode. Because of the mutual repulsion among neighboring electrons, the beam requires further collimation below the anode. This additional collimation is controlled with a focus coil that allows variation of the distance from the gun to the workpiece. A deflection coil deflects the reconverging beam to a designated location on the workpiece.

A high vacuum is needed in the region where the electrons are emitted and accelerated, both to protect the emitter from oxidation and to prevent interference with the electrons while they are still at low velocity. Therefore, the electron-gun housing is pumped and maintained at a vacuum of 10^{-5} torr. The workpieces are contained in an enclosure under a vacuum of approximately 5×10^{-2} torr. An intermediate vacuum level provides short evacuation times and higher production rates. Treating at one atmosphere does not require any evacuation time.

In electron-beam heat treating, the energy exchange is simply a matter of the electrons in the beam transferring their kinetic energy to the atomic structure of the target material in the

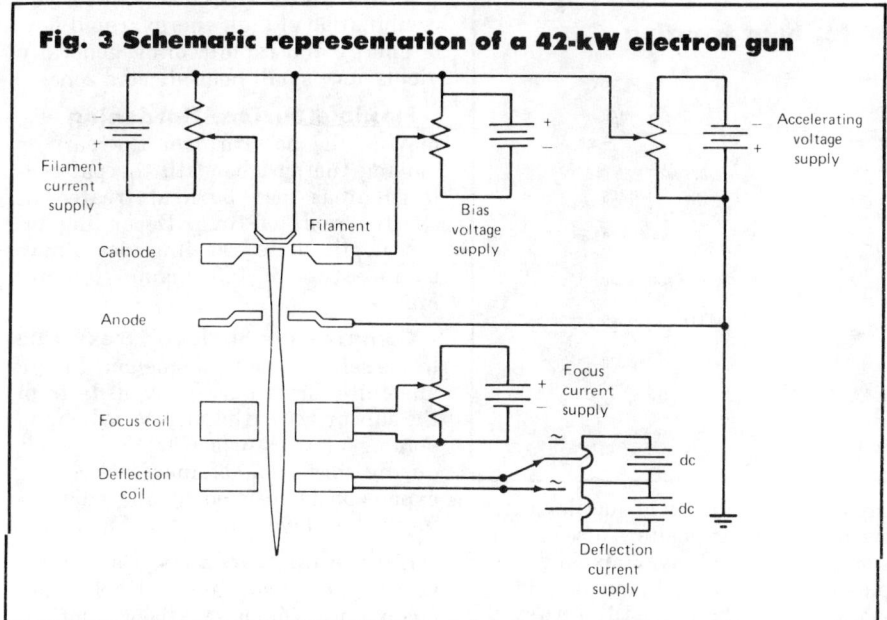

Fig. 3 Schematic representation of a 42-kW electron gun

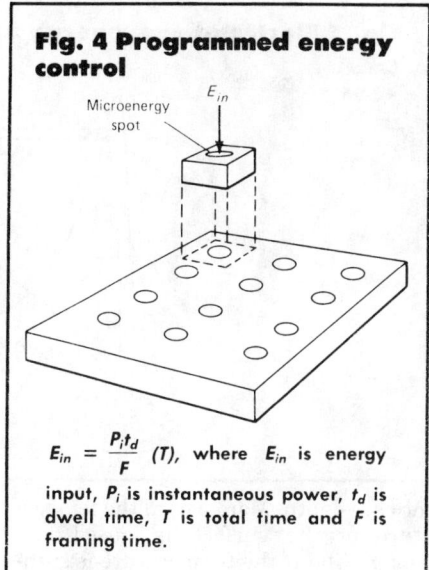

Fig. 4 Programmed energy control

$$E_{in} = \frac{P_i t_d}{F}(T),$$ where E_{in} is energy input, P_i is instantaneous power, t_d is dwell time, T is total time and F is framing time.

form of heat. The electron beam, when sharply focused for welding, is capable of impingement power densities on the order of 10 MW/cm^2 (65 MW/in.^{2}). Because this powerful concentration of energy is easily controllable in power magnitude, power density and beam position, it is well suited for surface hardening as well. These power densities are much too high for nondestructive heat treating, however. Destructive heat treating in this context refers to controlled remelting of ferrous and nonferrous materials.

An energy concentration of 3.1 kW/cm^2 (20 kW/in.^{2}) is more suitable for selective heat treating. To reduce the beam energy to this level, a single electron beam is programmed through a group of discrete beam positions referred to as a raster pattern.

Programmable Raster Pattern. The advancement that made electron-beam heat treating practical was the development of a programmable dedicated computer-control system coupled with electron-beam equipment. The control system generates a raster pattern by magnetically maneuvering the electron beam accurately from one programmed location to another while controlling energy input at each location, as shown in Fig. 4.

The power of the electron beam is distributed over the surface to be hardened at points spaced so that no "hot spots" exist when a defocused beam is applied. These locations may be spaced as required to obtain the necessary power-density distribution. The electron beam is positioned successively at each specific location, dwelled for an adjustable time interval (20×10^{-6} s minimum) and then translated at high speed to the next coordinate point. Patterns that conform to the contours of each surface to be hardened are thus produced. These patterns are repeated throughout the entire heat treating cycle, typically at a rate of 500 times per second.

Electron-beam power is also controlled as a function of time. Initial power density is sufficient to heat the surface rapidly to an austenitizing temperature just below the melting point of the ferrous alloy being hardened. A thermal gradient is generated from the surface to the interior, and the depth of the gradient increases as a function of time. As the thermal gradient gets deeper, the surface temperature tends to rise. To keep the surface temperature just below the melting point, the temperature rise is countered by decreasing the surface power density with respect to time, using the computer capability to adjust electron-beam power rapidly.

The austenitizing temperature is held just below the melting point of the alloy to take advantage of rapid carbon migration. Typical minimum times required for austenitization of the structure and dissolution of the iron carbide (Fe$_3$C) are on the order of $\frac{1}{3}$ to $\frac{1}{2}$ s. In this length of time, material to a depth of only about 0.25 mm (0.010 in.) can be heated sufficiently so that it will be hardened when the electron-beam en-

ergy is turned off and the workpiece allowed to self-quench. Soaking for longer times prepares deeper layers for hardening.

When the desired depth of carbon-bearing austenite is prepared, the electron-beam power is turned off. Heat flow to the interior of the metal continues until the temperature becomes equalized throughout the workpiece. The rate usually is high enough to exceed the critical cooling rate and to convert the austenite to hard martensite. A typical relationship among these various factors is shown in Fig. 5.

Raster-Pattern Control. Although part configuration must be compatible with the electron-beam process, as indicated in Fig. 2, flexibility is possible through use of static or traveling raster patterns.

Static patterns are raster patterns whose geometric shapes may be varied infinitely. The patterns are displayed on the part with no relative movement between the part and the pattern.

Traveling patterns are used when heat treating of large areas is required. A static pattern of some geometric shape is displayed on the part, and motion is applied either to the raster pattern or to the part.

Control of Case Depth

Hardening depth is simply a function of time. Because the heat applied to the surface travels through the material by conduction, the longer the surface is maintained at a temperature above the critical value, the deeper the heat will penetrate, and thus the greater will be

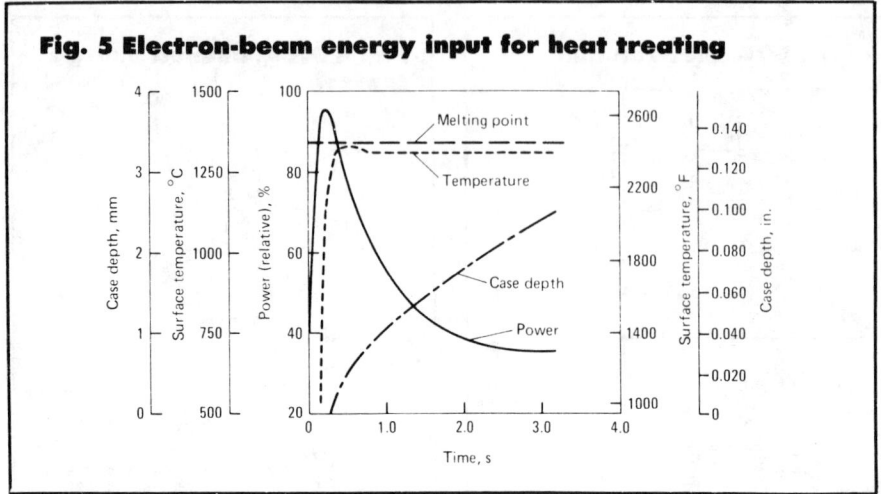

Fig. 5 Electron-beam energy input for heat treating

the case depth. Moreover, if the temperature of the surface is raised fast enough and if the temperature is sufficiently high to enable the carbon to go quickly into solution, it is possible to obtain a reasonably deep case with very little heat-affected zone.

Quench Media

To complete the surface-hardening process, the area heated must be quenched rapidly to develop full hardness. This limits the majority of electron-beam applications to those materials that are self-quenching. There must be sufficient mass beneath the surface being hardened to conduct the heat from the surface to the core of the workpiece rapidly. The transfer of heat must be rapid enough to prevent formation of pearlite and to reduce the surface temperature below the M_s (martensite-start) temperature before bainite transformation begins. For carbon steels, eight units of mass generally are required for each unit of case at the desired hardness.

In such a self-quenching process, quenching of the austenite occurs from the inside to the outside, which means that the inner layers harden first. Be-cause transformation of ductile austenite to hard martensite causes expansion, the inner layer expands and plastically extends the surface layers while the surface is still austenitic. When the surface material finally transforms and expands, it generates moderate biaxial compressive forces in conjunction with the hard, virtually unyielding underlayer. This is an ideal residual-stress condition for fatigue resistance, similar to the benefits derived from controlled shot peening.

Advantages of Electron-Beam Heat Treating

Electron-beam heat treating is a precise process that allows the heat treater control over every aspect of the flow of energy into a workpiece for selective surface hardening.

Low Energy Usage. Because the flow of energy is very rapid and the area being heated is held to a selective minimum, the core and the surface area not heat treated will not be adversely affected.

Minimal Part Distortion. When energy flow is controlled and rapid and is restricted to highly selective areas, minimum distortion may be realized.

Small Heat-Affected Zone. The combination of high energy, rapid flow of energy and rapid cooling generally yields very small heat-affected zones.

Flexible Surface Hardening. By moving the pattern over the part or moving the part beneath the pattern, large areas may be heat treated in short periods of time. Depending on case depth, these traveling speeds may be in excess of 150 cm/min (60 in./min).

Compressive Surface Stress. Due to the self-quench phenomenon, ductile austenite forms hard martensite from the subsurface to the surface. When the volume of the surface layer expands during formation of martensite, this expansion is resisted by the adjacent unyielding layer.

Maximum Hardness. Case hardness attained by means of the electron-beam process follows textbook relationships of hardness and carbon content, as illustrated in Fig. 1. The ultimate hardness achieved by this process may be slightly greater than that obtained by conventional heat treating methods because of the high efficiency of self-quenching, which is more effective and much faster than quenching with external quenchants.

Short Heat Treating Time. An accelerated electron beam delivers high-level energy to the surface being modified. Due to this high energy level and to the relatively low thermal conductivities of ferrous materials, heating times for electron-beam heat treating normally are on the order of 0.5 to 2.5 s for static patterns. For most applications, heating times range from 0.5 to 1.0 s.

Minimal Workpiece Oxidation. Carbon-bearing materials heat treated in an electron-beam system may be protected in a vacuum environment and thus may be treated without surface discoloration.

SECTION IV
Powder Metal Parts

Sintering

By the ASM Committee on Sintering*

SINTERING is the process by which loose or compressed powders are bonded by heating at temperatures below the melting points of the major constituents. Densification may or may not occur. If powders of two or more different metals are heated together to a sufficiently high temperature, alloying may take place simultaneously with sintering. Sometimes a liquid phase forms and assists in consolidation, or a compact may be sintered for a short time and then infiltrated with a molten metal of lower melting point.

The processes operative in sintering include vapor and/or liquid transport, diffusion, and plastic flow. The predominant process is diffusion. A preform, usually a "green" compact, becomes a sintered part in a series of continuous stages. As the temperature increases, the interparticle contacts increase in size and strength. As diffusion progresses, neck formation occurs, which causes the pores to become rounded. With continued sintering, densification increases as a result of volume diffusion. Ultimately, pores coalesce. The total sintering process causes strength, ductility, and thermal and electrical conductivity to increase. With increased temperature, approaching the melting point, porosity becomes isolated and theoretical density is approached asymptotically.

A common production method used to achieve higher densities is re-pressing and resintering. Re-pressing closes up the larger pores mechanically, and fresh bonds are formed during resintering. The improvement obtainable by these operations is illustrated by the following example.

Example. Re-pressing and Resintering for Increased Density (Fig. 1). Densities of compacts 25 mm (1 in.) in diameter and 25 mm high were measured after pressing at pressures varying from 4.2 to 8.4 tonnes/cm^2 (1 tonne = 1 Mg, or 1 metric ton), or 30 to 60 tons/in.^{2}, and sintering for 1 h in dissociated ammonia at 1120 °C (2050 °F). The relation of compacting pressure and density, after sintering, is shown graphically in Fig. 1. Also shown in Fig. 1 is the increase in density that is effected by re-pressing at 7.0 tonnes/cm^2 (50 tons/in.^{2}) and resintering for 1 h in dissociated ammonia at 1120 °C (2050 °F).

Recent technology has been developed that utilizes high-temperature sintering to achieve densification as well. If near-theoretical density is required, other processes, such as hot pressing or re-pressing, hot isostatic pressing, hot forging and super solidus sintering, are employed.

Sintering Furnaces

The burn-off chamber of a sintering furnace used for sintering ferrous preforms is usually controlled to heat the preforms to temperatures from 500 to 800 °C (930 to 1470 °F). It is important that all lubricants, including zinc stearate, stearic acid and waxes, be volatilized and expelled from the furnace before the preforms enter the high-temperature section, and both the flow of gas and the time of heating should be sufficient to ensure that this is done. If lubricants pass into the sintering section, they will be decomposed, and the liberated products may adversely affect the sintering process, the parts and the furnace.

The sintering section of the furnace may be refractory lined or fully muffled. In small furnaces it is cheaper to use a muffle to obtain gas tightness. In larger furnaces it is more economical to make the shell gastight and eliminate the muffle. Production sintering furnaces are supplied throughout with a protective atmosphere and are divided into: (a) a burn-off section, which serves

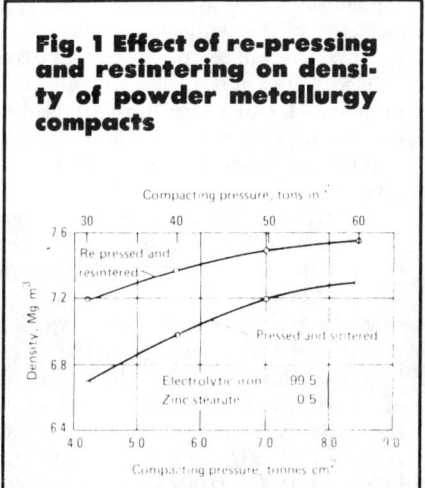

Fig. 1 Effect of re-pressing and resintering on density of powder metallurgy compacts

*Kenneth H. Moyer, *Chairman*, Product Development Engineer, Hoeganaes Corp.; J. Howard Beck, President, BTU Engineering Corp.; A. J. Craig, Jr., Chief Metallurgist, Homelite; Donald Dyke, Engineering Manager, Sintered Specialties; Donald Grendon, Sales Engineer, Drever Co.; Erhard Klar, Manager Research and Development, Glidden Metals, SCM Corp.; George Otto, Maytag Co.; Thomas Sibley, Manager, P/M Applications, Air Products & Chemicals, Inc.; Sang-Kee Suh, Senior Research Engineer, Ford Motor Co.

also for preheating; (b) a high-temperature heating section; and (c) a cooling section. These sections have typical length ratios of 1:1:2 or 1:1:3.

The high-temperature heating section must be long enough to allow sufficient time for the preforms to heat up to temperature and enough soak time at temperature for adequate sintering. Multiple-control zones are used to obtain suitable temperature gradients.

Large sintering furnaces usually are constructed with a gastight shell and electrical-resistance heating elements or gas-fired radiant tubes exposed to the heating chambers. Muffles generally are not used in larger furnaces because they are expensive to purchase and to maintain, and also introduce thermal losses. Muffle construction, however, is widely used when the dew point of the atmosphere must be kept below about −40 °C (−40 °F). By using high-purity alumina refractories, muffle-free furnaces with exposed molybdenum heating elements have been operated successfully at low dew points. Full muffle furnaces have one further advantage: they purge faster, because there is no porous brickwork in the chamber and because less purge gas is required.

The cooling section often begins with a short, insulated zone in which the preforms cool slowly enough to avoid thermal shock and to allow for carbon restoration. This is followed by a cooling section that may be a long, water-jacketed extension or a shorter, forced-convection cooling system. Automatic control of the temperature of the cooling water is most desirable. If the temperature of the walls should fall below the dew point of the protective gas, condensed water may collect on the workpieces and cause staining. If cooling time is insufficient, the sintered parts will oxidize when they emerge into the air.

Preforms commonly are conveyed through the furnace by mesh belts, roller hearths, pusher mechanisms or walking beams.

Mesh-belt conveyors typically handle a nominal loading of 480 Pa (10 lb/ft^2) at 1120 °C (2050 °F). Stretching of the alloy belt limits the length of the furnace and the size of the furnace load. It is desirable to keep the temperature below 1150 °C (2100 °F) when using a mesh belt. Because each end of this type of furnace usually remains open during operation, consumption of the protective gas is high and ample gas capacity must be provided. Flame curtains or nitrogen baffles may be used to prevent oxygen intrusion. A variable-speed drive gives flexibility in adjusting time and temperature cycles.

Roller-hearth furnaces are similar in arrangement to mesh-belt furnaces except that, in place of the belt, a series of driven rollers is fitted along the entire length of the hearth. These rollers are spaced to support and carry trays loaded with preforms. Maximum operating temperature is limited by the properties of the alloy rolls and usually is about 1120 °C (2100 °F). Depending on roll spacing, loads four to seven times greater than can be handled by a mesh-belt furnace of equal length can be conveyed on a roller hearth. Furnace doors are provided and are opened only to charge or discharge the trays. Thus, consumption of protective gas is less, and heat losses are lower, than for mesh-belt furnaces, because the atmosphere flow may be diminished.

Pusher-type furnaces are suitable for sintering of preforms that are too heavy to be carried by a mesh belt or that require sintering temperatures greater than 1150 °C (2100 °F). With this type of equipment, preforms are fed into the furnace on trays, which are advanced by mechanical pushers. Alternately, for small batches, the trays may be pushed through manually.

Walking-beam furnaces are used for high-temperature sintering. These furnaces can move heavy loads at high temperatures. The moving mechanism is capable of four basic motions: up, forward, down and reverse. The up motion lifts the tray from the shelves; the forward motion advances the trays; the lowering motion unloads the trays from the beam, and the reverse motion returns the empty beam to the original position. Normal operation involves temperatures up to 1400 °C (2550 °F). Doors at both ends are closed except during loading and unloading, thus reducing gas consumption.

Vacuum furnaces are also used for high-temperature sintering. They may be mechanized to provide continuous production of parts. Parts are protected by the absence of reactive gases, and furnaces may be backfilled with an inert or protective gas. Burn-off generally is accomplished in a separate conventional furnace. Heating is done by means of resistance heating elements. Cooling can be accelerated by backfilling with a nonreactive gas.

Sintering Atmospheres

Protective atmospheres are used in powder metallurgy (a) to prevent oxidation and reduce oxides, (b) to control carbon contents of iron and iron alloy preforms, and (c) to flush volatilized lubricants from the furnace.

Oxidation and decarburization of iron preforms are caused by oxygen, water vapor and carbon dioxide when present in excess proportions with respect to hydrogen and carbon monoxide contents. Iron oxides are reduced by hydrogen, carbon monoxide and carbon. Carburization is caused by carbon monoxide and by hydrocarbons such as methane.

Copper and bronze preforms are susceptible to general oxidation and to scaling or discoloration by oxygen. These compacts are not adversely affected by hydrogen, carbon monoxide or carbon. Selective attack on zinc in brass compacts is caused by carbon dioxide, oxygen, sulfur and water vapor.

Vacuum is used mainly for sintering preforms of stainless steels and tool steels; soft magnetic materials; and refractory metals such as tantalum, titanium, zirconium and uranium—all of which react with most of the usual protective gases, including hydrogen. Vacuum is also being used to an increasing extent for sintering of conventional ferrous materials at high temperatures.

When the moisture content of any atmosphere must be kept very low (as in sintering of alloys containing chromium), the furnace must be operated and maintained with special care, to eliminate all leakage or back-diffusion of air that would contaminate the furnace atmosphere. One factor is often overlooked; the dew point of the gas fed into the furnace may be different from the moisture content of the gas in contact with the workpieces, owing to higher oxygen content within the porous compacts. Under vacuum, care must be exercised to prevent the pressure from dropping below the vapor pressure of the alloy constituents, which can result in depletion.

The atmospheres most commonly used for sintering are: hydrogen, dissociated ammonia, nitrogen-base exothermic gas, purified rich exothermic gas, endothermic gas, and vacuum. Each is discussed individually in the sections that follow.

Table 1   Characteristics of sintering atmospheres for powder metallurgy products

Atmosphere	Typical dew point, °C	Gas characteristic(a) at sintering temperature for:												Carbon steel	Stainless steel	Relative cost per unit volume(b)
		Al	Cu	Brass	Bronze	Ni	Ag	Mo	W	Fe	Fe-Cu	Fe-C	Fe-Cu-C			
Hydrogen:																
Liquid	−75	R	R	R	R	R	R	Y	Y	R	R	...	...	...	Y	9–20
Bulk gas	−70	R	R	R	R	R	R	Y	Y	R	R	...	...	...	Y	20–35
Steam-methane	−40 to −50	R	R	R	R	R	R	Y	Y	R	R	...	...	...	Y	9–14
Nitrogen base, with:																
Endothermic enrichment	−20 to −10	X	R	R	R	R	R	...	C2	C2	C2	C2	C2	C2	...	1.5–6(e)
Hydrogen enrichment	−70	Y(d)	R	R	R	R	R	R	R	N	N	N	N	N	R	1.7–7(e)
Methanol enrichment	−20 to −10	X	R	R	R	R	R	...	C2	C2	C2	C2	C2	C2	...	1.6–6.5(e)
Ammonia-base:																
Dissociated NH₃	−40 to −50	R	R	R	R	R	R	R	R	N	N	...	...	N	R	3.3–7.2
Burned NH₃, rich	+20 to +30(c)	R	R	...	R	R	R	...	...	D3	D3	...	...	...	...	2.3–5.1
Exothermic gas:																
Rich, saturated	+20 to +30	X	R	...	R	R	R	...	...	D3	D3	...	...	...	...	1
Medium rich, saturated	+20 to +30	X	...	...	R	R	R	...	...	...	...	...	...	...	...	0.9
Purified exothermic gas:																
Rich	−40	R	R	R	R	R	R	...	...	C1	C1	C1	C1	C1	...	1.5–2.2
Medium rich	−40	R	R	R	R	R	R	...	...	C1	C1	C1	C1	C1	...	1.5–2.2
Endothermic gas:																
Rich, dry	−20 to −10	X	R	R	R	R	R	...	C3	C3	C3	C3	C3	C3	...	1.6–3.2
Fairly rich, dry	−5 to 0	X	R	R	R	R	R	...	C2	C2	C2	C2	C2	C2	...	1.5–3.1
Medium rich, saturated	+20 to +30	X	R	...	R	R	R	...	...	D1	...	...	...	...	...	1.5–3
Lean, saturated	−20 to −30	X	R	...	R	R	R	...	...	D3	D3	...	...	...	...	1.5–2.5

(a) R, reducing; Y, recommended; C1, mildly carburizing; C2, carburizing; C3, strongly carburizing; N, neither carburizing nor decarburizing; X, not recommended; D1, mildly decarburized; D2, decarburizing; D3, strongly decarburizing. (b) Costs are approximate and relative to rich, saturated exothermic gas. (c) Dew point may be reduced by refrigeration or by absorbent-tower dehydration. (d) Nitrogen with no enriching gas is recommended. (e) Price range includes on-site plant production or liquid tanker deliveries.

The characteristics and applications of gas atmospheres for sintering are summarized in Table 1.

Hydrogen provided as a liquid or in bulk gaseous trailers is the most economical source for most sintering facilities. For volumes larger than 170 000 to 340 000 m³ (6 to 12 million ft³) per month, hydrogen may be produced at lower cost by steam methane reforming.

The explosiveness of hydrogen-oxygen mixtures demands that hydrogen be handled with extreme care. Its high thermal conductivity (seven times that of air) helps in increasing the rate of heat transfer in the heating and cooling chambers. Its low density causes it to diffuse rapidly outward and can allow back-diffusion of air through small openings and cracks, which may result in contamination of the atmosphere.

Hydrogen delivered as a liquid typically contains less than 0.001% impurities, with an oxygen content of less than 0.0002%. Bulk gaseous hydrogen usually has less than 0.05% impurities and less than 0.0005% oxygen. This high purity makes hydrogen suitable for sintering of stainless steels, carbides, tungsten and other refractory metals, magnetic materials, superalloys and other metals in which high reducing potential and no nitriding is desired.

Dissociated Ammonia. Cracked or dissociated ammonia is made by passing ammonia gas (from large cylinders or tanks) over a heated catalyst, and consists of a mixture of 75% hydrogen and 25% nitrogen by volume. It is dry and can be used as a substitute for pure hydrogen in nearly all sintering applications, including stainless steel, iron, brass, copper and tungsten. Its use, particularly for sintering molybdenum and ferrous materials, is sometimes avoided because of the danger of nitriding by traces of undissociated ammonia, which are nearly always present and which dissociate on contact with hot metal. This residual ammonia may be removed almost completely by passing the gas through water (and subsequently drying it), through activated alumina or through a molecular sieve.

Exothermic gas is made by burning natural gas, propane or a similar hydrocarbon gas in a refractory-lined combustion chamber with controlled amounts of air.

Rich or medium-rich exothermic gases are the atmospheres most commonly used for sintering. The richest gas is formed from a 6-to-1 ratio of air to natural gas and contains about 14% hydrogen, 10% carbon monoxide, 1% methane, 5% carbon dioxide and 70% nitrogen. It has a dew point approximately 5 °C (9 °F) above that of the cooling water and is useful for sintering preforms of copper, bronze, silver, iron, and iron-copper. It usually is strongly decarburizing at the sintering temperature to ferrous metals. A medium-rich gas, made by reacting air with natural gas in a ratio of 6.75 to 1, has been used for sintering nonferrous compacts. The principal advantages of exothermic gas are low flammability and low cost (it is the most economical gas available). Exothermic gas also may be usefully employed where removal of lubricant is desired.

Like all gases derived from hydrocarbons, exothermic gas may have a carburizing or a decarburizing effect on ferrous parts, depending on the carbon content of the work and on temperature.

Purified Rich Exothermic Gas. By removal of water and carbon dioxide from exothermic gas, a stable and mildly reactive gas with a dew point below −45 °C (−50 °F) may be produced. This gas is called purified rich exothermic gas. It has virtually no carburizing or decarburizing effect on iron-graphite compacts. Purified rich exothermic gas is used for sintering preforms of iron, iron-copper, iron-carbon and iron-copper-carbon; and also for copper infiltration of iron or iron-carbon compacts.

Endothermic gas is made in an externally heated catalytic chamber by reacting a hydrocarbon with an amount of air insufficient to support combustion. The ratio of air to natural gas can be varied from 4.5:1 to 2.4:1, depending on the desired carbon potential of the atmosphere. The most common air:gas ratio is approximately 2.4:1, which produces gas of the following composition: 20% CO, 38% H_2, traces of CO_2 and CH_4, rem N_2. Dew point for practical operation is between -5 and $+5$ °C (23 and 41 °F). Traditionally, endothermic generators have provided low-cost carbon-controlled atmospheres. Depending on the air:gas ratio, endothermic gas can be either carburizing or decarburizing; therefore, it is used to sinter ferrous-base preforms and most nonferrous preforms.

Nitrogen-base atmospheres are made by mixing cryogenically produced nitrogen with one of several enriching gases. Liquid storage vessels are the most economical source for most sintering facilities. For larger requirements, on-site plants provide gaseous nitrogen at lower cost.

The atmosphere generally consists of 75 to 95% nitrogen, so that advantage may be taken of nitrogen's low dew point. Gaseous hydrogen, dissociated ammonia, endothermic gas or methanol is used for enrichment. Hydrogen and dissociated ammonia provide reducing atmospheres with neutral carbon potentials. Endothermic gas and methanol produce atmospheres with medium carbon potentials. Methanol is provided as a liquid that dissociates in the furnace to form essentially carbon monoxide and hydrogen. Hydrocarbons can be added to any of the atmospheres to increase carbon potential. With the proper enrichment gas, nitrogen-base atmospheres can be used for sintering and infiltrating iron, iron-carbon, iron-copper and iron-copper-carbon compacts, and for sintering nonferrous compacts. Stainless steels and refractory metals can be sintered when nitriding is not critical. For some applications, such as soft magnetic materials or highly corrosion-resistant stainless steels and some refractory metals, ni-

Table 2 Typical sintering temperatures and holding times for various metals and prealloyed powders

| Material | Temperature | | Time, min |
	°C	°F	
Aluminum	570–650	1060–1200	5–25
Bronze	760–870	1400–1600	10–20
Brass	760–940	1400–1725	10–45
Copper	900–1010	1650–1850	12–45
Iron, iron alloys	1010–1425	1850–2600	8–45
Nickel	1010–1150	1850–2100	30–45
Stainless steel	1095–1315	2000–2400	30–60
Alnico magnets	1205–1300	2200–2375	120–150
Ferrites	1205–1480	2200–2700	10–30
90% tungsten, 6% nickel, 4% copper	1345–1595	2450–2900	12–20
Tungsten carbide	1425–1480	2600–2700	20–30
Molybdenum	2050	3730	120 approx
Tungsten	2345	4250	480 approx
Tantalum	2400	4350	480 approx

trogen absorption occurs and should be avoided. The principal advantages of nitrogen-base atmospheres are safety and elimination of generating equipment.

Vacuum is finding increasing application as an environment for protection of materials that are reactive at elevated temperatures. It is beneficial for sintering soft magnetic preforms, stainless steels, tantalum, titanium, uranium, zirconium, refractory metals, and superalloys that react with hydrogen, carbon or nitrogen. Care must be taken to prevent alloy depletion, which can occur if the vapor pressure of an element is higher than the pressure in the vacuum vessel. Further, backfilling with reactive gases for cooling purposes can cause detrimental reactions.

Sintering Practice

Time and temperature cycles must be carefully chosen to develop the properties required in sintered preforms. The major causes of low density or low strength are insufficient temperature or insufficient time at temperature. Of the two, low temperature usually is the more critical condition. When the preform is composed of a mixture of two or more constituents, longer time at the sintering temperature is required for diffusion than is required for sintering a similar prealloyed preform.

Invariably, the optimum values of time and temperature for sintering any given compact are determined empirically. Within practical limits, higher sintering temperatures result in better

properties and are more effective than longer times. The limits of temperature are defined by the available equipment. Temperature is also limited by the increased degree of difficulty of controlling dimensions at higher temperatures. As sintering is prolonged, the measurable increase in density continues at a lower rate, until a point is reached beyond which the diminishing returns do not justify the expense of further sintering. The ranges of time and temperatures most widely used are given in Table 2.

Alloying in Iron-Base Compacts. Carbon is the most common alloying addition to iron. It usually is added, as graphite, to the original blend of powders so that alloying will occur during sintering, but it may be added from a carburizing furnace atmosphere. The reactions that occur between iron and graphite and among iron, graphite and the furnace atmosphere are fairly rapid at the sintering temperature. For this reason, it is necessary that the atmosphere be maintained at the desired carbon potential or that restoration of the carbon potential be made prior to cooling. Carbon gradients may result from improper control of carbon from the atmosphere. Final carbon content is determined by the amount of graphite added, by the carbon potential maintained and by control of restoration.

Copper often is alloyed with iron in the sintering furnace. It has a solid solubility in iron of 8 to 9% at normal sintering temperatures. As cooling occurs, some precipitation hardening is likely, depending on cooling rate. Additions of copper cause compacts to expand, which often makes it more difficult to control dimensions. For consistent compact size, it is necessary to maintain close control over raw materials and over sintering temperature, atmosphere and time.

Carbon is also alloyed with iron copper preforms. The solubility of copper in iron is lowered by the addition of carbon, and the presence of copper decreases the sensitivity of compacts to variations in the sintering atmosphere. Iron-copper-carbon alloys have good mechanical properties, and dimensions of compacts made from these materials are generally easier to control than are those of similar compacts made from iron and copper.

To obtain increased strength and density, postsintering operations such as re-pressing and resintering may be employed. When the part is to be re-pressed and resintered, it generally is not given a thorough sintering until the last pass through the furnace. The initial pass (or passes, when resintering is done more than once) is done mainly to anneal the compact and burn off the lubricant. It is desirable that alloying be kept to a minimum during initial sintering so that the compact will not be strengthened sufficiently to resist compaction in subsequent re-pressing operations. As the density of the compact approaches that of a solid material and interconnecting porosity disappears, the reaction of the compact to the atmosphere also approaches that of solid material.

Prealloyed powders provide a means of obtaining preforms with structures more uniform than those of preforms produced by diffusing particles of the constituent metals. Additional alloying may be achieved through mixing of elemental or other

Table 3 Nominal compositions and sintering temperatures for some prealloyed metal powders

Composition	Sintering temperature			
	Hydrogen-base atmosphere		Nitrogen-base atmosphere	
	°C	°F	°C	°F
Copper-base alloys				
90 Cu, 0.5 P, 9.5 Zn	940	1725	...	...
90 Cu, 1.5 Pb, 8.5 Zn	900	1650	900	1650
90 Cu, 10 Zn	900	1650	900	1650
85 Cu, 15 Zn	900	1650	900	1650
78.5 Cu, 1.5 Pb, 0.3 P, 19.7 Zn...........	900	1650	...	...
78.5 Cu, 1.5 Pb, 20 Zn	900	1650	900	1650
70 Cu, 0.3 P, 29.7 Zn	870	1600	...	...
70 Cu, 30 Zn	900	1650	870	1600
60 Cu, 40 Zn	...	...	815	1500
64 Cu, 18 Ni, 18 Zn	980	1800	980	1800
64 Cu, 1.5 Pb, 18 Ni, 16.5 Zn	870	1600	940	1725
70 Cu, 10 Ni, 20 Zn	940	1725	...	...
Iron-base alloys				
17 Cr, 12 Ni, 2.5 Mo, rem Fe	1210	2210	...	...
12.5 Cr, rem Fe	1210	2210	...	...
50 Fe, 50 Ni..........................	1175	2150	...	...
97 Fe, 3 Si	1260	2300	...	...
1.8 Ni, 0.5 Mo, add 0.6 graphite, rem Fe........................	1135	2075	...	...
1.75 Ni, 1.5 Cu, 0.5 Mo, rem Fe.........	1110	2030	...	...
0.4 Ni, 0.7 Mo, 2 Cu, rem Fe	1120	2050	...	...
1 Mo, 0.4 Mn, rem Fe..................	1120	2050	...	...

intermetallic powders to achieve more complex alloys. Table 3 lists nominal compositions of some prealloyed metal powders, and gives sintering temperatures for these materials.

Liquid-Phase Sintering. The oldest and most important example of liquid-phase sintering is production of 90Cu-10Sn bronze for self-lubricated bearings. At the sintering temperature, the elemental tin is liquid, which allows lower sintering temperatures (815 to 895 °C; 1500 to 1640 °F), and shorter sintering times (5 to 10 min) than those required for prealloyed

bronze. Powder producers supply lubricated bronze premixes with graded particle size distributions, which allow better control of dimensional change (growth) during sintering. The self-lubricated bearings are sized after sintering for further control of dimensions.

Liquid-phase sintering may be successfully employed for other alloys as well. These include iron-copper, iron-phosphorus, aluminum, iron-sulfur, tungsten carbide–nickel, tungsten carbide–cobalt and tungsten carbide–iron alloys.

Guidelines for heat treating powdered metal parts

How are P/M parts hardened? This article explains effects of part density, alloys, and quench media, offers guidelines for in-house and commercial processing, and highlights new technology.

by HOWARD FERGUSON

Heat treatment of P/M parts has in the past been considered a secondary operation. The nature of hardening was not well understood; and in most cases its application was to provide a file-hard surface for wear resistance with little regard for other properties. With the advent of high-compressibility powders, heat treatment is being re-evaluated in a more serious manner.

As density approaches theoretical, compressive residual stress patterns which approximate those in wrought steels are created. This has been shown to improve such properties as fatigue strength, impact strength, and hardenability.

Density

Although many variables affect the heat treat response of P/M parts, none has as strong an influence as density. The response of a metal to heat treatment depends on its heat conductivity, which in turn is dependent upon surface area.

Powder metal parts have both external and internal surface areas due to contained porosity. By going higher in density, we are essentially decreasing internal surface area and changing the heat conductivity of the material. This characteristic can be correlated with the weight-to-volume ratio of a metal.

In wrought materials where we have the highest weight-to-volume ratio, heating and cooling rates are fast. This is why wrought steels can be satisfactorily hardened at relative-

H. Ferguson is manager of the Lansing (Michigan) Division of Lindberg Heat Treating Co.

ly low carbon contents.

In low-density P/M materials with a relatively low weight-to-volume ratio, slow heat dissipation inhibits hardenability, and slack quenching or shallow hardening results. This is why in many instances a low-density P/M part of eutectoid carbon composition tends to form a case when quenched. Low-hardenability allows pearlite to transform at a certain depth, which etches to a different color upon polishing. This appears as a case, although carbon content re-

mains the same throughout the cross-section.

Case depth

To determine the effect of density on case depth, slugs with a 1-inch (25.4-mm) outside diameter were pressed at densities ranging from 6.0 to 6.8 g/cm³ from pure iron powder. A slug was machined from SAE C1018 bar stock to show a comparison with wrought steel. The slugs were carbonitrided at 1600°F (870°C) for 30 minutes and oil quenched.

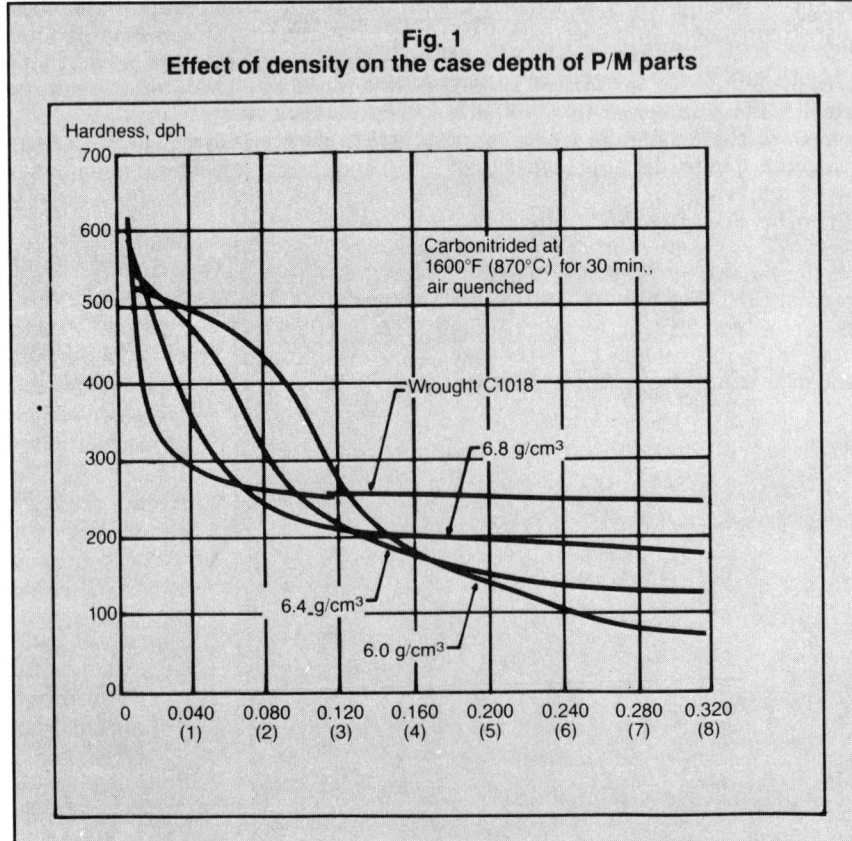

**Fig. 1
Effect of density on the case depth of P/M parts**

Results for carbonitrided P/M parts compared with results obtained with wrought steel.

Reprinted from Heat Treating, May 1984, 34-39, © 1984 Fairchild Publications, with permission from the author

Cross sections were mounted and microhardness traverses were made to determine effective case depth **(Fig. 1)**.

As can be seen, the lower we go in density, the greater the case depth becomes. This is due to the greater amount of porosity, which allows for deeper gas penetration. However, surface hardness decreases with de-creasing density. This occurs because of the poor heat conductivity of the material. Pores act as insulators that inhibit heat dissipation. You will notice, however, that core hardness increased with density. In this area, the carbon level is essentially the same, and this shows only the effect of density.

By studying this graph, you can see why it would be difficult to achieve maximum physical properties in low-density parts. As stated before, the purpose of case hardening is to get a hard, wear-resistant surface and a ductile core. With the deep carbon penetration at the low densities, the core properties would be approxi-mately the same as the case properties with the normal cross sections in P/M parts today.

With a fast quenchant, enough sur-face hardness can be gained in the low-density parts for wear resistance; but for improved fatigue and impact life, the higher densities have to be used to provide this sharp transition zone between case and core.

This fact becomes apparent when we look at the microstructures of heat treated slugs with increasing densities **(Fig. 2)**. The slugs were carbonitrid-ed at 1600°F (870°C) for 30 minutes, oil quenched, and subsequently drawn at 400°F (205°C) for 1 hour. Carbon has penetrated deeply into the low-density slugs.

Upon examining these micro-structures at higher magnifications, we noticed that the martensite formed in the low-density slugs was quite coarse, and some bainite had precipitated at the grain boundaries. This indicated that carbon potential was low and the critical cooling rate had been exceeded.

Also, the 7.0 g/cm³ slug showed a coarse martensite but no trans-formation products. Wrought C1018 had a fine, consistent martensitic structure. Rapid diffusion at the lower densities prevents the forma-tion of a steep carbon gradient necessary for effective case hard-ening. Without this gradient, we do not get the high compressive stresses on the surface of the part which gives the material its high fatigue and im-pact resistance.

From **Fig. 3** we see that as density increases, we have a proportional de-crease in carbon penetration up to 6.8 g/cm³. Above this density, carbon penetration drops off rapidly, which means that steeper carbon gradients exist and optimum case-core proper-ties can be attained.

Hardenability

In this test, slugs were made from a eutectoid carbon premix similar in chemistry to SAE 1080. Modified Jominy bars were pressed, sintered, and reheated to 1600°F (870°C) for

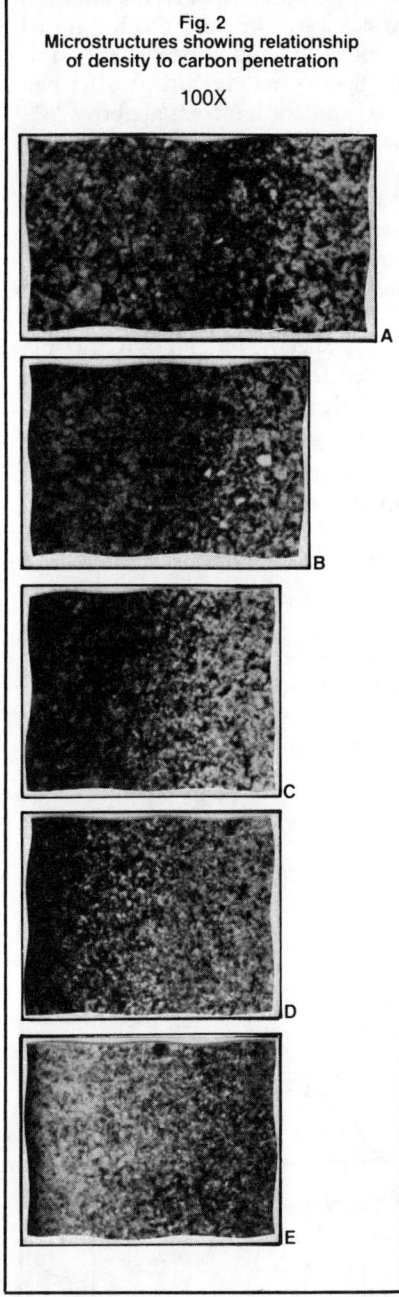

Fig. 2
Microstructures showing relationship of density to carbon penetration

100X

(A) density 6.0 g/cm³; carbon penetration 0.044-in. (1.1-mm); **(B)** density 6.4 g/cm³; car-bon penetration 0.030-in. (0.76-mm); **(C)** den-sity 6.8 g/cm³; carbon penetration 0.022-in. (0.55-mm); **(D)** density 7.0 g/cm³; carbon penetration 0.012-in. (0.30-mm); **(E)** density 7.87 g/cm³ (wrought) C1018; carbon penetra-tion 0.004-in. (0.10-mm).

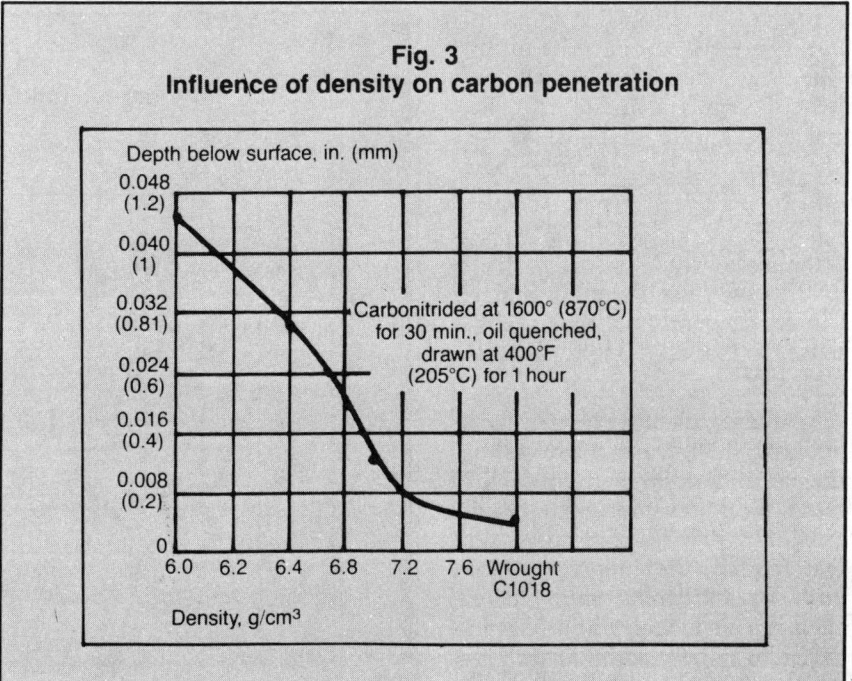

P/M parts are compared with wrought steel.

30 minutes in a nitrogen atmosphere to prevent oxidation.

Bars were then end-quenched in a water column on one flat of the bars. Hardness readings were taken every 0.1-inch (2.5 mm) from the quenched end and plotted on a graph. Results are compared with those for wrought C1080 drill rod (Fig. 4).

This graph dramatically illustrates the poor heat conductivity of low-density parts. As-quenched surface hardness decreases with density, and depth of hardening drops off. The wrought slug reached its maximum hardness at approximately 0.1-inch (2.5 mm) below the surface because the fast quench caused an excessive amount of austenite to be retained on the surface. This lowers the hardness of the material. Peak hardness denotes the depth at which the micro-structure is composed of 100% martensite.

Alloy content

We know that by combining carbon with iron we gain tensile strength in proportion to the added carbon content. In wrought steel, this ratio is maintained up to approximately 1.4% C.

In P/M steels, ultimate strength is reached at 0.8 to 0.9 C. Above this range a network of cementite forms at the grain boundaries, which embrittles the material and destroys its transverse rupture strength. As we add alloying agents, optimum carbon content is lowered. These elements tend to reduce the critical cooling rate of the steel, which moves the TTT curve to the right. As alloy content is increased, carbon content is decreased to maintain optimum physical properties.

In sintered metals, the two most common alloying agents are carbon and nickel. Addition of copper increases both hardness and tensile strength in the sintered condition. Heat treated, the copper increases depth of hardening but reduces toughness and elongation. The influence of copper additions in relation to combined carbon content is shown in Fig. 5.

Carbon-copper

As carbon content increases, strength increases proportionately up to a certain maximum, then begins to decrease with increasing carbon. In the straight iron-carbon material, this peak occurs at approximately 0.65% as heat treated. In the as-sintered condition, peak strength occurs at 0.8 C.

By adding copper, we increase our strength level and reduce the amount of combined carbon required for maximum strength. However, there is a maximum copper content over which strength begins to level off. This content seems to vary with density. At high densities—6.8 g/cm³ and above—2% Cu appears to be maximum. Over this, strength remains relatively constant; and dimensional change becomes erratic. In the 6.2-6.4 g/cm³ range, a 5% Cu content seems to be maximum for optimum heat treated properties.

Although blending copper powder with iron powder in a premix decreases toughness, the opposite is true when copper is infiltrated into a straight iron-carbon premix. Here, high density is obtained by the copper filling porosity, which provides high strength and hardness while maintaining good toughness upon heat treatment.

Nickel-carbon

Use of nickel in powder pre-mixes has increased greatly since the advent of low-cost, high-compressibility powder. It exerts more influence on heat treat properties than any other alloying agent blended with iron. It provides the maximum impact and fatigue strength besides increasing hardenability.

In Fig. 6 we can see that increasing nickel content significantly increases tensile strength. But this benefit begins to taper off when the addition exceeds 2%. The same result is noted in elongation and impact energy. Impact energy actually drops off when nickel content is increased above 2%. For this reason, most nickel steel used in production contains only 2%. Economics also dictates this maximum addition. However, when requirements call for maximum impact energy and fatigue strength, along with high surface hardness, our plant usually specifies a 2Ni-0.5C premix pressed to a 7.0-7.2 g/cm³ density, sintered, and heat treated.

There are some alloy premixes on the market that contain both nickel and copper. Small additions of nickel to copper enhance hardenability, but impact energy and elongation drop off considerably.

Fig. 7 (p. 38) shows the influence of copper and nickel additions on case depth upon carburizing at 1560°F (850°C) for 2 hours. By adding nickel to the copper premix, there

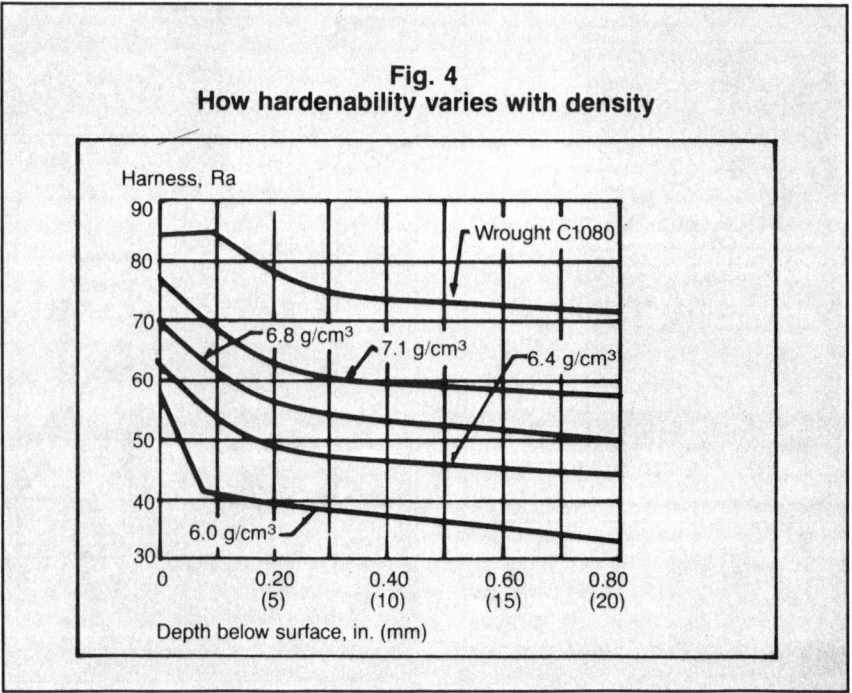

**Fig. 4
How hardenability varies with density**

Harness, Ra

Wrought C1080

6.8 g/cm³ 7.1 g/cm³ 6.4 g/cm³

6.0 g/cm³

Depth below surface, in. (mm)

Results for wrought C1080 steel compared with P/M parts

is practically no increase in surface hardness; however, case depth and hardenability increase significantly. Although copper reduces impact energy, many fabricators use it with nickel because it supposedly increases the diffusion rate of nickel into iron.

The new P/M materials are prealloy powders. Alloy materials are added to the melt prior to atomization. These powders are specifically designed for heat treat response. A commonly used prealloy corresponds to the SAE 4600 series in chemistry. The 4600 prealloy contains 2 Ni and 0.5 Mo.

The 4600 grade shows exceptionally good hardenability. It is being used in developing the hot forging process for P/M preforms. As the P/M forging technique is developed, it will be used in much greater quantities.

Processing variables

Temperature is the most important variable. It must be high enough to completely austenitize the material so it can be quenched out to a fully martensitic structure. In P/M, this temperature is also affected by density. As density decreases, the critical cooling rate increases because of poor heat conductivity.

Normally, to obtain a fully martensitic microstructure at a density below 6.2 g/cm³, a quenching temperature exceeding 1600°F (870°C) should be used.

Time at temperature is another factor to consider. Low-density parts require less time to reach heat than those of higher density because of the rapid heat penetration through the porosity.

Next to temperature, atmosphere is the most important consideration. Most heat treating is done in an endothermic atmosphere which contains approximately 20 CO, 35 H_2, 0.3 CH_4, balance N_2. Usually, other gases are added to adjust carbon potential to meet requirements for a part. Additive gases are normally methane and dissociated ammonia. Depending on the process, these additions range from 15% to 25% for methane and 2% to 5% for ammonia.

Carbonitriding is the most commonly used case hardening method because it lowers the critical cooling rate and increases hardenability. However, the amount of ammonia added is more critical for P/M parts than it is for wrought types. Contained porosity allows for much faster diffusion of the nitrogen atoms. This has the effect of retaining austenite upon quenching, which lowers surface hardness and causes dimensional instability. For this reason, carbonitriding is normally carried out within the 1500° to 1575°F (815° to 855°C) temperature range.

Carburizing is normally applied when deeper case depths are wanted. In this instance, only methane is added to the generator gas to increase carbon potential into the 0.80 to 0.90 range. For this reason, this process

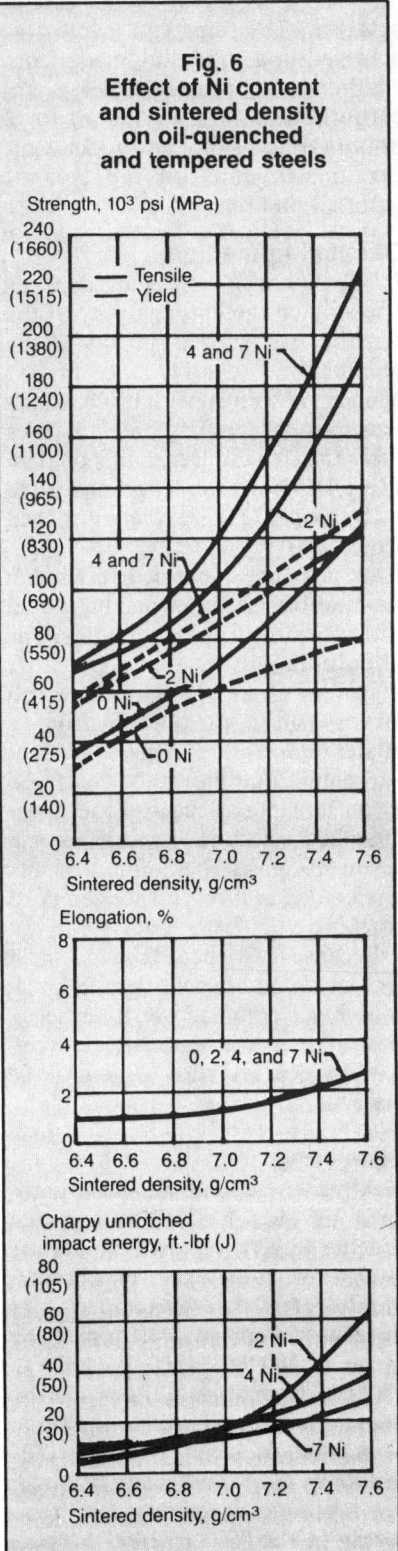

**Fig. 6
Effect of Ni content and sintered density on oil-quenched and tempered steels**

Specimens were quenched from 1600°F, then tempered for 30 minutes at 400°F.

Fig. 5
Effect of carbon-copper additions on heat treat strength

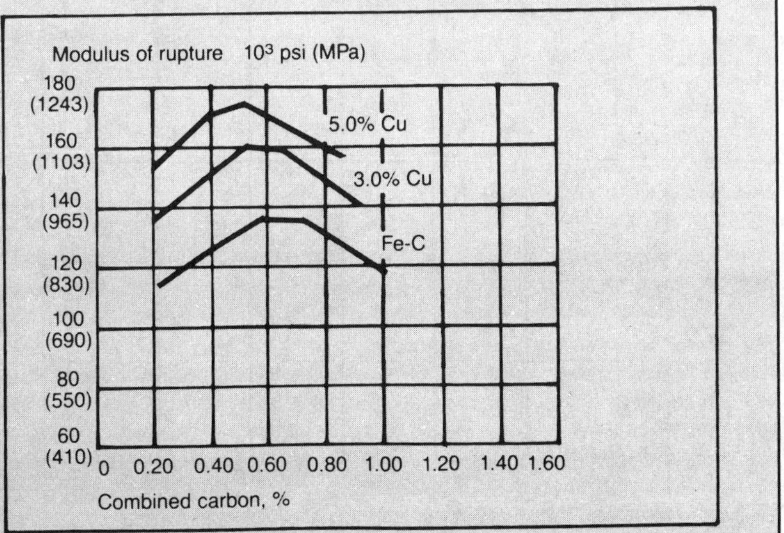

Graph shows influence of copper additions in relation to combined carbon content.

Source: Heat Treating, May 1984, 34-39

requires less time. With wrought SAE 4620, a case depth of 0.050-inch (1.28 mm) can be attained by carburizing at 1700°F (925°C) for 6 hours at temperature. The same case depth was achieved in a P/M slug made from 4600 pre-alloy pressed to 6.8 g/cm³ carburized at 1575°F (855°C) for 2 hours. Here again the proper temperature depends on the type of material and density.

Quenching medium

The extent of hardening depends not only on the hardenability of the steel but also on the severity of the quenchant. Ideally, the fastest quench is obtained with brine, which increases the cooling rate to three times that for oil. Depending on density, a P/M part can absorb up to 3% oil by weight (**Fig. 8**) when quenched from 1600°F (870°C).

A part that contained this much water or brine after quenching would corrode within a matter of hours. The disadvantage of oil, however, is that it is less severe than brine, which means you must quench from a higher temperature to get equivalent hardening. But this can be construed as an advantage because the lower thermal gradient between surface and center decreases the amount of distortion and reduces the possibility of cracking.

Because P/M heat treating must include an oil quench, desired hardenability is obtained by proper selection of the time-temperature cycle, part density, and alloy content of the material.

Tempering

After a steel part has been hardened by quenching, it is common practice to give it a low-temperature anneal for stress relief. This is commonly referred to as tempering or drawing. The normal tempering range is 300° to 1200°F (150° to 650°C), depending on the properties desired. If as-quenched hardness is to be maintained, a 300° to 375°F (150° to 190°C) temper is usually required. For optimum fatigue strength, tempering in the 800° to 1000°F range (425° to 540°C) is required.

As tempering temperature in-creases, hardness, tensile, and yield strength decrease proportionately, while impact energy and elongation increase.

Guidelines for in-house heat treating

The porous nature of P/M parts make them rather difficult to heat treat. Certain precautions are necessary to assure both the quality of the parts and the safety of the equipment and operator.

The following recommendations are suggested when processing parts in-house:

1. If parts are to be heat treated, do not rustproof after sinter as is

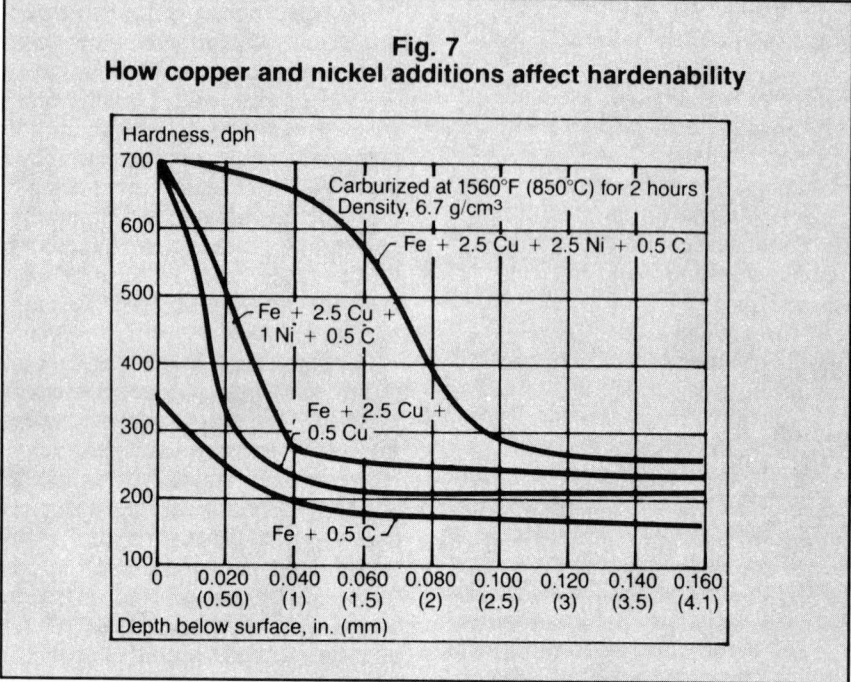

Fig. 7
How copper and nickel additions affect hardenability

When nickel is added to the copper premix, hardenability increases significantly.

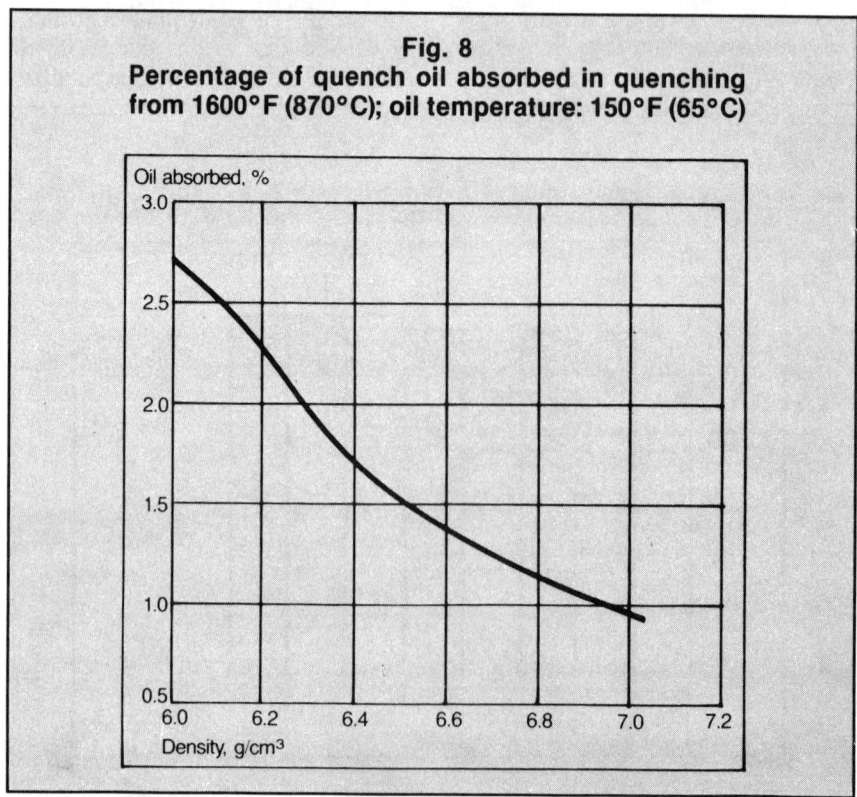

Fig. 8
Percentage of quench oil absorbed in quenching from 1600°F (870°C); oil temperature: 150°F (65°C)

Depending on density, a P/M part can absorb up to 3% oil by weight.

commonly done. These are oil-base products which will decompose upon heating and disrupt atmosphere control in the furnace.

2. If machining is to be done prior to heat treat, this same precaution applies.

3. Equip your tempering furnace with suitable exhaust capability to prevent oil vapor ignition. This can occur at temperatures above 400°F.

4. Adhere to precise process control specifications based on hardenability results of properly sintered parts. Parts should pass a sinter structure inspection prior to heat treat.

Guidelines in selection of commercial heat treating

1. Don't shop for lowest price.

2. Visit your heat treat vendor and evaluate process control.

3. Provide vendor with process to follow and request incoming inspection on sintered parts.

4. Review critical dimensions. Request dimensional inspection of each H. T. lot.

5. Explain apparent hardness measurement and request hardness checks to be made using one scale only. The published Rockwell conversion chart does not apply to P/M parts.

6. Review specific heat treat procedures such as:

A. Load size and weight if batch operation. My rule of thumb: 1 pound P/M parts per 4 gallons quench oil.

B. Belt speed and loading if continuous operation.

C. Quench temperature and time at heat.

D. Method of fixturing.

E. Transfer time to quench.

F. Oil temperature and agitation control.

G. Necessity of temper.

7. Request lab certification of all H. T. lots.

New alternatives to conventional heat treating

Most applications calling for heat treat are for improved wear resistance. Because of the notch sensitivity of hardened P/M parts, a uniform improvement in such properties as fatigue or impact resistance cannot be assured.

Conventional heat treatments such as carburizing, carbonitriding or neutral hardening rely on transformation to a martensitic microstructure for improved properties. These transformations occurring within a porous structure create significant dimensional distortions which in many applications require finish grinding to bring to print tolerance.

Over the past year, the Lansing Division of Lindberg Corp. has developed a low-temperature gas nitrocarburizing treatment which imparts a file hard surface to P/M parts using temperatures below 1100°F and without the need of a quenchant. This is a proprietary process we call "Lindure."

There are several significant advantages of this process when applied to P/M parts:

1. No structural transformations occur. Distortion is held to a minimum. Many parts evaluated show no change from sinter sizes.

2. Without the need of a quench, the total porosity is available for oil impregnation. This is important for parts designed to provide self-lubrication.

3. The process develops a hard compound layer of iron nitride on the surface and to a depth below the surface depending on percent of interconnected porosity.

4. This treatment allows the sintered structure to remain intact, which in many cases provides improved fatigue and impact properties compared to conventional hardening treatments.

An application of this process is shown in **Fig. 9.** This is a P/M rotor currently used in automotive oil pumps. Density of the part is 6.7 gm/cc.

The prior process was mold, sinter, oil quench from 1550°F and temper at 300°F. This caused distortion in the lobes, and a finish grind operation was necessary to return the lobe spacing to print tolerance of .0004 tir.

The Lindure process eliminated the need for finish grinding and improved wear resistance due to a lower coefficient of friction at the surface.

The microstructure of the surface and core of the P/M parts is shown in **Fig. 10.**

Another process being developed at Lindberg that is finding application in P/M parts is ion-nitriding. This process utilizes an ion exchange from an anode to the cathode work piece in a partial vacuum. File hard surfaces can be attained at temperatures as low as 850°F.

Both of these treatments are now being looked at as cost-effective alternatives to conventional heat treating by eliminating the need for final machining. Close tolerance control and file hard surfaces are achieved by both processes. **HT**

Fig. 9	Fig. 10

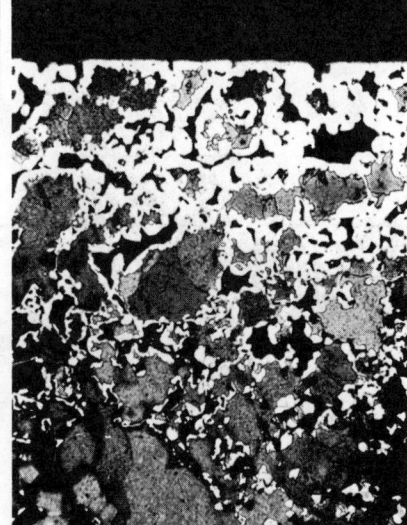 

P/M oil pump rotor: surface epsilon iron nitride extends to .015-in. depth.

Core structure of P/M rotor: density of part is 6.7 g/cm³.

Microstructures, quality control of P/M parts heat treated under various atmospheres

The quality of steel P/M parts sintered and heat treated under various atmospheres is examined. While all heat treat atmospheres proved acceptable, nitrogen sintering seemed to improve heat treat uniformity.

by P. JOHNSON, W.Q. JUDGE, and H.S. NAYAR

Because the P/M industry has grown up with endothermic atmospheres, green parts are conventionally designed around the use of endo; consequently, sintering specifications have gotten "married" to endo. With the introduction of alternative atmospheres for both sintering and heat treatment,[1,2] there is a need to reexamine the processes; the role of each component in furnace atmospheres needs to be understood, and rationales for using various atmospheres need to be established.

In a previous study on carbon control during sintering,[3] the core carbon was found to decrease a little more under nitrogen-based atmosphere than under endo. **Fig. 1** indicates a more striking difference in the surface carbon levels. Under endo the part decarburizes in the high heat (line A); it then recarburizes in the slow cool to a maximum, as shown by B. Depending on the balance between the decarburizing and recar-

W.Q. Judge is a metallurgist with Engineered Sinterings and Plastics, Watertown, Conn. H.S. Nayar is manager, Particle Technologies Group, for Airco Industrial Gases Division of the BOC Group, Murray Hill, N.J. Johnson, who was with Airco when this paper was written, is now senior metallurgist for National Standard Co., Niles, Mich.

This paper was originally presented at the Metal Powder Industries Federation P/M conference in New Orleans in 1983. It was published in Progress in Powder Metallurgy, vol. 39, and is copyrighted by the MPIF.

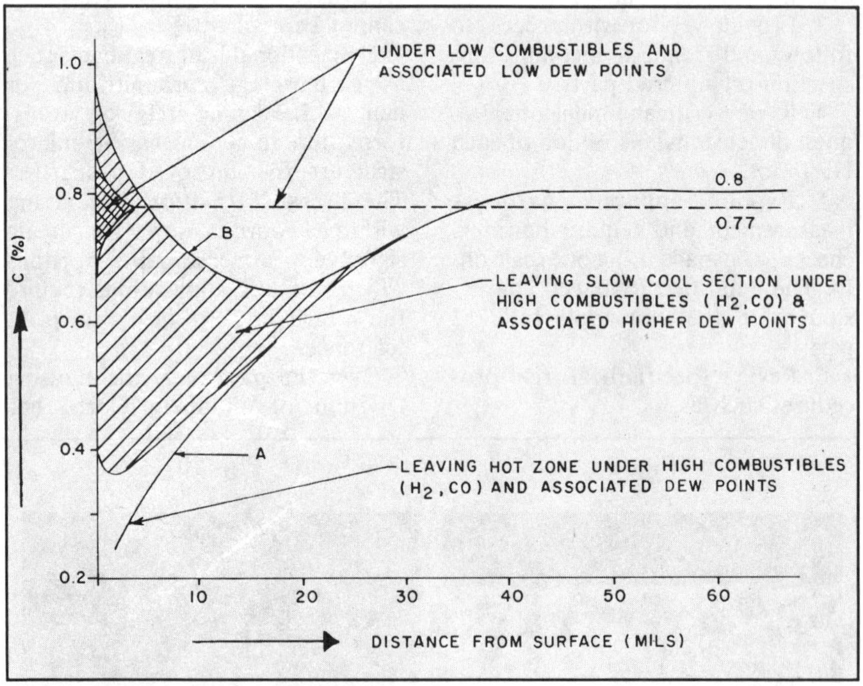

Fig. 1: Effects of endo and nitrogen-based atmospheres on surface and core carbon during sintering

Table I Green part specifications		
	High-carbon parts	**Low-carbon parts**
Prealloyed 4600	99.4%	99.7%
Hydrogen loss	.18	.18
Carbon	.01	.01
Sulfur	.019	.019
Apparent density	3.03	3.03
Flow for 50 grams	25.6	25.6
Graphite	.6%	.3%
Atomized Acrawax "C"	1.0%	1.0%
Nominal diameter	2-1/16"	2-1/16"
Nominal thickness	0.7"	0.7"
Average density	6.94	6.95

All photos are of part surface unless indicated otherwise.

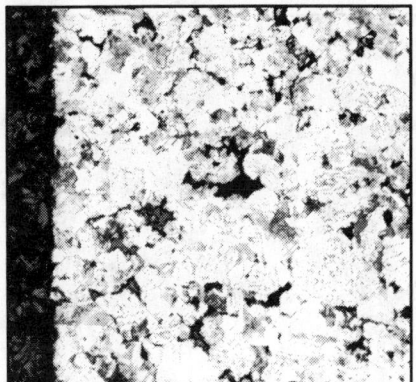

Fig. 2: High-carbon part sintered in endo

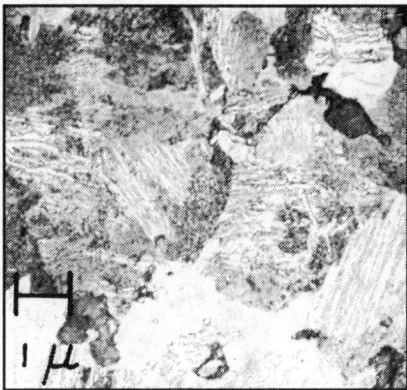

Fig. 3: High-carbon part sintered in endo (higher magnification)

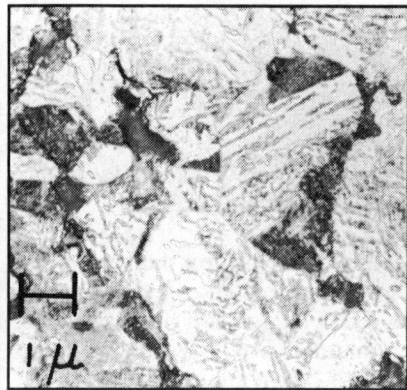

Fig. 4: High-carbon part sintered in endo (core)

Fig. 5: High-carbon part sintered in N$_2$-based

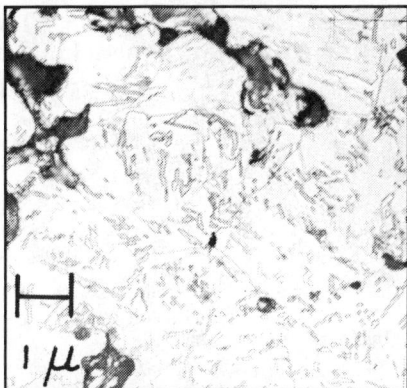

Fig. 6: High-carbon part sintered in N$_2$-based (core)

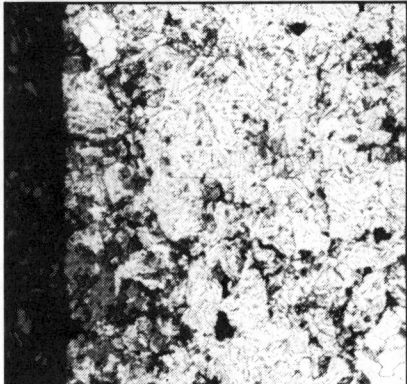

Fig. 7: Low-carbon part sintered in endo

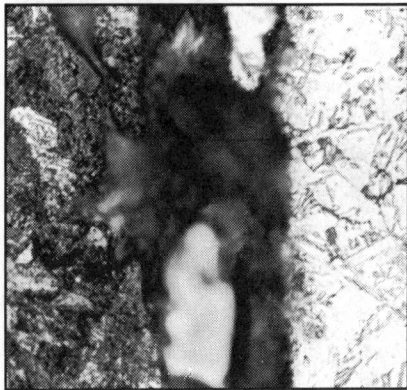

Fig. 8: Low-carbon parts sintered in endo (left) and N$_2$-based (right)

burizing, the carbon can vary over the limits shown, as shaded. Under nitrogen, the surface carbon variations are much less, as the crosshatched area indicates.

The present study reports the effects of various atmospheres on the microstructure and properties of sintered and heat treated P/M parts. It also explores the use of hardness readings for quality control.

Sintering: procedures

Parts in these studies were made from prealloyed 4600 with two different levels of graphite added. The powder specifications and green part properties are shown in **Table I**.

A 6-inch mesh belt metal muffle furnace was used for all sintering runs. The electrically heated preheat was run at 793°C (1460°F) set temperature to give a part temperature of 680°C (1256°F). The gas-fired high heat was set at 1171°C (2140°F) to give a maximum sintering temperature of 1140°C (2084°F). This furnace has a typical slow-cool section followed by two water-cooled sections. The belt speed was 1.8 inches/minute for about 20 minutes residence at sintering furnace.

Endothermic and nitrogen-based atmospheres were used to sinter both high- (.6% added graphite) and low- (.3% added graphite) carbon parts. Endo was introduced entirely in the slow cool, as is typical practice. For the high-carbon parts, atmosphere dewpoint was 2° to 4°C (35-40°F), and for the low-carbon parts, it was 4° to 7°C (40-45°F). The total atmosphere flow was 500 cfh, and no enrichment spike was used.

Under the nitrogen system, both the high- and low-carbon parts were processed in one atmosphere composition. No flow changes were made because of the carbon level differences. Furnace zoning[3] was used to optimize the metallurgical action. The entrance of the furnace was fitted with a burner to eliminate smoking. The total atmosphere flow was, again, 500 cfh, with about 4% hydrogen. This produced a hot-zone dewpoint of −42° to −37°C (−44° to −35°F). "Wet" nitrogen was used in the pre-heat to assist in lubricant removal. Dry nitrogen/hydrogen was used in the slow cool, along with a small amount of natural gas for carbon content. Dry nitrogen was used in the water cool section to exclude air and help in cooling the parts. This atmosphere contains no carbon monoxide.

The parts were marked prior to sintering to indicate the front top surface. After every 10 parts, two standards were placed on the belt to obtain the data presented here. The furnace was loaded continuously for a minimum of five hours to give about two full belt loadings.

Results

Fig. 2 is a typical micrograph of a high-carbon part sintered in endo. The structure is ferrite, carbides and unresolved pearlite near the surface. The pearlite formation is most pronounced on the bottom, and is very slight on the top. Even at 500X, **Fig. 3**, the pearlite is still unresolved. The core structure consists of ferrite and carbides, as seen in **Fig. 4**. The nitrogen-sintered parts shown in **Fig. 5** show no signs of pearlite formation, and as shown in **Fig. 6**, have a very uniform structure from surface to core.

The low-carbon parts sintered in endo show a structure similar to that of high-carbon parts. **Fig. 7**, again, shows some pearlite near the surface, although much less than in the high carbon parts. The surface structure of both the endo and nitrogen sintered parts are shown in **Fig. 8**. **Figs. 9** and **10** for the low-carbon parts sintered in nitrogen-based show no signs of pearlite and a uniform structure from surface to core.

Changes in dimension for both high- and low-carbon parts sintered under both endo and nitrogen are summarized in **Table II**. These are presented in terms of mil (.001-inch) per inch. The endo-sintered parts show a greater change in diameter, but a smaller change in thickness. In all cases, the deviations from the size of the green parts are smaller under the nitrogen system. Weight loss data shows a greater loss in weight for the nitrogen parts. Again, the *deviation* in that weight loss is less under nitrogen.

For the sintered low-carbon parts, we see a greater shrinkage (both in diameter and in thickness) in the

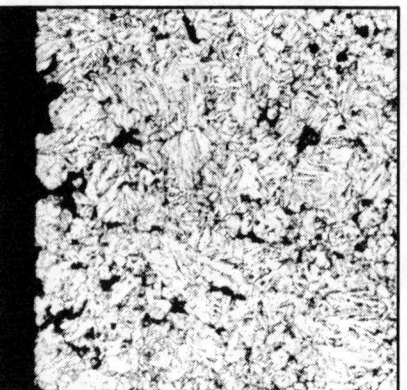

Fig. 9: Low-carbon part sintered in N_2-based

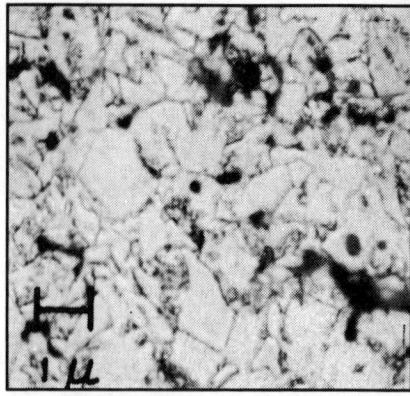

Fig. 10: Low-carbon part sintered in N_2-based (core)

Table II
Comparison of dimensional changes/deviations on sintering

High-carbon parts

Property	Units	Endo	Nitrogen
Diameter 12-6	Mils/in	−2.693/.093	−2.431/.072
3-9	Mils/in	−2.731/.081	−2.426/.062
Thickness	Mils/in	−4.538/.748	−5.644/.515
Weight	%	−1.323/.135	−1.395/.010

Low-carbon parts

Property	Units	Endo	Nitrogen
Diameter 12-6	Mils/in	−2.507/.091	−2.722/.093
3-9	Mils/in	−2.493/.099	−2.746/.078
Thickness	Mils/in	−5.358/.704	−5.697/.810
Weight	%	−1.145/.010	−1.194/.013

Table III
Average chemistries/deviations as sintered

Endo atmosphere

Part	Carbon (%)	Oxygen (ppm)	Nitrogen (ppm)
High-carbon			
surface	0.537/.023	735/92	162/13
core	0.438/.010	587/19	133/5
Low-carbon			
surface	0.413/.023	641/25	150/5
core	0.230/0	601/28	143/4

Nitrogen atmosphere

Part	Carbon (%)	Oxygen (ppm)	Nitrogen (ppm)
High-carbon			
surface	0.502/.015	450/10	208/3
core	0.400/0	469/10	219/2
Low-carbon			
surface	0.298/.005	394/21	231/11
core	0.190/0	480/19	230/5

Table IV
Average hardness as sintered (R_B/deviation)

Location	High-carbon parts Endo	Nitrogen	Low-carbon parts Endo	Nitrogen
Top	72.3/2.3	70.4/2.3	68.9/1.8	65.5/1.7
Bottom	78.7/2.1	65.8/2.4	65.4/2.1	60.7/2.5
12:00	75.4/1.3	70.5/1.3	70.2/1.1	63.8/1.0
3:00	74.2/1.2	71.6/0.9	71.5/1.2	65.7/1.1
6:00	75.6/1.4	70.3/1.0	70.9/1.0	64.1/1.3
9:00	73.0/1.7	72.8/1.0	70.2/1.2	66.8/1.1
Average	74.9/2.3	70.2/2.4	69.5/2.2	64.4/2.1

Table V
Average atmosphere composition during heat treating

Heat treat atmosphere	Dewpoint (°F)	High-carbon parts, neutral-hardened		
		CO_2 (%)	CO (%)	CH_4 (%)
Endo	45	0.56	17.6	1.2
Endomix	28	0.44	17.0	0.8
(N_2 + methanol)				
Natural gas/water	54	0.40	16.8	7.7
Nitrogen/natural gas	−24	0.001	0.4	1.2
		Low-carbon parts, carburized		
Endo	30	0.28	18.0	1.2
Endomix	22	0.18	16.5	0.7
Natural gas/water	36	0.17	16.0	5.7

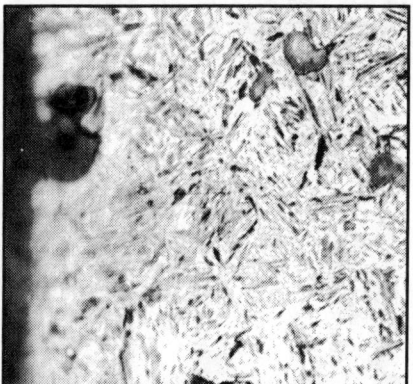

Fig. 11: High-carbon part sintered & heat treated in endo

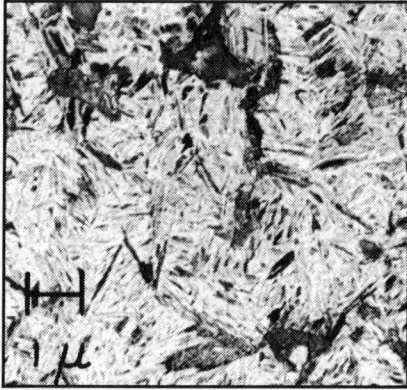

Fig. 12: High-carbon part sintered & heat treated in endo (core)

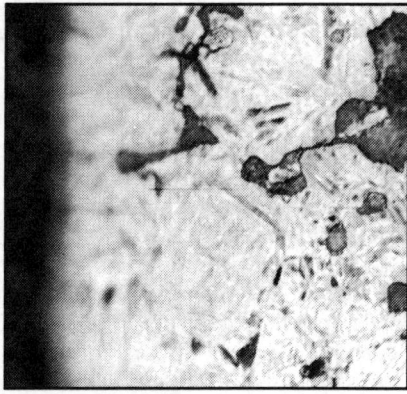

Fig. 13: High-carbon part sintered in N_2-based, HT'd in endo

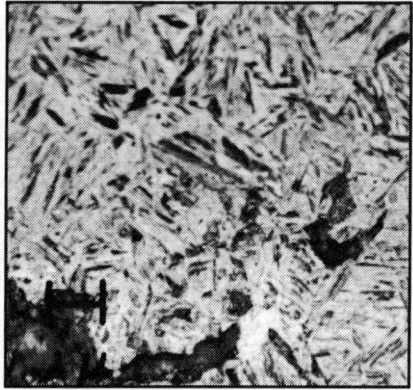

Fig. 14: High-carbon part sintered in N_2-based, HT'd in endo (core)

nitrogen-sintered than in the endo-sintered parts. From the dimensional parameters, it appears that with the nitrogen system there is less deviation from the mean. This is consistent with the microstructures, which show more uniform structure in the nitrogen-sintered part than in those sintered under endo. The weight loss is greater in the nitrogen-sintered low-carbon parts.

The average chemistries for both high- and low-carbon parts sintered in endo and nitrogen are given in **Table III**. This data indicates a greater carbon loss in nitrogen-sintered parts. In fact, low-carbon parts picked up surface carbon when sintered in endo. Both the high- and low-carbon parts sintered in the nitrogen-based system lost about 0.04% more carbon. This is consistent with the previous work.[3] Even though the core carbon loss is lower

for the low-carbon parts, about .08% compared to .17%, the percentage change is much closer. The endo parts lost about 0.25% of the starting carbon, and the nitrogen parts lost about 0.35% of the starting carbon.

Oxygen levels are higher in the endo-sintered parts, but the nitrogen levels are higher in the nitrogen-sintered parts. The sum of the oxygen and nitrogen is lower in the nitrogen-sintered parts than in the parts sintered under endo. Again, if we look at the deviation, the nitrogen-sintered parts show a smaller variation.

Sintering quality control

Rockwell B hardness readings have been proposed for quality control of sintered and heat treated P/M parts.[4] According to the technique, the RB hardness were taken on the top, bottom, front (12:00), back (6:00), right (3:00) and left (9:00) of the standard parts after sintering. The endo parts are harder. The greatest variation is seen between the top and the bottom of each sample. The higher hardness of the endo parts appeared to be related to the pearlite formation seen in **Figs. 2** and **3**. The greater variation between the top and the bottom might be related to the cooling rates. The top of the sample is cooled by radiation and convection/conduction to the gaseous atmosphere. The bottom cools by these same mechanisms, but also by conduction through the belt to the furnace jacket. Therefore, the bottom seems to cool slightly faster than the top or the sides. This conclusion is supported by the near equality of the top and side hardnesses.

Another feature in the hardness measurements is the appearance of a double peak. This is most apparent for the bottom hardness in the high-carbon, nitrogen-sintered parts. It was seen in all parts for the top and bottom. In the top, it appears as a greater deviation in hardness. This effect has been traced to a slight difference in as-pressed density between the two surfaces. Since the part orientation was initially selected randomly, about equal numbers of samples have the lower density surface on the top and on the bottom. The actual density variation has been measured by slicing a part in half. One surface has a density of 6.94 gm/cc, while the other is 6.87 gm/cc. This is a relatively small density variation, and yet it shows up as about a 4-point difference in RB reading. The average hardness values and their standard deviations are shown in **Table IV**.

Heat treatment: procedures

After being sintered, these same parts were heat treated to study the effects of various heat treat atmospheres on nitrogen- and endo-sintered parts. The high-carbon parts were neutral-hardened, and the low-carbon parts were carburized to about 0.8% surface carbon. For neutral hardening, atmospheres of endo, nitrogen/methanol (Endomix),[5] natural gas/water,[6] and nitrogen/natural gas[7] were used. For carburizing, only the first three of these were used, as nitrogen/natural gas does not produce controlled carburizing to any appreciable extent.

Endo was produced in an endo generator from natural gas and air. The catalyst was heated to 999°C (1830°F), and the output was controlled by monitoring the dewpoint and making manual changes in the ratio. Endomix was produced by dissociating methanol according to the reaction: $CH_3OH \rightarrow$

$CO + 2 H_2$ + traces of H_2O, CO_2 and CH_4.

This takes place in the furnace hot zone and, when mixed with the correct amount of nitrogen, produces atmospheres of 20% CO, 40% H_2, and 40% N_2. Natural gas can also be reacted with water in the furnace hot zone to produce an atmosphere with higher levels of CO and H_2 and reduced levels of nitrogen. This atmosphere requires a tighter furnace than Endomix or endo, and shows a greater variation in CO level. It also puts greater restrictions on the furnace design and construction. As an example, some catalyst must be present in the hot zone.

All three of these atmospheres can be controlled with dewpoint, CO_2 or

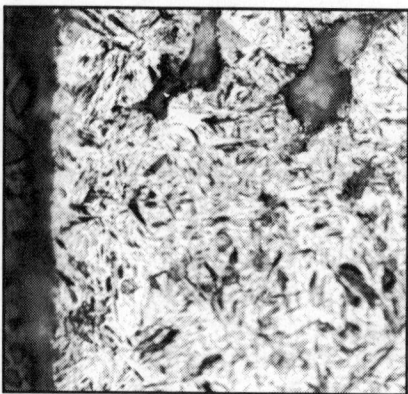

Fig. 15: High-carbon part sintered in N_2-based, HT'd in Endomix

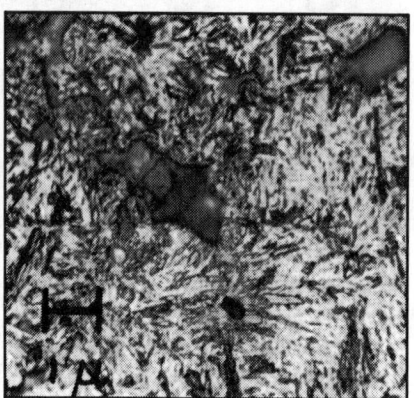

Fig. 17: High-carbon part sintered in N_2-based, HT'd in N_2/nat. gas

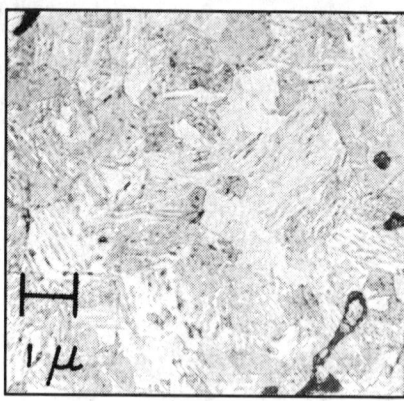

Fig. 19: Low-carbon part sintered & HT'd in endo (core)

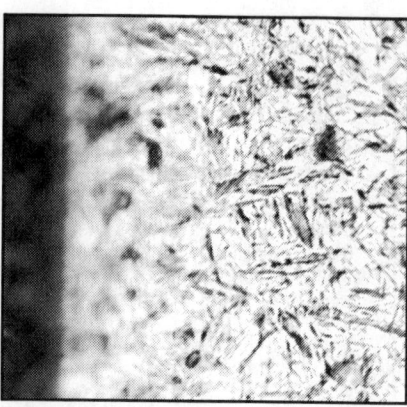

Fig. 16: High-carbon part sintered in N_2-based, HT'd in nat. gas/water

Fig. 18: Low-carbon part sintered & HT'd in endo

other common control instruments. The atmosphere produced from nitrogen/natural gas does not require any control. It cannot carburize and is used for neutral or, more correctly, inert, hardening. The nitrogen displaces the air in the furnace and the natural gas removes the last traces of air. This atmosphere is also non-flammable.

The furnace used was an L&N Tricarb integral-quench batch fur-

nace. This is a gas, radiant-tube-fired furnace and is rated at 600 pounds/hour gross weight. The quench is Park AAA operated at 65.6°C (150°F). Loads were made up of several test pieces plus about 100 pounds of dummy stock. The neutral hardening cycles were run at 843°C (1550°F) for 45 minutes at heat. The carbon level was controlled to provide an atmosphere neutral to a .6% carbon, except in the nitrogen/natural gas atmosphere. This latter atmosphere was run with no control. Carburizing cycles were one hour at 871°C (1600°F), with a carbon setpoint equal to a .8% carbon. The average atmosphere compositions are given in **Table V**.

Results

Figs. 11 and **12** show the surface and core structure of parts sintered in endo and neutral-hardened in endo as a reference. The structure is martensitic both at the surface and in the core. A part sintered in nitrogen and neutral-hardened in endo looks

Fig. 20: Low-carbon part sintered in N₂-based, HT'd in endo

Fig. 21: Low-carbon part sintered in N₂-based, HT'd in Endomix

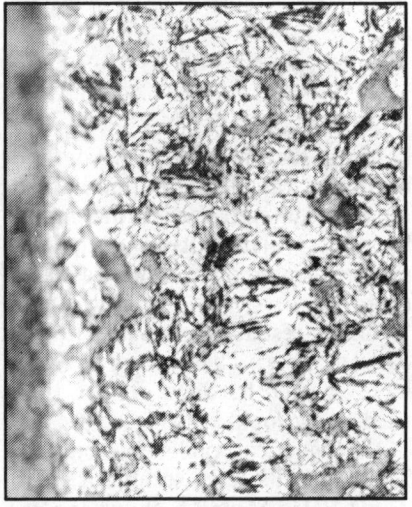

Fig. 22: Low-carbon part sintered in N₂-based, HT'd in nat. gas/water

Table VI
Change in diameter/deviation on heat treating and tempering for parts sintered in endo and N₂-based atmosphere

Heat treat atmosphere	Change in diameter/deviation (mils/inch)			
	High-carbon		Low-carbon	
	Endo	Nitrogen	Endo	Nitrogen
Endo	.80/.03	.60/.03	−.17/.13	−.05/.09
Endomix	.71/.17	.59/.02	−.17/.08	−.14/.07
Natural gas/water	.59/.06	.77/.07	.08/.12	.16/.02
Nitrogen/nat. gas	.29/.09	−.10/.09		

Table VII
Change in thickness/deviation on heat treating and tempering for parts sintered in endo and N₂-based atmosphere

Heat treat atmosphere	Change in thickness/deviation (mils/inch)			
	High-carbon		Low-carbon	
	Endo	Nitrogen	Endo	Nitrogen
Endo	−.29/.61	1.15/.62	2.73/.004	2.01/.20
Endomix	−1.93/1.11	.36/.10	1.87/.81	1.87/.0006
Natural gas/water	−.93/.10	−.36/.30	2.37/.10	2.87/.41
Nitrogen/nat. gas	−.07/1.11	.86/.20		

Table VIII
Average surface hardness/deviation as heat treated

Heat treat atmosphere	Hardness (R_B) High-carbon parts, neutral-hardened	
	Endo-sintered	Nitrogen-sintered
Endo	111.48/.99	111.56/.50
Endomix	111.29/.60	110.94/.61
Natural gas/water	111.02/.70	110.67/.78
Nitrogen/natural gas	110.54/.75	110.46/.87
Hardness (R_B) low-carbon parts, carburized		
Endo	113.23/.83	112.79/.78
Endomix	113.77/.59	113.46/.59
Natural gas/water	113.56/.92	113.50/.68

the same, **Figs. 13** and **14**. As **Figs. 15, 16,** and **17** show, parts sintered in nitrogen and then heat treated in Endomix, natural gas/water, and nitrogen/natural gas all show structures similar to endo-processed parts.

Figs. 18 and **19** show a low-carbon part sintered in endo and carburized in endo. The surface structure is similar to **Fig. 11** of the high-carbon parts, but it does show a slightly higher carbon level. The core shows a structure of ferrite and pearlite with some carbides. The nitrogen-sintered parts carburized in endo (**Fig. 20**), in Endomix (**Fig. 21**), and in natural gas/water (**Fig. 22**) all show structures similar to the parts sintered and carburized in endo.

The changes in the diameter and thickness on heat treating are shown in **Table VI** and **Table VII**. Again, the slightly smaller variation for the parts sintered in nitrogen seems to carry through the heat treatment.

Heat treat quality control

The use of RB readings has been proposed as a quality control tool.[4] **Fig. 23** shows the expected RB reading as a function of P/M part density at the RC60 equivalent hardness band. The parts in this study have densities of 6.94 to 6.95, and they should show RB readings of 105 to 113 to achieve the desired RC60 particle hardness.

All of the high-carbon parts, as heat treated and tempered, have

Source: Heat Treating, May 1985, 34-40

hardness readings well in the desired hardness range. The endo- and nitrogen-sintered parts show similar hardness distribution, but the nitrogen-sintered parts show a better uniformity.

Similar results are seen for the low-carbon parts. These parts actually have greater hardness values than needed, as would be expected because of the higher carbon levels produced. These results are summarized in **Table VIII**. **Table IX** shows the actual carbon levels for both high- and low-carbon parts after heat treatment.

Summary

P/M parts sintered in endo and nitrogen-based atmospheres were examined. These same parts were then heat treated in several different atmospheres. We've seen from the microstructures and the deviations in measured physical properties that:

Table IX Carbon levels in heat treated parts				
	Neutral-hardened			
	Endo-sintered		**Nitrogen-sintered**	
Heat treat atmosphere	**Surface**	**Core**	**Surface**	**Core**
Endo	0.60	0.48	0.60	0.51
Endomix	0.56	0.45	0.56	0.48
Natural gas/water	0.60	0.47	0.61	0.50
Nitrogen/natural gas	0.56	0.43	0.61	0.47
	Carburized			
Endo	0.75	0.41	0.75	0.30
Endomix	0.69	0.26	0.71	0.27
Natural gas/water	0.79	0.28	0.80	0.32

• The nitrogen sintering system gives better part uniformity,

• Different heat treat atmospheres show little difference in microstructures,

• All heat treat atmospheres give acceptable results,

• Greater uniformity in sintering (nitrogen-based) shows greater uniformity in heat treatment.

The greater uniformity of parts produced under a nitrogen-based system should be even more evident in normal production, where an endo generator can change as fast as the weather changes. **HT**

REFERENCES

1) H. S. Nayar and J. D. Drew, "Heat Treating Atmospheres: Feedstock Makeup, Properties," *Heat Treating*, July and September 1980.

2) H. S. Nayar, "Endo Vs. Nitrogen: Some Technical Considerations," *Heat Treating*, March and April 1980.

3) H. S. Nayar, "The Concept of Furnace Zoning: Its Use in Developing Highly Effective Sintering Atmospheres," presented at International Powder Metallurgy Conference, Florence, Italy, June 1982. Associazione Italiana di Metallurgia, Italy.

4) W. Q. Judge, "Evaluation of Heat Treated Powder Metal Parts," 7th Heat Treat Conference/Workshop, Chicago, Illinois, May 9-12, 1983.

5) B. J. Sheehy, "Nitrogen-Methanol Carburizing Atmospheres," *Metal Progress*, September 1981.

6) U.S. Patent 4,049,473—Assigned to AIRCO, Inc.

7) P. Johnson and H. Nayar, "Natural Gas and Its Nitrogen Alternative For the Metal Working Industries," Summary of Paper, Sixth Annual UMPR-PNR Conference on Energy, Oct. 16-18, 1979.

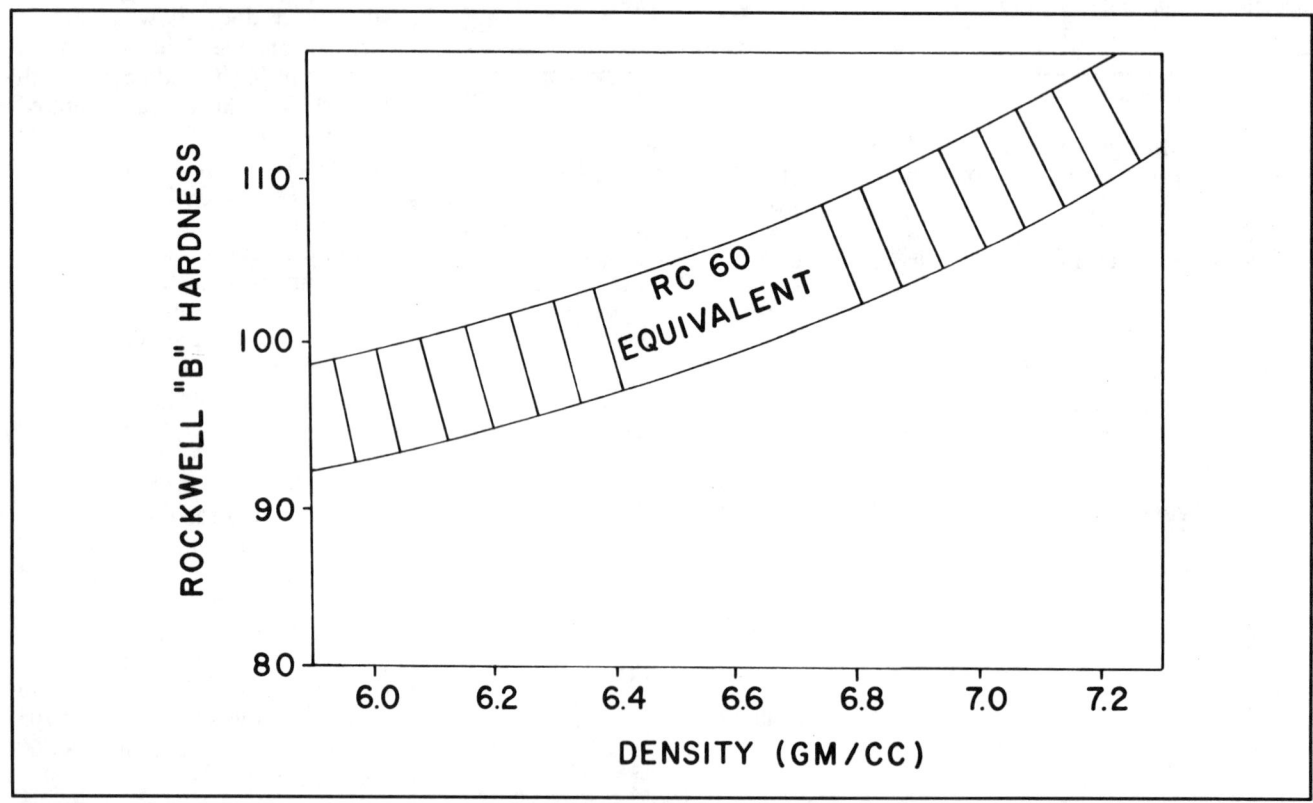

Fig. 23: Rockwell hardness (RB) as a function of P/M parts' density for heat treated parts showing a particle hardness of RC 60.

The Application of CBN Tools to Machine PM Parts

The machining of sintered iron-base components has for some time created problems for the PM industry because of the poor tool life when employing conventional cutting tool materials such as cemented carbides. The following article by Nobuo Furukawa of the Sintered Alloy Division of Sumitomo Electric Industries Ltd, in Itami, Japan, describes the successful application of cubic boron nitride tools for the machining of various high-strength sintered ferrous components.

PM parts are used in many important components of automobiles, agricultural machines and copying machines. In order to increase the number of applications of PM parts, we have been developing high-strength materials, wear-resistant materials and a new brazing technique. While these materials are not usually easily machinable, we have improved their machinability by adding other elements. In addition, we have recently developed a new CBN tool, coating tool and cermet tool, among others, which we use in the machining of PM parts.

NEW CBN TOOLS

The CBN content, binder, colour, Vicker's hardness and result of the tool life test for wrought steel are shown in Table 1. From these data it is very easy to conclude that BN100 and other CBN grades are suitable for continuous cutting and BN200 is suitable for intermittent cutting. Several types of CBN tools were also examined in the machining of PM parts. The valve-seat-ring material which is used in automotive diesel engines was selected for this machining test because of its poor machinability (Table 2). Usually tool life is measured in terms of length of flank wear (VB), which we determined by microscope. The results of these machining tests are shown in Figs. 1 to 3. The following results were obtained:

(1) The flank wear of BN100 is smaller than that of BN200 and the carbide tool.

(2) The tool life of BN100 under wet cutting conditions is longer than that under dry conditions.

(3) The tool life of BN100 is not greatly affected by the cutting velocity.

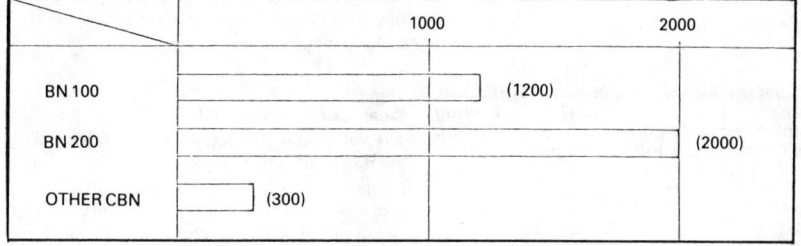

	BN 100	BN 200	OTHER CBN
C.B.N. content	High	Low	High
Binder	Special ceramics	Special ceramics	Metal
Colour	Black	Brown	Black
Vicker's Hardness	4000~4500	3000~3500 0	4000~4500
Toughness	2	1	3

Number of times tool was used until chipping occurred (Intermittent cutting)

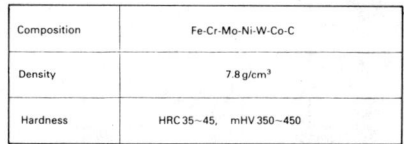

		1000	2000
BN 100		(1200)	
BN 200			(2000)
OTHER CBN	(300)		

TABLE 1 Comparison of CBN tools. **Work:** SKD-11, V=100m/min, d=0.5mm, f=0.1mm/rev.

Composition	Fe-Cr-Mo-Ni-W-Co-C
Density	7.8 g/cm³
Hardness	HRC 35~45, mHV 350~450

TABLE 2 Part specifications for the machining test

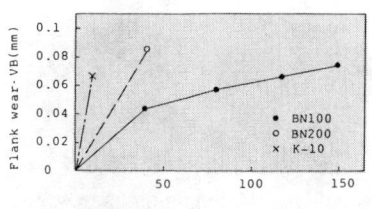

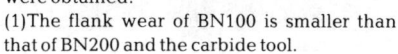

FIG.1 Comparison of CBN tools and Carbide tool. **(Tool:**SNG 432; **Cutting Conditions:**V=64m/min, d=0.2mm; f=0.5mm/rev., Dry)

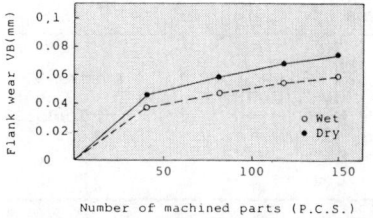

FIG.2 The effect of cutting liquid. **(Tool:**SNG 432, BN 100; **Cutting Conditions:**V=100m/min., d=0.2mm; f=0.1mm/rev.)

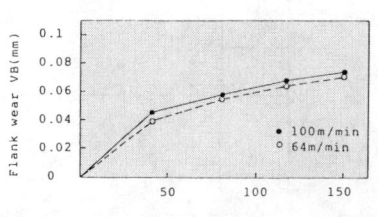

FIG.3 The effect of cutting speed. **(Tool:**SNG 432, BN 100; **Cutting Conditions:**d=0.2mm, f=0.1mm/rev. Dry)

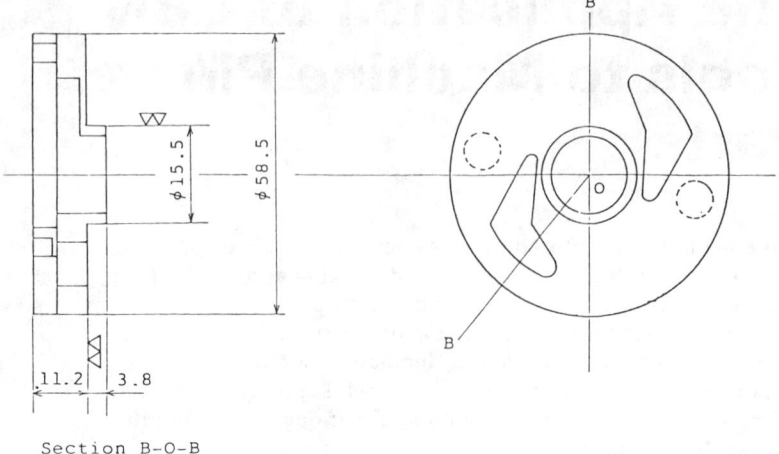

Section B-O-B

FIG.4 Side plate for powder steering pump. (----;Brazed line; Fe-Cu-C, Density=6.5g/cm³; HRB65~80)

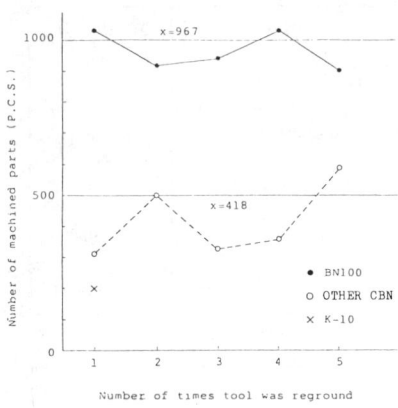

*FIG.5 Comparison of tool life. (**Work:**Side plate (Brazed part); **Cutting Conditions:**V=100m/min., d=0.2mm; f=0.05mm/rev.; **Tool:**Rake angle=10°; Nose Radius=0.2mm; Dry)*

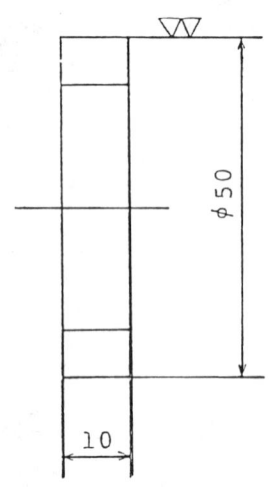

*FIG.6 Valve seat ring for diesel engine. (Fe-Cr-Mo-Ni-W-Co-C, Density=7.8g/cm³; **Hardness:**HRC 35~45)*

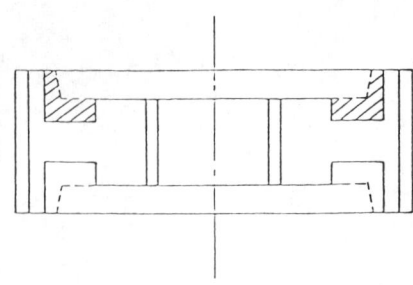

FIG.7 Clutch hub for manual transmission made from Distalloy AE+C

APPLICATION OF CBN TOOLS

Brazed Part

The side plate shown in Fig. 4 is used in the power steering pump of automobiles. By using a newly developed brazing technique we were able to produce this complicated shape. From the point of view of machinability, however, this side plate is poor, because it has hard brazed portions and two holes. Using this part, we performed a tool life test using BN100, other CBN and K-10. As the data of Fig. 5 indicates, the tool life of BN100 is higher than that of other CBN or the carbide tool.

Wear Resistant Material

The valve-seat-ring shown in Fig. 6 is used in the diesel engine of automobiles. As the hard components of W, Cr, Mo, Cr and Co are dispersed in this part, it is very easily concluded that the machinability of this part is very poor. Part specification and cutting conditions are shown in Fig. 6 and Table 3. The tool life of BN100 is approximately the same level as that of other CBN but in the case of the carbide tool, tool life is only a few pieces.

	Total number of machined parts (P. C. S)
BN 100	1396
OTHER CBN	1213

*TABLE 3 Tool life test of CBN tools for VSR. (**Work:**VSR; **Cutting Conditions:**V=73/min., d=0.3mm; f=0.18mm/rev.; **Tool:**rake angle=10°, Nose radius=0.3mm; Wet)*

Distalloy AE

The clutch hub, shown in Fig. 7, is used in the manual transmission of automobiles. As high strength and high wear resistance are needed for this part, Distalloy AE is used for the material. It is well known that the machinability of Distalloy AE is not good. The result of a machining test of several tools is shown in Table 4. BN200 is the best tool for the machining of this part, but due to considerations of cost and the speed with which the tool can be changed, the cermet tool of throw-away type was decided best for this machining operation.

	Tool life (PCS/reg.)	cost
BN 200	200	high
H-1 (Carbide tool)	30	low
T-12A (Cermet tool)	120~150	low
T-12A Cermet tool, throw away type	120~150	low

*TABLE 4 Tool life test. **Work:**Clutch hub; **Cutting Conditions:**V=120/min; f=0.1mm/rev., Wet)*

FIG.8 Gears used in the clutch system of copying machines

MACHINING OF HEAT TREATED PM PARTS

Continuous Cutting

The gears which are shown in Fig. 8 are used in the clutch system of copying machines. As these parts are mainly used in the spring type, one-way clutch, high wear resistance and closed tolerance are needed. Usually we use heat treatment and grinding operations to solve these problems. However, we cannot adopt the grinding operation for this part because its maximum corner radius (shown in Fig. 9) is only 0.3mm. Using BN100 we have been producing these parts without any tool trouble.

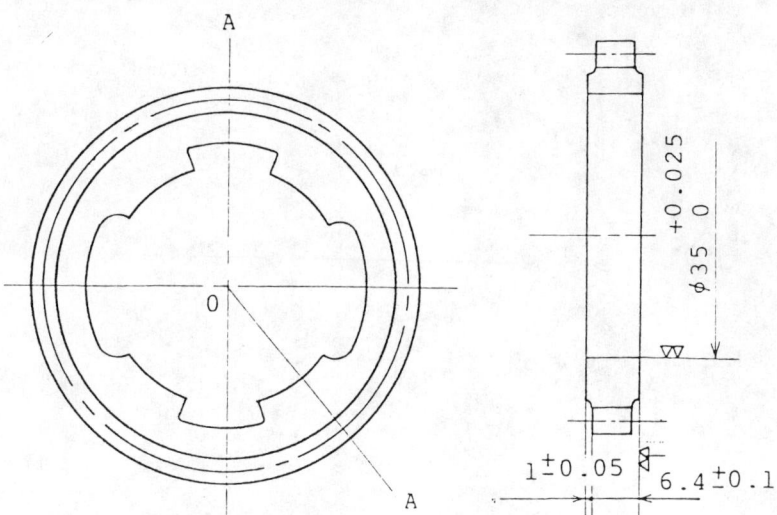

FIG.10 Gear used in injection pump of diesel engine. (Fe-Ni-Cu, Density=7.1g/cm^3; HRC 34~40 (Heat treated)

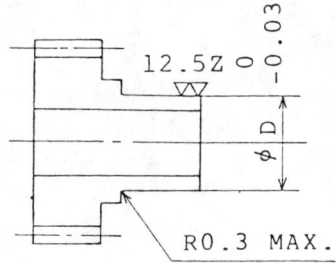

FIG.9 Gear used in copying machines. (Fe-Cu-C, Density=6.8g/cm^3; Hardness HRA 62~72

Intermittent Cutting

The gear which is shown in Fig. 10 is used in the fuel injection pump of diesel engines. The inside diameter and the side surface are ground in order to achieve close tolerance and good surface roughness. Because grinding is expensive, however, our customer plans to change its machining operations from grinding to turning. There are many technical problems in carrying out this plan, such as the selection of the tool material and the attainment of a close tolerance of the inside diameter, good surface roughness and flatness.

We established the surface roughness after the turning operation was the same level as that of grinding.

	Total number of machined parts (P. C. S.)
BN-100	450
OTHER CBN	400

TABLE 5 Tool life for hardened gear. (**Work:**gear (Heat treated part); **Cutting Conditions:**V=120m/min.; d=0.1mm; f=0.1mm/rev.; **Tool:**Brazed tool; Rake angle=10°; Nose radius=0.8mm. Wet)

Heat Treating
Nonferrous Metals

Principles of Heat Treatment of Nonferrous Alloys

By Charlie R. Brooks, University of Tennessee

THE PRINCIPLES which govern heat treatment of metals and alloys are applicable, of course, to both ferrous and nonferrous alloys. However, in practice there are sufficient differences to make it convenient to emphasize as separate topics the peculiarities of the alloys of each class in their response to heat treatment. For example, in nonferrous alloys, eutectoid transformations, which play such a prominent role in steels, are seldom encountered, so that the principles associated with time-temperature-transformation diagrams and with martensite formation are not emphasized in this review (they are covered in another section of this handbook). On the other hand, the principles associated with chemical homogenization of cast structures are applicable to many alloys in both classes.

Examination of the heat treatments used for nonferrous alloys reveals that a wide variety of processes are employed. However, because the process of diffusion underlies nearly all heat treatments, the concepts of diffusion are summarized first in this article. Annealing after cold working is a very important heat treatment for nonferrous alloys, and this topic is discussed next. Then the subject of homogenization annealing is reviewed, because it is an important heat treatment for as-cast structures. The process of precipitation, and the hardening that accompanies it, are described next, because these phenomena are especially important in Al-base alloys (and also in some Mg-, Cu- and Ni-base alloys). Then, to illustrate the formation of structures in which two phases are present in comparable quantities (e.g., Ti-base alloys, some Cu brasses, etc.), the heat treatments of a specific type of Cu-Zn alloy are examined. Finally, references are listed which provide additional information on the principles of heat treatment of nonferrous alloys.

DIFFUSION IN METALS AND ALLOYS

In heat treatment of metals and alloys, the rate of structural changes is usually controlled by the rate at which the atoms in the lattice change position. Thus, when cold worked copper is annealed and softens, or an aluminum-base alloy is aged, we are interested in how the atoms move relative to each other so as to bring about the observed changes in properties. The movement of the atoms involved here is called diffusion, and it is this process of diffusion which is examined in this section.

Diffusion in Pure Metals (Self-Diffusion). Atoms in a lattice at finite temperature are not static, but are vibrating in three dimensions around the normal atom position, usually the lattice site. Thus, consideration arises as to whether these atoms, by some mechanism, can exchange positions with each other and thereby move through the lattice. Such movement of the atoms of a pure metal is termed self-diffusion, and it is usually detected by experiments in which a thin layer of a radioactive atom is placed on the surface (e.g., by plating) of the same metal which is not radioactive and then the sample is given an annealing treatment at sufficient temperature and for sufficient time to allow diffusion. Because the difference between the radioactive and nonradioactive atoms is in the nuclear structure, and not in the valence electrons which are related to bonding, it is assumed that the radioactive atoms move through the lattice by the same mechanism and at the same rate as do the nonradioactive atoms. Thus, the movement of the radioactive atoms, which can be followed by a suitable radioactivity detector, reflects the type of movement the atoms in the metal undergo.

Such an experiment is illustrated schematically in Fig. 1. The radioactive layer is depicted as only two atoms thick, whereas it will really be much thicker (e.g., 1 mm). The sequence of time from 0 to t_3 shows increasing amounts of radioactive atoms (closed circles) moving into the lattice of the nonradioactive atoms (open circles), and simultaneously the lattice sites of the radioactive atoms are occupied by the nonradioactive atoms. The amount of radioactivity is measured as a function of depth into the sample from the surface, giving the profiles shown at the bottom of the figure.

Vacancies. The movement of atoms in the lattice, as depicted in Fig. 1, can be conceived to occur by several mechanisms. For example, at any instant of time, it is possible that the nearest two neighboring atoms have vibrated in directions so that space is left around the two atoms, allowing them to exchange positions simultaneously. Such an event is depicted in Fig. 2(a). It is clear that the two atoms which exchange positions must move, to some extent, the neighboring atoms in order to pass each other during the exchange process. It may also be possible for four atoms to vibrate at some instant so that they move cooperatively in a ring, allowing all four to move simultaneously to new neighboring positions, as depicted in Fig. 2(b).

Although mechanisms such as those just suggested probably occur in some alloys, in most

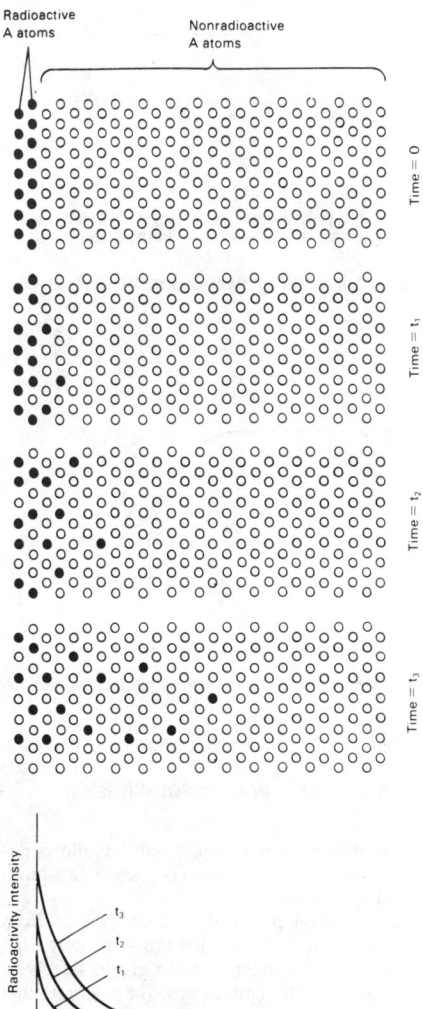

Fig. 1. Schematic diagram showing self-diffusion in a pure metal (radioactive atoms represented by filled circles)

In (a), two atoms move simultaneously to exchange positions. In (b), four atoms move cooperatively to rotate simultaneously to move to new positions.

Fig. 2. Schematic representation of two possible diffusion mechanisms

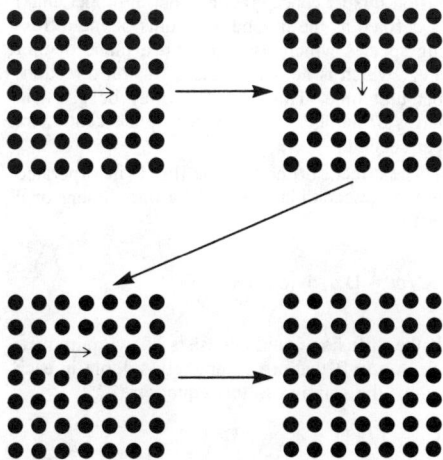

The vacancy moves to the new positions with time as shown by the small arrows. The large arrows show the changes with time.

Fig. 3. Schematic depiction of diffusion by vacancy movement

Reprinted from Metals Handbook, Desk Edition, 28.64-28.78, © 1985 American Society for Metals

metals and alloys diffusion occurs by vacancy movement. An unoccupied normal atom position in the crystal structure (usually a lattice site) is a vacancy. The presence of vacancies in a lattice at equilibrium is a consequence of a balance between the energy required to form the vacancies ΔH and the entropy ΔS created by their presence. Thus, there is an equilibrium concentration which minimizes the free energy change ($\Delta G = \Delta H - T\Delta S$).

If a vacancy exists in a lattice, then it requires much less energy for an atom to change positions than in the mechanisms depicted in Fig. 2. An atom has only to move into the vacancy, with much less energy. Such movement is shown in Fig. 3. It is to be noted that the diffusion occurs by rather random movement of the vacancies throughout the lattice.

Diffusion in Alloys (Chemical Diffusion). When two metals (or alloys) are placed in contact, atoms will begin to migrate across the contacting interface. Such diffusion of unlike species is called chemical diffusion, and is illustrated schematically in Fig. 4. (For the process to occur as shown in Fig. 4, the metals have to be soluble in each other; otherwise, when sufficient amounts of one metal diffuse into the other to reach a concentration corresponding to the solubility limit, precipitation of a second phase occurs.) The chemical diffusion depicted in Fig. 4 actually occurs by vacancy diffusion.

Fick's Laws of Diffusion. The mathematical relation that connects the concentration of the diffusing species with distance is Fick's law, a phenomenological equation which fits well most diffusion data. Fick's first law states that the diffusion flux, J (in one-dimensional diffusion), is given by

$$J = -D(dC/dx)$$

where C is concentration and x is distance. D is a constant at a given temperature, but may be concentration-dependent; it is called the diffusivity or diffusion coefficient. Figure 5 illustrates the relation between these terms and the concentration profile associated with chemical diffusion, such as illustrated in Fig. 4. Figure 6 shows data typical of those obtained by machining thin layers from a diffusion couple and analyzing each for the amount of the metals present. The diffusion flux (if concentration is put in proper units) is defined as the number of atoms of the diffusing species which pass through a plane of unit area, which is normal to the diffusion direction, per unit time. Thus, the flux may be given in terms of number of atoms per square centimetre per second.

The effect of time t on the flux is incorporated in Fick's second law (again, for one-dimensional diffusion):

$$dC/dt = D\ d/dx(dC/dx) = D\frac{d^2C}{dx^2}$$

If the diffusion couple consists of two pure metals A and B that are completely soluble in each other, the solution to this equation is

$$C_A = 1/2\ [1 - \phi\ (x/2\ Dt)]$$

where ϕ is the Gauss error function and C_A is the concentration of A at distance x from the original interface. (Similar expressions are obtained for different starting conditions — e.g., an alloy coupled against a pure metal, etc.) To extract D,

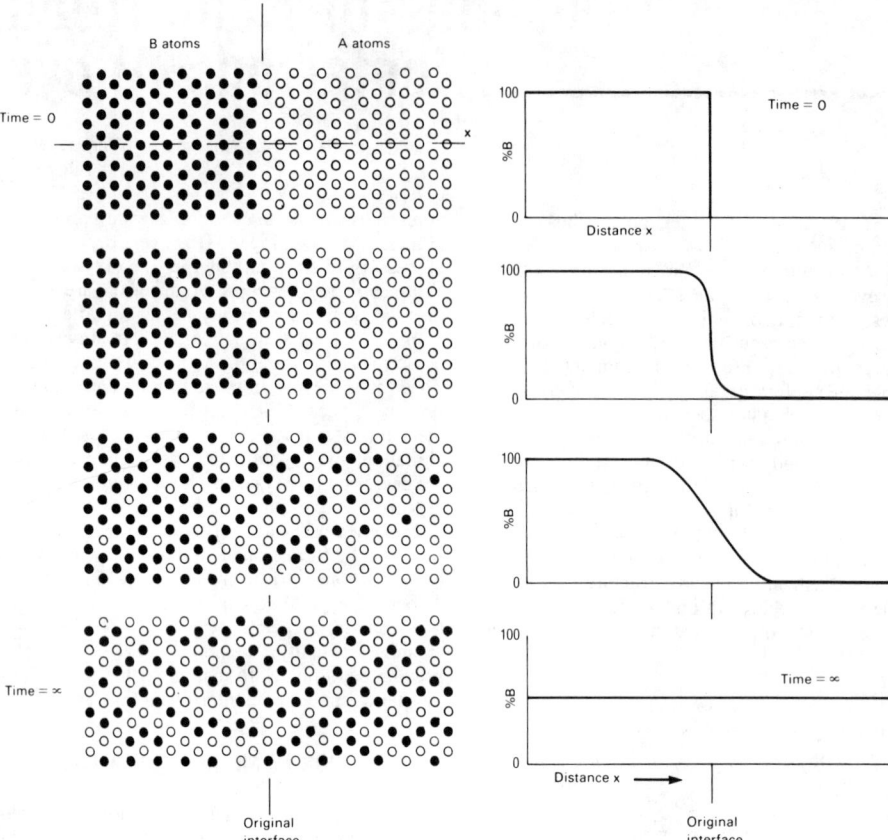

The diffusion couple is made up of pure B (filled circles) and pure A (open circles). As time progresses, mixing on the two sides occurs. At infinite time, complete mixing has been achieved, with the chemical composition being identical on both sides.

Fig. 4. Schematic illustration of chemical diffusion involving two different metals

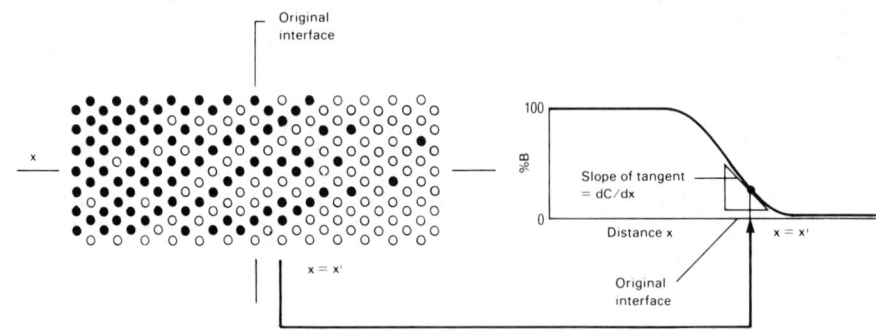

Flux of atoms across the plane at x = x' is the number of atoms crossing a plane 1 cm square per unit time (s) and is proportional to the gradient dC/dx at that location (x = x'):
$$J = -D(dC/dx)$$
The proportionality constant is the diffusivity or diffusion coefficient. The negative sign is required to make the flux positive to be physically realistic, as the gradient dC/dx is negative.

Fig. 5. Illustration of the meaning of the terms in Fick's first law of diffusion

then, for a given diffusion time t at a given distance x, the value of C_A is obtained (e.g., read from Fig. 6). This allows a value of ϕ to be obtained. Then, error function tables are used to determine the argument of ϕ — that is, to determine a value for (x/2 Dt). Then D is obtained.

Such a procedure should yield the same value of D no matter what value of x is chosen. However, it is found that D will usually vary, meaning that it is a function of composition. In this case, the equation to use is

$$dC/dt = d/dx(D\ dc/dx)$$

The solution is more complicated, but allows determination of the diffusion coefficient as a function of composition.

An important practical relation evolves from the solution to Fick's second law — namely, that the time-distance relation for a given concentration C is $x^2 \cong Dt$. This means, for example, that during a homogenization treatment designed to remove the effects of dendritic segregation (coring), the time is proportional to x^2, where x is approximately the dendritic arm spacing.

Temperature Dependence of the Rate of Diffusion. The dependency of the rate of diffusion on temper-

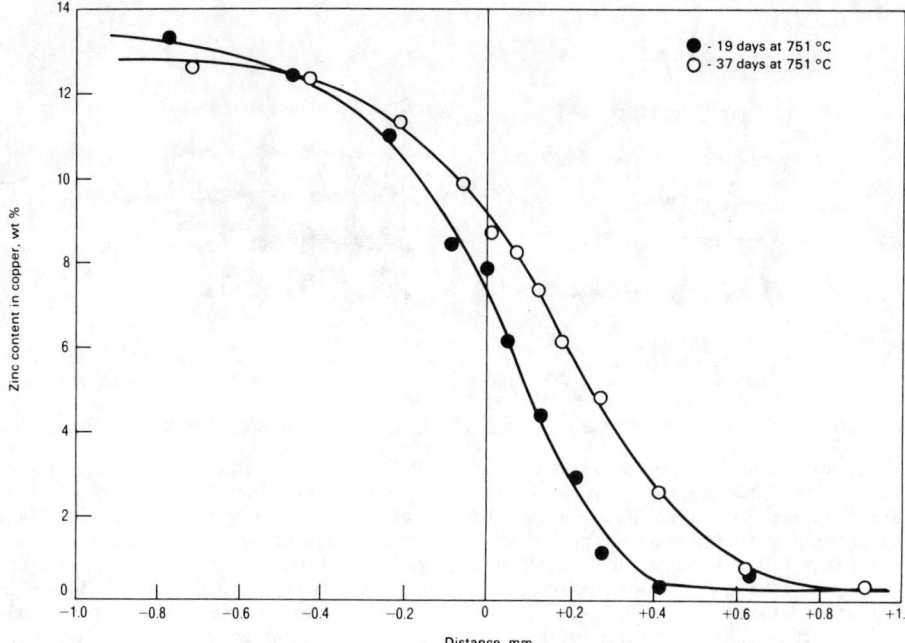

Each point represents the chemical analysis of a thin layer machined from the sample. Adapted from F. N. Rhines and R. F. Mehl, *Trans AIME*, Vol 128, p 185 ff, 1938.

Fig. 6. Concentration profile data typical of metals obtained from a diffusion couple, which in this case was copper:zinc

ature is found to be exponential, which is not surprising, because many rate reactions obey such a dependency. Thus, D is given by

$$D = D_0 e^{-B/T}$$

where D_0 and B are constants, and T is absolute temperature. Theoretical treatments show that this should be written as

$$D = D_0 e^{-Q/RT}$$

where R is the ideal gas constant and Q is the activation energy for the diffusion process. Q reflects the energy required to move an atom over a barrier from one lattice site to another; the barrier is associated with the requirement that the atom must vibrate with sufficient amplitude to

break the nearest neighboring bonds in order to move to the new locations.

The values of D_0 and Q shown in Table 1 typify those found in metals. The equation above for the temperature dependency of D predicts that log D plotted versus 1/T should be a straight line, and Fig. 7 shows some typical linear results for metals and alloys.

The exponential temperature dependence is important in heat treating. It shows that the rate

of change in processes which are diffusion-controlled will increase greatly with an increase in temperature. Thus, an increase in temperature of 10 K will approximately double the rate of the process.

Intrinsic Diffusion Coefficients. If the original interface of the diffusion couple is identifiable, then experiments show that the location where half of the diffusing species will have moved from one side to the other does not coincide with the original interface. This is sometimes referred to as the *Kirkendall effect*, and is taken as strong experimental evidence of the vacancy mechanism of diffusion in metals. Darken showed that the relation between the measured diffusion coefficient (as described above) and the intrinsic diffusion diffusivities of the individual atom species (for a binary system of atoms A and B) is

$$D = C_A D_A + C_B D_B$$

Here C_A and C_B are the mole fractions of A and B, respectively, and D_A and D_B are the intrinsic diffusivities of A and B, respectively. D_A and D_B are concentration-dependent.

Interstitial Diffusion. If the solute atom is sufficiently small, it will locate in an interstice between the larger solvent atoms, forming an interstitial solid solution. Diffusion of interstitial atoms occurs, not by a vacancy mechanism, but by the atoms jumping from one interstitial site to another. (Fick's laws still apply.) As the interstitial solute atom increases in size, the activation energy increases (Table 1), showing that it becomes more difficult for the atom to move between the solute solvent atoms to a neighboring interstitial site. In general, the activation energy for interstitial diffusion is less than that for substitutional diffusion.

Grain-boundary diffusion. Experimental studies have shown that diffusion along grain boundaries, along the core of dislocations and on free surfaces is considerably more rapid than diffusion through

Table 1. D_0 and Q values for diffusion in various substitutional and interstitial solid solutions

Solute	Solvent (host structure)	D_0, cm²/s	Q, calories per mole
Substitutional diffusion			
Cu	Cu	0.78	50 500
Cu	Sn	0.11	45 000
Cu	Ni	1.92	68 000
Ni	Cu	1.1	53 800
Cu	Al	0.647	32 270
Zn	Cu	0.73	47 500
Pb	Pb	0.887	25 500
Ti	Ti	0.000358	31 200
Al (4%)	Cu	0.0455	39 500
Zn (24-29%)	Cu	0.095	35 000
Interstitial diffusion			
H	Cu	10^{-2}	10 000
O	Cu	10^{-3}	46 000
C	Ti	0.00302	20 000
O	Ti	1	40 000
H	Ta	...	6 000
C	Ta	0.0061	38 520
N	Ta	0.0056	37 840
O	Ta	0.0044	25 450

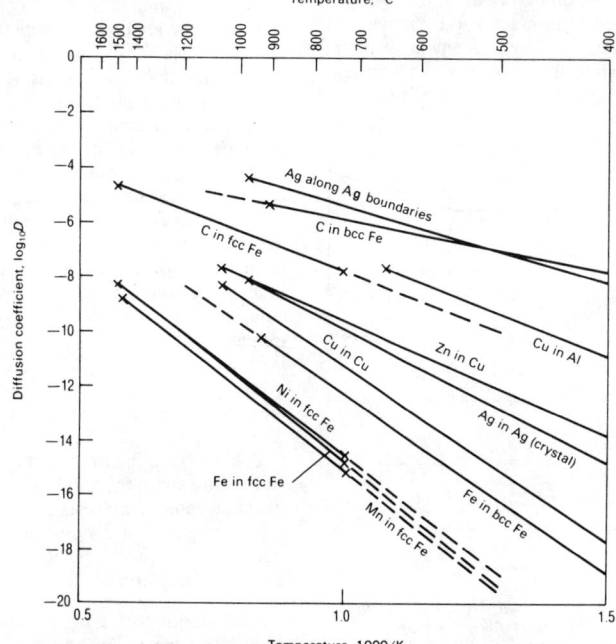

The straight lines are prominent and commonly found. From L. H. Van Vlack, *Elements of Materials Science*, 2nd Ed., Addison-Wesley Publishing Co., Reading, MA, 1964.

Fig. 7. Plots of log D versus 1/T for several metals

Source: Metals Handbook, Desk Edition, 28.64-28.78

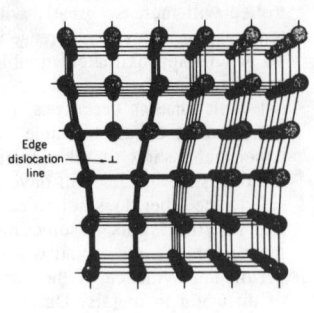 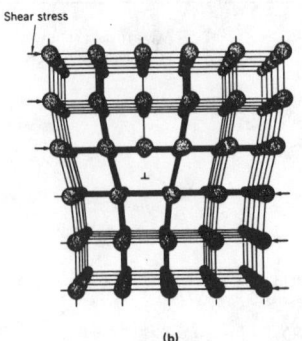 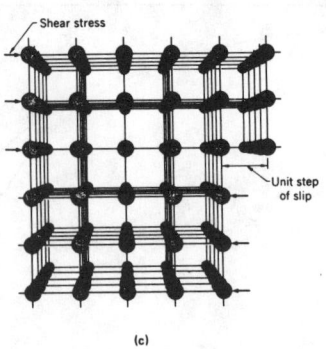

(a)   (b)   (c)

(a) An edge dislocation in a crystal. (b) The dislocation has moved one lattice spacing due to the shearing force. (c) The dislocation has reached the edge of the crystal and produced unit slip. Adapted from A. G. Guy, *Essentials of Materials Science*, McGraw-Hill, New York, 1976.

Fig. 8. The motion of an edge dislocation and the production of a unit step of slip at the surface of the crystal

the interior of a crystal. Of particular interest here is grain-boundary diffusion, which influences precipitation and phase changes at the boundary. The data in Fig. 7 for self-diffusion in silver show that the grain-boundary diffusivity is several orders of magnitude greater than bulk diffusion. Also, as temperature decreases, bulk diffusion becomes slower and grain-boundary diffusion becomes more important.

ANNEALING OF COLD WORKED METALS

Dislocations. Plastic deformation in metals and alloys occurs primarily by relative movement or slip of blocks of material on specific crystallographic planes (slip planes) and in certain directions (slip directions). (Plastic deformation in metals can also occur by twinning. However, in the brief treatment here, this mechanism will not be discussed.) This occurs not by movement of regions of the crystal as a whole, but by movement of successive dislocations. A dislocation is a lattice defect (either edge or screw) which is present in even well-annealed metals as a consequence of prior processing. Dislocations play a central role in plastic deformation because less energy is required to produce slip by movement of the dislocations than by movement of entire regions of a crystal past each other. This process is illustrated in Fig. 8 for an edge dislocation.

Obviously, millions of dislocations must repeat this process in order to generate visually obvious shape changes. This is possible, however, because the dislocations, which are present in the metal prior to plastic deformation, create other dislocations by a multiplication mechanism during plastic deformation.

In hexagonal close-packed crystals, the prominent slip plane is the close-packed (001) plane, and the slip directions in this plane are the close-packed directions, of which there are three nonparallel, identical choices. Thus, this crystal structure exhibits three slip systems. In the face-centered cubic structure, the slip plane is also the close-packed plane {111}. However, in this system there are four types of nonparallel {111} planes. In each plane there are three possible slip directions (⟨110⟩ type), and hence 12 slip systems. In the body-centered cubic structure, the slip plane is of the {110} type (also the most closely packed plane in this system), and the slip directions are of the ⟨111⟩ type, of which there are three in each plane. Thus, the body-centered cubic structure also has 12 slip systems. The types of slip plane and slip direction are sensitive to temperature, and in some alloys other slip systems are activated when temperature changes.

Effect of Cold Working on Properties and Microstructure. The multiplication of dislocations on several slip systems upon plastic deformation leads to their interaction with each other, and this restricts their movement, so that further deformation requires an increase in external load. Thus, the material work (or strain) hardens. This effect is illustrated in Fig. 9, which shows the strengthening induced by deformation in rolling of pure copper, and of copper-zinc solid-solution alloys, at 25 °C. Plastic deformation such that strengthening or hardening occurs is called *cold working;* plastic deformation such that work hardening does not occur is called *hot working.* (Alternative definitions are given below, under "Hot Working.") Note that these definitions have no particular attachment to room temperature.

Cold working increases hardness, yield strength and tensile strength, and lowers ductility. It also increases electrical resistivity because the increasing density of dislocations scatters the electrons. Fig. 10 illustrates the effects of cold working on several properties.

Cold working of a metal causes distortion of grains, and the specific nature of this distortion depends on the type of deformation (e.g., rolling, swaging, etc.). If the plane of observation is parallel to the rolling direction, the grains will appear elongated in the rolling direction. Also

observed in the microstructure are parallel striations within the grains, the density of which increases with the amount of deformation. These striations are actually rows of etch pits, or etched grooves, where the etchant has removed metal preferentially at surface locations at which the dislocations emerge. Such striations are sometimes called *deformation bands.* In metals and alloys which show annealing twins (mainly face-centered cubic metals, such as copper and brass), the twins, originally appearing as straight lines crossing (or nearly crossing) the grains, become bent, distorted and fragmented. All of these microstructural features of cold worked metals are illustrated in Fig. 11.

Recovery, Recrystallization and Grain Growth. In shaping of metals and alloys by cold working, there is a limit to the amount of plastic deformation attainable without fracture. However, proper heat treatment prior to reaching this limit restores the metal or alloy to a structural condition similar to that prior to deformation, and then

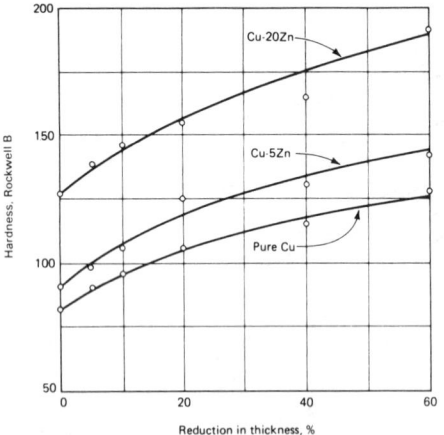

From Charlie R. Brooks, *Heat Treatment, Structure and Properties of Nonferrous Alloys*, American Society for Metals, Metals Park, OH, 1982.

Fig. 9. The effect of plastic deformation (by rolling at 25 °C) on hardness of pure copper and two Cu-Zn solid-solution alloys

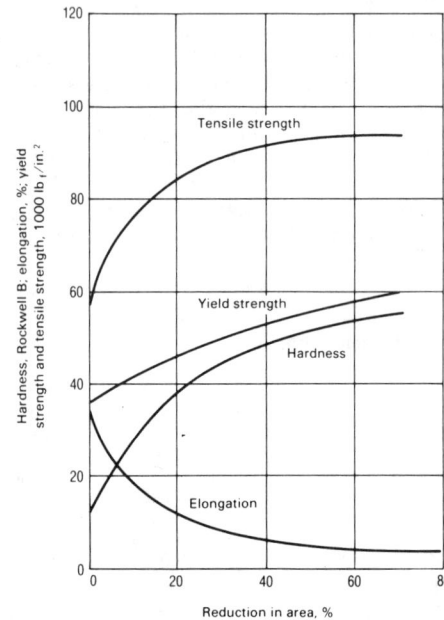

Adapted from R. A. Wilkins and E. S. Bunn, *Copper and Copper Base Alloys*, McGraw-Hill, New York, 1943.

Fig. 10. The effect of cold working (by rolling at 25 °C) on the tensile mechanical properties and hardness of oxygen-free, high-conductivity (OFHC) copper

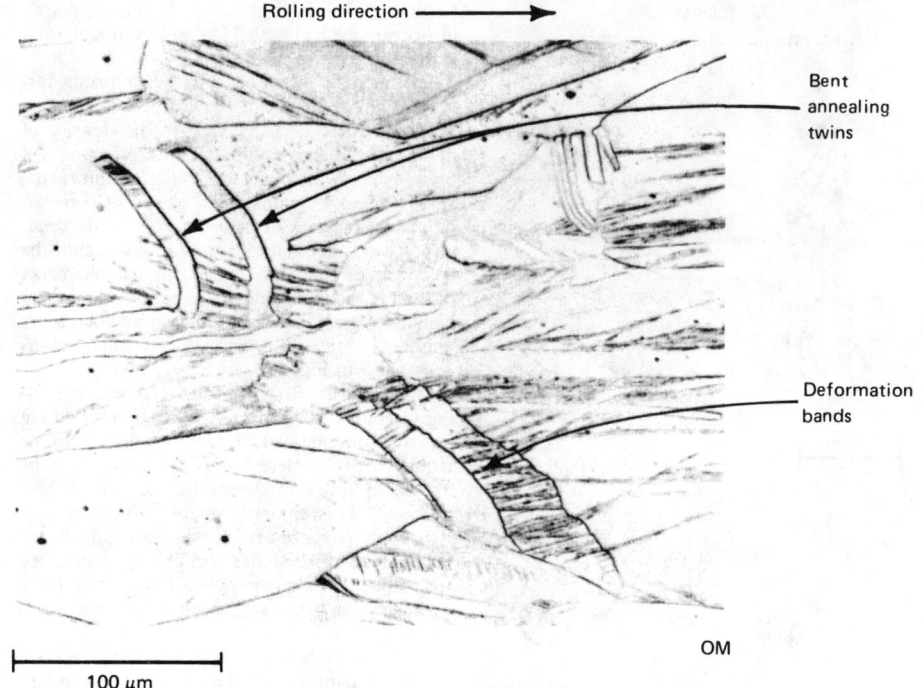

Rolling direction ➞

Bent annealing twins

Deformation bands

OM

|← 100 μm →|

From same source as Fig. 9.

Fig. 11. The microstructure of a Cu-15Zn alloy cold rolled at 25 °C to a 40% reduction in thickness, showing deformation bands and bent annealing twins revealed by etching the polished surface

decrease, followed by a continued, but gradual, decrease. The data shown in Fig. 12(a) are for a fixed temperature. A similar result is obtained by annealing samples for a fixed time at increasing temperatures, as shown in Fig. 12(b).

The stage of annealing for short times or at low temperatures wherein the hardness remains constant, or increases slightly, is called the *recovery* region. Here the dislocations undergo movement by thermal activation, being rearranged into arrays somewhat more stable and more difficult to move than in the cold worked, unannealed condition, and hence cause a slight increase in hardness. In this period, such rearrangement allows some properties to attain their values prior to cold working, and hence is referred to as recovery. One such property is electrical resistivity, as illustrated in Fig. 13. The cellular arrangement of the dislocations, compared with that of the cold worked condition, in-

additional cold working can be conducted. This type of heat treatment is called *annealing,* and in this section some of the principles involved and the effects which occur are summarized.

Because cold working produces an increasing concentration of lattice defects (e.g., dislocations), the energy of the crystals is increased. Thus, there is a thermodynamic driving force for the metal to undergo changes which will return

it to the original, low-energy condition. The rates of these changes depend on the mechanisms involved, and are sensitive functions of temperature and alloy.

The changes in strength that occur during annealing are illustrated by the hardness data in Fig. 12(a). The hardness (and the yield and tensile strengths) initially remains approximately constant (or increases slightly), then shows an abrupt

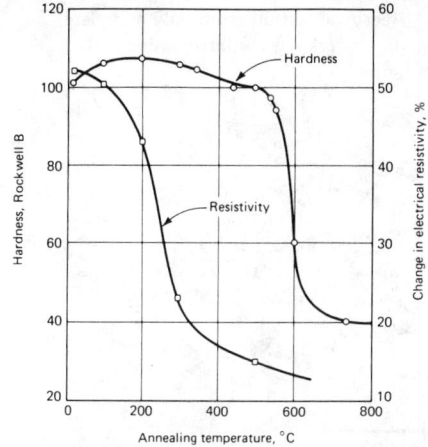

The metal had been cold worked at 25 °C almost to fracture. Annealing time, 1 h. Adapted from J. E. Wilson and L. Thomassen, *Trans ASM,* Vol 22, p 769, 1934.

Fig. 13. Effect of annealing temperature on hardness and electrical resistivity of nickel

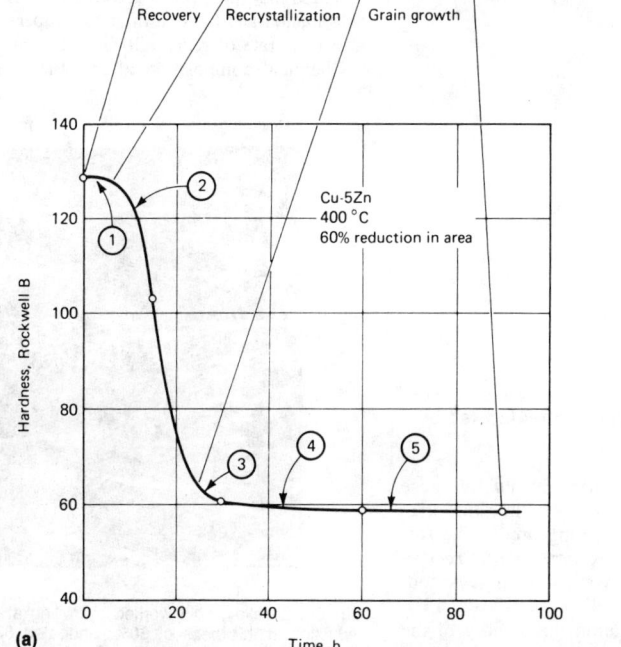

(a)

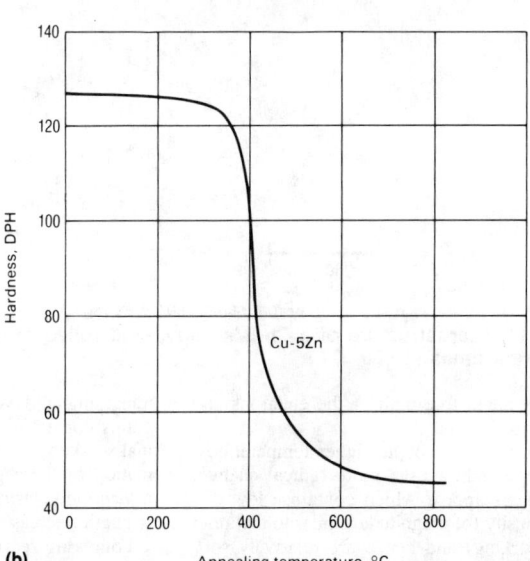

(b)

Part (a) from same source as Fig. 9.

Fig. 12. (a) Effect of annealing time at fixed temperature (400 °C) on hardness of a Cu-5Zn solid-solution alloy cold worked 60%. (b) Effect of annealing temperature at fixed time (15 min) on hardness of a Cu-5Zn solid-solution alloy cold worked 60%.

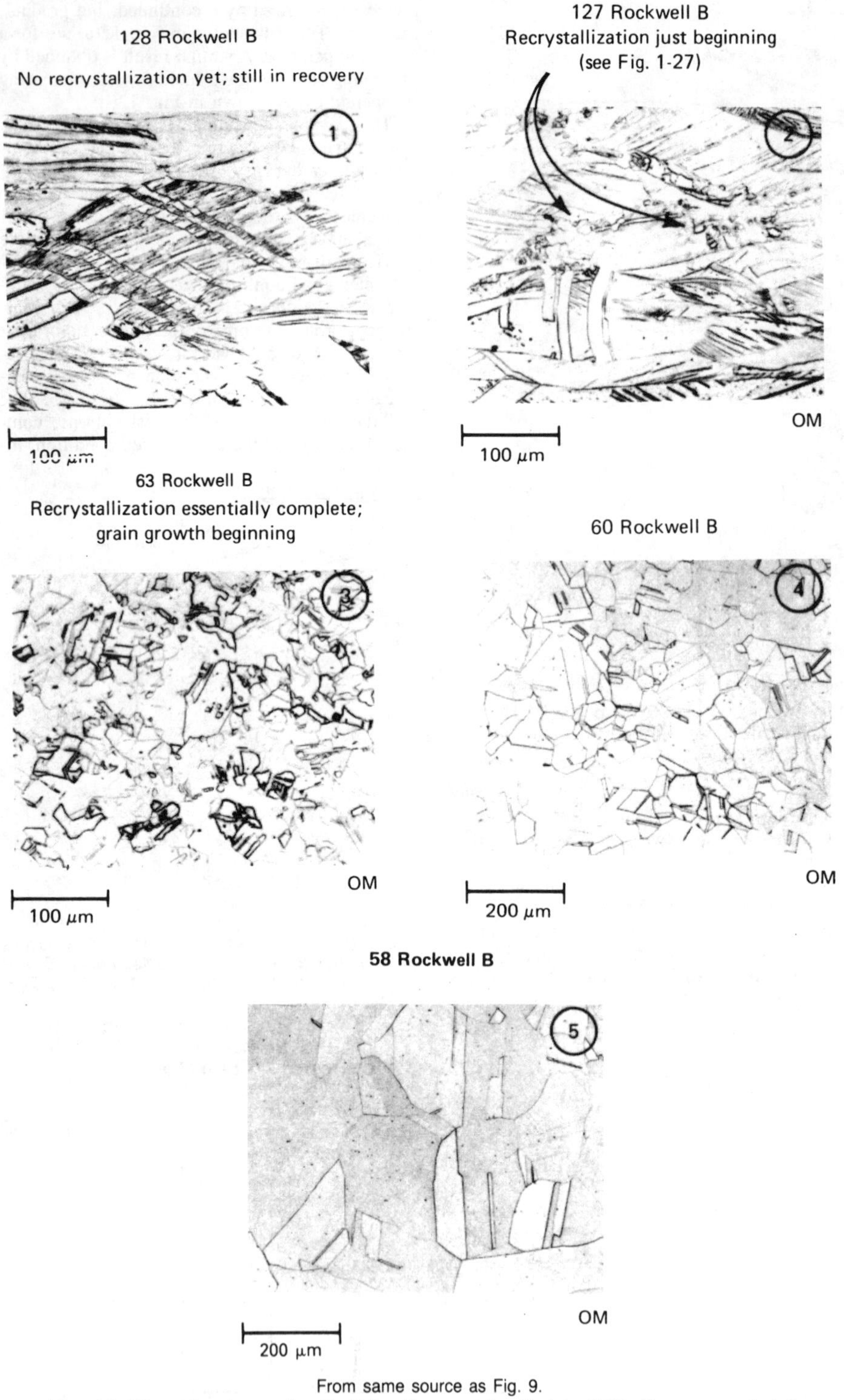

128 Rockwell B

No recrystallization yet; still in recovery

1

⊢ 100 μm ⊣

127 Rockwell B
Recrystallization just beginning
(see Fig. 1-27)

2

⊢ 100 μm ⊣   OM

63 Rockwell B
Recrystallization essentially complete;
grain growth beginning

3

⊢ 100 μm ⊣   OM

60 Rockwell B

4

⊢ 200 μm ⊣   OM

58 Rockwell B

5

⊢ 200 μm ⊣   OM

From same source as Fig. 9.

Fig. 14. Microstructure of a Cu-5Zn alloy, cold rolled to 60%, then annealed for different times at 400 °C

strength decreases as grain size increases, during this period the hardness decreases, although only gradually (Fig. 13).

The microstructural changes which occur during annealing are illustrated in Fig. 14. During recovery, there is a decrease in the density of deformation bands, although this effect is not prominent. When recrystallization commences, small, equiaxed grains begin to appear (see micrograph 2 in Fig. 14, and Fig. 15) in the structure. These continue to form and grow until the cold worked matrix is consumed, which marks the end of the recrystallization period and the beginning of grain growth. Further annealing causes only an increase in grain size (see micrographs 3, 4 and 5 in Fig. 14).

Factors Affecting Recrystallization. Because annealing of cold worked metals is usually carried out to soften the material, the temperature and time required to complete recrystallization must be known in order to determine the proper heat treatment. It is common to refer to the *recrystallization temperature* as an indicator of the temperature at which the metal must be annealed for softening. (This temperature can be taken to be that which gives any specified amount of recrystallization.)

Several factors affect the value of the recrystallization temperature. Two of the most important are annealing time and amount of prior cold work. Figure 16 illustrates the effect of annealing time. The longer the time at a given temperature, the farther the metal progresses in the annealing process. Thus, if a metal just commences recrystallization at 200 °C in 15 min, then it may be completely recrystallized in 30 min.

The effect of the amount of prior cold work is illustrated in Fig. 17. Increasing amounts of plastic deformation increase the concentration of lattice defects (e.g., dislocations) and make the metal more thermodynamically unstable. Hence, recrystallization occurs at lower temperatures, or in shorter times, the greater the amount of cold work. Although this is the main effect, it is to be noted that the type of deformation, the rate of deformation and the deformation temperature also affect the rate of recrystallization.

Chemical composition affects the recrystalli-

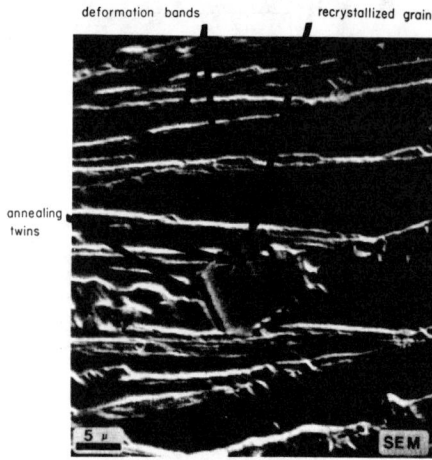

Cu–5% Zn alloy, cold worked by rolling at 20 °C to reduction in thickness of 60%; annealed 60 min at 350 °C. From same source as Fig. 9.

Fig. 15. High-magnification scanning electron micrograph showing a small recrystallized nucleus

creases the mean free path of the electrons and lowers the resistance.

After longer times or at higher temperatures, the structure undergoes a more radical change. Small crystals appear which contain a low dislocation density (of magnitude similar to that prior to cold working) and hence are relatively soft. These crystals nucleate in regions of high dislocation density, and thus in the microstructure appear at or near deformation bands. With time, these nuclei grow, and more nuclei form in the

remaining cold worked matrix. Eventually, these grains contact each other (at that time the original worked material has disappeared). The formation of these grains is referred to as *recrystallization*. During this recrystallization period, strength decreases drastically (Fig. 12 and 13).

Following recrystallization, the energy of the alloy is reduced further by a decrease in the grain-boundary area by grain growth. Thus, the long-time or high-temperature region of the annealing curve is referred to as *grain growth*. Because

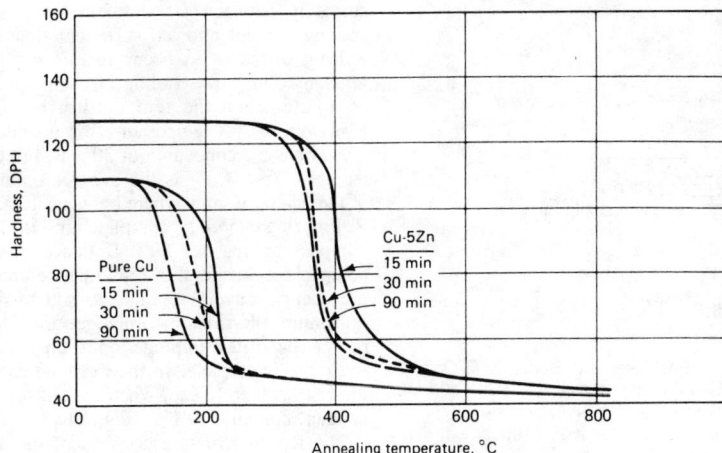

Both materials were originally cold rolled at 25 °C to 60% reduction in thickness. From same source as Fig. 9.

Fig. 16. Illustration of effect of annealing time on the annealing process in pure Cu and a Cu-5Zn alloy

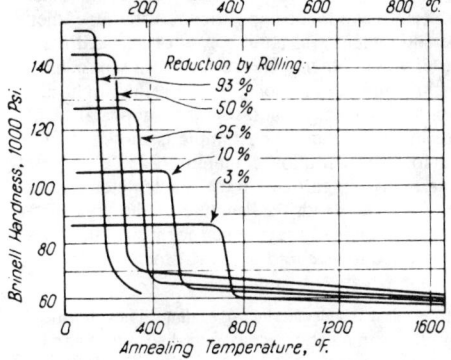

From G. Sachs and K. R. Van Horn, *Practical Metallurgy,* American Society for Metals, Metals Park, OH, 1951.

Fig. 17. Illustration of effect of amount of cold working on the annealing process for pure copper

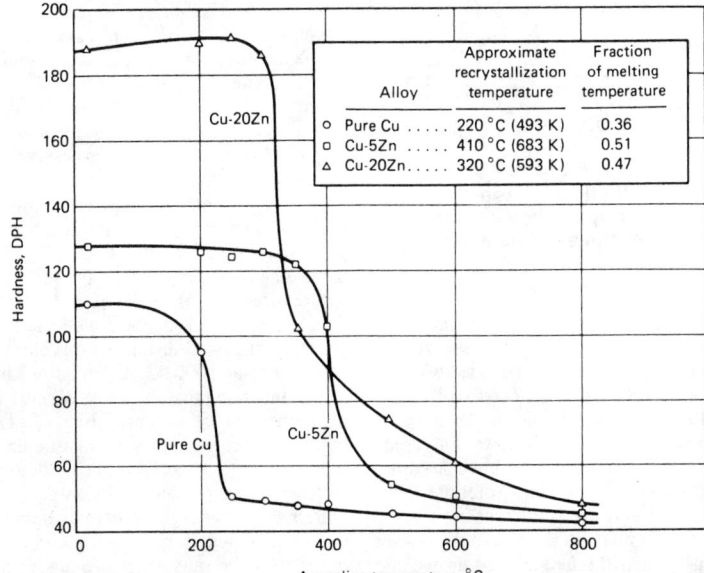

The alloys were originally cold rolled at 25 °C to 60% reduction in thickness. The recrystallization temperatures listed are based on the inflection point of each curve. From same source as Fig. 9.

Fig. 18. Illustration of effect of Zn content of Cu-Zn solid-solution alloys on the annealing process

zation process, and here a distinction must be made between solid-solution alloys and multiphase alloys. In many alloys containing second-phase particles, the presence of such particles favors formation of recrystallization nuclei and thus lowers the recrystallization temperature. In solid-solution alloys, even quite small amounts of solute can have potent effects on the recrystallization temperature. For example, addition of 0.05% Ag to copper will increase the recrystallization temperature from about 140 to about 340 °C. Thus, because silver only slightly lowers the electrical conductivity of copper, this alloy is used in applications which require the alloy to be cold worked for strength but in which slight heating may occur, and stress relaxation and recrystallization must be prevented.

If the solubility is sufficiently high to allow considerable solute concentration, the recrystallization temperature may decrease. This is illustrated in Fig. 18 for Cu-Zn alloys. This effect is expected to be related to the influence of zinc on the atom mobility in Cu-Zn alloys, and indeed the activation energy Q for diffusion in these alloys increases slightly with additions of up to 10% Zn, then decreases considerably with additions from 10 to 20% Zn.

It is useful here to note a rule of thumb—that the recrystallization temperature is approximately 0.3 to 0.6 of the absolute melting point. In the case of Cu-Zn solid-solution alloys, addition of zinc to copper lowers the melting point, and thus the recrystallization temperature will decrease for high zinc contents (e.g., 20 to 30%) (see Fig. 18).

Because recovery, recrystallization and grain-growth processes all involve atom movement, it is expected that the rates of these processes will depend on temperature in the same functional relation as does diffusion—that is, the rate is proportional to $e^{-Q/RT}$, where Q is the activation energy for the particular process. Thus, we may take as an approximation that the time required at a given temperature for recrystallization to commence (or for any given amount of recrystallization to be attained) will be inversely proportional to this exponential expression. Using as typical activation energies those given for diffusion, it is found that a decrease in temperature of 10 °C may increase by a factor of two the time required for recrystallization to commence.

Abnormal Grain Growth. The recrystallization process referred to in the preceding discussions is sometimes called *primary recrystallization,* to distinguish from other situations which lead to unusually large grains on annealing. Under conditions of very high amounts of plastic deformation and high annealing temperatures, abnormally large grains can develop following primary recrystallization: this is called *secondary recrystallization.* Such behavior is favored by the presence of grain-growth inhibitors, such as insoluble particles (e.g., inclusions). Abnormally large grains can also form if the metal has received a critical, but small, amount of deformation (e.g., about 10% or less) prior to annealing. In this case, primary recrystallization does not occur, but a few grains with less deformation than neighboring grains grow relatively rapidly at the expense of the cold worked grains. This effect is also called *germinative grain growth.*

Hot Working. Alternative definitions of cold working and hot working to those given previously can now be presented. *Cold working* is plastic deformation such that recrystallization does not occur within a reasonable time. *Hot working* is plastic deformation at or above a temperature at which recrystallization occurs in a rather short time. Thus, if the deformation temperature is sufficiently high, the metal cannot be cooled rapidly enough, even in a short time, to prevent recrystallization. This rather "spontaneous" recrystallization is depicted in Fig. 19.

HOMOGENIZATION OF CASTINGS

One of the most important commercial heat treatments is homogenization of castings. Such a treatment is used prior to mechanical processing of the cast ingot, and it is often used even when an object is cast into essentially the final shape. The temperatures and times used depend on the diffusion rate and the starting structure (the latter dictates the concentration gradients and the diffusion path). To understand how this enters into the situation, it is important to know how solidification occurs in alloys, and especially how

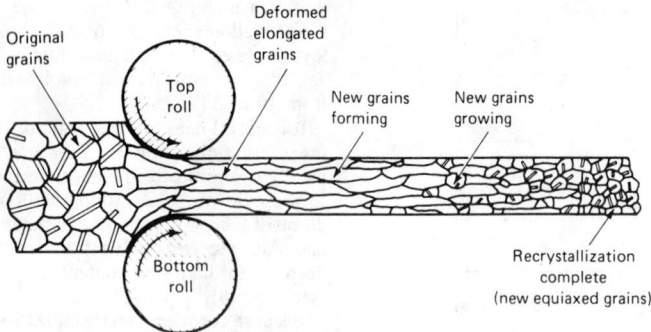

From R. A. Grange, in *Fundamentals of Deformation Processing*, ed. E. A. Backofen, J. J. Burke, L. F. Coffin, N. T. Reed and V. Weiss, Syracuse Univ. Press, Syracuse, N.Y., 1964, as adapted from J. M. Camp and C. B. Francis, *The Making, Shaping and Treating of Steel*, 5th Ed., U.S. Steel Corp., Pittsburgh, 1940.

Fig. 19. Schematic illustration of the change in grain structure on hot rolling

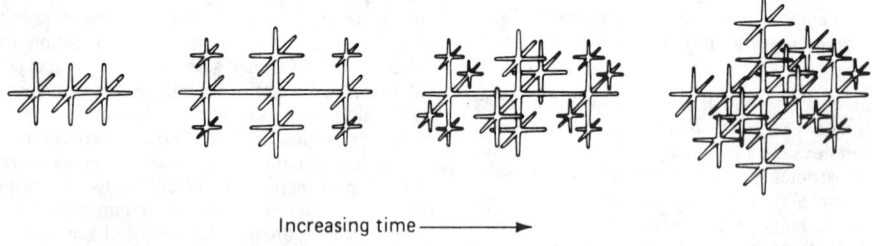

Increasing time ———▶

Adapted from P. S. Hurd, *Metallic Materials*, Holt, Rinehart and Winston, New York, 1968.

Fig. 20. Schematic illustration of a dendritic crystal forming in a liquid

chemical segregation develops during solidification.

Dendrite Formation. In metals and alloys, the crystals which form in the liquid during freezing generally have a configuration consisting of a main branch with many appendages. A crystal of such a morphology is called a *dendrite* ("fern-like"), and its formation is illustrated schematically in Fig. 20. During freezing, many crystals form, usually on the cold sidewalls of the mold, but also in the center of the casting. As these dendritic crystals grow, they eventually become large enough so that impingement occurs. Then the remaining liquid freezes, with a boundary formed between the differently oriented grains. The original dendritic pattern may not be apparent by observation of only the geometry of the grain boundaries outlining the grains.

Coring. In solidification of most alloys, chemical segregation intrinsically accompanies dendrite formation. To see how this develops, consider a hypothetical alloy whose phase diagram is that shown in Fig. 21. On slow cooling of a liquid alloy containing 30% B, crystallization commences at temperature T_0. The chemical composition of this crystal will be 10% B. As cooling continues, the crystal grows in size (as a dendrite). The phase diagram shows that the equilibrium composition of the crystal must follow the solidus line (line abc in Fig. 21). Thus, the crystal continuously changes its chemical composition, approaching 30% B as the temperature approaches that of completion of freezing, T_2. At T_2, the metal consists of crystals each containing uniformly 30% B. Note that the center of each crystal corresponds to the original nucleus, which had only 10% B when freezing commenced. Thus, on cooling, as the dendrites increased in size, from each layer frozen onto the crystal some B atoms must move throughout the crystal, including some to the center, to maintain the chemical composition uniformly at the value

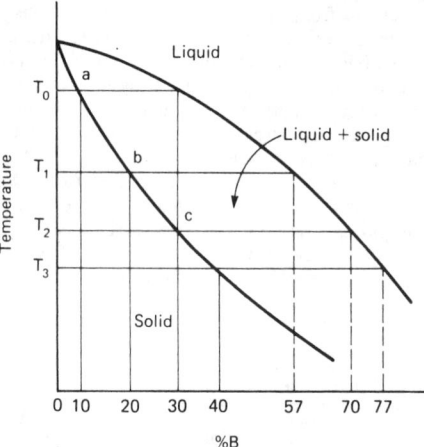

Adapted from same source as Fig. 9.

Fig. 21. Hypothetical phase diagram of system A-B, showing the composition of the solid as a 30% B alloy freezes

given by the solidus at any given temperature.

Clearly, such atom movements require finite time, and the question naturally arises as to what deviations from equilibrium will occur if the alloy is cooled rapidly from the liquid. A simplified picture of what occurs is as follows. On rapid cooling, the first crystals to form have a composition of 10% B. As these grow, the interface between the liquid and the solid crystals maintains the chemical composition given in the phase diagram. Thus, when the crystal has grown as the temperature has decreased from T_0 to T_1, the outside of the crystal will have a composition of 20% B. However, due to the rapid cooling, the center of the crystal will still be 10% B. Between the center and the outside, the composition varies

smoothly between 10 and 20% B. The rapid cooling has not allowed sufficient time for significant diffusion to occur in this composition gradient. On slow cooling, freezing would be complete when the temperature reached T_2. However, at this temperature the outside of the crystal has a composition of 30% B, but the center only 10% B. Thus the average composition of the crystal is somewhere between 10 and 30% B. Freezing cannot be complete until the average composition reaches 30% B (since this is the composition of the alloy), and hence undercooling occurs. Layers continue to add to the dendrite, until the sidearms impinge, and finally all of the dendrites impinge on each other, and freezing is complete. In the example used here, the last layer to freeze, when the sidearms make contact, contains 40% B (Fig. 21).

The frozen structure consists of dendrites in which the central regions of the main branch and of the sidearms contain about 10% B, and the regions where the sidearms met on completion of freezing contain about 40% B. If an etchant is used for which the rate of attack of the metal is sensitive to this compositional difference, then certain regions will be dissolved or attacked more readily than others. The surface then will consist of low and high regions, which reflect light differently, causing contrast in the appearance of the microstructure. An example is shown in Fig. 22 for a Ni-Cu alloy containing 30% Cu. Note that at low magnification the uneven etching has revealed the dendritic structure of the crystals.

This chemically segregated, dendritic structure is referred to as *cored*, and the process of its formation is called *coring*.

Chemical Homogenization Annealing. The chemical gradients in a dendritically cored structure can be reduced to an acceptable level by annealing at a sufficiently high temperature for a sufficient time. The rate of diffusion is given by an appropriate solution to Fick's law. As an approximation, the required time is $x^2 \cong Dt$, where x is the distance between the regions of low and of high concentration in the dendrite cell, which is one-half of the cell size. As an example, in Fig. 22(d) the cell size is approximately 40 μm, so x = 20 μm. Taking $D = 2 \times 10^{-10}$ cm²/s at 1000 °C for a Ni-30Cu alloy, then the required homogenization time is about 6 h. At 1100 °C, $D = 10^{-9}$, and the required time is 1 h. Obviously, higher temperatures lower considerably the required time, but other factors, such as excessive oxidation, must be considered.

If an ingot with a cored, cast structure, such as that shown in Fig. 22, is reduced in thickness 50% by rolling, then the dendritic cells will (on the average) be elongated in the rolling direction but reduced in thickness 50% in the through-thickness direction of the rolled plate. Thus the effective diffusion distance x becomes about 10 μm. Then, at 1000 °C, the required homogenization annealing time becomes about 1 h, instead of the 6 h for the as-cast structure. This points out the advantage in processing of coupling a homogenization anneal with plastic deformation to remove coring present in the as-cast structure.

In many alloy ingots, there also occurs gross, or ingot, segregation, where the chemical composition of the outside of the ingot may be different from that along its centerline. Here the final liquid freezes, and rejection of solute elements (frequently impurities) from the advancing front of the freezing crystals, in which they have a lower solubility, results in a region rich in these ele-

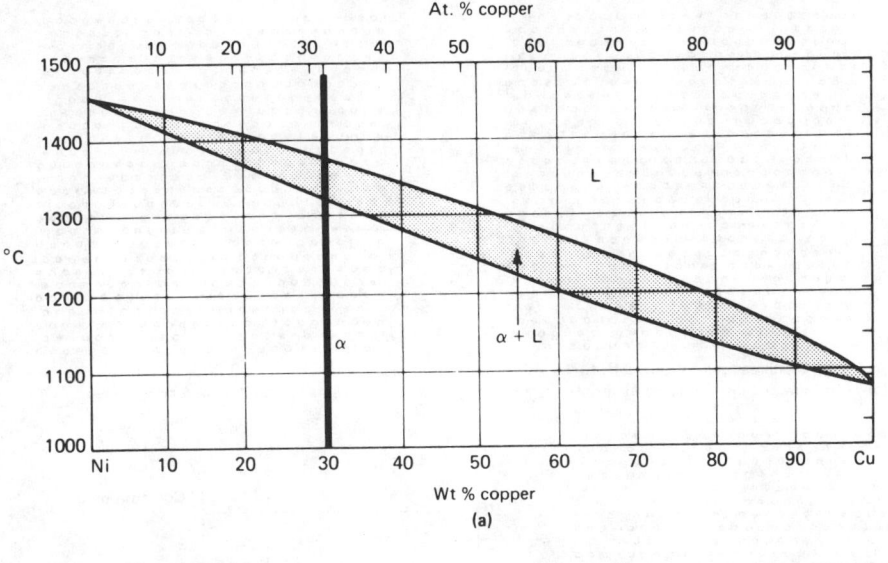

(a)

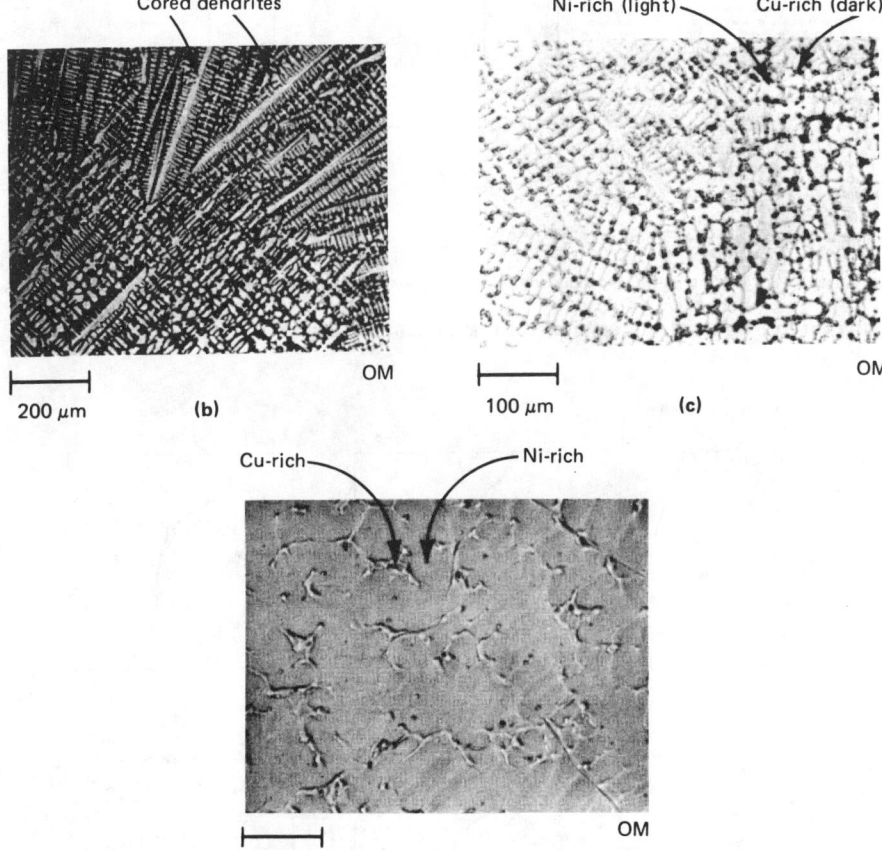

Cored dendrites

OM

200 µm   (b)

Ni-rich (light)   Cu-rich (dark)

OM

100 µm   (c)

Cu-rich   Ni-rich

OM

40 µm   (d)

(a) Ni-Cu phase diagram. (b, c and d) The microstructure at increasingly higher magnifications. Note that the dendrite cells are approximately 40 µm across (d). From same source as Fig. 9.

Fig. 22. The Ni-Cu phase diagram, and the microstructure of a Ni-30Cu alloy that has been cooled rapidly from the liquid, developing a nonequilibrium cored structure

ments near the center. However, a calculation similar to that above shows that the diffusion distance in this type of chemical inhomogeneity is much too great to be reduced appreciably by homogenization annealing.

In many commercial nonferrous alloys, the as-cast structure will not only be cored, but also will contain nonequilibrium, second-phase particles. In such systems, on slow cooling, when freezing

is complete a single-phase solid will be present (as described above). However, on rapid cooling, in which coring occurs, the liquid composition may increase to the value of the eutectic before freezing is completed. Then this liquid freezes to a solid eutectic structure. The microstructure then consists of a dendritically cored matrix containing small regions of multiphase, eutectic solid. These regions will dissolve on

proper solution heat treatment, and thus will be removed along with the coring.

PRECIPITATION HARDENING HEAT TREATMENTS

In designing alloys for strength, an approach often taken is to develop an alloy in which the structure consists of particles which impede dislocation motion dispersed in a ductile matrix. The finer the dispersion, for the same amount of particles, the stronger the material.

Such a dispersion can be obtained by choosing an alloy which, at elevated temperature, is single phase, but which on cooling will precipitate another phase in the matrix. A heat treatment is then developed to give the desired distribution of the precipitate in the matrix. If hardening occurs from this structure, then the process is called *precipitation hardening*. It is to be noted that not all alloys in which such a dispersion can be developed will harden. However, in this section attention is placed on systems which do harden if the precipitation process is properly controlled.

Solution Heat Treatment. A prerequisite to precipitation hardening is the ability to heat the alloy to a temperature range wherein all of the solute is dissolved, so that a single-phase structure is attained. This is shown schematically in Fig. 23 for a 10% B alloy in a hypothetical system A-B. Heating above the solvus temperature T_2 for this alloy, and holding in the α range for sufficient time, will form the single phase α. This is the required *solution heat treatment*. This structure is then retained at ambient temperature by cooling rapidly (e.g., water quenching) from the α range to prevent the precipitate from forming. The structure is supersaturated with respect to the solute, and hence is unstable.

The Process of Precipitation. After quenching from the α region (Fig. 23), precipitation is achieved by reheating the alloy below the solvus (T_2 in Fig. 23) at a suitable temperature for a suitable time. During this time, at localized regions (e.g., grain boundaries), the precipitates nucleate. Because these precipitates have a higher solute content than the matrix, the region in the matrix surrounding them is reduced in solute content. This forms a concentration gradient such that the solute atoms diffuse from the adjacent matrix toward the particles, allowing the precipitates to continue to grow. The rate of growth is diffu-

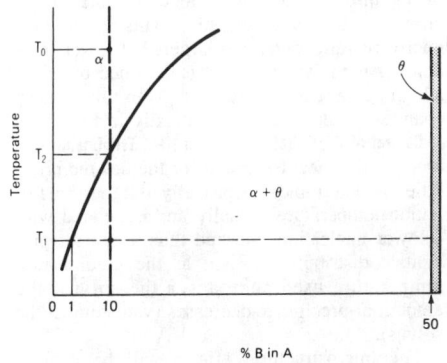

The decreasing solubility of B in α with decreasing temperature allows an alloy containing 10% B to be single-phase at high temperature (i.e., above T_2) but two-phase at low temperature (T_1). Adapted from same source as Fig. 9.

Fig. 23. Hypothetical phase diagram of system A-B

Source: Metals Handbook, Desk Edition, 28.64-28.78

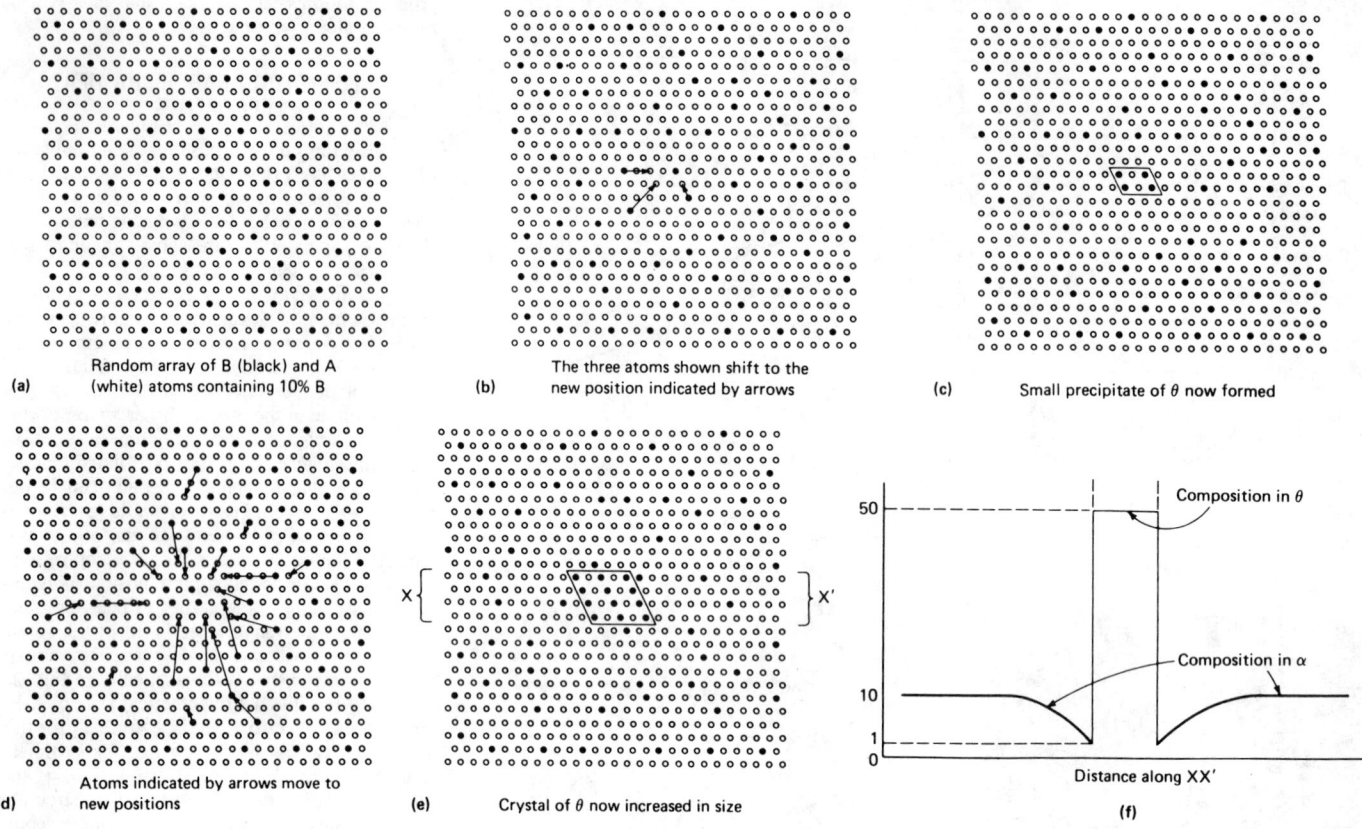

(a) Random array of B (black) and A (white) atoms containing 10% B

(b) The three atoms shown shift to the new position indicated by arrows

(c) Small precipitate of θ now formed

(d) Atoms indicated by arrows move to new positions

(e) Crystal of θ now increased in size

(f) Composition in θ / Composition in α / Distance along XX'

Time is increasing from (a) to (e), but at (e) equilibrium is not yet attained. In (f) is shown the concentration profile through the precipitate in (e). From same source as Fig. 9.

Fig. 24. Schematic illustration of formation of a precipitate in a supersaturated matrix

sion-controlled and is given by an appropriate solution to Fick's law. The precipitation process is depicted schematically in Fig. 24. Here the precipitate contains 50% B (see Fig. 23).

The maximum amount of precipitate which can form is given by the equilibrium amount, which can be calculated from a mass balance (lever rule). Once this equilibrium amount of precipitate has been attained, then further change in the precipitates is caused by the tendency for the system to reduce the precipitate/matrix interfacial area. Thus, with time at a given aging temperature, the smaller precipitates dissolve, with the solute diffusing through the matrix to contribute to the growth of the larger particles. This results in a microstructure containing larger, but fewer, particles. An equivalent effect is obtained by using a high aging temperature for a given time. These changes are depicted schematically in Fig. 25.

Control of Precipitation Through Heat Treatment. The precipitation heat treatment for the desired properties is determined empirically. Higher precipitation temperatures usually are associated with a lower nucleation rate and thus a coarser precipitate distribution. Also, as the precipitation temperature used approaches the solvus, the amount of precipitate decreases (vanishing at the solvus).

The microstructural effects will be demonstrated by referring to the aging of an Al-5Cu alloy. Figure 26 shows that this alloy must be solution heat treated at temperatures between 500 °C (solvus) and 575 °C (solidus). If this alloy is quenched from 545 °C (after 1 h), then aged for 12 h at 400 °C, the structure obtained will be that shown in Fig. 27(a). The precipitates are fine

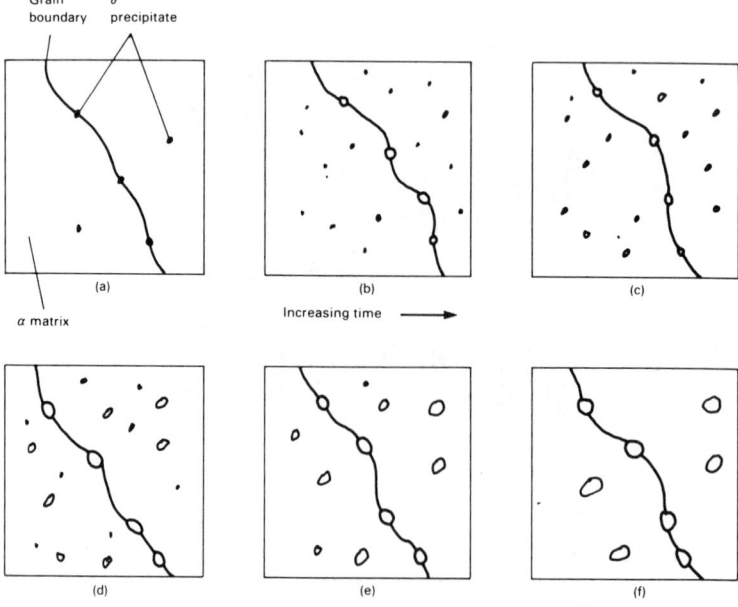

Fig. 25. Schematic illustration of formation of Θ precipitates in the α matrix (a and b) and their coarsening (c through f)

and evenly distributed, and are about 1 μm in size. If an aging temperature of 300 °C is used (for 12 h), then the structure in Fig. 27(b) is obtained. It can be seen that this higher aging temperature produced a somewhat coarser distribution of Θ than that at 300 °C.

In the Al-5Cu alloy (and in most other precipitation-hardenable alloys), the precipitation process is not as simple as that depicted schematically in Fig. 24. Instead, formation of the equilibrium precipitate (Θ in the Al-Cu alloy) is preceded by formation of one or more nonequi-

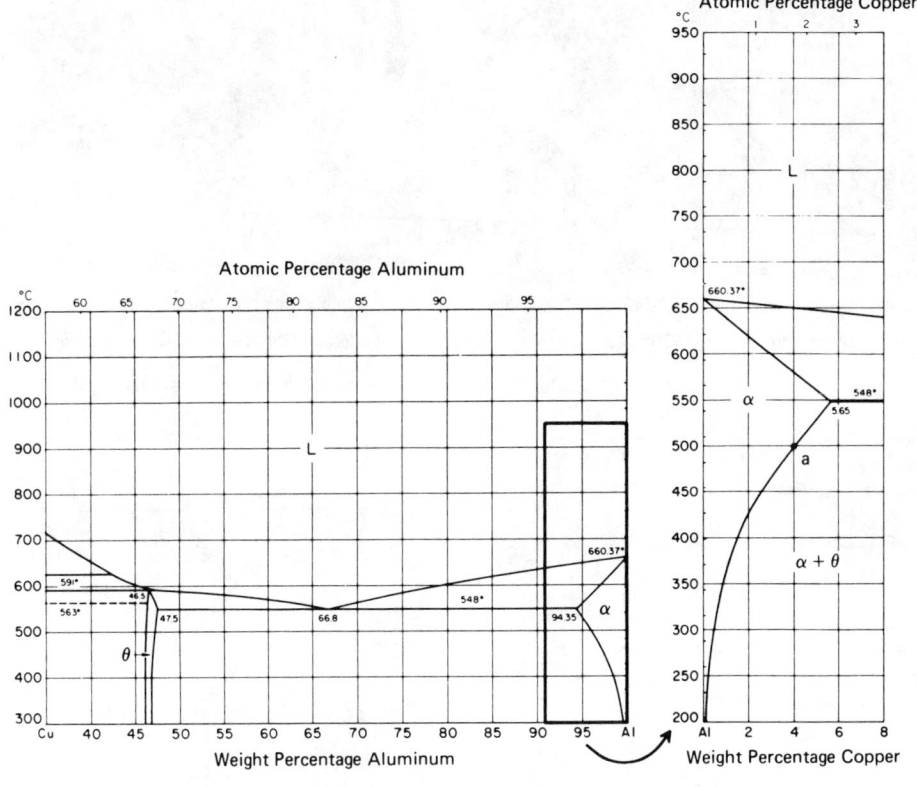

Atomic Percentage Aluminum

Atomic Percentage Copper

Adapted from same source as Fig. 9.

Fig. 26. The Al-rich end of the Al-Cu phase diagram, showing the 5% Cu line

librium configurations or precipitates. For an Al-4.6Cu alloy, this is shown in Fig. 28. In the earliest stage, Cu-rich zones form (called *Guinier-Preston Zones*), followed by two metastable precipitates (Θ'' and Θ'), before the equilibrium Θ appears. Note how fine these metastable phases are. In Fig. 28, the Θ'' particles are approximately 0.01 μm in size, corresponding to particles about 50 atoms in size.

Precipitation hardening. The strengthening which occurs during aging of an Al-4Cu alloy is illustrated in Fig. 29. Note that the maximum hardness is about double that in the as-quenched (supersaturated α) condition. Also note that the maximum hardness does not correspond to formation of the equilibrium Θ phase, but to the metastable, transition phases, which form in a considerably finer distribution than does Θ (compare Fig. 28 and 27).

The effect of temperature and time on aging is illustrated by the data in Fig. 30. As pointed out previously, the higher the precipitation tempera-

ture, the lower the maximum hardness, because less precipitate forms as the solvus temperature is approached. However, the higher the temperature, the higher the rate of precipitation, and hence the maximum hardness is attained in less time.

In most commercial precipitation-hardenable alloys, the rate of precipitation is low at ambient temperature, although sufficiently rapid to bring about measurable hardness changes in a reasonable time, as shown in Fig. 30 for aging at 30 °C. If hardening occurs at or near ambient temperature, it is termed *age hardening;* aging at other temperatures is called *precipitation hardening*.

Commercial alloys usually contain multiple elements, so that the required heat treating temperatures cannot always be deduced from an examination of related binary phase diagrams. In many alloys which contain mainly two alloying additions, the ternary phase diagram can be used as a guide for establishing the required heat treatments. For example, consider the Al-base alloy 2024. It contains approximately 4% Cu and 1% Mg, with lesser amounts of Mn, Si, Fe, Cr and Zn. Considering the alloy to be an Al-Cu-Mg ternary alloy, and using 4% Cu and 1% Mg to represent the average concentrations of these elements, then the Al-rich end of the ternary phase diagram can be used to illustrate the required heat treatments. This is shown in Fig. 31. The liquidus is about 650 °C, the solidus 570 °C and the solvus 500 °C. Thus, the solution annealing temperature must be between 500 and 570 °C. When consideration is given to the allowable range of the amounts of Cu and Mg in the 2024 alloy, then the solution annealing range is narrowed. Avoiding heating above the liquidus is of particular importance, because this will allow formation of small regions of liquid in the structure, which on cooling form compounds, and can lead to problems in achieving desired properties. The specification for the solution annealing temperature for alloy 2024 is 488 to 499 °C, only an 11 °C spread. Thus, if 494 °C is used, only a deviation of about ±5 °C is allowed.

Aging of alloy 2024 must be carried out below the solvus, about 500 °C. The response to aging for this alloy is typified by the data in Fig. 32.

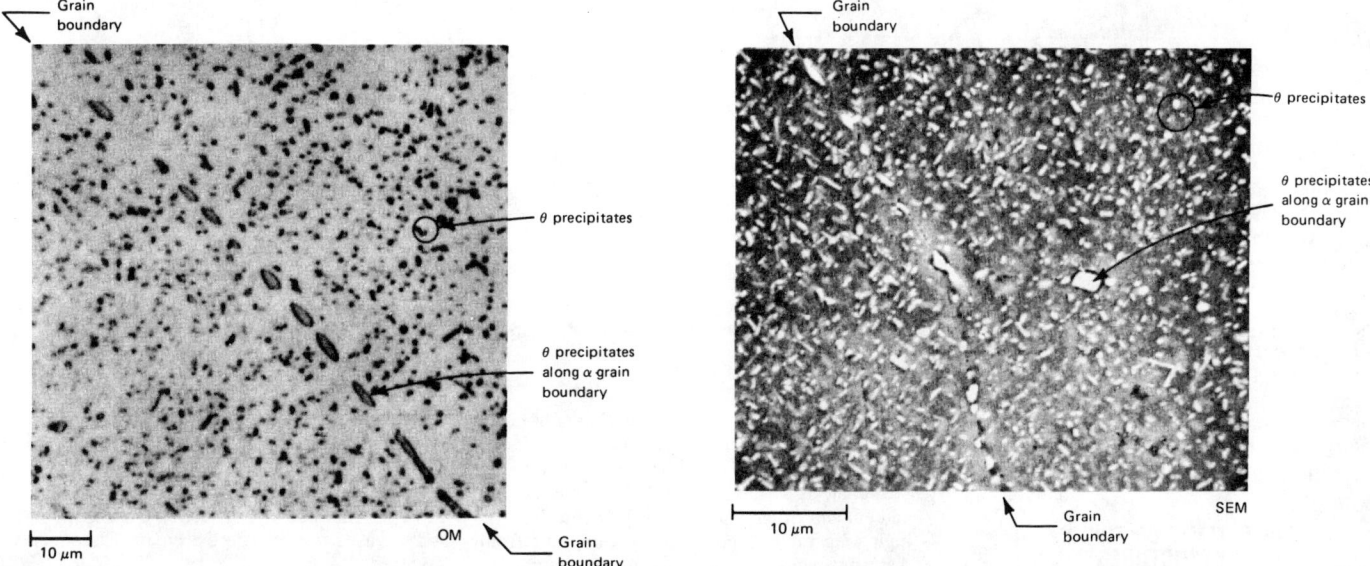

(a) is an optical micrograph, and (b) is a scanning electron micrograph. From same source as Fig. 9.

Fig. 27. Microstructure of Al-5Cu alloy heated for 1 week at 545 °C, cooled rapidly to 25 °C, then held 12 h at (a) 400 °C and (b) 300 °C

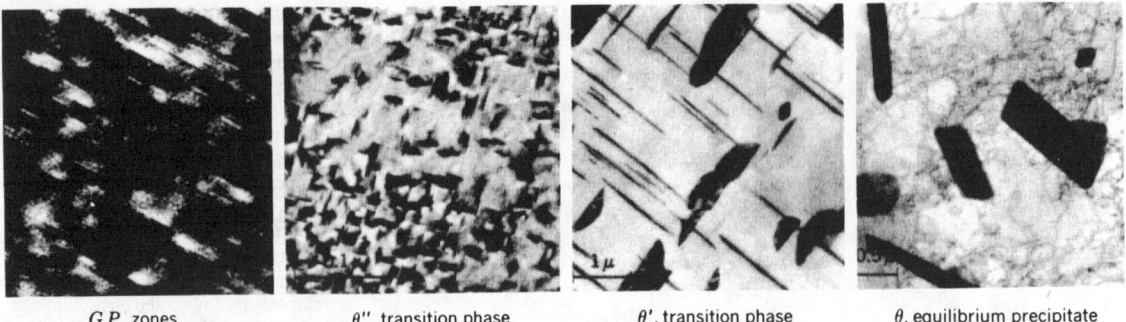

| G.P. zones | θ'', transition phase | θ', transition phase | θ, equilibrium precipitate |

The micrograph at far right shows θ precipitates similar in size to those shown in Fig. 27(b). Adapted from *Introduction to Materials Science,* by A. G. Guy, McGraw-Hill, New York, 1979.

Fig. 28. Transmission electron micrographs of precipitates formed in an Al-4.6Cu alloy with increasing aging time (left to right)

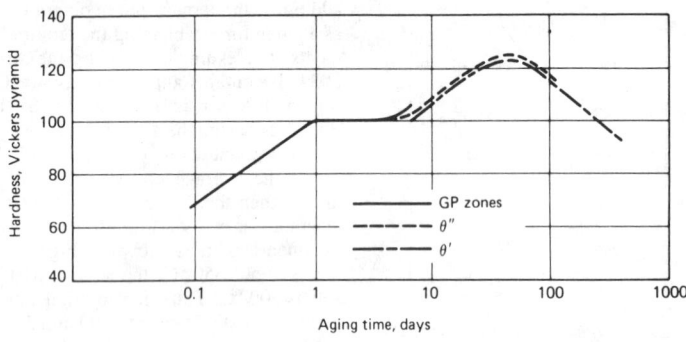

Compare this curve to the structures shown in Fig. 28. Adapted from J. M. Silcock, T. J. Heal, and M. K. Hardy, *J Inst Metals,* Vol 83, 1953, p 239.

Fig. 29. Hardness curve for an Al-4Cu alloy showing the relationship between the various precipitates formed and the hardness on aging at 130 °C

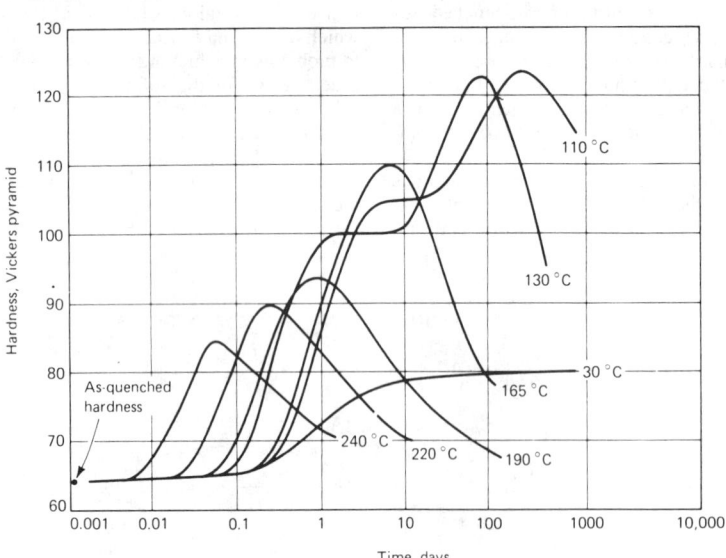

The alloy was solution annealed for at least 48 h at 520 °C, then cooled quickly (water quenched) to 25 °C. Adapted from H. K. hardy, *J Inst Metals,* Vol 79, 1951, p 321.

Fig. 30. Hardness as a function of aging time for an Al-4Cu alloy

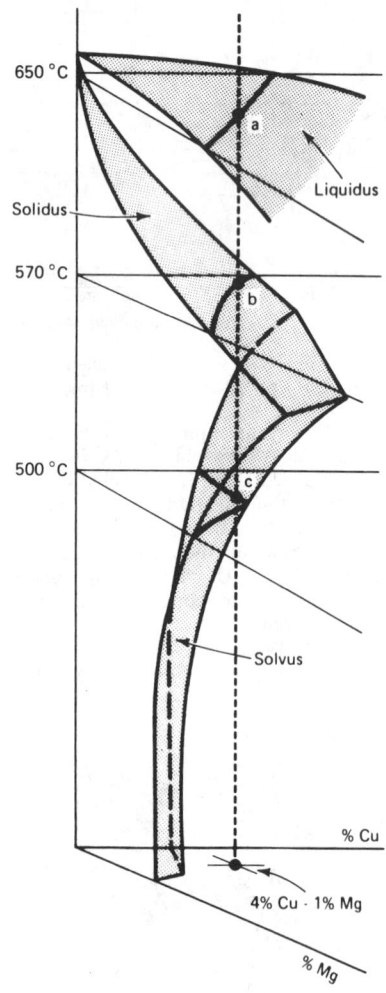

These temperatures are based on experimental observations. Adapted from same source as Fig. 9.

Fig. 31. Schematic illustration of the Al-rich end of the Al-Cu-Mg phase diagram, with the liquidus, solidus and solvus shown for a 4Cu-1Mg alloy

Note that there is a range of combinations of temperature and time which will give about the same optimum mechanical properties.

DEVELOPMENT OF TWO-PHASE STRUCTURES

In some nonferrous alloys (e.g., Ti-base alloys and high-zinc Cu-Zn alloys), the desired struc-ture consists of a mixture of two phases of comparable quantity (unlike the two-phase structures developed in precipitation hardening, where the precipitate is in the minority). The morphology and amount of each are varied by control of the high temperature used and the cooling rate from that temperature. The preferred microstructure can be quite complex, and the required treatment differs considerably for different systems, so that a systematic treatment of the principles involved is difficult. Instead, in this section a specific alloy will be used to illustrate the types of treatments involved.

In the Cu-Zn system, alloys containing about 40% Zn serve as the basis for some commercial alloys (e.g., Muntz metal and naval brass). The

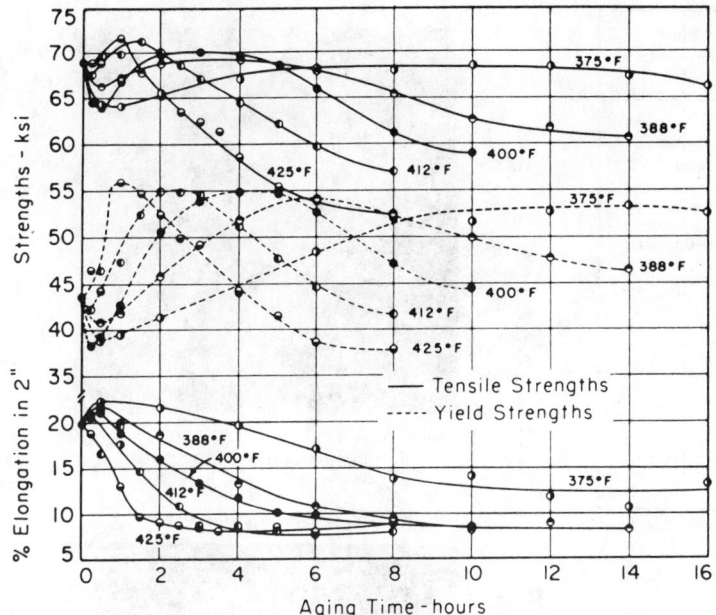

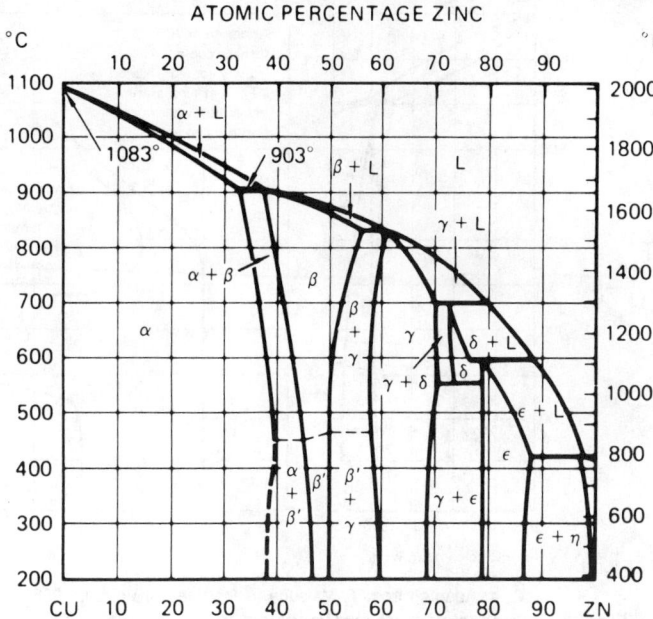

The initial condition was the natural aged state (temper T4). From W. A. Anderson, in *Precipitation from Solid Solution*, American Society for Metals, Metals Park, OH, 1958.

Fig. 32. Effect of aging time and temperature on mechanical properties of Al-base alloy 2024

The β phase is body-centered cubic; the β' phase is an ordered structure based on this arrangement.

Fig. 33. The Cu-Zn phase diagram

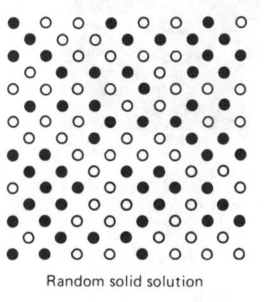

Random solid solution

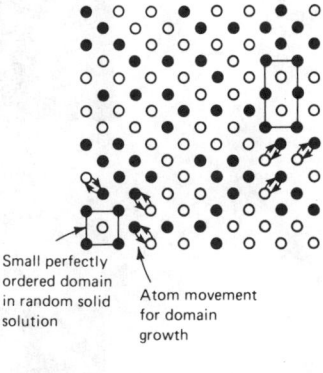

Small perfectly ordered domain in random solid solution

Atom movement for domain growth

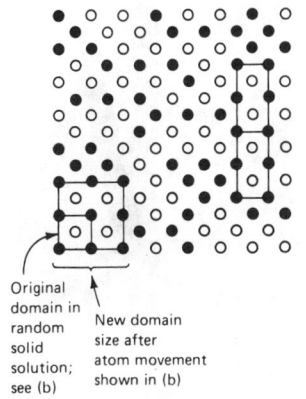

Original domain in random solid solution; see (b)

New domain size after atom movement shown in (b)

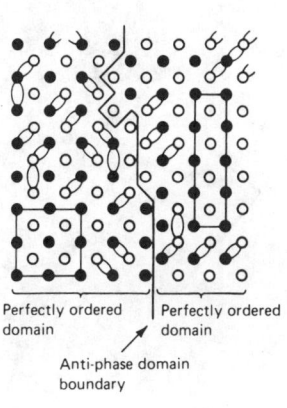

Perfectly ordered domain

Perfectly ordered domain

Anti-phase domain boundary

(a) (b) (c) (d)

Adapted from same source as Fig. 9.

Fig. 34. Schematic illustration of a possible mechanism for the formation of ordered β' from the disordered β in Cu-Zn alloys

Cu-Zn phase diagram (Fig. 33) shows that the alloys of interest are in the region of α and β phase stability. The β phase is body-centered cubic, with the Cu and Zn atoms located at random on the lattice sites.

On cooling to temperatures below the dashed line (about 450 °C), the Cu and Zn atoms take specific relative positions on the sites, forming an *ordered structure*, or a *superlattice*. This phase is denoted β' in Fig. 33. If the composition is exactly 50 at.% Zn, then the ordered structure is based on a body-centered cubic cell with Zn atoms at the center and Cu atoms on the corners (or vice versa).

The formation of an ordered structure from a disordered matrix of the same basic lattice involves the localized exchange of atom positions (via the vacancy mechanism) to the desired structure. This process is depicted schematically in Fig. 34. It can be seen that an ordered region grows by atoms at the β/β' interface, taking on

the arrangement of the ordered β' region. When two interfaces from neighboring regions meet, the arrangement of atoms may be out of sequence (out of phase). Such an interface is called an *antiphase boundary,* and the enclosed regions are called *domains*. The properties of the β' ordered structure depends on the degree of perfection (correctness of relative atom location) within the domains and on the domain size, both of which depend on the temperature and time involved in forming β' from β.

These alloys in the β' form are not suitable for commercial use, because this structure is brittle. However, alloys in which the β' phase coexists with the ductile α phase are useful. The Cu-40Zn alloy can be heat treated at high temperature so that it is all β. The structure developed at lower temperatures depends on the heat treatment, because this controls precipitation and formation of the α phase. If the alloy is cooled slowly from 800 °C, the phase diagram (Fig. 33) shows that

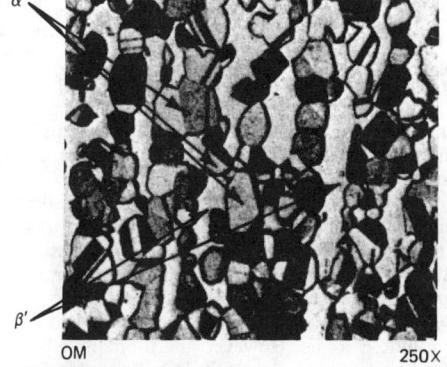

The clear, white regions are the β', and the dark and gray regions showing annealing twins are α. Adapted from D. K. Crampton, *Metal Progress,* Vol 46, 1944, p 276.

Fig. 35. Typical microstructure of annealed Muntz metal (Cu-40Zn)

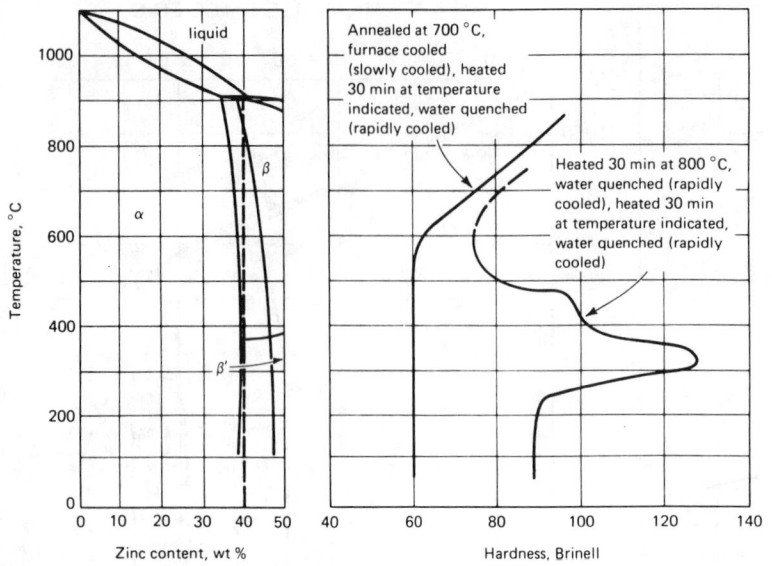

All β′

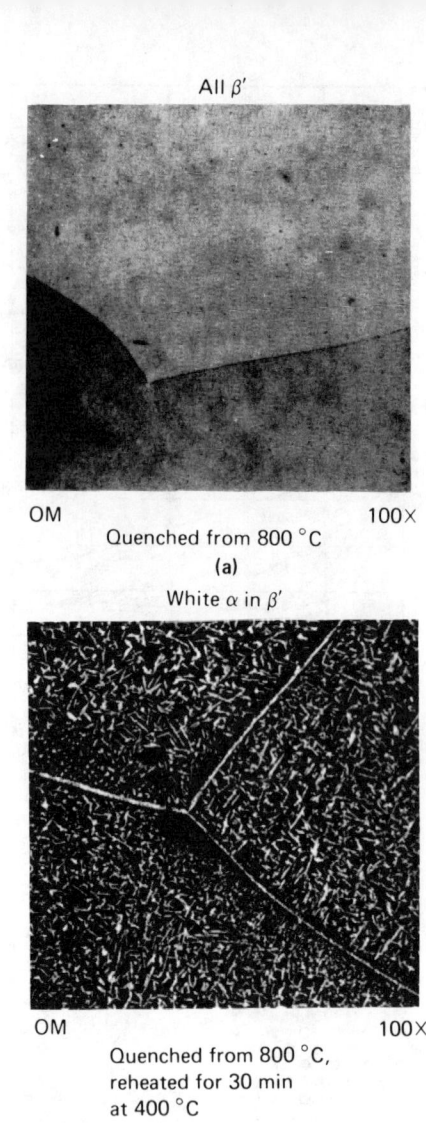

OM   100×

Quenched from 800 °C

(a)

White α in β′

Quenched from 800 °C,
reheated for 30 min
at 400 °C

(b)

White α in β′

OM   100×

Quenched from 800 °C,
reheated for 30 min
at 600 °C

(c)

Adapted from T. Matsuda, *J Inst Metals*, Vol 39, 1928, p 67.

Fig. 36. Influence of heat treatment on the hardness at 25 °C of a Cu-40Zn alloy

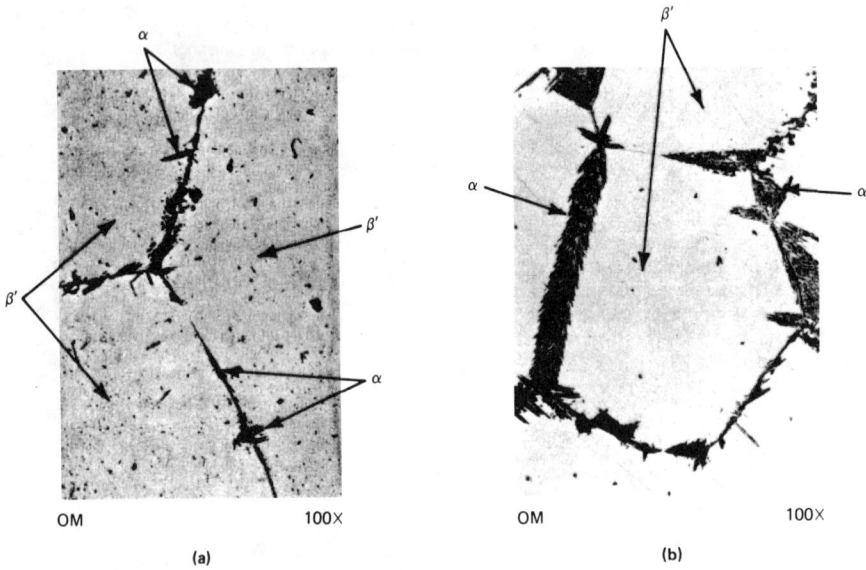

Even rapid cooling has not prevented some α from forming. (a) Cu-40Zn alloy quenched into ice water from 825 °C. Adapted from *Engineering Physical Metallurgy*, by R. H. Heyer, Van Nostrand Reinhold, 1939, used with permission of Brooks/Cole Publishing Co. (b) Quenched Muntz metal. From *Metals Handbook*, 8th Ed., Vol 7, American Society for Metals, Metals Park, OH, 1972.

Fig. 37. Microstructures typical of Cu-40Zn alloys cooled rapidly from the β region to 25 °C

The higher reheating temperature gives a coarser structure, and hence a softer material. Adapted from C. H. Samans, *Metallic Materials in Engineering*, MacMillan Co., New York, 1963.

Fig. 38. Microstructures of Cu-42Zn alloy quenched from the β region, then reheated to develop an α precipitate structure

at 25 °C the alloy should consist of approximately equal amounts of α and β′. Figure 35 shows a typical microstructure.

One of the curves in Fig. 36 shows that the amount of β′ influences hardness. The alloy was cooled slowly from 700 °C, where it was mostly β, to 25 °C, then reheated to temperature for 30 min, followed by rapid cooling. On heating at 800 °C, the structure is all β, and on rapid cooling little α forms. However, the β orders to β′, giving a hardness around 90 HB. Reheating for 30 min in the lower temperature range (25 to 500 °C) is not sufficient to affect significantly the originally slowly cooled structure, and the hardness remains constant. In this temperature range, the structure consists of approximately equal

amounts of α and β′. However, as the temperature increases from 500 °C, 30 min is sufficient time to allow the equilibrium amounts and α and β to form. Thus, as the temperature increases, increasing amounts of β and decreasing amounts of α are present at temperature, giving increasing amounts of β′ on cooling rapidly to 25 °C, and hence a rise in hardness.

If the Cu-40Zn alloy is cooled rapidly to 25 °C after sufficient holding (e.g., 30 min) above about 750 °C, a structure of essentially all β′ is obtained. Often some α is observed to have formed in the β grain boundaries, and the morphology will vary somewhat depending on the exact cooling rate. Usually, the α is present as "needles" emanating from the boundaries, with a clear

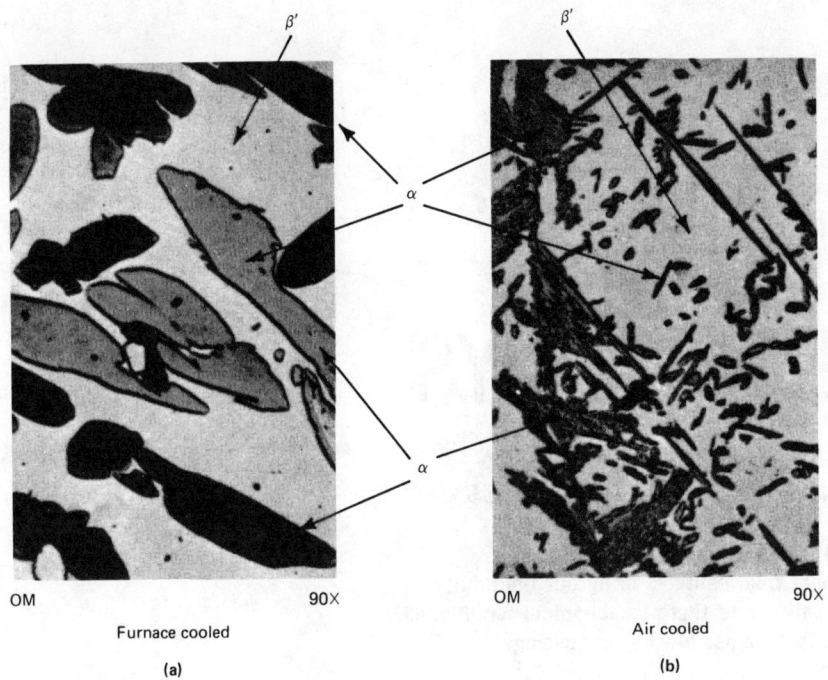

OM   90×   OM   90×

Furnace cooled   Air cooled

(a)   (b)

Adapted from R. F. Mehl and G. T. Marzke, *Trans AIME,* Vol 93, 1931, p 123.

Fig. 39. Microstructures of a Cu-43Zn alloy after cooling from 700 °C, the β region, showing effect of cooling rate on structure of α crystals

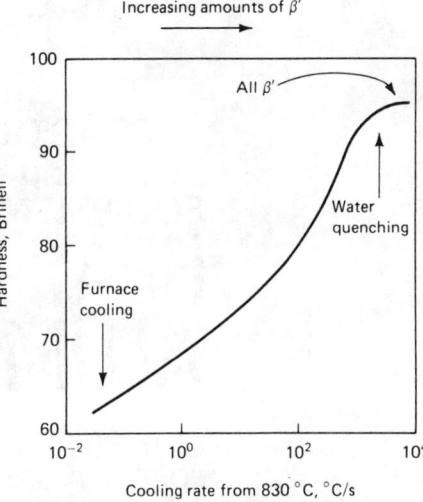

Adapted from T. Matsuda, *J Inst Metals,* Vol 39, 1928, p 67.

Fig. 40. Effect of cooling rate from the β region on hardness of a Cu-40Zn alloy

crystallographic relation between the α and the β′ in which it has formed. Figure 37 shows two examples.

On reheating β′ in the intermediate temperature range, the morphology of the α formed will vary depending on the exact heat treatment. Also, reheating will influence the change in the ordered structure. Both changes affect properties, and the hardness can be increased considerably by judicious treatment. In Fig. 36 are shown hardness data for a Cu-40Zn alloy after reheating for 30 min following an initial treatment of quenching from 800 °C. Supposedly the maximum hardness obtained by treatment around 300 °C is caused by formation of a fine α precipitate and some changes in the ordered β′ phase. The types of microstructures obtained by such heat treatments are illustrated in Fig. 38 for a Cu-42Zn alloy. In this alloy the zinc content is sufficiently high to

completely suppress any α formation on rapid cooling from β, giving at 25 °C only β′ (Fig. 38a). Reheating for 30 min at 400 °C gives a fine α precipitate on the β′ grain boundaries, and a fine intercrystalline precipitate of α (Fig. 38b). Reheating for 30 min at a higher temperature, 600 °C, gives a coarser α structure (Fig. 38c).

If the rate of cooling from the α region is quite low (several hours to 25 °C), then α nucleates at a high temperature at which the nucleation rate is low, and the α crystals grow relatively large as few crystals nucleate. This gives a rather coarse structure, typified by Fig. 39(a). As the cooling rate increases, the nucleation rate increases, but the individual α crystals do not have time to grow large before the temperature becomes too low for significant growth to continue. This gives a finer structure (see Fig. 39b) and increases strength. Eventually, the cooling rate becomes sufficient

to suppress formation of α altogether, giving a structure entirely of highly unstable β′ at 25 °C. Figure 40 illustrates the influence of cooling rate from β on hardness.

SELECTED REFERENCES

Physical Metallurgy Principles, 2nd Ed., by R. E. Reed-Hill: Van Nostrand Reinhold, New York, 1973

Fundamentals of Physical Metallurgy, by J. D. Verhoeven: Wiley, New York, 1975

Structure and Properties of Engineering Alloys, by W. F. Smith: McGraw-Hill, New York, 1981

Structure and Properties of Alloys, by R. M. Brick, R. B. Gordon and A. Phillips: McGraw-Hill, New York, 1965

Heat Treatment, Structure and Properties of Nonferrous Alloys, by C. R. Brooks: American Society for Metals, Metals Park, OH, 1982

An Introduction to the Solidification of Metals, by W. C. Winegard: Institute of Metals, London, 1964

Principles of Solidification, by B. Chalmers: Wiley, New York, 1964

Precipitation Hardening, by J. W. Martin: Pergamon, New York, 1968

Aluminum Alloys: Structure and Properties, by L. F. Mondolfo: Butterworths, Boston, 1976

PRACTICES and EQUIPMENT for HEAT TREATING ALUMINUM ALLOYS

By DAVID S. THOMPSON, OGLE R. SINGLETON, ROBERT D. McGOWAN, and GRANT E. SPANGLER

The alloys listed in the Metal Progress Data Sheet on p. 86 require rather sophisticated heat treatments to bring out their full engineering potential. Heat treating and thermomechanical techniques, as well as the commercial equipment needed for processing, are discussed in detail in this article.

THERE ARE TWO prerequisites for a heat-treatable alloy to be considered age or precipitation hardenable: 1. Solid solubility of major alloying elements shall decrease with decreasing temperature. 2. Guinier-Preston (GP) zone solvus shall be sufficiently high to be able to form GP zones in a reasonable time.

Three major systems produce practical heat-treatable aluminum alloys — Aluminum Assn. designations for wrought alloys are used:

Al-Cu	2XXX series (2219, 2021, 2014, 2024)
Al-Mg-Si	6XXX series (6063, 6061)
Al-Mg-Zn	7XXX series (7005, 7039, 7075, 7178, 7079)

Other systems include Al-Ag, Al-Zn, Al-Mg, Al-Mg-Ag, or Al-Si, but none is practical for reasons such as economics, limited hardening effects, or inability to nucleate homogeneous precipitation.

Mr. Thompson is director, alloy development; Mr. Singleton is research engineer, fabrication technology; and Mr. Spangler is director, metallurgy, Metallurgical Research Div., Reynolds Metals Co., Richmond, Va. Mr. McGowan is metallurgical engineer, Aluminum Mill Products Div., American Metal Climax Inc., Morris, Ill.

General Characteristics

The degree of strengthening due to precipitation hardening is much greater than that due to solid solution hardening. The 7000 series (Al-Zn-Mg alloys) have the highest mechanical properties because of the high volume fraction of $MgZn_2$ available for precipitation. The hardening effect depends on the volume fraction and also the size of the precipitates. GP zones are the most effective hardeners because of their small size and large numbers. Generally, it is desired to age to maximum hardness. Tempers are designated T5, T6, or T8 (see box). In certain circumstances, overaging may enhance the stress-corrosion resistance; and this temper is designated T7.

Many factors can influence precipitation and resultant properties such as: quench conditions, deformation after quenching, minor addition elements, and aging practice.

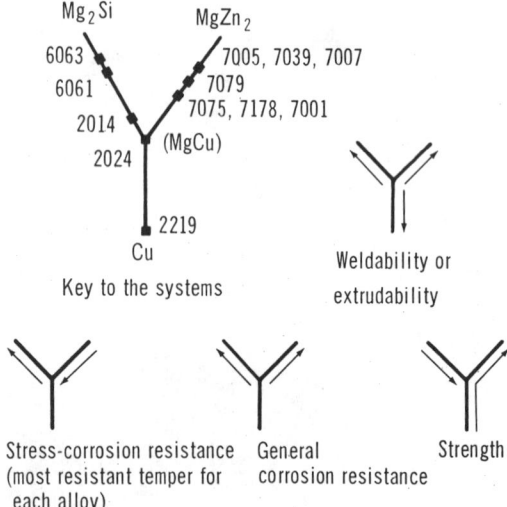

Fig. 1 — Diagrams illustrate general properties of heat-treatable aluminum alloys. Arrows indicate direction of improving properties, but not magnitude. No arrow indicates no change.

Reprinted from Metal Progress, September 1970, 78-83, © 1970 American Society for Metals

It is interesting to note that a compositional connection exists between the three major systems and their subgroups. This connection is magnesium and to a lesser extent copper, as shown in Fig. 1. The relative positions of some principal alloys are shown on this diagram, although distances are not proportional to composition.

The Data Sheet on p. 86 includes a rating of the heat-treatable alloys according to corrosion resistance (general and stress) and weldability. In many applications, alloy selection depends upon the best compromise of these properties. Poor general corrosion resistance is probably the least detrimental property since the effect of corrosion can be minimized by coating or by cladding° with an alloy more anodic than the core.

The Process of Heat Treatment

Heat treating is a time-temperature process with or without cold working of the alloy. A typical cycle is shown in Fig. 2. Vital features are the choice of solution heat treatment temperature, quench rate, and aging practice. Aging can consist of holding at ambient temperature (natural aging) until substantially stable properties are achieved, or holding at an elevated temperature (artificial aging) to achieve stable properties in a shorter time. Other details of the cycle may include cold working which can have important effects (discussed later).

The specific details of most heat treatment practices are contained in the military specifications for aluminum alloys (MIL-H-6088D, Amendment 2, Dec. 23, 1968).

Solution Heat Treatment — The primary objective of solution treatment is to obtain as complete a solid solution of the alloying elements as

___

°Cladding alloys include the conventional low-strength aluminum alloys such as 7072 and 1230 as well as the high-strength, age-hardenable alloys such as 7011 and 6003.

How Aluminum Assn. Designates Tempers of Heat-Treatable Alloys

-W Solution Heat Treated: An unstable temper applicable only to alloys which spontaneously age at room temperature after solution heat treatment. Specific only when period of natural aging is indicated (-W ½ hr).

-T Thermally Treated to Produce Stable Tempers Other than -F, -O, or -H: Applies to products which are thermally treated, with or without supplementary strain hardening, to produce stable tempers. The -T is always followed by one or more digits. Numerals 1 through 10 indicate one specific sequence of basic treatments, as follows:

-T1 Partially Solution Heat-Treated and Naturally Aged to a Substantially Stable Condition: Applies to products partially solution heat treated by an elevated-temperature rapid-cool fabrication process (casting or extrusion).

-T2 Annealed (cast products only): Indicates a type of annealing treatment to improve ductility and increase dimensional stability.

-T3 Solution Heat Treated and Cold Worked: Applies to products cold worked to improve strength, or in which effect of cold work in flattening or straightening is recognized in applicable specifications.

-T4 Solution Heat Treated and Naturally Aged to a Substantially Stable Condition: Applies to products not cold worked after solution heat treatment, but in which effect of cold work in flattening or straightening may be recognized in applicable specifications.

-T5 Partially Solution Heat Treated and Artificially Aged: Applies to products artificially aged after an elevated-temperature, rapid-cool fabrication process, to improve mechanical properties and/or dimensional stability.

-T6 Solution Heat Treated and Artificially Aged: Applies to products not cold worked after solution heat treatment, but in which effect of cold work in flattening or straightening may be recognized in applicable specifications.

-T7 Solution Heat Treated and Stabilized: Applies to products stabilized to carry them beyond point of maximum hardness, providing control of growth and/or residual stress.

-T8 Solution Heat Treated, Cold Worked, and Artificially Aged: Applies to products cold worked to improve strength, or in which effect of cold work in flattening or straightening is recognized in applicable specifications.

-T9 Soluton Heat Treated, Artificially Aged, and Cold Worked: Applies to products cold worked to improve strength.

-T10 Partially Solution Heat Treated, Artificially Aged, and Cold Worked: Applies to products artificially aged after an elevated-temperature, rapid-cool fabrication process and then cold worked to improve strength.

A period of natural aging at room temperature may occur between or after the operations listed for tempers -T3 through -T10. Control of this period is exercised when it is metallurgically important

Additional digits may be added to designations -T2 through -T10 to indicate a variation in treatment which significantly alters the characteristics of the product:

-TX51 Stress Relieved by Stretching: Applies to products stress relieved by stretching the following amounts after solution heat treatment:

Sheets and Plates: 1½ to 3% permanent set

Rods, Bars, and Shapes: 1 to 3% permanent set.

Applies to sheets and plates and rolled or cold finished rods and bars. These products receive no further straightening after stretching.

Applies to extruded rods, bars, and shapes only when designation is subdivided as follows:

-TX510: Applies to extruded rods, bars, and shapes which receive no further straightening after stretching.

-TX511: Applies to extruded rods, bars, and shapes which receive minor straightening after stretching to comply with standard tolerances.

-TX52 Stress Relieved by Compressing: Applies to products stress relieved by compressing after solution heat treatment to produce a nominal permanent set of 2½%.

-TX53 Stress Relieved by Thermal Treatment: The following two-digit -T temper designations have been assigned for some wrought products heat treated by the user:

-T42: Applies to some alloys solution heat treated by the user.

-T62: Applies to some alloys solution heat treated and artificially aged by the user.

Source: Metal Progress, September 1970, 78-83

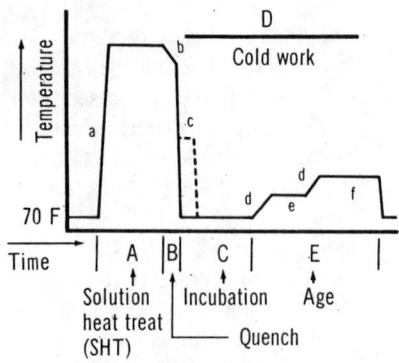

Fig. 2—The principal and secondary features of heat treatment processes for aluminum include (a) heat-up to SHT temperature (850 to 1,000 F); (b) quench delay (cooling between leaving SHT oven and quench); (c) interrupted or step quench (unusual); (d) heat-up during age; (e) first step age (200 to 400 F); (f) second step age (200 to 400 F).

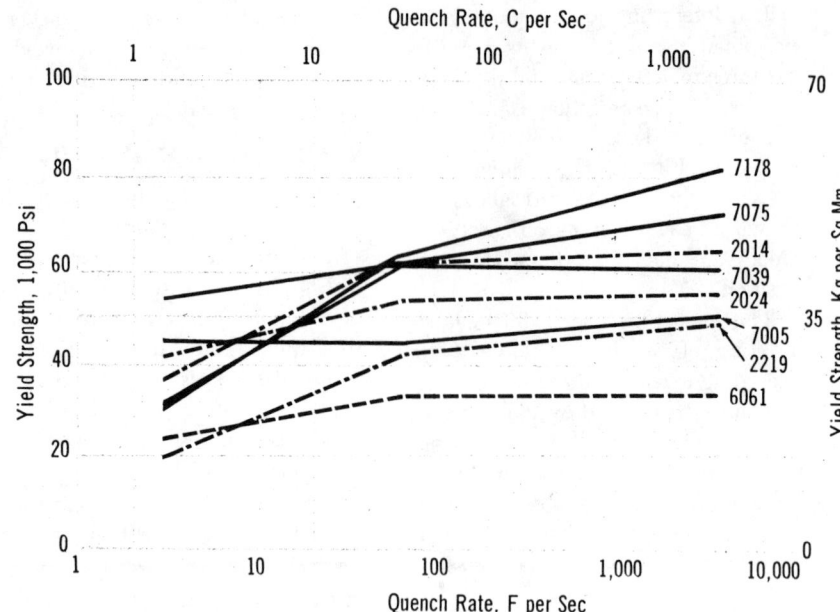

Fig. 3—These data show the effect of quench rate on yield strength of various aluminum alloys after aging to T6 temper.

possible without producing melting or undesirable recrystallization. This can be achieved by either mill (or press) quenching or by a separate (formal) solution heat treatment cycle.

Mill quenching comes directly after hot forming (rolling or extrusion) and is carried out at a temperature above the solvus of the alloy. This process can be less expensive than a separate heating cycle.

However, there are two drawbacks: control of temperature during hot working and hot working characteristics above the solvus are poor. Consequently, this process is only suitable for dilute alloys with a large temperature difference between solvus and solidus.

A separate, formal solution heat treatment is generally required to achieve maximum property levels. Holding time is controlled by the metallurgical structure of the material and can vary from as low as 5 min to as long as 12 to 16 hr. Commercial practices usually do not produce complete solid solution, and recently it has been shown that extended solution treatment times can lead to increased properties, where recrystallization or grain growth presents no problems.

Quench Delay — Military specification MIL-H-6088D details the maximum quench delays permitted for high-strength alloys. This specification is really aimed at controlling the minimum temperature of the metal immediately prior to quenching to avoid premature precipitation.

Quenching — After solution heat treating, quenching is required to retain both the solid solution and a sufficient supersaturation of vacancies for effective aging. Too slow a quench can lead to: 1. Precipitation of large equilibrium precipitates which contribute little to strength. 2. Increased precipitation in grain boundaries which may reduce ductility or corrosion resistance. 3. Annealing out of vacancies which would subsequently alter the precipitation kinetics.

The maximum quench rate at which these symptoms appear will vary from alloy to alloy. If this critical quench rate is high (100 to 10,000 F per sec), the alloy is quench sensitive. If the rate is low (0.1 to 1 F per sec), the alloy is quench insensitive.

The influence of quench rate on mechanical properties of various alloys in the T6 temper is shown in Fig. 3. The more highly alloyed materials are the most sensitive. It is well known

that chromium (or the unrecrystallized structure which chromium promotes) greatly increases quench sensitivity of the 7000 series alloys, particularly those with high zinc, magnesium, and copper contents.

One factor almost universally overlooked is that of optimizing the aging practice for slowly quenched material. In Fig. 4, aging curves show that protracted aging of very slowly quenched Al-Zn-Mg alloys can yield high properties. Also, the beneficial effect of incubation at room temperature prior to aging is shown. These data suggest that poor quenching primarily lowers the vacancy concentration which in turn lowers the maximum temperature for homogeneous nucleation and also slows the diffusion of solute atoms to GP zones. Note that if only a typical T6 aging practice had been used (say 16 hr at the aging temperature used in Fig. 4), then virtually no age hardening would take place for the slow-cooled specimen with no incubation. Thus, a false idea of the quench sensitivity of this alloy would have been formed. A similar, less pronounced effect is observed in some higher-strength alloys.

Mechanical properties are generally improved by specifying the fastest possible quench rate. For other properties, the reverse can be true. With copper-free 7000 series alloys, for example, a slow quench rate is particularly beneficial to the stress-corrosion resistance. Also, residual stresses and distortion are generally minimized with a slow quench.

Thus, the selection of a suitable quench medium may depend upon several factors. Cold water, the fastest quench, is also the most convenient and economical. Hot water on heavy sections minimizes residual stresses and distortion of parts.

New quenchants may further reduce quenching stresses. Such stresses are not necessarily caused simply by speed of quench but are related more directly to the change in heat transfer coefficient with surface temperature and from point to point on the surface due to the formation of steam films. One method of breaking down such steam films is the technique called "superquenching," utilizing high pressure and high volumes of coolant.

Another approach is to deposit a thin insulating film on the surface of the part to even out the heat transfer to the quenchant and so increase the over-all speed of quenching in a boiling fluid. The vapor-film mode of cooling is also effective. With liquid nitrogen as the quenchant, a stable vapor-film exists to below room temperature. Distortion in such quenches is low; however, the cost and the low cooling power generally limit use to complex sheet parts.

It is also possible to reduce residual stresses through the use of hot nonaqueous coolants such as oil, liquid metals, or fused salts. Generally, these quenchants have such a high boiling point that vapor phase cooling is not obtained.

These high-temperature quench media are used in the process known as isothermal quench aging (IQA). Although IQA has shown definite advantages, industry has been reluctant to take on the problems such as fire hazards, postquench cleanup, and bath maintenance generally associated with nonaqueous quenchants.

Stress Relief — Stress relief is generally done by stretching; how-

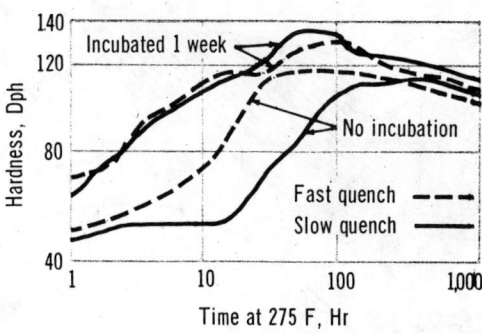

Fig. 4 — Here is the effect of quench rate and room temperature incubation prior to rapid heating to 275 F on aging curve of a 5 Zn, 1 Mg, 0.1 Cr alloy.

ever, some reduction can be achieved by roller leveling. Surface residual stress can be reduced by shot peening or planishing.

Thermal treatments above about 425 F will dimensionally stabilize and stress relieve parts, although these treatments result in lower mechanical properties. Wrought products are stress relieved only partially during artificial aging treatments for maximum hardening. For example: residual quenching stresses are reduced about 10% for 7075 alloy aged at 240 F and by about 30% for 2014 aged at 320 F.

A method especially suitable for relief of residual stresses in complex castings or forgings is that of "upquenching." The method involves strain reversal which is accomplished by reversing the quenching process — that is, the part is cooled to cryogenic temperatures, then up-quenched with steam. The method can result in tensile stresses on the surface, however, which may be undesirable.

Incubation — Incubation is merely a portion of the aging cycle and if no elevated-temperature aging is to be used, is synonymous with natural aging.

In the case of Al-Zn-Mg-Cu alloys, T6 mechanical properties can vary with incubation time prior to aging. No satisfactory explanation has been offered for this effect, though it is almost certain that reversion is involved. The use of a slow heating rate between room temperature and the aging temperature or the use of a two-step aging practice can eliminate these variations.

In the case of the 6000 series alloys, increased incubation leads to a continual decrease in final properties

for highly alloyed material or to an increase for dilute alloys. In these alloys, interruption of the quench at a temperature above the GP zone solvus or brief high-temperature treatments above the GP zone solvus can retard the influence of incubation by annealing out vacancies at the high temperature.

Postquench Working — As was pointed out above, working is useful for the relief of quenching stresses. Postquench working can also serve other purposes including simple straightening, flattening, increasing strength, improving corrosion resistance, reducing incubation effects, and providing nucleation sites for the later stages of precipitation.

In alloys 2021, 2024, and 2219, the T8 temper (1 to 8% cold work, followed by artificially aging) results in markedly increased strength and stress corrosion resistance. Cold work also accelerates the aging process. These effects may be obtained by any method of cold working, including explosive shocking which can be applied so that no dimensional change takes place. In most cases, cold working results in the nucleation of fine precipitates on the dislocations introduced during plastic deformation. Alloy 2014 is one exception.

In the Al-Mg$_2$Si system, cold work can be added before aging, between steps of a two-step age, or after aging with increasing improvement in mechanical strength.

In the instance of Al-Zn-Mg alloys, even 50% cold work prior to aging produces little change in final mechanical properties. The aging process is merely accelerated.

The combination of low-temperature aging, cold working, and final

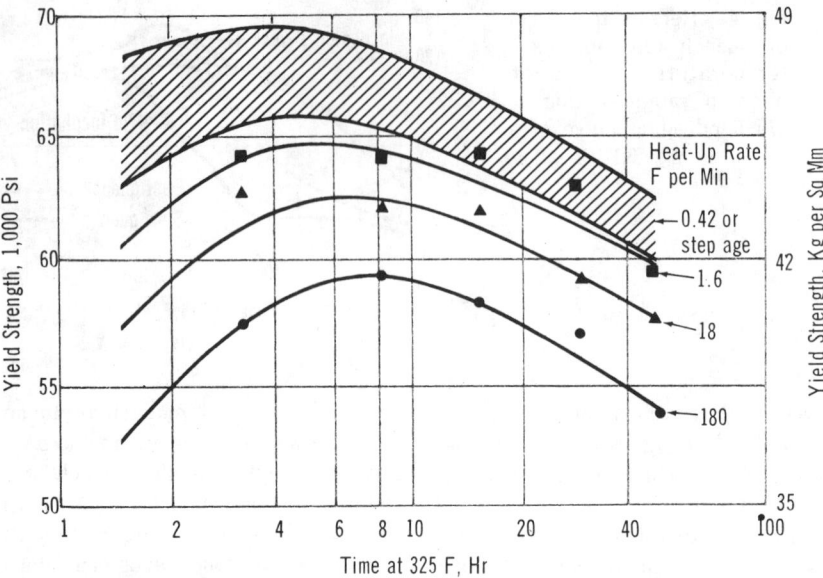

Fig. 5 — The influence of rate of heating to aging temperature on aging curves for 7075 plate (2 in. thick) is depicted by these curves.

aging at a higher temperature substantially increases strength without loss of elongation for 7000 alloys. Cold working after complete aging leads to similar increases in strength but elongation is greatly reduced. Thus, an age-cold work-age cycle appears to offer the best improvement in strength and ductility.

Aging — Isothermal aging can be used up to the GP zone solvus, provided the heating rate is not too rapid. To age above the GP zone solvus, a two-step practice or a slow heating rate must be specified. Approximate ranges for the GP zone solvus of the main systems are:

2000 series	350 to 410 F
6000 series	350 to 425
7000 series	230 to 325

In general, the highest mechanical properties are achieved at the lowest practical aging temperature. However, the recently developed stress-corrosion-resistant temper (T73 for 7075) and exfoliation-resistant tempers (T76 for 7075 and 7178), require some overaging. To do this in reasonable times, it is necessary to go to aging temperatures above the GP zone solvus. Hence, either a two-step aging practice (with the first step below the GP zone solvus) or a slow heat-up cycle must be adopted.

The beneficial effects of these practices on mechanical properties are shown in Fig. 5. Optimum properties are obtained with either a two-step aging practice or a heat-up rate of less than approximately 1 F per min. Peak properties using these aging practices are in fact close to those obtained using a normal T6 aging practice. To attain the stress-corrosion resistance of the T73 tempers, however, it is necessary to age beyond the peak. Typical maximum heating rates for fully loaded commercial aging ovens are in the range 0.3 to 1.5 F per min. Therefore, in the plant, two-step aging may be unnecessary. As a precaution, though, the hold at a low temperature is specified for these tempers.

Commercial Heat Treating

Two basic furnace types for solution heat treating are the salt bath and the gaseous atmosphere furnace. Salt baths offer high heat capacity, fast-heat-up rates, and uniform temperature distribution. The major user of the salt bath furnace is the aircraft industry because of the complex shapes of castings, extrusions, and formed wrought products treated. Disadvantages are related to cost, dragout, and the corrosive nature of the medium. In addition, it is necessary that the piece to be heat treated be clean to avoid contamination of the salt and be free of moisture to avoid explosion.

Handling of parts, especially heavy sections, can be a problem. The nature of the salt bath furnace requires that parts be lifted upward to clear the top during transfer to a quench tank. A safe distance must be maintained between the bath and the quench tank due to the incompatibility of molten salt and water. Naturally, this distance cannot be too far because the temperature drop could negate the heat treatment. Transferring is usually done manually or with the aid of an overhead hoist. Only through the use of sophisticated handling equipment does the salt bath lend itself to high production rates. Dragout from the furnace combined with the corrosive nature of salt dictates that an effective rinse be employed after quenching.

Most aluminum sheet and plates are heat treated in either air or controlled-atmosphere furnaces. Controlled atmospheres may be achieved by purging with bottled gas such as dry nitrogen or argon. However, for large-scale operations, the controlled atmosphere is obtained through the combustion of air and natural gas. It is desirable to eliminate oxygen and water vapor from the atmosphere because both are highly reactive with aluminum and can cause stains, blisters, and high-temperature oxidation (HTO). Inasmuch as water vapor is one of the products of combustion, it is necessary to separate it from the other gases, and this can be done by cooling. A proper mix of air and natural gas will yield, after drying, an atmosphere composed mainly of nitrogen and carbon dioxide with less than 0.3% O_2.

Both air and controlled-atmosphere furnaces may utilize electricity, gas, or oil as the heat source. However, direct firing is rarely employed with gas and oil due to the detrimental effects of the products of combustion, especially sulfur dioxide. These fuels are fired through radiant tubes, and the gas is circulated past these tubes. With all atmosphere furnaces, rapid circulation of air is essential for uniform temperature distribution and this is provided by fans located within the furnace. Proper design and baffling is also required to eliminate hot and cold spots.

There are two basic types of atmosphere furnaces. In one, the metal is held stationary within the furnace while in the other the metal is propelled through a long, multizoned unit — coils are handled this way. Cut-to-length sheets and plates can be processed by both types of furnaces.

In a vertical heat treatment furnace, metal is supported vertically during the thermal operation. This type is normally erected directly over a quench tank, and access is through a sealable opening in the bottom. Metal is lifted upward by a hoist. After sufficient time at temperature, the furnace bottom is opened, and the load is plunged into the quench tanks.

Vertical furnaces are relatively small in comparison with continuous types. Typical over-all dimensions are 10 ft by 10 ft by 30 ft. Production rates are also relatively low due to the sequence of operations (load, heat, soak, quench, and unload). The furnace has the advantages of simple design, very few zones, and the ability to monitor temperature readily.

In a continuous heat treatment furnace, metal is propelled through a series of zones at a predetermined speed which is a function of the alloy itself, furnace heat input, gas velocity, and furnace length.

In a continuous furnace for cut-to-length sheets and plates, the metal is heat treated in the horizontal position. It is supported on and moved along by a series of drive rolls or cables. The load enters at one end of the furnace, is heated and soaked, and then, at the opposite end, quenched and unloaded. The drive system in the last few zones, separate from the balance of the furnace, provides for the rapid exit of metal into the quench system. A series of water sprays floods the load on both top and bottom surfaces.

The length of the furnace depends to a large extent upon the type of product to be heat treated — the lighter the gage, the shorter the furnace. A furnace 60 ft long is sufficient for the heat treatment of 0.010 to 0.125 in. sheet gages. For plates as thick as 6 in., furnaces 200 ft long

may be required. Typical line speeds are in the range of 10 to 15 ft per min for sheets. Line speeds for plates range from 3 in. per min to 6 ft per min.

The same type of furnace is used in the continuous heat treatment of coiled sheet, but it requires more sophisticated auxiliary equipment than one treating cut-to-length material. Faster line speeds are necessary because a part of the coil is always entering the quench. The unwind and take-up reels must be synchronized with the line speed of the furnace, and provisions must be made for the rapid changing and attachment of succeeding coils. This is achieved with accumulator towers at both ends of the furnace.

A recent development is "air flotation." The force needed to propel the coil is provided by "pinch" rolls on the entry and exit sides of the furnace. The sheet in the furnace is supported by high velocity air streams impinging on the sheet surfaces. The streams are recirculated past burner tubes, also serving as a heat source. The high velocity required to support the sheet also provides rapid heating rates. Because relatively fast line speeds can be attained (in excess of 100 ft per min), a furnace of this type is capable of high production rates. It also has the advantage of eliminating contact marks.

Flattening Operations

One of the possible consequences of quenching is severe distortion of the metal. Usually, the lighter the gage, the more distortion encountered. Therefore, a flattening operation is normally employed on heat-treated metal.

Cut-to-length material is flattened by stretching, rolling on a planishing mill, or by reverse flex leveling. Material in coils is flattened by tension or roll flex leveling or both and by light reductions on a cold rolling mill (½ to 6%, depending on final temper).

Applications for Heat-Treated Alloys

While the 6000 series of heat-treatable alloys are applied in a wide

range of products, largely as extruded shapes, the principal applications of the 2000 and 7000 series alloys are in the aircraft and aerospace industries. Modest inroads are being made by titanium alloys, beryllium, and composite materials, but aluminum is still the principal structural material in both aircraft and missiles. Even in the new giant transport and passenger planes, as well as the supersonic *Concorde*, designers working with materials engineers have chosen aluminum as the major structural material. The alloys that are popular in these aircraft are 2014, 2024, 7075, and 7079 — 2618 is used in the *Concorde* for its better elevated-temperature properties.

The aluminum industry and government research agencies are actively working to improve the general and stress-corrosion resistance and mechanical properties of present alloys and tempers. One important demand in the new large aircraft is for heavier sections. Aircraft design has moved from riveted structures made from thin sheets and extrusions to integrally stiffened skins machined from heavy plate and large extrusions. Forgings up to 8 in. thick are being made and heat treated.

In our laboratory and others, development work is aimed at reducing quench sensitivity in the highest-strength 7000 alloys to maintain high properties in thick sections. One new alloy meeting some of the above requirements is 7080, which is currently being evaluated.

The more recently developed medium-strength, weldable 7000 series alloys, such as 7004 and 7005, have promising futures. They are particularly suited to the ground transportation field, where high strength is not as critical as it is in aircraft, but where a medium-strength, readily weldable material can be effectively employed. The major limitation lies in poor stress-corrosion resistance in the short transverse direction. However, this problem can be avoided by recognizing it during design and construction. These alloys have been successfully applied in trucks, buses, trains, LP gas tanks, and portable bridges. ◉

Properties, Characteristics, and Applications

Alloy Designation (a)	Nominal Composition, %	Product Forms (b)	Weldability (c)	Typical Annealed Properties — Tensile Strength, 1,000 Psi	Yield Strength, 1,000 Psi	Elongation, %	Solution Treatment Temperature, F	Temper (d)	Corrosion Resistance — General (e)	(f) Stress
Al-Cu										
2011	5.5 Cu, 0.4 Bi, 0.4 Pb	BW	—	—	—	—	975	T3	B	B
								T8	B—	A
2219	6.3 Cu, 0.3 Mn, 0.1 V, 0.18 Zr, 0.06 Ti	BEFST	A	25	11	18	995	T31, T351	B	B
								T62	B	B
								T81, T851	B—	A
								T87	B—	A
2021	6.3 Cu, 0.3 Mn, 0.12 Cd, 0.05 Sn, 0.1 V, 0.18 Zr, 0.06 Ti	S	A	—	—	—	—	T8	B—	A
Al-Cu-Mg										
2017	4.0 Cu, 0.5 Mg, 0.7 Mn	BW	B	26	10	22	935	T4, T451	B	B
2117	2.6 Cu, 0.35 Mg	BW	—	—	—	—	935	T4	B	—
2018	4.0 Cu, 0.7 Mg, 2.0 Ni	F	B	—	—	—	950 (j)	T61	B	B
2218	4.0 Cu, 1.5 Mg, 2.0 Ni	F	B	—	—	—	950 (j)	T72	B	B
2618	2.3 Cu, 1.6 Mg, 1.0 Ni, 1.1 Fe, 0.07 Ti	FSE	B	—	—	—	985 (j)	T61	B	B
2024	4.4 Cu, 1.5 Mg, 0.6 Mn	BESTW	B	27	11	21	920	T3	B	B
								T4, T351	B	B
								T6, T651	B	B
								T81, T851	B—	A
								T86	B—	A
Al-Cu-Si-Mg										
2014	4.5 Cu, 0.8 Si, 0.5 Mg, 0.8 Mn	BEFST	B	27	14	18	935 (k)	T4, T451	B	B
								T6, T651	B	B
Al-Si-Mg-Cu										
4032	12.2 Si, 1.1 Mg, 0.9 Cu, 0.9 Ni	F	—	—	—	—	950 (k)	T6	B	—
Al-Mg-Si-Cu										
6951	0.6 Mg, 0.35 Si, 0.28 Cu	S	A	16	6	30	985	T6	A—	A
6061	1.0 Mg, 0.6 Si, 0.25 Cu, 0.20 Cr	BEFSTW	A	18	8	27	985	T4, T451	A—	A
								T6, T651	A—	A
								T913	A—	A
6262	1.0 Mg, 0.6 Si, 0.25 Cu, 0.1 Cr, 0.5 Pb, 0.5 Bi	BW	—	—	—	—	1,000	T651	A—	A
								T9	A—	A
6066	1.1 Mg, 1.3 Si, 0.9 Cu, 0.8 Mn	BEFT	A	22	12	18	990	T6, T651	B	A
Al-Mg-Si										
6101	0.6 Mg, 0.5 Si	BET	A	—	—	—	—	T6	A	A
6003	1.2 Mg, 0.7 Si	S	—	—	—	—	—	—	—	—
6151	0.6 Mg, 1.0 Si, 0.25 Cr	F	A	—	—	—	960	T6	A	A
6053	1.2 Mg, 0.65 Si, (ratio) 0.25 Cr	BFW	—	16	8	35	970	T6, T651	A	A
6063	0.7 Mg, 0.4 Si	{ ET	A	13	7	—	970	T1	A	A
								T4	A	A
								T5	A	A
								T6	A	A
6463	Lower impurity limits than 6063							T832	A	A
Al-Zn-Mg-Cu										
7001	7.4 Zn, 3.0 Mg, 2.1 Cu, 0.25 Cr	BET	—	37	22	14	870	T6	B	B—
								T75	B—	B
7049	7.7 Zn, 2.5 Mg, 1.6 Cu, 0.15 Cr	F	—	—	—	—	—	T73	B—	A
7075	5.6 Zn, 2.5 Mg, 1.6 Cu, 0.25 Cr	BEFSTW	B	33	15	16	880 (k)(l)(m)	T6	B	B—
								T76	B—	B
								T73	B—	A
7175	Lower impurity limits than 7075	F	—	—	—	—	—	T736	B—	A
7178	6.8 Zn, 2.7 Mg, 2.0 Cu, 0.25 Cr	ES	B	33	15	15	875 (m)	T6	B	B—
								T76	B—	B
7079	4.3 Zn, 3.3 Mg, 0.6 Cu, 0.18 Cr, 0.2 Mn	{ EFS	B	33	15	16	830	T6	B	B—
7179	Lower impurity limits than 7079									
X7080	6.0 Zn, 2.3 Mg, 1.0 Cu, 0.4 Mn	F	—	—	—	—	—	T7	B—	A
Al-Zn-Mg										
7005	4.5 Zn, 1.4 Mg, 0.13 Cr, 0.4 Mn, 0.15 Zr	{ EST	A	27	10	23	880 (n)	T5	A	B—
7004	4.2 Zn, 1.5 Mg, 0.45 Mn, 0.15 Zr									
X7007	6.5 Zn, 1.8 Mg, 0.15 Cr, 0.15 Zr	S	B	—	—	—	880	T7	A—	B—
7011	4.8 Zn, 1.3 Mg, 0.12 Cr, 0.2 Mn	S	—	—	—	—	—	T6	A	—
7039	4.0 Zn, 2.8 Mg, 0.2 Cr, 0.3 Mn	S	A	36	15	20	850	T61	A—	B—

(a) Aluminum Assn., New York. Alloy systems in bold face.

(b) B, bars or rods; E, extrusions; F, forgings; S sheets or plates; T, tubing; W, wire.

(c) Almost all aluminum alloys are arc weldable in the simplest terms; problems can arise when trying to weld for high joint efficiencies or "in restraint." A, easy to weld; B, less weldable (application may demand development of welding procedures and special design).

(d) See p. 79 for explanation of temper designations.

(e) A, good outside service potential, unprotected, even in sea coast and industrial areas; B, outside protection generally required with special attention to edges, pockets, seams. Minus sign indicates less resistance than that of others in that group.

(f) A, no service failures known (in some instances, failures of laboratory tests have occurred); B, some service failures have been reported or short transverse stresses should be minimized and environmental behavior checked.

# of Heat-Treatable Aluminum Alloys

Typical Properties at Room Temperature				Typical Properties at —320 F (h)			Typical Properties at 300 F (i)			
Tensile Strength, 1,000 Psi	Yield Strength, 1,000 Psi	Elonga-tion, %	Hardness-Bhn (g)	Tensile Strength, 1,000 Psi	Yield Strength, 1,000 Psi	Elonga-tion, %	Tensile Strength, 1,000 Psi	Yield Strength, 1,000 Psi	Elonga-tion, %	Applications
55	43	15	95	70	50	20	28	19	25	Screw machine products
59	45	12	100	72	51	15	28	20	24	
52	36	17	—	—	—	—	—	—	—	Welded tanks, engine parts.
60	42	10	—	73	49	16	45	33	17	
66	51	12	120	83	61	15	49	40	17	
69	57	10	—	—	—	—	—	—	—	
73	63	9	—	90	77	10	—	—	—	Experimental: cryogenic uses in welded applications.
62	40	22	105	80	53	28	40	30	15	General structurals, fittings, screw machine products.
43	24	27	70	56	33	30	30	17	20	Rivets.
61	46	12	120	75	56	15	45	40	12	Cylinder heads, pistons.
48	37	11	95	—	—	—	—	—	—	Cylinder heads, turbine engine parts.
64	54	10	—	78	61	12	50	44	14	Aircraft frames; engine parts.
70	50	18	120	85	62	18	55	45	11	Structurals (particularly aircraft), screw machine products, hard hats.
68	47	20	120	84	61	19	45	36	17	
69	57	10	—	84	68	11	45	36	17	
70	65	7	—	85	78	8	55	49	11	
75	71	5	—	92	85	5	54	48	11	
62	42	20	105	75	50	20	—	—	—	General structurals, hydraulic cylinders;
70	60	13	135	84	72	14	40	35	20	drill pipe, cryogenic vessels, shotgun receivers.
55	46	9	120	66	48	11	37	33	9	Pistons, fittings.
39	33	13	82	—	—	—	—	—	—	Core for brazing sheet.
35	21	23	65	45	25	25	30	21	25	General structurals including marine, transportation and highway products, tread plate, stadium seats, towers.
45	40	15	95	60	47	22	34	31	20	
67	66	10	—	—	—	—	—	—	—	
45	40	17	—	60	47	22	34	31	20	Screw machine products.
58	55	10	120	74	67	14	38	37	14	
57	52	12	120	—	—	—	—	—	—	General structurals.
32	28	15	71	43	33	24	21	19	20	Electrical distribution, busbar, motor components.
—	—	—	—	—	—	—	—	—	—	Cladding alloy for 2014.
48	43	17	—	57	50	20	28	27	20	Machine parts, marine structurals, fuses.
37	32	13	—	—	—	—	25	24	13	Rivets.
22	13	20	42	34	16	44	21	15	20	Irrigation and construction pipe,
25	13	22	—	—	—	—	—	—	—	utility structurals, optical frames,
27	21	12	60	37	24	28	20	18	20	trim (architectural and bright trim).
35	31	12	73	47	36	24	21	20	20	
42	39	12	95	—	—	—	—	—	—	
98	91	9	160	—	—	—	—	—	—	Missile structurals.
86	78	9	—	—	—	—	—	—	—	
77	69	11	—	—	—	—	—	—	—	Experimental, heavy, aerospace forgings.
83	73	11	150	102	92	9	31	27	30	Wing skin plates, structurals in aircraft, screw machine products, nose cones, tubes.
76	65	11	—	—	—	—	—	—	—	
73	60	11	—	92	72	14	31	27	30	
77	70	11	—	—	—	—	—	—	—	Premium forgings.
88	78	11	160	106	94	5	31	27	40	Aerospace structurals.
83	73	11	—	106	89	10	31	27	40	
78	68	14	145	92	80	12	33	28	37	Aerospace structural, heavy forgings, deep submersible hulls.
68	58	11	—	—	—	—	—	—	—	Experimental, heavy, aerospace forgings.
56	47	14	—	70	55	14	—	—	—	Weldable structurals, mobile bridges, dump trucks.
73	67	12	—	93	81	12	—	—	—	Experimental, weldable plates.
55	45	13	—	—	—	—	—	—	—	Cladding for high-strength composite with 7075, 7178, 7079.
60	50	14	123	83	59	15	45	39	20	Armor plate, unfired pressure vessels, cryogenic, structurals.

(g) 500 kg load, 10 mm ball.
(h) Data are for specimens equilibrated at temperature.
(i) Data are for specimens tested after prolonged heating, usually 10,000 hr at temperature.
(j) Quench forgings and rings in boiling water.
(k) Quench forgings and rings in water at 140 to 180 F.
(l) Temperatures up to 920 F may be used for sheets less than 0.051 in. thick.
(m) Use 870 F for extrusions.
(n) Slow quench rate (about 1 F per sec between 750 and 500 F) favors stress corrosion resistance.
Source: Reynolds Metals Co.

Heat Treating Titanium and Its Alloys

By JOHN A. BURGER
and DEAN K. HANINK

Alpha grades are annealed and stress relieved, while alpha-beta
and beta alloys can be strengthened by solution treating and aging.
Because the metal is easily contaminated by atmosphere gases, inert gases
and vacuum are often used to protect parts during heat treatment.

APPLICATIONS OF TITANIUM and its alloys range from critical missile and aircraft components (Fig. 1) to corrosion resistant racks for plating. Eight grades account for 85 to 90% of all applications. Within the group are three grades of unalloyed titanium and five alloys, Ti-5Al-2.5Sn, Ti-6Al-4V, Ti-8Al-1Mo-1V, Ti-6Al-6V-2Sn, and Ti-13V-11Cr-3Al. The most used grade, Ti-6Al-4V, accounts for about 60 to 65% of all production.

Types of Titanium Alloys

Metallurgical changes dictate division into three classes, alpha, alpha-beta and beta.

In alpha types, the beta phase transforms completely to the alpha phase as the alloy cools through the transformation temperature range. For all practical purposes, these alloys cannot be strengthened by heat treatment.

Alpha-beta alloys are strengthened by solution

Mr. Burger is section chief, Materials and Processes Specifications, and Mr. Hanink is chief, Materials Laboratories, Allison Div., General Motors Corp., Indianapolis.

treating and aging. Solution treatment consists of (1) heating to temperatures approaching, but below, the beta transus; (2) holding for a time; and (3) cooling rapidly to retain the beta phase to room temperature. The next step, aging the alloy between 900 and 1100 F, precipitates particles of alpha titanium which add strength.

There are only a few beta alloys. Containing beta stabilizers of vanadium and columbium, plus tantalum or molybdenum (or both), they remain 100% beta when cooled (even in air) from above the beta transus temperature (about 1325 F). Solution treating has about the same effect on them as annealing.

Treatments for Common Alloys

In this report, we cover Ti-5Al-2.5Sn (alpha), Ti-6Al-4V (alpha-beta) and Ti-13V-11Cr-3Al (beta) and discuss typical heat treatment practice and effects on properties. Table I lists recommended heat treating practices for these and other titanium alloys. Table II presents stress relieving data. For sheets, short times and high temperatures are

Reprinted from Metal Progress, June 1967, 70-75, © 1967 American Society for Metals

Fig. 1 — Ring spacer, dome, and pressure vessel (all Ti-6Al-4V) for the Apollo program typify the application of titanium alloys in aerospace vehicles. They are usually strengthened by heat treatment at 1750 ±15 F for 1 hr, quenched rapidly in circulating water, aged at 1000 to 1150 ±10 F for 4 hr. The pressure vessel was welded and stress relieved at 1000 F for 4 to 8 hr.

usual. Bars and forgings are treated for longer times at lower temperatures. Times and temperatures are specified by customers, depending on configuration, properties required, processing, sheet size and thickness, plate and forging cross section.

However, even though annealing time and temperature are primarily dictated by customer specifications, type of furnace equipment, part configuration and usage, and maximum annealing response desired (reduction in area, percentage of elongation) must also be considered. The most effective annealing response is at higher temperatures. If sheet stock is annealed before forming, maximum formability is required. Also, if only an air atmosphere is available, an annealing temperature of 1550 F would be used for a short time (10 to 15 min) to keep surface contamination and grain growth to a minimum. Oxide is removed mechanically or chemically. With a vacuum furnace, effective annealing response can be obtained by holding at 1450 F for 30 to 60 min. Time at heat is not critical in vacuum, but economy enters the picture

Source: Metal Progress, June 1967, 70-75

Table I — Heat Treating Titanium and Its Alloys

Grade	Beta Transus Temperature, F	Forms	Annealing Treatment	Solution Treatment and Aging
Alpha Grades				
Ti (99.0%)	1690	Sheets, bars, forgings	1250 to 1300 F, 2 hr, air cool	Alpha alloys usually not strengthened by aging
Ti-5Al-2.5Sn	1900	Sheets, bars, forgings	1325 to 1550 F, 10 min to 4 hr, air cool	
Ti-5Al-2.5Sn (low O_2)	1910	Sheets, bars, forgings	1325 to 1550 F, 10 min to 4 hr, air cool	
Ti-5Al-5Sn-5Zr	1815	Sheets, bars, forgings	1650 F, 1/2 to 4 hr, air cool	
Ti-7Al-12Zr	1825	Sheets, bars, forgings	(1) 1600 to 1650 F, 1/2 to 4 hr, air cool (low strength anneal) (2) 1300 F, 1 hr, air cool (recommended for high strength)	
Ti-8Al-1Mo-1V	1900	Sheets	1400 to 1450 F, 8 hr, furnace cool (simple anneal)	
		Sheets	1400 to 1450 F, 8 hr, furnace cool + 1400 to 1450 F, 1/4 hr, air cool (duplex anneal for maximum notch toughness)	
		Sheets	1400 to 1450 F, 8 hr, furnace cool + 1850 F, 5 min, air cool + 1375 F, 1/4 hr, air cool (triplex anneal for notch toughness and creep resistance)	
		Bars, forgings	1650 to 1850 F, 1 hr, air cool + 1100 F, 8 hr, air cool (duplex anneal for bars and forgings)	
Alpha-Beta Grades				
Ti-4Al-4Mn	1700	Sheets, bars, forgings	1300 F, 2 to 4 hr, furnace cool	1400 to 1500 F, 1/2 to 2 hr, water quench and age 800 to 1000 F, 8 to 24 hr, air cool
Ti-6Al-4V	1820	Sheets, bars, forgings	1300 to 1550 F, 1 to 8 hr, slow cool to 1050 F, air cool	1500 to 1750 F, 5 min to 1 hr, water quench and age 1000 to 1150 F, 1 to 12 hr, air cool
Ti-679	1730	Bars and forgings	1650 F, 1 to 2 hr, air cool	1650 F, 1 hr, water quench and age 930 F, 24 hr, air cool
Ti-6Al-4V (low O_2)	1820	Sheets, bars, forgings	1300 to 1550 F, 1 to 8 hr, slow cool to 1050 F, air cool	Not recommended
Ti-6Al-6V-2Sn-1(Fe,Cu)	1735	Bars, forgings	1300 to 1400 F, 1 to 2 hr, air cool	1600 to 1675 F, 1 hr, water quench and age 900 to 1100 F, 4 to 8 hr, air cool
Ti-7Al-4Mo	1840	Bars, forgings	1450 F, 1 to 8 hr, slow cool to 1050 F, air cool	1650 to 1750 F, 1/2 to 1 1/2 hr, water quench and age 900 to 1200 F, 4 to 16 hr, air cool
Beta Grades				
Ti-1Al-8V-5Fe	1525	Sheets, bars, forgings	1250 F, 1 hr, furnace cool to 900 F, air cool	1375 to 1425 F, 1 hr, water quench and age 925 to 1000 F, 2 hr, air cool
Ti-13V-11Cr-3Al	1325	Sheets, bars, forgings	Same as solution treating	(1) 1400 to 1500 F, 1/4 to 1 hr, water quench (or air cool) and age 900 F, 2 to 96 hr, air cool (2) 1450 F, 1/3 hr, air cool+cold roll+ 800 F, 24 hr, air cool (cold worked material)

if long times and high temperatures are involved.

If a large, thick forging is annealed and stock is removed after heat treatment, the practice for carbon and alloy steels can be used. You hold 1 hr per in. of cross section in an air-atmosphere furnace at 1325 F. With data nomographs, you estimate the depth of contamination from the combination of time and temperature exposure to air. Recommended practice for bars and forgings: anneal at 1325 to 1550 F for 4 hr and air cool; stress relieve at 1000 to 1200 F for 15 min to 4 hr and air cool.

Annealing Procedures

The alpha alloy, Ti-5Al-2.5Sn, is annealed for 10 min to 4 hr in the 1325 to 1550 F range — shorter

Table II — Stress Relieving Titanium Alloys

Grade	Time, Hr	Temperature, F
Commercial titanium	1/4 to 4	900 to 1100
Ti-5Al-2.5Sn	1/4 to 4	1000 to 1200
Ti-5Al-2.5Sn (low O_2)	1/4 to 4	1000 to 1200
Ti-8Al-1Mo-1V	1/2 to 4	1100 to 1200
Ti-5Al-5Sn-5Zr	1/2 to 4	1100 to 1200
Ti-7Al-12Zr	1/2 to 4	1000 to 1100
Ti-8Mn	1/2 to 2	900 to 1100
Ti-4Al-3Mo-1V	1/2 to 2	1000 to 1100
Ti-6Al-4V	1 to 12	1000 to 1100
Ti-6Al-6V-2Sn	1 to 8	900 to 1200
Ti-679	5 to 10	900 to 950
Ti-7Al-4Mo	1 to 8	900 to 1300
Ti-4Al-4Mn	1 to 2	1000 to 1300
Ti-13V-11Cr-3Al	1 to 4	900 to 1200
Ti-1Al-8V-5Fe	1 to 2	1000 to 1100

exposure times are at higher temperatures. Temperatures above 1600 F are generally avoided because they promote excessive oxidation and grain growth. Cooling can be fast or slow; its rate has little effect on properties.

Typical properties of an alloy annealed at 1600 F are 143,200 psi yield strength, 146,000 psi tensile strength, 16.7% elongation and 43.4% reduction in area. Stress relieving (which is normally needed) is in the 1000 to 1200 F range for 15 min to 4 hr. Air cooling follows.

In treating Ti-6Al-4V (the alpha-beta alloy), the temperature range for annealing is 1200 to 1400 F. By solution annealing and quenching from higher temperatures in the alpha-beta zone (of the phase diagram), the alloy is conditioned for strengthening by subsequent aging.

The standard annealing temperature for bars and forgings of Ti-6Al-4V is 1300 ±25 F. Hold at heat for 2 hr, and cool in air. To anneal sheet, strip and plate stock, heat to 1325 ±25 F, hold at heat for 1 hr, cool at a rate not faster than 50 F per hr to below 800 F, then cool conventionally to room temperature. Higher annealing temperatures (1300 to 1550 F), long times (2 to 8 hr), and faster cooling rates raise toughness with some decrease in strength. Lower annealing temperatures are recommended for most applications.

With higher temperatures, response increases because more beta appears and its alloy content changes. As the solution temperature is raised, the vanadium content is lowered below the limit at which beta remains on quenching. Then beta transforms partially to alpha prime (martensite), forming a structure which is retained by quenching rapidly in water. Time, temperature, and rate of cooling must be selected to obtain a mixed alpha-beta structure.

For Ductility, Solution Treat

Treaters develop maximum ductility and best formability by solution treating near 1500 F. Solution treating at 1750 F, followed by water quenching and aging, produces maximum strengths (Fig. 2). The closer to the beta transus temperature, the lower the elongation. In industry, elongation required at these strengths is 8% minimum. In addition, 1750 F is largely an industry standard, especially for rocket motor cases. This temperature reduces distortion and growth, which can occur especially on large parts above 1800 F. Also, allowance is made to age longer than 8 hr if additional aging is required as in stress relieving. At 1750 F, strength is still on the upgrade.

Fast quenching is needed. Strength is effectively lowered when the solution treated workpiece is not quenched within 10 sec, as shown in Fig. 3. (This is perhaps the most important item to know when heat treating alpha-beta alloys for high strength.)

Because heat treatable titanium alloys show shallow hardenability (due to low thermal conductivity), we do not expect full strength in sections much over 1 in. thick. As noted, the strength of solution treated Ti-6Al-4V is improved by aging between 900 and 1100 F for 1 to 24 hr. Higher aging temperatures result in overaging and loss in strength; ductility is improved only slightly.

Aging temperatures are determined by composition (especially interstitial levels) and tensile properties required, if the cross section, solution temperature and quench remain constant. Times vary with cross section of the part and processing sequence.

For example, 4 hr of aging is standard for bars and forgings. Times for sheets and assemblies vary similarly, lasting 1 to 4 hr. Solution treatment and aging of rough machined parts are followed by more machining, then welding and stress relief. Stress relief lasts another 4 hr. If reworking is needed for a weld or if a section must be replaced (and rewelded), another stress relief is required at the same temperature range as aging. Aging and stress relief can extend up to 24 hr (accumulative) with little effect on tensile properties. Longer times are a convenience in processing.

Heating at 1000 to 1200 F for ½ to 1 hr, followed by air cooling, stress relieves Ti-6Al-4V. To reduce development of oxidation, times and temperatures should be low as possible. Tests show that heating at 1200 F for 8 hr completely relieves stresses with some oxidation, while heating at 1000 F for the same period removes only 50% of the stresses—but produces no oxidation. Time is again primarily accumulative as required by processing, and also as an economy factor. However, after 10 hr, an improvement of about 5000 psi is realized in the relief of residual stresses from 900 to 1200 F. Up to 24 hr, an additional 2000 to 5000 psi rise is obtained. Because cost usually becomes a factor, 8 hr is standard.

Heat Treating Beta Alloy

In processing Ti-13V-11Cr-3Al, fast or slow cooling from the annealing temperature retains the beta phase structure. Annealing and solution treatment are done between 1300 and 1900 F. However, long periods at high temperatures lower ductility by causing excessive grain growth. High strengths are developed by solution treating above 1400 F, water quenching (or air cooling), and aging at 800 to 1000 F. Even thick sections respond to hardening when they are aged.

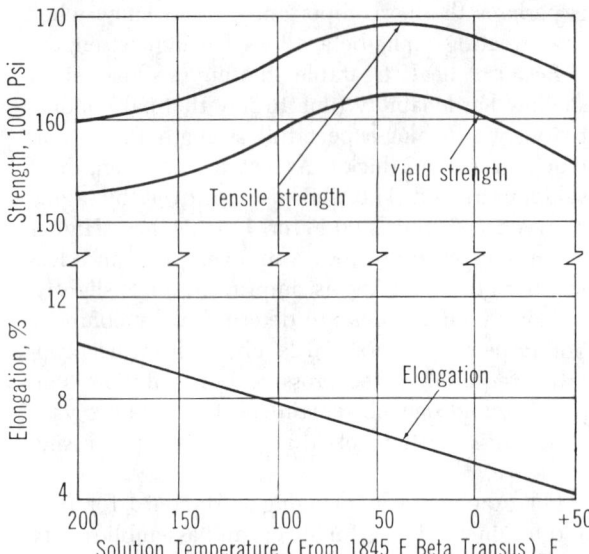

Fig. 2 — Varying the solution temperature of Ti-6Al-4V forgings produces different combinations of strength and ductility. Aging at 1100 F for 8 hr brought out the indicated properties.

Fig. 3 — If water quenching is delayed more than 10 sec after solution treatment (at 1750 F here), Ti-6Al-4V loses strength—especially after aging.

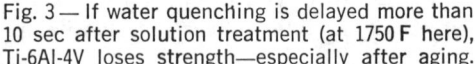

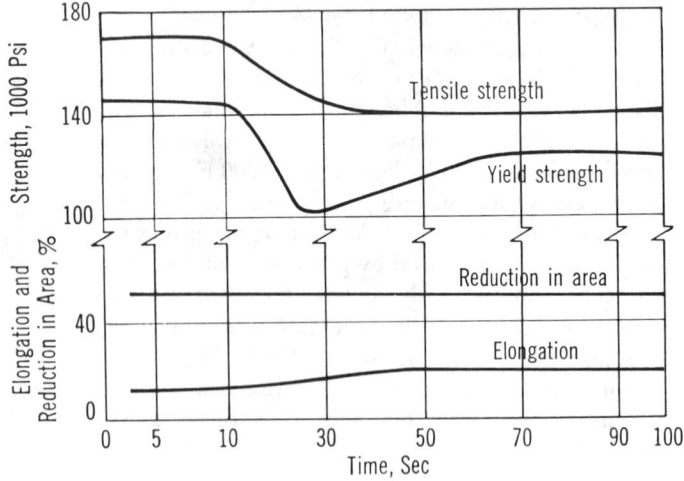

Normal cooling from the annealing temperature retains the beta-phase structure. So annealing Ti-13V-11Cr-3Al is the same as solution treating it. Any shop will do the heat treatment specified by a customer. Most treatments depend on the application and processing. The producer of this alloy recommends solution treatment of 1425 ±25 F for 30 min for sheets, with interstage annealing of 1375 ±25 F for fabricated sheet parts for high tensile strength after aging. Above 1550 F, grain growth proceeds rapidly.

Aging boosts strength. Normally, this beta alloy is aged at 800 to 1000 F for 20 to 100 hr. You can treat it up to 500 hr at 1000 F without overaging, and maximum response develops around 900 F. Cold or warm working speeds aging response, producing higher strengths in aged material.

Stress relieving is normally carried out by annealing 1 to 4 hr at 900 to 1200 F — for shorter times, you go to higher temperatures. Less relief is realized during subsequent aging at 900 F. There is no loss in tensile properties in material held at 900 F up to 2 hr. However, the degree of stress relief for ¼ to 2 hr jumps from 50 to 75%. At 1000 F for ¼ hr, the percentage of relaxation is 80%; for 2 hr, it is 95% with 2 to 5% loss in aged properties. Times are not fixed. They vary with the amount of stress relief and properties desired.

Contamination Problems

Titanium, which is chemically active at elevated temperatures, reacts with almost all elements, including those in furnace atmospheres — carbon, oxygen, nitrogen and hydrogen. It also reacts with, and is contaminated by, most gaseous compounds, such as CO, CO_2, H_2O, NH_4, and many volatile organic materials. Gaseous elements not only form compounds at the surface, but also penetrate the metal lattice, causing a drop in ductility and toughness. Further, contaminating gases (with the exception of hydrogen) cannot be removed by vacuum heat treatment.

In general, strength rises while ductility and toughness decline — toughness is the property most seriously affected. Embrittlement effects (especially those promoted by hydrogen) are another matter. Certain alloys can tolerate more interstitial elements than others. The degree varies with temperature and the alloy.

Titanium resists many corrosive environments because a thin film of natural oxide slowly forms with heating. Since the film does not thicken significantly until the temperature exceeds 1000 F, oxidation during service (always at lower temperatures) is usually not a problem. However, at higher temperatures — and 1400 to 2000 F is the heat treating range — oxidation is rapid. More important, gas diffuses into the metal, hardening the surface and causing embrittlement.

The only solution is to remove the contaminated layer, and its depth will vary with time and temperature. Tests reveal that at 600 to 1000 F, no film appears; up to 1700 F, the film is less than 2 mils in thickness; at 1800 F, it is 2 mils; at 1900 F, 4 mils; at 2000 F, it grows to 14 mils.

Controlling Contamination

After bar stock and forgings are heat treated, sufficient stock is removed to eliminate air contamination and quenching distortion. However, pickup of hydrogen is critical. If more than 200 ppm is present, the gas embrittles the alloy, lowering its impact and notched tensile strength. Prevention precedes heat treatment: remove fingerprints, mill stencils, oils, greases, and other foreign matter with chloride-free cleaning solutions.

Nearly finished and finished machined parts and titanium sheet present serious problems. Unlike steel, titanium cannot be protected from oxidation by conventional atmospheres (such as cracked ammonia, carbon monoxide or hydrogen). Inert gases like argon and helium will give protection if they are highly purified and free of moisture.

Because high purity gas furnaces are not generally available for commercial heat treating, we suggest: solution treat the alloy for as short a period as possible, using the best inert atmosphere economically available or treating under vacuum. In the inert atmosphere approach, follow with pickling or light belt grinding to remove the contaminated skin.

If your furnaces also heat treat steels, they should be purged for several hours with the atmosphere you select. The step prevents contamination from residual gas in crevices and refractories. In the instance of air-blown furnaces that also heat treat ferrous alloys, the operator should purge them with large volumes of air (150 cu ft per min for 4 hr) before starting the cycle.

Titanium parts can also be protected by encapsulating them in low carbon steel containers. Seam weld the envelope around the titanium part, and include a fitting to evacuate air. Another method: provide two fittings through which a continuous purge of inert gas can circulate in the container. In either situation, the envelope protects the titanium from oxygen and hydrogen. At the same time, the cooling rate is not seriously affected.

Proprietary coatings, which also prevent contamination, are applied to sheets, bars and precision forgings. It is generally believed that coatings help (some are commercially available), but they do not eliminate the problem. For this reason, each treatment must be considered with respect to allowable depth of contamination and in terms of the compatibility and usefulness of the coating.

One manufacturer employs a proprietary coating during precision forging of titanium compressor blades. After finish forging, he uses a final polishing operation to remove any oxide that may have resulted from a leak in the coating.

Another manufacturer forged a precision compressor wheel coated with his proprietary coating, primarily to reduce the amount of contamination and machining time. Approximately 0.015 to 0.025 in. stock was removed. (Reduction of titanium oxide on the surface lowers cutting tool wear; the oxide is hard — above Rockwell C 50 — and brittle.)

Furnace and Equipment Types

Heating and heat treating equipment that handles steel is satisfactory if precautions are taken to prevent contamination. Vertical gantry furnaces are popular in processing large parts. Both electric or gas fired, they operate to 2100 F and have quench tanks below floor levels. The hot furnace travels (on a track) to the quench tank, and lowers the charge into the bath in less than 5 sec.

Although vacuum heat treatment is normally more expensive than conventional furnace procedures, it is ideal in many applications. Such treatment becomes more economical because problems associated with contamination and cleanup are eliminated — especially with machined and thin sheet structures requiring high temperature exposure and close dimensional tolerances. Also, the only practical way of lowering hydrogen content is by vacuum treatment.

Heat Treating Nickel-Base Superalloys

By JOHN A. BURGER
and DEAN K. HANINK

Cast and wrought alloys are solution treated and age hardened
to strengthen them for service at high temperatures. They are
subject to intergranular oxidation at heat treatment temperatures
and must be protected from contamination by atmospheres during processing.

BECAUSE OF ITS face-centered cubic crystallographic structure, nickel displays a high degree of ductility and toughness at elevated to cryogenic temperatures. The metal can be cast, welded and brazed. Its mechanical properties are altered by hot working and thermal treatment. Nickel's ability to alloy readily with other metals encourages its selection as the basic component of many casting and wrought alloys (Table I).

These rather complex alloys primarily combine strength with corrosion resistance and have superior resistance to creep and rupture at elevated temperatures. All are commonly used in missile and aerospace applications (Fig. 1).

In these grades, copper, molybdenum and chromium, singly and in combination, enhance specific corrosion resistance. Various combinations of chromium, tungsten, molybdenum, columbium and cobalt impart heat resistance and strength at elevated temperatures. Titanium, aluminum, columbium, silicon, magnesium and beryllium make the material age hardenable.

Effects of Heat Treatment

Castings respond to thermal treatment, and are normally heat treated before going into service. Mechanical properties of wrought alloys (generally used as machined forgings or components formed from sheet metal or machined bar stock) are altered by hot working as well as by cold deformation. Likewise, they respond to heat treatment, with corresponding changes in mechanical properties. For example, age hardening usually improves room-temperature mechanical properties significantly, and it produces tremendous increases in resistance to creep and rupture at elevated temperatures.

Annealing the Alloys

To soften these alloys, heat at a high temperature for a time and cool rapidly to prevent the precipitation of intermetallic compounds and carbides. (Table II gives treatments for 19 alloys.) The procedure recrystallizes the metal and raises ductility. It is effective for wrought material that has been work hardened by deep drawing, bending or other forming operations. For many of the hardenable alloys, annealing cycles are the same as those for solution treating. Annealing and stress relieving practices vary considerably, and experience with

Mr. Burger is section chief, Materials and Processes Specifications, and Mr. Hanink is chief, Materials Laboratories, Allison Div., General Motors Corp., Indianapolis. This is the second in a series of three articles on heat treatment. Part I in the June issue dealt with titanium and its alloys; Part III in August will cover precipitation hardening stainless steels.

Reprinted from Metal Progress, July 1967, 61-66, © 1967 American Society for Metals

Fig. 1 — The turbine disk, with blades, is made of Inconel 713C, while the other component, a rough-machined turbine wheel, is of Waspaloy, Hastelloy X, and Inconel 718. Parts have been heat treated as shown in Table II.

treatments of parts for known requirements usually indicates that processors have to modify factors shown in Table II.

Water quenching is generally preferred. However, we suggest slower cooling (by hot salt, oil or air) for heavy complex sections of Waspaloy, René 41, Udimet 500, Udimet 700 and Inconel X-750 as being adequate and more practical. Rapid cooling in air is usually suitable for strip and sheet parts.

Strengthening Is Needed

Strengthening requires solution treating, then aging. In many instances, the solution treating temperature will depend on the mechanical properties desired. To produce optimum creep and creep-rupture properties, use a higher solution temperature; to develop optimum short-time tensile properties at elevated temperatures, use a lower solution temperature. Compromises in mechanical properties can, of course, be achieved by intermediate solution temperatures. Table II lists typical solution treating cycles — a variation of ±25 F in listed temperatures is generally satisfactory.

Be sure to cool below about 1000 F rapidly enough to prevent precipitation in the intermediate temperature range. For sheet metal parts of most alloys, rapid air cooling will suffice. Molten salt, oil or water quenching is usually satisfactory for heavier sections.

An alternative solution treatment for Inconel 718 is 1950±25 F for 1 hr and air cool; the aging treatment is the same as that in Table II.

For Inconel X-750 bars and forgings, use the 2100 F solution treatment, stabilizing at 1625 F; or solution treat at 1900 F for ½ hr, air cool, and age 20 hr at 1300 F without stabilizing.

For Nimonic 80A, the listed treatment (Table II) for long-time and short-time properties is standard. For greater creep extension at fracture, solution treat at 1975 F for 8 hr, transfer directly to another furnace heated to 1550 F, hold at 1550 F for 24 hr, furnace cool, and age.

The Table II treatment gives optimum short-time tensile properties for René 41. For optimum stress-rupture properties, heat the alloy at 2150 F for 2 hr and air cool, age at 1650 F for 4 hr. and air cool.

An alternative treatment for high temperature stabilization of Udimet 700 is solution treating at 2150 F for 4 hr, air cooling, heating at 1975 F for 4 hr, air cooling and aging at 1550 F for 24 hr, air cooling, heating at 1400 F for 16 hr, and air cooling.

For optimum stress-rupture properties, use listed

Table I — Composition of Nickel-Base Alloys

Alloy	C	Cr	Co	Mo	Ti	Al	Fe	Other
Wrought								
Hastelloy R-235	0.12	15	—	5	2.5	2.0	10	—
Hastelloy B	0.03	—	—	28	—	—	5	0.30 V
Hastelloy C	0.06	15.5	—	16	—	—	5	0.20 V, 4 W
Hastelloy W	0.08	5	—	24.5	—	—	5.5	0.30 V
Hastelloy X	0.10	22	1.5	9	—	—	18	0.60 W
Inconel 600	0.04	15	—	—	—	—	7	—
Inconel 700	0.13	15	30	3	2.2	3.2	1	—
Inconel 702	0.04	15.5	—	—	0.65	3.25	0.4	—
Inconel 718	0.04	19	—	3	0.8	0.6	18	5.2 Cb
Inconel X-750	0.04	15	—	—	2.5	0.9	7	1 Cb
Nimonic 75	0.12	20	—	—	0.40	—	5 max	—
Nimonic 80A	0.08	20	1	—	2.25	1.25	3 max	—
Nimonic 90	0.08	20	18	—	2.50	1.50	3 max	—
René 41	0.09	19	11	10	3	1.5	2	0.005 B
Udimet 500	0.10	18	18	4	3	2.9	0.5	0.004 B
Udimet 700	0.10	15	19	5	3.5	4.5	0.5	0.03 B
Waspaloy	0.10	19.5	13.5	4	3	1.3	0.75	0.0045 B, 0.06 Zr
Cast								
GMR 235	0.15	15.5	—	5.25	2.0	2.0	10	0.05 B
Hastelloy B	0.10	—	—	28	—	—	5	—
Hastelloy C	0.08	16.5	—	17	—	—	5	4.5 W, 0.40 V
Hastelloy X	0.10	22.0	1.5	9	—	—	18	0.60 W
Inconel 713	0.12	13	—	4.5	0.6	6	0.5	2 Cb+Ta
Inconel 713LC	0.06	12	1.5	4.5	0.6	6	0.3	2 Cb+Ta
IN-100	0.15	10.5	14	3.0	4.8	5.4	—	1 V, 0.015 B, 0.06 Zr

Table II — Typical Heat Treating Cycles for Nickel-Base Superalloys

Alloy	Annealing Temperature, F	Annealing Time, Hr	Solution Treating Temperature, F	Solution Treating Time, Hr	Intermediate Aging Temperature*, F	Intermediate Aging Time, Hr	Final Aging Temperature*, F	Final Aging Time, Hr
Hastelloy R-235	1975	1	1975	1	—	—	1400	16
Hastelloy B	2150	1	2150	1/2	—	—	—	—
Hastelloy C	2225	1	2225	1	—	—	—	—
Hastelloy W	2150	1	2150	1	—	—	—	—
Hastelloy X	2150	1	2150	1	—	—	—	—
Inconel 600	1900	1/4 to 1/2	2050	2	—	—	—	—
Inconel 625	1700 to 1900	1	2000 to 2200	1	—	—	—	—
Inconel 700	2200	2	2160	2	—	—	1600	4
Inconel 718	1750	1	1750	1	1350	8	1325	8
Inconel X-750	1900 to 2000	1/2 to 3/4	2100	2	1550	24	1300	20
Nimonic 80A	1975	2	1975	2	—	—	1300	16
Nimonic 90	1975	2	1975	2	—	—	1300	16
René 41	1975	2	1975	1	—	—	1400	16
Udimet 500	1975	4	1975	4	1550	24	1400	16
Udimet 700	2075	4	2050 to 2150	4	1600 +1800	8 4	1200 +1400	24 8
Waspaloy	1850	4	1975	4	1550	24	1400	16
Inconel 713	Usually not annealed	—	2100 to 2150	2	—	—	1700 to 1825	4 to 16
Inconel 713LC	Usually not annealed	—	2100 to 2150	2	—	—	1700 to 1825	4 to 16
IN-100	Usually not annealed	—	2100 to 2150	2	—	—	1700 to 1825	4 to 16

*A variation of ±25 F from the stated temperature is satisfactory.

treatment for Waspaloy. Should you desire optimum short-time tensile properties, however, solution treat at 1875 F, and follow with the same aging treatment.

Age Hardening These Grades

Many nickel-base alloys can be age hardened (to develop high strength) by heating at 1200 to 1650 F after solution treating. Such strengthening (also called precipitation hardening) occurs because different phases are dispersed and dissolved in the matrix at the solution treatment temperature. This structure is retained when the material is cooled rapidly. Aging at an intermediate temperature precipitates the dissolved constituent, strengthening the alloy. Selection of aging conditions depends primarily on the mechanical properties desired. Many of the alloys may require double aging. The first treatment, also referred to as stabilizing, is usually carried out at a higher temperature, such as 1500 to 1650 F.

Treatment at about 1550 F, followed by reheating between 1300 and 1400 F, results in optimum creep and rupture properties. When higher ductility is required, the 1550 F aging treatment is omitted (if service is not to be in the 1500 to 1600 F range).

The solution treating temperature may be changed for any one of several reasons. The development of desired properties often requires tailored aging. For example, to obtain maximum stress-rupture properties in René 41, solution treat at 2150 F for 2 hr and age 4 hr at 1550 F. However, for maximum room temperature strength, the solution temperature is lowered to 1950 F and aging is changed to 16 hr at 1400 F.

Factors that influence the selection of aging temperature include type and number of precipitating

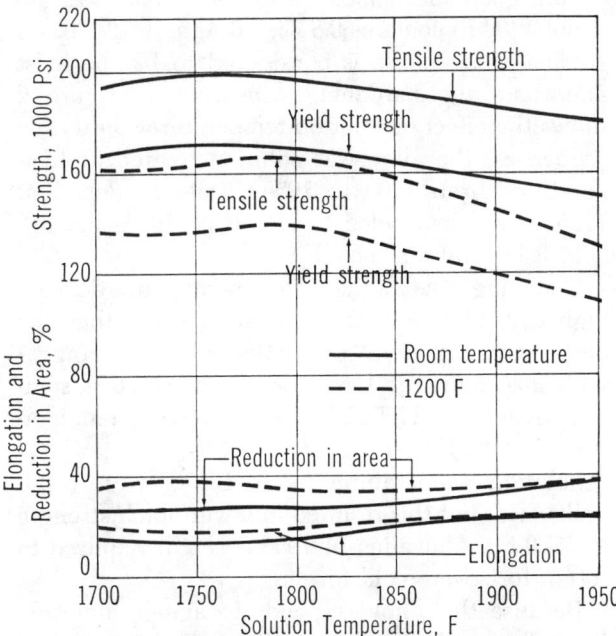

Fig. 2 — Varying the solution temperature for Inconel 718 affects strength and ductility. After being solution treated, specimens were aged at 1325 F for 8 hr, cooled to 1150 F (at 100 F per hr) held 8 hr, and air cooled.

Fig. 3 — Long aging times are needed for annealed Inconel 718. The highest hardness, Rockwell C 39 to 40, was developed by holding at 1250 F for at least 20 hr.

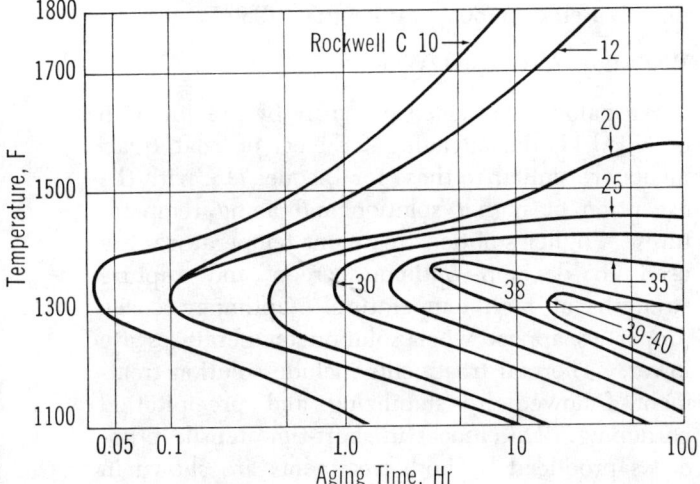

Fig. 4 — When heat treated, Inconel 718 retains strength to about 1000 F and then weakens rapidly. Ductility remains steady to 1200 F, then rises.

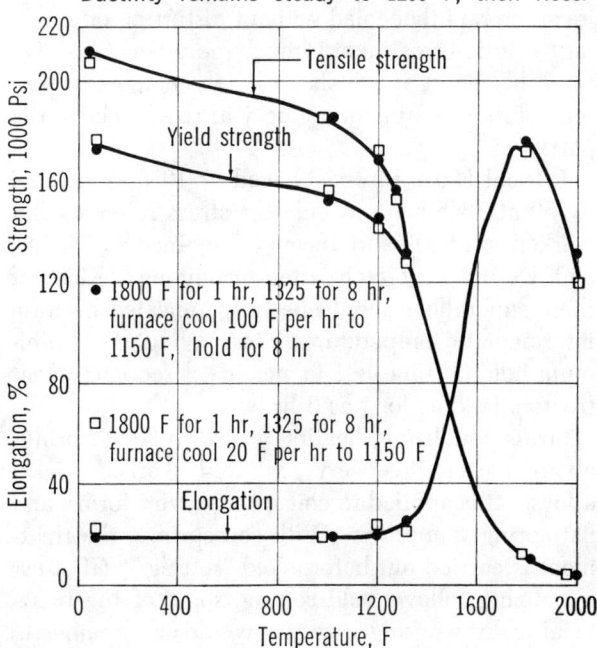

phases available, anticipated service temperature, precipitate size, and the combination of strength and ductility desired. Table II lists typical aging cycles for wrought nickel-base alloys.

Air cooling is generally recommended after aging. An exception is Inconel 718. After final aging, cool in a furnace at 100 F per hr to 1150 F, hold for 16 hr, then cool in air.

Stress Relieving

To stress relieve these alloys, heat to a suitable temperature, hold long enough to reduce residual stresses, then cool slowly to minimize development of new stresses. This treatment promotes dimensional stability and minimizes distortion that otherwise would occur if internal stresses are subsequently released during machining and welding or during service. It also prevents the material from cracking during welding by removing previously introduced internal stresses, increases its ability to withstand stresses imposed during subsequent forming operations without cracking, and precludes cracking during storage or service.

Many hot or cold worked nickel alloys are generally stress relieved in the 900 to 1300 F range. Time and temperature vary and usually must be determined by trial and error.

Avoid stress relieving the age hardenable, nickel-base alloys as much as possible because it may reduce their response to aging, promote brittleness or impair corrosion resistance. If required (for example, following the welding of a restrained joint), one or two courses can be taken. First, if the welded part or assembly is rather simple (so that it can be rapidly cooled without distorting or cracking), solution treatment is much more beneficial. On the other hand, if the shape is too complex to permit solution treatment, aging will stress relieve the part.

Inconel 600 is stress relieved at 1650 F; Inconel X-750 at 1625 F (used only for stress relieving hot worked grades); and Inconel 713, Inconel 713LC, and IN-100 at 1750 F after machining. Most of these superalloys should be fully annealed because intermediate temperatures cause aging. The minimum holding time is 1 hr per in. of section. Heat treaters hold as long as 3 hr.

Stress equalizing, heating at 500 to 900 F, brings about partial "recovery" in cold worked nickel alloys. It is applied to coil springs, wire forms, and flat spring stampings. With coil springs, the treatment is carried out before cold "setting." (If stress equalizing follows cold setting, some of the beneficial cold working stresses would be removed.)

Treatments of this type are not usually carried out on age hardenable alloys unless they have been cold worked after age hardening.

Effects of Specific Treatments

Thermal treatments influence mechanical properties, as we will demonstrate with Inconel 718 and Waspaloy.

Inconel 718 (a precipitation hardenable alloy having good mechanical properties from −423 to about 1200 F) contains 0.8 Ti, 0.6 Al and 5.2 Cb. A gamma-prime phase is considered to be the most important age hardening constituent. Figure 2 shows the effect of solution temperatures on tensile properties; the alloy was solution treated at temperatures from 1700 to 1950 F, age hardened at 1325 F for 8 hr, cooled to 1150 F (at 100 F per hr), held 8 hr, and air cooled.

Changing the aging treatment can develop higher yield strengths. Our tests show that the best combination of properties seems to appear with final aging at 1150 F on sheet which is solution treated at 1700 F. (The stabilizing temperature is 1325 F.)

Solution temperatures up to 1950 F were originally used, but this resulted in lower notch strength at 1200 F. Annealing at 1950 F is still required to soften for severe cold forming.

Because this alloy responds to aging sluggishly (Fig. 3), welding and annealing can be achieved without rapid hardening during heating and cooling. Figure 4 shows how mechanical properties at high temperatures are influenced by heat treatment.

Stress rupture life is also affected by the solution temperature. For instance, the life of Inconel 718 at 1300 F at a stress of 75,000 psi was tripled (from 45 to 136 hr) by raising the solution temperature from 1750 to 1950 F. However, maximum stress rupture life is obtained for notched specimens solution treated at 1750 F, rather than 1950 F.

Waspaloy Also Used Widely

Waspaloy is hardened primarily by precipitation of Ni(Al,Ti) during aging. Effects of heat treatments are similar to those for Inconel 718, with the exception of higher solution and aging temperatures. Higher solution treatment temperatures are used to develop optimum creep and rupture strengths at high temperature. Optimum tensile properties appear when solution temperatures are lower. Thermal treatments include solution treatment followed by stabilizing and precipitation hardening. Differences in short-time tensile properties produced by both treatments are shown in

Fig. 5. The high solution temperature (1975 F) is recommended for optimum creep and rupture properties.

Watch Out for Contamination

Nickel-base alloys resist oxidation well at normal (1500 to 1800 F) service temperatures, depending on the alloy. However, at temperatures used for solution treating, these alloys are susceptible to intergranular oxidation, an effect that lowers their resistance to thermal fatigue.

If the atmosphere has a carburizing potential, carbon is picked up and forms a stable carbide (TiC). This action removes titanium from solid solution, preventing normal precipitation hardening in the surface layers. Titanium nitride (TiN) can be formed in the same manner due to nitrogen contamination.

All exposed surfaces of parts should be kept free of dirt, fingerprints, oil, grease, forming compounds, lubricants, and scale. Avoid, also, lubricants or fuel oils that contain sulfur-bearing compounds because they corrode metal surfaces.

Furnace Equipment

Batch heating for annealing or solution treating is usually done in conventional equipment such as box, roller hearth and standard muffle furnaces. They may have provisions for purging, preheating and quenching if the high temperature compartment is supplemented by other chambers.

Nickel-base alloys are commonly aged in the 1200 to 1650 F range in rotary or box furnaces, with or without atmosphere, again depending on processsing sequence. The usual operating temperature tolerance is ±25 F for wrought alloys and ±15 F for casting alloys. Continuous furnaces are seldom used because of the long aging cycles.

Salt baths are not recommended because chloride in the bath can react with the alloy surface during long-time immersions required for aging. Vacuum and hydrogen furnaces are employed for heat treating some alloys. To cool the parts, the vacuum retort is pressurized with an inert gas that provides conductive cooling.

Protective atmospheres are used in annealing or solution treating if oxidation cannot be tolerated. Otherwise, alloys can be solution treated in air or in the normal mixture of air and combustion products found in gas-fired furnaces.

A lean, dilute exothermic atmosphere is relatively safe and economical. Any surface scale which forms can be removed by pickling or by mechanical means. Endothermic atmospheres are not recom-

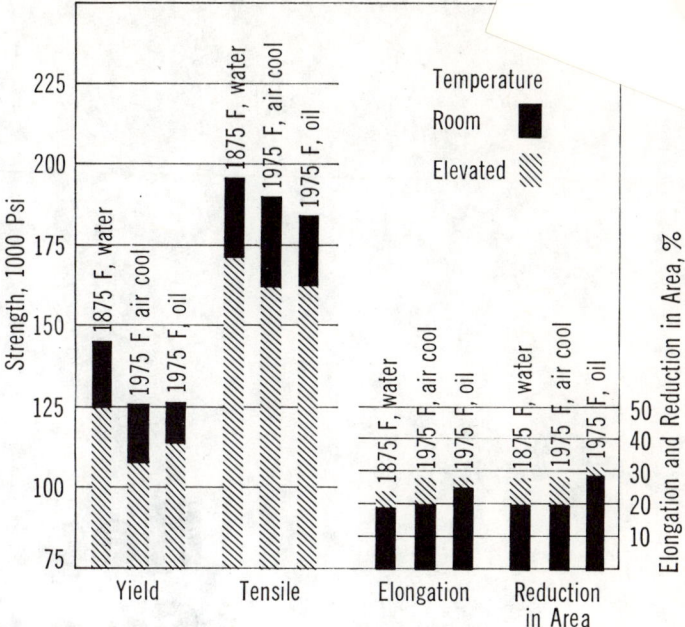

Fig. 5 — Strengths and ductilities of Waspaloy at room temperature and 1000 F can be varied by using different solution treating temperatures and quenchants. After being quenched, specimens were aged at 1550 F for 4 hr, air cooled to room temperature, heated to 1400 F, held 16 hr and air cooled.

mended. If no oxidation can be tolerated, use dry argon (dew point, −60 F or lower) in a sealed retort or furnace chamber. A purge of at least ten times the volume of the retort is recommended before placing the retort in the furnace. Also, the argon must be kept flowing continually during and after the treatment until the parts have cooled nearly to room temperature, to prevent an oxide film from forming.

Alloys containing stable-oxide formers (such as aluminum and titanium) must be bright annealed in a vacuum or in an inert gas such as argon (dew point, −60 F or lower). Dry hydrogen (dew point, −60 F or lower) can be used instead of dissociated ammonia, providing that residual hydrocarbons such as methane are limited to 50 ppm to prevent carburizing.

Air is the most common atmosphere for aging. The smooth, tight oxide layer that is formed is usually acceptable on the finished part. However, if the oxide layer must be minimized, employ a lean exothermic gas (air-gas ratio, about 10 to 1). Gases containing hydrogen and carbon monoxide for aging cycles are dangerous because of the explosion hazard below 1400 F.

Heat Treating of Special-Purpose Alloys

HEAT TREATING procedures for several special-purpose alloys used in military and aerospace applications, and in other products where special properties are required, are discussed in this article. Included are procedures for treating depleted uranium, (DU) zirconium, tantalum, niobium and alloys of these metals. Also presented are lists of references for expanded study of these metals and their alloys.

Heat Treating of Depleted Uranium and Its Alloys

By the ASM Committee on
Heat Treating of Depleted Uranium*

The major applications of depleted uranium (DU) and its alloys are those for which density is an important, if not an overriding, consideration (at 18.7 g/cm³, unalloyed depleted uranium is one of the densest of all elements). Among these applications are kinetic energy penetrators for military use, aircraft and missile counterweights, radiation shielding, gyrorotors and ballast. Dilute alloys containing 0.75 wt%

Ti or 2 wt% Mo are used in production of kinetic energy penetrators (the largest single application of DU) because superior mechanical properties can be developed in these alloys, and their corrosion resistance can be improved, by heat treatment.

Most unalloyed DU is produced in cast form and is used without heat treatment and with or without machining. However, requirements for rolled and heat treated DU may increase significantly in the future.

Phase changes in DU and its alloys are accompanied by significant changes in volume. Volumetric shrinkage from the high temperature gamma phase to the low-temperature phase is 1.8%. Changes in linear dimensions are influenced by preferred orientation. For a random orientation, linear shrinkage is 0.6%. Because shrinkage cannot be predicted accurately, these dimensional changes make it impossible to machine (DU) to final dimensions before heat treatment. In rough machining for heat treatment, an envelope at least 0.4 to 0.5 mm (0.015 to 0.020 in.) thick should be allowed for final machining.

Metallurgical Characteristics of Depleted Uranium

In processing of uranium, several considerations must be addressed: hydrogen embrittlement and hydride formation; ready oxidation in air, attack by hot water; and dissolution by acids. In addition, fine particles of uranium metal are pyrophoric and can ignite spontaneously at room temperature.

The metal uranium is obtained from uranium hexafluoride (UF_6) tailings from the uranium enrichment process that provides U-235 uranium for the nuclear industry. In typical production of depleted uranium, UF_6 tailings are reduced to uranium tetrafluoride (UF_4), called "green salt", which is further reduced to derby uranium metal by a thermite-type bomb reduction with magnesium metal. Calcium metal has also been used to reduce UF_4 to derby metal. Typical derby chemical analysis ranges (in ppm) are 5 to 50 Cu, 8 to 40 Al, 30 to 150 Fe, 10 to 50 Ni, 10 to 100 Si, 1 to 10 Mg, 10 to 50 C, 15 to 40 O, 8 to 40 N and 4 to 18 H. Minor high-vapor-pressure contaminants can be re-

*Arthur L. Geary, Senior Metallurgist, Nuclear Metals, Inc.; Nicholas C. Jessen, Jr., Group Leader, Nuclear Division, Union Carbide Corp.; Allen B. Townsend, Development Consultant, Union Carbide Corp.

Table 1 Typical chemical and gas analyses for vacuum-induction-melted DU

Element	Average analysis, ppm
Carbon	32
N_2	3
O_2	14
Bulk H_2	0.16
Aluminum	10
Copper	10
Iron	30
Manganese	8
Nickel	4
Lead	5
Silicon	40

moved by volatilization during subsequent vacuum melting operations, while the other listed elemental contaminants remain. These impurities should be monitored and controlled to ensure expected metallurgical response to processing. Typical chemical and gas analyses for vacuum-induction-melted depleted uranium are shown in Table 1. In general, hardness varies directly, and ductility varies inversely, with impurity content.

Phase Changes. Chemically and metallurgically, DU is identical with natural uranium. Transformation from the alpha phase (orthorhombic) to the beta phase (tetragonal) occurs at 662 °C (1224 °F). The beta phase is stable up to 773 °C (1423 °F), where it transforms to the high-temperature gamma phase (body-centered cubic). The latter is stable to the melting point, 1132 °C (2070 °F). The gamma phase in unalloyed DU cannot be retained to room temperature by commercial quench rates.

Grain Size and Orientation Control

The grains of cast DU or of DU worked in the gamma region are quite large, typically 2 to 3 mm (0.08 to 0.12 in.) in diameter. Large grain sizes are undesirable for material that is to be worked because they result in rough machined surfaces and variations in mechanical properties. Grain size can be refined considerably by multiple beta quenching. In workpieces more than 25 mm (1 in.) thick, however, refinement is limited to an outer layer of grains because the cooling rates at greater depths are too low. Generally,

the rim of fine-grain material is sufficiently thick to produce a smooth surface after subsequent working.

Beta treatment is also used as a heat treatment for DU worked in the high alpha temperature range. Beta treatment consists of heating the DU into the beta range, holding for a suitable time and cooling at a rapid rate. A common temperature range is 720 to 730 °C (1330 to 1350 °F). The purpose of this heat treatment is to eliminate the preferred orientation that develops during working.

The final alpha grain size is insensitive to temperature within the beta range and to variations in holding time. Times from 1 min to 1 h at temperatures greater than 700 °C (1290 °F) have no effect on final alpha grain size. Cooling rate, however, has a significant effect. Water-quenched material has a significantly finer grain size, and the grains have rough scalloped edges. Air-cooled grains have more uniform boundaries. Extremely low cooling rates, such as those obtained in a furnace-cool cycle, produce large alpha grains.

High cooling rates such as those achieved by water quenching produce high residual stresses in beta-treated material. In thin sections, these residual stresses can produce appreciable plastic deformation in the alpha phase. This alpha phase can be recrystallized, and sometimes grain refined, by an anneal in the alpha range. Alpha annealing after beta treatment will not produce recrystallization at the center of a thick section, because the center does not cool rapidly enough for sufficient straining of the lattice to occur.

When DU is water quenched from the beta phase, high stresses develop due to the combination of (a) volume contraction during beta-to-alpha transformation and (b) the radial temperature gradient. These stresses are compressive at the surface and tensile at the center. The tensile stresses are high enough to produce failure near the centerline. Large numbers of repetitive quenches from the beta phase can produce sponginess, cracks or holes in the center section of the workpiece.

Grain growth is extremely sensitive to orientation as well as to differences in metal purity, to prior deformation and to heat treatments that affect the dispersions of contaminant second phases. Those contaminants that are in solution tend to delay recrystallization and often result in mixed, or incom-

plete, recrystallized structures. Those elements having uranium compounds that show limited solubility, thereby existing as inclusions, do not delay recrystallization appreciably.

The amount of work that exists in uranium metal prior to heat treatment has an important effect on final grain size. A 1 to 2% strain in uranium constitutes critical strain. Recrystallization of material with this amount of strain results in very large grains. Consequently, a plate prepared for forming operations is produced with 10 to 15% warm work to ensure that no areas of critical strain exist in the final wrought product. As the amount of work in the metal to be annealed is increased, the temperature needed for uniform recrystallization is lowered. For most formed parts made from relatively pure material, recrystallization will not occur below 400 °C (750 °F) with annealing times of up to 10 h.

Cold Working

Alpha uranium is slightly softer than steel and is considered to be malleable and ductile. Alpha uranium is readily worked at room temperature; however, directionality and texture persist because of a pronounced anisotropy. Aside from the complication of directionality, the tensile strength of uranium can be greatly enhanced by cold working, as shown in Table 2; hardness also can be increased significantly, as shown in Table 3.

Annealing

Annealing of cold worked DU is similar to that in other metals. The first stage is recovery, in which there is a slight decrease in hardness, a small decrease in electrical resistivity and a pronounced sharpening of x-ray line shape. Recovery is followed by recrystallization. The variation of recrystallization temperature as a function of cold work is shown in Fig. 1 for an annealing time of 1½ h. Recrystallization begins at 400 °C (750 °F) and is complete at 450 °C (840 °F) in material cold worked 90 to 94%. Light cold working (about 4%) causes recrystallization to begin at 525 °C (975 °F), but recrystallization is not complete after 1½ h at 600 °C (1110 °F).

The grain size of cold worked and annealed DU depends upon a variety of factors: annealing time and temperature; amount and homogeneity of the cold work strain; cold working temperature; and volume, size and dispersion of inclusions. Cold or warm working of DU often results in a banded or duplex structure; this persists as a duplex grain size after recrystallization. The average grain size for material rolled to 50% reduction at 300 °C (570 °F) is illustrated in Fig. 2. This shows the effect of annealing temperature (annealing time, 1 h) on materials of various purities. Grain size is about 0.01 to 0.015 mm (0.0004 to 0.0006 in.) for impure uranium and as large as 0.04 mm (0.0015 in.) for high-purity metal.

Table 4 presents typical mechanical properties of uranium in various conditions. These data are not precise values and are intended for use only as guidelines for selection of heat treatments.

The following heat treating procedure can be followed to produce fine-grain material. It is best to start with a relatively pure material that has very small amounts of inclusion-producing impurities. Following a hot breakdown of the rolling ingot at 630 °C (1165 °F), rolling operations are performed at 300 to 400 °C (570 to 750 °F). After 40 to 60% warm work, the material is given a short recrystallization anneal (about 30 min per inch of thickness) at 630 °C. The rolling stock is cooled from the annealing temperature to below 400 °C and given further warm work. This process is repeated as often as final stock thickness will allow.

If the rolled stock is to be used in subsequent forming operations, the final rolling procedure should leave 15 to 20% warm work in the plate. Forming operations should be carried out warm at temperatures not exceeding 375 °C (705 °F). A final anneal at 630 °C (1165 °F) for 30 min per inch of thickness (minimum, 6 to 8 min) will produce parts with a grain size of ASTM 6 to 10, depending on the amount of warm rolling possible. Intermediate anneals at

temperatures below the suggested 630 °C (1165 °F) will help develop the finer grain size. Test specimens will be required to establish the times necessary at these lower temperatures to effect complete recrystallization.

Alpha recrystallization annealing will not remove the anisotropy of the crystal structure produced by the rolling and forming processes. The stock for most forming operations is produced by "square rolling", or by giving the rolled plate essentially equal reductions in the longitudinal and transverse directions. This process produces

a plate with relatively uniform properties in the plane of the plate. Equal reductions taken at 45° to the standard longitudinal and transverse directions, will produce even more uniform forming stock.

Cast Uranium. The grain size of cast uranium is difficult to define because the large cast grains all have well-organized substructures (see Fig. 3). The cast microstructure and mechanical properties can be improved by beta heat treating (see Fig. 4). In this process, the casting is heated to about 740 °C (1365 °F), water quenched, and

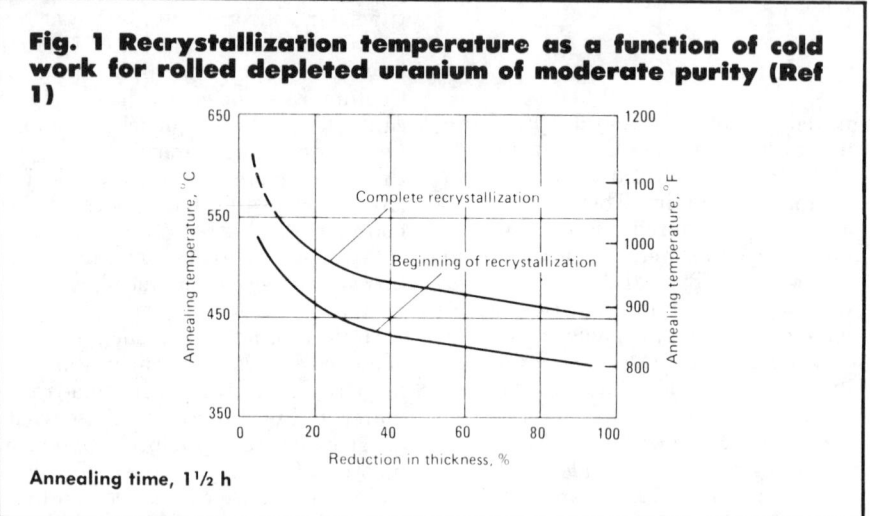

Fig. 1 Recrystallization temperature as a function of cold work for rolled depleted uranium of moderate purity (Ref 1)

Annealing time, 1½ h

Table 2 Typical mechanical properties of DU as functions of amount of cold work

Cold work, %	Ultimate tensile strength		Tensile yield strength(a)		Compressive yield strength		Elongation(b), %	Hardness(c), HV
	MPa	ksi	MPa	ksi	MPa	ksi		
0(d)	1060	154	375	54	405	58	15	294
15	1140	165	525	76	500	72	17	352
25	1190	173	600	87	575	83	14	354
40	1280	186	660	96	605	87	13	359
55	1360	197	905	131	690	100	11	397

(a) At 0.2% offset. (b) In 50 mm or 2 in. (c) 1-kg load. (d) This material is highly directional and was not beta heat treated.

Table 3 Hardness data for cold-worked uranium rod

Cold work, %	Hardness									
	HR15N		HR30N		HR30T		HR45T		HV(a)	
	Average	Range(b)	Average	Range	Average	Range	Average	Range	Average	Range(c)
0........	72	71 to 74	43	42 to 44	79	77 to 82	68	67 to 69	294	281 to 308
15........	73	71 to 74	46	42 to 49	81	79 to 82	72	69 to 73	352	335 to 366
25........	75	74 to 76	47	43 to 49	81	79 to 83	73	71 to 74	354	348 to 361
40........	76	75 to 77	50	47 to 52	83	81 to 85	73	69 to 75	359	339 to 376
55........	78	76 to 79	56	55 to 58	85	83 to 86	79	78 to 80	397	376 to 423

(a) 1-kg load. (b) Results of five indentations. (c) Results of eight indentations across the diametral cross-section of the bar

Fig. 2 Average grain size of depleted uranium as a function of annealing temperature (Ref 2)

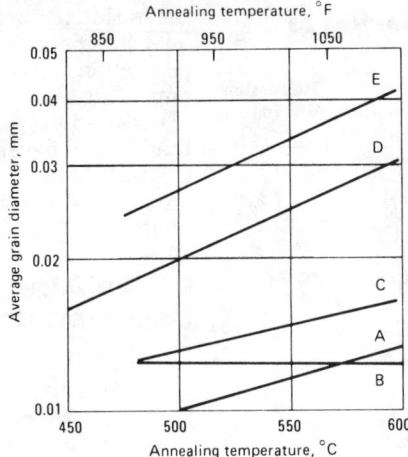

One-hour anneals after 50% reduction by rolling at 300 °C (570 °F). A and B are impure metal; C, D and E are high-purity uranium in decreasing order of submicroscopic inclusion content

then given by an alpha anneal. Grain refining occurs because of the presence of small levels of uranium-iron and uranium-silicon compounds. These compounds are put into solution by the 740 °C (or higher) beta heat treatment, kept in solution by the water quench, and then precipitated by the alpha anneal. If the alpha-anneal temperature is low, the precipitation of the compounds is fine and well dispersed. After this fine precipitate dispersion is achieved, a second beta quench can be more effective as a grain refining step. Two beta quenches and two alpha annealing cycles will produce the desired grain structure in stock as thick as 32 mm, or 1.25 in. (see Fig. 3).

The specific details of the recommended beta heat treatment—such as temperature, time, quenching procedure and furnace conditions—are dictated by the final metallurgical condition desired. When beta heat treating is followed by water quenching, the uranium lattice undergoes heavy strain as

Fig. 3 Typical grain structure of cast depleted uranium

a result of the beta-to-alpha transformation. This transformation can occur by either diffusion or martensitic mechanisms, depending on the severity of quenching. Transformation from beta to alpha is a diffusion reaction at low cooling rates, a martensitic reaction at high cooling rates, and a mixed diffusion and martensitic at intermediate cooling rates. In general, to relieve the residual stresses induced by beta quenching and to precipitate a fine dispersion of secondary compounds, an annealing temperature of 575 °C (1065 °F) is used. Typical tensile properties that can be developed by various heat treatments in cast and wrought uranium are shown in Tables 5 and 6, respectively.

Dilute Alloys of Depleted Uranium

Dilute alloys that are heat treated in larger quantities are DU-0.7 wt% Ti and DU-2 wt% Mo. Both are used as cores in kinetic energy penetrators. The ability of these alloys to age harden is related to the fact that titanium and molybdenum have extended solid solubility in the high-temperature gamma phase and essentially complete

Table 4 Typical room-temperature mechanical properties of uranium in various conditions

Fabrication history	Yield strength(a)		Tensile strength		Elongation(b), %	Reduction in area, %
	MPa	ksi	MPa	ksi		
Cast	207	30	448	65	5	10
Gamma extruded	172	25	552	80	10	12
Beta rolled	207	30	586	85	12	...
					20	
Alpha extruded, 600 °C (1080 °F)	207	30	621	90	15	
Alpha rolled at:						
300 °C (570 °F)	759	110	1172	170	7	14
500 °C (930 °F)	414	60	897	130	20	...
600 °C (1080 °F)	276	40	759	110	20	...
Annealed, after rolling at:						
300 °C (570 °F)	345	50	759	110	5	...
500 °C (930 °F)	276	40	690	100	15	...
Beta treated after alpha rolling:						
Water quenched	241	35	586	85	10	12
Slow cooled	207	30	414	60	7	...

(a) At 0.2% offset. (b) In 50 mm or 2 in.

Fig. 4 Grain of cast depleted uranium refined by beta quenching

Table 5 Typical tensile properties vs heat treating methods for cast uranium

Heat treating methods	Tensile strength MPa	ksi	Yield strength(a) MPa	ksi	Elongation(b), %	Reduction in area, %	J-integral J/mm²	in.·lb/in.²	Tearing modulus	Charpy impact energy J	ft·lb
As cast	420	61	205	30	6	...	...	...	...	...	...
Vacuum heat treated, 640 °C (1184 °F), 1 h	450	65	215	31	5	...	...	...	...	...	...
Vacuum heat treated, 650 °C (1202 °F), 2 h, then 630 °C (1196 °F), 24 h	565	82	185	27	13	...	...	...	...	...	...
Salt annealed	450	65	215	31	8	...	...	...	...	...	...
Beta quenched, vacuum annealed	785	114	295	43	22	17	0.034	192	35	~14(d)	~10(c)
							0.016	90	11	~7(c)	~5(d)

(a) At 0.2% offset. (b) In 50 mm or 2 in. (c) 21 °C. (d) 54 °C

Table 6 Typical tensile properties vs heat treating methods for wrought uranium

Method of heat treating	Tensile strength MPa	ksi	Yield strength(a) MPa	ksi	Elongation(b), %	Reduction in area, %
Vacuum heat treated	800	116	271	39	31	28
Salt annealed or short vacuum heat treated	655	95	272	39	12	12
Vacuum arc melt, vacuum heat treated	780	113	217	31	49	...
Vacuum heat treated plate	835	121	273	40	40	...
Salt annealed	885	128	213	31	20	...

(a) At 0.2% offset. (b) In 50 mm or 2 in.

Table 7 Cooling rates for DU-0.75Ti in various quench media

Media	Quench rate °C/s	°F/s
Flowing argon	3.8	6.8
Conventional or soluble oil	38–40	68–72
0.05% PVA(a)	80	145
Water	98	175
10% brine	190	340

(a) Polyvinyl alcohol

insolubility in the low-temperature alpha phase. On rapid quenching, the gamma transforms martensitically to supersaturated alpha prime. A fine dispersion of intermetallic compound develops during subsequent aging at temperatures above 300 °C (570 °F).

Figure 5 shows the microstructure of water-quenched 36-mm- (1.4-in.-)diam U-0.75Ti bar. The outside of the bar (Fig. 5a) has transformed completely to lenticular alpha prime. The fineness of this structure increases with increasing quench rate. The grain boundaries visible in the structure are those of the prior gamma grains. The U-0.75Ti alloy is shallow hardening, however. At the center of the bar, transformation to alpha phase plus U_2Ti starts at prior grain boundaries (Fig. 5b). Similar structures are found in 18-mm- (0.7-in.-) diam bar oil quenched from the gamma phase. Aging to peak hardness produces no detectable change in structure.

Solution Treating and Aging. Heat treating for improved hardness and mechanical properties in dilute DU alloys consists of solution treating in the gamma-phase temperature range of 800 to 850 °C (1470 to 1560 °F), quenching to room temperature, and aging in the alpha temperature range. Alternatively, interrupted quenching in a molten metal or salt bath held at the appropriate temperature can be used.

The time at the solution-treating temperature is not critical. Times as short as 2 to 5 min produce a completely gamma structure. Longer solution-treating times are generally used—typically, ½ to 1 h. Excessively long times should be avoided because they lead to large gamma grain sizes.

An important consideration in the selection of conditions for gamma solution treatment is the hydrogen level required in the final product. Hydrogen is detrimental to ductility in U-0.75Ti and must be maintained at 1 ppm or less to ensure high ductility in heat treated parts. These levels have been achieved consistently in extruded 36-mm (1.4-in.) -diam rod by vacuum outgassing for 2½ h at 850 °C (1560 °F) at 10^{-5} torr. Unalloyed uranium and uranium alloys are sensitive to hydrogen and for maximum material properties require extensive outgassing. The literature should be consulted before selecting conditions for these alloys.

Quenching. Table 7 gives the rates at which DU-0.75Ti cools when quenched in various media. The test slugs used for measuring these rates were 22 mm (0.875 in.) in diameter by 21 mm (0.845 in.) long. Cooling rates for other DU alloys should be similar. Except for DU-0.75Ti, which is cooled by very slow argon gas quenching, response to subsequent aging at 350 °C (660 °F), as determined by hardness measurements, was independent of quench rates above 40 °C/s (72 °F/s). Because DU-Ti alloys are shallow hardening, higher quench rates are needed to achieve uniform hardening response in larger-diameter bars or thicker plates.

Small-diameter bars and plates can be plunge quenched, but larger diameter bars (greater than 19 mm, or 0.75 in.) develop centerline voids if plunge quenched. These voids pose a particularly serious problem. Once they are formed, there is no easy way to heal them. Void formation is related to the stresses caused by the large volume change associated with the gamma–to–alpha prime transformation and high radial thermal gradients. Centerline voids can be minimized by end quenching—that is, by lowering

Source: Metals Handbook, 9th Edition, Vol. 4, 777-789

Fig. 5 Microstructure of solution treated and quenched U-0.75Ti

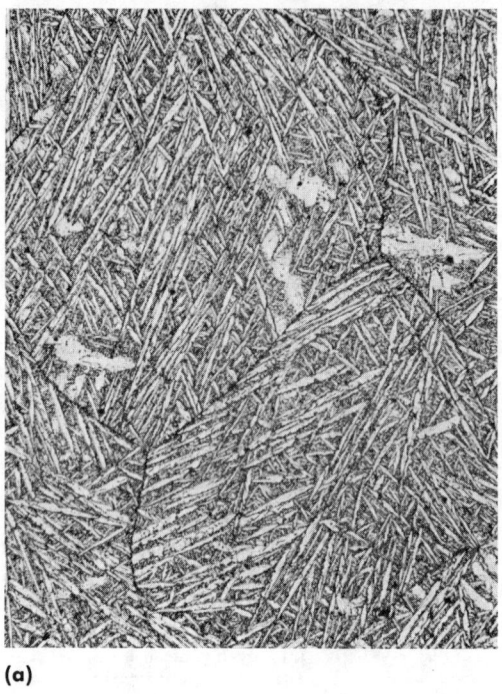

(a)

(b)

Bar 36 mm (1.4 in.) in diameter end quenched into water at 455 mm/min (18 in./min). Chrome-acetic electroetch. Magnification, 100×. (a) Edge. (b) Center

the bars, end on, at a controlled rate, into the quench media. Bars 36 mm (1.4 in.) in diameter have acceptable levels of centerline voids when quenched in this way, 18 at a time, into circulating water at 455 mm/min (18 in./min). The number and size of centerline voids, as detected by ultrasonic techniques, are substantially lower in bars end quenched at 255 mm/min (10 in./min). The aging response of the more slowly quenched bars is identical with those quenched at 455 mm/min (18 in./min).

Aging Results. Hardness and strength levels achieved on aging of dilute DU alloys are illustrated in Fig. 6, 7 and 8. The hardness curves for U-0.75Ti apply equally well to oil-quenched and water-quenched material (Fig. 6). The scatter band is caused by nominal differences in titanium and trace element contents of the alloys; iron and copper, even at low levels, contribute to the hardening response. Silicon is reported to retard hardening.

The heat treater has a wide selection of time-temperature options to achieve specified combinations of hardness and strength. For example, U-0.75Ti can be hardened to 45 HRC by any of the following treatments: 16 h at 380 °C (720 °F); 5 h at 400 °C (750 °F); or 1¾ h at 420 °C (790 °F). For production runs, conditions are selected to optimize equipment utilization.

Metastable High Alloys

Heat treatment of DU alloys with 4.0 wt% or more molybdenum or equivalent is similar to that for dilute alloys. The treatment starts with a gamma-phase solution treatment of 1 h at about 800 °C (1470 °F). This is followed by either (a) quenching to room temperature followed by aging in the alpha region or (b) quenching directly to the aging temperature and holding to achieve the desired properties. Special care must be taken in heat treating large sections. Cracking has occurred in billets of U-6 wt% Nb 205 mm (8 in.) in diameter (or larger) that were water quenched from 800 °C (1470 °F). Refer-

ences should be consulted for details regarding specific high alloys.

Processing and Equipment

Generally, three basic furnace designs are used for heating or heat treating of unalloyed uranium: molten baths, inert-atmosphere furnaces and vacuum furnaces. The type of furnace chosen depends primarily on desired final properties and material quality.

Uranium has been heated in molten lead for 50 h at 350 °C (660 °F) with no appreciable reaction. Longer periods, and/or temperatures of 800 to 1000 °C (1470 to 1830 °F), however, have caused uranium to be completely penetrated. Molten salts are the most common heating media used in industry for preheating uranium prior to fabrication operations and for final heat treatment. Table 8 shows corrosion results for six heat treating mixtures of salts. The main disadvantage of molten salt baths is the potential for hydrogen

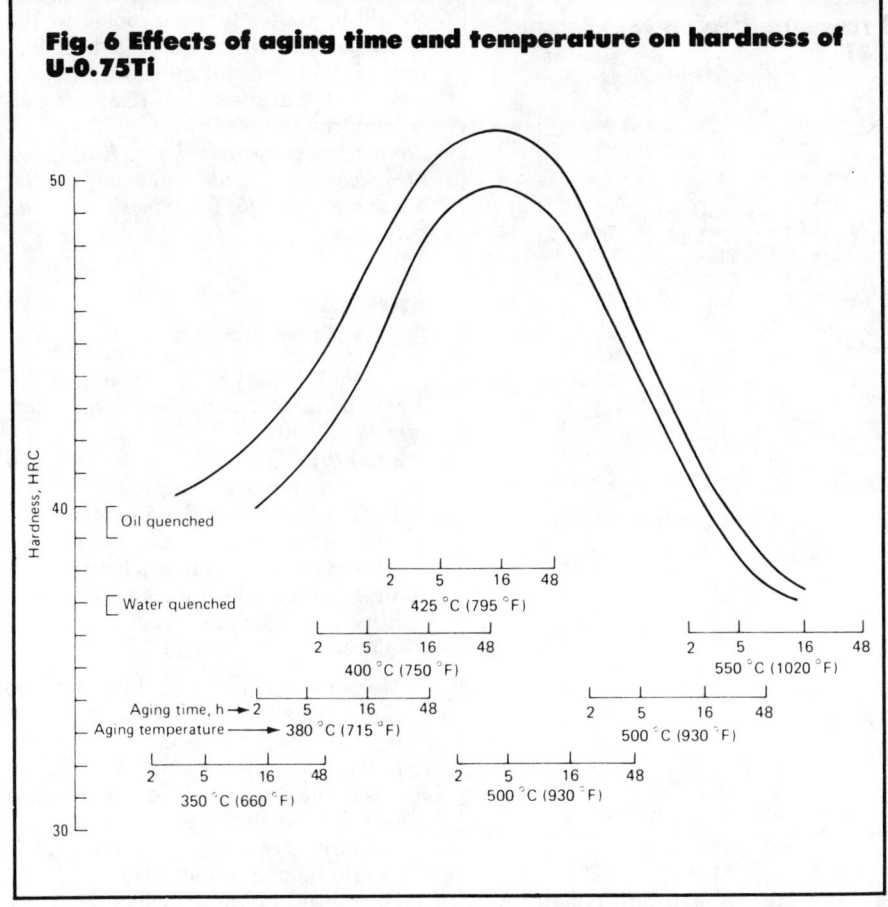

Fig. 6 Effects of aging time and temperature on hardness of U-0.75Ti

Table 8 Corrosion of uranium in molten salts at 595 °C (1100 °F)

Salt mixture	Time, h	Observed attack
44Na$_2$CO$_3$-30K$_2$CO$_3$-26Li$_2$CO$_3$	½ to 2	No corrosion
	4	Surface pitting
74K$_2$CO$_3$-26Li$_2$CO$_3$	½ to 2	No corrosion
	4	Pitted
47Na$_2$CO$_3$-32K$_2$CO$_3$-21Li$_2$CO$_3$	½ to 2	No corrosion
	2	Pits beginning
	4	Badly pitted
53K$_2$CO$_3$-46.6Li$_2$CO$_3$	½ to 1	No corrosion
	2	Surface pits
20NaOH-30K$_2$CO$_3$-50Na$_2$CO$_3$	½ to 4	Surface etching
47Na$_2$CO$_3$-47K$_2$CO$_3$-6Li$_2$CO$_3$	½	Scaling, 0.24% weight loss
	2	Scaling, 1.0% weight loss
	4	Scaling, 1.3% weight loss

contamination. Consequently, any planned use of molten salt baths for heating of uranium should include a design for removing the hydrogen from the bath, such as sparging the bath with CO$_2$ gas. Hydrogen, which is par-ticularly deleterious to uranium, drastically reduces its tensile elongation, as shown in Fig. 9.

Furnace Atmospheres. No appreciable attack on uranium occurs in dry furnace atmospheres of helium, argon, carbon monoxide, carbon dioxide or hydrocarbon gases at temperatures up to 500 °C (930 °F). However, relatively slight amounts of water vapor in any of these gases can cause extensive corrosion. Uranium reacts with water, in either liquid or vapor form, to produce UO$_2$ and hydrogen. The UO$_2$ spreads over the entire surface area as a black powder and, depending on the amount of exposure and purity of the metal, can result in excessive pitting and surface cratering. Thus, in heat treating of uranium in atmospheres of commercial inert gases, furnaces should be equipped with gas line dryers to dry the gas thoroughly before it is used.

Vacuum Treatment. Vacuum heat treating of uranium provides the best over-all environment for obtaining maximum tensile properties and high-quality metal surfaces. The principal advantage of vacuum-furnace heat treating is the potential for removing hydrogen from the metal.

Figure 10 shows the time required for vacuum heat treating uranium stock of different thicknesses to achieve maximum ductility. This illustration assumes an initial hydrogen concentration within the part of 2 ppm, which is typical of material processed through carbonate preheating baths, and it can be used for cast material.

A vacuum of 300 torr limits oxidation to an acceptable level while maintaining the hydrogen at its initial level. A vacuum of 10^{-5} torr is needed for significant lowering of the hydrogen level.

Cleaning. Surfaces of parts to be heat treated should be free of moisture, grease and cutting lubricants. Heavier oxides that form on DU and its alloys can be removed by pickling for about ½ h in a 1:1 mixture of nitric acid and water at 25 °C (75 °F). Copper, which is often used as cladding during fabrication, can also be removed with 1:1 nitric acid.

CAUTION: U-Nb and U-Zr alloys can produce explosions during pickling in nitric acid. The problem can be eliminated by adding 1 to 2 vol % of hydrofluoric acid to the pickling solution.

Quenching. The oil-quench media can be contained within the vacuum chamber. Following solution treating, the furnace is backfilled with argon or helium. The work is then transferred to a position directly over the quench tank and lowered at a controlled rate into the quench media.

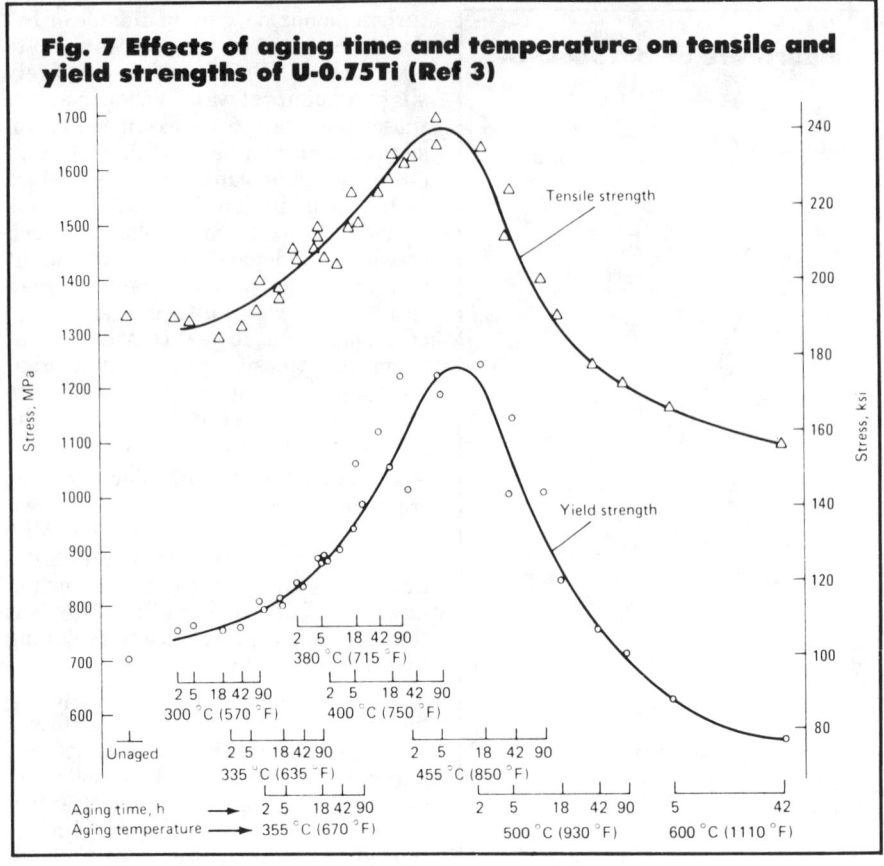

Fig. 7 Effects of aging time and temperature on tensile and yield strengths of U-0.75Ti (Ref 3)

The high vapor pressure of water precludes incorporation of water-base quench media into the furnace. Specially designed bottom-loading furnaces have been used successfully with separate quench tanks. The sequence of events is as follows. At the end of solution treating, the quench tank is moved under the furnace. The furnace is then backfilled with argon or helium, the power turned off, and the bottom door removed. The load is dropped rapidly to a position 50 to 75 mm (2 to 3 in.) above the water level and then is dropped at a controlled rate until all of the workpiece is submerged.

Fixtures. Neither DU nor its alloys are especially strong at gamma-solution treating, temperatures. Low strength coupled with a high density places special requirements on fixturing to provide proper support and thus prevent sagging. If possible, rods and irregular-shape pieces should be placed with the long axis in a vertical position and supported every 150 mm (6 in.); spans for horizontal pieces should be limited to 75 to 100 mm (3 to 4 in.).

Figure 11 shows a fixture used for solution treating of bars 36 mm (1.4 in.) in diameter by 380 mm (15 in.) long. This basket is made of Inconel 600. Copper shims are used at the points of contact between the DU and the Inconel because these two materials react at solution treating temperatures to produce a DU-Fe eutectic composition with a melting point of 725 °C (1335 °F). Molten-metal attack would result in extensive local wastage of the DU and welding to the fixture.

Salt Baths. If a salt bath is used for solution treating, the times should be kept as short as possible, not only to minimize hydrogen pickup but also to limit corrosive attack. Pieces exposed to molten chloride salts ($BaCl_2$, KCl and NaCl eutectic) and molten carbonate salts (35% Li_2CO_3 and 65% K_2CO_3) at 730 °C (1350 °F) have lost 0.08 mm (0.003 in.) of thickness in 1 h. Such attack would be significantly more severe at 850 °C (1560 °F).

Aging treatments can be carried out in recirculating inert gas furnaces, lead baths, lead-tin baths or molten salt baths. Because the aging reactions are temperature sensitive, the temperature should be controlled to within +5 °C (+9 °F) and preferably to within +2 °C (+4 °F). Following aging, the pieces should be either furnace cooled to 100 °C (212 °F) or water quenched. Oxidation of DU is exothermic, and the heat generated can make the reaction self-sustaining. *Rods 18 mm (0.7 in.) in diameter have reached red heat when they were not water quenched after being aged at 425 °C (795 °F) in a lead pot.*

Examples of Heat Treatment

The following procedures are examples of heat treatment used to meet certain specifications.

Example 1. A DU-0.75Ti alloy is to be heat treated to a hardness of 44 to 52 HRC. The heat treated material is to be machined to form a kinetic energy penetrator with a cylindrical body and a conical nose. Mechanical properties and hydrogen level are not as follows: The procedure is specified.

- Machine alpha-extruded bar stock to the approximate dimensions, allowing a 0.4-mm (0.015-in.) envelope for finish machining.
- Assemble premachined blanks in a rectangular basket with the grid-support structure; the nose ends should be pointed upward.
- Degrease to remove residual cutting fluid.
- Place the baskets in a vacuum furnace with an integral oil quenching tank. Solution treat ½ h at 850 °C (1560 °F) and at a pressure of less than 300 torr.
- Oil quench.
- Degrease.
- Age 2 h at 440 °C (820 °F) in an inert-gas recirculating furnace.
- Furnace cool.

Example 2. Bars of DU-0.75Ti alloy, 36 mm (1.4 in.) in diameter, are to be heat treated to the following specifications: hardness, 38 to 44 HRC; minimum 0.2% yield strength, 725 MPa (105 ksi); minimum elongation 12%; and maximum hydrogen content, 1 ppm. The procedure is to:

- Cut extruded bar stock to length.
- Pickle in 1:1 nitric acid to remove copper sheath.
- Rinse and air dry.
- Place rods vertically in a basket (see Fig. 7).
- Solution treat 2½ h at 850 °C (1560 °F) in a vacuum of 5×10^{-5} torr, or better.

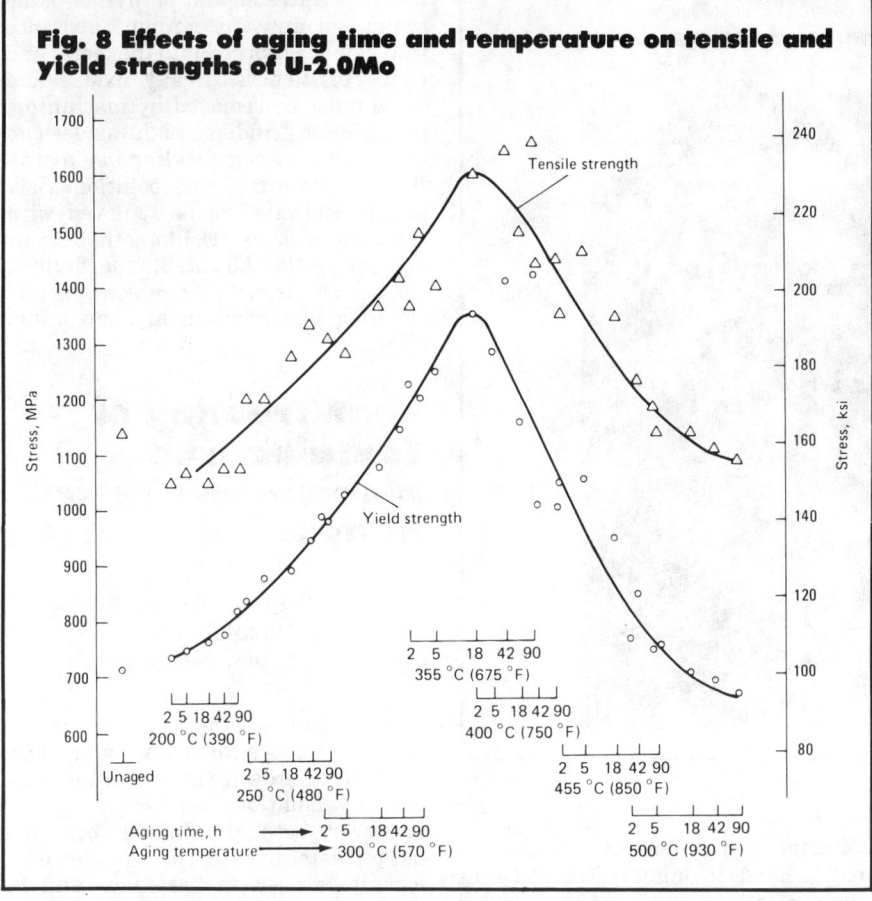

Fig. 8 Effects of aging time and temperature on tensile and yield strengths of U-2.0Mo

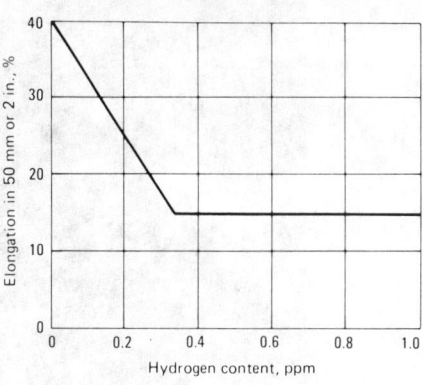

Fig. 9 Elongation vs hydrogen content for wrought depleted uranium

Fig. 10 Time required to achieve maximum ductility in depleted uranium plate of various thicknesses under a vacuum of 10⁻⁴ torr

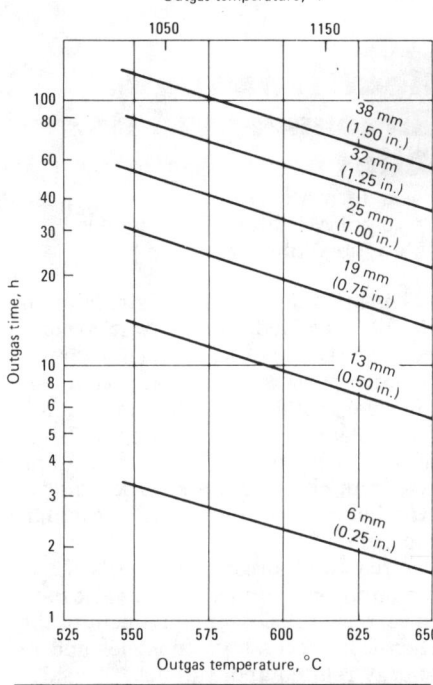

- Quench into circulating water at 455 mm/min (18 in./min).
- Air dry.
- Age 16 h at 350 °C (660 °F) in an inert gas recirculating furnace.

Licensing and Health and Safety Requirements

Possession of more than 15 lb (6.8 kg) of depleted uranium in any form requires a license from the U.S. Nuclear Regulatory Commission. Title 10, Part 40, of Federal Regulations describes the steps necessary and the requirements to obtain such a license. In addition, all other local, state and federal regulations also are effective as applicable.

The greatest potential source of contamination in the heat treating area is uranium oxide. The area should be isolated from the remainder of the plant, and everyone entering should be required to wear disposable protective footwear. Smoking and eating should be restricted.

The toxicity of depleted uranium if it enters the blood stream, may result in poisoning similar to that caused by lead, arsenic, mercury or any other heavy metal.

REFERENCES

1. "Recrystallization of Cold-Rolled Uranium", by E. E. Hayes: U. S. Atomic Energy Commission Report TID-2501, 1949
2. Recrystallization and Grain Growth in Uranium, by E. S. Fisher: in Reactor Technology and Chemical Processing, Vol 9 of *Proceedings of the International Conference on the Peaceful Uses of Atomic Energy*, United Nations, 1956
3. The Effect of Aging on the Mechanical Behaviors of U-0.75 wt. % Ti and U-2.0 wt. % Mo, by K. H. Eckelmeyer and F. J. Zanner: *Journal of Nuclear Materials*, Vol 62, No. 1, Oct 1976, p 37-49

SELECTED REFERENCES

- *Nuclear Reactor Fuel Elements: Metallurgy and Fabrication*, edited by A. R. Kaufmann: Interscience Publishers, a division of John Wiley and Sons, New York, 1962.
- *Physical Metallurgy of Uranium Alloys*, edited by J. J. Burke *et al*: Brook Hill Publishing Co., Chestnut Hill, MA, 1976

Fig. 11 Fixture for solution treating and end quenching of 18 depleted uranium bars

metal surfaces should be free of oxide scale, soil and any foreign substance that can be absorbed into the metal and cause brittleness. Heavy oxides and scale must be removed by machining, abrasion or grinding. Light oxides can be removed by acid pickling in a hydrofluoric acid-nitric acid solution. Soil, grease and oils can be removed with solvents such as trichloroethylene or acetone. ANSI/ASTM Standard B614 can be used as a guide for descaling and cleaning of zirconium and zirconium alloys.

Heat Treating of Tantalum and Niobium and Their Alloys

By the ASM Committee on Heat
Treating of Tantalum
and Niobium and Their Alloys*

At the present time, tantalum is primarily used as a capacitor material for electronic hardware and niobium as an alloying addition to other materials. However, both are finding increased use by themselves and as base elements for alloys used in aerospace and in applications where resistance to chemical attack is important. Niobium-base alloys are widely used as superconductors.

Tantalum and niobium are sister metals, not only in the sense that they are often found together in nature, but also because their like chemical and metallurgical properties require similar procedures for melting, fabrication and heat treating. Both are chemically active, ductile metals that are readily embrittled by pickup of a few hundred parts per million of O_2, N_2, C or H_2. Surface oxidation of both metals occurs in air above 300 °C (570 °F), and the oxidation rate increases with increasing temperature. Hydrogen embrittlement occurs if either metal is cathodic in a galvanic couple or is exposed to a hydrogen atmosphere during cooling from elevated temperature.

To prevent pickup of these four interstitial elements during heat treating, it

*Robert E. Droegkamp, Manager, Research and Development, Fansteel Metals, Inc.; Louis (Ned) E. Huber, Jr., Kawecki-Berylco Industries, Inc., Division of Cabot Corp.; R. Terrence Webster, Principal Metallurgical Engineer, Teledyne Wah Chang.

Heat Treating of Zirconium and Its Alloys

By R. Terrence Webster
Principal Metallurgical Engineer
Teledyne Wah Chang

The major application of zirconium is in the fuel cladding for water-cooled nuclear reactors and in ancillary reactor core parts such as water channels and fuel rod spacers. The major alloys for nuclear applications are the Zircaloys, which are a series of zirconium, tin, iron, chromium and nickel alloys. Another alloy used is zirconium-niobium.

Zirconium, zirconium-tin alloys and zirconium-niobium alloys are also used in corrosion resistant equipment in the chemical processing industries and in energy related applications.

Annealing Temperatures

For most of the zirconium alloys in industrial applications, strength cannot be increased by heat treatment. Consequently, the only heat treatments required are full annealing and stress-relief annealing.

Zirconium-niobium alloys can be heat treated to higher strength, but they normally are used in the annealed condition. The heat treatments for zirconium and zirconium alloys are as follows:

- Full recrystallization anneal—700 to 785 °C (1290 to 1445 °F) for 1 h per inch of thickness
- Stress relief anneal—480 to 595 °C (895 to 1100 °F) for 1 h per inch of thickness.

To avoid excessive grain growth in highly cold worked thin sheet and foil, full annealing at 595 °C (1100 °F) for 6 h may be desirable.

Processing

The nuclear reactor-grade alloys are annealed in vacuum (at pressures from 10^{-2} to 10^{-5} torr) or in an inert-gas atmosphere of argon or helium. This is done to prevent formation of an oxide layer on the surface, which can be removed only by mechanical means such as abrasion or machining.

The commercial alloys can be annealed in air because the oxide layer is fully corrosion resistant and need not be removed. Prior to heat treating, the is essential to use appropriate surface-cleaning procedures, furnace maintenance and operational practices.

Both metals first can be purified by electron-beam melting in a vacuum of 10^{-5} torr or better. Alloying can be done by either electron-beam or vacuum-arc melting, the latter method being chosen for better control when alloying elements tend to be volatile.

Mill products also are made from powders by resistance or induction sintering of cold isostatically compacted bars in a vacuum of 10^{-5} torr.

Conventional fabrication procedures are used; if heavy sections are heated in air, all surface contamination must be removed by machining or grinding and pickling before annealing.

It is strongly suggested that contact be made with the primary fabricated metal supplier for discussion and review of plans and procedures prior to any thermal treatment of tantalum or niobium or their alloys. Numerous expensive failures have resulted from attempts to perform thermal treatment attempts in air, in atmospheres with inadequate inert gas protection or in inadequate vacuum systems.

Cleaning Procedures

To avoid contamination of tantalum and columbium by interstitial elements and metallic contaminants, it is mandatory that the material be chemically clean before it is subjected to any heating operation such as annealing or welding. Cleaning and degreasing present no special problems (conventional methods and materials may be used, although hot caustics must be avoided). First, thorough degreasing is carried out using a detergent or solvent. Degreasing is followed by chemical etching, typically with a mixture 60 HNO_3, 20 HF and 20 H_2SO_4 (volume percentages); hot and cold water rinses in distilled water; and spot-free drying. The etching solution may be strengthened by HF additions, or weakened by water additions, to achieve the amount of stock removal necessary to ensure cleanness of the metal surface. One company eliminates H_2SO_4 because some evidence indicates that it can contribute to weld embrittlement. Nitric acid should always be present, however, because it prevents hydrogen pickup during pickling.

Elevated-temperature forgings will have an oxygen-contaminated outer layer, and this must be removed from all surfaces by machining or grinding before acid pickling.

Annealing Practice

Tantalum and niobium and their alloys are most often used in the fully recrystallized condition to achieve the best fabrication response, although some applications require stress relieved or cold worked properties. The recrystallization temperature is so highly dependent on purity, amount of cold work and prior history, that current practice is to anneal pilot samples to ensure that the correct temperatures are used. Time at temperature is typically 1 h.

Table 9 can be used as a guide for choosing pilot temperatures. Materials given heavy fabrication reductions will recrystallize to finer grain sizes at lower temperatures than will those given lighter fabrication reductions. The recrystallization annealing temperature is also somewhat dependent on interstitial purity. For example, pure tantalum containing 200 ppm oxygen requires a higher recrystallization annealing temperature than pure tantalum containing less than 50 ppm oxygen.

A typical annealing sequence is:

- Visually verify material cleanness
- Load, using tantalum, tantalum alloy or molybdenum fixtures for support, or tantalum foil for protection, as required. Tantalum, niobium and certain of their alloys exhibit low yield strengths at the necessary annealing temperatures. Sticking to fixtures and other pieces can cause damage, especially with tantalum because of its high density.
- Pump down
- Leak rate check
- Power on to temperature
- Time as required at temperature
- Power off
- When temperature drops below 1000 °C (1830 °F), backfill to 15 mm Hg with industrial high-purity (99.995% min) argon or helium.
- Before removing load from furnace, allow to cool to below 200 °C (390 °F), which can require from 3 to 15 h depending on furnace size and mass of load.

This sequence is not intended to be a detailed procedure for annealing these materials. This sequence information is intended to create awareness of the difficulties and risks of heat treating these materials to avoid repetition of past costly errors. The major risk is loss of vacuum at temperature resulting in the extremely costly destruction by oxidation not only of the parts being heat treated but also of the shielding and resistance elements.

Furnaces

Tantalum and niobium and their alloys are easily contaminated during

Table 9 Annealing temperatures for tantalum and niobium and their commercial alloys

Alloy designation	Nominal alloy additions, %	Annealing temperature			
		Stress relief		Recrystallization	
		°C	°F	°C	°F
Tantalum alloys					
Ta	None	850	1560	1000–1250	1830–2280
Ta	None(a)	1000	1830	1200–1350	2190–2460
FS63	2.5W, 0.15Nb	1000	1830	1200–1300	2190–2370
FS61	7.5W(a)	...	...	1400–1550	2550–2820
FS60	10W	1100	2010	1300–1600	2370–2910
T111	8W, 2Hf	1100	2010	1400–1650	2550–3000
T222	9W, 2.4Hf, 0.01C	1100	2010	1400–1650	2550–3000
Niobium alloys					
Nb	None	800	1470	900–1200	1650–2190
FS80	1Zr	875–1150	1610–2100	1150–1250	2100–2280
SNb 291	10Ta, 10W	1000	1830	1150–1200	2100–2190
Nb 752	10W, 2.5Zr	...	...	1300–1400	2370–2550
C 129Y	10W, 10Hf, 0.1Y	900	1650	1150–1250	2100–2280
FS85	28Ta, 11W, 0.8Zr	1150	2100	1300–1400	2370–2550
C103	10Hf, 1Ti, 0.7Zr	...	...	1250–1375	2280–2510

(a) Powder metallurgy; all other compositions are vacuum melted.

annealing, and special care must be exercised in furnace selection, cleanness of work and annealing practice used. Normally used are cold-wall radiant-heated furnaces with refractory metal heater elements, primary heat shields, permanent hearth materials and support fixtures. These furnaces operate at vacuums of 10^{-4} torr or better and have an acceptably low leak rates.

Vacuum Leak Control. Leak-rate control tends to be the key to successful heat treating of tantalum and niobium alloys, especially with products having high surface-to-volume ratios, such as low gage wire, tube and strip. A suggested maximum leak rate is 0.1 μm/min. This expression includes chamber volume and pressure rise per unit of time, which is the only meaningful measurement.

A leak rate is only meaningful in a stabilized system—that is, one that has pumped for a period of time and is no longer outgassing. It is defined as the difference in pressure (torr) for one second in a volume of one litre.

For example, a 1200-litre vacuum chamber is isolated by closing the high-vacuum valve. The pressure rise is observed and an increment of rise is timed by stopwatch. The chamber in this example rose 2×10^{-3} torr in 5 min. The rate is calculated as:

$$\frac{\text{Pressure change} \times \text{chamber volume}}{\text{Time}} =$$

$$\frac{2 \times 10^{-3}}{300} \times 1200 =$$

$$0.8 \times 10^{-2} \text{ torr l/s}$$

where pressure change is measured in torr; chamber volume in litres; and time in seconds. This rate is acceptable within the suggested limit of 10^{-2} torr l/s.

Hot-wall argon-atmosphere furnaces have been used, but adsorbed gases and metals on hot furnace walls are more likely to cause contamination. Argon must be free of H_2 and have a dew point below -50 °C (-60 °F).

Furnaces must be clean and usually must not be used for other operations or other metals unless a given practice has been found to be satisfactory. Furnaces previously used to perform brazing operations should be avoided.

Good practice dictates that furnaces be heated to a temperature 100 °C (180 °F) above the annealing temperature in the empty condition, to remove adsorbed gases.

Furnace qualification is often a customer requirement and usually includes a limit on the amount of allowable contamination as measured by increased hardness or interstitial content.

The trend toward bonding tantalum and niobium to other metals presents special problems that must be carefully reviewed from a metallurgical standpoint. The recommendations is Table 9

probably do not apply to most clad materials.

SELECTED REFERENCES

- *Tantalum and Niobium, Metallurgy of the Rarer Metals -6,* by G. L. Miller: Butterworths Scientific Publications, London, 1959
- *Columbium and Tantalum,* by Frank T. Sisco and Edward Epremian: John Wiley & Sons, Inc., London, 1963
- "The Engineering Properties of Tantalum and Tantalum Alloys", by F. F. Schmidt and H. R. Ogden: DMIC Report 189, Sept 13, 1963
- "The Engineering Properties of Columbium and Columbium Alloys", by F. F. Schmidt and H. R. Ogden: DMIC Report 188, Sept 6, 1963
- *Metals Handbook,* Vol 2, 8th ed., "Heat Treating, Cleaning and Finishing", 1979, p 269–270
- *Metals Handbook,* Vol 3, 9th ed., "Properties and Selection: Stainless Steels, Tool Materials and Special-Purpose Metals", 1980, p 321–325
- Postheating of Cb-1Zr and T111 (Ta + 8W + 2Hf) Weldments: NASA-Lewis Specification No. RM–5, June 1971
- Interactions of Refractory Metals with Active Gases in Vacua and Inert Gas Environments, by H. Inoye: in *Refractory Metals Alloys, Metallurgy and Technology,* edited by I. Machlin, R. T. Begley and E. D. Weisert, p 165

Heat Treating of Titanium and Titanium Alloys

By the ASM Committee on Titanium and Titanium Alloys*

TITANIUM AND TITANIUM AL-LOYS are heat treated for the following purposes:

- To reduce residual stresses developed during fabrication (stress relieving)
- To produce an optimum combination of ductility, machinability, and dimensional and structural stability (annealing)
- To increase strength (solution treating and aging)
- To optimize special properties such as fracture toughness, fatigue strength and high-temperature creep strength.

Various types of annealing treatments (single, duplex, beta and recrystallization annealing, for example), and solution treating and aging treatments, are imposed to achieve selected mechanical properties. Stress relieving and annealing may be employed to prevent preferential chemical attack in some corrosive environments, to prevent distortion (a stabilization treatment) and to condition the metal for subsequent forming and fabricating operations.

Response of titanium and titanium alloys to heat treatment depends on the composition of the metal. Unalloyed titanium is allotropic. Its close-packed hexagonal structure (alpha phase) changes to a body-centered cubic structure (beta phase) at 885 °C (1625 °F), and this structure persists at temperatures up to the melting point.

With respect to their effects on the allotropic transformation, alloying elements in titanium are classified as alpha stabilizers or as beta stabilizers. Alpha stabilizers, such as oxygen and aluminum, raise the alpha-to-beta transformation temperature. Nitrogen and carbon also are alpha stabilizers, but these elements usually are not added intentionally in alloy formulation. Beta stabilizers, such as manganese, chromium, iron, molybdenum, vanadium and niobium, lower the alpha-to-beta transformation temperature and, depending on the amount added, may result in retention of some beta phase at room temperature. Alloying elements such as zirconium and tin have essentially no effect on the alpha-to-beta transformation temperature. Based on the types and amounts of al-

loying elements they contain, titanium alloys are classified as alpha, near-alpha, alpha-beta or beta alloys. Near-alpha alloys are alloys with predominantly alpha stabilizer, plus limited beta stabilizers (normally, 2% or less).

Alpha and near-alpha titanium alloys can be stress relieved and annealed, but high strength cannot be developed in these alloys by any type of heat treatment. The commercial beta alloys are, in reality, metastable beta alloys. When these alloys are exposed to selected elevated temperatures, the retained beta phase decomposes and strengthening occurs. For beta alloys, stress-relieving and aging treatments can be combined, and annealing and solution treating may be identical operations.

Alpha-beta alloys are two-phase alloys and, as the name suggests, comprise both alpha and beta phases at room temperature. These are the most common and the most versatile of the three types of titanium alloys. Phase compositions, sizes and distributions can be manipulated by heat treatment within certain limits to enhance a spe-

---

*Walter Herman, *Chairman*, Technical Director, Viking Metallurgical Corp.; Roger V. Carter, Manager, Metals Technology, Boeing Commercial Airplane Co.; William H. Heil, Supervisor, Quality Assurance, Timet; Ralph J. Kotfila, Lead Engineer—Technology, McDonnell Aircraft Co.; Charles J. Scholl, Chief Product Metallurgist, Wyman-Gordon Co.

cific property or to attain a range of strength levels.

Not all heat treating cycles are applicable to all titanium alloys, because the various alloys are designed for different purposes. Alloys Ti-5Al-2Sn-2Zr-4Mo-4Cr (commonly called "Ti-17") and Ti-6Al-2Sn-4Zr-6Mo are designed for strength in heavy sections; Ti-6Al-2Sn-4Zr-2Mo for creep resistance; Ti-6Al-2Cl-1Ta-1Mo and Ti-6Al-4V-ELI for resistance to stress corrosion in aqueous salt solutions, and for high fracture toughness; Ti-5Al-2.5Sn for weldability; and Ti-6Al-6V-2Sn, Ti-6Al-4V and Ti-10V-2Fe-3Al for high strength at low-to-moderate temperatures.

Stress Relieving

Titanium and titanium alloys can be stress relieved without adversely affecting strength or ductility. Stress-relieving treatments decrease the undesirable residual stresses that result from (a) nonuniform hot forging deformation from cold forming and straightening, (b) asymmetric machining of plate (hogouts) or forgings, and (c) welding and cooling of castings. Removal of such stresses helps maintain shape stability and eliminates unfavorable conditions, such as the loss of compressive yield strength commonly known as the Bauschinger effect.

When symmetrical shapes are machined in the annealed condition, employing moderate cuts and uniform stock removal, stress relieving may not be required. Compressor disks made of Ti-6Al-4V have been satisfactorily machined in this manner, conforming with dimensional requirements. In contrast, thin rings made of the same alloy could be machined at a higher production rate to more stringent dimensions by stress relieving 2 h at 540 °C (1000 °F) after rough machining.

Separate stress relieving may be omitted when the manufacturing sequence can be adjusted to employ annealing or hardening as the stress-relieving process. For example, forging stresses may be relieved by annealing prior to machining. Large, thin rings have been effectively processed with minimum distortion by rough machining in the annealed state, followed by solution treating, quenching, partial aging, finish machining and final aging. Partial aging relieves quenching stresses, and final aging relieves

Table 1 Recommended stress-relief treatments for titanium and titanium alloys

Parts can be cooled from stress relief by either air cooling or slow cooling

Alloy	Temperature °C	°F	Time, h
Commercially pure Ti (all grades)	480 to 595	900 to 1100	1/4 to 4
Alpha or near-alpha titanium alloys			
Ti-5Al-2.5Sn	540 to 650	1000 to 1200	1/4 to 4
Ti-8Al-1Mo-1V	595 to 705	1100 to 1300	1/4 to 4
Ti-6Al-2Sn-4Zr-2Mo	595 to 705	1100 to 1300	1/4 to 4
Ti-6Al-2Cb-1Ta-0.8Mo	595 to 650	1100 to 1200	1/4 to 2
Ti-0.3Mo-0.8Ni (Ti Code 12)	480 to 595	900 to 1100	1/4 to 4
Alpha-beta titanium alloys			
Ti-6Al-4V	480 to 650	900 to 1200	1 to 4
Ti-6Al-6V-2Sn (Cu + Fe)	480 to 650	900 to 1200	1 to 4
Ti-3Al-2.5V	540 to 650	1000 to 1200	1/2 to 2
Ti-6Al-2Sn-4Zr-6Mo	595 to 705	1100 to 1300	1/4 to 4
Ti-5Al-2Sn-4Mo-2Zr-4Cr (Ti-17)	480 to 650	900 to 1200	1 to 4
Ti-7Al-4Mo	480 to 705	900 to 1300	1 to 8
Ti-6Al-2Sn-2Zr-2Mo-2Cr-0.25Si	480 to 650	900 to 1200	1 to 4
Ti-8Mn	480 to 595	900 to 1100	1/4 to 2
Beta or near-beta titanium alloys			
Ti-13V-11Cr-3Al	705 to 730	1300 to 1350	1/12 to 1/4
Ti-11.5Mo-6Zr-4.5Sn (Beta III)	720 to 730	1325 to 1350	1/12 to 1/4
Ti-3Al-8V-6Cr-4Zr-4Mo (Beta C)	705 to 760	1300 to 1400	1/6 to 1/2
Ti-10V-2Fe-3Al	675 to 705	1250 to 1300	1/2 to 2
Ti-15V-3Al-3Cr-3Sn	790 to 815	1450 to 1500	1/12 to 1/4

stresses developed during finish machining.

Table 1 presents combinations of time and temperature that are used for stress relieving titanium and titanium alloys. The ranges in both time and temperature indicate that more than one combination may yield satisfactory results. The higher temperatures usually are used with shorter times, and the lower temperatures with longer times, for effective stress relief. During stress relief of solution-treated and aged titanium alloys, care should be taken to prevent overaging to lower strength. This usually involves selection of a time-temperature combination that provides partial stress relief. The parts, in bulk or in fixtures, may be charged directly into a furnace operating at the stress-relief temperature. If a part is mounted in a massive fixture, a thermocouple should be attached to the largest part of the fixture.

Figure 1 illustrates the effects of stress relieving Ti-6Al-4V at five temperatures ranging from 260 to 620 °C (500 to 1150 °F) for periods of time ranging from 5 min to 50 h.

The rate of cooling from the stress-relieving temperature is not critical. Uniformity of cooling is critical, partic-

ularly in the temperature range from 480 to 315 °C (900 to 600 °F). Oil or water quenching should not be used to accelerate cooling, however, because this can induce residual stresses by unequal cooling. Furnace or air cooling is acceptable.

Stress-relieving treatments must be based on the metallurgical response of the alloy involved. Generally, this requires holding at a temperature sufficiently high to relieve stresses without causing an undesirable amount of precipitation or strain aging in alpha-beta and beta alloys, or without producing undesirable recrystallization in single-phase alloys that rely on cold work for strength.

Stress relieving of the more highly alloyed alpha-beta compositions, and of beta alloys, should be done using a thermal exposure that is compatible with annealing, solution-treating, stabilization or aging processes.

There are no nondestructive testing methods that can measure the efficiency of a stress-relief cycle other than direct measurement of residual stresses by x-ray diffraction. No significant changes in microstructure due to stress-relieving heat treatments can be detected by optical microscopy.

Weldments. The temperatures used for stress relieving complex weldments of alpha or alpha-beta alloys should be near the high ends of the ranges given in Table 1. Complex weldments may be defined as those having multiple welds in complex configurations, possibly involving combinations of machine and manual welding. In complex weldments made with commercially pure titanium, Ti-5Al-2.5Sn alloy or Ti-6Al-4V alloy, more than 70% of the residual stress is relieved during the first hour at temperature. Simple weldments of commercially pure titanium often are used without stress relief.

Annealing

Annealing of titanium and titanium alloys serves primarily to increase fracture toughness, ductility at room temperature, dimensional and thermal stability, and creep resistance. Many titanium alloys are placed in service in the annealed state. Because improvement in one or more properties generally is obtained at the expense of some other property, the annealing cycle should be selected according to the objective of the treatment. Common annealing treatments are:

- Mill annealing
- Duplex annealing
- Triplex annealing
- Recrystallization annealing
- Beta annealing.

Recommended annealing treatments for several alloys are given in Table 2. Mill annealing is a general-purpose treatment given to all mill products. It is not a full anneal, and may leave traces of cold or warm working in the microstructures of heavily worked products (particularly sheet). Duplex and triplex annealing alter the shapes, sizes and distributions of phases to those required for improved creep resistance or fracture toughness. Both recrystallization and beta annealing treatments are used to improve fracture toughness. Beta annealing is done at temperatures above the beta transus of the alloy being annealed.

Straightening, sizing and flattening may be combined with annealing by use of appropriate fixtures. The parts, in bulk or in fixtures, may be charged directly into a furnace operating at the annealing temperature.

Either air or furnace cooling may be used, but the two methods may result

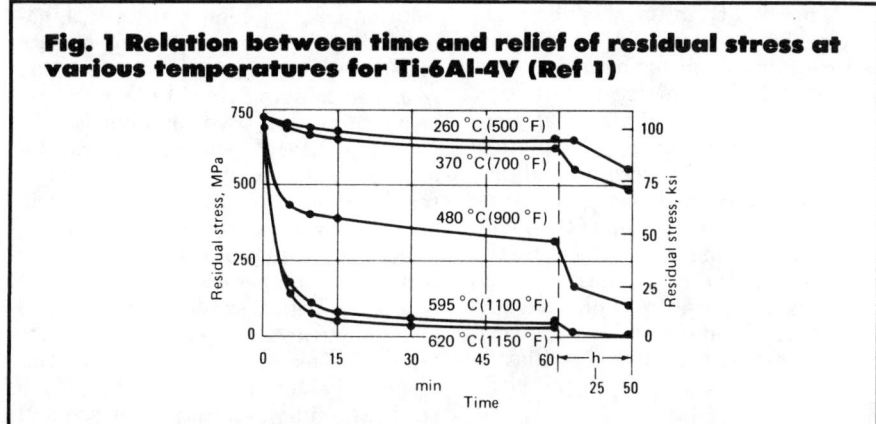

Fig. 1 Relation between time and relief of residual stress at various temperatures for Ti-6Al-4V (Ref 1)

Table 2 Recommended annealing treatments for titanium and titanium alloys

Alloy	Temperature °C	°F	Time, h	Cooling method
Commercially pure Ti (all grades)	650 to 760	1200 to 1400	1/10 to 2	Air
Alpha or near-alpha titanium alloys				
Ti-5Al-2.5Sn	720 to 845	1325 to 1550	1/6 to 4	Air
Ti-8Al-1Mo-1V	790(a)	1450(a)	1 to 8	Air or furnace
Ti-6Al-2Sn-4Zr-2Mo	900(b)	1650(b)	1/2 to 1	Air
Ti-6Al-2Cb-1Ta-0.8Mo	790 to 900	1450 to 1650	1 to 4	Air
Alpha-beta titanium alloys				
Ti-6Al-4V	705 to 790	1300 to 1450	1 to 4	Air or furnace
Ti-6Al-6V-2Sn (Cu + Fe)	705 to 815	1300 to 1500	3/4 to 4	Air or furnace
Ti-3Al-2.5V	650 to 760	1200 to 1400	1/2 to 2	Air
Ti-6Al-2Sn-4Zr-6Mo	(c)	(c)	. . .	. . .
Ti-5Al-2Sn-4Mo-2Zr-4Cr (Ti-17)	(c)	(c)	. . .	. . .
Ti-7Al-4Mo	705 to 790	1300 to 1450	1 to 8	Air
Ti-6Al-2Sn-2Zr-2Mo-2Cr-0.25Si	705 to 815	1300 to 1500	1 to 2	Air
Ti-8Mn	650 to 760	1200 to 1400	1/2 to 1	(d)
Beta or near-beta titanium alloys				
Ti-13V-11Cr-3Al	705 to 790	1300 to 1450	1/6 to 1	Air or water
Ti-11.5Mo-6Zr-4.5Sn (Beta III)	690 to 760	1275 to 1400	1/6 to 1	Air or water
Ti-3Al-8V-6Cr-4Zr-4Mo (Beta C)	790 to 815	1450 to 1500	1/4 to 1	Air or water
Ti-10V-2Fe-3Al	(c)	(c)	. . .	. . .
Ti-15V-3Al-3Cr-3Sn	790 to 815	1450 to 1500	1/12 to 1/4	Air

(a) For sheet and plate, follow by 1/4 h at 790 °C (1450 °F), then air cool. (b) For sheet, follow by 1/4 h at 790 °C (1450 °F), then air cool (plus 2 h at 595 °C or 1100 °F, then air cool, in certain applications). For plate, follow by 8 h at 595 °C (1100 °F), then air cool. (c) Not normally supplied or used in annealed condition (see Table 3). (d) Furnace or slow cool to 540 °C (1000 °F), then air cool.

in different levels of tensile properties. For example, air cooling of Ti-6Al-6V-2Sn from the mill-annealing temperature results in lower tensile strength than that obtained by furnace cooling. If distortion is a problem, the cooling rate should be uniform down to 315 °C (600 °F).

Stability. In alpha-beta titanium alloys, thermal stability is a function of beta-phase transformations. During cooling from the annealing temperature, beta may transform and, under certain conditions and in certain alloys, may form the brittle intermediate phase omega. A stabilization annealing treatment is designed to produce a stable beta phase capable of resisting further transformation when exposed to elevated temperatures in service. Alpha-beta alloys that are lean in beta, such as Ti-6Al-4V, can be air cooled from the annealing temperature without impairing their stability. Furnace (slow) cooling may promote formation of Ti₃Al, an ordering reaction that can degrade resistance to stress corrosion. Slight increases in strength (up to 34 MPa, or 5 ksi) can be gained in Ti-6Al-4V and in Ti-6Al-6V-2Sn by cooling from the annealing temperature to 540 °C (1000 °F) at a rate of 56 °C/h (100 °F/h).

To obtain maximum creep resistance and stability in the near-alpha alloy Ti-8Al-1Mo-1V and Ti-6Al-2Sn-4Zr-2Mo, a duplex annealing treatment is employed. This treatment begins with solution annealing at a temperature high in the alpha-beta range, usually 28 to 56 °C (50 to 100 °F) below the beta transus for Ti-8Al-1Mo-1V and 19 to 56 °C (35 to 50 °F) below the beta transus for Ti-6Al-2Sn-4Zr-2Mo. Forgings are held for 1 h (nominal) and then air or fan cooled depending on section size. This treatment is followed by stabilization annealing for 8 h at 595 °C (1100 °F). Final annealing temperature should be at least 56 °C (100 °F) above the maximum anticipated service temperature. Maximum creep resistance can be developed in Ti-6Al-2Sn-4Zr-2Mo by beta annealing or beta processing.

Straightening During Annealing. It may be difficult to prevent distortion of close-tolerance thin sections during annealing. Straightening of bar to close tolerances, and flattening of sheet, present major problems for titanium producers and fabricators. Because of springback and resistance to straightening at room temperature,

it is necessary to employ elevated-temperature forming. At annealing temperatures, many titanium alloys have creep resistance low enough to permit straightening during annealing. With proper fixturing, and in some instances judicious weighting, sheet-metal fabrications and thin, complex forgings have been straightened with satisfactory results. Again, uniform cooling to below 315 °C (600 °F) can improve results.

Various jigs and processing techniques have been proposed for annealing titanium in a manner that will yield a flat product. "Creep flattening" and "vacuum creep flattening" are two such techniques. Creep flattening consists of heating titanium sheet between two clean, flat sheets of steel in a furnace containing an oxidizing or inert atmosphere. Vacuum creep flattening is used to produce stress-free flat plate for subsequent machining. The plate is placed on a large, flat ceramic bed that has integral electric-heating elements. Insulation is placed on top of the plate, and a plastic sheet is sealed to the frame. The bed is slowly heated to the annealing temperature while a vacuum is pulled under the plastic. Atmospheric pressure is used to creep flatten the plate.

Solution Treating and Aging

A wide range of strength levels can be obtained in alpha-beta or beta alloys by solution treating and aging. The origin of heat treating responses of titanium alloys lies in the instability of the high-temperature beta phase at lower temperatures. Heating an alpha-beta alloy to the solution-treating temperature produces a higher ratio of beta phase. This partitioning of phases is maintained by quenching; on subsequent aging, decomposition of the unstable beta phase occurs, providing high strength. Commercial beta alloys, generally supplied in the solution-treated condition, need only be aged.

After being cleaned, titanium components should be loaded into fixtures or racks that will permit free access to the heating and quenching media. Thick and thin components of the same alloy may be solution treated together, but the time at temperature (soaking time) is determined by the thickest section. For most alloys, the rule is 20 to 30 min per inch of thickness, to get the required temperature, followed by the required soak time.

Time/temperature combinations for solution treating are given in Table 3. A load may be charged directly into a furnace operating at the solution-treating temperature. Although preheating is not essential, it may be used to minimize distortion of complex parts.

Solution Treating. To obtain high strength with adequate ductility, it is necessary to solution treat at a temperature high in the alpha-beta field, normally 28 to 83 °C (50 to 150 °F) below the beta transus of the alloy. If high fracture toughness or improved resistance to stress corrosion is required, beta annealing or beta solution treating may be desirable. A change in the solution-treating temperature of alpha-beta alloys alters the amount of beta phase and consequently changes the response to aging (see Table 4). Selection of solution-treating temperature usually is based upon practical considerations such as the desired level of tensile properties and the amount of ductility to be obtained after aging.

Because solution treating involves heating to temperatures only slightly below the beta transus, proper control of temperature is essential. If the beta transus is exceeded, tensile properties (especially ductility) are reduced and cannot be fully restored by subsequent thermal treatment. The beta transus temperatures for commercial alloys are listed in Table 5.

Beta alloys normally are obtained from producers in the solution-treated condition. If reheating is required, soak times should be only as long as necessary to obtain complete solutioning. Solution-treating temperatures for beta alloys are above the beta transus; because no second phase is present, grain growth can proceed rapidly.

Quenching. The rate of cooling from the solution-treating temperature has an important effect on strength. If the rate is too low, appreciable diffusion may occur during cooling, and decomposition of the altered beta phase during aging may not provide effective strengthening.

For alloys relatively high in beta-stabilizer content, and for products of small section size, air or fan cooling may be adequate; such slow cooling, where allowed by specified mechanical properties, is preferred because it minimizes distortion. Beta alloys generally are air quenched from the solution-treating temperature.

Water or a 5% brine or caustic soda solution is preferred for quenching

Table 3 Recommended solution treating and aging (stabilizing) treatments for titanium alloys

Alloy	Solution temperature °C	°F	Solution time, h	Cooling rate	Aging temperature °C	°F	Aging time, h
Alpha or near-alpha alloys							
Ti-8Al-1Mo-1V..................	980 to 1010(a)	1800 to 1850(a)	1	Oil or water	565 to 595	1050 to 1100	...
Ti-6Al-2Sn-4Zr-2Mo	955 to 980	1750 to 1800	1	Air	595	1100	8
Alpha-beta alloys							
Ti-6Al-4V	955 to 970(b)(c)	1750 to 1775(b)(c)	1	Water	480 to 595	900 to 1100	4 to 8
	955 to 970	1750 to 1775	1	Water	705 to 760	1300 to 1400	2 to 4
Ti-6Al-6V-2Sn (Cu + Fe)...........	885 to 910	1625 to 1675	1	Water	480 to 595	900 to 1100	4 to 8
Ti-6Al-2Sn-4Zr-6Mo	845 to 890	1550 to 1650	1	Air	580 to 605	1075 to 1125	4 to 8
Ti-5Al-2Sn-2Zr-4Mo-4Cr	845 to 870	1550 to 1600	1	Air	580 to 605	1075 to 1125	4 to 8
Ti-6Al-2Sn-2Zr-2Mo-2Cr-0.25Si	870 to 925	1600 to 1700	1	Water	480 to 595	900 to 1100	4 to 8
Beta or near-beta alloys							
Ti-13V-11Cr-3Al	775 to 800	1425 to 1475	¼ to 1	Air or water	425 to 480	800 to 900	4 to 100
Ti-11.5Mo-6Zr-4.5Sn (Beta III)......	690 to 790	1275 to 1450	⅛ to 1	Air or water	480 to 595	900 to 1100	8 to 32
Ti-3Al-8V-6Cr-4Mo-4Zr (Beta C)	815 to 925	1500 to 1700	1	Water	455 to 540	850 to 1000	8 to 24
Ti-10V-2Fe-3Al	760 to 780	1400 to 1435	1	Water	495 to 525	925 to 975	8
Ti-15V-3Al-3Cr-3Sn	790 to 815	1450 to 1500	¼	Air	510 to 595	950 to 1100	8 to 24

(a) For certain products, use solution temperature of 890 °C (1650 °F) for 1 h, then air cool or faster. (b) For thin plate or sheet, solution temperature can be used down to 890 °C (1650 °F) for 6 to 30 min, then water quench. (c) This treatment is used to develop maximum tensile properties in this alloy.

Table 4 Variation of tensile properties of Ti-6Al-4V bar stock with solution-treating temperature

Solution-treating temperature °C	°F	Room-temperature tensile properties(a) Tensile strength MPa	ksi	Yield strength(b) MPa	ksi	Elongation in 4D, %
845	1550	1025	149	980	142	18
870	1600	1060	154	985	143	17
900	1650	1095	159	995	144	16
925	1700	1110	161	1000	145	16
940	1725	1140	165	1055	153	16

(a) Properties determined on 13-mm (½-in.) bar after solution treating, quenching and aging. Aging treatment: 8 h at 480 °C (900 °F), air cool. (b) At 0.2% offset

alpha-beta alloys, because these quenchants provide cooling rates necessary to prevent decomposition of the beta phase obtained by solution treating, to provide maximum response to aging. The need for rapid quenching is further emphasized by short quench-delay-time requirements. Depending on the mass of the sections being heat treated, some alpha-beta alloys can only tolerate a maximum delay of 7 s, whereas more highly beta-stabilized alloys can tolerate quench delay times of up to 20 s. For example, the effect of quench delays on Ti-6Al-4V bar is shown in Fig. 2.

Less sensitive to delayed quenching are alloys such as Ti-6Al-2Sn-4Zr-6Mo and Ti-5Al-2Sn-2Zr-4Mo-4Cr, in which fan air cooling develops good strength through 100-mm (4-in.) sections.

Section size influences effectiveness of quenching and, in turn, response to aging. The amount and type of beta stabilizer in the alloy determine depth of hardening or strengthening. Thick sections exhibit lower tensile properties unless the alloy is highly alloyed with beta stabilizers. The practical significance of section size for some alloys is shown in Table 6. The effects of quenched section size on the tensile properties of Ti-6Al-4V alloy are illustrated in Fig. 3. (For additional data, see Tables 2 and 3 on page 526 of Volume 1 of the 8th Edition of this Handbook.)

Aging. The final step in heat treating titanium alloys to high strength consists of reheating to an aging temperature between 425 and 650 °C (800 and 1200 °F). Aging causes decomposition of the supersaturated beta phase retained on quenching. A summary of

aging times and temperatures is presented in Table 3. The time/temperature combination selected depends on required strength.

Aging at or near the annealing temperature will result in overaging. This condition, called solution treated and overaged, or STOA, is sometimes used to obtain modest increases in strength while maintaining satisfactory toughness and dimensional stability.

Although the aged condition is not necessarily one of equilibrium, proper aging produces high strength with adequate ductility and metallurgical stability. Heat treatment of alpha-beta alloys for high strength frequently involves a series of compromises and modifications, depending on the type of service and on special properties that are required, such as ductility and suitability for fabrication. This has become especially true where fracture toughness is important in design and strength is lowered to improve design life.

During aging of some highly beta-stabilized alpha-beta alloys, beta transforms first to a metastable transition phase referred to as omega phase. Retained omega phase, which produces brittleness unacceptable in alloys heat treated for service, can be avoided by severe quenching and rapid reheating to aging temperatures above 425 °C (800 °F). Because a coarse alpha phase forms, however, this treatment might not produce optimum strength proper-

Table 5 Beta transformation temperatures of titanium alloys

Alloy	Beta transus °C, ±15	°F, ±25
Commercially pure Ti, 0.25 max O_2	910	1675
Commercially pure Ti, 0.40 max O_2	945	1735
Alpha and near-alpha alloys		
Ti-5Al-2.5Sn	1050	1925
Ti-8Al-1Mo-1V	1040	1900
Ti-6Al-2Sn-4Zr-2Mo	995	1820
Ti-6Al-2Cb-1Ta-0.8Mo	1015	1860
Ti-0.3Mo-0.8Ni (Ti code 12)	880	1615
Alpha-beta alloys		
Ti-6Al-4V	1000(a)	1830(b)
Ti-6Al-6V-2Sn (Cu + Fe)	945	1735
Ti-3Al-2.5V	935	1715
Ti-6Al-2Sn-4Zr-6Mo	940	1720
Ti-5Al-2Sn-2Zr-4Mo-4Cr (Ti-17)	900	1650
Ti-7Al-4Mo	1000	1840
Ti-6Al-2Sn-2Zr-2Mo-2Cr-0.25Si	970	1780
Ti-8Mn	800(c)	1475(d)
Beta or near-beta alloys		
Ti-13V-11Cr-3Al	720	1330
Ti-11.5Mo-6Zr-4.5Sn (Beta III)	760	1400
Ti-3Al-8V-6Cr-4Zr-4Mo (Beta C)	795	1460
Ti-10V-2Fe-3Al	805	1480
Ti-15V-3Al-3Cr-3Sn	760	1400

(a) ± 20. (b) ± 30. (c) ± 35. (d) ± 50

Fig. 2 Effects of quench delay on tensile properties of Ti-6Al-4V bar (Ref 2)

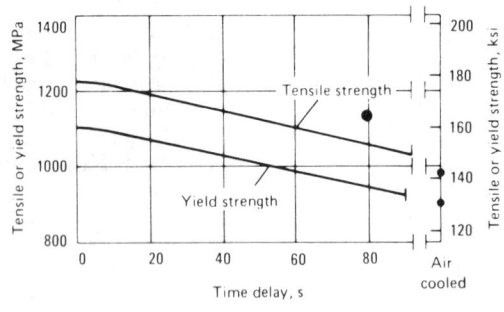
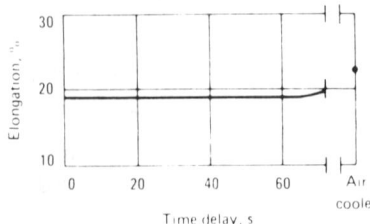

Bar, 13 mm (½ in.) in diameter, was solution treated 1 h at 955 °C (1750 °F), water quenched, aged 6 h at 480 °C (900 °F) and air cooled.

---

ties. An aging practice that ensures that aging time and temperature are adequate to carry out any omega reaction to completion usually is employed. Aging above 425 °C (800 °F) generally is adequate to complete the reaction.

The metastable beta alloys do not require solution treatment. Final hot working, followed by air cooling, leaves these alloys in a condition comparable to a solution-treated state. In some instances, however, solution treating at 790 °C (1450 °F) has produced better uniformity of properties after aging. Aging at 480 °C (900 °F) for 8 to 60 h produces tensile strengths of 1.10 to 1.38 GPa (160 to 200 ksi). Aging for times longer than 60 h may provide higher strengths, but will decrease ductility and fracture toughness if the alloy contains chromium and titanium-chromium compounds are formed. Short aging times can be used on cold worked material to produce a significant increase in strength over that obtained by cold working. Use of beta alloys at service temperatures above 315 °C (600 °F) for prolonged periods is not recommended, because the loss of ductility caused by metallurgical instability is progressive.

Other Special Thermal Treatments. Certain physical properties, such as notch strength, fracture toughness and fatigue resistance, can be enhanced in some alloys by special thermal treatments. Three such treatments are given below:

- *Solution treating and overaging of Ti-6Al-4V:* Heat 1 h at 955 °C (1750 °F), water quench, then 2 h at 705 °C (1300 °F), air cool. Advantages: improved notch strength, fracture toughness and creep strength at strength levels similar to those obtained by regular annealing.
- *Recrystallization annealing of Ti-6Al-4V or Ti-6Al-4V-ELI:* Heat 4 h or more at 925 to 955 °C (1700 to 1750 °F), furnace cool to 760 °C (1400 °F) at a rate no higher than 56 °C/h (100 °F/h), cool to 480 °C (900 °F) at a rate no lower than 370 °C/h (670 °F/h), air cool to room temperature. Advantages: improved fracture toughness and fatigue-crack-growth characteristics at somewhat reduced levels of strength.
- *Beta annealing of Ti-6Al-4V, Ti-6Al-4V-ELI and Ti-6Al-2Sn-4Zr-2Mo.* Ti-6Al-4V or Ti-6Al-4V-ELI: Heat 5 min to 1 h at 1010 to 1040 °C (1850 to 1900 °F), air cool to 650 °C (1200 °F)

Table 6 Relation of tensile strength of solution treated and aged titanium alloys to size

Alloy	13 mm (½ in.) MPa	ksi	25 mm (1 in.) MPa	ksi	50 mm (2 in.) MPa	ksi	75 mm (3 in.) MPa	ksi	100 mm (4 in.) MPa	ksi	150 mm (6 in.) MPa	ksi
Ti-6Al-4V	1105	160	1070	155	1000	145	930	135	...	...	...	...
Ti-6Al-6V-2Sn (Cu + Fe)	1205	175	1205	175	1070	155	1035	150	...	...	...	...
Ti-6Al-2Sn-4Zr-6Mo	1170	170	1170	170	1170	170	1140	165	1105	160	...	...
Ti-5Al-2Sn-2Zr-4Mo-4Cr (Ti-17)	1170	170	1170	170	1170	170	1105	160	1105	160	1105	160
Ti-10V-2Fe-3Al	1240	180	1240	180	1240	180	1240	180	1170	170	1170	170
Ti-13V-11Cr-3Al	1310	190	1310	190	1310	190	1310	190	1310	190	1310	190
Ti-11.5Mo-6Zr-4.5Sn (Beta III)	1310	190	1310	190	1310	190	1310	190	1310	190	...	...
Ti-3Al-8V-6Cr-4Zr-4Mo (Beta C)	1310	190	1310	190	1240	180	1240	180	1170	170	1170	170

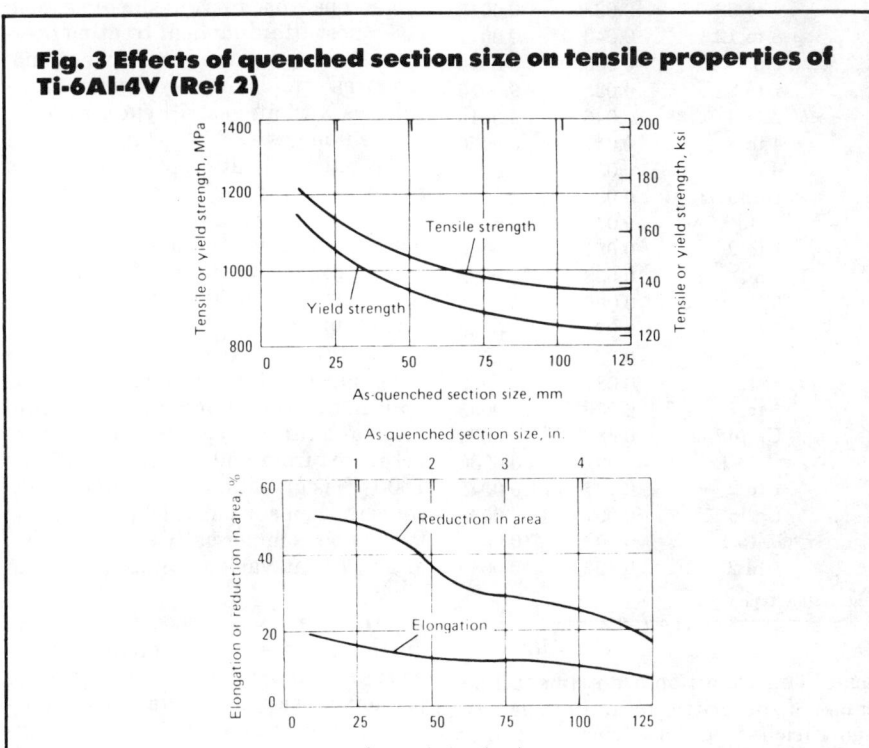

Fig. 3 Effects of quenched section size on tensile properties of Ti-6Al-4V (Ref 2)

at a rate of 85 °C/min (150 °F/min) or higher, then 2 h at 730 to 790 °C (1350 to 1450 °F), air cool. Advantages: improved fracture toughness, high-cycle fatigue strength and resistance to aqueous stress corrosion. Ti-6Al-2Sn-4Zr-2Mo: Heat ½ h at 1020 °C (1870 °F), air cool, then 8 h at 595 °C (1100 °F), air cool. Advantages: improved creep strength at elevated temperatures as well as improved fracture toughness.

Post Heat Treating Requirements. Titanium reacts with the oxygen, water and carbon dioxide normally found in oxidizing heat treating atmospheres and with hydrogen formed by decomposition of water vapor. Unless the heat treatment is performed in a vacuum furnace or in an inert atmosphere, oxygen will react with the titanium at the metal surface and produce an oxygen-enriched layer commonly called "alpha case". This brittle layer must be removed before the component is put into service. It can be removed by machining, but certain machining operations may result in excessive tool wear. Standard practice is to remove alpha case by other mechanical methods or by chemical methods, or by both.

Oxidation rates of commercial titanium alloys vary, and Table 7 can be used as a guide to determine how much metal should be removed. Temperature and total time at temperature must be known. One method to check for complete removal of alpha case is to etch the component with a solution composed of 18 g of ammonium bifluoride per litre of water (2.4 oz/gal). The presence or absence of alpha case is detected by the difference in etching characteristics: light gray shows the presence of alpha case; dark gray indicates its absence. If the component has been machined, such as a forging, the ammonium bifluoride treatment must be preceded by etching in a solution consisting nominally of 5% HF, 30% min HNO_3, balance water. For other mill products, such as plate, microexamination of representative samples removed from the plate is commonly used.

Small amounts of hydrogen (100 to 200 ppm) can be tolerated in titanium alloys with the specific limiting amount determined by the type of alloy. High hydrogen content can lead to premature failure of a component. Hydrogen pickup occurs not only during heat treatment but also during pickling or chemical cleaning operations used to remove alpha case. The amount of hydrogen pickup can only be determined by chemical analysis. If high hydrogen content is found, vacuum annealing is required. A typical vacuum annealing cycle consists of heating at or close to the annealing temperature for 2 to 4 h in a vacuum of not less than 10 μm.

Hardness testing is not recommended as a nondestructive method of checking the efficiency of heat treatment. The correlation between strength and hardness is poor. Whenever verification of a property is required, the appropriate mechanical test should be used.

Contamination During Heat Treatment

Before being subjected to any thermal treatment, titanium components should be cleaned and dried. Caution:

Table 7 Minimum metal removal after thermal exposure of titanium alloys

Heat treating temperature °C	°F	Time at temperature, h	Minimum stock removal per surface(a) mm	in.
480 to 593	900 to 1100 .... Up to 12		0.005	0.0002
594 to 648	1101 to 1200 .... Up to 4		0.008	0.0003
		4 to 12	0.015	0.0006
649 to 704	1201 to 1300 .... Up to 1		0.013	0.0005
		1 to 8	0.020	0.0008
		8 to 12	0.025	0.0010
705 to 760	1301 to 1400 .... Up to 1		0.025	0.0010
		1 to 4	0.036	0.0014
		4 to 8	0.038	0.0015
		8 to 12	0.043	0.0017
761 to 787	1401 to 1450 .... Up to 1		0.030	0.0012
		1 to 2	0.038	0.0015
		2 to 4	0.046	0.0018
		4 to 8	0.051	0.0020
		8 to 12	0.056	0.0022
788 to 815	1451 to 1500 .... Up to ½		0.036	0.0014
		½ to 1	0.041	0.0016
		1 to 2	0.051	0.0020
816 to 871	1501 to 1600 .... Up to ½		0.058	0.0023
		½ to 1	0.066	0.0026
		1 to 2	0.076	0.0030
872 to 898	1601 to 1650 .... Up to ½		0.058	0.0023
		½ to 1	0.081	0.0032
		1 to 2	0.089	0.0035
899 to 926	1651 to 1700 .... Up to ½		0.086	0.0034
		½ to 1	0.091	0.0036
		1 to 2	0.107	0.0042
927 to 954	1701 to 1750 .... Up to ½		0.097	0.0038
		½ to 1	0.107	0.0042
		1 to 2	0.122	0.0048

(a) Values shown are typical; actual values may vary with alloy type.

Do not use ordinary tap water in cleaning titanium components. Oil, fingerprints, grease, paint and other foreign matter should be removed from all surfaces. Cleaning is required because the chemical reactivity of titanium at elevated temperatures can lead to its contamination or embrittlement and can increase its susceptibility to stress corrosion. After cleaning, parts should be handled with clean gloves to prevent recontamination. If a component is to be sized, straightened or heat treated in a fixture, the fixture also should be free of any foreign matter and loosely adhering scale.

Titanium is chemically active at elevated temperatures and will oxidize in air. However, oxidation is not of primary concern in heat treating of titanium, although it may be a problem in sheet-forming operations. Oxygen pickup during heat treatment results in a surface structure composed predominantly of alpha phase and causes formation of

scale. This condition is detrimental because of the brittle nature of the oxygen-enriched alpha structure, which also is very abrasive to either carbide or high speed steel machine tools. At 955 °C (1750 °F), the alpha structure can extend 0.2 to 0.3 mm (0.008 to 0.012 in.) below the surface and must be removed.

An antioxidant spray coating may be applied to clean sheet-metal parts in order to minimize oxygen pickup. Such coatings work effectively at temperatures up to about 760 °C (1400 °F), but their use does not fully eliminate the need for removing the surface structure after heat treating.

The danger of hydrogen pickup is of greater importance than that of oxidation. Current specifications limit hydrogen content to a maximum of 125 to 200 ppm, depending on alloy and mill form. Above these limits, hydrogen embrittles some titanium alloys, thereby reducing impact strength and notch

tensile strength and causing delayed cracking.

Hydrogen Pickup. With the exceptions of high vacuum, salt baths and chemically inert gases such as argon, all heat treating atmospheres contain some hydrogen at temperatures used for annealing titanium. Hydrocarbon fuels produce hydrogen as a by-product of incomplete combustion, and electric furnaces with air atmospheres contain hydrogen from breakdown of water vapor. However, because small amounts of hydrogen can be tolerated in titanium and because inert media are expensive, most titanium heat treating operations are performed in conventional furnaces employing oxidizing atmospheres with at least 5% excess oxygen in the flue gas.

An oxidizing atmosphere serves in two ways to reduce hydrogen pickup: it reduces the partial pressure of hydrogen in the surrounding atmosphere, and it provides the titanium with a protective surface oxide that retards hydrogen pickup.

Oxidation rates of titanium alloys vary considerably. A comparison of the scaling rates of commercially pure titanium and titanium alloys in air at temperatures from 650 to 980 °C (1200 to 1800 °F) is given in Fig. 4. Table 8 indicates the measurable thickness of oxide formed on commercially pure titanium after ½ h at various temperatures in air.

Nitrogen is absorbed by titanium during heat treatment at a much slower rate than oxygen and thus does not present a serious contamination problem. Dry nitrogen has been used successfully as a lower-cost protective atmosphere for heat treating of titanium forgings that are to be fully machined after treatment. If absorbed in sufficient quantities, however, nitrogen forms a hard, brittle compound.

Carbon monoxide and carbon dioxide decompose in the presence of hot titanium and produce surface oxidation.

Chlorides. Titanium alloys are subject to stress corrosion when parts with high residual stress are exposed to chlorides at temperatures above 290 °C (550 °F). Salt from fingerprints, and the chlorides contained in some degreasing solutions, may cause stress-corrosion cracking at temperatures above 315 °C (600 °F). Although this phenomenon is readily produced in laboratory testing, and is known to occur during heat treatment, hot-salt cracking has not

Fig. 4 Scaling rates of titanium, titanium alloys and stainless steel in air at various temperatures

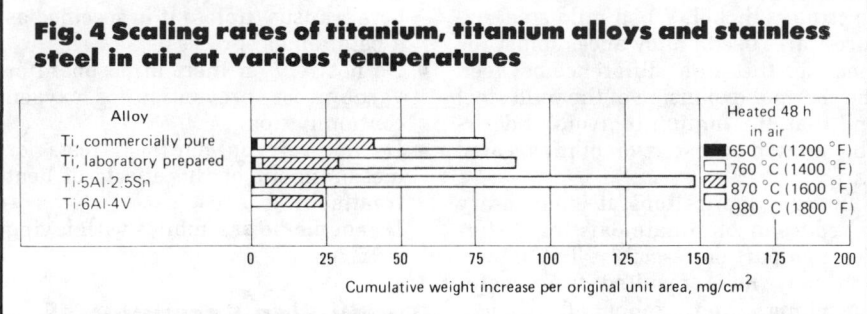

Cumulative weight increase per original unit area, mg/cm²

Table 8 Thickness of oxide on commercially pure titanium heated for ¹/₂ h in air

Temperature		Measurable thickness	
°C	°F	mm	in.
315	600	None	
425	800	None	
540	1000	None	
650	1200	<0.005	<0.0002
705	1300	0.005	0.0002
760	1400	0.008	0.0003
815	1500	<0.025	<0.001
870	1600	<0.025	<0.001
925	1700	<0.05	<0.002
980	1800	0.05	0.002
1040	1900	0.10	0.004
1095	2000	0.36	0.014

been a significant problem in service. Care is required during thermal processing to ensure freedom from chloride contamination.

Growth During Heat Treatment

Solution treating of large parts requires allowances for growth during heat treatment. The growth due to heating may be retained after cooling, and this growth may be increased either by longer holding times at solution temperature or by lower heating rates. Table 9 gives examples of net growth of Ti-6Al-4V specimens heated to 955 °C (1750 °F).

Furnace Equipment and Accessories

Atmospheres. An oxidizing atmosphere should be maintained during any thermal treatment of titanium. Furnaces normally operated with exothermic atmospheres, endothermic cracked-ammonia atmospheres or hydrogen atmospheres, because of the danger of hydrogen pickup, should be thoroughly "burned out" before being used for processing of titanium. If dimensions, shape or size do not permit removal of scale by subsequent pickling or machining, antioxidant coatings suitable for use to 760 °C (1400 °F) can be employed to minimize contamination. A vacuum or an inert gas such as argon also can be used.

Furnaces. Titanium usually is annealed or stress relieved in conventional furnaces constructed for annealing of steel. These furnaces are electric, gas fired or oil fired, in order of decreasing popularity. The temperature-control equipment for these operations should have an accuracy of ±5.5 °C (±10 °F) and should be capable of controlling and recording the desired temperature within ±14 °C (±25 °F), except where

control within ±8 °C (±15 °F) is required by MIL-H-81200.

Vacuum annealing furnaces are of either the cold-wall or the hot-wall type and may be heated by gas or electricity. Cold-wall electric vacuum furnaces are used most commonly with titanium. Maximum furnace operating temperature depends upon the heating elements and radiation shields, but usually these furnaces are designed for a maximum temperature of 980 °C (1800 °F) and are adequate for all titanium alloys. Hot-wall electric furnaces and gas-fired vacuum furnaces have been used in production. When the furnace employs a metallic retort, operating temperatures are held below 980 °C (1800 °F); higher temperatures can be achieved with ceramic retort tubes.

Laboratory vacuum annealing furnaces usually are operated at pressures of 0.1 μm or less, whereas production furnaces are designed to operate at pressures of 0.5 to 3.0 μm.

Vacuum annealing is expensive, and generally it is used only when: (a) a reduction in hydrogen content is required, (b) further hydrogen contamination is prohibited or (c) allowances that can be made for stock removal are insufficient to permit surface contamination resulting from annealing in air. Hydrogen outgassing at 705 °C (1300 °F) and below is so slow that its cost may be prohibitive. A temperature of 730 °C (1350 °F) is recommended as a minimum, and temperatures from 760 to 790 °C (1400 to 1450 °F) are preferred. At a temperature of 760 °C, removal of 100 ppm of hydrogen from 13- to 25-mm (¹/₂- to 1-in.) sections of Ti-6Al-4V alloy required approximately 2 h at a pressure of <10 μm. Actual time at temperature may vary widely depending on the capacity of the furnace to maintain a vacuum.

Solution-treating equipment can vary from a simple furnace with accurate temperature control and a water-

Table 9 Effect of heating rate and time at 955 °C (1750 °F) on growth of Ti-6Al-4V

Test conditions: 50-mm (2-in.) specimens were taken in the longitudinal direction (except where otherwise indicated) from material annealed 2 h at 705 °C (1300 °F) and air cooled. No growth was observed in specimens tested during annealing.

Mill heat(a)	Heating rate °C/min	°F/min	Holding time(b) h	Net growth(c), %
A	3.3	6	0	0.27
B	3.3	6	0	0.22
A	3.3	6	1	0.60
B	3.3	6	1	0.49
A	3.3	6	2	1.00
B	3.3	6	2	0.90(d)
B	10	18	1	0.32
B(e)	10	18	1	0.35

(a) Beta transus temperatures (determined metallographically) were 990 °C (1810 °F) for heat A and 1015 °C (1860 °F) for heat B. (b) All specimens water quenched after holding for time indicated. (c) As determined by Leitz-Wetzler dilatometer. (d) Calculated from curve. (e) Specimen taken in transverse direction.

quench tank to specialized installations for treating complex parts. Electrically heated furnaces are preferred because they minimize hydrogen pickup, although fuel-fired furnaces with slightly oxidizing conditions or with muffles that protect the metal from combustion products have been used successfully. Resistance and induction heating also have been used to reduce heating times and to minimize contamination during solution treatment. Accuracy of temperature-control equipment should be within ±2.8 °C (±5 °F), and the desired temperature should be controlled within ±14 °C (±25 °F).

To reduce distortion in long, thin

products such as sheet or extrusions, in hollow cylinders and in long forgings during immersion quenching, parts often are suspended vertically in an electrically heated drop-bottom furnace. In addition, weights usually are attached to the bottom ends of sheet to improve flatness during heating and to facilitate lowering of the sheet into the quench tank.

Quenching Media. Because rapid cooling is required after solution treating of most titanium alloys, either water or a 5% brine or caustic soda solution is most widely used as the quenching medium. Low-viscosity oil with a high flash point has been used effectively in vertical immersion quenching of sheet to reduce distortion. Quenching oils used with steel provide rapid cooling to 370 to 425 °C (700 to 800 °F), and these oils are satisfactory. Their use, however, should be limited to thin sections to avoid degradation of strength compared to that obtained by water quenching from the same solution temperature. Various concentrations of glycol in water will produce quench rates between those of water and those of oil.

Aging Furnaces. Because they do not involve combustion by-products, furnaces of the electrical-resistance type are preferable for aging titanium and its alloys. Retorts, however, may be used with oil-fired or gas-fired furnaces to avoid contamination. Aging furnaces normally are equipped with internal fans to promote circulation of air or other atmosphere throughout the work zone. Temperature-control equipment should be accurate to ±1.1 °C (±2 °F) and should be capable of controlling temperature within ±8 °C (±15 °F).

At normal aging temperatures of 480 to 595 °C (900 to 1100 °F), a protective atmosphere is not required. Aging in air produces a superficial scale that can be removed easily by mechanical or chemical means (this scale also may be left in place, because it does not affect properties).

Fixtures. In fixturing titanium components or assemblies to prevent distortion, the thermal-expansion characteristics of both the titanium alloy and the fixture itself must be considered. Ideally, both the alloy and the fixture will have equivalent thermal expansion characteristics within the intended aging-temperature range. Mild steel is commonly used because it is low in cost and can be made reasonably resistant to oxidation at aging temper-atures through use of coatings such as electroless nickel. When mild steel fixtures are used, allowances must be made for the slight difference between the thermal expansion of the mild steel and that of titanium to avoid undesirable growth or distortion of the treated part.

In some applications, it is necessary to reduce or eliminate existing distortion in a part or assembly. This distortion may have resulted from water quenching, from relief of residual stresses during machining, from stresses induced by welding, or from uncontrollable springback after forming. Proper fixturing during aging can be used to minimize such distortion. Fixtures also must guard against sagging; for example, Ti-6Al-4V has a tendency to sag at 955 °C (1750 °F) during solution heat treating. Because titanium alloys exhibit creep behavior within the normal range of aging temperatures, it is possible to fixture and "creep form" components or assemblies to desired shape. Parts also may be sized by fixturing during aging.

Summary of Practice. Key considerations in heat treating of titanium and its alloys—practices that are to be followed and those that should be avoided—are summarized below.

- Provide sufficient stock for post-treatment metal-removal requirements (contaminated metal removal).
- Clean components, fixtures and furnaces prior to heat treatment. (Caution: Do not use ordinary tap water in cleaning of titanium components.)
- Use temperature controls with an upper cutoff to prevent temperature from exceeding beta transus.
- Charge cold components into furnaces operating at the required temperature.
- Stack and support components to allow free access of heating and quenching media.
- Observe quench-delay requirements to ensure hardening response during aging.
- Review property requirements and select optimum heat treating procedure.
- Review strength requirements and select proper aging cycle.
- Remove alpha case after all heat treating is complete.
- Check for the presence of hydrogen after all processing is complete.
- Do not nest components.
- Do not allow temperature to exceed beta transus (unless it is specified as a beta anneal process).
- Do not rely on inert atmosphere or vacuum for prevention of oxygen contamination.
- Do not rely on hardness tests for measurement of the effects of heat treatment.
- Do not pickle assemblies with faying surfaces.

Production Examples of Heat Treating Processes

The examples that follow describe applications of heat treating processes to specific titanium parts and assemblies, and indicate typical relationships among heat treating and other production operations.

Example 1 (Alpha Alloy Weldment). Because alpha alloys are not hardenable, heat treatment of welded compressor cases made of alpha alloy Ti-5Al-2.5Sn was limited to annealing and stress relieving. After the subassemblies had been formed, machined and stress relieved, the cases were assembled, using manual and mechanized inert-gas-shielded tungsten-arc welding and resistance welding. The completed assemblies were stress relieved for 1 h at 620 °C (1150 °F) in an electric muffle furnace with air atmosphere. No fixturing or protective coating was used. After being stress relieved, the cases were descaled by grit blasting and light pickling in a nitric-hydrofluoric acid solution.

Alpha-Beta Alloy Weldments. Development of maximum properties in alpha-beta alloy weldments requires solution treating followed by rapid quenching and then aging. However, if reduced strength is acceptable or higher ductility is required, workpieces may be annealed only, after welding.

Example 2 (Wing Rib). After being inert-gas-shielded tungsten-arc welded with Ti-6Al-4V filler rod, a wing rib made of annealed Ti-6Al-4V sheet was solution treated for 1 h at 900 °C (1650 °F) in an air muffle furnace and water quenched. The part was protected by a glass coating during heating. (The coating, which consisted of hydrous borax, boric acid and aluminum hydrate in a volatile carrier, was sprayed on the sheet at a dry weight of 269 to 323 g/m², or 0.88 to 1.06 oz/ft².) After being quenched, the part was inserted in an aging fixture and forced into the desired shape. It was aged in the fixture

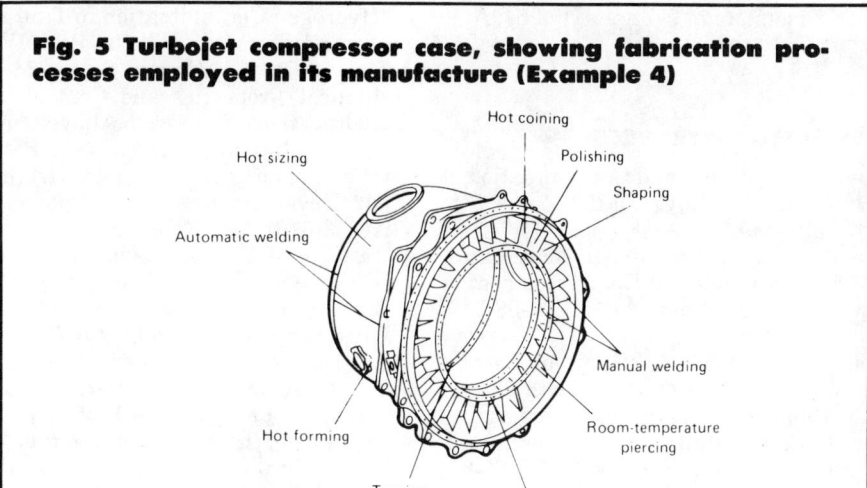

Fig. 5 Turbojet compressor case, showing fabrication processes employed in its manufacture (Example 4)

for 4 h at 540 °C (1000 °F) in an air muffle furnace. Sufficient relaxation and relief of stresses occurred during aging to bring the part into conformity with dimensional requirements. After aging, the welds had an ultimate tensile strength of 1090 MPa (158 ksi), a yield strength of 985 MPa (143 ksi), and 6% elongation in 12.7 mm (½ in.).

Example 3 (Pressure Vessel). A spherical pressure vessel 610 mm (24 in.) in diameter and 6.4 mm (¼ in.) in wall thickness was fabricated from 19-mm- (¾-in.-) thick hemispherical forgings of Ti-6Al-4V. The hemispheres were rough machined and tack welded to a special frame mounted on the lid of a cylindrical electric furnace. The rough machined hemispheres were then solution treated for 1 h in argon at 900 °C (1650 °F), water quenched and finish machined. Furnace loading and unloading were accomplished by use of an overhead crane, and the parts were quenched within 6 s after being removed from the furnace. The two hemispheres were joined by mechanical inert-gas-shielded tungsten-arc welding, using Ti-6Al-4V titanium filler metal and a reinforced weld area. Following welding, the vessel was aged for 6 h in a circulating-air furnace at 540 °C (1000 °F). No fixturing or protective coating was used. Cleaning after aging consisted of pickling in nitric-hydrofluoric acid.

Example 4 (Turbojet Compressor Case). Figure 5 shows a 760-mm-(30-in.-) diam jet-engine compressor case that was fabricated from Ti-5Al-2.5Sn. The shell of the case was roll formed into two half-round segments after the ports had been punched. After rolling, the ports were deep drawn in resistance-heated dies at 595 °C (1100 °F). After being loaded in stainless steel fixtures, the half-round segments were stress relieved in an air muffle furnace for 1 h at 620 °C (1150 °F). The ports were then sized to final dimensions while the curvature of the segments was maintained with a fixture. After the ports had been drawn and the segments stress relieved, all components were cleaned by grit blasting and pickling. The half-round segments then were welded by the inert-gas-shielded tungsten-arc method, and the resulting weldments were stress relieved for 1 h at 620 °C without fixturing.

The shrouds were brake formed to a radius of 2.5 *t* at room temperature. Between brake-forming operations, the shrouds were resistance heated in air at 595 °C (1100 °F) for 5 min. Following brake forming, the shrouds were wrap formed to their final diameters at 120 °C (250 °F) in heated dies to reduce springback. They were then stress relieved 1 h at 620 °C (1150 °F) in air, and air cooled. No fixturing was necessary for maintaining tolerances.

The vanes were fabricated from machined bar stock and sheet. The annealed material was welded to form hollow vanes and then hot coined at 650 °C (1200 °F) in dies to flatten the parts and relieve welding stresses. Following coining, the vanes were cleaned and machined to an airfoil contour.

The final assembly was made by joining the various stress-relieved components by fusion welding. After assembly, the entire compressor case was stress-relieved (without fixturing) in an air furnace for 1 h at 620 °C (1150 °F) and air cooled. Then the assembly was cleaned by grit blasting and pickling.

Wing section heat treating practices and applications are discussed in Examples 5 and 6.

Example 5. Skins of Ti-6Al-4V were stretch formed at room temperature in the fully annealed condition and solution treated in air for 30 min at 900 °C (1650 °F). After being water quenched, the distorted parts were placed in a steel aging frame and clamped to make them conform to the desired configuration. The skins and fixtures were aged 4 h at 540 °C (1000 °F) in a muffle furnace. Sufficient relaxation occurred during aging so that the skins were within drawing tolerances.

Example 6 (T-section extrusions). Ti-6Al-4V wing sections (1.22 m or 48 in. long) for spars were hung vertically on a rack for solution treatment in an air muffle furnace at 900 °C (1650 °F). The rack and the parts were then rapidly quenched in water. Straightening of the distorted (primarily, bowed) extrusions was accomplished by use of fixtures during aging. The parts were subjected to a 50% overbend by insert spacers and aged 4 h at 540 °C (1000 °F) before being machined. This aging cycle was repeated after each of the three machining operations required. Heat treating was performed without protective atmosphere. After the solution treatment and the first aging treatment, the parts were pickled in nitric-hydrofluoric acid. Pickling removed the contaminated surface layer and increased tool life in machining.

Example 7 (Rocket-Motor Cases). In fabrication of rocket-motor cases made of Ti-6Al-4V, the individual components were machined to a maximum wall thickness of 12.7 mm (0.500 in.), or to 7.62 mm (0.300 in.) wherever possible, allowing at least 1.3 mm (0.050 in.) of stock for cleanup after solution treating. Only cylindrical components were placed in fixtures for the solution treating operation. Solution treating was performed by holding the parts for 2 h at a temperature 19 to 33 °C (35 to 60 °F) below the beta transus (aim: 28 °C, or 50 °F, below beta transus) in a bottom-loading gantry furnace and then quenching them rapidly in a violently agitated 3% solution of sodium hydroxide.

Tensile-test coupons representative of all parts in the load were heated and quenched with the work. Before the parts themselves were aged, the test coupons were aged for 8 h at each of four temperatures—480, 510, 540 and 565 °C (900, 950, 1000 and 1050 °F)—to determine the optimum aging temperature in terms of desired mechanical properties. Then the parts were fixtured (a predetermined load was applied to promote creep forming), aged 8 h at optimum temperature, and air cooled. Use of a protective atmosphere was not required.

After being aged, the components were machined to final dimensions prior to welding, removing surface material that had been oxidized and contaminated during solution treating. Components were then welded into the final assemblies without preheating or postheating. The completed assemblies were stress relieved for 2 h at 480 °C (900 °F) and air cooled.

REFERENCES

1. "Titanium Alloy Handbook", by R. A. Wood and R. J. Favor: Report MCIC-HB-02, Battelle Memorial Institute, Columbus, OH, Dec 1972
2. "Properties and Processing Ti-6Al-4V": Timet, Apr 1980
3. How to Descale Titanium, by A. E. Durkin: *Materials and Methods,* Vol 38, Oct 1953, p 107-109
4. Properties and Structure of Titanium After 30-Min. Heating at 1200 to 2000 °F, by E. Walden and L. A. Dixon: *Metal Progress,* Vol 64, Aug 1953, p 88-89

SELECTED REFERENCES

- The Oxidation and Contamination of Ti and Ti Alloys: DMIC Memo 238, July 1968
- Production Techniques for Extruding, Drawing and Heat Treatment of Titanium Alloys: AFML-TR-68-349, Dec 1968
- "Influence of Metallurgical Factors on the Fatigue Crack Growth Rate in Alpha-Beta Titanium Alloys", by J. C. Chesnutt, A. Thompson and J. C. Williams: AFML-TR-78-68, Rockwell Science Center
- "Improvement of Reliability and the Mechanical Properties of Titanium Alloy Forgings", by T. Gurganus and G. Hall: AFML-TR-75-211, Alcoa Technical Center
- "Improved Manufacturing Methods for Producing High Integrity More Reliable Titanium Forgings", by R. Sparks and J. Long: AFML-TR-73-301, Wyman Gordon, Worcester, MA, Feb 1974
- *Titanium Science and Technology,* by R. I. Jaffe and H. M. Burte: Plenum Press, 1973
- "Residual Stresses, Stress Relief and Annealing of Titanium and Titanium Alloys", by D. J. Maykuth: DMIC Report S-23, July 1968
- Heat Treating Titanium and Its Alloys, by J. A. Burger and D. K. Hanink: *Metal Progress,* Vol 91, No. 6, June 1967, p 70-75
- "Hydrogen Contamination in Titanium and Titanium Alloys; Part IV: The Effect of Hydrogen on the Mechanical Properties and Control of Hydrogen in Titanium Alloys", by D. N. Williams *et al:* report from Battelle Memorial Institute to Wright Air Development Center, Contract AF 33(616)-2813, Mar 1957
- "Scaling of Titanium and Titanium Alloys", by H. W. Maynor, Jr., B. R. Barrett and R. E. Swift: report from University of Kentucky to Wright Air Development Center, issued as WADC Report 54-190, Part I, Mar 1955, and Part II, June 1955
- "A Study of the Air Contamination of Three Titanium Alloys", by J. E. Reynolds, H. R. Ogden and R. I. Jaffee: Titanium Metallurgical Laboratory Report 10, Battelle Memorial Institute, Columbus, OH, 1955
- Kinetics of the Reaction of Titanium with Oxygen, Nitrogen and Hydrogen, by E. A. Gulbransen and K. F. Andrew: *Journal of Metals,* Vol 1, 1949, p 741-748
- "Hydrogen in Titanium and Titanium Alloys", by D. N. Williams: Titanium Metallurgical Laboratory Report 100, Battelle Memorial Institute, Columbus, OH, May 1958
- An Experimental and Thermodynamic Investigation of the Hydrogen-Titanium System, by A. D. McQuillan: *Proceedings of the Royal Society* (London), Vol A204, Dec 1950, p 309-323
- Vacuum Degassing, by C. B. Griffith and M. W. Mallett: *Vacuum Metallurgy,* Vol 147, 1954

Heat Treating of Lead and Lead Alloys

By William B. Hampshire
Technical Service/Metallurgy
Lead Industries Association, Inc.

LEAD normally is considered to be unresponsive to heat treatment. Yet, some means of strengthening lead and lead alloys may be required for some applications. Lead alloys for battery components, for example, can benefit from improved creep resistance in order to retain dimensional tolerances for the full service life. Battery grids also require improved hardness to withstand industrial handling.

The absolute melting point of lead is 327.4 °C (621.3 °F). Therefore, in applications in which lead is used, recovery and recrystallization processes and creep properties have great significance. Attempts to strengthen the metal by reducing the grain size or by cold working (strain hardening) have proved unsuccessful. Lead-tin alloys, for example, may recrystallize immediately and completely at room temperature. Lead-silver alloys respond in the same manner within two weeks.

Transformations that are induced in steel by heat treatment do not occur in lead alloys, and strengthening by ordering phenomena, such as in the formation of lattice superstructures, has no practical significance.

Despite these obstacles, however, attempts to strengthen lead have met with some success.

Solid-Solution Hardening

In solid-solution hardening of lead alloys, the rate of increase in hardness generally improves as the difference between the atomic radius of the solute and the atomic radius of lead increases.

Specifically, in one study of possible binary lead alloys, it was found that the following elements, in the order listed, provided successively greater amounts of solid-solution hardening: thallium, bismuth, tin, cadmium, antimony, lithium, arsenic, calcium, zinc, copper and barium. Unfortunately, these elements have successively decreasing solid-solution solubilities, and thus the most potent solutes have the most limited solid-solution hardening effects. Within the midrange of this series, however, are elements that, when alloyed with lead, produce useful strengthening.

A useful level of strengthening normally requires solute additions in excess of the room-temperature solubility limit. In most lead alloys, homogenization and rapid cooling result in a breakdown of the supersaturated solution during storage. Although this breakdown produces coarse structures in certain alloys (lead-tin alloys, for example), it produces fine structures in others (lead-antimony alloys, for example). In alloys of the lead-tin system, the initial hardening produced by alloying is quickly followed by softening as the coarse structure is formed.

In lead-antimony alloys, at suitable solute concentrations, the structure may remain single phase with hardening by Guinier-Preston (GP) zones formed during aging. At higher concentrations, and in certain other systems, aging may produce precipitation hardening as discrete second-phase particles are formed.

Alloys that exhibit precipitation hardening typically are less susceptible to overaging and thus are more stable with time than alloys hardened by GP zones. Lead-calcium and lead-strontium alloys have been observed to age harden through discontinuous precipitation of a second phase—Pb_3Ca in lead-calcium alloys and Pb_3Sr in lead-strontium alloys—as grain boundaries move through the structure.

Solution Treating and Aging

Useful strengthening of lead can be attained by adding sufficient quantities of antimony to produce hypoeutectic lead-antimony alloys. Small amounts of arsenic have particularly strong effects on the age-hardening response of such alloys, and these effects are enhanced by solution treating and rapid quenching prior to aging.

An investigation (Ref 1) was conducted on the effects of additions of 0.15% arsenic on the age-hardening behavior of five hypoeutectic lead-antimony alloys. Accurately weighed quantities of commercially pure lead, antimony and arsenic were melted under a nitrogen atmosphere to produce the alloys listed below:

Alloy	Antimony, %	Arsenic, %
200	2.0	. . .
215	2.0	0.15
400	4.0	. . .
415	4.0	0.15
600	6.0	. . .
615	6.0	0.15
800	8.0	. . .
815	8.0	0.15
1000	10.0	. . .
1015	10.0	0.15

To minimize segregation during solidification, the melts, each weighing about 100 g (3.5 oz), were chilled rapidly by casting into a shallow horizontal steel boat. Pieces weighing about 15 g (0.5 oz) were cut from the castings and were used as hardness-test specimens after being subjected to one of two heat treatments. One group of specimens, designated group A, was air cooled from the liquid state and quenched in ice water as soon as solidification was complete. Another group, designated group B, was solution treated for 4 h at 250 °C (480 °F) and then quenched in ice water.

In view of the rapid age hardening of the arsenical alloys, it was desired that the hardness tests be commenced as soon after quenching as possible, and one minute was chosen as a reasonable period of time. For group B specimens, this was easily achieved by solution treating the specimens, which had previously been ground with 600 grade paper, in a stream of nitrogen in a vertical tube furnace. The specimens then were dropped into ice water and transferred after 20 s to water at room temperature. Hardness of these specimens was measured with a Vickers hardness tester using a 2.5-kg load.

Because the specimens in group A could not be prepared for hardness testing before they were quenched, a different procedure was necessary to

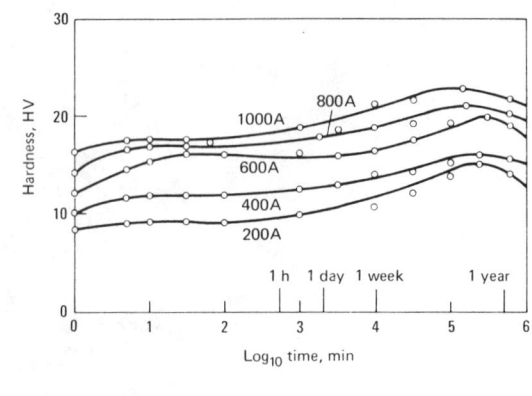

Fig. 1   Age hardening of Pb-Sb alloys, solidified and water quenched

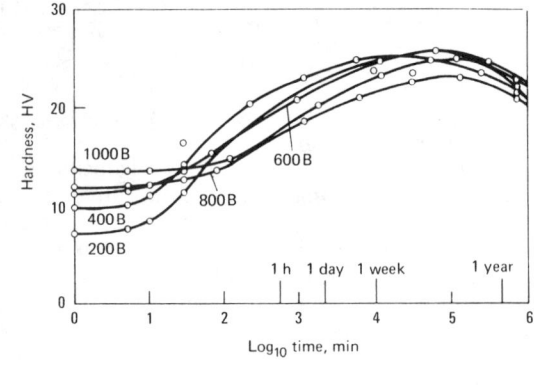

Fig. 2   Age hardening of Pb-Sb alloys, solution treated 4 h at 250 °C (480 °F) and water quenched

permit testing to commence after one minute. These specimens were remelted in small flat-base cylindrical cups punched from thin aluminum foil. After quenching, the foil was peeled from the specimen and the hardness measurement was made on the flat surface (after the top surface had been quickly smoothed on a file to ensure proper seating in the test machine).

Cooling curves were obtained for the specimens in group A by means of Chromel-Alumel thermocouples, which were thinly sheathed with mild steel. After remelting, the thermocouple was removed before the beginning of solidification. Quenching then was timed to follow complete solidification on the basis of the previous cooling curves. The group A specimens were cooled at a rate of 1 °C/s (1.8 °F/s) to solidification

at 232 °C (450 °F). These specimens showed evidence of surface segregation.

Hardness tests were continued for up to two years, with the specimens being stored at ambient temperature, which for the majority of the time was between 20 and 22 °C (68 and 72 °F). To investigate variations in hardness, tests were made on some of the specimens using a Reichert microhardness tester (5.3-g load, applied for 10 s).

Test Results. Results of hardness testing showed that lead-antimony alloys of commercial purity demonstrate significant age hardening, particularly after solution treating, as shown in Fig. 1 and 2. In the quench-cast alloys of group A, the hardness at any time increased with antimony content. For the solution-treated specimens of group

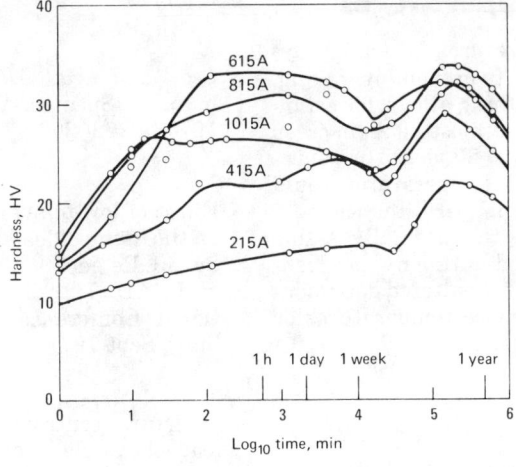

Fig. 3 Age hardening of Pb-Sb as alloys, solidified and water quenched

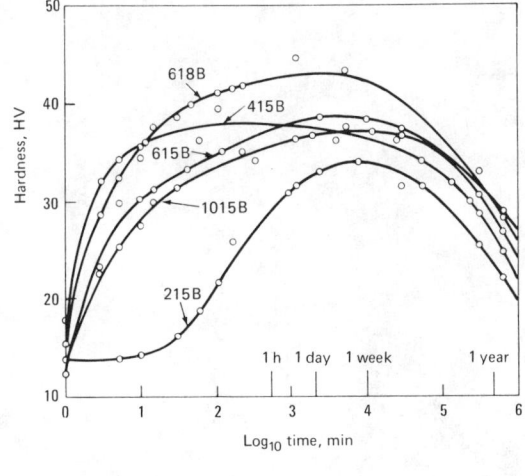

Fig. 4 Age hardening of Pb-Sb as alloys, solution treated 4 h at 250 °C (480 °F) and water quenched

B, aging was more effective for alloys with lower antimony contents.

Figures 3 and 4 show that both the rate and the extent of hardening are increased by addition of 0.15% arsenic. As shown in Fig. 4, increases in hardness of the solution-treated hypoeutectic alloys are pronounced in the first 10 min.

Hardness Stability. For any given alloy, both heat treatments result in hardnesses after 1 min and after 2 years that are somewhat comparable, as indicated in Fig. 1 through 4. For most of the two-year period, the solution-treated specimens were hard-er than the quench-cast specimens. Other investigations have also shown that alloys cooled slowly after casting are always softer than quenched alloys. As shown in Fig. 3, the alloys with 2 and 4% antimony harden comparatively slowly, and the alloy containing 6% antimony appears to undergo optimum hardening.

Application. To reduce the antimony contents of the positive plates in lead-acid storage batteries, because of antimony's detrimental effect on charge retention, there has been a trend toward replacing eutectic alloys with a Pb-6Sb-0.15As alloy. Battery grids made of this arsenical alloy will age harden slowly after casting and air cooling. However, storing grids for several days constitutes unproductive use of floor space and results in undesirable interruptions in manufacturing sequences.

Although large-scale solution treatment of battery grids might be difficult to justify economically or to achieve without some distortion, quenching of grids cast from arsenical lead-antimony alloys offers an attractive alternative method of effecting improvements in strength. The suitability of quench-cast grids can be assessed by comparing the values given in Fig. 3 with the hardness level that battery grids require in order to withstand industrial handling (about 18 HV, the hardness of the eutectic alloy). The alloy containing 2% antimony clearly does not respond sufficiently well to be considered as a possible alternative. The 4% antimony alloy, however, attains a hardness of 18 HV after 30 min, and the alloys that contain 6, 8, and 10% antimony could be handled almost immediately. Furthermore, the values given in Fig. 3 for hardness after two years are superior in all instances to those for air-cooled alloys of similar composition. Hardness curves decline steadily after two years, and full evaluation of these alloys for use in the battery industry would require battery performance tests.

Dispersion Hardening

Another mechanism for strengthening of lead alloys involves elements that have low solubilities in solid lead, such as copper and nickel. Alloys that contain these elements can be processed so that no homogenization results; most of the strengthening that occurs is developed through dispersion hardening, with some solid-solution hardening taking place as a secondary effect. The resulting structure is more stable than those developed by other hardening processes. Dispersion strengthening also has been achieved through powder metallurgy methods in which lead oxide, alumina or similar materials are dispersed in pure lead.

Fabrication

Although alloy selection is important, care must be taken in fabrication as well. Castings should be cooled rapidly to a temperature below that at which the structure breaks down, or a coarse structure will be obtained. Age-hardening alloys should be extruded at a temperature above the breakdown temperature, and extrusions should not be allowed to cool slowly. Rolled alloys often are processed at insufficient temperatures; when this occurs, homogenization after rolling is required if age hardening is to produce a beneficial response.

Cold Storage

Cold storage has been shown to improve the response of lead-antimony alloys to age hardening. Cooling a homogenized Pb-2Sb alloy to -10 °C ($+15$ °F) and holding for one or two days prior to room-temperature aging results in increases in both the rate of age hardening and the maximum hardness attained. This behavior has been explained as the result of a reduction in the mobility of quenched-in free vacancies and a consequent reduction in their annihilation. The process allows the vacancies to form complexes with solute atoms, and these complexes improve the efficiency of nucleation during aging.

Service Temperatures

Service temperatures for lead alloys must be kept low to prevent overaging. Some cable-sheathing alloys, for example, have retained most of their creep resistance for up to 20 years, but exposure to elevated temperatures could have reduced this performance substantially. Even the normally stable age-hardened lead-antimony and lead-calcium alloys can be altered detrimentally by high service temperatures or excessive working.

REFERENCE

1. The Effect of Arsenic on the Age-Hardening of Lead-Antimony Alloys, by J. D. Williams: *Metallurgia,* Vol 74, No. 443, Sept 1966, p 105–108

SELECTED REFERENCES

For additional information on heat treatment of lead and lead alloys, refer to the following sources and their bibliographies:

- *Lead and Lead Alloys,* by W. Hofmann: Springer-Verlag, Berlin, 1970, p 262–267

- Effect of low-temperature treatment on the aging of lead antimony alloys, by J. J. Regidor *et al:* paper presented at LEAD '71, The Fourth International Conference on Lead, Hamburg, Sept 1971

- Structural control of non-antimonial lead alloys via alloy additions, heat treatment and cold work, by R. D. Prengaman: paper presented at Pb80, The Seventh International Lead Conference, Madrid, May 1980

Vacuum Methods

Five Decades of Vacuum Furnace Progress

By WILLIAM C. DIMAN

Vacuum furnaces have evolved from simple pot-type chambers used mainly for outgassing electronic components to systems capable of most heat treating operations. Part I of a two-part article.

WITHIN 50 years, heat treating furnaces and processes have developed from the introduction of controlled atmosphere environments through the creation of vacuum technology to the vacuum carburizing process, which unites atmosphere and vacuum into a single furnace system.

Progress began back in 1925 when Carl I. Hayes invented the Certain-Curtain process for atmosphere furnaces, which entails use of a gas/air mixture to form a protective atmosphere in an electric furnace. Vacuum techniques came along shortly thereafter (in 1927), but their acceptance for heat treatment did not begin to emerge until the mid 1960's due to a variety of causes:

First, because the industrial furnace manufacturers' market centered on the use of exothermic and endothermic atmospheres, most engineering and research time was devoted to improvements in these areas.

Second, the degree of vacuum was stated in "inches of mercury," "millimeters of mercury," "microns," or "torr," tending to create an aura of mystery.

Third, and probably the most important, vacuum equipment was costly. For management to open its purse strings to purchase any equipment, it demanded a genuine need for the equipment and a reasonable period of amortization. The furnace, in short, had to pay for and support itself.

The Early Years — The few vacuum furnaces available in the twenties were simple, top-loading, water-jacketed potlike vessels having low floor-to-floor production rates. First used by radio valve manufacturers, these prototype systems were later adopted by manufacturers of aircraft parts because the equipment was suited to their needs. The materials they processed required annealing, bonding, and degasification rather than hardening, which necessitates quench rate capability.

Mr. Diman is director of sales, C. I. Hayes Inc., Cranston, R. I.

Heat treaters, however, were concerned with quenching rates. Some time passed before their needs were eventually met by vacuum furnaces equipped with integral oil quenching tanks (Fig. 1). This system demonstrated a wide range of heat treating capability at a low cost per unit weight of charge. Furthermore, steel surfaces remained unaffected metallurgically and in appearance. Next came improvements such as the triple-quench vacuum furnace (water, oil, and inert gas), the conveyor vacuum furnace, and vacuum carburizing.

The use of graphite has aided these developments. To explain, detrimental oxidizing and decarburizing influences have been virtually eliminated in the neutral environment provided by fibrous graphite linings. In addition, vacuum heat treating systems have been beneficial in the area of ecology by eliminating the need for venting waste atmosphere and by reducing power requirements.

With the introduction of fibrous graphite felt, graphite cloth, and solid graphite for use in thermal insulation and heating elements, the vacuum furnace became competitive with controlled atmosphere heat treating furnaces. Today's systems handle every heat treating chore with the exception of some surface treatments such as carbonitriding and nitriding — and these processes are being actively researched.

What Furnace Does — The function of the heat treating furnace, whether it provides controlled atmosphere or vacuum, is to produce a metallurgically correct product by annealing, normalizing, hardening, or other heat-activated process. This is no easy task because any metal or alloy responds according to its individual characteristics. Unless steel parts are heat treated as specified, they can distort, crack, pit, and develop a soft skin, among other things. In some members of the stainless steel family, for instance, carbides can emerge in the grain boundaries during heat

treatment, changing the entire nature of their performance.

Steel mills have gone to considerable effort to place information on design limitations and recommended heat treating processes in the hands of design engineers and metallurgists. When these recommendations are followed exactly, metallurgical expectations are consistently achieved.

As an example of useful information supplied by the steel mills, consider isothermal transformation curves, visual aids that illustrate the cooling rates that must be used to harden given grades properly. The time indicated for the steel to cool from the austenitizing temperature into the martensitic temperature range while still remaining to the front, or nose, of the curve (thus staying completely austenitic) clearly indicates whether slow cooling, forced convection, or liquid quenching is necessary (Fig. 2).

The only remaining factor is control of the protective atmosphere — or lack of atmosphere as within a vacuum furnace. For the latter, this chore is relatively simple, being efficiently handled by elaborate pumping systems plus accurate vacuum measuring and pressure control gages.

Atmosphere Furnaces Develop — The previously mentioned Certain-Curtain has the following features: controlled volumes of gas and air are introduced into a small combustion chamber located under the hearth of a box-type furnace (Fig.

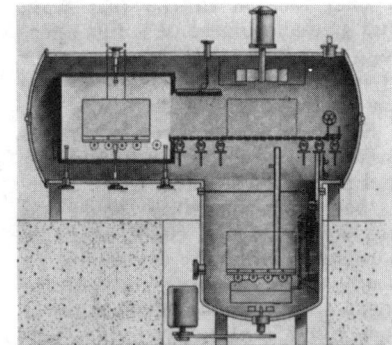

Fig. 1 — Diagram of vacuum furnace with integrated oil quenching bath shows heating chamber and tank. Work is heated in the chamber at the right, and is then conveyed above the tank where it is quenched.

Reprinted from Metal Progress, September 1975, 91-92, 94, 96, © 1975 American Society for Metals

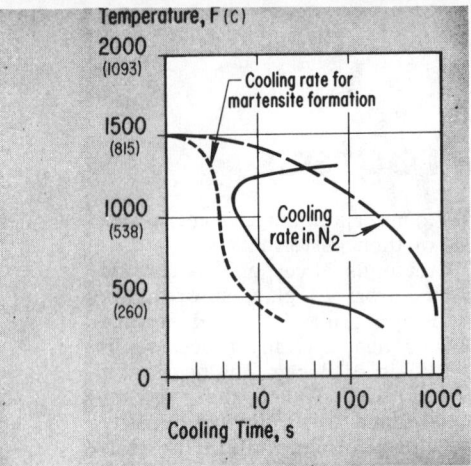

Fig. 2 — Time-temperature-transforma-tion diagram for AISI 52100 (1.06 C, 0.33 Mn, 0.32 Si, 1.44 Cr) demonstrates cooling rate of nitrogen and cooling rate needed to assure a predominately martensitic structure in the quenched part.

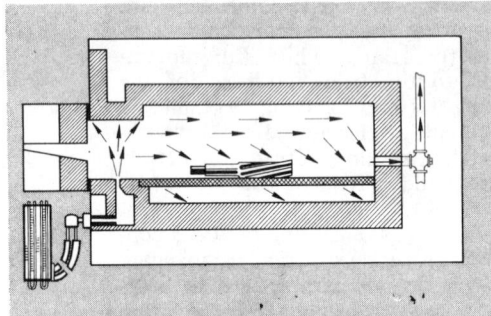

Fig. 3 — In Certain-Curtain furnaces, gas and air flow into the chamber under the hearth, through the hearth, and out of the rear end.

3). The combusted product consists of a gas containing given amounts of CO, CO_2, H_2, and trace amounts of other gases. The protective atmosphere flows upward through a slotted opening in the hearth above the combustion chamber, enveloping the work in the heating chamber. The gas then flows out through a vent located in the rear wall, and a small volume of it fills openings, sealing the loading door against intrusion of room air into the heating chamber.

Such an atmosphere minimizes decarburization, scaling, burning, and pitting of steel surfaces. High-carbon, high-chromium steels, and high-speed steels could now be hardened without developing a heavy scale, which is difficult to remove and results in depletion of carbon to a greater depth than indicated by the scale thickness.

When the early Certain-Curtain furnaces were installed, they were accompanied by a field engineer, who set up the furnace, heated the chambers, and then analyzed the atmosphere within the heating chamber under various gas-air ratios. Once this was done, a chart was left with the furnace operator to enable him to select the proper gas-air setting for each type of steel.

Recirculating Furnaces — Following this advance came the recirculator system (Fig. 4), which partially removes water vapor products of combustion from the precombusted atmosphere. In the process, CO was increased and CO_2 was decreased, resulting in an atmosphere which would not decarburize any steel then available at any temperature. The system is basically an endothermic generator consisting of a positive pressure pump, a purifying chamber, a condenser bottle, and a special arrangement for introducing circulated gases into the back of the furnace.

The recirculator was attached to the integral atmosphere combustion chamber. When the recirculator was in operation, gases from the combustion chamber were drawn down and through it to be dried and purified. Only after purification was complete did the gases enter the furnace heating chamber.

These gases were introduced into the rear of the chamber against a labyrinth baffle, which acted as a catalyst to complete the gas reaction. The baffle also functioned as a chamber for reheating the cooled gases, which were now ready to enter the furnace as a protective atmosphere.

Because the pump circulates several times the volume of gases being burned per hour as well as those which are introduced, gases in the heating chamber continually recirculate during heat treating. In the recirculation method, the hot gases already in the furnace heating chamber combine with the fresh gas-air mixture in the combustion chamber. This action supplies heat to the fresh mixture, aiding its combustion and assuring a high degree of precombustion.

For example, if the volume of gases being handled by the recirculating system was three times the volume of fresh gases being introduced, the combustion chamber received hot gases equal to double the volume of the oncoming cold gases to be burned — a decided help in burning the cold mix.

At the same time, the system reduced the water vapor content coming from two sources — the gases from the combustion chamber, and the reaction of reheating and carbon dioxide removal — before re-introducing the mixture into the heating chamber.

Endothermic Generator Appears — The basic recirculator system was augmented by placing a horizontal catalyst-filled retort beneath the

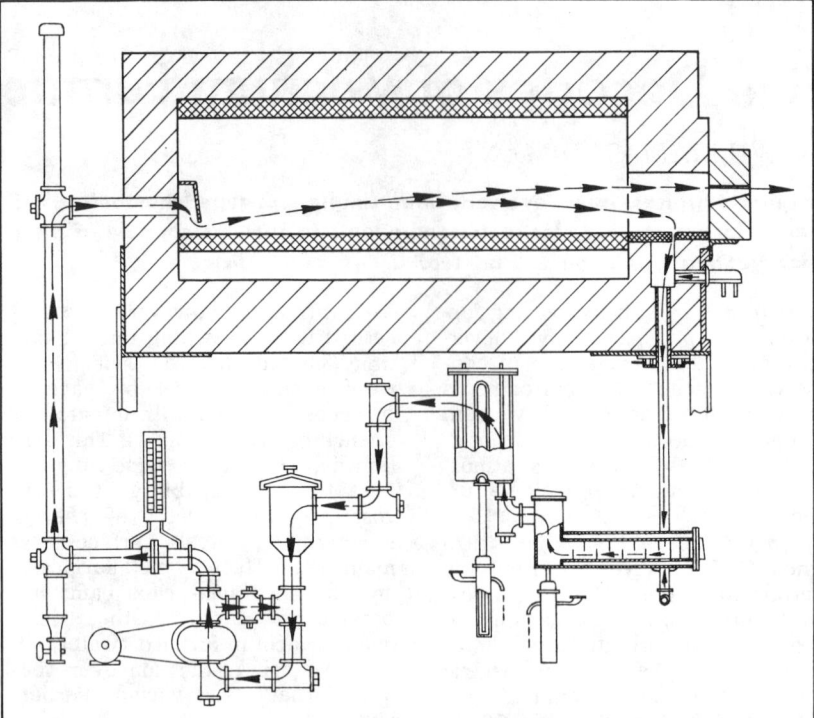

Fig. 4 — In the recirculation system, water vapor products of combustion are removed from the precombusted atmosphere by drying and purification.

heated floor of the furnace. Then a controlled mixture of gas and air could be directed into the pre-combustion chamber, through this retort, and into the furnace, through the rear wall where it contacted a labyrinth baffle cast of nickel.

Within the chamber, this controlled atmosphere produced an environment suitable for heat treating hot-work steels and high-speed steels. Properly handled, the furnace could even heat treat carbon steels without decarburizing them.

The next step was to place this catalyst-filled retort in its own controlled heating environment, thus creating an endothermic generator, which could be maintained at the optimum temperature, 1850 F (1010 C). Through proper sizing, the generator could be built big enough to supply several furnaces simultaneously.

Atmosphere generated by endothermic units had constant compositions, and flow to each individual furnace was enriched as required to produce the proper environments for the particular alloy being heat treated. Since the same atmosphere was produced by the generator, the same environment could be continually duplicated within the heating chamber of the furnace, whether it be a box-type, roller hearth, or walking-beam installation.

Furthermore, a central control room could be set up so that the metallurgist could monitor each furnace from one spot. The entire task could, in fact, be reduced to push-button operation.

Modern endothermic generators are usually equipped with the latest in gas control equipment to provide a constant caloric content. Such controls compensate for varying environmental conditions within the furnace insulation and correct for gas inconsistencies which may be introduced by the local supplier.

Vacuum Furnaces — Unlike atmosphere equipment, a vacuum furnace does not contend with, or need to compensate for, CO_2, CO, H_2O, O_2, or H_2. Furthermore, it does not have to be held at temperature under protective atmosphere during its idle periods. It is a demand-type unit, usable either around the clock or during single shifts.

Other characteristics are ability to heat to temperature at either extremely rapid rates, or heat at rates adjusted to produce optimum preheating conditions. Vacuum furnaces can also heat unequal masses of steel to a uniform temperature via multiple plateau programming, which is important in bonding.

The primary difference between endothermic atmosphere furnaces and vacuum furnaces is in pressure. Atmosphere furnaces function with a positive pressure atmosphere, while a vacuum furnace functions with a partial pressure environment.

When a vacuum heat treating system is purchased, the buyer should be aware that this partial pressure atmosphere furnace is not an absolute vacuum. Nor is it a mystical tool that will harden a particular steel without ever heating it above its Curie point, or heat and drastically quench a dovetail cutter of high-speed steel without ever having the tool crack.

Three Styles — Vacuum furnaces are available in three basic models: top loading (pit-type); bottom loading (bell-type); and horizontal loading (box-type). Each type may be stretched or compressed, equipped with internal doors, integral water and oil quench systems, or any combination of these. They can also include automatic conveyor or pusher systems.

Early vacuum furnaces, as mentioned earlier, were simple cold-wall types used exclusively for outgassing metals. Use of backfill gas in these early furnaces enabled operators to speed cooling, thus shortening turnaround time. The need for a safer, more economical approach to heat treatment of stainless steel encouraged development of forced-convection backfill cooling in vacuum furnaces.

The cooling rate needed to harden a type 400 stainless was achieved by backfilling the vacuum furnace with an inert gas, and employing an internal fan to force this gas through the work load, the gas being cooled in turn by water-cooled walls of the furnace. Use of such innovations as finned tubing, positioned between the shields and the furnace shell, improved the rate of cooling further.

Unfortunately, the radiation shields and heating elements were being heated and cooled at the same rates as the work and work tray, which created a maintenance problem. Because of constant expansion and contraction stresses, these parts distorted and cracked in service.

While heat treaters accept replacement of a tray or fixture as a conventional operating cost factor, they could not tolerate such replacement for costly shields and heating elements. One logical method of eliminating this problem was to stop quenching the heating chamber.

Thus furnaces were provided with water-jacketed cooling sections adjacent to heating zones. Fans and acceleration duct systems in these sections cooled hot loads rapidly. This new design was beneficial for two reasons — it reduced maintenance costs and produced faster quenching rates because only the work was being cooled.

One does not have to be a mathematician to realize that the backfill quenching rate of a 200 lb (90 kg) load is considerably faster than the backfill quenching rate of a 300 to 400 lb (136 to 181 kg) load under identical conditions. (The additional 100 to 200 lb [45 to 90 kg] represents the weight of the radiation shields and heating elements.) Employing the dual chamber furnace design, we found that we had adequate cooling capacity to handle backfill quenching of a heavy, dense, solid mass of material.

Source: Metal Progress, September 1975, 91-92, 94, 96

Electric Vacuum Furnaces for Heat Treating Steel

By CHRISTOPHER F. MASTERS
The Kinetic Co.
Greendale, WI 53129

Editor's Note: *How hardening and carburizing processes can be implemented for many steels in electric vacuum furnaces is described in this article. It is from a paper presented during the 13th Biennial National Electric Heating Conference, sponsored by The Electrification Council in Cincinnati, Ohio, Feb. 6-8, 1978. Included is a discussion of the constructional features of these vacuum furnaces, emphasizing the various components and cycles which are needed in order to successfully carry out the required processes. Limitations as well as advantages are explained.*

Two of the many types of heat processes which steel undergoes in order to fully develop its properties are hardening and carburizing.

Hardening Process

The hardening process in itself is quite simple. The steel is first heated up to its austenitizing temperature which is anywhere from 1500°F to 2300°F, depending on the grade of steel involved, and then quenched or cooled rapidly. The process has several requirements if it is to be successful. First, the austenitizing temperature must be controlled to within very narrow limits, typically 25°F for commercial work. Next, the heating must be done in such a way so as to limit the scale formation on the steel surface. Finally, and this is the most difficult part of the process, the steel must be cooled quickly enough to develop martensite, the hard phase of steel. Depending on the grade of steel involved, there is anywhere from a few seconds to an hour to cool the steel from its austenitizing temperature to room temperature.

As far as vacuum heat-treat furnaces are concerned, a steel which requires a slow quench to harden is much easier to process than one which requires a relatively fast quench. A "simple" vacuum furnace is limited by the speed at which it can quench its load of steel. Fortunately for the development of vacuum furnaces, most of the steels which exhibit superior wear resistance and strength properties can be hardened by a process which incor-

porates a slow quench. These slow quenching or readily hardenable steels are very highly alloyed, and are used to produce a tremendous variety of superior industrial parts ranging from small drills and taps to huge forming dies used in the automotive industry. The section size of the steel part being hardened also contributes to its hardenability. Obviously, it is much easier to cool down a thin section than a thick section.

The other problem which arises in the hardening of steels is the protection of the steel surface against scaling (oxidation), carburization or decarburization. Most steel parts which have to be hardened are first machined as close as possible to their final size prior to hardening because the steel in the soft or annealed state can be machined by relatively inexpensive operations such as milling, drilling, turning, etc. After hardening, the machining in most instances has to be by grinding, which is very slow in comparison. Furthermore, the removal of any scale is difficult, for it acts differently than the base metal to any machining process. Therefore, any hardening process should preserve the original surface conditon as much as possible.

When steel is heated to its austenitizing temperature as part of the hardening process, air must be excluded to prevent the scale formation or decarburization. Ways to do this are to surround the steel with an inert atmosphere of nitrogen, argon, helium, an "endothermic" atmosphere which consists of a mixture of air and natural gas which has been reacted by externally supplied heat, or by simply excluding all the air and not replacing it with anything else, i.e., vacuum.

Vacuum Furnace Hardening

With this brief introduction into some of the background problems and requirements of a furnace which can be used to harden steel, the constructional details of a "simple" vacuum furnace can now be presented. Referring to Figs. 1 and 2, the center of the fur-

(Continued on page 26)

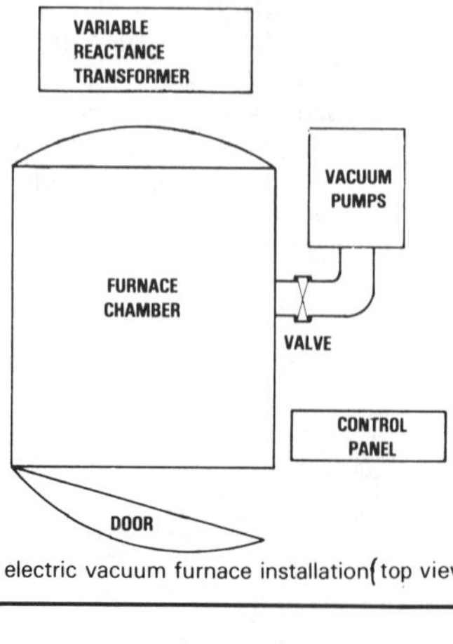

Fig. 2 Interior cross section of electric vacuum furnace

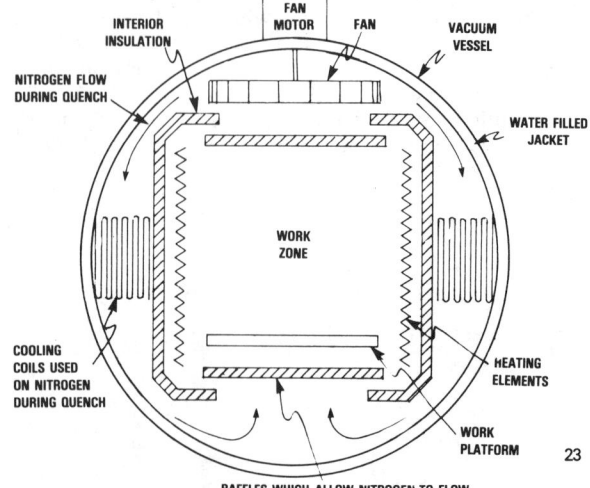

VARIABLE REACTANCE TRANSFORMER

VACUUM PUMPS

FURNACE CHAMBER

VALVE

CONTROL PANEL

DOOR

Fig. 1 Layout of electric vacuum furnace installation(top view).

FAN MOTOR

INTERIOR INSULATION

FAN

VACUUM VESSEL

NITROGEN FLOW DURING QUENCH

WATER FILLED JACKET

WORK ZONE

COOLING COILS USED ON NITROGEN DURING QUENCH

HEATING ELEMENTS

WORK PLATFORM

BAFFLES WHICH ALLOW NITROGEN TO FLOW

23

nace is the work zone, where the steel to be hardened is located. Surrounding it, either on the top or the bottom or sides, are the heating elements, which are electrically powered by means of a variable reactance transformer. The transformer, in conjunction with an electronic controller which adjusts its power output according to the temperature read inside the furnace by a thermocouple, supplies the necessary heat in a stepless fashion which allows exceptionally fine temperature control. This heating system and sophisticated electronic controls together with high quality of insulation completely surrounding the work zone allows empty furnace temperatures up to 2500°F to be controlled to within ±5°F.

To vacuum heat treat work accurately, it is necessary to monitor its temperature directly with a thermocouple. Instrumentation for reading this work thermocouple can easily be incorporated into the instrumentation necessary to control the furnace.

The following typical cycle has evolved for heat treating in a vacuum furnace. The steel load is placed in the furnace, which is then sealed, and pumped down to a vacuum. The vacuum level required is in the vicinity of 100 microns of mercury (100 millitorrs). This level is readily attained and maintained in an industrial environment with simply a mechanical pump and a blower. Power is applied to the heating elements and the load is uniformly heated to its austenitizing temperature. Quenching is accomplished by turning off the vacuum pumps (by means of a large valve on the pumping line) and quickly backfilling with an inert gas such as nitrogen to a pressure just less than atmospheric. The internal fan then blows the nitrogen past the internal heat exchangers and over the load, cooling it down to room temperature. The water jacket, which is part of the furnace vacuum chamber, also acts as a heat exchanger and helps in cooling down the load. Furthermore, the jacket prevents the exterior of the furnace from rising above the cooling water temperature even though the interior of the furnace may be at its maximum temperature.

This typical cycle allows the work to enter and leave the furnace at room temperature and thus its surface after hardening is just as clean as before hardening. The furnace will not clean up a scaled work surface because the vacuum levels are not low enough nor the temperatures high enough to cause any significant vaporization. In fact, should any scale be present on the work surface at the beginning of the hardening cycle, it would be diffused deeper into the work.

The basic advantage of a vacuum furnace is its inherent simplicity. The cycle is completely automatic with the addition of a few timers, relays, and solenoids. Inside the furnace the only moving part is the fan; outside there are the vacuum pump and some solenoid valves. The electronic controls are all well developed; repair consists of diagnosing the problem and simply replacing the defective assembly. There is nothing explosive associated with the furnace. It is completely powered by electricity. The heating elements and furnace proper are shut down when not in use, although the vacuum pump should be left running. The furnace is clean. Because of its water jacket it can be used in an air conditioned room in the summer, and in the winter its cooling water can be used to heat the building in which it is located.

However, there are some negative aspects. A simple vacuum furnace, as compared to a muffle furnace with a controlled atmosphere, can easily cost several times as much in initial capital outlay. Yet operating costs may be comparable, inasmuch as a vacuum furnace consumes fuel only during the heating part of the cycle. Furthermore, other economies may be evidenced in the overall manufacturing costs of the part being hardened in a vacuum furnace. For example, the part can be machined to closer

tolerances and finish ground easier than comparable parts hardened in other types of furnaces. Therefore, for the particular hardening process overall costs must be considered.

A fundamental limitation to the utilization of a "simple" vacuum furnace is the minimum cooldown time for the steel load in the furnace. The furnace becomes much more useful with faster cooldown rate because it allows more lightly alloyed (and cheaper) steels and thicker cross section loads to be properly hardened.

The most obvious way to decrease the cooldown time is to include an oil quench unit as part of the furnace. The oil quench unit is located inside of the vacuum chamber and the work load is physically moved, while under vacuum, from the heat zone into a quench tank. The cooldown time for a load with such a unit is on the order of seconds. However, the inclusion of an integral oil quench in a vacuum furnace can easily double its original cost. Furthermore, many mechanical parts have to function properly in a vacuum and in the presence of temperatures which often exceed 2000°F.

Progress is being made in improving the cooldown rate of the "simple" vacuum furnace through improvements in the heat exchangers and internal baffles. Different inert gases, such as helium and argon with higher thermal transfer rates, have been utilized. Designers have even increased the pressure of the nitrogen in the furnace during cooldown to greater than that of the atmosphere. By increasing the total mass of nitrogen present in the furnace, the rate at which heat is transferred from the furnace load to the nitrogen can be increased.

Considerations in Tempering

Subsequent tempering is much simpler to perform than the original hardening process, for the temperatures required are much lower. Furthermore, there is no requirement for the cooling speed.

However, there are two problems associated with using vacuum furnaces for the subsequent tempering operation. Work tempered in a conventional furnace may not scale enough at the lower temperatures involved to economically justify the surface condition resulting from vacuum tempering. However, for overall practicality the entire manufacturing process needs to be examined.

The second problem involves temperature control. All heating inside a vacuum furnace is via radiation. At temperatures below about 800°F the radiation heat transfer rate (which is proportioned to the absolute temperature raised to the fourth power) is so slow that the control system cannot react properly. The result is that the furnace cannot be used to heat loads to such low temperatures.

Benefits in Hardening

The "simple" vacuum furnace today has evolved to the point where it has become almost irreplaceable for the hardening of highly alloyed, thin cross section loads. Benefits in such applications stem from fuel economy, work surface integrity, dependabilty, cleanliness, and simplicity of operation.

Carburizing Process

In order for steel to absorb carbon, it must be heated up to a relatively high temperature, then brought in contact with a carbonaceous material, and finally held at that temperature for a time sufficiently long for the carbon to penetrate into the surface to the required depth. At the temperatures used in commercial carburizing, i.e., 1600°F to 1900°F, the depth to which carbon will penetrate is approximately proportional to the absolute temperature raised to the seventh power times the square root of the time the work is held at that temperature. This means that the carbon will penetrate about 2½ times as deep into the work for the same process time if 1900°F is used instead of 1600°F. Or conversely, the process will take about 6½ times as long to accomplish the same depth of carburization at 1600°F as compared to 1900°F. However, the higher operating temperatures can drastically shorten the life of the furnace and the supporting fixtures used. So the exact temperature used represents a trade off between the cost of the furnace and fixtures and the length of time they have to last.

Vacuum Furnace Carburizing

The simplest form of vacuum carburizing consists of utilizing a vacuum furnace as the source of heat required for the process, and utilizing natural gas as the source of carbon. The process consists of heating the steel load in a vacuum furnace as previously described, then while the load is still at heat and in the furnace, injecting natural gas into the furnace until the pressure inside is about one-half atmospheric. There is more than enough total carbon present at these pressures to cause carburization. The rate of the carburization is principally controlled by the temperatures utilized, which are approximately the same as that used in conventional carburizing.

The principal problem faced with vacuum furnace carburization is the constant rejuvenation of the gas right at the surface of the steel. Fresh gas must constantly be brought in contact with the steel surface. This can be accomplished by including a fan inside the furnace which will constantly circulate the gas. Or it can be done by evacuating the furnace and refilling it with fresh gas several times during the carburizing process.

Because in a vacuum carburizing furnace the heating method does not supply any of the carbon required and the carbon addition does not supply any heat, process segments can be combined in a very straightforward manner. For example, in Fig. 3 a typical carburizing cycle is shown. The steel is first heated to the carburizing temperature in the stabilizing portion of the cycle, so named because its purpose is to insure that the entire load reaches one uniform temperature. Next the natural gas is added, which is the source of carbon for the steel.

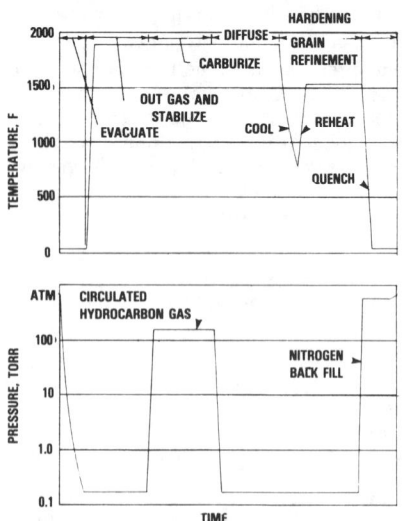

Fig. 3 Typical vacuum carburizing cycle, with temperature and pressure as functions of time.

The third portion of the cycle is the diffusion of the carbon into the steel interior, thereby effecting a thicker case or carbon depth. The unique thing about the diffusion portion in a vacuum furnace is the relative ease with which it can be accomplished. In a controlled atmosphere furnace the atmosphere must be blended just right to avoid any decarburization or futher carburization of the steel surface.

The last portion of the cycle is for grain refinement. At the high temperatures and long times needed for proper carburization and diffusion, the steel grains tend to agglomerate, which could cause inferior mechanical properties in the finished steel part. By lowering the steel temperature below its critical temperature, and then raising it back above the critical temperature the steel grain structure becomes refined.

For the carburized steel to develop its full wear or strength potential, it has to be hardened. This can be accomplished with a simple vacuum furnace only if the steel is highly alloyed. Or alternatively, the furnace can be equipped with an integral oil quench, which allows practically any steel to be hardened by quenching it from the grain refinement temperature. If necessary, the steel can always be hardened in a subsequent operation which includes a fast quench.

The equipment required to perform the vacuum carburizing process consists of a vacuum furnace previously described but modified to include the necessary timers, solenoids, and temperature controls to automatically control the furnace. Also included must be some means to circulate the gas inside the furnace.

Summary of Advantages and Disadvantages

The primary advantage of vacuum carburizing is the exceptional repeatability and uniformity which results when similar loads are processed.

Other advantages of the process are: (1) it is very clean with the parts coming out of the furnace just as shiny as initially charged; (2) the furnace itself is also clean and does not give off tremendous quantities of heat to the surrounding area; (3) the amounts of natural gas required are miniscule compared to that required by conventional carburizing equipment; (4) there is nothing explosive associated with vacuum carburizing.

The disadvantage of the process is that the initial capital cost of the equipment can easily be several times that of conventional equipment. However, the cost of electricity to operate a vacuum carburizing furnace should be considerably less than that for a conventional furnace. However, pros and cons for each application must be evaluated individually.

The ultimate future of vacuum carburizing may be dependent upon price of natural gas. If its price doubles or triples in the near future, then vacuum carburizing which uses electricity as its fuel, may become the most economical way to perform gas carburizing.

Conclusion

The application of vacuum furnaces to the hardening and carburizing of steel represents a new technology which started in the 1960's and which is having an ever greater impact on commercial heat treating practices. The cleanliness, repeatability, straight forwardness of operation, safety, and ability to shut the equipment completely down between cycles are important advantages. Also, these furnaces are versatile. A vacuum furnace which can harden can also anneal, braze, and sinter. A vacuum carburizing furnace can nitride, by using ammonia as the carrier gas, and carbonitride by using a mixture of natural gas and ammonia. And the most versatile feature of these furnaces is that they operate on electricity, which is available everywhere in the world.

Bibliography

G.A. Roberts, J.C. Hamaker, Jr., A.R. Johnson, *Tool Steels*, American Society for Metals, 1962

ASM Committee, *Carburizing and Carbonitriding*, American Society for Metals, 1977

ASM Committee, *Metals Handbook, Vol. II*, American Society for Metals, 1964

Heat Treating in Vacuum Furnaces and Auxiliary Equipment

VACUUM HEAT TREATING is a relatively new development in metallurgical processing. Vacuum heat processing consists of thermally treating metals in heated enclosures that are evacuated to partial pressures compatible with the particular metals and processes. Vacuum is substituted for the more commonly used protective gas atmospheres during either part or all of the heat treatment. Furnace equipment used in vacuum heat treating differs widely in size, shape, construction and method of loading.

Although originally developed for processing of electron-tube materials and refractory metals for aerospace applications, vacuum furnaces are now employed in brazing, sintering, heat treating and diffusion bonding of metals. Vacuum furnaces also are used for annealing, carburizing, heating and quenching, tempering, and stress relieving. Furnaces for vacuum heat treating are equipped for workloads ranging from several pounds up to 100 tons, and heated working chambers range in size from 0.03 m³ (1 ft³) up to hundreds of cubic feet. Although most vacuum furnaces are batch-type installations, continuous vacuum furnaces with multiple zones for purging, preheating, high-temperature processing and cooling by gas or liquid quenching also are used.

VACUUM MEASUREMENTS

A theoretical or ideal vacuum is an empty space that does not contain either vapors, particles, gases or other matter and consequently has no atmospheric pressure. Because this condition does not exist, even in outer space, an ideal vacuum cannot be achieved. Therefore, a manufactured vacuum is expressed in relative terms of pressure compared with the standard atmospheric pressure surrounding the earth. Standard atmospheric pressure at sea level, 45° latitude and 0° C has the following values: 760 mm Hg, 760 000 μ or μm Hg, 29.921 in. Hg, or 14.696 psi.

FURNACE EQUIPMENT

Although conventional atmosphere furnaces can be adapted for vacuum heat treating by adding a vacuum-tight retort connected to a suitable pumping system, furnace equipment developed especially for vacuum heat treating is generally used.

COLD WALL FURNACES

Cold wall furnace units consist of a water-cooled vacuum vessel maintained near ambient temperature during high-temperature operations. Consequently, because the operating temperature does not affect the strength of the vessel material, large units can be constructed for use at high operating temperatures.

In the cold wall design, the water-cooled vacuum vessel contains and supports the internal insulation, the electrical heating elements and the hearth on which the workload rests. The vacuum acts as (a) a substitute for the normal heat treating atmosphere to protect the workload; (b) an insulating medium in the furnace, because the thermal conductivity of a vacuum is essentially zero; and (c) an effective protective coating around the heating elements, the heat shields and the supporting hearth.

Bottom-Loading Furnaces. As shown in Fig. 6, bottom-loading furnaces are stationary and elevated well above floor level. The bottom descends to floor level for ease of loading. The work is loaded on trays that are placed on the hearth by a fork lift when the bottom is in the lowered position. Such furnaces are built to handle large, heavy loads and are cooled rapidly by a high-velocity internal or external circulating gas system.

Horizontal-Loading Furnaces. A box-type horizontal-loading furnace consists of a gastight cylindrical shell with circular convex end plates. In some designs, both of the end plates are hinged to permit easy access to the furnace interior.

Cross-sectional views of a three-chamber oil-quench furnace are shown in Fig. 7. The front chamber is equipped with internal cooling coils and a circulating fan for accelerated gas cooling. The center chamber is the heating chamber, which can be sealed at both ends during the heating cycle by internal moving heat shields and doors equipped with O-rings. The third chamber contains the oil quench and the vertical transport system required to immerse the work in the circulated quenching oil.

HEATING ELEMENTS

Resistance heating and induction heating are the two most common methods of heating within cold wall furnaces. When vacuum furnaces are heated inductively, a graphite cylinder is used as a susceptor; the graphite is heated by induction and radiates the heat to the work inside the cylinder. When heating is provided by the more common resistance elements, the heat transfer is also completed by radiation; therefore, the active heating surface should be large enough to effect a rapid transfer of heat.

Resistance heating elements operating in a vacuum do not require oxidation-resistant properties equal to those required in oxidizing atmospheres. Because, to improve operating efficiency, resistance heating elements are heated to higher temperatures than are the elements used in conventional furnaces, resistance heating elements require low vapor pressures to ensure long life. Materials meeting these requirements are:

- Refractory metals, such as tungsten, molybdenum and tantalum
- Pure solid graphite in the form of bar, rod or tube
- Pure graphite cloth woven from fine filaments of pyrolyzed graphite
- Chromium-nickel elements for operating temperatures below 980 °C (1800 °F).

Refractory Metals

Tungsten is capable of withstanding higher operating temperatures than the other refractory metals. As a heating-element material, it is used as a thin sheet or as sections of woven wire screen.

Molybdenum, in the form of solid rod, strip or thin sheet, is the most widely used metallic heating-element material. Molybdenum in sheet form is normally preferred because its electrical power density (watts per square inch of radiating surface) is low compared with that of cylindrical rod, resulting in lower operating temperatures and thus longer service life.

Solid Graphite Heaters

All metals lose some strength when heated, whereas crystalline carbon in the form of graphite increases in strength as the temperature increases. Pure graphite in the form of flat bar and rod is less expensive than other high-temperature metallic resistors. Graphite also has a much lower heat expansion coefficient and is more resistant to thermal shock than most metallic materials, and has a high melting point and a low vapor pressure; thus, it is an excellent choice for a vacuum furnace heating-element material.

Graphite Cloth Heaters

A third type of material used for vacuum heating elements is a cloth composed of fine graphite fibers. This material is made from rayon cloth pyrolyzed at high temperature to convert the carbon in the rayon to crystalline graphite. The cloth is strong and very flexible. It can be cut with ordinary scissors to the desired size and shape. Because the cloth is flexible, the supporting sys-

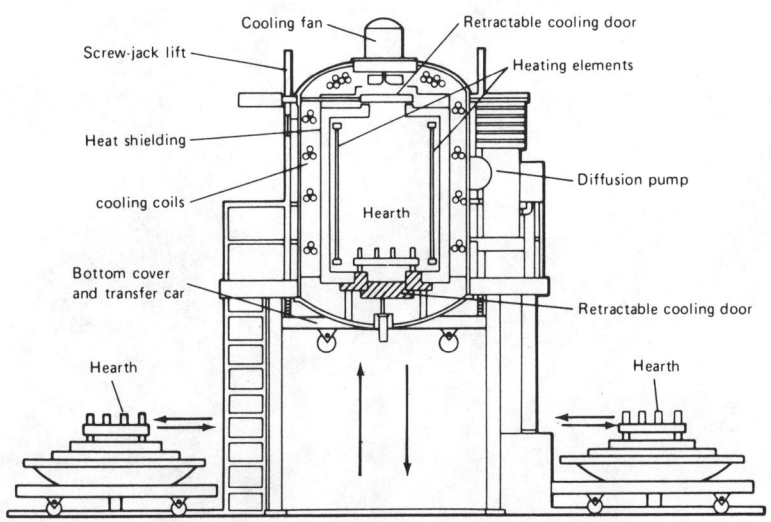

Fig. 6. Bottom-loading cold wall vacuum furnace

Reprinted from Metals Handbook, Desk Edition, 28.35-28.36, © 1985 American Society for Metals

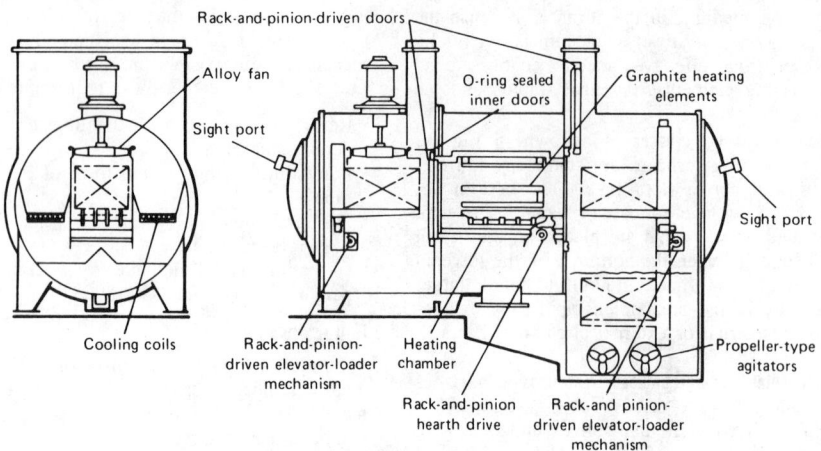

Fig. 7. Three-chamber vacuum oil quench furnace

A schematic diagram of a three-nozzle vapor-diffusion pump is shown in Fig. 8. Vapor from a liquid held in a closed boiler heated at the bottom is forced upward inside the boiler. The vapor passes quickly through a narrow circumferential opening in the nozzles at a downward angle. Molecules of gas that stray from the vacuum chamber above the pump toward the vapor jet streaming from the nozzles encounter the downward-directed stream of heavy molecules. The over-all effect is to compress the gas molecules and force them downward to a point where they can be removed by the mechanical forepump.

tem can be simplified considerably. The ends usually are clamped in graphite electrodes.

PUMPING SYSTEMS

Vacuum vessels are evacuated by various types of pumping systems that depend, to a great extent, on the pressure range needed for processing. An adequate vacuum pumping system must attain the specified pressure and must have sufficient capacity to handle the processing gas load, not only at the ultimate pressure but at all intermediate pressures during the pumpdown cycle. Pumping systems are usually divided into two subsystems: the roughing pump and the high-vacuum pump. For certain requirements, a single pumping system is sufficient for the entire range and cycle. Pumps usually are classified as mechanical pumps or diffusion pumps.

Mechanical pumps operate on the fluid-flow principle and are primarily positive-displacement pumps with suitable seals to permit operation at low pressures. Piston pumps or rotary blowers in various pumping-speed ratings are available. Vacuum-system levels down to 25 μm Hg can be obtained with oil-sealed rotary mechanical pumps.

Diffusion Pumps. For pumping at a vacuum-system level below 10^{-3} torr, a vapor-diffusion pump generally is used. Pumping action is directed by a high-velocity stream of heavy molecules in the form of a pump fluid, usually oil.

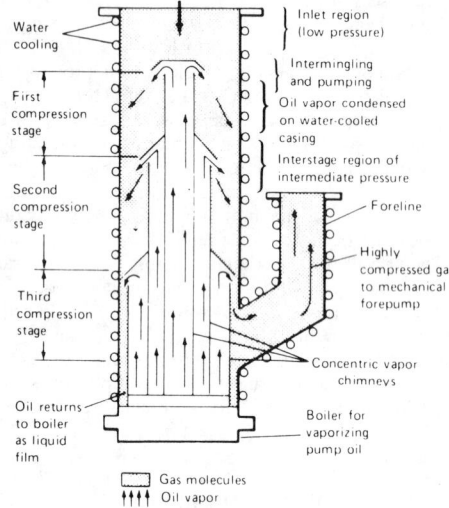

Fig. 8. Oil vapor-diffusion pump

SECTION VII
Salt Bath Processing

Processes and Furnace Equipment for Heat Treating Tool Steels

By the ASM Committee on Heat
Treating of Tool Steels*

MOLTEN SALTS of various compositions are well adapted to all operations in the heat treatment of tool steels. For tools that cannot be ground after hardening, or for tools that require the best possible surface condition and the maintenance of sharp edges, salt bath heating provides the best results. Table 1 lists various salt bath compositions and processing temperatures for heat treating tool steels. The salt bath method of hardening tools—particularly high speed steel tools—has greatly expanded with increased use of molybdenum steels. With correct operating conditions, tools can be heat treated without carburization, decarburization and scaling. The surface will be fully hard with minimum distortion. Three salt baths are generally used—preheating, high-temperature, and quenching baths. The function of the quenching bath is to equalize the temperature as well as to ensure a clean surface after heat treatment.

Salt Baths

Most tools heat treated in salt baths are fully hard from surface to core regardless of section thickness. Because salt baths provide temperature uniformity in preheating, high-temperature heating and in quenching, distortion and residual stresses are minimized.

Tools that are heat treated in molten salt baths are heated by conduction, the molten salt provides a ready source of heat as required. Although steels come in contact with heat through the tool surfaces, the core of a tool rises in temperature at approximately the same rate as its surface. Heat is quickly drawn to the core from the surface. Salt baths provide heat at a rate equal to the heat absorption rate of the total tool. Convection or radiation heating methods are unable to maintain the rate of heating to reach equilibrium with the rate of heat absorption. The ability of a molten salt bath to supply heat at a rapid rate accounts for the uniform, high quality of tools heat treated in salt baths. Heat treating times are also shortened; for example, a 25-mm- (1-in.-) diam bar can be heated to temperature equilibrium in 4 min in a salt bath, while 20 to 30 min would be required to obtain the same properties in convection or radiation furnaces.

Salt baths are the most efficient method of heat treating tool steels;

*W. James Laird, Jr., *Chairman,* Vice President—Marketing, Research & Development, Upton Industries, Inc.; Carl J. Oxford, Jr., Vice President—Technology, National Twist Drill & Tool Division, Lear-Siegler, Inc.; Percy Rawcliffe, Manager, Metallurgical and Physical Laboratories, Morse Cutting Tools, Division, Gulf & Western Manufacturing; Carl Reichel, Drill & End Mill Division, Inc.; TRW, Inc.; Ronald Spitzer, Chief Metallurgist, Bearings Division, TRW, Inc., Daniel S. Zamborsky, Corporate Metallurgist, Warner & Swasey Co.

Table 1 Typical compositions and recommended working temperature ranges of salt mixtures used in heat treating tool steels

Salt mixture No.	Composition, %						Melting point		Working range	
	Barium chloride	Sodium chloride	Potassium chloride	Calcium chloride	Sodium nitrate	Potassium nitrate	°C	°F	°C	°F
Austenitizing salts (high heat)										
1............	98-100	...	...	...	...	...	950	1742	1035-1300	1895-2370
2............	80-90	10-20	...	...	...	...	870	1598	930-1300	1705-2370
Preheat salts										
3............	70	30	...	...	...	...	335	635	700-1035	1290-1895
4............	55	20	25	...	...	...	550	1022	590-925	1095-1700
Quench and temper salts										
5............	30	20	...	50	...	...	450	842	500-675	930-1250
6............	...	...	...	...	55-80	20-45	250	482	285-575	545-1065

about 93 to 97% of the electric power consumed with a salt bath operation goes directly into heating. In atmosphere furnaces, 50% of the energy consumed goes for heating, and the remaining 50% is released up the furnace stack as waste. Tool steels that are heat treated in molten salts typically are processed in ceramic-lined furnaces with submerged or immersed electrodes containing chloride-base salts.

Immersed-Electrode Furnaces

Ceramic-lined furnaces with immersed (over-the-side) electrodes have greatly extended the useful range and capacity of molten salt equipment when compared with externally heated pot furnaces. The most important of these technical advantages are

- The electrodes can be replaced without bailing out the furnace.
- Immersed electrodes allow more power capacity to be put into the furnace, thus increasing production.
- Immersed electrodes permit easy start-up when the bath is solid. A simple gas torch is used to melt a liquid path between the two electrodes, thus allowing the electrodes to pass current through the salt to obtain operating temperatures.

Immersed electrode furnaces are not as energy efficient as submerged-arc furnaces, however. The area in which the immersed electrodes enter the salt bath allow additional heat loss, through increased surface area. As can be seen in Table 2, the surface area of the salt bath (A) in the submerged-electrode furnace is smaller than the surface area plus the immersed electrodes (A + B) in the immersed electrode furnace. However, a good cast ceramic and fiber insulated cover placed over the bath and electrodes will reduce surface radiation losses up to 60%.

The immersed-electrode furnace has a pot made of high-temperature fireclay brick, surrounded on five sides by approximately 12.7 cm (5 in.) of castable and insulating brick. Figure 1 is a schematic drawing of an immersed-electrode furnace with interlocking tiles and removable electrodes. The removable electrodes enter the furnace from the top, and a seal tile is located in front of the electrodes to protect them from exposure to air at the air-bath interface. This protection helps prolong electrode life. Table 2 compares service life of electrodes and refractories for four basic furnace designs.

Over-the-top electrodes are usually built with laminated cold legs, and water cooling is always required. A typical life expectancy for electrodes operating in such a furnace at 840 °C (1550 °F) is approximately 6 months to 2 yr for over-the-top electrodes, compared with 4 to 8 years for submerged electrodes.

Submerged-Electrode Furnaces

Submerged-electrode furnaces have the electrodes placed beneath the working depth for bottom heating. Figure 2 is a schematic of a typical submerged-electrode furnace. Many submerged-electrode furnaces are designed for specific production requirements and are

Fig. 1 Salt bath furnace for neutral heating

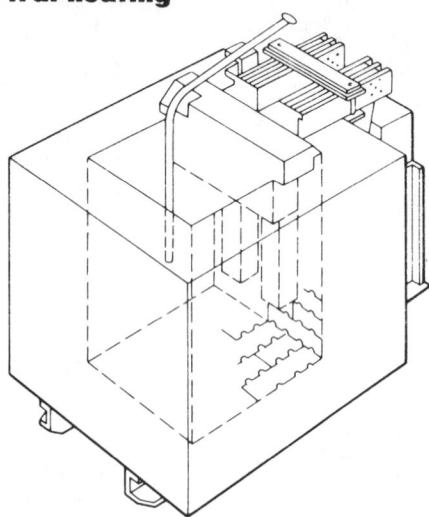

Furnace features a ceramic pot and over-the-top (immersed) electrodes.

equipped with patented features, which offer certain economical and technical advantages:

- *Maximum work space with minimum bath area:* The electrodes do not occupy any portion of the bath, so that only the alloy comes in contact with the salt. Bath size is consequently smaller, and electrode life increases many times over by incorporating unidirectional wear and eliminating excessive deterioration at the air-bath interface.
- *Circulation-convection currents:* Bottom heating provides more uniform bath temperatures and bath move-

Table 2 Service life of electrodes and refractories

| Operating temperature | | Service life, years | |
°C	°F	Electrodes	Refractories
Submerged-electrode furnace: Furnace A			
535-735	1000-1350	10-20	10-20
735-955	1350-1750	4-8	4-8
955-1175	1750-2150	3-4	3-4
1010-1285	1850-2350	1-3	1-3
Immersed-electrode furnaces			
Furnace B			
535-735	1000-1350	2-4(a)	4-5
735-955	1350-1750	1-2(a)	2-3
955-1175	1750-2150	½-1(a)	1-2
1010-1285	1850-2350	¼-½(a)	1-1½
Furnace C			
535-735	1000-1350	2-4(a)	4-5
735-955	1350-1750	1-2(a)	2-3
955-1175	1750-2150	½-1(a)	1-2
1010-1285	1850-2350	¼-½(a)	1-1½
Furnace D			
535-735	1000-1350	2-4(a)	4-5
735-955	1350-1750	1-2(a)	2-3
955-1175	1750-2150	½-1(a)	1-2
1010-1285	1850-2350	¼-½(a)	1-1½

Furnace A   Furnace B

Furnace C   Furnace D

Note: Service life estimates are based on the assumption that proper rectification of chloride salts is being done, as well as routine unit maintenance and care. (a) Hot leg only

ment through the use of natural convection currents.

- *Triple-layer ceramic wall construction:* The temperature gradient through the wall causes any salt penetrating the wall to solidify before it can penetrate the cast refractory material that forms the center portion of the wall construction. This design requires from 5 to 8% of the initial salt charge to fill the ceramic pot. By comparison, in some designs, 140 to 150% of the initial charge is needed to seal the ceramic walls of furnaces

built with two layers of ceramic brick, backed up and supported by a steel plate. Salt penetrates the ceramic walls of any furnace and distorts the geometry of the walls. Reducing the amount of salt allowed to penetrate the ceramic walls aids in maintaining dimensions and in promoting longer furnace life.

- *Electrode placement:* Enclosing the electrode in a clear rectangular box free of any protruding obstructions eliminates potential hazards to operating personnel during cleaning. Any sludge formed in the furnace is removed easily by operating personnel.

Frame Construction. A typical submerged-electrode furnace is made of brick and ceramic material assembled, regardless of size, in a rigid, self-supporting welded steel frame. This frame consists of supporting channels or beams welded to the underside of a heavy steel plate that forms the frame base. To this base are welded lengths of heavy angle iron around the outside and on top of the plate. These pieces are notched to permit welding of the heavy angle-iron posts to the plate and vertical sides of the base-plate angle iron. Lengths of heavy angle iron are welded similarly to the top of the posts. When required, additional vertical reinforcing members are welded between the bottom and top pieces of angle iron, and prestressed horizontal members also are used to ensure that the refractory material cannot move after the furnace has been brought to operating temperature.

Brick Construction. Three types of refractory materials commonly are used in submerged-electrode furnaces.

Submerged-electrode furnace liners are constructed with 23-cm (9-in.) thick high-temperature burned bricks. Consisting of approximately 42% alumina and 52% silica, the brick material is used in standard brick sizes such as 6 by 11 by 23 mm (2½ by 4½ by 9 in.) and in various brick shapes such as straights, flat backs and splits. The bricks are laid with a high-quality air-setting mortar that resists abrasion, erosion and chemical attack by chloride, fluoride and nitrate/nitrite salts. The mortar offers sufficient wear and corrosion resistance to be economically used with some salts containing cyanide. For straight cyanide or carbonate salts, a welded steel pot is used.

The outer wall of the salt bath fur-

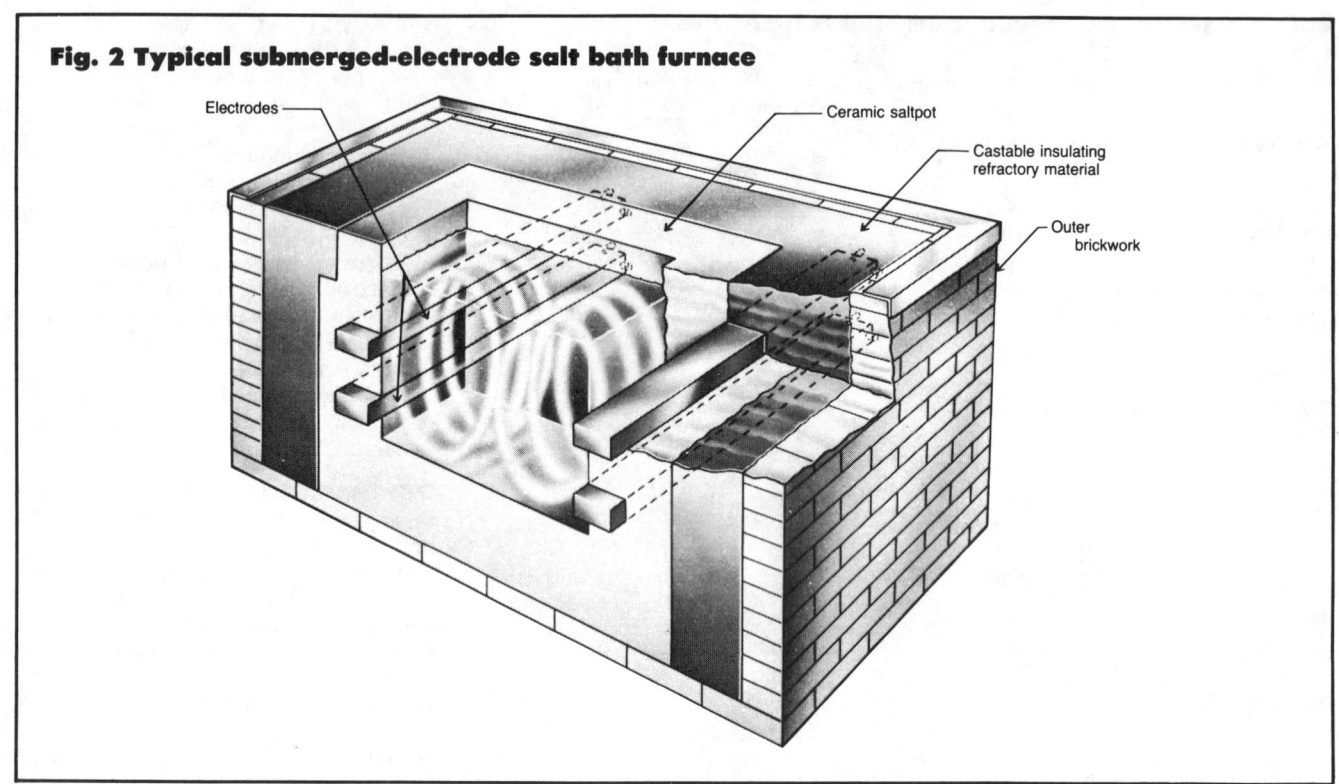

Fig. 2 Typical submerged-electrode salt bath furnace

Electrodes

Ceramic saltpot

Castable insulating refractory material

Outer brickwork

nace is made of a second-quality fire-brick with the same dimensions as brick used for the liner. The important qualities of this brick are the strength of material and uniformity of size and shape.

The inner castable wall is constructed with a mixture of refractory cement and aggregate that is poured between the liner and outer wall to form a 50-cm (9 ½-in.) thick monolithic wall structure. This dimension provides a temperature gradient sufficient to cause the salt to freeze in the wall, thus making the wall self-sealing. With this design, salt penetration into the wall amounts to less than 8% of the bath volume. The maximum temperature of the outside wall during furnace operation is 60 °C (140 °F).

Electrode Construction. The electrodes used in submerged-electrode salt bath furnaces vary widely in size and shape, depending on the geometry of the furnace and power requirements. All of the electrodes are located near the bottom of the bath and are built into the wall so that only one face of the electrode is in contact with the salt. This placement leaves the bath area free of obstruction for ease of cleaning and eliminates the possibility of touching the electrodes with the work.

Alloy electrodes are made by welding a 1612.9-mm² (2½-in.²) alloy material to a mild steel backing, or by welding a 127-by-127-mm (5-by-5-in.) alloy material directly to the mild steel shank.

The durability of typical electrode and ceramic components of submerged electrode furnaces is described in Table 1.

Automatic Heat Treating of Tool Steels

Figure 3 illustrates three different heat treating arrangements for production heat treatment of tool steels. Table 3 gives relative process times and tem-

Table 3 Relative process times and temperatures for automated heat treating of tool steels

Process stage	Operating temperature		Total time in furnace(a)
	°C	°F	
First preheat	650-870	1200-1600	X
Second preheat	760-1035	1400-1900	X
High heat	1010-1285	1850-2350	X
Isothermal quench	535-705	1000-1300	X
Air cool	Room temperature		6X, 12X, 24X
Wash, hot water	80-95	180-200	6X
Rinse, hot water	80-95	180-200	X

(a) See Table 4 for drill sizes and times in the high heat indicated by an "X" in this table.

peratures for heat treating, and Table 4 gives process times for twist drills. The systems are equipped for cycles ranging from less than 1 min to 10 min. The parts are suspended on tong-type fixtures and are carried through the process by a chain conveyor on carrier bars. To facilitate rapid transfer of the tool steels, rotary transfer arms are placed between the preheat and the high heat units and between the high heat and the quench units. Transfer-arm placement is chiefly governed by production rate; however, transfer arms are always required between the high heat and the quench units to satisfy metallurgical conditions. The lines

Table 4 Time cycles for heat treating twist drills

Diameter		Time
mm	in.	
2.54-4.77	0.100-0.188	1 min 30 s
4.80-8.08	0.189-0.318	1 min 40 s
8.10-12.90	0.319-0.508	1 min 50 s
12.91-18.24	0.509-0.718	2 min 0 s
18.26-23.32	0.719-0.918	2 min 20 s
23.34-38.10	0.919-1.500	2 min 40 s
102-mm (4-in.) diam cups		6 min
64-mm (2½-in.) diam end mills		7 min
76-mm (3-in.) diam end mills		10 min

Pieces in high heat on smaller diameters

2.54 mm (0.100 in.) = 160 pieces/tong = 480 pieces in bath = 1.2 kg (2.6 lb)
4.77 mm (0.188 in.) = 85 pieces/tong = 255 pieces in bath = 3.4 kg (7.65 lb)
6.50 mm (0.256 in.) = 63 pieces/tong = 188 pieces in bath = 5.5 kg (12.3 lb)
8.08 mm (0.318 in.) = 25 pieces/tong = 75 pieces in bath = 3.8 kg (8.6 lb)
12.90 mm (0.508 in.) = 16 pieces/tong = 48 pieces in bath = 8.2 kg (18.2 lb)

Table 5 Ranges of endothermic-atmosphere dew point for hardening tool steels(a)

Furnace dew point; AGA class 302 atmosphere

Steel	Furnace temperature(b)		Dew point range	
	°C	°F	°C	°F
W2, W3	800	1475	7 to 12	45 to 55
S1	925	1700	4 to 7	40 to 45
S2	870	1600	4 to 16	40 to 60
O1	800	1475	7 to 12	45 to 55
O2	775	1425	7 to 12	45 to 55
O7	855	1575	−4 to 2	25 to 36
D2, D4	995	1825	−7 to −1	20 to 30
D3, D6	955	1750	−7 to −1	20 to 30
H11, H12, H13	1010	1850	1 to 7	35 to 45
T1	825	2350	−18 to −12	0 to 10
M1	1205	2200	−15 to −12	5 to 10
F2, F3	830	1525	−5 to 1	23 to 34

(a) For short times at temperature. (b) Approximate midrange of austenitizing temperatures for the specific types of tool steels

also have areas above the furnaces to accommodate air cooling of the tools. In special cases, lines will be made with a station for an isothermal nitrate quench after the neutral salt quench. This additional stage allows rapid reduction of the temperature of the tools and reduces the air cooling time from 24 times to 6 times the time at the high heat temperature. **Caution: if as little as 600 ppm of nitrate salts are allowed to enter the high heat furnace, extreme surface damage can be done to the tool being heat treated.**

Rectification of Salt Baths

Neutral salts used for austenitizing steel become contaminated with soluble oxides and dissolved metals during use, resulting from a reaction between the oxide layers present on fixtures and workpieces and the chloride salts. Because the buildup of resulting oxides and dissolved metals renders the bath oxidizing and decarburizing toward steel, the bath must be rectified periodically.

Baths of salts such as salt mixtures No. 1 and 2 in Table 1 can be rectified with silica, methyl chloride or ammonium chloride. The higher the temperature of operation, the more frequent the need for rectification. Baths in which the electrodes protrude above the surface require daily rectification with either ferrosilicon or silicon carbide. Baths operated above 1080 °C (1975 °F) require rectification once daily or more. During rectification of a bath, the silica combines with the dissolved metallic oxides to form silicates. Although these silicates settle out as a viscous sludge that can be removed, sufficient soluble silicates can remain to cause the bath to become decarburizing. If the bath is not rectified, it becomes more viscous than water.

Methyl chloride bubbled through the bath or the submerging of ammonium chloride pellets in a perforated cage in the bath are more effective methods of rectifying salt baths. The ammonium chloride pellets react with the oxides to regenerate the original neutral salt without sludge formation or bath thickening. To remove dissolved metals from high-temperature baths, graphite rods are introduced at operating temperature. The graphite reduces any oxides to metal, which adheres to the rod. The metal can be scraped off and the rod reused.

To control the decarburizing tendency of high-temperature baths, test specimens frequently should be hardened by quenching in oil or brine. A file-soft surface indicates a need for more rectification. This test may be supplemented by analysis of the bath. High-heat baths containing in excess of 0.5% barium oxide are likely to be decarburizing to steel.

One method of rectifying austenitizing baths such as salt mixtures No. 2 and 3 of Table 1 is as follows:

- Add 57 g (2 oz) of boric acid for each 45 kg (100 lb) of salt, after every 4 h of operation.
- Insert a 76-mm (3-in.) graphite rod into the bath for 1 h for every 4 h of operation.

Atmospheres

In selecting an atmosphere that will protect the surface of tool steel against addition or depletion of carbon during heat treatment, it is desirable to choose one that requires no adjustment of composition to suit various steels. An ammonia-base atmosphere (AGA class 601) meets this requirement and has the advantage of being sufficiently reducing to prevent oxidation of high-chromium steels. In the range of dew points generally found in this gas, −40 to −50 °C (−40 to −60 °F), there is no serious depletion of carbon, because the decarburizing action is slow and any loss of carbon at the surface is partially replaced by diffusion from the interior. For applications in which high superficial hardness is important, a carburized surface can be obtained by the

Fig. 3 Process designs for automated salt bath furnaces for heat treating high speed tool steels

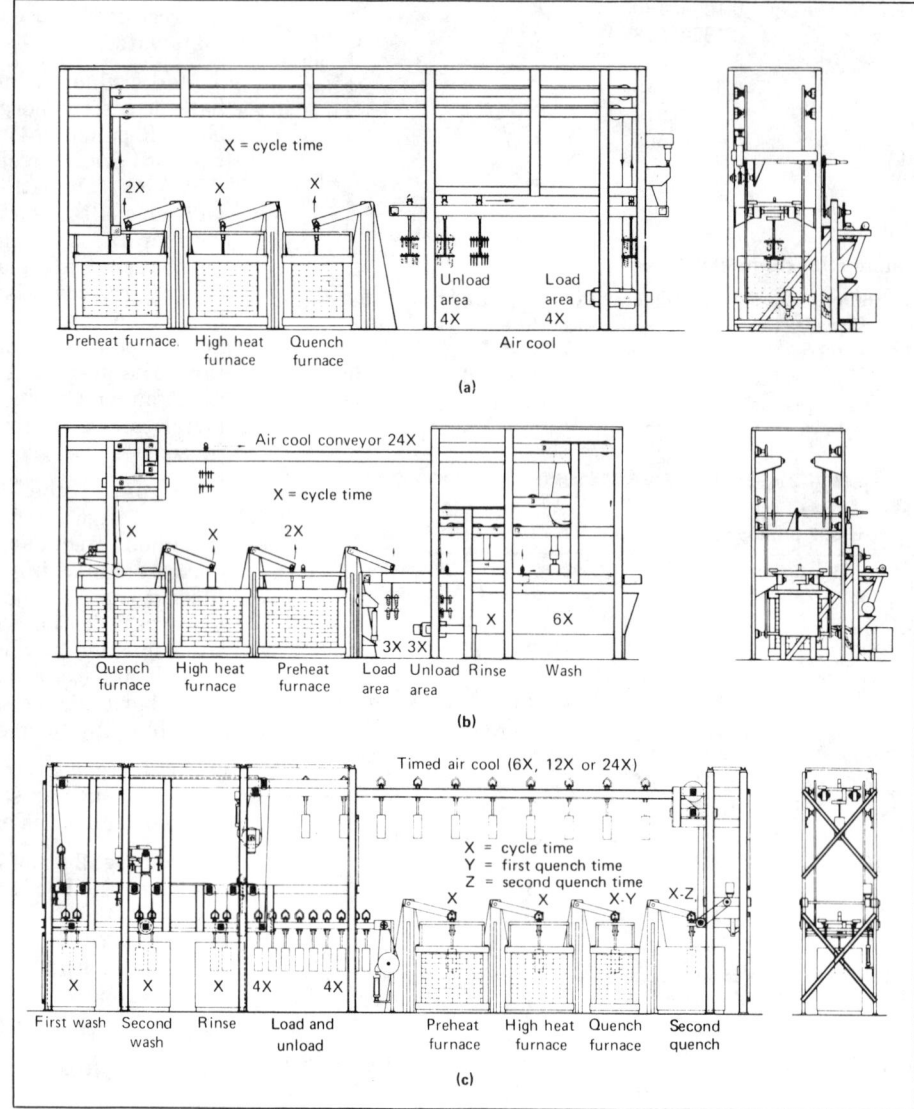

Installations can be custom designed to meet specific customer requests. (a) Does not include wash and rinse. (b) Similar to (a), but includes wash and rinse operation necessitating relocation of load and unload operations. (c) Similar to (b), but includes second quench and a variation in wash cycles as specified by customer.

addition of about 1% methane to the atmosphere. Although ammonia-base atmosphere costs more than endothermic gas, this seldom becomes important because tool treating furnaces generally are comparatively small and therefore require a correspondingly small quantity of gas.

Endothermic-base atmospheres are often used for the protection of tool steel during heat treatment. Suggested ranges of dew point for an AGA class 302 endothermic atmosphere when used for hardening some common tool steels are listed in Table 5. Relatively short heating times for hardening small tools allow treatment to be carried out with the theoretical carbon balance of the atmosphere varying over a rather wide range. However, for the hardening of large die sections, the particular composition of the die steel being treated requires careful control of the atmosphere if carburization or decarburization is to be avoided during the relatively long heat treating cycle.

Salt Bath Equipment

By W. James Laird, Jr.
Vice President-Marketing
Research & Development
Upton Industries, Inc.

SALT BATHS are used in a wide variety of commercial heat treating operations, including cyaniding, liquid carburizing, liquid nitriding, austempering, martempering and tempering applications. Salt bath equipment is well adapted to heat treatment of tool steels, as well as to treatment of nonferrous alloys. Advantages of using salt bath equipment include thermal control and rapid heating rates. Applications of the various furnace designs and auxiliary equipment to specific heat treating processes are described in other articles in this volume.

Externally Heated Furnaces

Externally heated salt bath furnaces may be fired by gas or oil, or heated by means of electrical resistance elements. A typical gas- or oil-fired furnace that is commonly used in liquid carburizing applications is shown in Fig. 4a in the article, "Liquid Carburizing and Cyaniding", in this Handbook. These furnaces are generally lower in initial cost than electrode or resistance heated furnaces and are simple to install and operate. To contain the molten carburizing salt, fuel-fired furnaces employ a steel or alloy pot, which may be either round or rectangular. Heat is applied by two or more self-cooling burners that fire tangentially

between the outer wall of the pot and the inner surface of the furnace lining. The hot gases are vented through a flue located near the top for atmosphere-type burners, or near the bottom for pressure-type burners and for atmosphere-type burners for which the flue is connected to a stack about 1 to 2 m (3.3 to 6.6 ft) high, to maintain negative pressure in the firing chamber. The combustion chamber is lined with firebrick and additional insulation is required. A steel casing completely surrounds all sides of the furnace housing and provides adequate safety in the event of pot failure.

Electrical resistance furnaces for neutral heating or liquid carburizing are less widely used than furnaces fired by gas or oil. (See Fig. 4b in the article, "Liquid Carburizing and Cyaniding".) These furnaces are heated by a series of resistance heaters surrounding the salt pot. For this reason, pot failure may result in the total destruction of the electrical heating elements. Operating temperatures below 900 °C (1650 °F) are used to prevent pot failure.

Salt pots are usually supported from a flange; consequently, pot size is limited by the strength of the flange material. Round pots for gas- or oil-fired furnaces normally range from 250 to 900 mm (9.8 to 35.4 in.) in diameter and from 200 to 750 mm (7.9 to 29.5 in.) in depth; they are about 10 mm (0.4 in.)

thick. Larger sizes have been built for special applications and have operated successfully. Pots larger than about 350 mm (13.8 in.) in diameter and 450 mm (17.7 in.) deep are rarely used for electrical resistance furnaces. Although it is physically possible to support the bottom of a large pot on a refractory pier, excessive temperature gradients may result.

Pots may be press formed from a single piece of low-carbon steel or iron-nickel-chromium alloy; a composition of Fe-35Ni-15Cr is usually preferred for the latter. Less expensive welded pots may be fabricated from either of these materials.

In a well-designed furnace, the life of round alloy pots will vary with maximum operating temperature about as follows:

Temperature		Service life, months
°C	°F	
840	1550	9 to 12
870	1600	6 to 9
900	1650	3 to 6
920	1690	1 to 3

In one installation, the placement of an additional control thermocouple in the combustion chamber to prevent the temperature of the chamber from exceeding 1095 °C (2000 °F) served to extend the life of high-temperature (HT) alloy pots to 2 years (previous life had been 6 months). Pot temperature was maintained at 900 °C (1650 °F) during a work week of 120 h (24 h/day, 5 days/week).

Temperature of the salt is measured and indicated by a thermocouple and suitable pyrometer. Operating within the range from 790 to 920 °C (1455 to 1690 °F), the externally fired furnace may vary as much as 10 °C (18 °F) above and below the set temperature when using on-off or high-low control systems. This is considered acceptable for many applications. Where closer control of temperature is required, a proportional control system, which will hold temperature variations to less than ±5 °C (±9 °F) should be used.

Design and Operating Factors. In the design of fuel-fired furnaces, ample space must be provided for combustion so that the flame will not impinge on the pot. If flame impingement is unavoidable, the pot should be rotated slightly at least once a week. Rotating the pot and using a sleeve reduce local deterioration in the region of flame impingement and prolong service life of the pot. The combustion-chamber atmosphere also has important effects on pot life. A system with a control range from high fire to low fire is preferable to an on-off system, because the latter allows air to enter the combustion chamber during the "off" portion of the cycle, thereby increasing the rate of sealing of the outer surfaces of the pot.

Electrical resistance heated furnaces should be equipped with a second pyrometer controller whose thermocouple is in the heating chamber. This will prevent overheating of the resistance elements, particularly during meltdown, when the thermocouple that controls the temperature of the main bath is insulated by unmelted salt. Because heating elements and refractories are severely attacked by salt, all salt must be kept out of the combustion chamber. For this purpose, a mixture of high-temperature refractory cement, with long-fiber asbestos for strength, may be used to seal joints where the pot flange rests on the retaining ring at the top of the furnace.

Externally heated pots should be started on low fire—low heat input—regardless of the method of heating. Once the salt appears to melt around the top, heat can be gradually increased to high fire to complete meltdown. *Excessive heating of the sidewalls or pot bottom during start-up may cre-*

ate pressures sufficient to expel salt violently from the pot. For added safety, the pot should be covered during melt-down, either with a cover or with an unfastened steel plate.

The waste heat of flue gases may be fed to an adjacent chamber and used to preheat work. Flue gases should always be visible to the operator. The presence of bluish white or white fumes at the vent are an indication that salts have entered the combustion chamber; prompt corrective action is required.

Advantages and Disadvantages. Because of the ease with which they can be restarted, externally heated furnaces are well suited to intermittent operations. Another advantage of furnaces of this type is that a single furnace can be used for a variety of applications by simply changing the pot for one containing the proper salt composition.

Externally heated furnaces have several characteristics, however, that limit their usefulness in certain operations. They are less adaptable to close and uniform temperature control because they dissipate heat by convection, creating temperature gradients in the bath. Also, the temperature lag of the thermocouple and the recovery time of the furnace may result in overshooting and undershooting of a desired temperature control point by 14 °C (25 °F). In addition to requiring an exhaust system for generated flue gases, externally heated furnaces may overheat at the bottom and sidewalls in restarting creating a pressure buildup in the thermally expanding molten salt that may cause an explosion. Finally, externally heated furnaces are seldom practical for continuous high-volume production, because of the limitations of pots with respect to size and maximum operating temperature. High maintenance cost is also a factor.

Immersed-Electrode Furnaces

Introduction of the immersed-electrode furnace greatly extended the useful range and capacity of molten carburizing baths. The electrodes can be removed and replaced without bailing the furnace. This design is also suitable for neutral heating, as well as for cyanide and noncyanide carburizing processes. The molten salt is contained in a steel or ceramic pot surrounded by suitable insulating materials, which separate it from an exterior casing or

framework of heavy-gage steel. The salt is heated by passing alternating current through it with immersed electrodes. As a result of the resistance built up to passage of current through salt, heat is generated within the salt itself. This heat is quickly dissipated by a downward stirring action created by the electrodes. The electrodes are attached by copper connectors to a transformer that converts the line voltage of the plant to a much lower secondary voltage (approximately 9 to 12 V) across the electrodes. Temperature is measured and automatically controlled by a system containing a thermocouple, pyrometer, relay and magnetic contactor. A typical immersed-electrode furnace is shown in Fig. 4c in the article, "Liquid Carburizing and Cyaniding", in this Handbook.

The depth of salt pots for immersed-electrode furnaces is usually limited to 0.6 m (2 ft) for metal pots; ceramic pot depth is nearly unrestricted. Furnaces with pots up to 4.5 m (14.8 ft) in length, and with a power input of 360 kW, are presently in operation. They have a capacity to heat a workload of about 320 kg/h (705 lb/h). In contrast, smaller units with salt pots having a work space measuring 230 by 180 by 350 mm (9 in. by 7 in. by 14 in.) deep, and with 15-kW power input, can heat about 22 kg/h (50 lb/h) to 920 °C (1690 °F).

Advantages and Disadvantages. Immersed-electrode furnaces do not require the use of iron-chromium-nickel alloy pots. Under normal operating conditions, the life of a carbon steel pot is usually 1 year or more. Carbon steel pots of welded construction, set into insulating brick but not cemented in place, give the following service life:

Maximum operating temperature		Service life, years
°C	°F	
840	1545	2 to 3
870	1600	1½ to 2
900	1650	1 to 1½
920	1690	1

These furnaces require minimum floor space and maintenance and can be used for all types of carburizing salts. Electrodes made of alloy steel should have an average service life equivalent to that indicated for steel pots in the above table. Worn electrodes can be replaced while the furnace is in operation.

Depending on the positioning of electrodes, temperature control to within

±3 °C (±5 °F) is easily obtained with immersed-electrode furnaces. Heat is generated within the bath, and overshooting is readily avoided. These furnaces lend themselves to mechanization and are suitable for high-volume production in the range of 815 to 1300 °C (1500 to 2370 °F).

Maximum pot size is not restricted. Pots may vary in length and width to suit requirements, and multiple pairs of electrodes can be installed to furnish the necessary heating capacity.

The immersed-electrode furnace is not recommended for intermittent operation. Depending on furnace size, reheating the salt charge may require a day or more. Pots are not intended to be interchangeable. Removal of the pot usually involves replacement of the surrounding insulation.

Steel Pot Furnaces

Some metal treating processes are performed in salt compounds that cannot be contained in a ceramic liner. For these applications, furnace manufacturers make use of a welded steel pot with immersed electrodes. This type of furnace is suitable for special applications such as case hardening in straight cyanide baths, tempering, and marquenching.

Construction. The steel pot has a sloped back wall, which produces a "bottom heating" effect resulting in better circulation and uniform temperature. This is accomplished by sloping the electrodes as shown in Fig. 1 and 2. As the current passes through the salt between the electrodes, the salt is heated, decreasing its density and causing it to rise toward the bath surface. Control of the rate of rise of the salt is effectively obtained by decreasing the distance from the electrodes to the steel pot. At the lower extremity of the electrode, the current enters the metal pot upon leaving the electrode to follow a shorter path to the other electrode. This arrangement ensures current flow through the salt along the entire electrode length. Due to the close proximity of the lower portion of the electrode to the pot, most of the heating is done in the lower part of the bath. This is the desired method of heating any liquid.

The metal pots are made of either plain steel or hot-dipped aluminized steel depending on the application. Thicknesses range from 12 to 38 mm (½ to 1½ in.). Reinforcing members for light plate, usually angular in shape, are welded around the top. Where

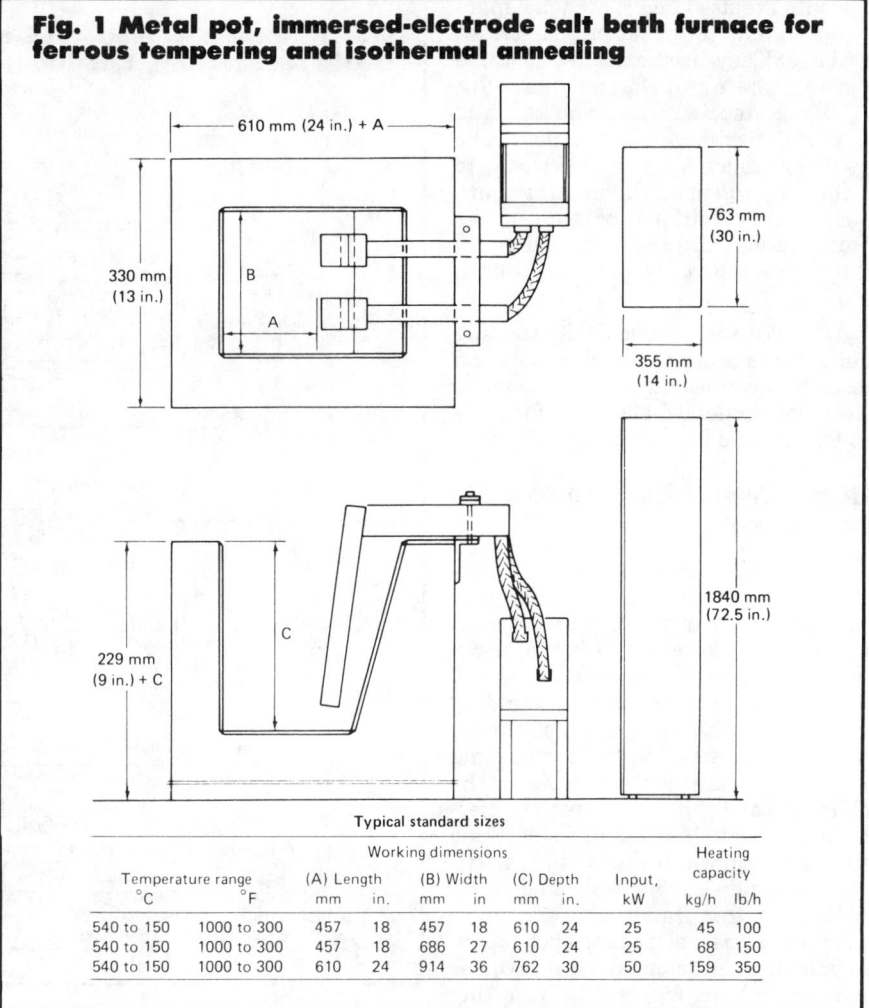

Fig. 1 Metal pot, immersed-electrode salt bath furnace for ferrous tempering and isothermal annealing

Typical standard sizes

Temperature range		Working dimensions						Input, kW	Heating capacity	
		(A) Length		(B) Width		(C) Depth				
°C	°F	mm	in.	mm	in	mm	in.	kW	kg/h	lb/h
540 to 150	1000 to 300	457	18	457	18	610	24	25	45	100
540 to 150	1000 to 300	457	18	686	27	610	24	25	68	150
540 to 150	1000 to 300	610	24	914	36	762	30	50	159	350

depth of the pot so requires, additional members are used at the midsection.

The pot is housed in an insulated 229-mm- (9-in.-) thick wall furnace with either a brick outside wall contained in a rigid welded steel frame or in a steel clad frame, depending on personal preference. In either type of construction, the frame is self supporting on a lattice formed by welding channels or beams to the underside of a steel base plate. The pot is supported on an insulated refractory pedestal.

Electrode Arrangement. Immersed electrodes are made of either mild steel or an alloy "hot" leg welded to a mild steel "cold" leg. As previously mentioned, these are shaped to follow approximately the slope of the pot wall. The portion of the electrode that crosses over the top of the salt bath and is connected to the plant power source is referred to as the "cold" leg. This is welded to the "hot" leg, or the portion of the electrode that is immersed in the

bath, with sufficient weld cross section to provide the necessary current conductor capacity (see Fig. 1). The shanks are drilled and tapped at the tinned terminal connection end for water cooling when necessary. If the latter is not required, the electrical connection is water cooled. Suitable clamping devices are used to facilitate electrode replacement.

- *Single-phase operation:* Several electrode arrangements can be used, depending on the size of the bath. If only two electrodes are required, they are normally positioned on the sloped wall side and at least 127 mm (5 in.) apart. Three electrodes are usually placed so that the center electrode, equal in size to two of the other electrodes, is used as a common conductor with equal current paths to each of the outer electrodes. More than three electrodes would be arranged in multiple groups.
- *Three-phase operation:* Three elec-

trodes are used and spaced in a manner similar to the spacing described above. They are connected to three single-phase transformers that have "Y" connected secondaries and "delta" connected primaries. The current flows from the electrodes to the pot, which is the neutral point. Several variations of three-phase connections are used, depending on the type of furnace and load requirements.

All accessories such as starting units, transformers, sludging tools and secondary connectors are the same for steel pot immersed-electrode furnaces as for ceramic furnaces.

Submerged-Electrode Furnaces

In a typical electrically heated submerged-electrode furnace, the frame is made of heavy angle iron with a steel plate at the base beneath the brickwork. The outer brickwork consists of hollow ceramic tile or common building brick. The salt pot is made of burned alumina firebrick. Castable insulating refractory fills the space between the sidewalls and the ceramic pot. An electrically heated submerged-electrode furnace is shown in Fig. 4d in the article, "Liquid Carburizing and Cyaniding", in this Handbook.

When salt is melted in the pot, it penetrates the refractory until it becomes cool enough to freeze. The resulting shell of solidified salt retains the liquid salt in the furnace. If the refractory develops a crack, bath temperature must be lowered to permit salt to solidify in the crack.

Water-cooled electrodes are in contact with liquid salt in the pot and are sealed in the refractory walls by frozen salt. Current travels between the electrodes, which are flush with the sidewalls. The path of current travel extends a few inches above the top of the electrodes.

Start-up and Shutdown. The submerged-electrode furnace can be started by adding molten salt from another furnace or by using a gas-fired torch or electric starting coil to melt a pool of salt that will wet both electrodes and provide molten salt for the current path. After the current path has been established in the molten salt between the electrodes, salt may be added to bring the bath up to working level. Additional salt will be required to maintain this level because a small

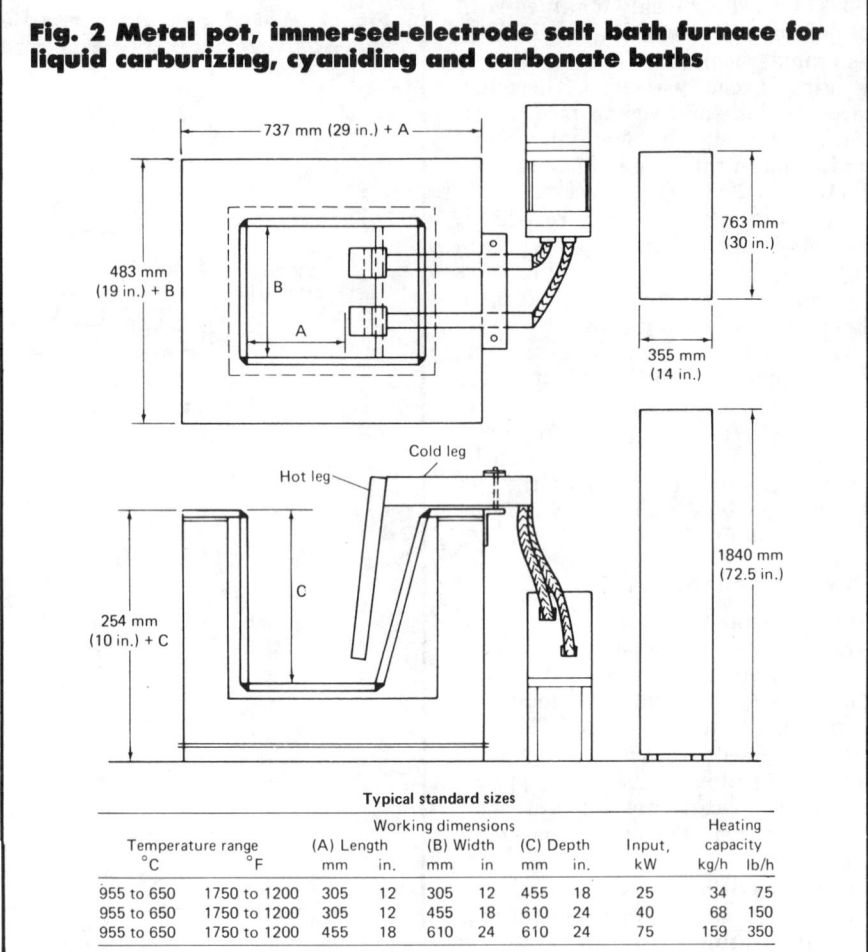

Fig. 2 Metal pot, immersed-electrode salt bath furnace for liquid carburizing, cyaniding and carbonate baths

Temperature range		Working dimensions						Input,	Heating capacity	
		(A) Length		(B) Width		(C) Depth				
°C	°F	mm	in.	mm	in	mm	in.	kW	kg/h	lb/h
955 to 650	1750 to 1200	305	12	305	12	455	18	25	34	75
955 to 650	1750 to 1200	305	12	455	18	610	24	40	68	150
955 to 650	1750 to 1200	455	18	610	24	610	24	75	159	350

amount, ≈5%, will seep into the brickwork and freeze.

If the furnace must be shut down, the molten salt should be bailed from the furnace before it freezes. However, if the salt is allowed to remain in the furnace, a resistance heated starting coil should be submerged in the bottom of the furnace while the salt is still molten. This coil remains in the frozen salt, and it is connected to the transformer leads to start up the furnace.

Advantages and Disadvantages. In common with immersed-electrode furnaces, submerged-electrode furnaces require minimum floor space and maintenance and are highly adaptable to mechanization.

Because the submerged-electrode furnace employs water to cool the electrodes and transformer, it may be operated at 50% overload without overheating the transformer, whereas the immersed-electrode furnace, being air cooled, should not be operated at an overload above 10%.

Because a ceramic pot is used, unexpected pot failure is rare with submerged-electrode furnaces, and the furnaces can be rebuilt on a planned schedule during annual shutdowns. In common with other electrical equipment, submerged-electrode furnaces are at a disadvantage where electric power rates are high, but this can be overcome to some extent by working the furnace in non-peak periods when lower power rates are applicable.

Because of the erosive effects on ceramic pots of water-soluble salts with high sodium carbonate or high sodium cyanide contents, submerged-electrode furnaces can be used only with low-cyanide, low-carbonate salts. Baths with high cyanide or carbonate salt require a modified basic brick. The furnace with modified brick and submerged alloy electrodes provides many years of service in noncyanide and cyanide operations. To increase furnace life, the furnace shown in Fig. 3 is recommended. This furnace has a modified basic

Fig. 3 Electrically heated submerged graphite electrodes

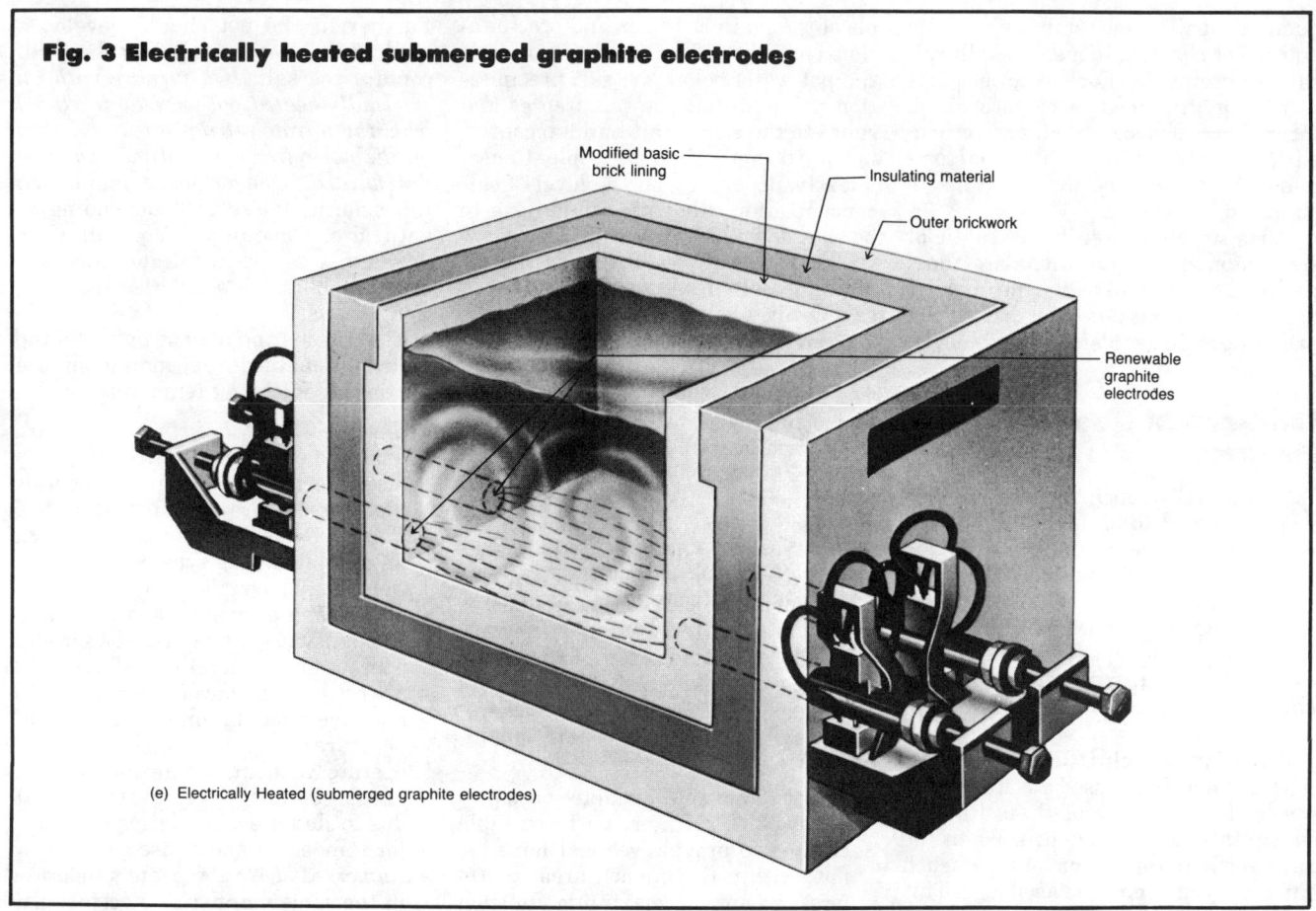

Modified basic brick lining

Insulating material

Outer brickwork

Renewable graphite electrodes

(e) Electrically Heated (submerged graphite electrodes)

brick lining for use with the basic carburizing salts. The alloy electrodes are replaced with continuing graphite electrodes. The electrodes are renewed as they become consumed without disconnecting them or even shutting off the power.

Automatic and Semiautomatic Lines

A fully automatic "jackrabbit" mechanism used for salt bath carburizing and hardening is shown in Fig. 6a in the article "Liquid Carburizing and Cyaniding", in this Handbook. The mechanism has synchronized, continuous chain conveyors that carry the work through the various operations. Work suspended from horizontal bars is moved through baths at the proper speed by a main conveyor. Transfer conveyors carry the work from bath to bath. Completed work is picked up by a third conveyor and is dried by warm air in the enclosed upper portion of the structure while it is being returned to the loading point. One operator loads and unloads the work.

This mechanism can be used only for work having similar requirements. It does not permit varying the time cycle of any one operation without affecting the cycles of the other synchronized operations.

Where part requirements are different, a semiautomatic mechanism may be used. (See Fig. 6b in the article, "Liquid Carburizing and Cyaniding".) Work is transferred between operations by an overhead monorail and is automatically advanced through the carburizing and tempering furnaces by means of a push-pull mechanism. This mechanism consists of two parallel beams with reciprocating push bars. Driven either hydraulically or electrically, the bars carry the dogs, which are spaced to advance the fixtures at the center of the furnace only a part of the stroke, while the end fixtures are advanced through the entire stroke. By closer spacing of the work at the center of the furnace and wider spacing at the ends, high productive capacity is achieved with ease of loading and unloading. A semiautomatic line of this type permits variations in the time cycling of any one operation without affecting other operations, and is less likely to require modification if the work requirements change.

A fully automatic programmable hoist carburizing line is shown in Fig. 6c in the article, "Liquid Carburizing and Cyaniding". The use of automated hoists makes possible the combination of austempering, martempering, and tempering or carburizing in one line. One or more hoists travel simultaneously back and forth automatically advancing the fixture carriers of work through the required stations.

The hoist movement is controlled by a solid state programmable control with functions that would normally require hundreds of relays, counters, switches and extensive wiring. Once programmed, the controller performs the desired commands and functions specified by the user. Time cycles, sequences, drills, and skips are easily entered or changed to meet metallurgical requirements. For instance, parts can be programmed to be carburized, air cooled, washed, rinsed, and returned for unloading. A push-button command then returns the program to standard processing.

Parts suitable for fully automatic or semiautomatic installations are those that can be fixtured by wiring, racking or placing in baskets and that do not present problems in either buoyancy or drainage.

Isothermal Quench Furnaces

Isothermal quench furnace systems were designed to eliminate the occurrence of chloride carry-over from the austenitizing bath to the quench bath, through salt separation and uniform vertical lamellar flow agitation. The three most common approaches to alleviating the salt concentration are chemical, temperature and gravity separation.

Chemical Precipitation. Chemical agents have been used to attempt to lower the solubility of the salts that precipitate into the quench. When the salts settle to the bottom of the quench tank, they are removed as sludge. This method offers little success in that the precipitate that forms is fine-textured and buoyant and thus tends to remain in suspension rather than precipitating out.

Temperature Precipitation. The elimination of carry-over salts has also been attempted by continuously pumping salt through a small auxiliary chamber whose temperature is maintained at a lower level than the main chamber. As the salt is processed through the auxiliary chamber, chlorides are continuously precipitated out.

Although this method appears practical, a fundamental error exists in its application. The salt is cooled by air blown through a space between the pot and outer shell of the precipitation chamber. Air is blown through this space to maintain the temperature lev-

els of the main chamber and precipitation chambers. The moving air cools the pot walls below the salt precipitation point so that the salt freezes and cakes to the sides. Salt buildup continues until the bath is unusable. Consequently, depending on the level of salt concentration, the bath would have to be shut down, possibly after only a few weeks of operation, to remove the remaining molten salt and chip away the caked salt.

Gravity Separation. This system of carry-over salt removal also uses a two-chamber design. The caking problem is eliminated by heavily insulating the pot walls at all points and using an internal air-water heat exchanger. Because the pot walls and salt are at the same temperature, there is no caking action. The chloride salts settle into an easily removable shallow pan at the bottom of the precipitating chamber, or if they are fine textured and buoyant, the salts float to the top of the tanks and are easily skimmed off.

The main advantages of two-chamber gravity equipment separation include:

- Easily removable variable speed propeller-type agitator with suitable baffling to provide vertical lamellar flow within the quench area, therefore, ensuring maximum quench power and minimum distortion
- Separate chloride precipitation chamber with adjustable weirs to maintain a low chloride level and subsequently high quenching power
- Easily removable internal heat exchanger to maintain quench temperature and precipitate chlorides
- Easily removable settling pan to ensure maximum efficiency in removal of chlorides
- Heavily insulated pot and precipitation chamber to eliminate salt caking on walls.

Furnace Heating. Generally, either gas or electricity may be used for heating isothermal quenching furnaces. When gas heating is desired, immersion tubes are recommended because the tubes are usually made of mild steel and provide long service life.

Further, if the pot should develop a leak, the insulation and outer shell will contain the salt. *If a furnace with an externally heated pot were to develop a leak, the nitrate-nitrite salt would drip on the hot refractory and may cause a fire hazard.* One or more immersion tubes normally are used, depending on bath size. Generally, they will have nozzle-mix sealed-in burners and will be available to FM or FIA specification.

Electric heating may be by one of the following methods depending on the maximum operating temperature:

- *Sheathed resistance strip heaters* are mounted externally to the side walls near the bottom. Maximum operating temperature is 425 °C (800 °F). They are easily removable through an insulated plug-type door. Protection against overshooting is achieved by locating a sensing device in close proximity to the heaters. The sensors operate directly on line voltage.
- *Sheathed resistance immersion heaters* have a maximum operating temperature of 425 °C (800 °F). They can operate without a transformer, but are susceptible to premature burnout due to sludge accumulation or operator tampering and abuse.
- *Immersed-electrode heaters* operate in the same manner as electrode pot furnaces for carburizing and tempering.

Furnace Construction. The pot is fabricated from firebox-quality steel plate double welded inside and out and properly supported to maintain its shape. Steel plate offers adequate resistance to chemical attack by the standard alkaline nitrate-nitrite salts at normal austempering and martempering temperatures. The pot is insulated with 102 to 152 mm (4 to 6 in.) of slab-type mineral insulation to prevent the chloride saturated nitrate salt from freezing to the sidewalls or bottom. The insulation is externally contained by a continuously welded outer steel shell. The shell is reinforced to ensure retention of the original shape and dimensions throughout its designed operating temperature range.

Furnaces and Furnace Equipment

Computerized Systems for Heat Treating

By Theodore K. Thomas
Business Unit Advertising Manager
Process Control Division
Honeywell Inc.

COMPUTER MONITORING AND CONTROL of heat treating processes have recently become common practices. Consequently, this article approaches the subject from three different perspectives: an introduction to computer systems and practices that simplifies the computerized approach and compares it with traditional control methods; a glossary of computer terms; and industrial examples of computerized systems utilized effectively in industry.

Although computer technology is ever-changing, several distinct trends are in evidence, including:

- A decrease in the physical size of computer components and equipment
- A decrease in the cost of equipment
- An increase in cost of programming, or software, to run the computer.

As an illustration of the dramatic change in the physical size of computer equipment over the period of its development, a typical contemporary hand-held calculator can be compared with ENIAC, the first successful electronic digital computer. ENIAC, which was built in 1946, occupied several rooms of a large warehouse at the University of Pennsylvania, used 18 000 vacuum tubes and required a power input of 140 000 W, and yet the modern hand-held calculator far surpasses ENIAC in computing power, speed and storage capacity.

The rising cost of energy, coupled with tighter requirements for quality control in the heat treatment of metals, parallel the development of computer technology. Rising energy costs are now offsetting computer costs. It is consequently economically feasible to consider the advantages of a digital computer over conventional instrumentation in the control of furnace process variables. A computerized system provides:

- Multiple-function capabilities
- Increased speed of measurement and response to process changes
- Higher precision of measurement and control
- Greater control flexibility
- Increased accuracy in recordkeeping and reporting
- The ability to communicate electronically between computers and other elements of computerized systems.

The application of computer technology as a means of increasing furnace operation efficiency and cost effectiveness foreshadows a more productive era of heat treating.

In addition to increased production and quality standards, many scheduling problems become manageable with computer control. Greater flexibility in the use of equipment to meet production requirements is realized, as well as the ability to schedule furnace operations effectively so as to capitalize on plant services and utilities. Improvements in furnace loading and unloading are easily matched to manpower availability.

Computerized monitoring leads to more accurate and efficient reporting of problems and irregular conditions that may occur in normal heat treating practice. Timely and effective reporting of alarm conditions adds greatly to energy savings and decreases downtime. The case studies that appear in the applications section of this article describe actual industrial applications in which computers have contributed greatly to the development of more efficient operations.

Computer Basics*

Computers operate by digital methods, rather than by the analog techniques traditionally associated with industrial instrumentation. An analog instrument handles data in terms of electronic or pneumatic signals that are proportional in size to the quantity of the variable being measured, such as temperature or flow. Digital logic electronically expresses the same quanti-

*Adapted with permission from Micro, Mini, and Mainframe Basics, by P. Masucci, *Instruments & Control Systems*, June 1977

ties in terms of binary digits, or bits, each with a value of either 0 or 1. With digital logic, immense amounts of information can be accumulated and processed within limited space to very high levels of accuracy and reliability.

A computer has three main parts: a central processing unit (CPU); a data-storage unit, or memory; and input/output (I/O) equipment. The central processing unit contains an arithmetic unit, where the actual computation takes place; a control unit that dictates to the arithmetic unit which programs are to be executed; and storage registers that accumulate the numbers calculated by the arithmetic unit in accordance with the predetermined or programmed directions of the control unit.

General Computer Classifications

Computers can be generally classified as microcomputers, minicomputers and mainframes.

Microcomputers. A microcomputer contains a microprocessor, which is a central processing unit (CPU) built on a silicon chip smaller than 1 cm². By itself, the microprocessor is of little use. However, when integrated with a memory chip, a clock chip, and provisions for interfacing with a power supply and with input/output (I/O) devices such as keyboards and visual display or readout units, it becomes a microcomputer. Most microcomputers have word lengths of 4 to 8 bits and memories that contain from 256 to 65 535 words, depending on application requirements. At present (1982), some microcomputers can access 8 million words.

At the sensor level, a microprocessor or microcomputer can increase system accuracy and the ease with which signals are transmitted. Whereas analog signals may suffer from variations in amplification or other losses when transmitted over long distances, digital data are transmitted in binary units (1's and 0's), which are less subject to error. A simple microprocessor or an even simpler LSI circuit in a digital converter can gather data with minimum system loss in signal strength and quality, and also can preprocess some of the data before sending it on in binary form. Nevertheless, a microprocessor or microcomputer is "dedicated" to (designed for) a specific task or group of tasks within an application. A digital watch, for example, contains a microcomputer dedicated to computation and

display of clock time; it cannot be used to figure income tax, although it has the computing power to do so. Similarly, a digital temperature controller contains a microcomputer dedicated to measurement, display and control of temperature over a given range, using a specific type of thermocouple or other temperature-sensing device. In an industrial control system such as a fuel-fired furnace, a microcomputer can sample and evaluate temperature at several different points in the furnace, compare the results with acceptable limits, and make necessary adjustments. It also can continuously measure and regulate the flows of fuel and of combustion air to maintain a preset ratio, which yields more energy-efficient control. Conventional analog instrumentation can do all this too, but is less accurate and involves equipment that is physically much larger and that requires more power.

A microcomputer (or a minicomputer) also can integrate several instruments by acting as a data logger, preprocessor or "smart front end" to a larger computer. Whereas large mainframes may have been used a few years ago for more complex control and measurement applications, microcomputers (and minicomputers) have found successful application as such components in larger systems approaches.

Most current applications of computer technology to heat treating involve microcomputers, although minicomputers are used for supervisory control of several furnaces—for example, where each furnace is separately controlled by a microcomputer system.

Minicomputers are distinctly different from microcomputers in two ways: (*a*) they are physically much larger, being mounted in metal racks and cabinets; and (*b*) they can handle several different applications. For example, a minicomputer can measure and control many separate variables of temperature, flow or vacuum simultaneously, and at the same time can perform the supervisory function of adjusting individual control setpoints in accordance with a mathematical "optimizing" model contained in its memory. To accomplish these functions, it requires more data-storage capacity and process-interface capability than a microcomputer, hence its greater physical size.

Minicomputers have word lengths of 12 to 16 bits and a minimum memory size of 4096 (4K) words. Minicomputers

are trending toward word lengths of 16 bits or more, for product speeds (production thruputs) greater than those obtainable with microcomputers.

Mainframes, or large-scale computers, usually have a word length of 32 bits or more and a memory capacity of 16 384 words or more. A large-scale computer usually is used for after-the-fact data processing and analysis. It also can supervise or monitor the operation of a minicomputer, which is, in turn, connected to either a furnace process or its instrumentation. Within this system or network structure, the large computer can either monitor or augment the smaller device's capabilities.

Memory Systems

The memory system is the region of the computer that stores program instructions and data for instant use. Memories are of either the read-only (ROM) or the read/write type. If the data or instructions in a computer's memory do not need to be altered, read-only memories are used, in which case the computer cannot alter the contents. An example of this type of content is temperature-emf (electromotive force) tables for thermocouples, which give the millivolt output generated at a particular temperature. Read/write memories, on the other hand, can be altered and accessed by the computer. Read/write memories are also known as random-access memories (RAM) because any data location within the memory can be accessed, or interfaced with, as easily as any other. Data such as values of furnace temperature sensed by a thermocouple are temporarily stored in this type of memory.

Memories are constructed with either a ferrite core or with semiconductors. Technically, core memory is a matrix whose elements are composed of tiny toroids, or doughnut-shape devices, made of ferrite materials. These elements are magnetized to store information. Semiconductor memory is a memory matrix composed of tiny semiconductor chip circuits. All core and semiconductor read-only memories retain data whether power is off or on. Such is not the case with semiconductor random-access memories. If the power source is interrupted, memory is lost. Random access memories can be protected by connecting batteries to the memory package for "battery backup" in the event of a power failure.

Semiconductor memory is usually faster, more compact and less expensive than core memory. In applications where volatility is a key factor, battery backup can be used. Most computers use a combination of semiconductor and core memory. The type of memory used with a computer depends in part on the application. For example, a microcomputer-based controller with a limited number of functions would probably incorporate a read-only memory as the main portion of memory and a small random-access memory for temporary storage of data.

In control applications, computers must connect either to a binary-type signal (either an open or closed relay), to an analog signal or in some cases, to a digital signal. The connection between the digital computer and the binary circuit may only involve the matching of voltage levels. If an instrument or valve has a digital output, connecting it to the computer again involves only voltage or current matching. If the control has an analog output, the analog signal from the sensor to the computer must be converted to digital form. The interface, in this case, is an analog-to-digital (A/D) converter that is placed between the computer and the analog control device. If the computer must send an analog signal to a control loop, a digital-to-analog (D/A) converter acts as the interface. The digital-to-analog converter constructs an analog signal from the digital information from the computer. Analog-to-digital and digital-to-analog interfaces are integral components of even very small processes because they are usually the only possible communication interfaces between processes and analog control devices.

Software

Software provides the computer with flexibility but also accounts for many of its problems. The problems are not necessarily due to the software, but to human error. These errors result because software usually is not written in English, but in a special language that the computer can understand and translate into the binary language it uses to execute the program commands. Software language may be "machine-like" or "English-like". Machine-like languages are called assembly or *low-level* languages, and English-like languages are called *high-level* or *compiled* languages.

Assembly languages are collections of mnemonics that refer to the exact functions a computer must perform. In

one particular assembly language, for example, CLA stands for "clear the accumulator", which is far easier to work with than the binary equivalent of 111110000000. Assembly languages sometimes are used with microcomputers because they conserve the microprocessor memory. Memory can be the largest hardware expense of a microcomputer system.

Whereas assembly languages are organized to match the way in which computers operate, high-level languages are organized more like spoken languages. High-level programs do not need to specify the individual steps of operations the computer performs; instead, they must only indicate what function to perform (for example, "add two fields").

While assembly languages require a special program to convert user code into binary code, high-level languages require a program to convert the high-level program into binary code. The special program, called either a compiler or an interpreter, occupies a significant portion of the memory system. This increases the size of memory required. Greater memory capacity is also required because high-level languages are not optimized for one specific machine. Consequently, they are not as efficient in their translation as are assembly languages.

Although high-level languages are not designed for specific machines, they are designed for specific functions. Of the popular languages, BASIC and FORTRAN were developed for math and scientific problem solving. FORTRAN, particularly FORTRAN IV, tends to have a greater number of features than BASIC, and thus is generally more difficult to learn. High-level languages are now in very common usage with computers of all sizes (microcomputers, minicomputers and mainframes); however, programs in these languages require greater memory capacity than programs in assembly languages. Most computers also have standard packaged operating systems written in assembly language or high-level language. With operating systems, users need to be concerned only with the application; the operating system handles file management, program location and programmers' "housekeeping" chores.

Computers have almost unlimited use. However, for a computer to provide the advantages of accuracy, flexibility and reliability, a potential user also

must investigate available software, memory requirements, storage, peripherals (typers, plotters and cathode-ray tubes) and, of course, interfaces (analog-to-digital and digital-to-analog converters, etc.) to ensure purchase of a system that is compatible with a broad range of needs, as well as to determine which system will afford the greatest possibility of flexibility and expansion. With the diversity of processor types, peripherals and software, it is possible to apply computer operations effectively to almost any control or instrumentation application.

Applications of Computerized Systems to Heat Treating Processes

Computerized control systems have not been utilized with heat treating furnaces in the past because of the high cost involved. Conventional instrumentation—such as recorder-controllers with thermocouple input, and indicating controllers with special reset options for hi-limit, or excess temperature control—has been much improved by solid-state circuitry, digital setpoint indexing, digital display, and other innovations designed to increase the accuracy, response time and reliability of operation. Computer systems have been viewed as expensive and complicated, especially by furnace builders who sell control packages. Therefore, updated versions of traditional and more familiar measurement and control instrumentation have been favored by both OEM's and users of heat treating furnaces, especially for use with single-chamber furnaces performing one or two operations.

However, in response to increased quality requirements and increased production quantities, in addition to higher energy costs, computerization provides a cost-effective and flexible system for controlling these variables. For example, proper operation of a multichamber vacuum furnace requires complex temperature ramps and soaks, mass flow measurement and control of atmosphere gas admission, varying vacuum levels between chambers, sequences of door interlocks, limit alarms, load and transfer adjustment sequences, and other functions within specific time limits. Under these conditions, the advantages of a digital electronic control system—that is, computerization—not only appear desirable, but are becoming mandatory. Microprocessor-based controllers can perform all of the control functions with electronic speed, a high degree of reliability and push-button ease. The digital-display capability presents setpoint and process-variable readings in a format that is difficult to misinterpret, and provides the flexibility of one- and two-decimal place read-out capacity.

Another advantage of the new digital instrumentation is its compactness, which saves panel space and cuts installation cost. Microprocessor-based controllers and programmers can interface and communicate with a minicomputer for performance monitoring or setpoint supervisory control. All of the functions, or relay logic, for interlock sequences and digital ("on-off") inputs can be provided by digital circuitry. These functions originate in a component called the programmable logic controller (PLC) that is provided as part of the modular hardware in a microprocessor-based furnace controller package.

In more advanced control systems, operation of a multiple-furnace facility can be monitored from one supervisory computer, while each furnace is controlled independently by its own microcomputer. This is the essential concept behind the term "distributed control", which is now widely used in discussions of computerized manufacturing applications.

Computerized Control of Carburizing Furnace Atmospheres*

Gas carburizing consists of treating the steel parts to be carburized at a suitable temperature and for the required length of time in a gaseous atmosphere to bring the carbon to the surface of the steel. Carburizing proceeds by the transfer of carbon to the surface of the steel and by the diffusion of carbon from the surface to the interior of the steel. Closer control over carbon potential is a primary means of attaining more reliable results in the carburizing process. Previously, single-component control systems had provided the only approach to this goal. With recognition of the fact that variations in furnace environment can limit the

*This case study was contributed by B. K. Gupta, F. W. Fraim, and V. Jayarama and is based on the article, How Microprocessors Can Give Close Carbon Potential Control, *Heat Treating*, July 1980. Information is based on the industrial experiences at Thermo Electron Corp., Holcroft Div.

accuracy of single-component systems, efforts were directed toward developing a multicomponent system for carbon control. A microprocessor-based multicomponent analyzing system was developed based on the calculation of actual carbon potential from several measured furnace atmosphere conditions. This system has demonstrated the ability to control and monitor furnace atmospheres as a cost-effective and efficient alternative to traditional instrumentation.

Monitoring and Control Systems. The extent to which various carbon compounds transfer carbon to steel is governed by the carbon potential of the surrounding atmosphere. Carbon potential may be defined as the percent of surface carbon attained by a steel specimen in a state of equilibrium with the surrounding atmosphere. The following reactions occur in a carburizing furnace atmosphere:

$$2\,CO \rightleftharpoons C + CO_2 \qquad \text{(Eq 1)}$$

$$CO + H_2 \rightleftharpoons C + H_2O \qquad \text{(Eq 2)}$$

$$CO \rightleftharpoons C + \tfrac{1}{2}O_2 \qquad \text{(Eq 3)}$$

where C is the carbon in solution in the steel. Each chemical reaction offers a basis for measuring and controlling the carbon potential of the furnace atmosphere.

CO_2-Base Control Systems. Equation 4 may be derived from Eq 1:

$$a_c = K_1 \cdot \frac{(P_{CO})^2}{P_{CO_2}} \qquad \text{(Eq 4)}$$

where a_c is activity of carbon; K_1 is an equilibrium constant, P_{CO} is partial pressure of CO in the atmosphere and P_{CO_2} is partial pressure of CO_2 in the atmosphere. Activity of carbon (a_c) is related to the carbon potential; K_1 is dependent only on temperature. If temperature and P_{CO} are constant, then the carbon potential can be measured and controlled on the basis of P_{CO_2} alone.

Dew Point–Base Control Systems. From Eq 2, Eq 5 can be derived:

$$a_c = K_2 \cdot \frac{P_{CO} \cdot P_{H_2}}{P_{H_2O}} \qquad \text{(Eq 5)}$$

If temperature, P_{CO} and P_{H_2} are constant, carbon potential can be measured and controlled on the basis of P_{H_2O} alone.

O_2-Base Control Systems. From Eq 3, Eq 6 can be derived:

$$a_c = K_3 \cdot \frac{P_{CO}}{(P_{O_2})^{1/2}} \qquad \text{(Eq 6)}$$

If temperature and P_{CO} are constant, carbon potential can be measured and controlled on the basis of P_{O_2} alone.

System Limitations. Each of the monitoring/control systems described above are based on the assumption that temperature remains constant. However, this is never the case in an actual production furnace. All of these systems also assume that a constant partial pressure of CO, or CO and H_2, exists in the atmosphere, which does not occur in actual practice. Errors result because of variations in temperature and in the partial pressures of CO and H_2. Figure 1 shows the errors in carbon potential that can result in a CO_2-base system due to such variations. Any system that could accommodate for such variations and make the necessary adjustments expediently would be more accurate in monitoring and controlling carburizing furnace atmospheres.

The Computerized Approach. The microprocessor-base multicomponent system operates on the reactions given in Eq 1 and 4, and it measures both CO and CO_2. In addition, it also measures furnace temperature. Consequently, all the variables affecting carbon potential are monitored. Having determined these variables, the computer control system, through its own internal microprocessor, computes the carbon potential according to:

$$\text{Carbon potential} = A + \frac{B}{T} \log a_c \qquad \text{(Eq 7)}$$

where A and B are constants, T is reaction (part) temperature and a_c is activity of carbon.

The computerized system also has the additional capability of measuring CH_4, although CH_4 is not used in the computation of carbon potential. The capability to measure CH_4 and limit its concentration in the atmosphere (through the use of a methane limit alarm) is necessary to maintain equilibrium conditions in the furnace and to minimize soot formation.

System Components. The multicomponent system is compact in size. The major components of the system are

- Sample input
- Calibration (zero or span) gas input
- Infrared system
- Thermocouple input
- Control switches (pushbutton or thumbwheel)

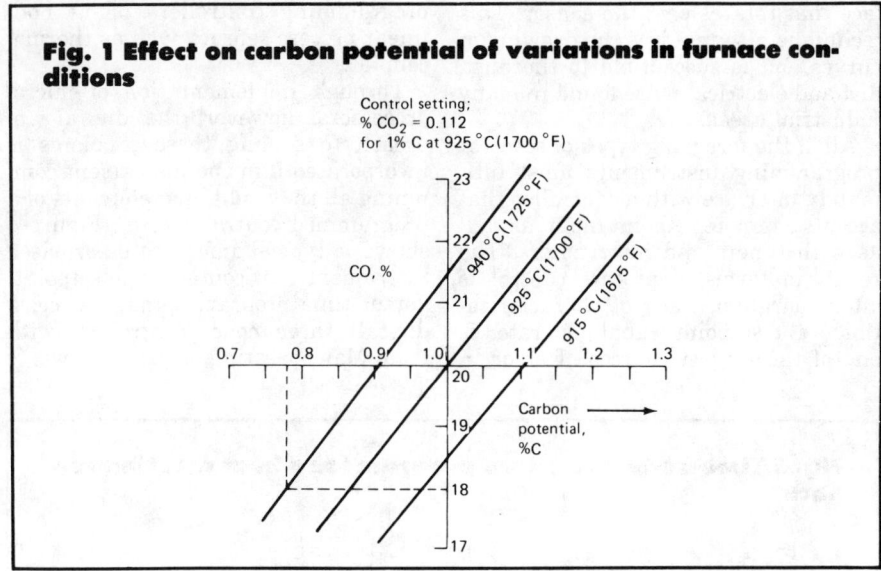

Fig. 1 Effect on carbon potential of variations in furnace conditions

Control setting;
% CO_2 = 0.112
for 1% C at 925 °C (1700 °F)

CO, %

Carbon potential, %C

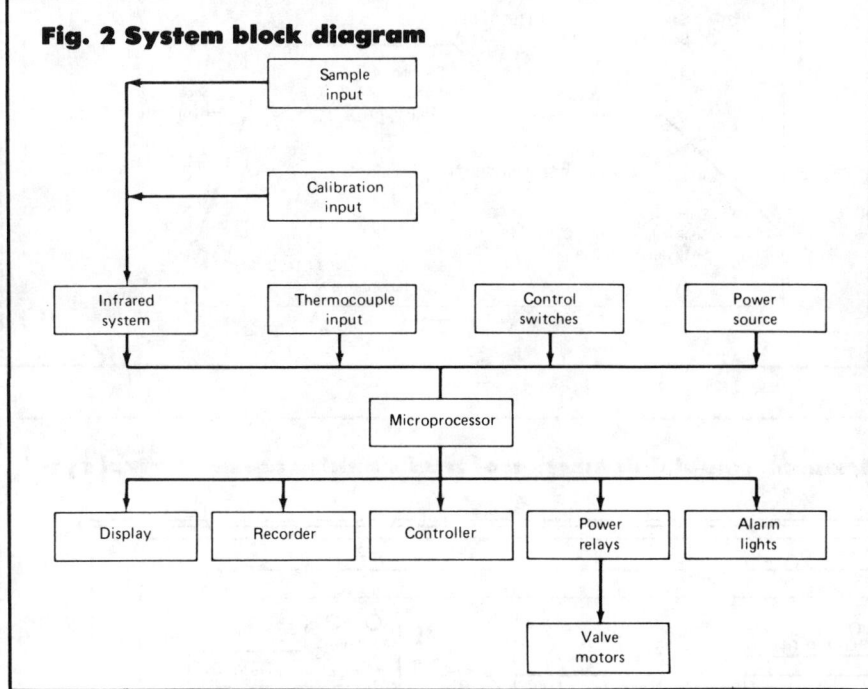

Fig. 2 System block diagram

- Sample input
- Calibration input
- Infrared system
- Thermocouple input
- Control switches
- Power source
- Microprocessor
- Display
- Recorder
- Controller
- Power relays
- Alarm lights
- Valve motors

- Power source and conditioning system
- Microprocessor
- Gas charge display
- Strip chart recorder
- Proportioning controller
- Power relays to operate valve motors
- Alarm lights.

The alarm lights are activated by automatic checks of the system to report conditions such as clogged filters or plugged sample lines. Figure 2 shows a schematic diagram of a multicomponent system. Flowmeters for the sample lines are connected to the compartments; one filter provides purge air and the other is for the sample going to the infrared analyzer system.

The computerized system is capable of monitoring and controlling six sample points, or zones of control. The unit gives the user a choice of basing the control on carbon potential, CO_2 or CH_4. For example, one or more sample points could control the atmosphere generator gas based on CO_2, while other points could control the furnace zones based on carbon potential or CH_4. All zones may be put on control, if desired. Some zones either may be on monitor or may be shut off. A choice of displays exists. The normal display shows CO, CO_2, CH_4, temperature and carbon potential. Displays may also include information on whether a zone is being monitored or controlled, and, if controlled, on what basis. A variety of other displays can be selected through use of the thumbwheel switches. A 24-point strip chart recorder shows CO, CO_2, temperature and carbon potential for each of the 6 zones of control. CH_4 is normally only on the display.

Infrared Analyzer. The infrared chamber is sealed and is continuously purged with clean, dry air. The drying is done by two heatless, regenerative-type, molecular sieve dryers, which also remove the CO_2 from the air. Controlled temperature and pressure conditions are maintained in the flow cell. The flow through the cell is also maintained at a constant level by holding a constant pressure drop across a fixed orifice; atmospheric pressure changes do not affect flow. In addition, the entire compartment is maintained at a constant temperature. These features provide the environmental stability that allows calibration adjustments to be kept to an absolute minimum.

Application of a Microprocessor-Based Control Programmer to Control of Time-vs.-Temperature Program*

Figure 3 shows a typical process profile for a heat treating furnace in which furnace temperature is controlled as a function of time in order to achieve the desired metallurgical conditioning of parts. This same type of program, however, could apply to applications for environmental chambers, weld stress relieving, or any application where process variables must be controlled as a function of time. To achieve this kind of programmed control, a separate set-point programming source, a controller, and several interrupters and timers are needed.

The types of programming sources available are limited in their ability to provide the degree of resolution and setpoint accuracies required to achieve

*This case study was provided by Theodore K. Thomas and is based on industrial experiences at Honeywell Process Control Div.

process control that consistently produces high-quality products.

The Cam Programmer has been used extensively in the heat treating industry for many years. This instrument utilizes a rotating plexiglas disk or cam, cut to conform to a specific program, or sequence of temperature rise, level and fall. The cam moves a mechanical follower arm that is spring-loaded against it. The arm in turn adjusts the setpoint of the furnace-temperature controller to maintain the program. The Cam Programmer has several inherent limitations. The most difficult problem that arises immediately is cutting the cam to the exact program desired. This is a time consuming task. Also, rates of rise (rates at which furnace temperature is increased or decreased) are restricted due to the inability of the mechanical follower arm to follow steep cam contours.

Another method of generating setpoint-versus-time profiles involves use of a photoelectric sensor that electronically follows a curve edge drawn on a rotating disk. Drawing the curve for each program is tedious, and dirt or dust in the operating area could ultimately blind the sensor at some critical point in the process.

Capacitive sensors have also been used in generating setpoint-versus-time profiles. These electronic sensing devices follow a line scribed on a surface that rotates past the sensor. This technique also requires the drawing of curves and is susceptible to the dust, dirt and electrical noise found in many industrial areas.

All of the foregoing types of setpoint programming instruments must ultimately interface with a controller that accepts a remote setpoint input, as well as with timers and interrupters. The result in terms of process control is often marginal. Lack of characterization of the setpoint signal generated is one of the largest sources of error in programming controllers using non-linear process sensors such as thermocouples.

Through implementation of microprocessors, however, the digital approach to solving these problems is incorporated into one instrument combining all the traditional elements of a programmed control system. Figure 4 shows a typical microprocessor-based instrument that contains the setpoint-versus-time program signal source, a digital three-mode controller with Auto-Man operating modes, and twelve

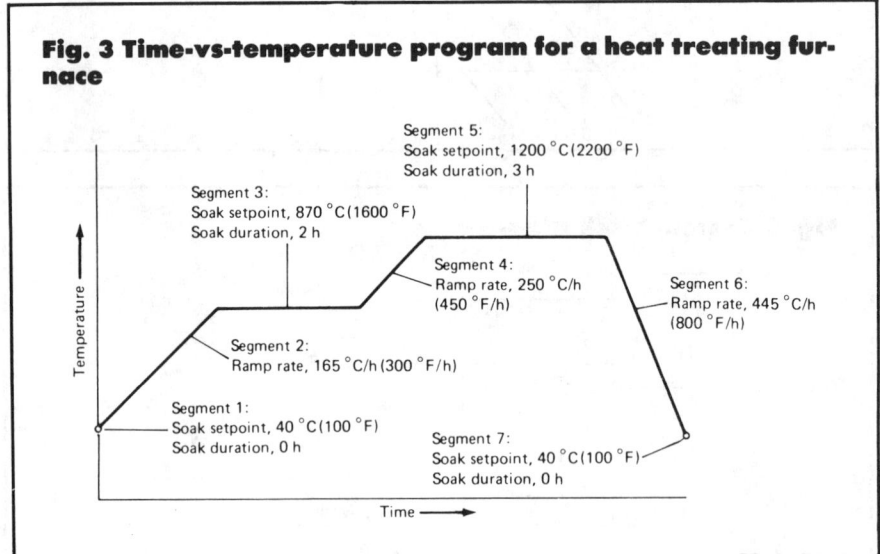

Fig. 3 Time-vs-temperature program for a heat treating furnace

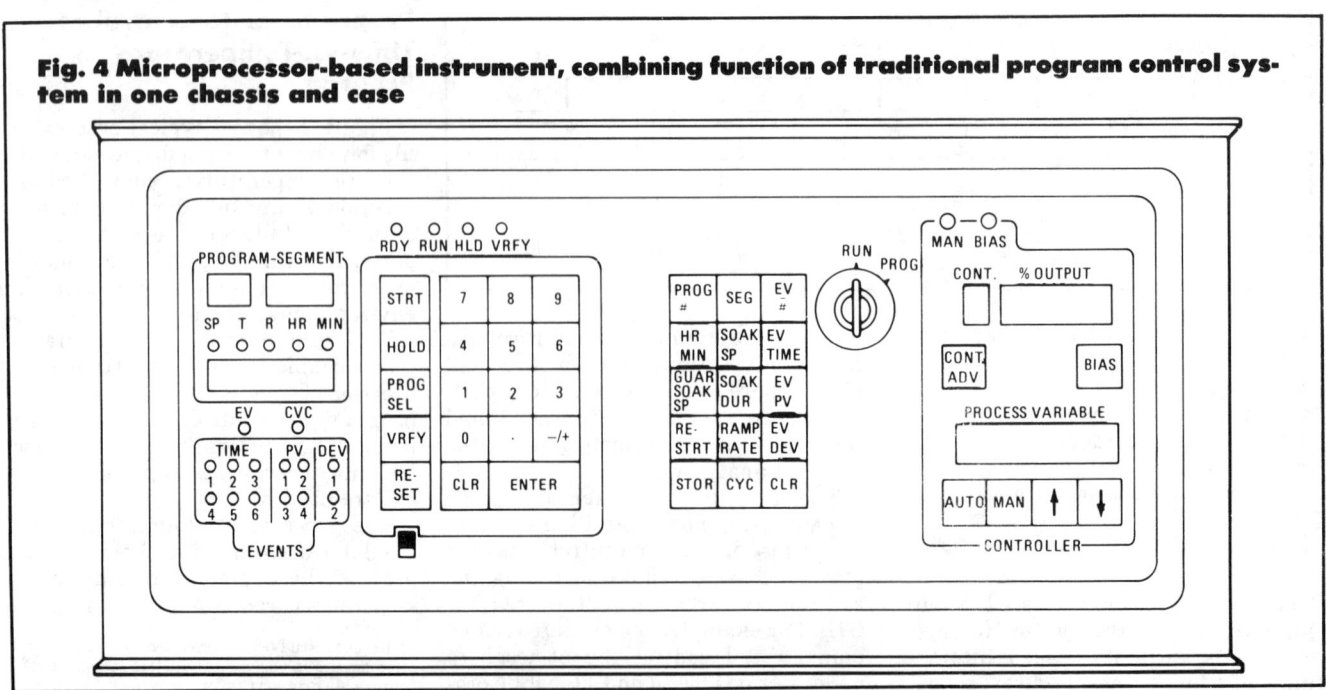

Fig. 4 Microprocessor-based instrument, combining function of traditional program control system in one chassis and case

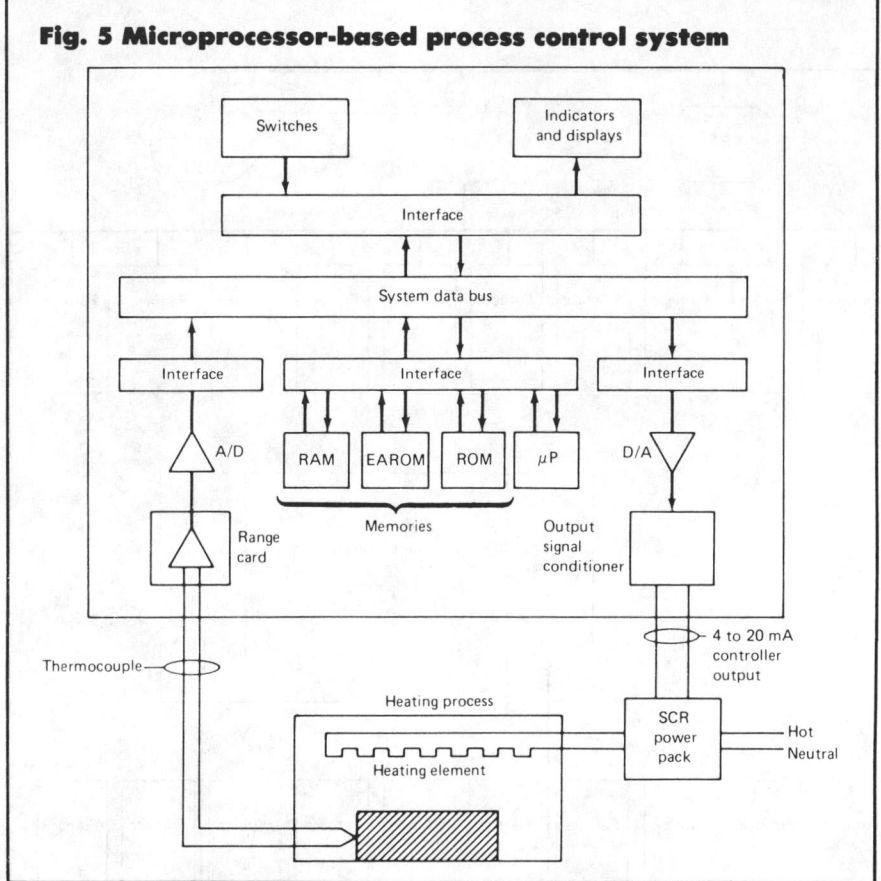

Fig. 5 Microprocessor-based process control system

A simplified description of the operation of the controller shown in Fig. 5 is as follows:

- The millivolt output from the thermocouple is conditioned and amplified in the range card to give an output of 0 to 2 Vdc.
- The analog 0-to-2-V signal is then changed to a 12-bit digital signal by the analog-to-digital converter.
- Every 300 ms, the microprocessor samples the output of the analog-to-digital converter and stores it in the RAM.
- The data in the RAM are compared with the thermocouple-versus-emf data in the ROM to determine the actual temperature being sensed by the thermocouple.
- The actual temperature is then sent to the operator's display and is also compared with the setpoint being generated by the program stored in the EAROM. The comparison and calculation of the control algorithm actually takes place in the RAM, which receives data from the ROM and EAROM and then stores the results in the EAROM.
- Every 300 ms, the microprocessor samples the results of the control algorithm and sends it to a 10-bit digital-to-analog converter whose output is 4 to 20 mA.
- The output can then be used directly to control the process or may be converted to either a time-proportioning or a position-proportioning control form if desired.

Use of Minicomputer for Job Scheduling, Temperature Control and Furnace-Pressure Control*

Production of parts for the aerospace industry includes a heat treating step after forging. Part failure could contribute to loss of life; consequently, it is extremely important that parts be heat treated properly and that accurate records be maintained concerning process control.

The critical close-tolerance heat treating is done in box furnaces, which are designed and operated to provide a high degree of temperature-uniformity. These furnaces are located in one building where two rows of furnaces face each other with oil and water quench tanks between them, as shown in Fig.

*This case study is based on industrial experiences at the Ladish Co.

programmable event switches. All operating data (setpoint, process variable, ramp rates, percent controller output) are displayed continuously by a seven-segment digital readout. Light-emitting-diode (LED) indicators show operating mode and event-switch status. Programs are entered through the use of a front panel keyboard. These programs are then permanently stored in nonvolatile memory which requires no battery backup. If a power loss occurs, the stored programs will not be lost; reprogramming is consequently not a requirement.

Figure 5 shows a process-control loop in which a microprocessor-based instrument is used to measure, indicate and control the temperature according to a preset schedule or program. Components required to complete the system are a temperature sensor and the final control device.

Three types of memory devices are depicted in the instrument shown in Fig. 5. The read-only memory (ROM) is the permanent memory; all of its information is "burned in" by the device

manufacturer and cannot be changed. The other two types, commonly referred to as random-access memory (RAM) and electrically alterable read-only memory (EAROM), are not permanent memories. Their information is changed, as required, when the system is operating. The main difference between the two temporary types of memories is that the RAM memory is volatile (that is, it loses its memory whenever the power is turned off or fails), whereas the EAROM type is non-volatile and does not require a battery backup to retain its memory when power is lost.

The less-expensive RAM type is used as a "scratch pad" to make temporary calculations before storing data that must be retained in EAROM. Data that must be retained include such items as:

- Temperature-versus-time programs
- Tuning constants for the algorithm
- Machine-state data, such as the current setpoint, controller output, and elapsed times.

Fig. 6 Furnace area layout

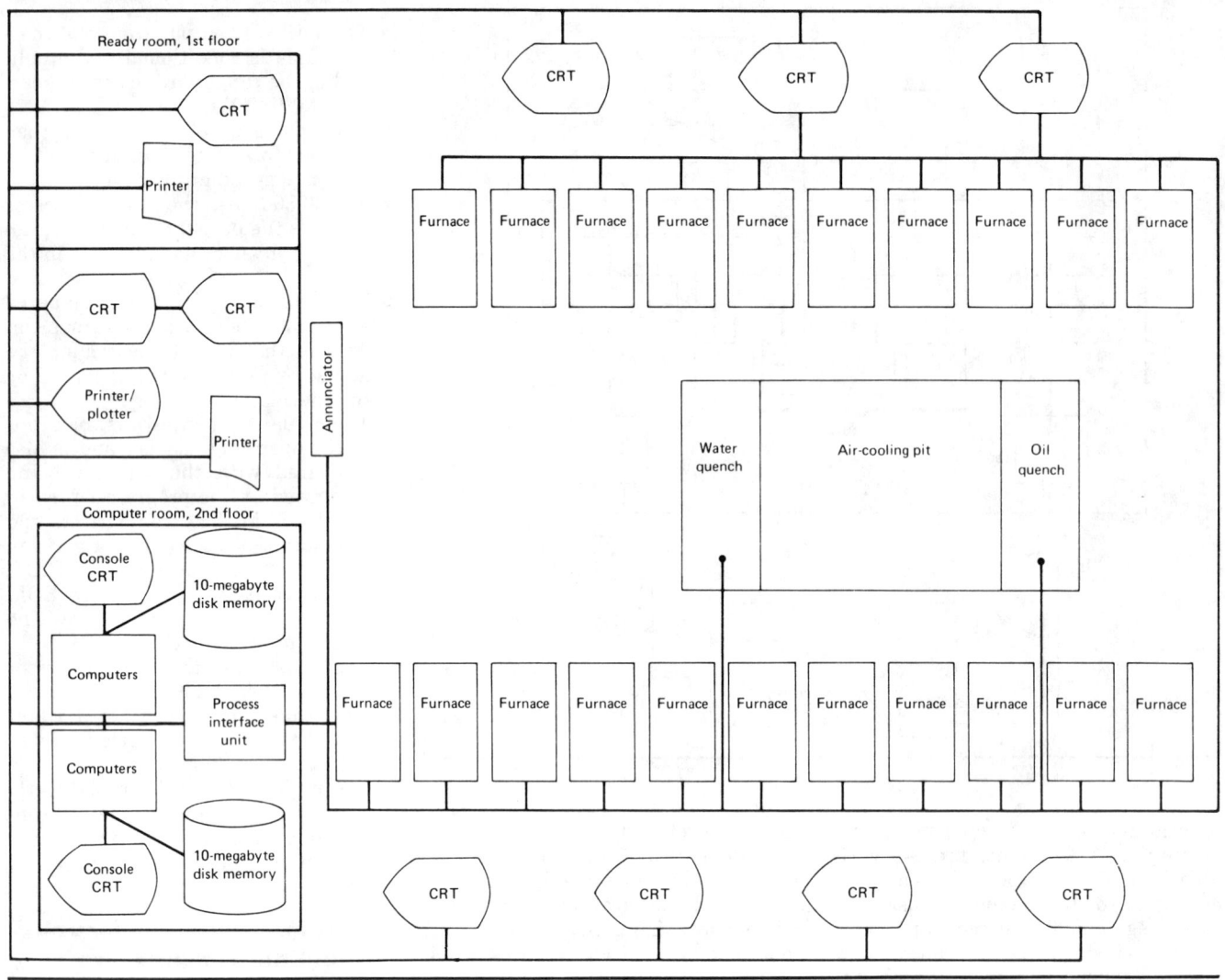

6. The furnaces are fired with excess air to meet temperature-uniformity requirements. There are five furnace thermocouples, one for control and four to indicate temperature uniformity, located in each furnace. In addition to these thermocouples, critical loads are thermocoupled in as many as nine locations and are monitored to ensure proper heat treatment. These furnaces are surveyed on a scheduled basis and are certified for very close temperature uniformity.

Computer Control. The traditional control system consisted of a controller, a circular chart recorder, and a four-point strip chart recorder for each furnace. An additional multipoint recorder was connected to the load ther-mocouples for critical loads. The increasing requirements for closer temperature controls, and improved recordkeeping, as well as the requirement of close tolerance quality, were the main reasons for computer control.

The main function of the computer is to provide temperature control and temperature records for the furnaces by monitoring up to 14 thermocouples in each furnace. The computer automatically adjusts the thermocouple readings for the deviation of the thermocouple wire so that all readings denote actual temperature, in either degrees Celsius or degrees Fahrenheit. The computer control system can be programmed to perform numerous functions, with several levels of response for each function. In a real-time system, the computer responds to events as they occur, with a predetermined set of priorities. If an event occurs that has top priority, such as furnace over-temperature, the computer will respond to that event first and then proceed to complete remaining lower priority commands.

System Operation. In setting up a load for heat treating, the operator enters data concerning the load, such as part and serial numbers, and information concerning the heat treating cycle, such as type of cycle, control temperature(s), heat-up rate, and length of cycle, on a cathode ray tube (CRT) terminal. Once an operator assigns a load to a furnace and instructs the computer that the furnace has been loaded, the computer executes the heat treating cycle. The system is flexible enough to allow operator changes (input) during the heat treating cycle.

The computer is programmed with several control schemes for the heat-up and hold portions of the heat treating cycle. The control scheme is automatically selected by the computer, based on information the operator has punched in at the CRT concerning the load. If an alarm condition occurs during the cycle, the computer will automatically adjust its control scheme until the problem has been corrected.

At periodic intervals, the computer will monitor and store the maximum and minimum readings of each thermocouple during the time interval. Alarm reports are also stored. At the end of each shift, all information entered by an operator concerning a load, as well as all temperature and alarm data, is transferred from disk to magnetic tape for storage as a permanent record. This data can be recalled at any time by replaying the tape.

At the end of each shift, the computer prints out a shift report showing what was heat treated during that shift. It can also generate other outputs upon operator request (such as any or all thermocouple readings, or expected completion time). Temperature-versus-time curves can be obtained using the graphics terminal and plotter. Operators are able to plot either the high and low readings for each thermocouple or the average of these readings, using an ordinate or abscissa of their choice.

Advantages of the System. Although the main purposes of the computer system are control of furnace temperature and keeping of temperature data, the system is capable of performing other functions, such as monitoring of furnace operating hours and the time until the next temperature-uniformity survey is due. The computer is also programmed to read gas meters in the plant. Through installation of additional furnace hardware, and input/output boards, the computer is able to control furnace pressure and to limit the amount of excess combustion air to the amount needed to maintain temperature uniformity. The software necessary for the computer to perform these functions is inserted in the computer before it is brought on line; however, it may be changed as heat treating needs and experience dictate.

In addition to providing improved temperature control and recordkeeping, the computer system also contributes to increased production. Many functions that are normally done at the discretion of the furnace operator are now done automatically. This in turn leads to increased energy savings.

Computer-Assisted Management System*

A commercial heat treating supplier is continuously faced with customer requests for different part configurations, use of different materials, and varying specifications, as well as close scheduling of jobs and accurate prediction of job-completion dates. The customer not only expects prompt delivery, but also high-quality parts.

One alternative to controlling costs and meeting customer service requirements is the computerized approach. The system chosen was patterned after the airline reservation system, because the goals and problems—finite capacity, generally unpredictable demand, and the need for fast delivery—were parallel. Expansion of the system to additional facilities is simple. Initial planning included the purchase of a mainframe with capabilities for handling many facilities in widely separated locations. To computerize a new plant, only a cathode ray tube, a printer and a phone line are needed to begin system operation.

A batch-type computer was in use to provide accounting, financial and managerial reports and analyses. The computer was an accepted, proven tool. The changeover from batch operations to a continuous on-line system required more modern computer hardware and a systems analyst familiar with on-line systems. Once the old accounting, financial and managerial programs were converted to the new computer, parallel systems were run for about two months. With the conversion completed, work efficiency was increased and the new computer used less space.

The on-line system was created to improve internal control of jobs progressing through the plant, to simplify paperwork, and to provide customers with fast, accurate information concerning order status. Status of jobs is maintained by heat treaters on the

*This case study was contributed by John D. Hubbard and is adapted from the article, Computer System at Hinderliter Heat Treating, *Metal Progress*, July 1980.

floor. A master job card travels with each batch of parts as it moves from work station to work station. Production information is entered on the master job card as a processing step is completed and is entered directly into the computer 24 h a day, 7 days a week.

When a customer inquires about a job status, the account number is keyed into the computer and every job the customer has in-house, including job number, purchase order number, number of pieces, weight, description, and date received, is displayed on the screen. Once the job in question is identified, the computer displays the shop routing in line-by-line detail of all operations the parts must proceed through to completion. Each individual furnace load is keyed to the actual furnace where it is being run or the exact location of that load, such as furnace number and receiving, inspection and shipping dates. From this information, the customer may be provided fast (2 min average) delivery information.

In addition to prompt and accurate customer service, shop routers are standardized; routers can be duplicated or custom built automatically. Invoices, acknowledgments and shipper forms are generated automatically, as well as special reports.

Glossary of Computer Terms

Access time: Amount of time required to access the contents of a memory location. This limitation is imposed by the speed of the memory circuitry.

Algorithm: Procedure used for performing a task

Backplane: Circuitry and mechanical elements used to connect the boards of a system

BASIC: Beginner's All-purpose Symbolic Instruction Code, oriented toward beginners rather than experienced programmers. Numerous incompatible versions exist.

Binary: Numbering system consisting of only 2 digits, either 0 or 1, as contrasted with the 10 digits, 0 to 9, of the decimal system. In electronics, the terms "binary", "two-state", and "digital" are synonymous.

BCD: Binary Coded Decimal, coding system in which each decimal digit from 0 to 9 is represented by 4 binary digits (bits)

Decimal digit	Binary code
0	0000
1	0001
2	0010
3	0011
4	0100
5	0101
6	0110
7	0111
8	1000
9	1001

Bit: Single binary digit. A bit may have two states; normally, a bit is considered "high" when its value is 1 and "low" when its value is 0.

Board: Card that contains circuitry for one or more specific functions, such as memory or interfacing

Bus: Circuitry in a backplane that allows transmission of electrical signals from one board to another

Byte: Group of 8 bits that are treated as a unit

Card frame: Enclosure that holds a system's boards in place

Chip: Integrated circuit

Clock: Device that generates electronic timing signals. Clock signals are often used to synchronize certain system operations.

COBOL: Common Business-Oriented Language, used primarily in business applications

Core memory: Type of memory that stores information on magnetically charged, doughnut-shape cores made of ferrite and lithium. Core memories have largely been superseded by semiconductor memories.

CPU: Central Processing Unit, the primary component of all computer systems. It is responsible for controlling system operations, as directed by the program it is executing.

Dedicated device: Device that is used exclusively for one function

DMA: Direct Memory Access, an arrangement where blocks of data can be transferred between main memory and a peripheral device (such as a disk drive) without processor intervention

EAROM: Electrically Alterable Read-Only Memory, a type of memory that combines the characteristics of random access memory (RAM) and read-only memory (ROM). It is nonvolatile (like read-only memory) but can be written into the processor (like RAM). EAROM, however, has a substantially longer writing time (currently about 2 ms versus 400 ns), as well as a limited number of writes (about 1 000 000) be-

fore the chip can no longer be reprogrammed.

EROM: Erasable Read-Only Memory, read-only memory that can be erased and reprogrammed. EROM is frequently erased through exposure to ultraviolet light. Also spelled EPROM.

Firmware: Part of a computer program that is incorporated, at least temporarily, as machine hardware—for example, instructions contained in a ROM

Fixed-point arithmetic: Arithmetic where the decimal point always remains at a predetermined position. Integer arithmetic is a type of fixed-point arithmetic, because the decimal point is always to the right of the mantissa.

Flag: Bit whose state signifies whether a certain condition has occurred

Flip-flop: Circuit that changes its logical state when signaled to do so by another device

Floating-point arithmetic: Arithmetic where the decimal point may occupy any position

Floppy disk: Component similar to a 45-rpm record made of flexible material and used for storing computer data

FORTRAN: Formula Translator, first high-level language. Emphasizes algebraic operations. Used primarily in scientific applications.

Gate: Circuit that performs a Boolean logic operation

General-purpose digital computer: A digital computer designed to solve a large variety of problems; that is, a computer that can be adapted to a large class of applications (as opposed to a computer that might be designed specifically to control a manufacturing process). A typical general-purpose digital computer consists of: (a) input/output (I/O) devices, which permit communications with the outside world; (b) memory, which stores data and instructions; and (c) central processing unit (CPU), which performs the arithmetic and data processing operations and provides the control that ties all of the subsystems together so that they operate in a fully automated manner.

Hard-wired logic: Group of solid-state logic modules mounted on one or more circuit boards and interconnected by electrical wiring. Logic control functions are determined by the way in which the modules are interconnected, as contrasted with a programmable controller or microprocessor, in which the logic is in program form.

Instruction: Group of bits that defines

a computer operation. An instruction may move data, do arithmetic and logic functions, control input/output devices, (typers, plotters or CRT's) or make decisions as to which instruction to execute next.

Intelligent device: Device that contains its own processor.

LSI: Large-Scale Integration, class of integrated circuits that contain the largest number of functions per chip. Microprocessors are LSI devices.

LED: Light-Emitting Diode, type of digital output display that is frequently used in calculators

Light pen: Input device used in conjunction with a video display. When the user touches the display screen with the light pen, the electronics associated with the pen will determine the coordinates of the point that the user touched. These coordinates will then be transmitted to the computer.

Line printer: Output device that prints an entire line of information at a time

Machine language: Binary code that can be directly executed by the processor, as opposed to assembly or high-level language

Mag tape: Magnetic tape, similar to that used by audio tape recorders, on which information can be stored in a computer-readable format

Mainframe: The computer itself, including the processor, main memory, input/output interfaces, and backplane

Main memory: Memory that the processor accesses directly, as opposed to peripherals such as disk and tape devices

Mass storage: Auxiliary or bulk memory, as opposed to main memory. Disk drives and tape drives are common mass storage devices.

Memory: Memory devices provide temporary and permanent storage of information. Permanent memories (ROM) instruct the microprocessor as to which logic operations to perform. Temporary memories contain information that the operator controls and has access to change.

Memory capacity: Number of bits that a memory can hold; for example, a 1K semiconductor memory can store 1000 bits (actually 1024 bits), and a 2K semiconductor memory can store 2000 bits (actually 2048 bits). Fixed memories usually contain instructions, and therefore their capacity is sometimes expressed as the number of words of a certain length that it can hold. For example, "256 × 4" means that the memory can store 256 4-bit words, which makes it a 1K memory (1024 bits). The same 1K memory could be a 128 × 8 memory. In either case, however, a 1K memory is purchases from the manufacturer, not a 256 × 4 or a 128 × 8 memory.

Microcomputer: Computer whose major sections—central processing unit (CPU), control, timing, and memory—are each contained on a single integrated (IC) chip, or at most, a few chips. In other words, a large-scale integrated (LSI) computer.

Microprocessor: Large-scale integrated (LSI) device that performs the functions of the central processing unit of a computer. It is called a microprocessor because of its extremely small size. Typically, it is contained on a single integrated chip (IC). In some cases, the microprocessor is made up of two, three or even more chips.

Minicomputer: Broad term describing any general-purpose digital computer in the low-to-moderate price range

Motherboard: Synonym for backplane

MTBF: Mean Time Between Failures, length of time for which a device can reasonably be expected to operate without malfunction

Multiplex: To combine two or more electrical signals into a single, composite signal. This may be done on a frequency basis (frequency-division multiplexing) or on a time basis (time-division multiplexing).

Off-line: Device that is not connected directly to its host computer. A keypunch is an example of an off-line device.

On-line: Device that is connected directly to its host computer

Parity check: An error-detection system in which an additional bit, the parity bit, is appended to each word or byte. Under even parity, the parity bit is 1 if there is an even number of 1's in the rest of the word. Under odd parity, the bit is 1 if there is an odd number of 1's in the rest of the word.

PASCAL: Advanced programming language, not an acronym

Peripheral: Unit, such as a communications terminal that is external to the system processor

Plotter: Hard-copy device that produces line drawings such as X/Y graphs. The coordinates of the points or lines to be plotted are normally supplied by the computer.

Port: Communication channel between a computer and another device

PCB: Printed Circuit Board, circuit board whose electrical connections are made through conductive material that is contained on the board itself, rather than with individual wires

PLA: Programmable Logic Array, device (usually an integrated circuit) containing a set of logic gates whose interconnections may be programmed

PROM: Programmable Read-Only Memory, type of read-only memory that can be programmed by the user. This programming usually requires special equipment.

RAM: Random-Access Memory, a read/write memory. A more strict definition of a RAM is a memory that stores information in such a way that each bit of information may be retrieved within the same amount of time as any other bit, as opposed to serial memory.

ROM: Read-Only Memory, memory in which information is stored permanently, such as a math function or a microprogram. An ROM is programmed according to the user's requirements during memory fabrication and cannot be reprogrammed.

Read/write memory: Memory whose contents can be continuously changed quickly and easily during system operation. It differs from a read-only memory (ROM), whose contents are fixed and not subject to change, and a reprogrammable ROM, whose contents can be changed but only periodically. A RAM is a read/write memory.

Real time: Pertains to the performance of a computation during the actual time that the related physical process occurs so that results of the computation can be used to guide the physical process

Register: Fast-access circuit used to store bits or words in a central processing unit (CPU). Registers play a key role in CPU operations. In most applications, the efficiency of the program is related to the number of registers.

Run time: Time at which the program is executed. Also, the amount of time required to execute the program.

Second source: Manufacturer who produces a product that is interchangeable with the product of another manufacturer

Semiconductor: Device (or material) with an electrical conductivity that lies between those of metal conductors and those of insulators. Integrated circuits, transistors and diodes are the most common semiconductor devices.

Semiconductor memory: Memory in which semiconductors are used as the storage elements, and characterized by low-to-moderate-cost storage and a wide range of memory operating speed, from very fast to relatively slow. Almost all semiconductor memories are volatile.

Serial memory: Memory whose contained data is accessible only in a fixed order, beginning at some prescribed reference point. Data in any particular location is not available until all data ahead of that location have been read. Such a memory is inherently slow compared with a random access memory (RAM).

Software: Coded instructions that direct the operation of a computer. A set of such instructions for accomplishing a particular task is called a program.

Solid state: Silicon or germanium semiconductor device, such as a diode, transistor or integrated circuit. May also refer to circuits, equipment or systems made from such devices.

Synchronous communication: Data transmission where the bits are transmitted at a fixed rate. The transmitter and receiver both use the same clock signals for synchronization.

Terminal: Device for communication with a computer. A typical terminal consists of a keyboard and a printer or video display.

Thermal printer: Hard-copy device that produces output on heat-sensitive paper

Volatile memory: Memory whose contents are irretrievably lost when operating power is removed. Practically all semiconductor memories are volatile.

Word length: Number of bits in a computer word. The longer the word length, the greater the precision (number of significant digits).

Furnace Control Instrumentation

Temperature Control

TEMPERATURE INSTRUMENTATION AND CONTROL SYSTEMS used in heat treating include temperature sensors, controllers, final control elements, measurement instruments and set-point programmers. A basic control loop includes a temperature sensor, a controller and a final control element.

BASIC CONTROL LOOP AND AUXILIARY DEVICES

The basic control loop is illustrated schematically in Fig. 1. Auxiliary devices used with this basic control loop include measurement instruments and set-point programmers (Fig. 2). The measurement instrument monitors the same temperature sensor as that used by the controller. The set-point programmer automatically varies the controller set point to provide a temperature cycle or temperature program in accordance with an established plan.

TEMPERATURE SENSORS

Thermocouples and resistance temperature detectors (also known as resistance thermometers) are the most important contact-type electrical temperature sensors used in the metals industry. Well over 90% of the sensors used in this industry are estimated to be thermocouples. A thermocouple is rugged, inexpensive and accurate; covers wide temperature ranges; and is fast in response. A resistance thermometer is more accurate and stable than a thermocouple. However, the resistance thermometer is more expensive, slower in response, and limited to lower temperatures, typically 540 °C (1000 °F).

Thermocouples

Thermocouples consist of two dissimilar wires that are metallurgically homogeneous. They are joined at one end, called the measuring or hot junction. The other end, which is connected to the copper wire of the measuring instrument circuitry, is called the reference or cold junction. The electrical signal output in millivolts is proportional to the difference in temperature between the measuring junction (hot) and the reference junction (cold). The different types of thermocouples, classified by their metallurgical compositions, have different output signal calibrations.

Thermocouples most commonly used in heat processing applications are listed in Table 1.

Resistance Temperature Detectors

Resistance temperature detectors are contact-type sensors. Their electrical resistance is proportional to temperature. Typical detector materials are platinum, copper and nickel. They are more stable and accurate than thermocouples, but even the platinum detectors have an upper temperature limit of approximately 540 °C (1000 °F), which reduces their usage in the metals industry.

Resistance temperature detectors are normally larger in size and slower in response than thermocouples. However, the new thin-film deposited detectors minimize this disadvantage, which characterizes conventional wire-wound detectors.

Noncontact Sensors

Radiation sensors are noncontact-type temperature sensors used with radiation pyrometers. One type of radiation pyrometer is the optical pyrometer.

Radiation sensors respond to radiant energy. They are classified as total-radiation (wide-band) or narrow-band types, depending on the width of the radiation wavelength band to which they respond. Total radiation sensors use thermal detectors, and narrow-band sensors use photoelectric detectors. Radiant energy from metals is characterized at lower temperatures as red hot (with longer wavelengths). At higher tempera-

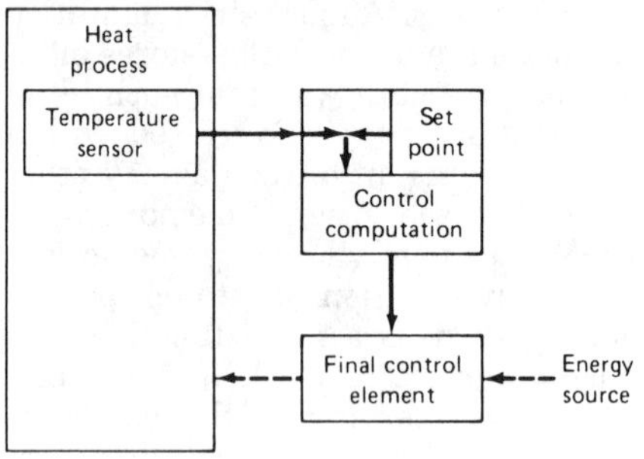

Fig. 1. Basic control loop

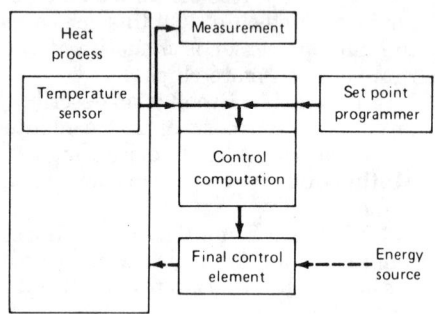

Fig. 2. Basic control loop with auxiliary devices

Table 1. Comparison of thermocouple types

Type	Usable temperature range °C	°F	Advantages	Restrictions
J (iron-constantan)	−185 to 870	−300 to 1600	Comparatively inexpensive; suitable for continuous service to 870 °C (1600 °F) in neutral or reducing atmospheres	Maximum upper limit in oxidizing atmosphere is 760 °C (1400 °F), due to the oxidation of the iron; protection tubes should be used above 480 °C (900 °F); protection tubes should always be used in a contaminating medium
K (nickel, chromium-nickel, aluminum)	−20 to 1370	0 to 2500	Suitable for oxidizing atmospheres; in higher temperature ranges, provides a more mechanically and thermally rugged unit than platinum or rhodium-platinum, and longer life than iron-constantan	Especially vulnerable to reducing atmospheres, requiring substantial protection when used
T (copper-constantan)	−185 to 370	−300 to 700	Resists atmosphere corrosion; applicable in reducing or oxidizing atmospheres below 315 °C (600 °F); its stability makes it useful at subzero temperatures; has high conformity to published calibration data	Copper oxidizes above 315 °C (600 °F)
E (nickel, chromium-constantan)	−185 to 870	−300 to 1600	Has high thermoelectric power; both elements are highly corrosion-resistant, permitting use in oxidizing atmospheres; does not corrode at subzero temperatures	Stability is unsatisfactory in reducing atmospheres
S (platinum, 10% rhodium-platinum) R (platinum, 13% rhodium-platinum)	−20 to 1480	0 to 2700	Usable in oxidizing atmospheres; provides a higher usable range than type K; frequently more practical than noncontact pyrometers; has high conformity to published calibration data	Easily contaminated in other than oxidizing atmospheres
B (platinum, 30% rhodium-platinum, 6% rhodium)	870 to 1650	1600 to 3000	Better stability than types S or R; increased mechanical strength; usable to higher temperatures than types S and R; reference-junction compensation is not required if junction temperature does not exceed 65 °C (150 °F)	Available in standard grade only; high temperature limit requires the use of alumina insulators and protection tubes; easily contaminated in other than oxidizing atmospheres

Reprinted from Metals Handbook, Desk Edition, 28.38-28.39, © 1985 American Society for Metals

tures, it is characterized as white hot (with shorter wavelengths).

Radiation sensors are used typically to control annealing furnaces as well as brazing and forge furnaces. They also are used for blast-furnace stoves, salt pots, checker bricks, rolling and strip mills, and induction-heating processes. One or more of the following conditions justify the use of noncontact sensors instead of contact sensors.

- Temperatures are too high for contact sensors.
- Work is moving too fast for detection by contact sensors.
- Required response rate is too fast for contact sensors.

MEASUREMENT AND CONTROL INSTRUMENTS

Measurement instruments measure the output signal of the temperature sensor and convert it to a temperature indication or recording in engineering units. Transmitters are used in some measurement systems to amplify and condition the temperature signal. The accuracy of the measurement depends greatly on the accuracy of the temperature sensor and the connecting leadwire. The accuracy of the measurement instrument is defined in its specifications under referenced conditions for its power supply, ambient conditions (temperature and humidity), electrical noise rejection and maximum source impedance. The accuracy of the transmitter has similar qualifications.

Measurement instruments are classified by their displays, analog or digital, and whether they are recording or nonrecording types. Analog displays include meters and motor-driven pointers. Analog strip-chart or round-chart recorders include analog temperature indication. Digital displays are available with or without digital printers. In general, digital equipment is more accurate than analog equipment, but specifications must be checked in either case.

Atmosphere Control

ALL METHODS of atmosphere control can effectively be divided into two groups: those involving control of the atmosphere-generating system, and those involving control of the atmosphere once it is inside the furnace. Both are important in maintaining a controlled condition throughout the heating process.

Most operations performed in the heat processing industry can be done under endothermic or exothermic protective atmospheres. Effective control of the generators that produce such atmospheres may be accomplished by means of certain types of controllers — namely, combustibles controllers, infrared CO_2 controllers and oxygen probes.

ANALYSIS AND CONTROL OF ENDOTHERMIC AND EXOTHERMIC ATMOSPHERES

Endothermic gas is used in heat treating furnaces as a protective atmosphere for hardening, stress relieving, carbon restoration and carburizing. The most recognized guides to endothermic generator operation are CO_2 and dew point. The ultimate guide is the relationship between CO_2 and dew point. Any deviation from this relationship indicates a generator problem, such as a leak or a carbon-coated catalyst. Figure 3 shows the proper carbon dioxide/dew point relationship.

Exothermic-type atmospheres are used where inert atmospheres are required. Applications are common in the metal treating industry for bright annealing of copper, annealing of motor laminations, processing of aluminum, and annealing of coils of wire and steel sheet.

OXYGEN PROBES

The oxygen probe is based in theory on a hot ceramic electrochemical cell. The probe will re-spond to oxygen, hydrogen, carbon monoxide, water and carbon dioxide and thus can determine the oxidization potential of a gas. The output of the oxygen probe is a direct measurement of the oxidation potential of the atmosphere at the process temperature of the furnace. Therefore, when the probe temperature is close to the furnace temperature, the response of the probe is a direct indication of whether the atmosphere will oxidize or reduce steel, provided that the composition of the atmosphere with regard to the proportions of carbon gases and hydrogen is known. Under such conditions, the probe will give a reliable indication of the oxidation/reduction situation for all furnace temperatures.

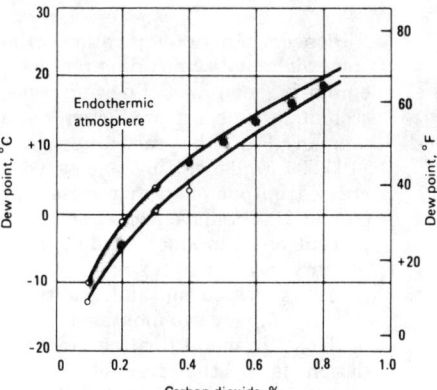

The generator in each plant was operated at a different temperature, all within the range from 1005 to 1095 °C (1840 to 2000 °F).

Fig. 3. Variation in the relation between dew point and carbon dioxide in generation of an endothermic atmosphere, as obtained from four plants

Source: Metals Handbook, Desk Edition, 28.38-28.39

Significance of Digital Control Programmers in Commercial Carburizing and Nitriding

By **GLENN RATLIFF**, President
    and
    WAYNE SAMUELSON, Vice President
    Shore Metal Treating
    Highland Heights, Ohio

Fig. 1 Exterior of Shore Metal Treating's 24,000 sq ft production facility in a Cleveland suburb. Company has doubled output capacity in three years with numerous production innovations.

Pioneers of high temperature carburizing and a new multi-step, high quality nitriding process, Shore Metal Treating is employing a number of new process methodology and control techniques to boost productivity and economy. To date, the company, Fig. 1, has raised gross annual sales per employee to $100,000 while cutting energy costs to 10% of gross; increased output of its five carburizing furnaces by between 50 and 100%; shaved scrap and rework to two-tenths of one percent of all product handled; and launched a line of proprietary heat treating products initially developed to upgrade its own output and quality.

The company's foundation is based on advancing the state of the art in metal treating. An important part of that foundation is in the area of more precise and responsive microprocessor-based process control.

Function of Digital Control Programmers

Already installed and operative are new digital control programmers (DCP) on one of five carburizing furnaces and one of two nitriding furnaces. The results thus far border on the dramatic in terms of both productivity and quality consistency.

The new instrumentation, Honeywell DCP 770133 triple programmer/triple controllers, combines in a single, easy-to-install unit variable set-point vs. time programmers with three mode controllers in each channel. Tighter control accuracy, increased resolution and repeatability of process are the major benefits over conventional cam type or curve follower programmers.

Fig. 2 Carburizing furnace of integral quench type is programmed controlled enhancing accuracy of the surface hardening process.

The digital control programmer can store up to nine separate programs consisting of up to 200 ramp or soak segments plus event switches. A guaranteed soak function allows the soak to proceed only when the process variable enters a selectable band around the set-point. Program entry and operation is via push buttons with digital LED seven segment displays in engineering units. Program storage is in non-volatile memory and is not affected by power failures. Key lock security is included to prevent unauthorized program tampering, and a self diagnostic program is permanently stored in the unit. The controller also implements bumpless transfer in both directions when switching from automatic to manual mode.

DCP Advantages in Carburizing

The microprocessor controlling the Ipsen carburizing furnace, Fig. 2, works with an Ipsen carbon probe which replaced an infra-red gas analyzer as the basic monitor of furnace atmosphere in the high volume, high temperature carburizing operation. The DCP 7700, Fig. 3, programs and controls three separate furnace functions and integrates their operation into a single unique control pattern. Furnace temperature, atmosphere and oil or air quenching temperatures are balanced, to produce what amounts to an unusual-

Fig. 3 Glenn Ratliff, pres. of Shore Metal Treating, activates fully automatic heat carburizing cycle via digital control programmer for carburizing furnace. New programmer stores up to nine separate heat treating programs, integrating control of furnace temperature, atmosphere and oil or air quench temperature programs.

Reprinted with permission from Industrial Heating, March 1982, 12-13, © 1982 National Industrial Publishing Co.

ly elongated carbon diffusion curve.

Control accuracy is greatly enhanced. With the old controller carbon depth could be controlled to ±.005 inches which provided satisfactory quality. But with the carbon probe-microprocessor system, it is possible to achieve case depth control to ±.0015 inches. Since it no longer is necessary to rely on basically manual control and because the control sequence can be duplicated, better quality is being obtained and, more importantly, in less time.

For example, if a customer previously required a .050/.060 inch guaranteed effective case, the deviation potential in the old control system would make it necessary to "shoot" for .060 inches. With the new controls and far less deviation, it is possible to aim for .053 inches and consistently end up with .052 - .054 inches. Thus, the reduction in overall furnace cycle is as much as 45-minutes (or 10-12%). In processing a large order, the faster turn around invariably means an extra furnace cycle per shift.

Another benefit is more economical use of manpower. Presently, the shop operates with one general foreman and one heat treating operator per shift.

Because of considerably improved quality and repeatability, customer complaints have been eliminated. One example of enhanced quality consistency is the current production of critical transmission bearing races for the new Ford Escort car.

The new process held total case between .048 - .054 inches over 130 loads initially. Every basket (load) run was sampled for hardness, depth of case and microstructure (see Fig. 4). Only two loads required corrective action. Using the old controls would have required some corrective action on 35 to 40 of those loads.

DCP Advantages in Nitriding

Although microprocessor controls on one of the company's two nitriding furnaces (Fig. 5) have just become operative, there are even greater potential benefits in efficiency.

Shore Metals already employs a self-developed and new state of the art three-to-four stage nitriding cycle. The multiple stage process significantly improves quality throughout the case depth range and limits finished product "growth" to as little as ±.0002 inches per linear inch.

With the microprocessor-based controllers and a hydrogen analyzer, the entire treating cycle — often as lengthy as 90 hr — can be controlled completely automatically. The DCP 7700 will react far more rapidly to gas deviations and will take the furnace up to temperature at the correct ramp rate every time to better control reproducibility and distortion. Once temperature has been reached, the soak can be controlled to within six seconds.

In addition, at any position during the soak cycle, Shore Metal will be able to control the percentage of disassociation of ammonia and thus provide unusually precise control of the nitrogen potential which, in turn, affects the final microstructure and related properties of the particular metal. When the cycle is completed, the new controller will automatically put the furnace on "blow-down" and will sound the alarm for unloading.

Considerably more labor was expended to operate the former control system. The operator was responsible for manual pipette samplings to determine the percentage of dissociation, elemental to the control of nitrogen in the furnace. The shortest time in which a thoroughly experienced operator could take accurate readings would be 10 minutes. Shore Metal checks readings twice every hour. Consequently, those readings are actually 10 minutes behind every hour of the process. In addition, the operator has to open and close valves in the correct proportions during the purge stage and then manually put the furnace on blow down.

With the hydrogen analyzer and the DCP, which also controls position proportioning motors, one can control pro-

Fig. 4 Wayne Samuelson, vice-pres. of Shore Metal Treating, checks microstructure of sample product in company's on-line testing lab. Mounted samples (on table in foreground) are made for every load of product the company processes. Hardness and case depth measurements are also made and filed to assure quality consistency.

Fig. 5 Pit type furnace of Lindberg design is used for nitriding, with microprocessor-based controller increasing the consistency of the multistage cycle and thus improving product properties.

portions within far more narrow bands. Proportioning is now controlled in stage one to ±1-1/2% and in stage two down to ±2%. The DCP actually will allow the control of proportions to as low as 1/4%. With manual control it's difficult to guarantee these percentages. Variations of as little as ±5% can compromise quality.

Further Enhancement of Production Efficiency

Among other refinements geared to increased production is the retrofit of independently fused circuits on the company's new, high-output, Ipsen two-zone furnace. This modification permits replacement of faulty microswitches without shutting down the furnace and facing a loss of production.

In addition, the company plans to replace the brickwork in its furnaces with fiber module insulation to save energy and cut the cost of furnace repairs and overhauls. Unless physically damaged, the 12x12 inch fiber modules will last indefinitely. More importantly, however, repairs in the event of damage can be handled in house, obviating the necessity of hiring expensive, specialized bricklaying labor.

Despite present progress in production efficiency the most important step is yet to come and that is in the area of automated control. All furnaces will eventually be equipped with microprocessor-based controllers. All of the controllers will be integrated into a computer with CRT display for true and highly viable central control, more effective, efficient scheduling, and faster, more positive response to widely varying customer requirements. ∎

Source: Industrial Heating, March 1982, 12-13

Heat-Resistant Alloys for Furnace Parts, Trays and Fixtures

By the ASM Committee on Trays and Fixtures*

TRAYS AND FIXTURES made of heat-resistant alloys are among the many parts used in industrial heat treating furnaces that operate at temperatures from 540 to 980 °C (1000 to 1800 °F). A complete list of these materials may be found in Vol 3 of this Handbook, together with room-temperature and high-temperature mechanical properties.

Basic Metallurgy and Product Forms

A partial list of typical products can be divided into two categories; parts that go through the furnaces and are, therefore, subjected to thermal and/or mechanical shock include trays, fixtures, conveyor chains and belts, and quenching fixtures. Parts that remain in the furnace with less thermal or mechanical shock, include support beams, hearth plates, combustion tubes, roller and skid rails, conveyor rolls, walking beams, rotary retorts, pit-type retorts, muffles, and drive and idler drums.

These heat-resistant alloys are supplied in either wrought or cast forms and, in some situations, may be a combination of the two. The properties and costs of the two forms vary, even though their chemical compositions are similar. Because there are many foundries and fabricators who are experienced in the design and application of these products, it is important to seek their advice when purchasing high-alloy parts.

There are five types of heat-resistant alloys listed in Vol 3 of this Handbook:

- Fe-Cr alloys
- Fe-Cr-Ni alloys
- Fe-Ni-Cr alloys
- Nickel-base alloys
- Cobalt-base alloys

The great majority of heat treating furnaces use only the second and third types, because the Fe-Cr alloys do not have sufficient high-temperature strength to be useful, and the nickel- and cobalt-base types (except Inconel) are generally too expensive except for very special applications. Therefore, this discussion will be limited to the use and properties of the Fe-Cr-Ni and the Fe-Ni-Cr grades.

Room-temperature mechanical properties have limited value in selecting materials or designing for high-temperature use, but they may be useful in checking the quality of the alloys. These properties are shown in Vol 3 and also may be found in ASTM specification A297. The useful high-temperature properties of these alloys are summarized in Table 1 for castings and Table 2 for wrought products. They include nominal composition, stress to produce a creep rate of 1% in 10 000 h, and stress to rupture in 10 000 h and

*Dennis M. Wagen, *Chairman*, Executive Vice President, Stanwood Corp.; Douglas J. Cleary, Vice President and General Manager, Stanwood Corp.; Francis Fahrenwald, Fahrenwald Consulting; Norman O. Kates, Vice President-Technology, Lindberg Corp.; James Kelly, Director of Technology, Rolled Alloys; Fred W. Klag, Vice President, The Alloy Engineering Co.; Arthur L. LaMasters, Marketing Director, Cleveland Alloy Casting Co.; Ross B. Shingledecker, Metallurgical Director, Manufacturing Services, Ladish Co.; John E. Stein, Vice President, Manufacturing, Resisto-Loy Company, Inc.; Donald J. Tillack, Technical Sales Engineer, Huntington Alloys, Inc.

Table 1 Composition and elevated-temperature properties of cast heat-resistant alloys

Grade	Approximate composition, % C	Cr	Ni	Temperature °C	°F	Creep stress to produce 1% creep in 10 000 h MPa	ksi	Stress to rupture in 10 000 h MPa	ksi	Stress to rupture in 100 000 h MPa	ksi
Fe-Cr-Ni alloys											
HF.......0.20 to 0.40	19 to 23	9 to 12	650	1200	124	18.0	114	16.5	76	11.0
			760	1400	47	6.8	42	6.1	28	4.0
			870	1600	27	3.9	19	2.7	12	1.7
			980	1800	...	...	...	...	...	...
HH ......0.20 to 0.50	24 to 28	11 to 14	650	1200	124	18.0	97	14.0	62	9.0
			760	1400	43	6.3	33	4.8	19	2.8
			870	1600	27	3.9	15	2.2	8	1.2
			980	1800	14	2.1	6	0.9	3	0.4
HK ......0.20 to 0.60	24 to 28	18 to 22	650	1200	...	...	...	...	...	...
			760	1400	70	10.2	61	8.8	43	6.2
			870	1600	41	6.0	26	3.8	17	2.5
			980	1800	17	2.5	12	1.7	7	1.0
Fe-Ni-Cr alloys											
HN ....0.20 to 0.50	19 to 23	23 to 27	650	1200	...	...	...	...	...	...
			760	1400	...	...	...	...	...	...
			870	1600	43	6.3	33	4.8	22	3.2
			980	1800	16	2.4	14	2.1	9	1.3
HT.......0.35 to 0.75	15 to 19	33 to 37	650	1200	...	...	...	...	...	...
			760	1400	55	8.0	58	8.4	39	5.6
			870	1600	31	4.5	26	3.7	16	2.4
			980	1800	14	2.0	12	1.7	8	1.1
HV ......0.35 to 0.75	17 to 21	37 to 41	650	1200	...	...	...	...	...	...
			760	1400	59	8.5	...	...	...	...
			870	1600	34	5.0	23	3.3	...	...
			980	1800	15	2.2	12	1.8	...	...
HX ....0.35 to 0.75	15 to 19	64 to 68	650	1200	...	...	...	...	...	...
			760	1400	44	6.4	...	...	...	...
			870	1600	22	3.2	...	...	...	...
			980	1800	11	1.6	...	...	...	...

Note: Some stress values are extrapolated.

100 000 h at temperatures of 650, 760, 870 and 980 °C (1200, 1400, 1600 and 1800 °F). A design stress figure commonly used for uniformly heated parts not subjected to thermal or mechanical shock is 50% of the stress to produce 1% creep in 10 000 h, but this should be used carefully and should be verified with the supplier.

In general, these materials contain iron, nickel and chromium as the major alloying elements. Carbon, silicon and manganese also are present and affect the foundry pouring and rolling characteristics of these alloys, as well as their properties at elevated temperature. Nickel influences primarily high-temperature strength and toughness. Chromium increases oxidation resistance by the formation of a protective coating of chromium oxide on the surface. An increase in carbon content increases strength.

All of the alloys commonly used in castings for furnace parts have essentially an austenitic structure. The Fe-Cr-Ni alloys (HF, HH, HI, HK and HL) may contain some ferrite, depending on composition balance. If exposed to a temperature in the range of 540 to 900 °C (1000 to 1650 °F), these compositions may convert to the embrittling sigma phase. This can be avoided by using the proper proportions of nickel-chromium, carbon and associated minor elements. Chromium and silicon promote ferrite, whereas nickel, carbon and manganese favor austenite. Use of the Fe-Cr-Ni types should be limited to applications where the temperatures are steady and are not within the sigma-forming temperature range. Transformation from ferrite to sigma phase at elevated temperature is accompanied by a change from ferromagnetic to nonferromagnetic material and from a soft to a very hard and brittle material.

All heat-resistant alloys of the Fe-Ni-Cr group are wholly austenitic and are not as sensitive to composition balance as the Fe-Cr-Ni group. They also contain large primary chromium carbides in the austenitic matrix and, after exposure to service temperature, show fine, precipitated carbides. This characteristic makes these alloys useful for carburizing trays and fixtures. They are considerably stronger than the Fe-Cr-Ni alloys and may be less expensive per part if the increased strength is considered when designing for a known load.

Life expectancy of trays and fixtures is best measured in cycles rather than hours, particularly if the parts are quenched. It may be cheaper to replace all trays after a certain number of cycles to avoid expensive shutdowns caused by wrecks in the furnace. Chains or belts that cycle from room temperature to operating temperature several times a shift will not last as

Table 2 Composition and elevated-temperature properties of wrought heat-resistant alloys

Grade	C	Approximate composition, % Cr	Ni	Other	Temperature °C	°F	Creep stress to produce 1% creep in 10 000 h MPa	ksi	Stress to rupture in 10 000 h MPa	ksi
Fe-Cr-Ni alloys										
309	0.08 max	17 to 20	34 to 37	· · ·	650	1200	48	7.0	· · ·	· · ·
					760	1400	14	2.0	· · ·	· · ·
					870	1600	3	0.5	10	1.45
					980	1800	· · ·	· · ·	3	0.5
310	0.08 max	24 to 26	19 to 22	· · ·	650	1200	63	9.2	· · ·	· · ·
					760	1400	17	2.5	· · ·	· · ·
					870	1600	9	0.3	13.5	1.95
					980	1800	· · ·	· · ·	4	0.6
Fe-Ni-Cr alloys										
330	0.08 max	17 to 20	34 to 37	· · ·	760	1400	25	3.6	30	4.4
					870	1600	13	1.9	12	1.8
					980	1800	3.5	0.52	4.5	0.65
330HC	0.4 max	17 to 22	34 to 37	· · ·	760	1400	47	6.8	54	7.8
					870	1600	18	2.6	18	2.6
					980	1800	5	0.7	5	0.7
333	0.08 max	24 to 27	44 to 47	3 Mo, 3 Co, 3 W	760	1400	43	6.2	65	9.4
					870	1600	21	3.1	21	3.1
					980	1800	6	0.9	7	1.05
800	0.1 max	19 to 23	30 to 35	0.15-0.60 Al, 0.15-0.60 Ti	760	1400	19	2.8	23	3.3
					870	1600	4	0.61	12	1.7
					980	1800	1	0.23	6	0.8
802	0.2 to 0.5	19 to 23	30 to 35	· · ·	760	1400	83	12.0	79	11.5
					870	1600	30	4.4	33	4.8
					980	1800	8	1.1	11.5	1.65
Ni-base alloys										
600	0.15 max	14 to 17	72 min	· · ·	760	1400	28	4.1	41	6.0
					870	1600	14	2.0	16	2.3
					980	1800	4	0.56	8	1.15
601	0.10 max	21 to 25	58 to 63	1.0-1.7 Al	760	1400	28	4.0	42	6.1
					870	1600	14	2.0	19	2.7
					980	1800	5.5	0.79	8	1.2

long as stationary parts that do not fluctuate in temperature. Parts for carburizing furnaces will not last as long as those used for straight annealing.

Finally, alloy parts represent a sizable portion of the total cost of a heat treating operation. Alloys should be selected carefully, designed properly and operated with good controls throughout to keep costs at a minimum.

Material Comparison for Heat-Resistant Cast and Wrought Components

The selection of a cast or fabricated component for furnace parts and fixtures depends primarily on the operating conditions associated with heat treating equipment in the specific processes, and secondarily on the stresses that may be involved. The factors of temperature, loading conditions, work volume, the rate of heating and furnace cooling or quenching need to be examined for the operating and economy trade-offs. Other factors, such as the furnace and fixture design, type of furnace atmosphere, length of service life, and pattern availability or justification, enter into the selection.

Some of the factors affecting service life of alloy furnace parts, not necessarily in order of importance, are alloy selection, design, maintenance procedures, furnace and temperature control, atmosphere, contamination of atmosphere or work load, accidents, number of shifts operated, thermal cycle and overloading. High-alloy parts may last anywhere from a few months to many years, depending on operating conditions. In the selection of the heat-resistant alloy for a given application, all properties should be considered in relation to the operating requirements to obtain the most economical life.

If either cast or wrought alloy fabrications can be practically used, both should be considered. Similar alloy compositions in cast or wrought form may have varying mechanical properties, different initial costs and inherent advantages and disadvantages. Castings are more adaptable to complicated shapes and fabrications to similar parts, but a careful comparison should be made to determine the over-all costs of cast versus fabricated parts. Initial costs, including pattern or tooling costs, maintenance expenses and estimated life are among the factors to be included in such a comparison. Lighter weight trays and fixtures will use less fuel in heating. Cast forms are stronger than wrought forms of similar chemical composition. They will deform less rapidly than wrought shapes but may

Table 3 Recommended materials for furnace parts and fixtures for hardening, annealing, normalizing, brazing and stress relieving

When more than one material is recommended for a specific part and operating temperature, each has proved satisfactory in service. Multiple choices are listed in order of increasing alloy content.

Retorts, muffles(a)		Radiant tubes(a)		Mesh belts, wrought	Chain link		Sprockets, rolls, guides, trays	
Wrought	Cast	Wrought	Cast		Wrought	Cast	Wrought	Cast
595 to 675 °C (1100 to 1250 °F)								
430	HF	430	HF	430	430	HF	430	HF
304		304		304	304		446	
							304	
675 to 760 °C (1250 to 1400 °F)								
304	HF	347	HF	309	309	HF	304	HF
347	HH	309	HH			HH	316	HH
309(b)							309	
760 to 925 °C (1400 to 1700 °F)								
310	HH	310(c)	HH	314	314	HH	310	HH
35-18(d)	HT(e)	35-18(d)	HK	35-18(d)	35-18(d)	HL	35-18(d)	HK
Inconel	HW(e)(f)	Inconel	HL			HT		HL,HT
925 to 1010 °C (1700 to 1850 °F)								
35-18(d)	HK	Inconel	HK	314	314	HL	310	HL
Inconel	HT		HL	35-18(d)	35-18(d)	HT	35-18(d)	HT
	HW		HT	Inconel	Inconel		Inconel	
1010 to 1095 °C (1850 to 2000 °F)								
35-18(d)	HK	Inconel	HL	35-18(d)	35-18(d)	HL	35-18(d)	HL
	HL		HX	80-20	80-20	HT	Inconel	HX
Inconel	HW							
	HX							
	NA22H							
1095 to 1205 °C (2000 to 2200 °F)								
Hastelloy X	HL	Inconel	HL	35-18(d)	35-18(d)	HX	Inconel	HL
Inconel	HU		HX	80-20	80-20			HX
	HX							

(a) Temperature gradients of 40 to 95 °C (100 to 200 °F) are assumed between heat-source side and work zone side of retorts, muffles and radiant tubes. (b) The stabilized grade 309S is recommended for applications involving mechanical or thermal shock. (c) Recommended for vertical mounting only. (d) A series of alloys generally of the 35Ni-15Cr type or modifications that contain from 30 to 40% Ni and 15 to 23% Cr and include RA-330, 35-19, Incoloy and other proprietary alloys. (e) HK or HL is recommended where greater strength is needed. (f) Recommended for applications requiring shock resistance, such as shaker hearths

crack more rapidly under conditions of fluctuating temperatures. Selection should be based on the practical advantages with all facts considered.

General Considerations. Both cast and wrought alloys are well accepted by the designers and users of furnaces requiring high-temperature furnace load-carrying components. There are certain advantages for each type of manufactured component; often, the compositions are similar, if the carbon and silicon levels in the castings versus the wrought material are ignored. In general, the specifications of the wrought grades have carbon contents below 0.25%, and many are nominally near 0.05% carbon. In contrast, the cast alloys have from 0.25 to 0.50% carbon. This difference has an effect on hot strength. The difficulty in hot working the higher carbon alloys accounts for their scarcity in the wrought series. Castings and fabricated parts are not always competitive; each product has advantages, which include:

Advantages of cast alloys

- Initial cost: A casting is essentially a finished product as cast; its cost per pound is frequently less than a fabricated item.
- Strength: Similar alloy compositions are inherently stronger at elevated temperatures than wrought alloys.
- Shape: Some designs can be cast that may not be available in wrought form; they also may not be economically fabricated if wrought material was available.
- Composition: Some alloy compositions are available only in castings; they may lack sufficient ductility to be worked into wrought material configurations.

Advantages of wrought alloys

- Section size: There is practically no limit to section sizes available in wrought form.
- Thermal fatigue resistance: Ductility of the fine grain microstructure of wrought alloys may promote better thermal fatigue resistance.
- Soundness: Wrought alloys normally are free of internal or external defects; they have smoother surfaces that may be beneficial in avoiding local hot spots.
- Availability: Wrought alloys are frequently available from stock in many forms.

Shape, complexity and number of duplicate parts (eventually affecting

Table 4 Recommended materials for parts and fixtures for carburizing and carbonitriding furnaces

When more than one material is recommended for a specific part and operating temperature, each has proved satisfactory in service. Multiple choices are listed in order of increasing alloy content.

| Part | Operating temperature of part(b) | |
	Wrought	Cast
Retorts(a)	...	HK
	35-18(c)	HT
	Inconel	
Muffles(a)	35-18(c)	HT
	Inconel	
Radiant tubes(a)	35-18(c)	HT
	Inconel	HU
		HX
Structural parts	35-18(c)	HT
	Inconel	
Pier caps, rails	35-18(c)	HT
	Inconel	
Tube supports	35-18(c)	HT
	Inconel	
Trays, baskets, fixtures (not quenched)	35-18(c)	HT
	Inconel	
	35-18(c)	HT(Nb)
	Inconel	HU
		HU(Nb)
Trays, baskets, fixtures (oil quenched)	35-18(c)	HT
	Inconel	HT(Nb)
		HU
		HU(Nb)
		HW

(a) Temperature gradients of 40 to 95 °C (100 to 200 °F) difference in temperature between heat-source side and work zone side of retorts, muffles and radiant tubes. (b) 815 to 1010 °C (1500 to 1850 °F). (c) A series of alloys generally of the 35Ni-15Cr type or modifications that contain from 30 to 40% Ni and 15 to 23% Cr and include RA-330, 35-19, Incoloy and other proprietary alloys

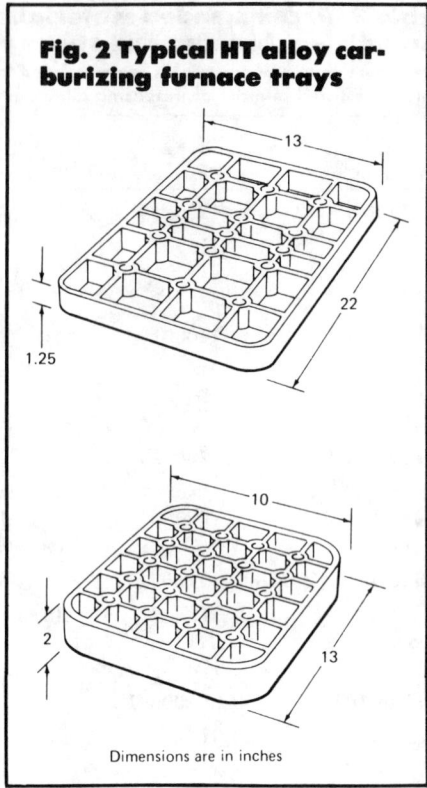

Fig. 2 Typical HT alloy carburizing furnace trays

Dimensions are in inches

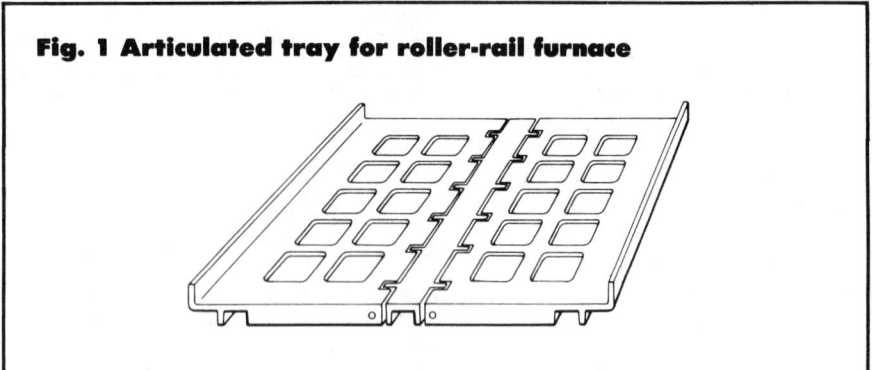

Fig. 1 Articulated tray for roller-rail furnace

cost) usually determine the choice between casting or fabricated part. Where section thickness and configuration permit, castings are usually cheaper. The cost of the metal per pound of casting is comparable to that of a fabricated part. Total cost of the fabrication usually would be higher because the cost of forming, joining and/or assembly must be added to the cost of material. However, when only one or two types of parts are to be made, the pattern cost precludes the use of a casting.

In energy-intensive heat treating industries, use of wrought fabrications enables fuel savings through reduced heat treating time cycles. At the present level of energy costs, wrought fabrications may be economically preferable because of improvements in thermal efficiency.

Fabrications are preferred for thin sections and for parts where less weight or greater heat transfer may be required. Where thick walls are necessary for strength, or where heavy loads are transported or pushed, the cost of fabricated sections may be prohibitive. Wrought materials have a greater degree of acceptance in fabricated baskets used under carburizing or carbonitriding conditions.

A factor that must be considered in evaluating castings and fabrications is the importance of good welding techniques, particularly for parts that are used in case hardening atmospheres. Castings have replaced fabricated products because of weld failures in multiwelded fabrications.

Although cast alloys exhibit greater high-temperature strength, it is possible to place too much emphasis on this characteristic in materials selection. Strength rarely is the only requisite and frequently is not the major one. More failures occur due to brittle fracture from thermal fatigue than do from stress rupture or creep. However, high-temperature strength is important where severe thermal cycling is required.

Specific Applications. Recommended applications for alloys for parts and fixtures for various types of heat treating furnaces, based on atmosphere and temperature, are summarized in Tables 3, 4 and 5. Where more than one alloy is recommended, each has proved adequate, although service life varies in different installations because of differences in exposure conditions.

Typical Applications

Trays and Grids. Many parts to be heat treated are irregular in shape and as such must be conveyed through the continuous heat treating furnaces or loaded and unloaded from the batch furnaces on grids or trays (Fig. 1). These trays or grids must withstand exposure to the same furnace conditions as the product and as such are subject to heating and cooling, detrimental and beneficial atmospheres, as well as compression and tensile loading. Heat-resistant alloys are used extensively for these parts, although there are instances where dispensible carbon or low-alloy steel fabricated trays are employed. In this instance, the choice is based on the economics of the particular situation, taking into account the cost of materials as well as the service life expected.

Two-thirds of the approximately 15 common heat-resistant alloy compositions find application in the heat treating industry. Of these, half are recommended for use in trays and grids. The particular alloy chosen should be selected on the basis of required strength at temperature, ductility and corrosion resistance.

Trays and grids made of alloying materials using approximately 10% nickel may find an application at furnace temperatures of 650 to 870 °C (1200 to 1600 °F), but as the use temperature goes up, for example, to 1040 to 1150 °C (1900 to 2100 °F), one would probably select an alloy with twice as much nickel. If the tray or fixture is to be subjected to the thermal shock of rapid heating and cooling, one would probably specify an even higher nickel content. The particular atmosphere surrounding the trays necessitates consideration of varied amounts of chromium addition to enhance resistance to oxidation or high-temperature corrosion. If trays are to be used in an atmosphere with very high sulfur, one would select an alloy with rather high chromium and moderate nickel. Some alloys contain relatively large amounts of silicon to fortify against carburization in carburizing applications (Fig. 2).

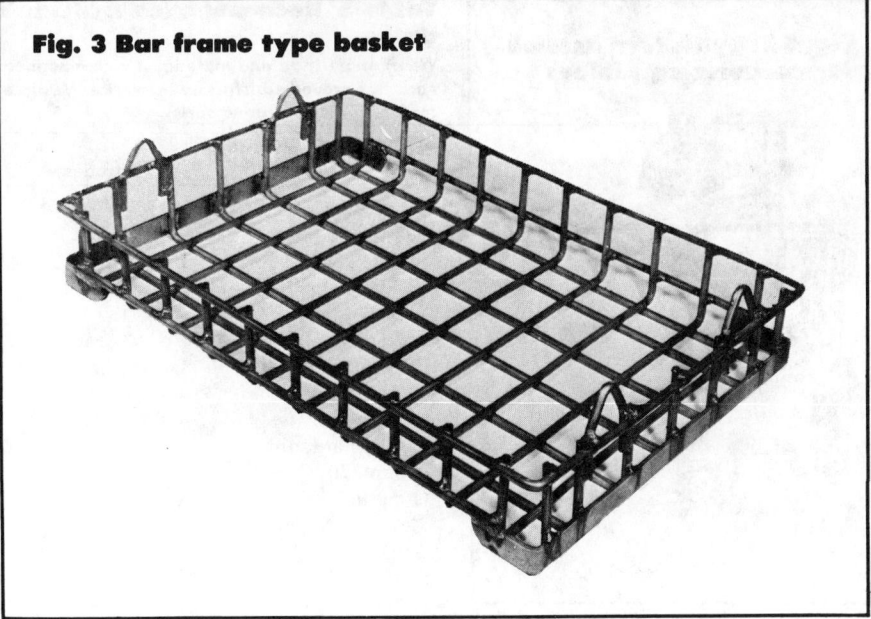

Fig. 3 Bar frame type basket

Fig. 4 Large pit furnace basket

Families of commercially available heat-resistant alloys provide sufficient selection so that an alloy that is optimized for each application and use can be specified. Alloy producers as well as vendors of trays and grids, both cast and fabricated, are an invaluable source of information regarding service life, design considerations and fabrication. Generally, a tray or grid should be of sufficient section size to provide reasonable service life under specified

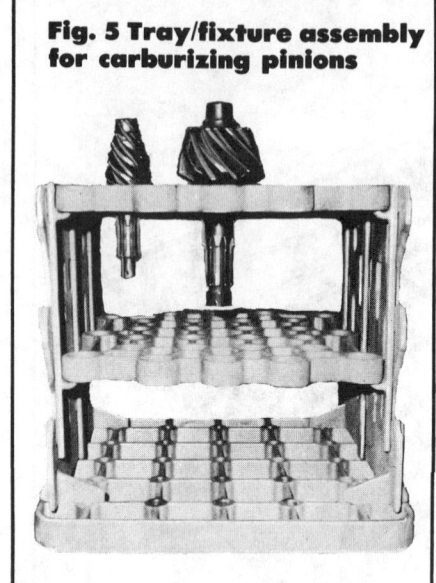

Fig. 5 Tray/fixture assembly for carburizing pinions

Table 5 Recommended materials for parts and fixtures for salt baths

Where more than one material is recommended for a specific part and operating temperature, each has proved satisfactory in service. Multiple choices are listed in order of increasing alloy content (except ceramic parts).

Process and temperature range	Electrodes	Pots	Thermocouple protection tubes
Salt quenching, 205 to 400 °C (400 to 750 °F)	Low-carbon steel	Low-carbon steel	Low-carbon steel, 446
Tempering, 400 to 675 °C (750 to 1250 °F)	Low-carbon steel, 446, 35-18(a)	Aluminized low-carbon steel, 309	Aluminized low-carbon steel, 446
Neutral hardening, 675 to 870 °C (1250 to 1600 °F)	446, 35-18(a)	35-18(a), HT, HU, Ceramic Inconel	446, 35-18(a)
Carburizing, 870 to 940 °C (1600 to 1720 °F)	446, 35-18(a)	Low-carbon steel(b), 35-18(a), HT	446, 35-18(a)
Tool steel hardening, 1010 to 1315 °C (1850 to 2400 °F)	Low-carbon steel(c), 446	Ceramic	446, 35-18(a), Ceramic

(a) A series of alloys generally of the 35Ni-15Cr type or modifications that contain from 30 to 40% Ni and 15 to 23% Cr and include RA-330, 35-19, Incoloy and other proprietary alloys. (b) Immersed electrode furnaces only. (c) Low-carbon steel is recommended for completely submerged electrodes only.

loading conditions. An overly heavy tray may prolong service life, but the added energy cost to heat the tray through each cycle may offset any cost savings realized through added life. It is sometimes possible to combine materials in trays to provide sufficient strength yet maintain minimum weight. For example, in an articulated tray used in an extremely long pusher furnace, the tray grid that is subject to the compressive force of the pusher bar is of a higher nickel content than the vertical load supports that must bear the compressive load on a per tray basis. This dual alloy tray represents a compromise among weight, cost and service life. In addition, service life is greatly affected by the tray cooling process, and in general, uniform section size throughout a tray is highly desirable to minimize stress during cooling.

All service conditions should be considered when selecting alloy for trays and grids. Unlike furnace structural parts, a tray is subject to alternate heating and cooling during each use cycle. The cooling can be rapid as in quenching work, or relatively slow as in furnace-cool applications. Selection of a proper alloy will ensure adequate service life if all service conditions are known and considered.

Baskets and Fixtures. In many situations, parts being heat treated are of a size that does not permit them to be loaded directly on a furnace hearth, tray or grid. They require some type of container, such as a basket; design of baskets varies because each product is developed for a specific application and loading and must function with a specific type of furnace equipment.

Baskets and fixtures can be produced from cast alloys or fabricated using wrought alloys. Fabrications are used in light to medium loading applications, intricate designs, complex shapes and generally with lighter metal sections. Typically, these include the bar frame type basket (Fig. 3) or corrugated box or shroud. In applications involving heavy loading and/or simple shapes and designs, cast alloys are commonly selected; typically, they are large pit furnace baskets (Fig. 4).

In specific applications, a part may require special positioning. This is accomplished by utilizing a fixture that is generally adaptable to an existing tray or grid or, in some instances, placed directly into a basket or container. These components can range from simple shapes, such as round, square, rectangular or fluted bars, to extremely intricate shapes. Figures 5, 6 and 7 are examples of such fixtures. Figure 5 is a tray/fixture assembly used for carbu-

rizing pinions; Fig. 6 was designed for heat treating lawn mower blades. Figure 7 was designed for heat treating shafts.

In most applications involving operating temperatures of 790 to 1010 °C (1450 to 1850 °F), the product is generally manufactured with a material having a nominal composition of 35Ni-15Cr, which provides a fully stable austenitic structure virtually free from any embrittling phases. In addition, it provides a reasonable cost/life ratio in applications involving endothermic, exothermic and inert atmospheres even with properly controlled enrichments of natural gas, air or ammonia, typically used for gas carburizing or gas carbonitriding. For quenching, a 35Ni-15Cr alloy provides acceptable life; however, in applications of severe quenching, higher nickel alloys or carbon steel may be considered, depending on the cost/life ratio of the product. If applications involve higher temperatures, excessive oxidation or carburization, consideration should be given to increase the Ni-Cr content of the alloy. For nitriding, higher nickel content provides the best cost/life ratio.

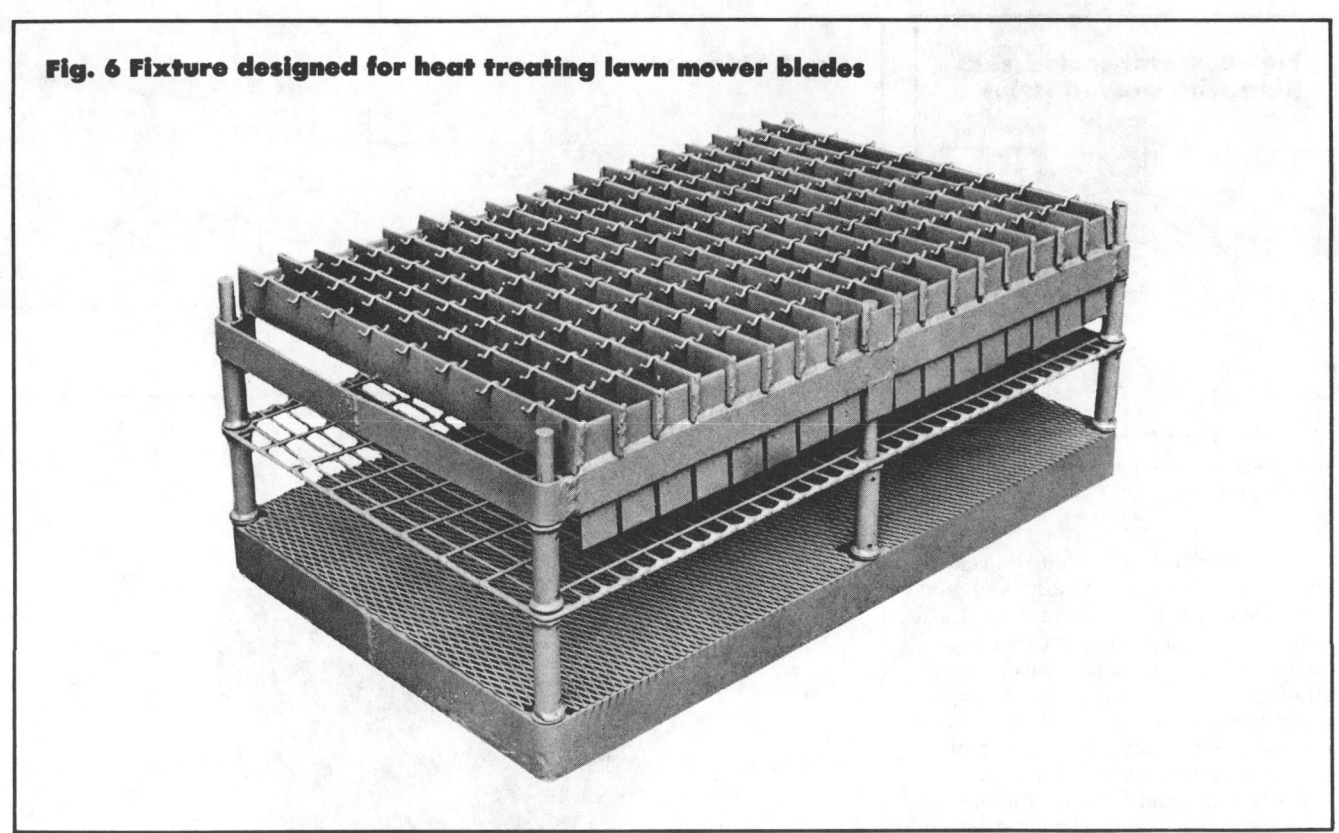

Fig. 6 Fixture designed for heat treating lawn mower blades

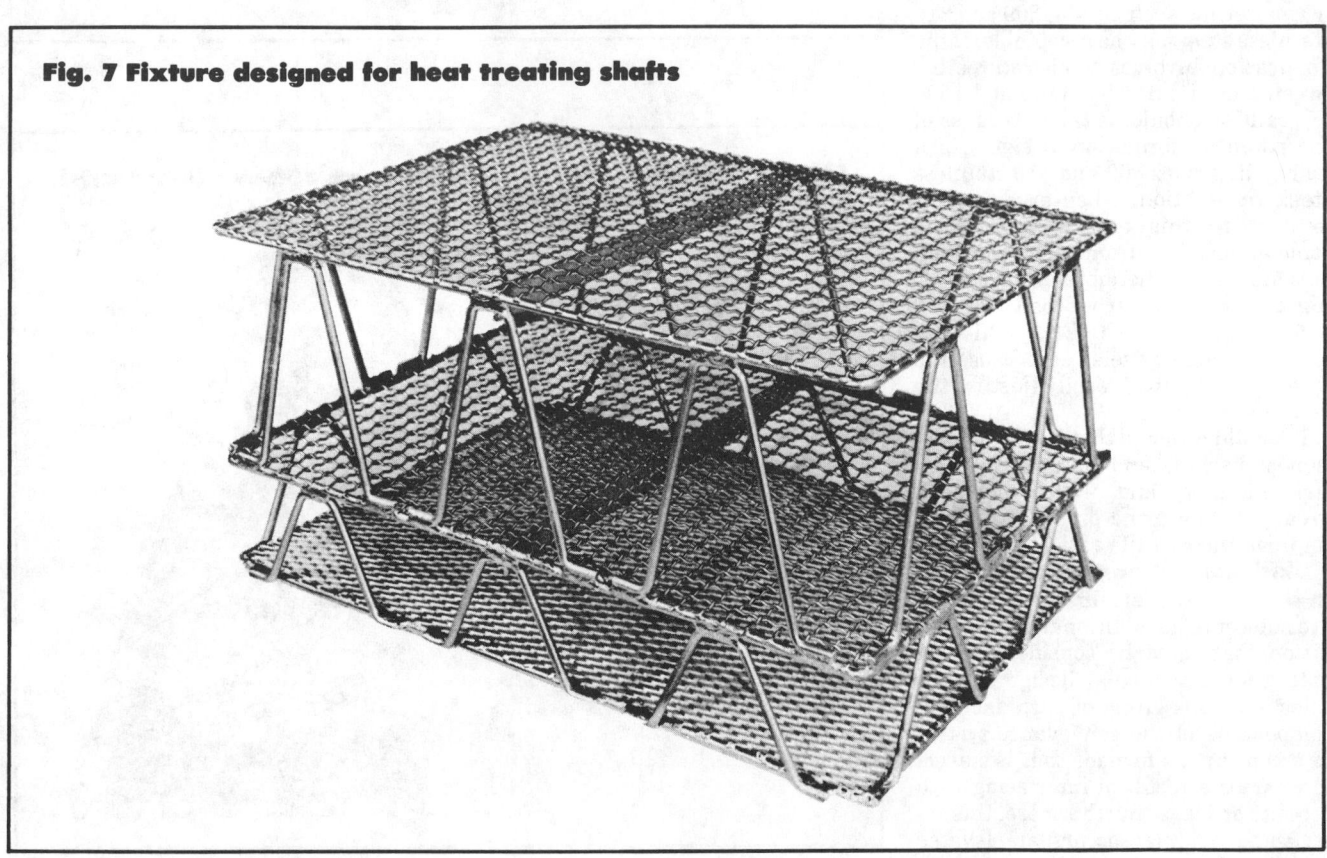

Fig. 7 Fixture designed for heat treating shafts

Source: Metals Handbook, 9th Edition, Vol. 4, 325-336

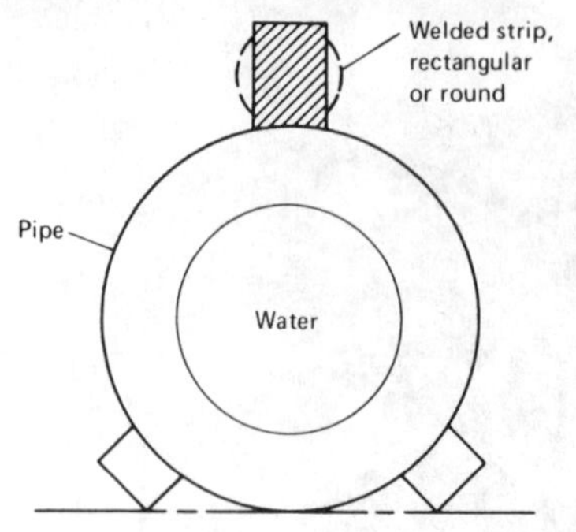

Fig. 8 Water-cooled skid pipe with welded strips

Welded strip, rectangular or round

Pipe

Water

Fig. 9 Thin-walled furnace roller

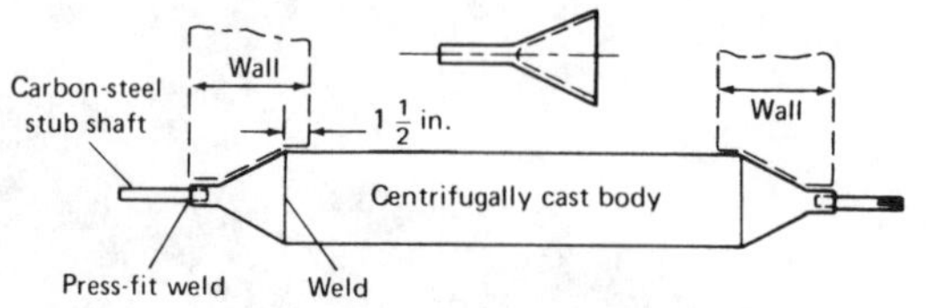

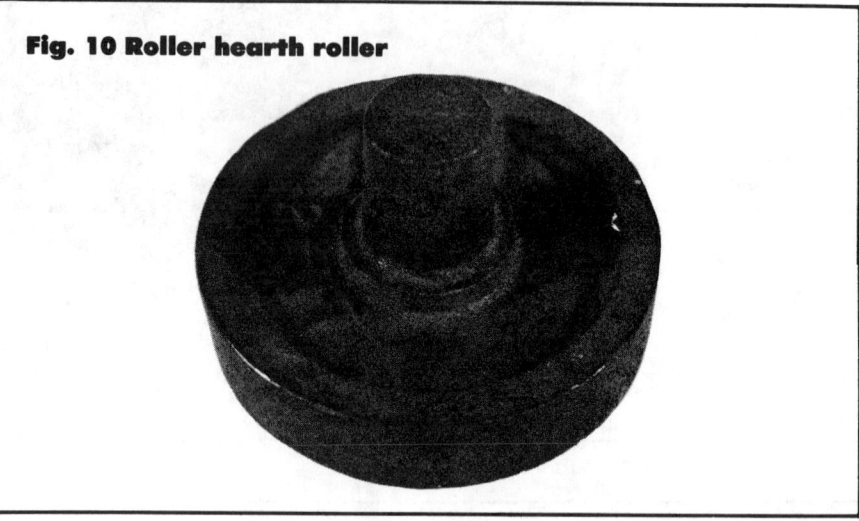

Carbon-steel stub shaft

Wall

$1\frac{1}{2}$ in.

Wall

Press-fit weld

Weld

Centrifugally cast body

Fig. 10 Roller hearth roller

In vacuum furnace applications, conventional heat-resistant alloys such as 35Ni-15Cr perform adequately. Caution should be taken to prevent vaporization of any element within these alloys. If a specific application has operating parameters that will not allow the use of a conventional alloy, molybdenum fabrications may be used.

For baskets and fixtures that may be restructured to lower temperature operations of 260 to 595 °C (500 to 1100 °F), materials such as 304, 309 and 310 stainless steel may be acceptable. If the application involves temperatures between 595 and 815 °C (1100 and 1500 °F), caution should be taken because of the potential formation of sigma, primarily in grades 309 and 310 stainless steel. In addition, when grade 304 is exposed to this temperature range, some embrittling from carbide participation results. Therefore, if the operating temperature is between 595 and 815 °C (1100 and 1500 °F), a 35Ni-15Cr alloy is generally selected because the extended life may easily justify the additional cost.

It should be noted that in the application of baskets and fixtures, periodic straightening and re-welding can greatly enhance the product's life and improve the cost/life ratio.

Skid Rails, Hearth Components and Rollers. Certain furnace parts are subject to an additional service condition that must be considered when opting for a particular design or alloy selection. This group of parts includes components of the conveyance system in a continuous furnace that is subject to wear as a result of interfacing with product or trays. Furthermore, this interfacing or wear occurs at elevated

Fig. 11 Conveyor belt assembled with 100-mm (4-in.) pitch and drive drum ready for installation

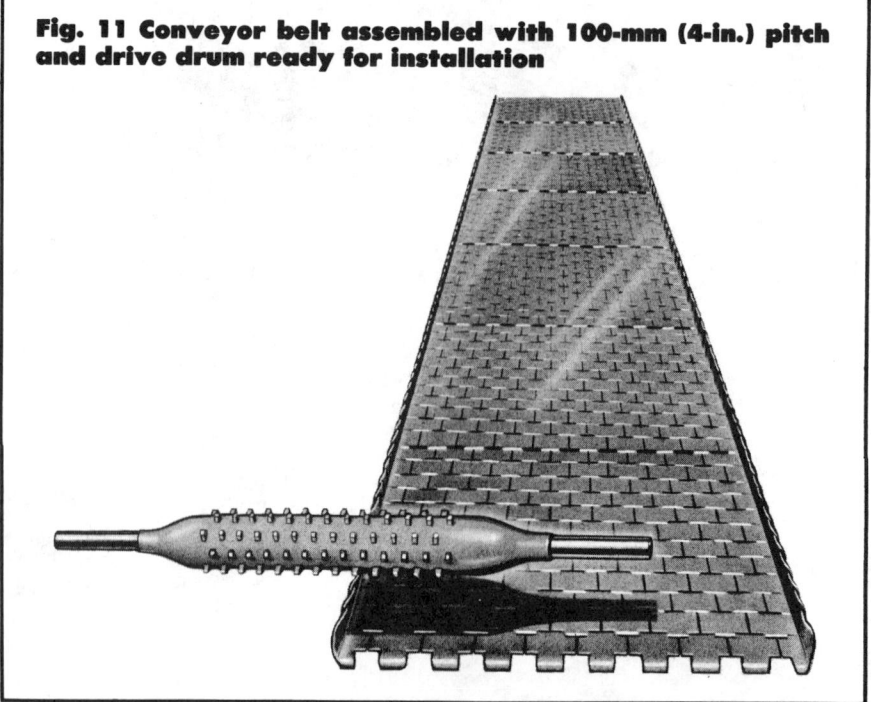

Fig. 12 U-shape radiant tube

Fig. 13 Straight radiant tube

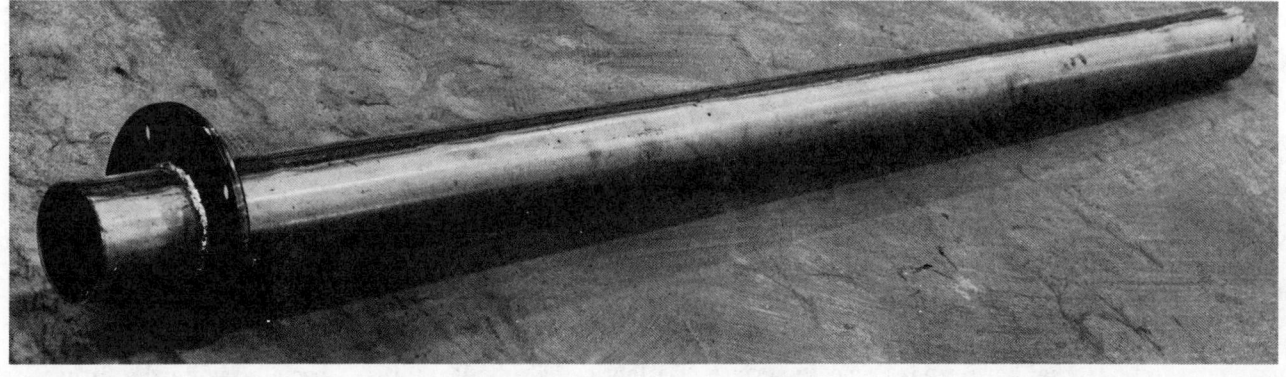

temperatures where alloy strength is diminished. Proper selection of an alloy for a specific high-temperature service involves consideration of many factors. One important factor is not to select the same composition for components that have sliding or rolling contact so as to minimize the possibility of galling or seizing. For example, when selecting an alloy to make skid rails (Fig. 8), one must consider (a) whether the rail will be cooled and the method of cooling, (b) whether adequate expansion space has been specified, (c) the amount of contact area present at the interface, and (d) how the rail will be supported and at what intervals. This implies that the design of skid rails and selection of the alloy is an integral part of furnace design; the same applies to rollers and hearth components.

Perhaps the greatest single factor affecting a roller in a heat treating furnace application is the actual bearing or roller support of the roller. In roller hearth furnaces, the rollers protrude through the furnace walls, and the roller bearings can operate in a relatively reduced ambient temperature (Fig. 9). However, in some roller-tray furnaces, the individual rollers operate within the furnace heated area and the roller spindle or shaft must rotate on a roller support without aid of a precise, lubricated bearing (Fig. 10). Hearth components usually are nonrotating or nonmoving parts and, in most situations, are well supported by refractory piers and/or ledges. They are almost always subjected to compressive loading, although they could on occasion be subject to lateral thrust and/or bending. When selecting an alloy for these applications, one should consider the elevated-temperature mechanical properties required for the anticipated loading.

Belting. Conveyor belts are used extensively in furnace designs for brazing, sintering and hardening of carbonitriding applications. The woven belts or mesh belting are commonly used for light duty loading, whereas the cast link belts are designed for heavy loading requirements. Figure 11 shows an assembled conveyor belt with a 100-mm (4-in.) pitch and the drive drum ready for installation.

When mesh belting is required for applications between 260 and 790 °C (500 and 1450 °F), medium carbon steel (grades 1040 to 1055) can be selected for application up to 540 °C (1000 °F). For higher temperatures, materials containing 1 to 5% chromium may be selected, or type 430 stainless is acceptable. Types 304, 309 and 316 stainless steel tend to be susceptible to carbide participation or the formation of sigma phase within this temperature range and, therefore, are not frequently selected. If stainless steel is required, type 347, which is stabilized with niobium and virtually free from carbide participation, may be selected.

Alloys commonly used for mesh belts in the temperature range of 790 to 1205 °C (1450 to 2200 °F) are 35Ni-15Cr,

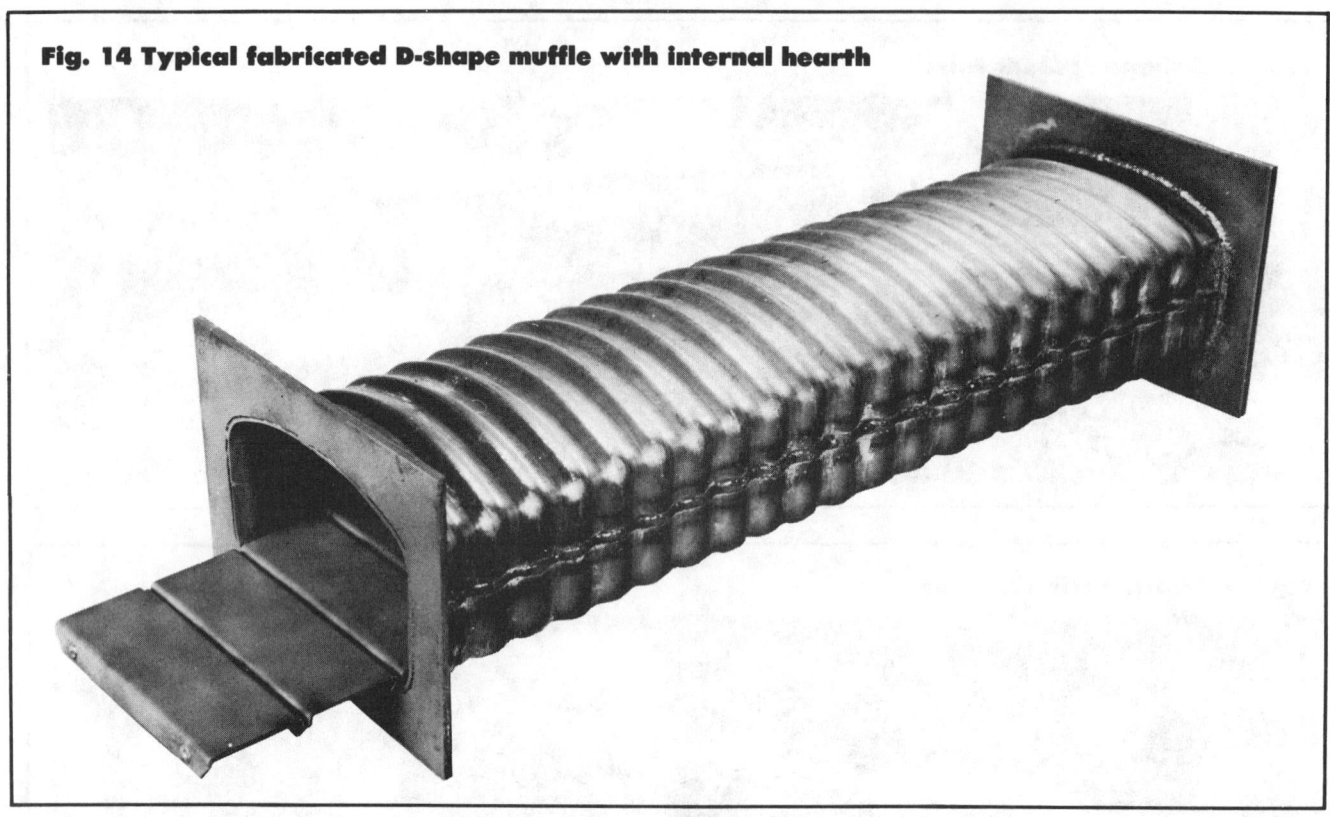

Fig. 14 Typical fabricated D-shape muffle with internal hearth

80Ni-20Cr and type 314 stainless steel. The selection of the proper alloy is based on temperature, atmosphere, possible process contaminants and cost/life ratios of the application. In addition to material selection, other key considerations for mesh belt applications are belt support, drive system, proper tension and control of side travel.

In applications involving heavier loading, the cast link belt is often used. These applications tend to be in the temperature range of 790 to 1095 °C (1450 to 2000 °F) and not in the low temperature range 260 to 790 °C (500 to 1450 °F). Materials, therefore, are similar to the high-temperature mesh belting alloys, except 35Ni-15Cr is the most popular alloy. It provides acceptable service in most conventional heat treating applications, such as hardening, gas carburizing and gas carbonitriding. The cast links generally are assembled using a wrought 35Ni-15Cr alloy with a higher carbon level. In the application of cast link belts, consideration should be given to support, drive systems, tension and side travel.

Radiant tubes can be manufactured from cast alloys or fabricated with wrought alloys and, in most applications, can be selected interchangeably depending on cost/life ratios. Fab-

rications may be selected because of the direct savings in fuel resulting from reductions in weight (fabricated tubes can weigh as much as one-quarter of the equivalent cast tubes). Also, the smooth surface of a fabricated tube is beneficial in avoiding focal points of concentrated or accelerated corrosion. The sound, smooth interior of the wrought tube permits optimum design stresses and helps to prevent the build-up of soot deposit. Figure 12 shows a typical U-shape radiant tube used in carburizing furnaces. Some furnaces use the straight radiant tube shown in Fig. 13.

A nominal 35Ni-15Cr alloy provides acceptable service life in most applications, and the material and proper application are basically the same as outlined in the discussion of baskets and fixtures. In addition to temperature and atmosphere, consideration should be given to tube design for proper expansion and contraction, support for horizontal mounting, and burner positioning to prevent flame impingement. These considerations as well as dissipation rates affect service life as severely as material selection.

Pots. Furnace design is the most important consideration in the selection of material for pots holding molten

lead or salt. Externally heated pots act as a muffle or barrier between the heating and work zones. This type of service is severe because of the great difference between outside and inside temperatures, especially while the furnace is being heated to the operating temperature, when the outside of the pot is subjected to maximum heat input and the lead or salt it contains is still solid.

When the furnace is heated by immersed or submerged electrodes, the pot is completely sealed from the outside air, and the inside of the pot is protected by the molten bath. A pot in this type of installation lasts much longer than an externally heated pot. Because of environmental reasons, salt operations, such as those using cyanide salts, have become extremely limited. The most popular operations remaining generally involve neutral salt and lead. The specific alloy selected for pots used in salt operations is directly related to salt composition.

Pots are available in both cast alloys or fabricated wrought alloys. However, the availability has become somewhat poor for cast pots, and therefore, fabricated pots are more widely used. Carbon steel pots can be used within a temperature range of 260 to 540 °C (500 to 1000 °F). For applications between 540 and 815 °C (1000 and 1500 °F), 309 stainless, 35Ni-15Cr and higher nickel alloys can be applied.

Electrodes. Choice of heat-resistant alloys used for electrodes depends chiefly on the type of furnace in which they are used. The most popular alloy for neutral salt pot electrodes is type 446 stainless steel. Immersed electrodes deteriorate rapidly along the line where the surface of the salt bath comes in contact with them. This is known as "air-line attack". Submerged electrodes, entering the bath through the side of the furnace, are never exposed to air and last much longer. This type of electrode is used only with ceramic pots.

Electrodes deteriorate badly at the salt line during the start-up period. Better service is obtained by maintaining them at a temperature just above the freezing point of the salt during short periods of inoperation. This practice not only prolongs the life of electrodes, but eliminates the tedious task of starting a cold bath. Very little power is required to hold a well-insulated, unused furnace at about 705 °C (1300 °F).

Retorts and muffles are used in heat treating furnaces to separate materials being heated from the products of combustion and, in some instances, to contain atmospheres that would otherwise escape through more porous containment vessels. In most situations, a muffle could be made either of metallic or nonmetallic materials. A typical D-shape muffle with internal hearth is shown in Fig. 14.

Muffles are treated as a separate category of HT alloy applications because an important and different set of constraints apply, in that the heat necessary to raise the inside of a muffle to the proper process temperature is applied from without. Materials and designs must be selected that will not only withstand the rigors of furnace temperature and corrosion conditions, but will, in addition, not act as a significant barrier to heat transfer. Designs must (a) provide for expansion and contraction, (b) be atmosphere tight, and (c) provide maximum area for radiating surfaces because most muffles do not include internal recirculation features. For this reason, many cast or fabricated muffles are corrugated in design. This design increases the radiating area while assisting in accommodating expansion and contraction as the muffle is cycled to and from operating temperature. Heat is transmitted to the inner wall radiating surface of a muffle by conduction, and in order to transfer heat, there must be a temperature drop across the wall of the muffle. The temperature drop is directly proportional to the thickness of the muffle wall. With heavy wall construction, the out-side temperature must be raised to effect a given temperature within the muffle. Muffle material should be selected to provide a balance between alloy content, which represents strength, cost and wall thickness.

Cost of any specific furnace part or fixture increases as the alloy content increases, although not necessarily in the same proportion as the base cost of the alloy. Some cost items will be about the same regardless of the type of alloy used.

To be meaningful, computations of cost for furnace parts and fixtures must be based on the number of hours of operation. In many instances, the more expensive alloys will prove more economical when computed on this basis. For example, service comparisons show that HU may be less expensive than HT for oil quenched carburizing trays, and HW may be cheaper than HT for oil quenched carburizing fixtures. Other examples, such as brazing belts, show that the alloy of lower initial cost also may be less expensive when judged by cost per service hour.

From a practical standpoint, even cost-per-service-hour data may be incomplete. Other factors should be considered for some components, notably: (a) labor cost of replacement, (b) loss of productivity during downtime, and (c) the possibility of damage to other components when failure occurs.

Production Systems

Improved fuel efficiency and temperature uniformity in heat treat furnaces are accomplished by an external high-volume recirculation system utilizing only 10 % excess air. Advantages over direct-fired combustion also include the ability to maintain temperature uniformity at variable production levels. Use of recuperators can result in 10 to 30 % fuel savings.

Heat treatment of oil country tubing

Max Hoetzl, Steel Mill Product Engineer, Surface Div., Midland-Ross Corp., Toledo, Ohio

A TYPICAL oil country tubing heat treating line, illustrated in Fig. 1, consists of an austenitizing furnace, quench units, a tempering furnace together with transfer tables, cooling beds and storage cradles.

After heating in the austenitizing furnace to the austenitic range of 1500 to 1700°F, tubing is either quenched or air cooled depending on the desired metallurgical properties. Quenching is usually performed in a series of spray rings at the exit end of the austenitizing furnace.

If the product is to be normalized, it is dummied through the quench units (without water quenching) onto cooling bed No. 1 and allowed to cool by natural convection. The tubing is then reheated between 800 to 1400°F in the tempering furnace where it is held at the desired temperature for 10 to 60 min, depending on metallurgical requirements. After tempering, the tubing is air cooled on cooling bed No. 2.

Although the austenitizing and tempering furnaces usually work in conjunction, the line layout shown in Fig. 1 also provides the flexibility to operate the furnaces separately.

This feature is facilitated by each furnace having its own inlet table and storage cradles. In addition, the tempering furnace can be charged from either side.

Furnace design

Heat treat furnaces are generally alloy-rail walking beam furnaces (Fig. 2). The furnace shown in Fig. 2 has three combustion chambers along its length and three zones across its width, which provide nine zones of furnace temperature, and fuel:air ratio control.

A typical furnace and tube temperature profile for an austenitizing furnace is shown in Fig. 3. The charge zone temperature is kept at a relatively low level to heat the tubes gradually, allowing tube stresses to be relieved slowly and evenly.

Tubes are heated uniformly through the walking beam motion of the furnace and will usually make two or more complete revolutions while in the furnace. Tube rotation and low charge zone temperature prevent tubes from bowing and becoming misaligned. The rotation also maintains tube straightness in the higher temperature zones, where the tube might sag if not rotated.

Fuel efficiency — The furnace illustrated in Fig. 2 is exhausted at the charge end. By exhausting from the lowest temperature zone, additional heat is extracted from the burner gases leaving the higher temperature zones. This reduces the fuel requirement in the charge zones and improves overall furnace efficiency.

To improve the efficiency further, a recuperator is installed in the flue between the furnace and the stack which

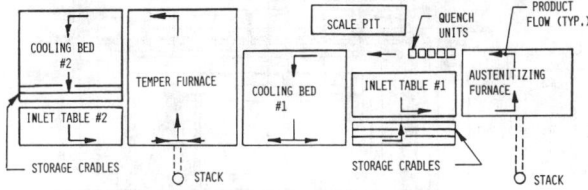

Fig. 1 — Typical oil country tubing heat treat line.

Fig. 2 — Austenitizing/tempering furnace for oil country tubing.

Courtesy Association of Iron and Steel Engineers, extracted from Iron and Steel Engineer magazine, October 1982, 36-38

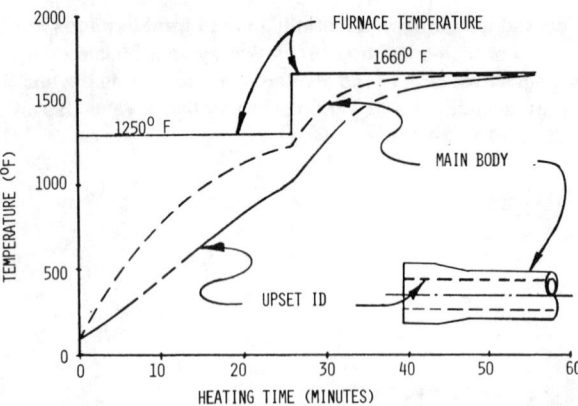

Fig. 3 — Typical furnace and tube temperature profile for austenitizing furnace.

transfers heat from the waste gases to the combustion air. By applying recuperation to an austenitizing furnace, fuel savings of 10 to 30% can usually be obtained. Actual savings are a function of the type of fuel, furnace exhaust temperature, preheat air temperature and the amount of excess air.

The relationship between the specific fuel consumption of an austenitizing furnace, excess air and combustion air preheat temperature is shown in Fig. 4. These data are based on natural gas firing and a furnace exhaust temperature of 1250°F. Fig. 4 shows that more fuel is used to heat a ton of steel with increasing excess air levels and decreasing combustion air temperatures. Fig. 4 also shows that preheated air has a leveling effect on the fuel efficiency at high excess air levels. With large amounts of excess air, a significant por-

tion of the fuel is used to heat the air to furnace operating temperatures. By using waste gases to preheat this air, less fuel is required.

For many years, excess air was used to achieve temperature uniformity in heat treat furnaces. The excess air lowered flame temperatures and reduced hot spots in the furnace. It also provided a high mass-flow of burner gases to even out furnace and tube temperatures. However, with high excess air levels, the fuel efficiency was poor.

To improve fuel efficiency and temperature uniformity, a new generation of heat treat furnaces is being built. These furnaces utilize high-volume recirculation systems for temperature uniformity, and combustion systems with only 10% excess air for fuel efficiency.

A side view of a 100-ton/hr austenitizing furnace with an externally recirculated charge chamber and a direct-fired discharge chamber is shown in Fig. 5. External recirculation systems, which incorporate a hot fan, are generally limited to temperatures of approximately 1500°F. (Because the discharge zone operates in the range of 1750°F, hot fans are not considered from a maintenance standpoint.) A charge end view of this furnace is shown in Fig. 6. It has four zones across its width with each zone having its own recirculating fan, fuel:air ratio controls and temperature controls. This furnace is installed at Algoma Steel, Sault Ste. Marie, Canada.

A side elevation of an externally recirculated tempering furnace is shown in Fig. 7. Since the maximum operating temperature is 1450°F, recirculating fans are used in all zones. The end view is identical to the austenitizing furnace (Fig. 6). This furnace is also located at Algoma Steel, and works in conjunction with the austenitizing furnace previously described.

An isometric view of an external recirculation system is illustrated in Fig. 8. Flue products are withdrawn from the furnace by the recirculating fan and then pass through a

Fig. 4 — Relationship between fuel consumption, excess air and combustion air temperature.

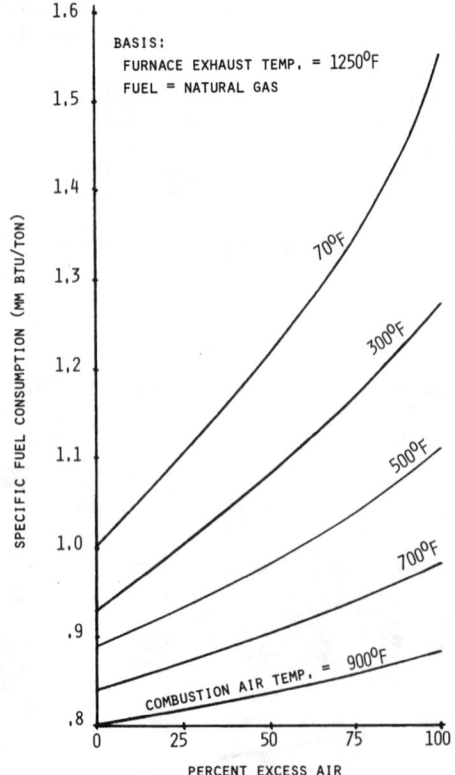

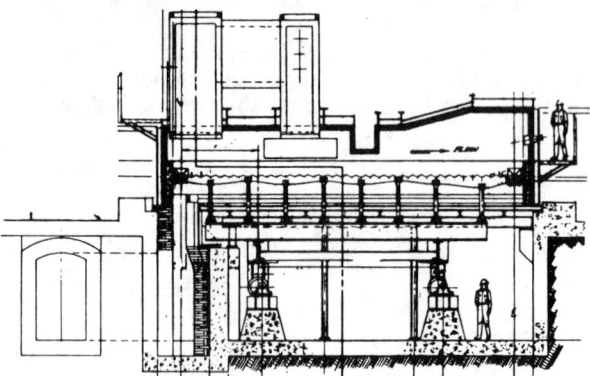

Fig. 5 — Walking beam austenitizing furnace (side view) with externally recirculated charge chamber and direct-fired discharge chamber.

Fig. 6 — Walking beam austenitizing furnace—charge end view.

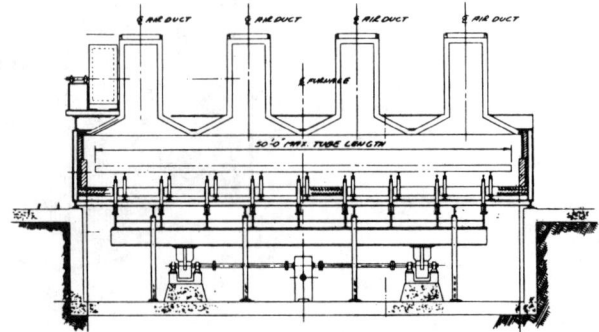

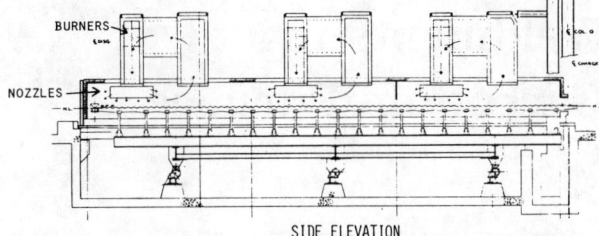

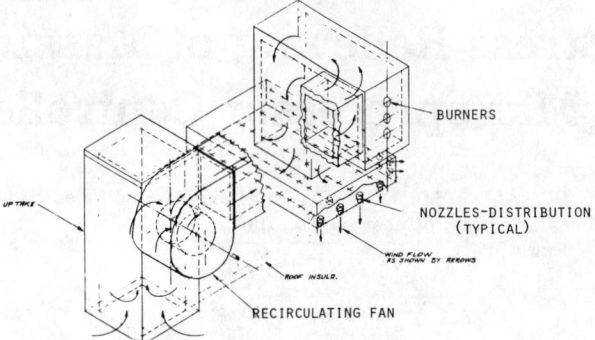

Fig. 7 — Externally recirculated walking beam tempering furnace.

Fig. 8 — External recirculation system.

combustion chamber. The burners, which operate with 10% excess air, heat the recirculated gases. The combined flow of hot gases are then distributed evenly through a series of plenums and nozzles. An added benefit of this system is the high rate of convection heat transfer, which is four to six times greater than in a direct-fired system. As a result, less floor space is required for a tempering furnace with external recirculation compared with a direct-fired furnace.

Product quality

With emphasis on increasingly deeper wells, the physical properties of oil well tubing have become more critical. Therefore, a consistently uniform product from the heat treat furnaces has become more important.

The external recirculation system has an advantage over the direct-fired combustion system because the mass flow of recirculated gases in an externally recirculated system is much greater than in a direct-fired system. As a result, the temperature drop in these gases (as they give up heat to the product) is less with external recirculation. This produces a more even furnace and tube temperature profile. Tube temperature uniformities of ±5°F are common from the tempering furnace at Algoma Steel, whereas large amounts of excess air are required to obtain ±10°F uniformity with a direct-fired furnace.

At less than rated production, the external recirculation system will maintain or improve temperature uniformity, because the mass flow of gases remains constant and the temperature drop is reduced with reduced production.

Temperature uniformity in the direct-fired furnace is less with reduced production because the mass flow of gases is related to burner input. As production is reduced, heat input and volume of burner gases are reduced. Consequently,

poorer heat and temperature distribution in the furnace result, which in turn, cause a wider temperature gradient in the tubing. To offset this problem, some direct-fired furnaces are operated with a constant amount of combustion air and only the fuel flow is decreased during reduced production. Although this maintains temperature uniformity, it sacrifices fuel efficiency.

An alternate approach is the use of high-velocity burners which create internal recirculation by the burner gases entraining many times their volume of furnace gases. The entrainment action lowers the flame temperature and causes the furnace gases to recirculate.

Although high-velocity burners improve temperature uniformity without the use of excess air, their recirculation effect is also limited by reduced production levels in the furnace. Where these conditions are frequent and temperature uniformity is critical, recirculating fans are recommended.

Summary

The rising cost of fuel and increases in demand for oil country tubing justify improvements in the tubing heat treat furnaces. Many options are available to improve furnace efficiency and product quality. Some of these options, such as recuperation and high-velocity burners, can often be applied to existing furnaces with a reasonable return on investment. Other options, such as external recirculation and the walking beam motion, should be considered in the original design. The tools for reducing specific fuel consumption and improving product quality are available. Ⓐ

Stress Relieving of Massive Weldments in Microprocessor Controlled Carbottom Furnace

By GERALD J. MacDONALD, Engineering and Sales Manager
Chesmont Engineering Co., Inc.
Eagle, Pa.

A microprocessor controlled carbottom stress relieving furnace at the Plate Fabricating Div. of Morris, Wheeler & Co., Trenton, N.J., is the key to successful production of large, heavy weldments. Known for their production of complex welded sections, the company fabricates components that include turbine cases, gear boxes, machine enclosures and various engineered products ranging up to 20 tons or more with wall thicknesses up to 12 in. (30.5 cm) thick. One of the principal products made is turbine castings, ranging from 3 in. to 4 in. (7.6 cm to 10.2 cm) thick, for the Delaval Turbine. Production of the massive weldments, Figs. 1 and 2, require highly skilled personnel, rigid quality standards and highly accurate stress relieving techniques.

Stress relieving specifications have become more and more stringent. Operator overtime and extreme care were necessary to meet customer specifications with existing fixed set point controls of a carbottom furnace. Many of the stress relieving cycles required 10 or more hours for heat-up, soak and controlled cooling. Therefore, as part of a $2.5 million expansion and renovation program a carbottom furnace was replaced with one that is microprocessor controlled and fully programmable (see Fig. 3).

During stress relieving heavy welded sections require close control to avoid the overheating of thinner sections. With the thin section used as the control point and the thick section also having to reach the stress relief temperature, obviously there is need for a patterned heat control system to cope with the nonuniformity of thicknesses.

Bill Turner, General Manager of the Plate Fabricating Div., expanded upon the reasons for the decision to install a modern computer-controlled furnace of special design.

(1) It would improve control, eliminating the need for manual adjustments and associated overtime labor, and provide hard copy documentation.

(2) It would improve quality and prevent any possibility of overheating or cooling stress in a critical area of a valuable weldment.

(3) It would provide dual fuel capability and reduce fuel costs. A fiber lining would save energy by reducing heating and cooling time for faster turn around.

(4) It would allow M-W to seek outside stress relieving work that requires "state of the art" furnace operation.

Fig. 2 High custom fabricated vessel is ready for shipment.

Fig. 1 Fabrication of heavy plate weldment.

Fig. 3 High velocity, dual fuel fired carbottom furnace under microprocessor control stress relieves wide range of loads of fabricated steel.

Special Furnace Design

The new furnace, designed by Chesmont Engineering Co., Eagle, Pa., was customized to handle Morris, Wheeler's stress relieving requirements. It was desired to handle a wide range of loads without special procedures and still maintain accurate stress relieving with uniformity. It was necessary to treat a variety of weldments, frequently with the same furnace cycle, and with the close control required this capability necessitates considerable flexibility. The furnace, as built, can heat loads from 2000 to 20,000 lb (907 to 9072 kg) from ambient to soak temperature at 100°F/hr (38°C/hr) and cool at 100°F/hour.

The furnace is top fired by five high velocity, dual fuel burners (Tempest II made by North American Manufacturing Co.). These burners were selected because their 13:1 turndown (high to low firing rate ratio) provides high degree of flexibility.

The microprocessor control system (MicRIcon 823, Fig. 4, made by Research Inc., Minneapolis, Minn.) was chosen for programming flexibility, alarm features, program storage and expandability for future furnaces. The operator merely determines the required profile (program) that meets the required specifications, enters the program into the memory, assigns alarms to prevent overheating and initiates the cycle.

The programmer carries out the cycle, including shutting down of the combustion system upon completion.

The furnace lining system, Fig. 5, is made up of both ceramic fiber (Kaowool 2200°F made by Babcock & Wilcox) and castable refractories for fuel efficiency and wear resistance. The fiber lining is anchored by 330 type stainless steel anchors that are highly resistant to oxidation and other thermal problems at elevated temperatures. The fiber lining consists of 6 in. of 4 lb/ft³ (15.2 cm of .06 g/cu m) density back-up material with a hot face lining of 2 in. (5 cm) thick, 8 lb/ft³ (0.13 g/cu m) density. The flues are vacuum formed ceramic fiber, each having an I.D. of 5 inches (12.7 cm). The lower sides of the furnace are lined with 2500°F (1371°C) insulating and dense castables.

The furnace car has both insulating and dense castable refractories rated to a maximum of 2500°F (1371°C) at the hot face. The car arch door seal consists of dense castable with 2% stainless steel fibers for superior resistance to erosion and spalling during thermal stress and shock associated with batch cycling processes.

Programmable Control System

The microprocessor/controller system for regulation of the new furnace is shown in Figs. 6 and 7. Temperature of weldment is sensed by a 1/8 in. (0.3 cm) O.D., type 316 stainless steel thermocouple (A) and is recorded. The microprocessor-based controller (B) senses the furnace temperature, compares it to the programmed set point and sends a proportional 4 to 20 ma signal to position the controller (C) which regulates the heating/cooling cycle in accord woth a "profile" or graph in the computer's memory, that is entered through the programmer keypad.

The position controller drives the modulation motor (D) clockwise (CW) or counterclockwise (CCW) and positions the air butterfly valve (E) in "proportion" to the 4-20 ma signal initially received by the controller. The air valve is characterized by 12 points of adjustment. Combustion air flow is "linearized" for optimum control; that is, 25% of stroke is 25% of air flow and 50% of stroke is 50% of air flow. The air flow signal is sensed by the gas valve (F) for proper gas flow — also in a linearized manner. The burners (G) are balanced (both fuel and air) for equal distribution of fuel input.

The variable ratio regulator (H) allows for stoichiometric

Fig. 4 Closeup view of microprocessor and position controller on control panel for dual-fuel carbottom furnace.

Fig. 5 Interior of carbottom furnace showing the ceramic fiber lining installed with type 316 stainless steel anchors, and dense castables at lower sides.

Fig. 6 Panel with microprocessor/controller system for programmed stress relieving of heavy weldments and custom fabrication in dual-fueled, carbottom furnace.

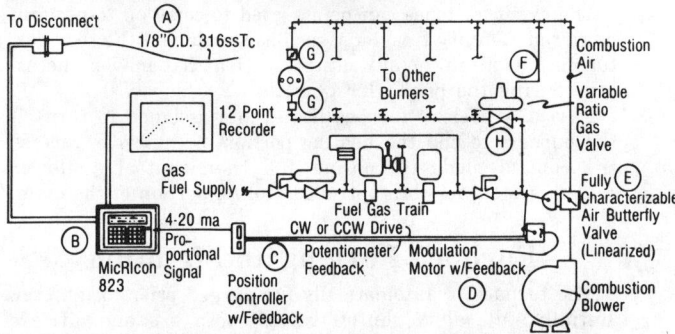

Fig. 7 Schematic diagram of microprocessor/controller system and combustion control in operation of carbottom furnace for stress relieving massive fabrications.

Source: Industrial Heating, December 1983, 24-26

or on-ratio firing at higher firing rates and leaner or 'tempered' flame temperatures at lower rates for soaks, as required. Products of combustion (POC) are always at a high volume to promote rapid recirculation of furnace atmosphere, thereby contributing to temperature uniformity of load.

The microprocessor control parameters, gain, reset and rate (3 mode or PID control) can be programmed to change as the cycle progresses. As the furnace responsiveness changes with temperature due to furnace wall radiation, burner recirculation characteristics and varying stack loss, this can be adjusted for and the control accuracy maintained through the full range of the furnace (0-2200°F) (–18 to 1204°C).

Memory and Storage Capability

There are 80 segments within the microprocessor memory. Each program or profile typically consists of ten segments, including a controlled heat-up, soak and controlled cooling. On the basis of utilizing ten segments, a minimum of eight separate programs can be stored, each with assigned alarm groups and event switches. (A digital cassette/printer can allow additional programs to be permanently stored and kept on file for future use).

Alarm Capabilities

To assure that costly weldments are not damaged during the heating process, a full selection of alarms can be programmed into the microprocessor system. High temperature deviation, low temperature deviation, high furnace temperature, low furnace temperature, and thermocouple burn out can be programmed to:

(1) Alert operating personnel by flashing on the programmer display;

(2) Hold the program until the system stabilizes itself;

(3) Abort the program to another segment;

(4) Abort completely and end the program;

(5) Engage an event switch or solid state relay (SSR) that can sound an alarm horn or energize any device that may be necessary in the particular situation.

All of these alarm conditions constitute an 'alarm group' and there are four independent groups that can be individually programmed. Any one alarm group can be assigned to any profile segment with instructions to alert, hold, abort, and/or engage an event SSR.

Morris, Wheeler's programs frequently assign alarm groups to hold the program should small temperature fluctuations occur, abort programs should a control thermocouple fail, and hold all programs entering the soak segment, until it is assured that the correct temperature is reached prior to the soak cycle beginning. This latter hold function guarantees the soak temperature as specified.

Event Switches

The event switches can be assigned to come on to perform more rudimentary functions, such as turning off the recorder, turning on a fan, or any other function that may be necessary during the particular cycle.

Morris, Wheeler's furnace has events assigned to turn off the burners to end the heating portion of the cycle, reverse the control mode to cooling and then control a blower cooling system for the controlled cooling portion of the cycle.

Furnace Purging and Ignition

The furnace is automatically pre-purged prior to ignition. Initially, all safety shutoff valves (both gas and oil) are proven closed by interlock switches. In addition, the fuel control valve is proven in low fire position prior to trial for ignition. The burners are ignited by gas pilots. Both pilots

and main flame are proven by the ultra-violet flame scanning system with a first-out annunciation indication. Should any pilot or main flame fail, the burner causing the failure is indicated, aiding greatly in any troubleshooting.

Typical Cycle

In a typical cycle a variety of steels for many applications will be stress relieved. These steels include A285 Grade C for pressure vessels, A515 Grade 70 for special pressure vessels, A516 Grade 70 for cryogenic vessels, A204 Grade B (carbon-molybdenum) and various grades of structural steels, including A588 (weathering types; that is Corten equivalents).

The cycle consists of heating the load to 600°F (316°C) (an intermediate stabilizing segment), heating up to 1150 to 1200°F (621 to 649°C) at the rate of 200° per hour, soaking for a specified period (dependent upon material section thickness), cooling to 900°F (482°C) or so (with fuel on) at 150°F (66°C) per hour, then cooling with air only below 900°F (482°C) at the rate of 150° per hour until 600°F (316°C) is reached again.

Up to 24, 1/8″ (0.3 cm) 315 stainless steel type K thermocouples may be employed for documentation of the stress relief treatment, each thermocouple being located at a strategic area. This documentation is filed with a copy forwarded to the customer.

At typical soaks of 1150°F (621°C) the results are ±2°F at each control point. A furnace temperature uniformity profile indicates ±15°F with top firing only. The high velocity burners create furnace atmosphere recirculation at temperatures below refractory radiation temperatures, thus accounting for good uniformity during lower heating-up temperatures.

Future Plans

The microprocessor control system has capacity for operation of seven additional furnaces in the future. Each furnace would merely require the addition of one input/output (I/O) card and a proportioning controller.

Plans call for a smaller furnace to handle minimum loads, and a larger, oil-fired furnace already has piping provisions for the future conversion to gas-firing combined with computerization.

Improved Overall Operations

The programmed furnace, the 'work horse' since its installation, has eliminated the need for monitoring the heat treat process. Now, hours after personnel have gone for the day, the furnace completes the process and even shuts itself off. The next morning, turn around can be immediate. Management of operations is improved with more flexible scheduling, improved furnace efficiency and enhanced quality.

The increasing demand for production of a quality product at a competitive price is a well known fact. At Morris, Wheeler & Co. they are taking steps towards continuing their long record of excellence by combining their expertise with the best of today's available technology.

Information Retrieval
For your guidance and filing convenience the following subjects are included in this article:

Key: microprocessor control, carbottom stress relieving furnace, turbine cases, gear boxes, ceramic fiber lining high velocity dual-fuel burners, insulating and dense castables, stainless steel fibers, controller, air butterfly valve, variable ratio regulator, digital cassette/printer, thermocouple, alarms, blower cooling system, safety shut-off valves (gas and oil), gas pilots, ultra-violet flame scanning system, pressure vessels (A285, A515 Grade 70), cryogenic vessels (A204 Grade B-C, Mo), type 316 stainless steel, type K thermocouples, proportioning controller

Heat Treating Steel Castings

Ray Pritchard
Larson Consolidated, Inc
Grafton, OH

Almost all steel castings produced today are heat treated prior to shipment. Heat treating is performed to obtain certain desirable properties of the cast steel. The basis of any heat treat process involves four fundamental decisions:

- the rate of heating;
- selection of a proper holding temperature;
- the holding time at the given temperature;
- and the rate of cooling.

Frequently it is necessary to repeat this sequence more than once in order to obtain the desired results. Many factors must be considered in exercising good judgment in determining the four above mentioned items. Certainly the steel chemistry, size and thickness, configuration and the required results of the castings are just as important, but let's look at these four items first.

Rate of heating is far less critical as far as end result than most people believe. However, many people still feel that consideration should be given to the following items:

- **Chemistry**—Some alloys are prone to cracking due to uneven stresses occurring when a relatively wide temperature variation exists between the interior and exterior surface of the casting. In most cases this is not true as the amount of heat travels rapidly through the casting.
- **Thickness**—A heavy section casting may crack upon rapid heating due to stresses caused by temperature variation. Again this is false in most cases.
- **Variations in section size**—Expansion varies for section size and can result in cracking. This is true because of the chance of gross differences in section size.

The control of the rate of heating is particularly important between room temperature and about 1200F in steel castings. Cracking is most prevalent at about 600F but may occur with slightly less risk up to about 1200F. Heating rates in the past have been specified for steel castings as low as 20F per hr. This is necessary only in extreme cases. The primary control factor for heating should be to obtain a reasonably uniformed temperature both in the furnace and in the casting.

Opinions relating to desired holding temperatures and time at temperatures vary. Generally 50-100F above the AC3 is accepted by the industry as correct, however there are some who feel that slightly higher temperatures may be needed for heavy section castings. In regards to holding times, for many years it was considered that holding times should follow the one hr/in. of section thickness rule of thumb. In recent work it has been shown that much less than this is required.

In order to understand the subject of cooling rates one must understand the results of the different cooling rates and practices.

Annealing is generally used on low carbon steels to produce a soft, easily machineable casting with high ductility and relatively low strengths. Castings are heated above the transformation temperature, held for the prescribed time and permitted to cool in the furnace below 800F to obtain a microstructure of ferrite and some pearlite.

The procedure for normalizing is basically similar to annealing, except the castings are removed from the furnace while at temperature and allowed to cool in the air. The microstructure again is pearlite and ferrite only more pearlite is obtained. Tensile strengths may vary as high as 100,000 psi and yields strengths will be higher than obtained by annealing. Microstructures and strength levels will vary with the rate of cooling, that is whether cooled in still air or with a fan. Care should be taken to insure that the castings cool uniformly to ensure consistent results.

Normalizing is frequently used because of its lower cost resulting from less time in the furnace and minimal quenching expenses. Frequently tempering is performed after normalizing. Quenching is a hardening treatment that is similar to normalizing except that the casting is cooled more rapidly, usually in water or oil. The fast cooling results in a microstructure that is predominantly martensite in alloy steels.

Although water quenching is preferred, oil is used frequently because it is a slower quench and minimizes cracking problems in some high carbon or complicated variable section sized castings. Quenching does create stresses and is normally followed by a tempering treatment.

Tempering and stress relieving are basically the same thermal treatment but for different reasons, though both relieve stresses. In these processes the casting is heated to a temperature below the transformation range, usually 500-1300F, held for a specified amount of time and cooled. Tempering is performed after quench hardening and many times after normalizing to relieve stresses and adjust mechanical properties to the desired level by modifying the metallurgical structure.

Tempering is time/C temperature related and is therefore best run at higher temperatures and short times to give maximum toughness. Care must be taken to avoid tempering any of the alloy steels in the temper embrittlement range which is generally from 800-1100F or reduced impact and mechanical properties will result.

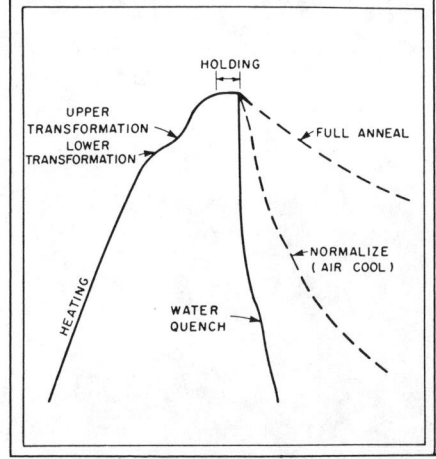

Fig. 4. This generalized sketch shows the steps in heat treating steel castings.**

Heat treatment is an important part of steel casting production because it develops the mechanical properties, however, selection of the right chemistry to be heat treated to meet the requirements is a must. There is a wide variety of chemistries that can meet many specifications if properly heat treated. Thus in any heat treating operation the first requirement is the selection of the proper alloy. This selection should be the best alloy for the job required. This takes reviewing the specified requirements, mechanical properties, impact of desired strength levels, the cost of the alloy and its heat treatment, etc.

The second thing is to make sure that the heat treat facilities are upgraded for maximum efficiency. Controlling heat treating temperatures by measuring them in the air of the furnace is not the way to get the most efficient applications of the furnace. Frequently by testing with thermocouples placed in the load of castings a correlation can be obtained between the readings of the oven thermocouple and what is actually happening with the castings. Also by putting thermocouples in test blocks of known dimensions an accurate picture of the heat castings are subjected to can be graphed.

Use a rapid rate of heating wherever possible. Slow heating just costs money

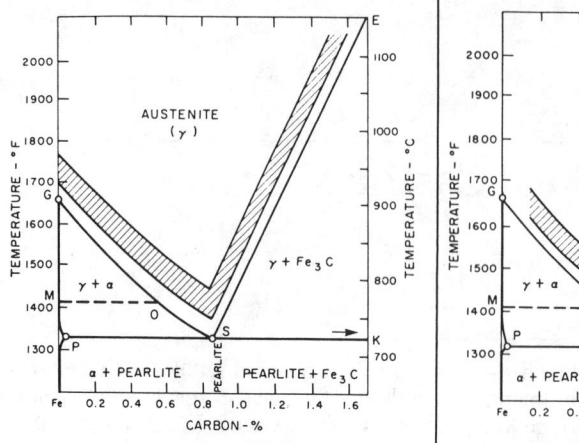

Fig. 5. Normalizing temperatures for carbon steels are illustrated here.**

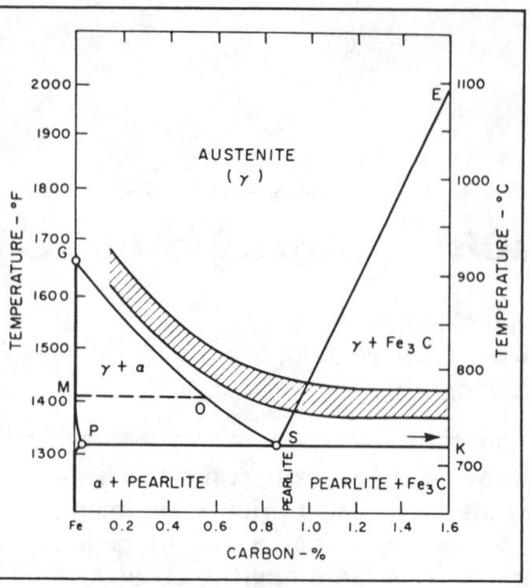

Fig. 6. Hardening temperatures for carbon steels are illustrated here.**

and unless you have a casting that is definitely one that is sensitive to cracking at high heating rates heat as rapid as possible. Holding temperatures should also be held to the minimum required as excess heat only increases costs. With the use of accurate thermocouples, excess heat will rarely be required.

Seldom is it necessary to hold any heat of steel for over an hour at temperature. To hold it longer than that is just a waste of time and may lead to poor practices where a dependence on that

heat treat time is required. Holding too long can also cause some grain growth which results in lower strength values.

Saving energy and furnace time is like money in the bank. Thus every effort should be made to heat to temperature rapidly and keep holding times to a minimum.

Since heat treating is a basic requirement of almost all steel castings, it is important that it be controlled. The proper selection of alloys and their heat treatment in an efficient furnace will result in a quality casting.

308

SECTION X
Fluidized Bed Process

Shown here is a gas fired fluidized bed furnace in action. Dry particles are fluidized by a stream of gas fed up through the bed. Rapid carburizing and heat transfer is achieved with good thermal efficiency and temperature uniformity.

Fluidized Bed Heat Treating

By Wally L. Bamford

MORE EXACTING metallurgical specifications call for greater precision in heat treatment operations. Modern processes require closer control of atmosphere, temperature level, and uniformity while complying with the need for reduced operating costs and acceptable environmental conditions for the operator. The use of fluidized beds in heat treatment is increasingly meeting these demands.

The major advantages of fluidized beds can be summarized as follows:

1. All types of ferrous and nonferrous materials may be treated.

2. High heat transfer rates can be achieved.

3. The atmosphere in the heating zone can be immediately adjusted to suit the requirements of the treatment.

4. There are no problems of fume or effluent disposal.

5. The units have high thermal efficiency and low fuel consumption. This results in low operating costs under service conditions.

6. As the furnace heating up time is relatively short, the bed can be shut down overnight without losing production time the following day.

7. Fluidized solids are nonabrasive, noncorrosive, and do not wet immersed objects.

Reprinted from Metal Progress, September 1981, 132-137, © 1981 American Society for Metals

Bed of Particles is Fluidized by Gas Flow

Fluidization involves making a bed of dry, finely divided particles (typically aluminum oxide in the current context) behave remarkably like a liquid. The individual particles become microscopically separated from each other by a moving gas fed up through the bed.

A bed of loose particles offers resistance to fluid flow through it. As the velocity of flow increases, the drag force exerted on the particles increases. When the fluid flow is upwards through the bed, the drag force will tend to cause the particles to rearrange themselves within the bed to offer less resistance to the fluid flow. Unless the bed is composed of large particles (1 mm [0.04 in.] mean diameter), the bed will expand (Fig. 1).

With further increase in the upward fluid velocity, the expansion continues and a stage will be reached where the drag forces exerted on the particles will be sufficient to support the weight of the particles. In this state, the fluid-particle system begins to behave like a fluid and it will flow under a hydrostatic head.

This is the point of incipient fluidization. The pressure drop across the bed will be equal to the weight of the bed although it is likely that this pressure drop will be exceeded just prior to the achievement of fluidization with gas fluidized systems because the residual packing and interlocking of particles within the bed must first be broken down (this is indicated by the hump in the stylized curve in Fig. 1 for bed pressure drop as a function of the fluid flow rate).

At the onset of fluidization the bed is more or less uniformly expanded. Beyond this point, however, the bed behavior is markedly different. The uniform expansion behavior is soon lost except with fine particles; the system becomes unstable and cavities containing few solids are formed. These look like bubbles of vapor in a boiling liquid. The value of the gas flow rate at which this happens depends on the properties of the fluidized solids, the design of the bed, and particularly on the type of gas distributor used. Over the flow range between incipient fluidization and the onset of bubbling, the bed is in a quiescent state. The bubbles are responsible for inducing particle circulation within the gas fluidized bed and it is this circulation which has a most important bearing on the advantageous heat transfer properties of gas fluidized systems. Most beds used for heat treatment are bubbling bed types.

At high fluid velocities, a point is reached where the drag forces are such that the particles become entrained within the fluid stream and are carried from the bed. Small particles tend to become entrained at lower fluid velocities than larger ones, and the way that bubbles burst at the surface of a gas fluidized bed and throw a spray of particles into the space above the bed can considerably affect the rate of loss of particles from a gas fluidized bed. The regime of fluidization lies between the extreme conditions of the packed bed and that of solids transport. Throughout this regime, the pressure drop across the bed remains approximately constant and sufficient to support the weight of the bed.

Although the properties of solid and fluid alone will determine the quality of fluidization (i.e., whether smooth or bubbling fluidization occurs), many factors influence the rate of solid mixing, the size of bubbles, and the extent of heterogeneity in the bed. These factors include bed geometry, gas flow rate, type of gas distributor, and internal vessel features such as screens, baffles, and heat exchangers.

Dense Phase — The most widely used fluidized bed for heat treatment is the dense phase type, although some units have been constructed[1] based on the dispersed phase bed with particle circulation for heat treatment of long thin metal parts such as shafts and plates. In a typical dense phase fluidized bed, the parts to be treated are submerged in a bed of fine solid particles held in suspension, without any particle entrainment, by an upward flow of gas.

Beneficial Features of Fluidized Beds

The use of fluidized beds for heat treatment operations has advantages and disadvantages as opposed to vacuum furnaces, protective gas atmosphere furnaces, and salt baths. A comparison of hardening and carburizing rates is shown in Table I.[2]

Heat Transfer — The high heat transfer coefficient of a fluidized bed, typically between 120 and 1200 W/m^2C (20 and 210 BTU/ft^2hF), is one of its most important properties. It enables products to be heated or cooled at speeds very close to those obtained in conventional salt or lead bath equipment.

The relative heating rates[3] of a 16 mm (0.6 in.) steel bar in a salt, lead, fluidized bed, and conventional furnace are illustrated in Fig. 2A; cooling rates for air, oil, water, and a fluidized bed are shown in Fig. 2B. Figure 3 shows further recovery rate data. The turbulent motion and rapid circulation rate of the particles, as well as the extremely high solid-gas interfacial area account for the good heating and cooling performance of the fluidized bed furnace.

Of all the parameters affecting the heat transfer coefficient in fluidized beds, the particle diameter has the greatest influence. Test data indicated that particle diameter should be as small as possible. However, below a certain size, electrostatic effects could cause problems.

Bed material and fluidization velocity of the gas also play a part. It was concluded that the governing physical property of the bed material is its density. There appears to be an optimum density for bed materials of around 1280 to 1600 kg/m^3 (80 to 100

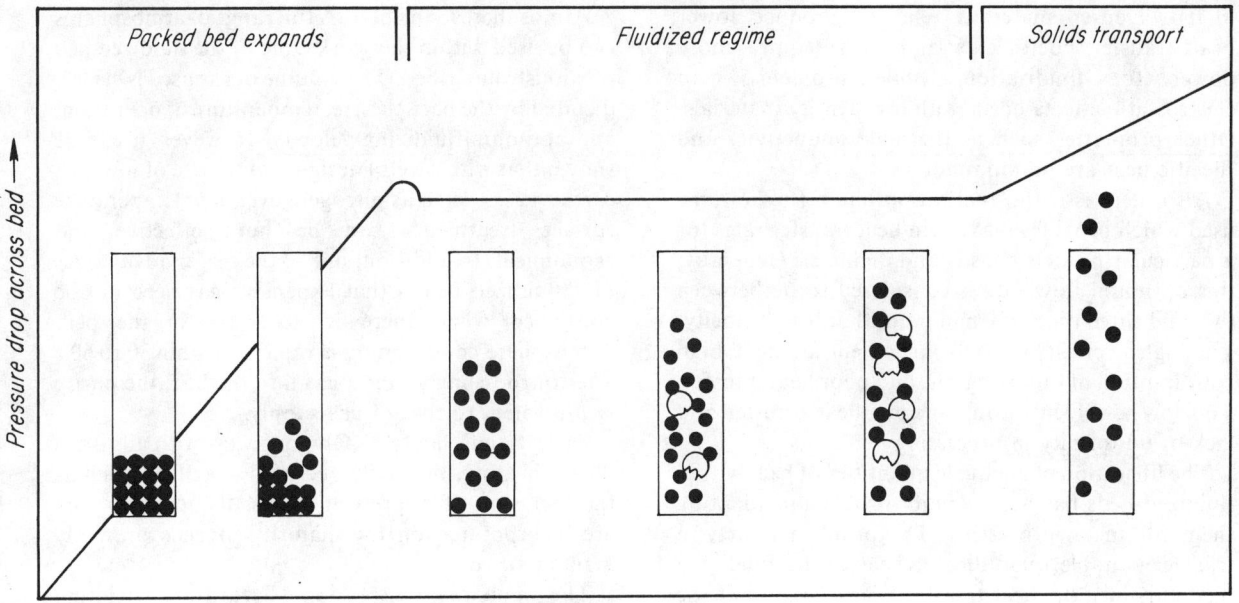

Fig. 1 — Schematic diagram of the fluidization process follows the response of the particle bed as gas flow rate increases. The bubbling seen toward the end of the fluidized regime is responsible for advantageous heat transfer properties.

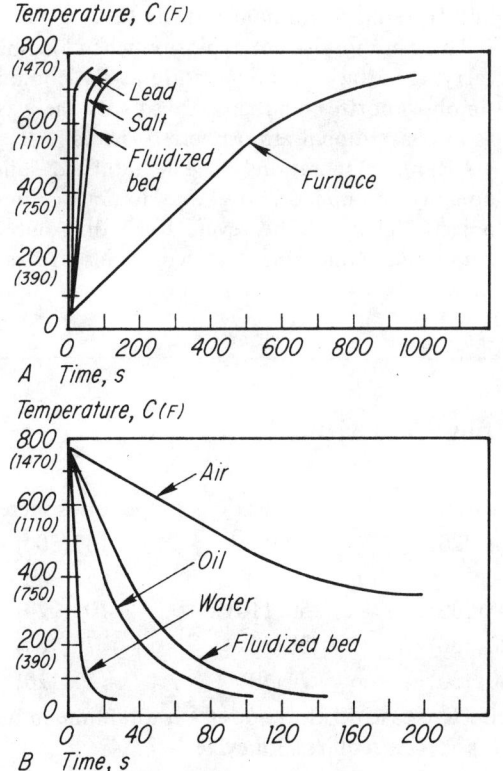

Fig. 2 — Relative heating rates (A) and quenching rates (B) are shown for different types of furnaces and different types of quench media. Steel bars with a diameter of 16 mm (0.6 in.) were treated.

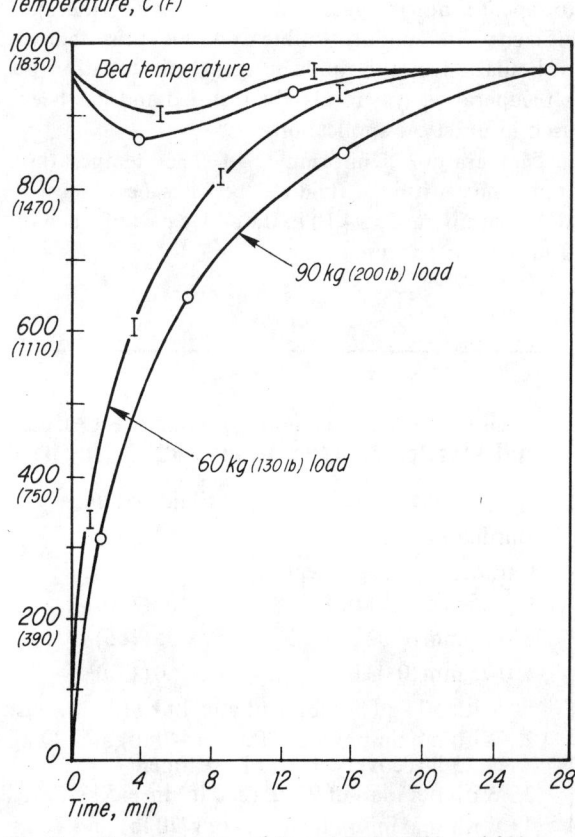

Fig. 3 — Recovery rates of 60 and 90 kg (130 and 200 lb) loads. The 25 mm (1 in.) in diameter steel parts were treated in a 0.3 m³ (10.6 ft³) fluidized bed.

lb/ft³). Denser materials tend to produce lower heat transfer coefficients and also require more power for fluidization, while problems with electrostatic effects occur with low density materials. Other properties such as thermal conductivity and specific heat are unimportant.

Also, it is essential that an optimum flow rate be used which provides maximum heat transfer rates for a particular particle density and diameter. Generally, this optimum flow rate is considered to be between two and three times the minimum fluidizing velocity. Too high a velocity leads to particle entrainment, high consumption of fluidizing gas, and poor heat transfer. Too low a velocity leads to poor heat transfer and lack of uniformity in processing.

The liberation of adequate quantities of heat within fluidized beds has been a major problem in adapting them for metal processing. The problem is first to transfer suitable quantities of heat to the fluidizing medium since the heat transfer characteristics of the bed itself are usually much more efficient than the transfer of heat to the fluidizing gas from the heat source.

In addition, the major part of the heat loss from any practical fluidized system is the heat content of the spent fluidizing gas. In instances in which thermal efficiency is unduly influenced by this factor, recirculation of the fluidizing gas or installation of a recuperative system may be justified and has been used in practical applications.

Temperature Uniformity — The temperature uniformity within a fluidized bed has been shown[4] to be within ±2 C (±4 F) in the active part of a bed of 3 m³ (108 ft³) volume.

Atmosphere Control — A full range of atmospheres can be used within the work zone of the fluidized bed previously described. The volume of gas used is clearly dictated by the particle size, temperature of operation, and optimum fluidizing velocity. However, it can be shown that with careful design and the use of low cost carrier gases, such as nitrogen, even low temperature surface treatments can be both effective and economical. In addition, one of the major advantages of a fluidized bed is that expensive gas need not be consumed while there is no work in the bed. Atmosphere conditioning is rapid — within 30 to 60 s after introducing an inert gas into the bed, the purity is equivalent to that of gas supply.

Operational Safety — Obviously, as with all forms of gas heating, normally accepted practices, such as the inclusion of nonreturn valves, flame traps, etc., are incorporated on the majority of beds presently manufactured.

The potential explosion hazard of igniting stoichiometric mixtures of gas and air is overcome by the use of premixing to noncombustible ratios and introducing secondary air into the mass of the fluidizing particles. In addition, the use of a flexible tile concept ensures that any failure of joints does not influence the performance of the bed.

There are no risks of explosion when loading parts carrying surface oil or moisture as the contaminant simply vaporizes, and is removed with the waste gas as in conventional atmosphere furnaces.

Cleaning Operations — The fluidized solids are nonabrasive, noncorrosive, and do not wet immersed objects. There is, however, some drag-out of the aluminum oxide since, as workloads are removed

Table I — Comparison of Furnace Outputs During Carburizing and Hardening Operations, kg/h (lb/h)

	Fluidized Bed	Vacuum	Salt Bath	Atmosphere
Hardening	150 (330)	144 (320)	70 (155)	137 (300)
Carburizing to case depth:				
0.25 mm (0.01 in.)	100 (220)	105 (230)	50 (110)	100 (220)
0.5 mm (0.02 in.)	75 (165)	80 (180)	38 (85)	75 (165)
0.75 mm (0.03 in.)	50 (110)[1]	60 (130)[2]	24 (50)[3]	55 (120)[4]

1. With 50 kg (110 lb) load and 10 kg (22 lb) basket, effective case depth requires 1 h total time in bed.
2. With maximum load 200 kg (440 lb) and 20 kg (44 lb) basket, requires 3 h cycle (1.25 h recovery + 1.75 h treatment).
3. With net load of 95 kg (210 lb) in 115 kg (250 lb) charge, requires 4 h total time in salt.
4. With maximum load 350 kg (770 lb) and 75 kg (165 lb) baskets, requires 5 h total time in furnace (1.75 h recovery + 3.25 h treatment).

from the fluidized bed, some collapse of the particles occurs on flat surfaces. It should be stressed that these particles can be re-used. They can be sieved and returned to the bed. It should also be noted that when parts already scaled or preoxidized are placed in a fluidized bed, particles will tend to adhere to the scale. However, these can be removed.

Defluidization/Bed Collapse — One of the common misconceptions about fluidized beds is that, because of their principle of operation, they are not very suitable for large solid parts with horizontal surfaces remaining stationary in the bed. With parts of this type, a cap of nonfluidized particles collects on the top of the horizontal surfaces and forms a thermal screen. This may appear to be a disadvantage but there are various methods of overcoming it, should it be desirable, which are normally designed into a fluidized bed. It should be noted that the higher the temperature of operation the greater the energy and agitation of the bed, and therefore, the less the likelihood of bed collapse taking place.

These methods are: movement of the parts being tested; introduction of additional agitation in the zone of the fluidized bed around the parts being treated, either by localized injection of fluidizing gas or careful design of the outline of the basket holding the parts; and increased fluidizing velocity.

Bed collapse can be turned to advantage of special heat treatments where, in the case of engineering parts of the shape described above, one area must be hard and tough while the rest must be soft and more ductile. In this case, after uniform heating, the product is submerged in a fluidized quenching bed, with the part to be hardened facing downwards. The top horizontal surface becomes covered with a cap of particles forming a thermal screen which retards the vigorous cooling caused by the fluidized bed.

Applications — Uses to date of fluidized bed furnaces include continuous wire annealing, treatment of tool steels such as AISI D2, marquenching of hot work tool steels, and heat processing of extrusion dies (see Fig. 4).

Thermochemical surface treatments appear particularly promising. Techniques such as nitriding, oxynitriding, carburizing, and FNC (ferritic nitrocarburizing) have been successfully tested in the gas fired fluidized bed.[5,6]

Internal Combustion, Gas Fired Fluidized Beds

A major development in the heating of fluidized beds occurred when an air-gas mixture was used for fluidization and ignited in the bed, generating heat by internal combustion (see Fig. 5).

The obvious advantage of this technique is that the bed is being fluidized by burning gases so that the heat is generated within the bed. In gas fired fluidized beds, if the vessel is well insulated, the heat-up rates from cold to 800 C (1470 F) can be typically 1 to 1.5 h. However, the limited applications due to atmosphere conditions within the bed constitutes a drawback of this method. Many attempts have been made to overcome the problems associated with this method of heating in order to take advantage of the tremendous heat-up and heat recovery rates.

In applying this technique to the heating of fluidized beds, the burner must be designed to meet the following criteria:

1. It must agitate the suspended particles to achieve the desired properties of heat transfer and uniformity of bed temperature.

2. It must deliver the required heat input to the combustion zone in the bed in order to achieve a recovery rate of 5 to 7 C/min (9 to 12 F/min) which is typical in internal combustion beds.

3. It must minimize contamination of the fluidizing gas where the gas is being used for atmosphere control.

4. It must occupy the smallest possible area of the fluidized bed so as to avoid undue waste of fluidizing gas. This is of particular importance when using expensive gases such as argon and helium.

A burner has now been developed which occupies

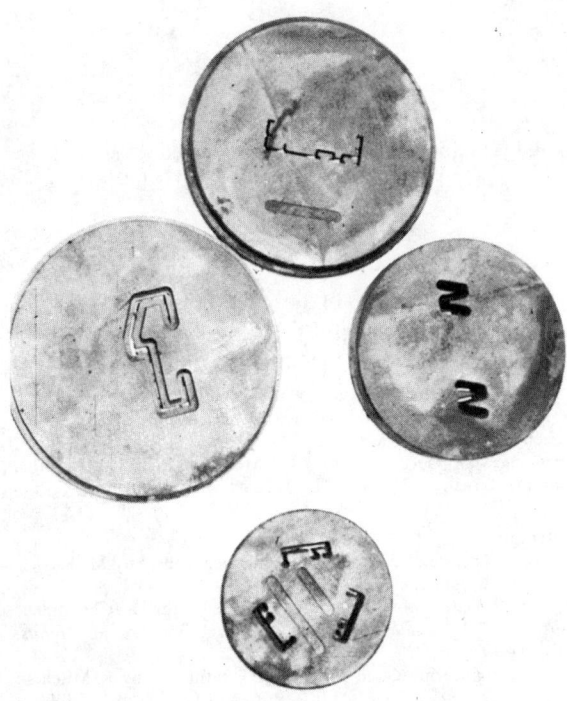

Fig. 4 — One example of parts heat treated in a fluidized bed are these extrusion dies. The furnace can be used for both ferrous and nonferrous parts.

Source: Metal Progress, September 1981, 132-137

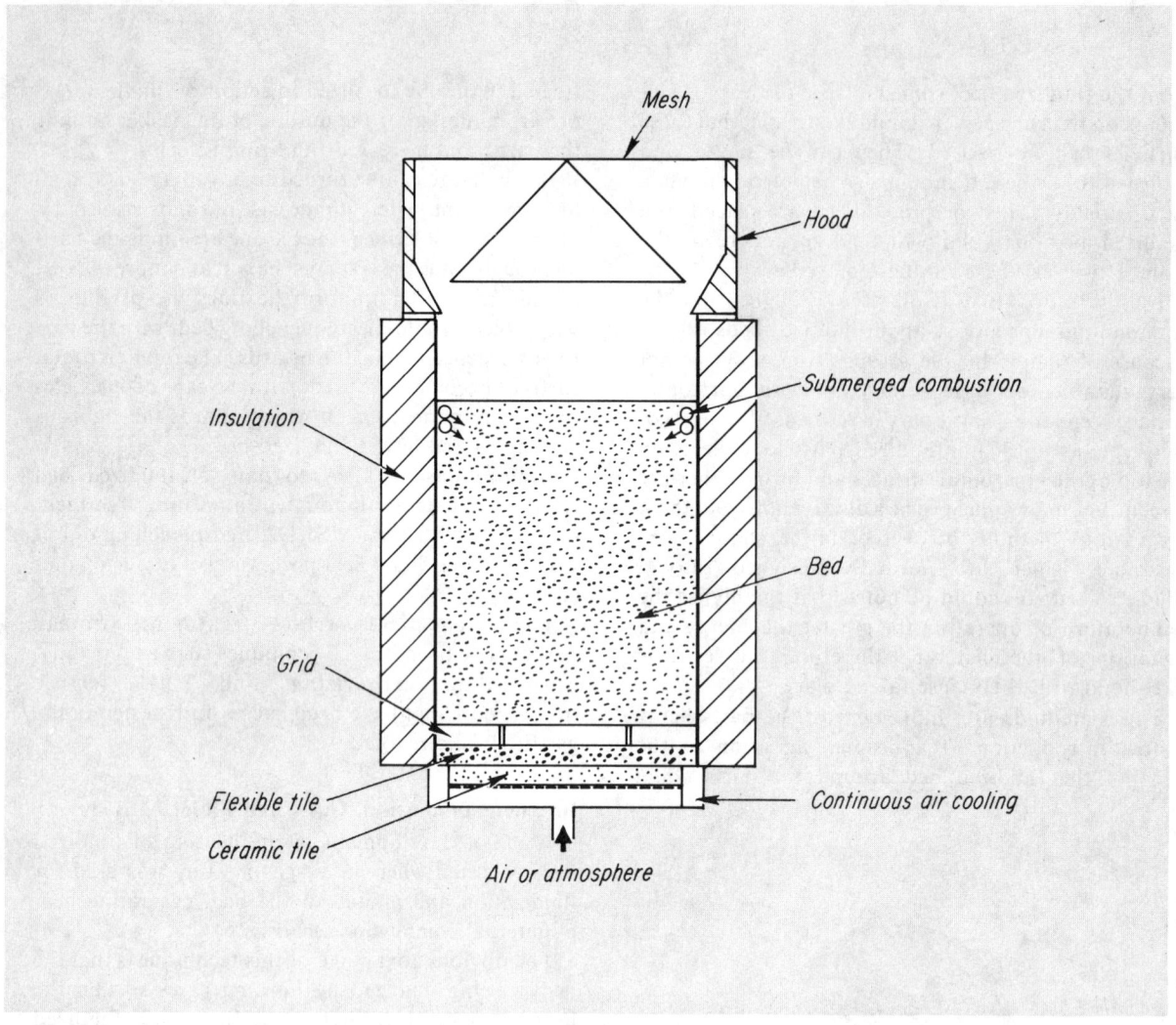

Fig. 5 — In this submerged combustion fluidized bed furnace, an air/gas mixture is ignited within the bed, and heat is generated by internal combustion. A temperature of 800 C (1470 F) can be achieved in 1 to 1.5 h.

only 10% of the open area of the bed and achieves a typical recovery rate in the temperature range 500 to 1000 C (930 to 1830 F) of 5 to 7 C/min (9 to 12 F/min). The burner operates on the principle of premixing air and gas to a noncombustible mixture (with propane the gas-air ratio is 1:9; with natural gas, 1:3). Final mixing with the secondary air in the bed is done in the immediate area of the fuel source tubes.

In practice, the particles in the immediate vicinity of the burner are circulated at a very high rate, and because the surface area available for heat transfer is high, the difference in temperature of the particles between entry and exit of the burner is small. This means that the temperature difference between the bed and the heated particles is low, resulting in a uniform bed temperature.

Utilizing the technique of submerged combustion, equipment has been constructed and is operating in the temperature range of 150 to 1100 C (300 to 2010 F). A typical midrange furnace has a heat-up rate of 5 to 8 C/min (9 to 14 F/min), and a temperature uniformity of ±5 C (±9 F). Recovery time of a 100 kg (220 lb) load to 950 C (1740 F) is 30 min.

For More Information: You are invited to contact the author directly by letter or telephone. Mr. Bamford is vice president, Can-Eng Manufacturing Ltd., 6800 Montrose Rd., Niagara Falls, Ontario, Canada L2E 6V5; tel: 416/356-1327.

References
1. "Heat Transfer Coefficients in Fluidized Beds," by M. Kovacs and J. C. Maunders: *BISRA Report*, PE/A/55/63.
2. "Controlled Atmosphere Fluidized Beds for the Heat Treatment of Metals," by R. W. Reynoldson: *Heat Treatment of Metals*, January 1977, p 14.
3. "An Engineering Concept of Heat Treatment," by E. Mitchell: *Heat Treatment of Metals*, Lliffe Books Ltd., 1963, p 107-149.
4. "Heat Treatment in Fluidized Beds," by M. J. Virr and R. W. Reynoldson: *Industrial Process Heating*, December 1972, p 18-22.
5. "Fluidized Bed Methods of Nitriding & Oxynitriding," by Z. Rogalski and H. Zowczak, 15th International Conference on Heat Treatment of Materials, Detroit, Mich., 6-8 May 1980, published by ASM, Metals Park, Ohio.
6. "Internally Fired Fluidized Bed Gas Carburizing," by J. J. Moore and G. G. Storey: *Metallurgic*, November 1979.

"FLUIDFIRE" FLUIDIZED BED FURNACES

By
RAJ TIWARI*

Introduction

The term "fluidisation" means attaining the properties similar to that of fluid. It is achieved when a stream of air-gas is passed through the bed of fine particles at certain velocity. When the air/gas velocity increases more than the static pressure of the bed, the bed suddenly fluidises and behaves like a liquid. This property has been used in developing the latest technique in Heat Treatment Furnace field.

In recent years the technology of heat treatment in fluidized beds has made significant advances. Fluidized bed heat treatment is now accepted as a normal operation.

The fluidized bed: As the name suggests, a fluidized bed is a container of graded particulate material which is kept in suspension by a uniformly distributed upward flowing gas stream. The minimum fluidization velocity MVF theoritically is a inverse function of the square of the particle diameter 'd' and linear function of the particle Mass 'P'

The Actual heat transfer rates obtained in the fluidized bed depends on the gas velocity and its properties such as particle size, density and thermal properties and on the furnace geometry.

The influence of gas velocity is such that there is an optimum fluidising velocity at which a maximum heat transfer co-efficient occurs. Increasing the fluidization velocity beyond this optimum value produces a slowly decreasing heat transfer co-efficient.

As for the particle size, the heat treatment transfer co-efficient decreases with increasing particle diameter.

In commercial designs a heat transfer co-efficient of between 40 and 120 Btu ft²h° F is achieved which is similar to salt bath performance but some 5 to 10 times better than with gas circulation furnaces.

Method of heating: There are two basic types internally or externally heated. In the internally gas fired units, fluidization of the media is effected by a near stoichiometric mixture of gas and air. This combustible mixture is ignited above the bed and quickly imparts its heat to the particles, which progressively heat up the incoming gases further down the bed, resulting in spontaneous combustion within the bed. Once 750°C has been exceeded combustion is completed within a narrow zone above the diffusion tile system at the base of the bed. The air and gas is mixed automatically to give normally a slightly gas rich mixture, thus maintaining an exothermic atmosphere in the heating zone of the bath. The temperature of the bath is automatically controlled via a proportional of integral controller, connected to motorised valve which meters appropriate amount of gas air mixture to the bed. Constant controlled temperature is achieved + 5°C.

In the externally heated units electric elements are controlled using a similar controller feeding three phase thyristor bank or gas burners by fully proportionating valves giving modulating operation. The fluidizing gas or gas/air is adjusted via a motorized valve to monitor optimum fluidization at all temperatures.

Furnaces: The furnaces have been designed as General Purpose Furnaces capable of being used to heat treat components in the temperature range 150°C to 1350°C.

Bath Furnaces: A range of bath furnaces has been available for a number of years and through for the heat treatment of high speed steel upto a temperature range of 12,50°C at 150 lb/hr. with a furnace work area of 400 mm dia by 450 mm deep.

Most popular type of furnace is PIT TYPE furnace, with a retort, supporting refractory, ceramic fibre insulation, mineral wool, hydraulic lifting lid, diffsion plate, gas inlets and work baskets.

Continuous Furnaces: The latest design of continuous furnace utilises a split roof construction allowing an overhead conveyor transport suspended baskets, billets or other components through the fluidized bed.

A current installation is for 1.5 ton/h. of rotary cultivator blades of various sizes, utilizing an internally fire fluidized bed for hardening the blades at a maximum temperature of 950°C. The operating efficiency of this line is of the order of 40% thermal efficiency calculated on a heat to steel basis.

Rotary Drum Furnace: The first furnace line of this design went into operation few years ago for hardening and tempering of lock washers at 850°C and 450°C respectively. At 800 lbh. the total gas consumption is 6.9 therm hour. The maximum capacity of the furnace is 1000 lb/h. The washers are quenched into oil from the hardening furnace and the line has a washer installed before the tempering furnace.

Similar lines have been used for Ball Bearing heat treatment.

Advantages

1 Remarkable temperature uniformity and accuracy (+ 5° in normal) resulting in minimal distortion factors.

2 High thermal mass contact with the work (3-4000 times greater than a conventional furnace).. This results in very little variation in temperature on loading.

3 High thermal exchange rate (5-10 times better than a convection environment). This means low fuel de-

---

* The author is with Simplicity Engineers P. Ltd., New Delhi 110 057.

mand, 35-50% of present consumption with a higher work throughput possibility.

4 The work leaves the furnace clean, therefore, post treatments to stop storage corrosion due to adherent salts are unnecessary.

5 No "Melting Heat" requirement results in a short heat up time (1-3 hours). The units can be shut down overnight without loss in production time the next day.

6 Safe operation, no toxic fumes, effluent disposal or explosion problems to worry about.

7 The furnace atmosphere is self generative and adjustable.

8 When compared to conventional furnaces it offers overall cost saving of approx. 35% plus.

PART II

Controlled atmosphere in fluidized bed furnaces

Fluidized bed furnaces fall, basically, into three categories, designed by their mode of heating.

1 Direct heated by internal combustion
2 Indirectly heated by internal combustion
3 Externally heated.

Although designs utilizing various combinations 1 & 2 have been used for surface treatments, the third type is the only one which offers a good degree of process control coupled with economic costs.

Method of Carburizing

Carburizing has always been the most popular application for fluidized beds. Earliest unit was developed in 1974.

The use of mixtures of hydrocarbon gas and air for carburizing has been found to be suitable in a fluidised bed furnace and results indicate a carbonizing rate 30% faster than in conventional carburising.

In addition, the diffusing rate of carbon in Austenite was found to be near to the theoretical value. The mechanism of the faster rate was attributed to higher carbon poential and the more favourable surface conditions than those found in conventional furnace.

Broadly the treoretical gas reaction can be compared with the classical hydrocarbon/oxygen reaction producing carburizing by dissolation of carbon monoxide at the steel surface i.e.

a) CH_4
   $C_3H_8 + O_2 \rightarrow CO_2 + H_2$
   C_4H_{10}

b) C(by thermal dissociation) + $H_2O \rightarrow CO + H_2$
   and
   C(by thermal dissociation) + $CO_2 \rightarrow 2\ CO$

c) $CO + H_2 \rightarrow C + H_2O$

Additional tests showed that very high surface carbon contents could be achieved in short periods of time.

It means free carbon is available in large volume. But it does not prevent carburising by sooting as in conventional gas carburing but promotes the reaction by a constant formation and removal upon the steel surface. This postulated a process of bombardment rather than surface diffusion.

Comparison of available atmospheres

There are at least three possible gas systems for fluidized bed carburising

1 Hydrocarbon gas + air
2 Hydrocarbon gas + Nitrogen
3 Hydrocarbon gas + Nitrogen/methanol.

Externally heated fluidized bed furnaces are used. Generally hydrocarbon gases methane or propane are metered through direct volumetric control system via. calibrated flowmeters.

The first two gas combinations are relatively easy to establish. But with Nitrogen methanol system much higher addition of hydrocarbon gas is needed.

To achieve 0—.70 mm case depth at 950°C following is the consumption :

Gas Mixture	Percentage
Methane + Air	75/25
Propane + Air	25/75
Methane + Nitrogen	60/40
Propane + Nitrogen	20/80
Methane + 50/50 Nigrogen/Methanol	25/75
Propane + 50/50 Nitrogen/Methanol	15/85

These ratios are for "Steady State" carburising throughout. But "Boost/Diffusion" technique can also be used.

Production Control of Atmospheres

Both the systems i.e. a) Direct volumetric control b) Continuous gas analysis can be successfully used with fluidised bed furnaces.

In Direct volumetric control system the flow regulation remains practically constant which makes proportional deviation control unnecessary. The use of a boost/diffuse technique was required for "deep" case depths over 1.5 mm.

For continuous monitoring of atmosphere, Zirconia probe oxygen sensor is more suitable than CO_2 — infrared analyser.

Other processes which can be carried out in fluidfire furnaces are:—

1 Nitriding
2 Nitro carburising
3 Stress Relieving
4 Bright Hardening
5 Bright Tempering
6 Bright Annealing

Effect of Temperature Gradient on the Quenching Power in Fluidized Beds

W. LUTY

The results presented in this paper indicate that the possibility exists, although not utilized so far, to increase the quenching rate in fluidized media and thereby to broaden the range of their application, by raising the difference between the temperature in the surface zone of the product being quenched and that in the fluidized medium. This may be relatively easily accomplished, in practice, by cooling the bed down to very low temperatures or, for certain grades of steel, by raising the austenitizing temperature. The above two measures may be also combined. In the present study, the medium was cooled down to −100 °C by fluidizing a corundum powder with nitrogen evaporated from the liquid state.

INTRODUCTION

The proper selection of the quenching rate during quench hardening is the factor that essentially affects the useful properties of the components of machinery and tools being quenched, as well as the hardening stresses and distortions produced in them. This is why the modern development of quenching techniques tends to increase the capacity of regulation and control of the quenching rate, so that it is better matched with the kinetics of phase transformations in undercooled austenite. Recent achievements in this field include addition of various substances to the mineral oils to stimulate the thermokinetic and other properties of the oil, and usage of some water-soluble polymeric compounds which permit a broad control of the quenching rate.

Besides the mineral oils and polymeric quenchants, gas media are more and more widely employed, especially after austenitizing in vacuum furnaces; also, powders of solid substances fluidized with gases are applied. In the literature on the application of fluidized media utilized as quenchants,

the quenching power of these media is generally described as lower than that of mineral oils, but much higher than the quenching power obtained with a gas stream. These general assessments of the quenching power and even the quenching curves given in some publications are usually for certain strictly specified conditions,[1] and are not sufficient for the engineer to select the quenching medium which will be most effective in hardening the machinery or tools at hand. To do this, one should have more precise data to describe the quenching rate, at least at the surface, as a function of the temperature of the products to be quenched. For most grades of steel, the quenching rate within the range of pearlitic transformation temperatures, *i.e.,* from 500° to 600 °C, is the most important. In earlier work,[2] the author has shown that a relatively low quenching rate of fluidized beds within the aforementioned temperature range (Figure 1) is the main restriction on the application of these beds in quench hardening. In this connection, it has been recognized necessary to perform further investigations aimed at overcoming the above barrier.

The studies made thus far on heat transfer in fluidized media were concerned with the relationship between the heat transfer coefficient, α, and such factors as the kind of bed material; the size of the solid particles; the kind of

W. LUTY, D.Sc., is Professor in the Institute of Precision Mechanics, Warsaw, Poland. This work was in part presented at the 2nd International Congress on Heat Treatment of Materials, Florence, Italy, September 20–24, 1982.

Reprinted from Journal of Heat Treating, December 1983, 108-113, © 1983 American Society for Metals

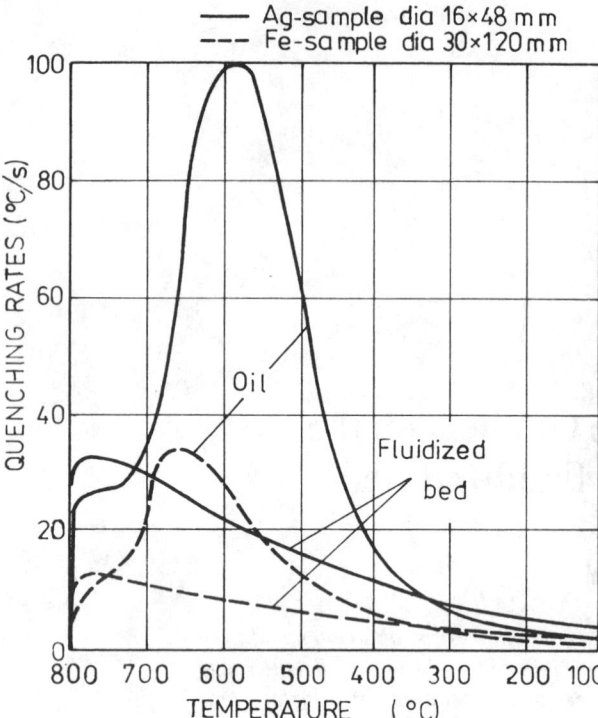

Fig. 1 — Quenching rate of a silver specimen and a low-carbon steel specimen in mineral oil and in a fluidized bed of corundum powder, as a function of the temperature of the specimen center.[2]

fluidizing gas; and the fluidization index, *i.e.*, the ratio of the actual velocity of the gas flow to the critical velocity.[3-6] As the parameter determining the fluidization index, the bed specific density in kg/m³ or g/cm³ may also be taken. The subject of the present study was the effect of another, easily controlled factor, the temperature of the fluidized medium or, more strictly, the temperature difference between the surface layer of the product being quenched and the bed, *i.e.*, the temperature gradient.

TEMPERATURE GRADIENT AS A QUENCHING RATE PARAMETER

The theoretical basis of the effect of the temperature gradient on the quenching rate is Newton's law. According to this law, the density of the heat flow received, q, is directly proportional to the difference in temperature between the surface of the body being cooled, t_p, and the cooling medium (in our case, the fluidized bed), t_z:

$$q = \alpha(t_p - t_z)$$

The proportionality coefficient, α, is called the heat transfer coefficient or heat absorption coefficient. It is obvious that the value of t_p decreases with time as the cooling process proceeds and depends on the thermal conductivity, λ, of the metal being cooled, whereas the heat flow across the cross-sectional area is the product of this conductivity and the

temperature gradient, dt/dx, in the surface layer of the metal. Newton's law may be thus also written in the form:

$$q = \lambda \, dt/dx = \alpha(t_p - t_z)$$

The above equation permits the value of α to be experimentally determined by measuring the temperature gradient in the product of known λ. However, for practical purposes of heat treatment, especially when evaluating the usefulness of the quenching medium to hardening of various grades of steel, it is more convenient to consider, instead of α, the quenching curves of temperature *vs* time or the curves of the quenching rate *vs* the temperature of the product being quenched. The plots of the temperature at the surface or, for instance, at the center of the product *vs* time, superimposed on the phase transformation (continuous or isothermal) diagrams known for a given grade of steel, enable us to assess the usefulness of the given quenchant to conventional quenching or austempering. Using the cooling curves given in Reference 2, together with the transformation diagrams for 20MoCr4 steel (SAE 4118) taken from the book by Wever *et al.*,[7] Figure 2 has been drawn as an example of how these curves may be utilized to determine the structures formed in carburized layers after they have been subjected to conventional quenching in a fluidized bed at a temperature of 20 °C or to austempering in the same medium at a temperature of 300 °C. From this figure, it follows that after conventional quenching in a cold fluidized medium the carburized layer can acquire a pure martensitic structure, while after austempering some amount of pearlite unavoidably occurs.

Critical quenching rates for carbon steels are 100° to 600 °C per second, depending on carbon content, whereas

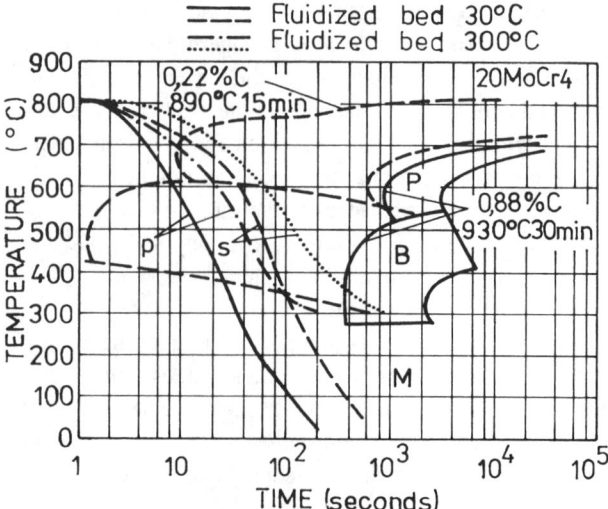

Fig. 2 — Quenching rate curves for the surface (*p*) and the center (*s*) of a low-carbon specimen of diameter 30 mm and length 120 mm in a fluidized bed of corundum powder, in comparison with a continuous TTT diagram for 20MoCr4 steel (AISI 4118) and for the same steel after carburization to 0.88 pct C.[2,7]

for alloy steels they vary from 0.1° to 100 °C per second.[6] Comparison of these data with the quenching rates obtained with fluidized media, especially within the temperature range from 600° to 500 °C, permits a rough evaluation of the possibility to use these media in hardening of various grades of steel. Obviously, an additional factor to be taken into account is the cross-sectional area of the product, since it also affects the quenching rate of the surface zone.

INVESTIGATION PROGRAM AND METHODOLOGY

The present study was concerned with the technological parameter of the primary difference in temperature between a silver cylindrical specimen of diameter 8 mm and length 24 mm, with a shielded thermocouple attached at its center, and the fluidized quenching medium. This primary temperature gradient may be raised either by lowering the temperature of the fluidized medium, by increasing the temperature of the specimen, or by combining both these effects. In the present study, only the temperature of the fluidized medium was changed. All other factors which also affect the quenching rate of the specimen were taken as constant, namely:

- the fluidized bed was composed of corundum (Al_2O_3) particles of practically uniform grain diameter of 100 μm;
- the gas used to bring the corundum grains into a fluidized state was commercially pure nitrogen obtained by evaporation from the liquid state;
- the nitrogen flow rate was controlled according to the temperature of the fluidized medium, so that the medium specific density was kept constant at 1.5 g/cm³, which could be easily accomplished by controlling the level of the bed in the chamber.

The primary temperature gradient was altered within the limits 1000° to 600 °C. The silver specimen was heated to a constant temperature of 900 °C. This means that the temperature of the fluidized bed was adjusted within the range from −100° to +300 °C. The thermocouple was attached at the geometrical center of the specimen. Since the specimen diameter was small (8 mm) and the thermal conductance of silver is high, one may, for practical purposes, assume that the temperature variations during quenching in the specimen as a whole correspond approximately to the temperature variations in the surface zone of the elements subjected to hardening.

The quenching rate of the specimen was measured with DIACPOT equipment designed by the Centre Technique des Industries Mechaniques (CETIM), France. The functional scheme of the equipment is shown in Figure 3 and the design diagram in Figure 4. This equipment has been gener-

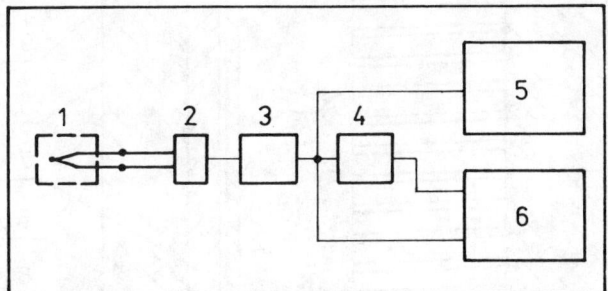

Fig. 3—Quenching rate measuring apparatus, DIACPOT, designed by CETIM, France: 1—specimen with thermocouple installed, 2—cold junction compensation, 3—amplifier, 4—differentiating circuit, 5—potentiometric recorder of the temperature–time function (XT), 6—quenching rate–temperature function recorder (XY).

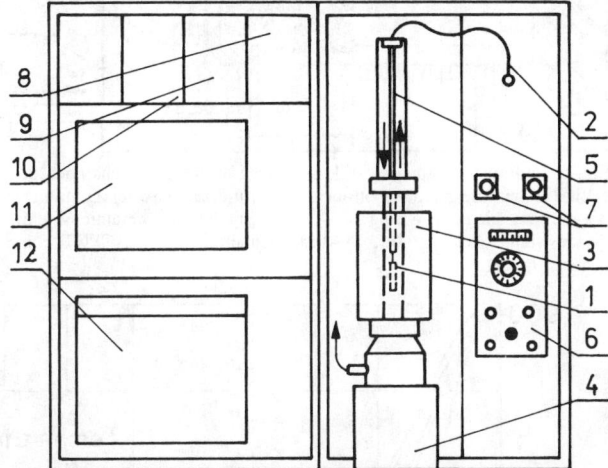

Fig. 4—DIACPOT apparatus, design diagram: 1—Ag cylindrical specimen, 2—shielded thermocouple, 3—electric furnace, 4—fluidizing apparatus (see Figure 5), 5—tube containing the thermocouple, 6—measurement and control panel, 7—push-buttons for controlling vertical motion of the tube 5, 8—potentiometer, 9—amplifier, 10—differentiation apparatus, 11—automatic recorder of the function XT, 12—automatic recorder of the function XY.

ally intended for studying liquid quenchants, especially quenching oils. In the present study, it has been adapted to study fluidized media. For this purpose, a special fluidizing apparatus was installed under the specimen heating furnace, the chamber of this apparatus being 90 mm in inner diameter and 160 mm in height. The apparatus arrangement is shown in Figure 5. The most important part of this apparatus is the porous bottom of the fluidizing chamber, which is a porous plate. The plate has been made of sintered powders of austenitic steel with grain diameters of 45 to 160 μm. The density of this sinter is about 3.5 g/cm³ and its porosity is 60 pct. Nitrogen flow density through the plate was about 40 m³/m²h at a pressure of 0.1 kPa beneath the plate, and about 120 m³/m²h at a pressure of 0.3 kPa. Nitrogen was supplied under the plate from a Dewar container through a spirally wound copper pipe installed within a thermostat.

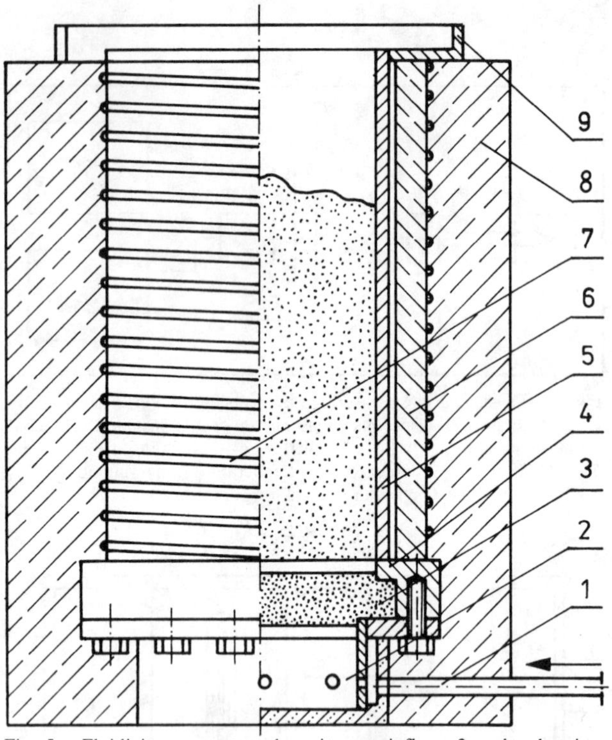

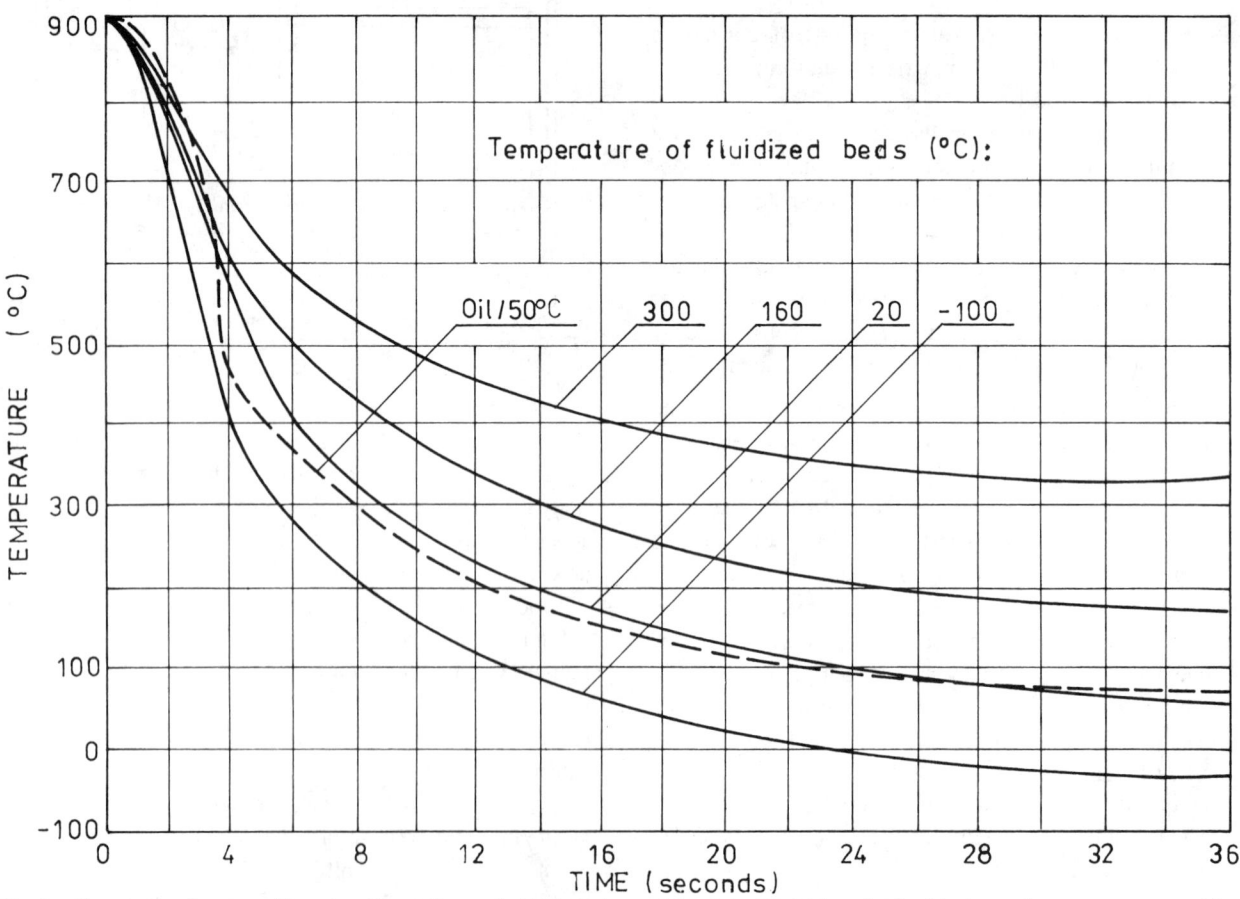

Fig. 5—Fluidizing apparatus: 1—nitrogen inflow, 2—chamber installed under the porous bottom of the fluidized powder container, 3 — porous plate, 4 — ring, 5 — steel cylinder, 6—ceramic shield, 7—resistance heating coil, 8—heat insulation, 9—DIACPOT connecting flange.

INVESTIGATION RESULTS AND THEIR INTERPRETATION

Figures 6 and 7 show the investigation results in the form of cooling curves plotted as the specimen temperature *vs* time and the quenching rate *vs* temperature. The first of these coordinate systems has the advantage that, in practice, one can easily find the quenching time necessary to cool the specimen down to a certain selected temperature, *e.g.,* the temperature of pearlitic transformation for a given grade of steel; also the quenching curve may here be easily drawn with time plotted on a logarithmic scale. In the latter case, the curve may be superimposed on the phase transformation diagram for undercooled austenite, as has been done in Figure 2. Of course, when steel is concerned, the curves obtained for a silver specimen may be treated only as the quenching curves for the surface zone. In Figure 6, a quenching curve for a conventional oil (low quenching rate) has also been drawn for comparison.

In Figure 7, the variations of the quenching rate as a function of the temperature of the specimen and as a function of the temperature of the fluidized medium are much more marked than in Figure 6. The basic difference in the character of these curves, in comparison with the curve for oil, is due to the phenomenon of strong heat absorption

Fig. 6—Temperature–time quenching curves for a silver cylindrical specimen of 8 mm by 24 mm in a fluidized bed at various temperatures and in a mineral oil of low quenching rate.

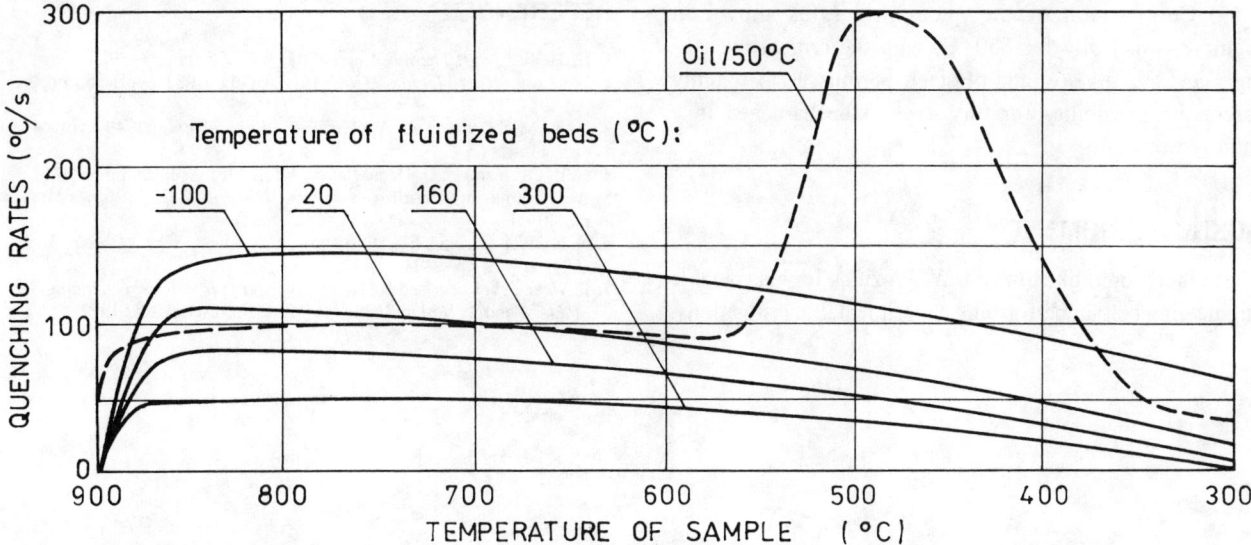

Fig. 7—Quenching rate of a silver cylindrical specimen sized at 8 mm by 24 mm, as a function of the specimen- and fluidized bed-temperature; the quenching rate curve in oil is given for comparison.

during boiling of the oil, which phenomenon obviously does not occur in fluidized media. This fact essentially reduces the quenching rate in fluidized media over the most important range of temperatures for many grades of steel, the temperatures of pearlitic transformation. On the other hand, the absence of the boiling phenomenon is advantageous with respect to the magnitude of hardening distortions, since no rapid changes in quenching rate occur.

The most essential result of our investigation, which determines the effect of the primary temperature gradient on the quenching rate at selected temperatures of 700° and 550 °C, is shown in Figure 8. Experimental points indicated in the diagram have been obtained from the curves recorded automatically in the system shown in Figure 7. A considerable rise in quenching rate with increasing primary temperature gradient, observed in the figure, is fully consistent with predictions based on the Newton law; this rise enables the quenching rate control to be performed over a relatively wide range. Comparing the quenching rate at 550 °C with the critical quenching rate for a given grade of steel, one can roughly estimate whether the steel can or cannot be hardened in a fluidized bed.

CONCLUSIONS

(1) The relatively low quenching power of fluidized media within the limits of pearlitic transformation, as compared to mineral oils, is by far the main restriction on the range of application of these media. However, if the primary temperature gradient is increased by lowering the temperature of the fluidized quenchant, *e.g.*, down to −100 °C, the

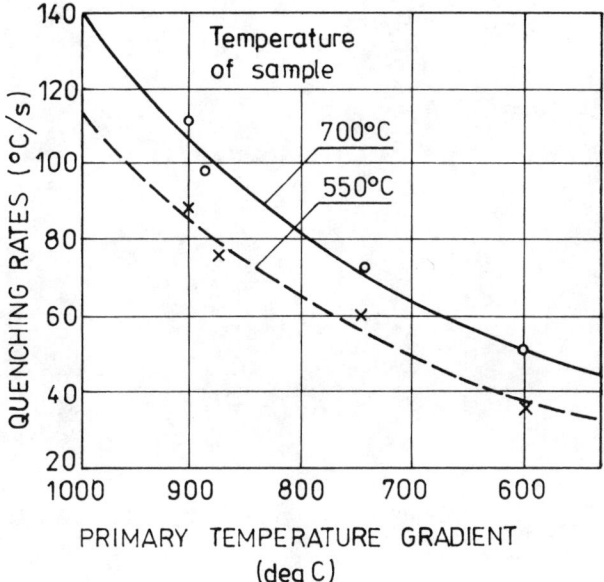

Fig. 8—Effect of the primary temperature gradient on the quenching rate of a silver cylindrical specimen sized at 8 mm by 24 mm in a fluidized bed, at the specimen temperatures of 700 °C and 550 °C.

quenching rate would be raised according to Newton's law, thus enabling steels of lower hardenability or of greater cross-sectional area to be hardened.

(2) Within the temperature range from the austenitizing temperature to about 550° to 600 °C (Figure 7), the quenching rate in fluidized beds is close to that in oil, and if the temperature of the fluidized medium is negative, it may be even greater.

(3) Comparison of the quenching rates shown in Figure 8, especially for 550 °C, with the critical quenching rate for a given grade of steel, permits us to roughly assess the possibility for this steel to be hardened in a fluidized medium.

ACKNOWLEDGMENT

The author would like to thank W. Woźniak for his valuable engineering help in performing experimental investigations.

REFERENCES

1. H. Hellio: *Trait. Therm. Mies.*, 1981, vol. 155, pp. 33–39.
2. W. Luty: *Hart.-Techn. Mitt.*, July-August 1981, vol. 36, no. 4, pp. 194–199.
3. H. S. Mickley and C. A. Thrilling: *Eng. Chem.*, 1949, vol. 41, pp. 1135–1147.
4. H. L. Qven and C. O. Dean: *Petrol. Eng.*, 1955, vol. 25, pp. 623–632.
5. A. J. Tamarin: *Metalloved. Term. Obrab. Met.*, 1968, vol. 3, pp. 10–11.
6. N. N. Warygin and E. I. Olshanov: *Metalloved. Term. Obrab. Met.*, 1971, vol. 6, pp. 2–11.
7. F. Wever, A. Rose, and H. Hougardy: *Atlas zur Wärmebehandlung der Stähle*, Band 2, Verlag Stahleisen M. B. H., Dusseldorf, 1972.

Fluidized Bed Heat Treating

and Its Applications in the Foundry

Carol A. Girrell

Procedyne Corp

New Brunswick, NJ

Fluidized bed furnaces have been found to be attractive for many heat treating processes because of their high, uniform rates of heat transfer and energy efficiency. This article covers the fundamentals of these furnaces and their specific application in the various heat treating processes.

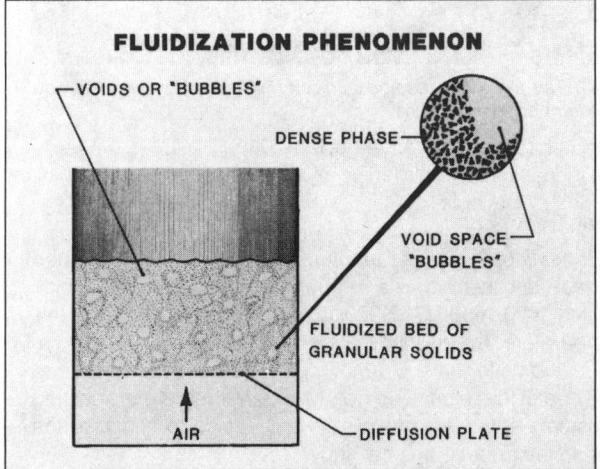

FLUIDIZATION PHENOMENON

VOIDS OR "BUBBLES"

DENSE PHASE

VOID SPACE "BUBBLES"

FLUIDIZED BED OF GRANULAR SOLIDS

AIR

DIFFUSION PLATE

Fig. 1. This diagram shows the basic geometry of a fluidized bed. Note the diffusion plate which allows air to travel through the bed of granular solids.

Over the last several years, fluidized beds have made a dramatic impact as furnaces for the heat treatment of metals. It has been proven that they are energy efficient, versatile and nonpolluting.

They have compiled an outstanding safety and performance record over a broad range of industrial applications. The use of nitrogen and nitrogen-based atmospheres, in addition to air or air-fuel mixtures as fluidizing gases, has enabled the movement of fluidized bed furnaces into all areas of heat treating including neutral and case hardening as well as annealing, tempering, stress relieving and preheating.

For the foundry industry, fluidized beds also show great promise in such applications as preheating for burning and welding, stress relieving on small and large castings and weldments, normalizing and hardening with or without protective atmospheres and tempering.

Although relatively new in heat treating, fluidized beds go back to the early 1920s and have been used extensively in the minerals, chemicals and petroleum refining industries. A general fluidized bed is illustrated in Fig. 1, which shows a chamber containing a particulate medium with a distributor plate across the bottom. The distributor plate diffuses air or other gases uniformly across the bottom of the chamber and upwards through the particles. This upward motion exerts a drag force on each particle which, when equal to the downward weight of the particle, causes the particles to separate, levitate and begin to move about. In this state, the particles are said to be fluidized.

This so-called fluidized bed exhibits remarkable liquid-like behavior. Objects placed on the bed will float or sink depending on their relative density. The surface of the bed remains horizontal even if the chamber is tilted; an opening in the side of the chamber will permit the solids to gush out just like liquid and if a divider is placed in the chamber, the levels will equilibrate and seek a common level as long as there is an opening in the partition. Furthermore, the bed maintains a constant pressure drop from bottom to top independent of the gas flow through the bed. When more gas flow than is required for minimum fluidization is passed through the bed, bubbles form within the medium which results in the ability to highly agitate the bed without mechanical means.

In this fluidized state, the bed has a high thermal conductivity and extremely excellent temperature uniformity much like molten liquids. Unlike molten liquids, however, it cannot boil, freeze, or vaporize nor will it explode or react when wet or oily parts are submerged in it.

A curve which most significantly defines the behavior of a fluidized bed for heat treating and heat processing applications is shown in Graph 1 which shows the pressure drop from the bottom to the top of a fluidized bed as a function of gas flow. There is an increase in pressure until the bed fluidizes and then an essentially constant pressure drop independent of gas flow once past minimum fluidization.

Graph 2 indicates that the surface heat transfer coefficient in a fluidized bed goes from near zero before minimum fluidization to a very high value once the bed has become fluidized. Heat transfer coefficients of 70-120 Btu/hr, per degree F, per sq ft are typically obtainable in a well fluidized bed. This is anywhere from 15-25 times more than natural convection coefficients and as much as 5-8 times more than would be obtained in a high velocity convection furnace. .

Even at high temperatures where radiation is significant, fluidized beds can be 2-3 times higher in heat transfer coefficients than conventional forced convection and radiation furnaces. In addition, a fluidized bed furnace is extremely

Fig. 2. A continuous wire fluidized bed heat treating furnace.

uniform in temperature throughout, with typical temperature differentials of no more than ±5-±10°F (±1.5-±3.0°C) over the working volume of the furnace, almost independently of size.

The particulate bed material is typically aluminum oxide, although olivine sand and other naturally occurring materials have also been used in a few special applications. These materials remain inert upon heating, and will not break down or melt even at very high temperatures. Because the material is physically and chemically stable, fluidized bed furnaces can be used over a wide range of temperatures.

Fluidized beds can also be used for removing heat from metal, as in quenching, cooling after shakeout and cooling following an isothermal anneal or temper. A primary benefit in using a fluidized bed is the uniformity of heat removal from the casting, reducing the chances of cracking and minimizing internal stresses, while enabling a rapid cool-down.

Fluidized bed furnaces can be heated in a variety of ways including internal and external electric resistance heaters, indirect gas-fired and direct gas-fired. In addition, these furnaces can be made in large sizes for the foundry industry.

Figure 2 shows a 35 ft long furnace for use up to 1400F (760C) where the fluidized bed is indirectly heated by electric resistance heaters on the outside of the retort. The design lends itself to indirect gas firing as well.

Figure 3 shows an electrically heated furnace, almost six ft in diameter by three ft deep. This unit is heated by electric resistance heaters immersed directly in the fluidized bed, and is good for temperatures up to 125F (675C). A unit similar to this, except rectangular in cross section (approximately 8 × 4 × 4 ft deep) is being used at a steel foundry for stress relieving and tempering of large castings, and a 4 × 4 × 5 ft

Fig. 3. This electrically heated fluidized bed is used in stress relieving and tempering operations.

deep unit for temperatures up to 200F (1100C) is presently being installed for hardening of alloy castings.

Heat Treatment

Annealing—Castings are submerged in the fluidized bed furnace and heated to a temperature above the Ac3 point, usually 1400-1650F (760-900C). After sufficient soaking at this temperature, the castings are cooled slowly so that a complete and uniform transformation occurs in the high temperature end of the pearlite range. Such treatment relieves casting stresses, refines the grain size and serves to minimize the dendritic structure in a casting.

While relatively simple, annealing is often time-consuming since it involves a slow cooling over the entire range from the austenitizing temperature to a temperature well below that at

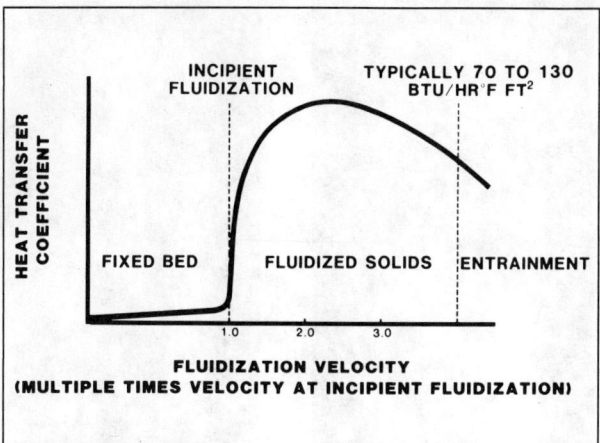

Graph 1. Fluidized bed pressure drop as a function of gas flow.

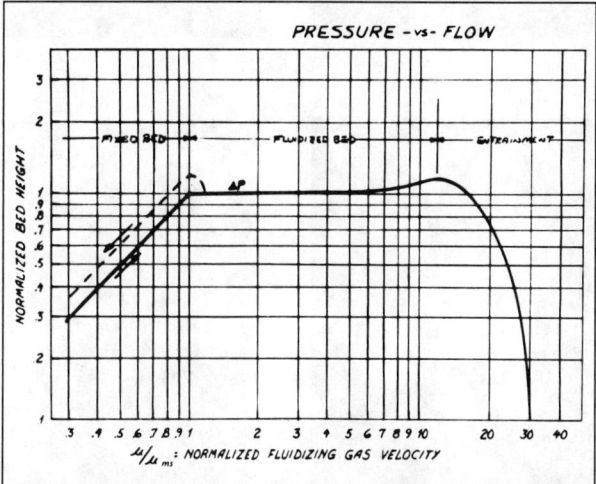

Graph 2. Surface heat transfer co-efficient as a function of fluidization velocity.

which transformation is complete; however, once transformation is complete a fluidized bed can be used to speed the rest of the cooling to room temperature, thereby minimizing cycle times. Fluidized bed furnaces for annealing can be of a batch type or continuous.

When batch equipment is used, the furnace is either shut down for cooling at the natural cool-down rate of the unit, or cooled at a controlled rate. When transformation is ensured and complete, the load of castings is removed from the high temperature unit and placed in an unheated fluidized bed for rapid cooling to room temperature. For continuous processing, parts are automatically transferred to beds at decreasing temperatures to provide the appropriate heating and cooling rates.

Both configurations of fluidized bed equipment (batch or continuous) are also ideally suited for isothermal annealing, a similar process, but one in which the cooling rates from the austenitizing temperature to the transformation temperature and from the transformation temperature to room temperature are not critical.

Normalizing—The normalizing treatment is similar to the annealing process, except that the castings are removed from the furnace at the end of the soaking period and quenched in either an unheated fluidized bed or other quench suitable to the material, such as oil or water. As internal stresses in the castings are not removed to the same extent as in the annealing processes, a draw or temper is frequently used

after normalizing. The draw or temper is also used to reduce the as-quenched hardness of the metal, when desired. Batch-type fluidized bed equipment, consisting of an austenitizing furnace, a quench and a tempering furnace, is most suitable for normalizing.

The indirectly heated fluidized bed furnaces are most often used, as the separation of the heat source from the bed material allows an inert atmosphere to be used to protect the surface of the castings from oxidation or decarburization.

Hardening, Austempering and Martempering—To develop optimum strength characteristics in certain cast alloys, these materials are sometimes treated to produce a martensitic or bainitic structure. The metal is heated to the austenitic region and quenched in either another fluidized bed, oil or water. Depending on the temperature of the quench and the rate of cooling to that temperature, martensite or bainite is produced. Austempering and martempering use heated quenches, so are less severe in terms of distortion and potential cracking; however, the rate of cooling to the quench temperature is much faster in austempering, producing similar distortion characteristics to straight hardening.

Fluidized bed furnaces can be used to replace the austenitizing furnace and often the quenching medium and usually produce less distortion than conventional furnaces and quenches. Again, the indirectly heated fluidized bed furnaces are preferred for the control of atmosphere chemistries.

Tempering and Stress Relieving—Tempering is principally confined to high-carbon and alloy steel castings to remove the internal stresses set up by quenching following hardening or normalizing. Stress relieving is performed on metals in the as-cast condition which have not been allowed to cool in the mold, to minimize internal stresses and reduce distortion. It is also used to relieve stresses following rough machining or flame hardening. Typical temperatures for stress relieving are 950-1200F (510-650C). Fluidized bed furnaces are good for these applications because of their temperature uniformity and high rate of heat transfer even below radiation temperatures, enabling shorter cycles and reduced distortion.

Other Applications—The uniform and rapid heating characteristics of fluidized bed furnaces can be utilized for other applications such as preheating for welding, burning, extrusion, rolling and forging. For those processes where drag-out of aluminum oxide particles may cause abrasion of dies or rolls, graphite is used as the bed medium.

The uniform heat removal characteristics of fluidized beds can also be used in such areas as cooling after shakeout, equalizing, stabilization and cooling after flame hardening.

Summary

Fluidized bed furnaces have been found to be useful for many foundry applications. Their inherent temperature uniformity, even in very large furnaces, provides for less distortion and cracking. The high rates of heat transfer, both in heating and cooling, allow optimization and reduction in cycle times. An additional energy saving benefit of fluidized beds is their ability to be turned on and off, rather than idled, with little loss of heat and no detrimental effects to the furnaces.

In an age where efficiency, optimization of processes and economy of energy are primary objectives, fluidized bed furnaces offer unique advantages over conventional equipment for the foundry. **mc**

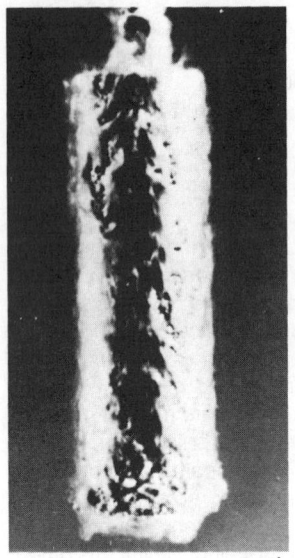
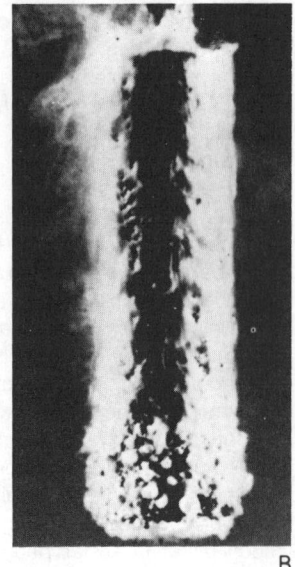
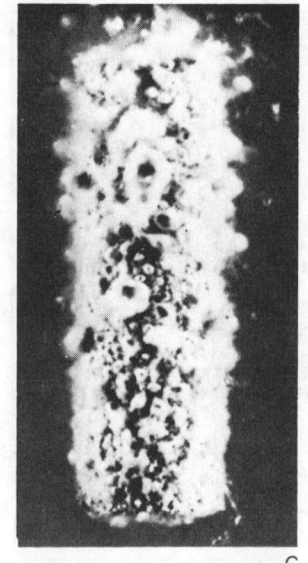
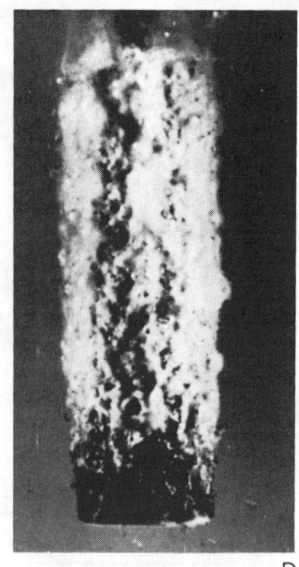

A | B | C | D

1. Photos show what happens when quenching with a polyglycol fluid. (A) At immersion, a film deposits on the surface of the hot metal. (B) After 15 sec the polymer begins activating. (C) After 25 sec the film is active at the corners of the workpiece. (D) At 35 sec the polymer coating is fully operative, permitting the cooling rate to increase. (E) The polymer redissolves after 60 sec to become a true solution again. (F) At the end of 75 sec the film is completely redissolved; heat is then removed by convection through the fluid.

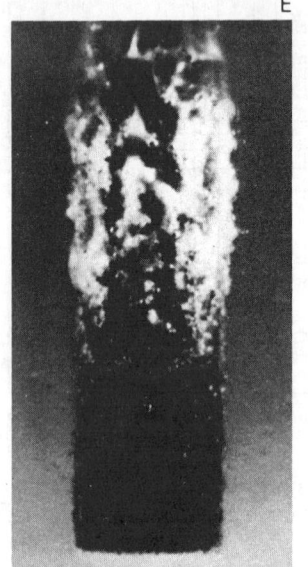
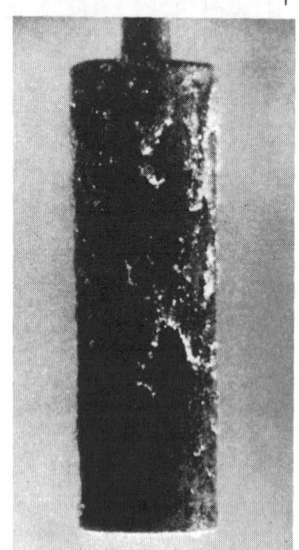

E | F

Heat treating jumps into synthetic fluids

Oil or water? You now have a third choice. Synthetic polymer quenching fluids are bridging the gap by offering more versatility and important savings in energy.

Three examples illustrate why commercial heat treaters are switching to synthetics:

• An air-brake manufacturer quenched SAE 1340 steel in a fast oil, but was unable to develop the required hardness. A water quench provided the hardness but cracked the shoes. A quench in boiling water eliminated the cracking, but caused a growth problem that required extra machining to meet specifications. Switching to an 11-percent-polyglycol quenching fluid provided the necessary hardness and avoided cracking or the need for secondary machining. In addition, the shoes could now be drilled prior to hardening, whereas in an oil or water

by Jack Hasson
E F Houghton & Co
Valley Forge, PA

quench, such drilling caused cracking.

• A forge shop specializing in plain carbon-steel forgings replaced two tanks, one of oil and one of water, with a single tank of 6-percent-glycol solution. Jobs quenched in the synthetic fluid range between crack-sensitive parts and those with borderline hardenability.

• A large midwestern heat treater does induction hardening, neutral and salt-bath hardening, carbonitriding and carburizing, austempering, martempering, and normalizing and stress relieving. The firm uses polyacrylate and polyglycol quenching fluids because of versatility, economy, and safety.

A plastic cushion

Polyglycol fluids are organic chemicals of high-molecular weight that dis-

Reprinted with permission from *Tooling & Production,* December 1983, 32-34, © 1983 Huebner Publications, Inc.

solve in water at room temperature.

But, when heated above 77 C, the polymer becomes insoluble and precipitates onto the work, going back into solution when the temperature falls below 77 C. This unusual property, called inverse solubility, accounts for desirable and unique cooling effects.

When heated steel is quenched in a polyglycol solution, cooling occurs in three distinct stages (**Figure 1** shows a series of time exposures during these stages).

Stage 1—The hot component is immersed, heating the quenching solution to above 77 C. A thick polymer layer, associated with the vapor stage, provides a period of slow cooling.

Stage 2—The polymer layer activates (boils), thereby permitting an increase in the cooling rate.

Stage 3—As the fluid approaches the inversion temperature, the polymer layer redissolves, and heat is removed from the quenched piece through convection in the liquid phase.

Ideal quenching speed is one that is as fast as possible, gives uniform hardening, and offers a cushion against cracking and distortion. Glycol polymers do just that, which is a boon to production heat treating.

The quenching speed of polyglycol is influenced by the temperature of the bath. A working range is between 35 C and 50 C, although up to 65 C can be tolerated in certain applications.

In addition, the cooling rate of a polyglycol quenching fluid can be varied by changing its concentration.* The thickness of the deposited film is a function of the concentration of the polymer in solution.

Solutions of 8 percent to 15 percent achieve quenching rates better than or as good as fast quenching oils, and are suitable for low hardenability applications where maximum mechanical properties are needed. Solutions of 15 percent to 30 percent provide cooling rates suitable for a range of through-hardening and case-hardening steels.

The slow quench

Polyacrylate aqueous quenching solutions, in contrast with the polyglycols, don't form a plastic film on the work material. An extended vapor phase and diminished heat extraction in the boiling range are responsible for the extremely

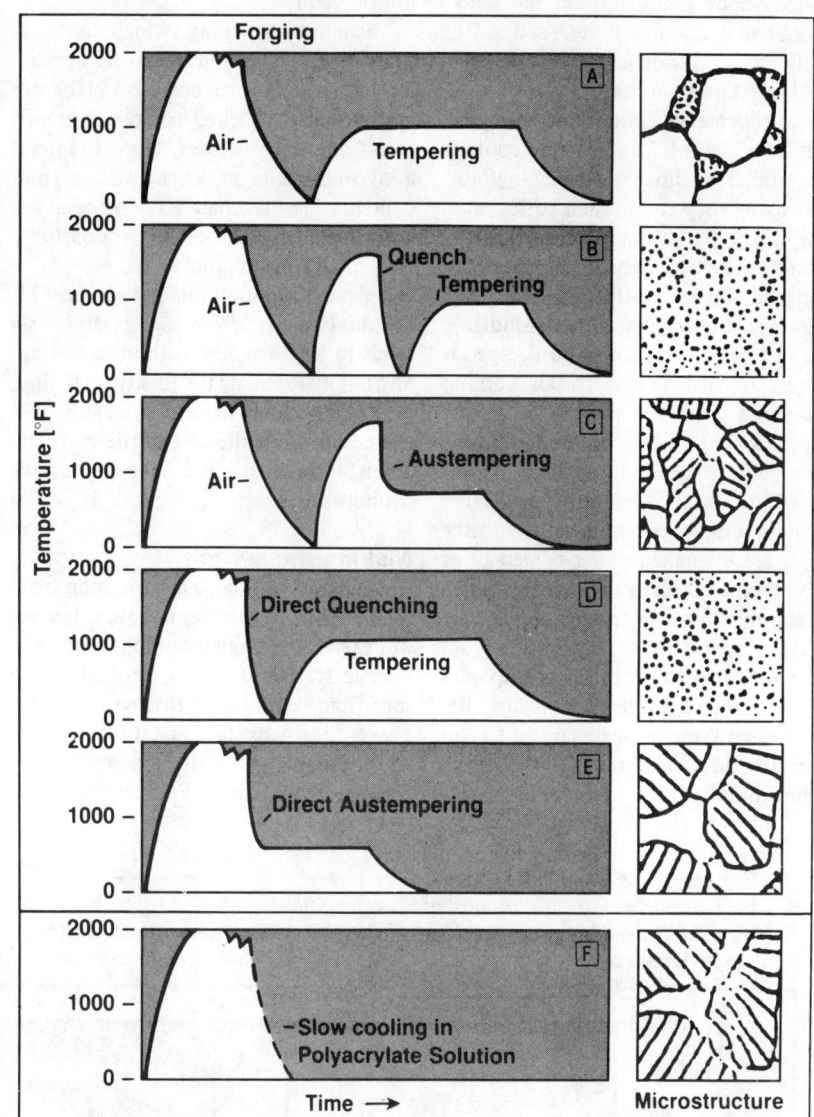

2. Many steel forgings are heat treated after forming operations to improve their machinability. Conventional treatments are shown here. (Note: The sample is carburizing steel that tends to smear when machined after forging.)

Every heat-treating method shown (except F) required additional amounts of energy obtained mainly from natural gas. With the rising price of gas, the opportunity to avoid tempering and austempering is tempting. The slow cooling rate that's possible when using polyacrylate solutions eliminates excess demands because it provides a good structure—pearlite plus about 20-percent large-grain ferrite—for critical machining.

Direct austempering, of course, will develop the same structure, and quenching and tempering produces spherodite—a better structure for forming and cutting. Simple tempering or annealing yields a bad working structure of ferrite with small amounts of high-carbon spherodite at the grain boundaries, which produces smearing during machining.

slow quenches possible with polyacrylates (their cooling curves approach a straight line because of the uniform heat extraction).

Soluble in water in all proportions, polyacrylates don't separate when heated. They are oil free, nonpoisonous, biodegradable, and nonflammable. They offer many cooling speeds, depending on the concentration and temperature of the solution.

A polyacrylate is not like glycol—it dissolves in water at any temperature. Polyacrylates are adaptable to a variety of programs, from normal hardening of ferrous metals of high or extremely high hardenability to cooling with a quench so slow that a martensitic structure will not form, **Figure 2**. This is possible because these solutions generate such a

* Concentration is easily monitored by measuring the refractive index of a fluid, then reading concentration from a calibration curve that plots the index against the concentration of glycol in solution.

durable vapor phase around the work material that cooling is retarded sufficiently to avoid transformation.

Polyacrylate solutions, agitated with high turbulence in a bath temperature of from 26.6 C to 51.7 C, generate cooling rates similar to those of quenching oils. This is why they can harden critical and crack-prone parts that have been traditionally quenched only in oil, martempering oils, or salt baths.

Studies at our technical center indicate that polyacrylates should be of special interest to firms presently air cooling carburized parts for partial machining before final hardening. Some carburized parts require grain refinement operations if the grain size grows too large during a long carburizing cycle, and the pieces are quenched or air cooled prior to hardening. This is no problem when using polyacrylate quenching fluids, **Figure 3**.

There is also good news for companies making continuous-cast steel slabs. Because these slabs must be cooled for inspection and surface repair, losses from scaling and decarburization take a heavy toll on yield.

To minimize scaling (which can consume 6 percent of the metal) and decarburization, plain carbon-steel slabs are often water quenched to speed up surface-finishing processes. Unfortunately, alloy-steel slabs or steels with carbon contents greater than 0.40 percent are not quenched because they will partially harden and often crack.

Polyacrylate solutions permit most alloy and high-carbon (even stainless) steels to be quenched without cracking. Savings by reducing the loss from scaling on these expensive alloys usually will more than cover the cost of the synthetic quenching fluid and the necessary equipment.

Making the switch

As might be expected, switching from one quenching medium to another requires preplanning and preparation. The change from water to a synthetic polymer fluid is probably the least difficult procedure. Adequate agitation is essential, though, and must be provided if it doesn't exist already. Painting the tanks should be unnecessary since most synthetics carry a corrosion inhibitor.

The most involved effort to switch from brine to a polymer is to scrub the quench tank thoroughly. Again, agitation may have to be increased.

Many conversions to polymer fluids are made to avoid the smoke and flames associated with oil quenches. Obviously, when making this change, it's necessary to carefully clean the entire quenching system. All residual oil must be removed.

Finally, it's necessary to determine if all strainers, drains, and other parts are chemically compatible with synthetic fluids.

Not every heat treater will pull the plug on the water, brine, or oil tank. But, where unfailing duplication of hardness and internal structure without cracking or deformation is needed, synthetic polymer quenching fluids are the solution.

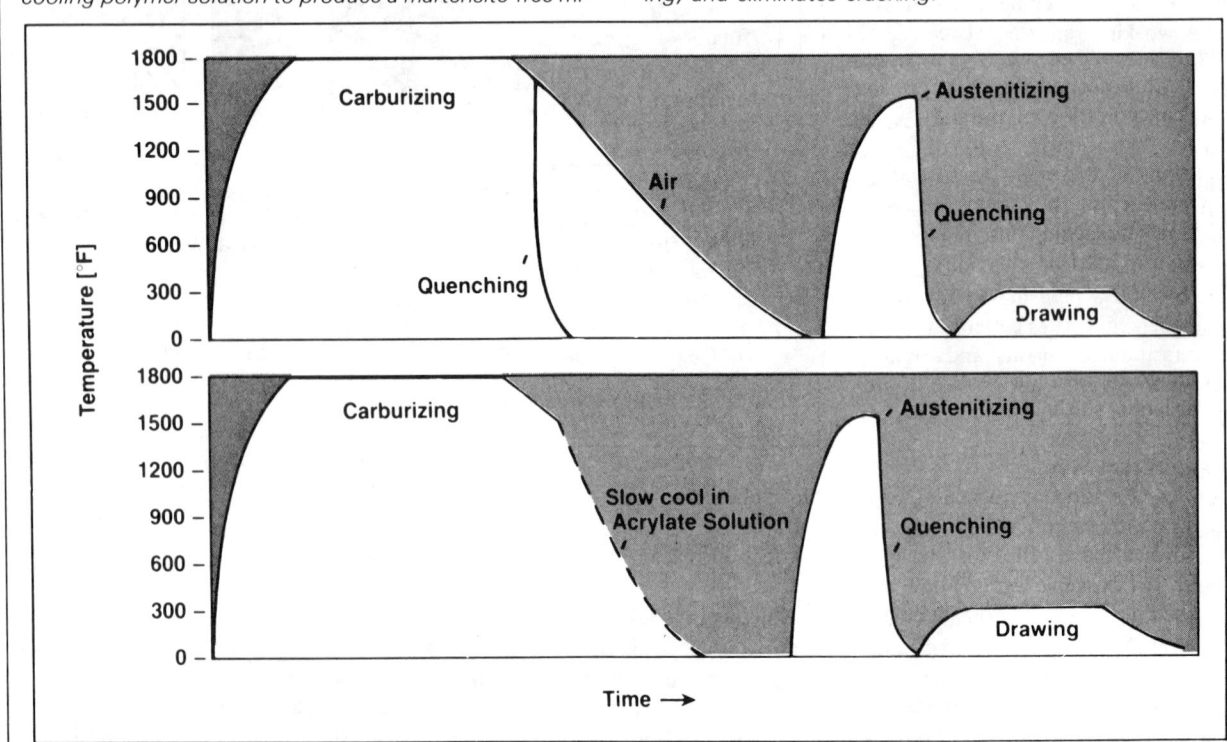

3. This comparison of heat treatments suggests that the first cooling cycle will be more efficient if done in a slow-cooling polymer solution to produce a martensite-free microstructure. Not only are the parts ready for working, but the polymer quench precludes scaling and decarburizing, and eliminates cracking.

Fluidized Bed Furnace Heat Treating Applications for the Die Casting Industry

Joseph E. Japka, Director, Metallurgy and Heat Treating Laboratory, Procedyne Corp., New Brunswick, New Jersey

The die casting industry is dependent upon precision engineered and manufactured tools, dies, and accessory steel tooling. All of the tooling is usually heat treated in numerous ways to provide the properties necessary for this industry. Procedyne Corp. of New Brunswick, New Jersey, manufactures fluidized bed furnaces for heat processing tooling economically. One furnace installation is capable of neutral hardening, carburizing, nitriding, annealing, and many other standard heat treating processes normally used by the tooling industry.

A significant property of a fluidized bed is the high heat transfer coefficient, which allows metal to be heated at speeds close to those obtained in lead or salt baths—without their inherent disadvantages. The heat transfer rate is about three times faster than the best available commercial furnace in the high-temperature range, and in the tempering range is five to eight times faster.

This rapid heating and the ability to turn the furnaces off or on at will without any furnace damage makes them ideal for intermittent as well as continuous usage.

Fluidized Bed Furnace Basics

A typical schematic drawing for a commercial installation used for general purpose heat treating is shown in Fig. 1. Full temperature control is possible from room temperature to the furnace's maximum allowable temperature of 1850°F. The furnace is built of an outer shell lined with heavy ceramic fiber blanket insulation which surrounds the electrically heated or gas fired heating chamber.

The heating chamber encloses a retort and diffusion plate constructed of heat-resistant alloys, which is the heart of the furnace. The patented diffusion plate is welded in the retort and contains a pattern of drilled and tapped drill screw orifices. The chamber at the bottom of the retort is fed by a gas atmosphere inlet pipe. At the top is a sand seal into which the lid with a burnoff vent is placed to close the retort during atmosphere operation. A thermowell with enclosed thermocouple is positioned at the side of the retort. Temperature is automatically controlled to within ±5 to ±10°F, depending on furnace size.

In the company's FLUIDHEARTH, furnace heating is indirect. Therefore, fluidization gas and velocities and compositions may be chosen to suit the process requirements. Since the bed material (generally, 80 or 120 mesh aluminum oxide) is inert even when heated, fluidizing gas composition can be chosen to provide the optimum chemistry for surface treatment.

The fluidized bed is quite liquid-like and can be stirred, bubbled, and agitated. The bed seeks its own level and has a pseudo density that determines whether objects float or sink

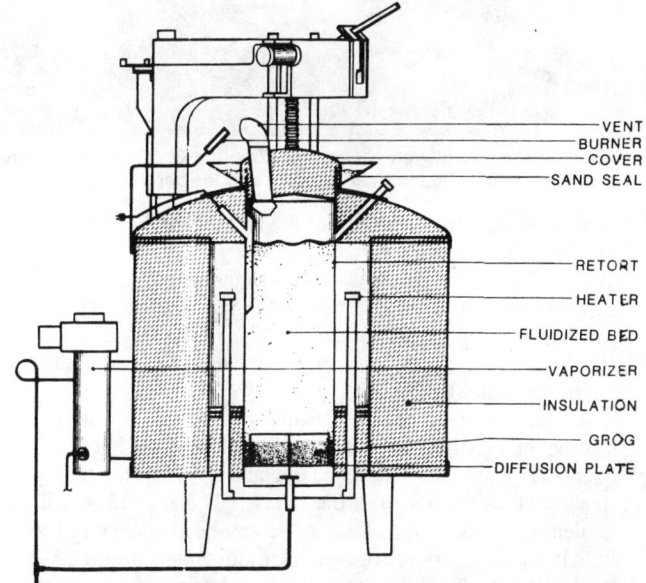

Fig. 1. Cross-section of FLUIDHEARTH heat treating furnace. Gas enters line, lower left. Vaporizer is used only with methanol for carburizing cycles.

into it. Although the particles which make up the bed are thermally nonconducting when in a nonfluidized state, when fluidized, thermal conductivity is acquired to a degree approaching that of a good heat transfer oil or molten salt or lead bath.

In operation, parts are held in baskets or racks, which may be either suspended above the retort bottom or rested on a load support frame which is built into the retort. Any atmosphere may be used, although the furnace is normally operated on compressed air, unless neutral hardening or other atmosphere heat treatments are necessary.

Neutral Hardening of Tool Steels

Almost all steels used for construction of tooling and dies in the die casting industry are heat treated. Metallurgically, steel hardening consists of raising the temperature of the steel to the austenite condition, usually over 1500°F, and quenching in a medium such as oil or water. This transformation of the steel microstructure from austenite to hard martensite is followed by a tempering or drawing treatment designed to increase toughness and reduce the hardness, so that the steel is in a condition suitable for use.

When this process is performed without any scaling or decarburization, it is known as neutral hardening. Fluid bed furnaces are ideal for this operation. The term *neutral* is

Fig. 2. System employed at Fox Steel Treating Co. L to r are the quench tank, fluidized bed hardening furnace, ambient fluidized bed quench, fluidized bed temper furnace, and the control system.

applied whenever the surface of the metal is to be protected from any change in the surface chemistry due to its exposure to heat. Typical hardening cycles for steels encounter temperatures from 1400 to 2200°F. Quenching or cooling methods produce the appropriate microstructure in the metal, using oil, water, brine, air, polymer quenchants, salt, or room-temperature fluidized beds.

In the majority of die casting industry applications, heating in the fluid bed furnace is followed by either oil quenching or ambient fluid bed quenching. Deep-hardening steels such as the hot work die steels H-11 and H-13, or air-hardening steels such as D-2 or A-2 can be quenched in the fluid bed using air or nitrogen as the fluidizing medium. Case hardening steels such as 8620 or carburized low-carbon steels, after heating, must be oil or polymer quenched, to achieve full hardness.

Use of a fluidized bed for hardening is similar to using a salt bath . . . but without the hazards. Parts are loaded in a basket or on a fixture for immersion in the fluid bed. Oily or wet parts can be processed in the fluid bed without worry of explosions or surface contamination. Since a fluidized bed is operable from ambient to its maximum operating temperature, any temperature is possible. Also, it can be turned off at any time when not in use and will retain much of its heat, adding to its excellent energy efficiency.

Neutral hardening is designed to prevent decarburization of the surface of steels and to prevent any other contamination, such as unwanted carburization. Prevention of carburization lessens the steel's susceptibility to heat checking. Decarburization may result in lower metal fatigue strength and lower wear resistance, and also results in a false hardness indication, when tested.

Proper use of an inert atmosphere avoids these problems. The normal atmosphere for neutral hardening in fluid beds is dry nitrogen, which may be obtained as a compressed gas or in bulk liquid form. The atmosphere is necessary only when neutral hardening is in progress. During other times, low-cost compressed air may be used for fluidizing the bed.

Typically, tool steels such as H-13 are neutral hardened in the following cycles: Parts are furnace loaded while the furnace is at a temperature of 1850 to 1950°F, using the nitrogen atmosphere. After furnace recovery, normally 20 to 30 min, the part may be soaked a relatively short time and

then quenched in the ambient fluid bed to room temperature. Using the inert atmosphere, no greater than 0.001 in. of decarburization takes place. The tools are then tempered in a fluid bed tempering furnace of similar construction using dry nitrogen in the range of 800 to 1000°F, depending upon the hardness desired.

The appearance of parts thus heat treated is clean and scale-free—but darkened slightly to a gray finish. For tight dimensional control, it usually is not necessary to finish-grind these parts a depth of more than 0.001 to 0.002 in. Another option with the fluid bed furnace, following the neutral hardening cycle, is to perform tempering and nitriding in one cycle, usually as the second of two tempering operations, as described in the next section.

Case Hardening Treatments in Fluid Beds

Case hardening treatments are designed to chemically impregnate either carbon or nitrogen or a combination of both into the surface of steel parts to develop a layer which is harder than the base material. (Procedyne's trade name for these processes is DYNA-CASE surface hardening treatments.) Case hardening may be subdivided further into deep case hardening and shallow case hardening. Examples of deep case hardening are carburization, which is normally performed at 1650-1850°F on steel to effect a hard layer between 0.010 and 0.100 in. deep. Shallow case hardening can be done at moderate temperatures from 1500 to 1650°F and is commonly known as carbonitriding.

Prior to the availability of fluid bed furnaces, case hardening often was done in cyanide-containing salt baths to provide a case depth 0.005 to 0.015 in. To achieve the hard case, the carbonitrided part must be quenched in oil or another quenchant. The need for quenching often causes problems in distortion, as a phase change occurs in the steel as it rapidly cools from 1500°F to room temperature.

An alternate shallow case hardening process is nitriding or nitrocarburizing, performed at a relatively lower temperature, avoiding any phase change, between 950 and 1100°F. This process often is performed on finished parts, as little dimensional change or surface damage occurs. The case thickness normally is between less than 0.001 up to 0.015 in., depending upon need.

It may be noted that heat treatment processes described

Fig. 3. Details of the furnace control systems. Box on left controls the temper furnace up to 1250°F. The center panel and atmosphere flowmeters control the hardening furnace, which can perform all cycles covered in the article.

using fluid bed furnaces with varying atmospheres and temperatures are performed with gas additions only. No salt or special chemicals are necessary in fluid bed furnaces. Furnaces designed for atmosphere treatments have an accessory panel containing gas flowmeters which are set for the different atmosphere heat treatments. Following is a brief description of each process as it is performed in the fluid bed.

Deep Case Carburization. This is performed in a fluid bed furnace by adding methanol as a liquid, which is vaporized in the furnace system and mixed with nitrogen and a small amount of natural gas to form a synthetic endothermic atmosphere within the furnace. Methanol is available in cylinders from commercial gas suppliers. Complete carbon potential control is possible.

Typically, deep case carburization may be performed from 1650 to 1850°F, from 1 to 8 hr. The parts are loaded in the furnace under a nitrogen atmosphere, and the carburizing atmosphere is turned on as soon as the parts are at temperature. Once the cycle of carburization is complete, normally the furnace is lowered to 1550°F and the parts are taken out and oil or polymer quenched, followed by a separate tempering cycle.

Carbonitriding. This is performed for an even harder shallow case on steel parts. Parts may be racked or loaded in a basket and are normally heated in the furnace from 10 min to 1 hr. Cases are available in depths approx. 0.005 to 0.015 in. Temperatures are 1500 to 1650°F, times are 10 min to 1 hr, and quenching is necessary.

Nitriding. Nitriding, metallurgically speaking, is the addition of nitrogen from raw ammonia into the surface of the material at temperatures between 950 and 1100°F. There are special steels, such as the Nitralloy family, often used for this purpose. A nitriding case usually is 0.001 to 0.015 in. deep and is commonly performed on finished machined metal parts as a final operation.

The extreme high hardness available from the nitrided surface is used to confer wear and fatigue resistance. As commonly practiced in the tooling industry, nitriding is a diffused case distinguished under the microscope by its etching characteristics. The time for successful nitriding often is quite long, varying from 4 to 48 hr for deep cases.

Nitrocarburizing. This surface treatment provides a very hard layer, 0.0005 to 0.002 in. deep, consisting of a combined nitrogen-carbon hard phase (which is sometimes referred to as *white layer*) on the surface. It is formed using mixtures of natural gas and raw ammonia in the fluid bed

furnace, from 0.5 to 2 hr, usually at temperatures between 1050 and 1150°F. Though this treatment provides a wear-resistant surface layer similar to nitriding, usually it is shallower. Many parts which are used in a sliding mode—such as ejector pins—can be processed in this manner in a relatively short time, compared with the longer times necessary for straight nitriding.

Steam Blueing. One additional surface treatment readily performed in a fluid bed furnace is steam treating, or blueing. An accessory to the furnace humidifies nitrogen or air before it enters the furnace for its fluidizing task. Immersing parts in this atmosphere usually is done between 650 and 1000°F for periods of 20 min to 1 hr and achieves a blue or black surface having some corrosion protection and an attractive cosmetic surface, which also retains lubricant.

It may be noted that the nitrocarburized case plus the steam blued surface coloration can be used for corrosion protection of steel components. It has been observed that this surface treatment provides at least 24-hr salt water corrosion protection with no rust formation.

Annealing. Occasionally, it is necessary to anneal tool steel components for remachining because of an error or a necessary tool change. This operation is readily performed in a fluid bed furnace using nitrogen as a protective atmosphere. The tool, made of H-13 or even D-2, is immersed in the fluid bed furnace at over 1600°F and held there until it is through heated and soaked, up to 0.5 hr. The furnace then is allowed to slow cool with a slight trickle of nitrogen atmosphere until the part cools to below black heat, i.e., below 1000°F.

At this point, the part may be removed from the furnace and cooled to room temperature. This annealing cycle often is done as an overnight operation where the furnace is merely shut off after the soak time, and the bed allowed to defluidize with just a trickle of nitrogen through it to prevent surface oxidation. The next morning, the part is removed, ready for further machining operations. The annealing cycle, when performed using dry nitrogen, is effective in avoiding surface decarburization or oxidation.

Combination Heat Treatments. Especially noteworthy is the versatility of fluid bed furnaces for combining heat treatments using only one furnace and an ambient fluid bed quench. In our facility, for example, H-13 steel dies have been heat treated to completion, including nitriding, in the following manner.

A finished machined die was hardened by heating under a nitrogen atmosphere to 1950°F in 20 min, and fluid bed quenched to room temperature. Then, it was tempered once at 1180°F for 2 hr, fluid bed quenched, and then tempered and nitrided 4 hr at 1050°F. No cleaning or intermediate operations were needed to finish the die, except for a final bright polish.

Case Histories

The fluid bed furnace system has shown uniform and dependable performance for die casting tooling. For example, one year ago, Detroit's Fox Steel Treating Co., a commercial heat treating company which specializes in tool and die work, purchased a FLUIDHEARTH system deep enough to handle 48-in.-long ejector pins, as shown Fig. 2 and 4.

According to VP Pat Fox, his customer used to fabricate 800 pins in order to guarantee that 400 would be usable after heat treating and straightening, a 50% rejection rate. "Any pin over 30 in. had to be laid down, to fit into an integral

Fig. 4. Milling cutters (cases at left) are nitrided. Note long ejector pins (foreground) awaiting heat treating.

quench furnace," he said, "and when it ran through the high heat and quench, it would warp like a pretzel. Half the pins would break in straightening."

The customer, Prospect Die and Mold (known as PDM), Roseville, Mich., fabricates its ejector pins of H-13, and the part is 90% completed before it goes to the heat treater. Now, when the parts are heat treated in the fluid bed, PDM makes 410 pins for an order of 400, a rejection rate of only 2.5%.

Fox carbonitrides the pins at 1650°F, and achieves a case depth of 0.015 to 0.020 in. The parts are then elevated in temperature to 1850°F, fluid bed quenched, and then tempered. Loads can vary from 250 to 600 lb., depending upon size, and the pins range from 1/32 in. dia. × 4 in. long up to 1 in. dia. × 48 in. long. Fox heat treats three to four loads per week and turns the work around in one day.

"Traditionally," Fox explained, "pins like these are made from prehardened stock and salt bath nitrided. Unfortunately for this application, a nitride case is brittle and has a tendency to chip in the mold. In the case of fluid bed heat treating, after we get our case depth, we go up to 1850°F to bring the core into solution, and then quench in the fluid bed. We have no problem with chipping, and less tendency of bending because the core is tougher."

Fox also carbonitrides H-13 cores and sleeves for PDM, preferring to run the parts in the fluid bed instead of in an integral quench furnace. Otherwise, the parts would need an oil quench to get the high hardness needed, which poses a hazard and requires that the parts be washed before tempering.

Fox also heat treats slides for Industrial Economy Tool Co., a Detroit mold builder. The slides are made of P-20 and H-13 steel, and are heat treated in a nitrocarburizing atmosphere at 1000°F and then fluid bed quenched. The customer sends 8 or 16 small sections weighing 3 to 4 lb. each, several times a month. Fox nitrocarburizes the parts to a case depth of 0.005 in. and returns them the same or next day.

"The customer used to have the parts nitrided somewhere else," according to Pat Fox, "but he likes the appearance of these better. They come out gun barrel blue and shiny. More important, however, is size stability, eliminating the necessity of additional fitting after the nitrocarburizing process."

Fox Steel Treating does all die casting tooling work in fluid beds (Fig. 2), including a 20 in. dia. × 48 in. deep case hardening furnace system with atmosphere capability for nitriding, carburizing, carbonitriding, and neutral hardening; a tempering furnace 20 in. dia. × 48 in. deep; and a fluid bed quench 24 in. dia. × 48 in. deep. Photographs of the Fox installation are shown in Fig. 2-4.

Fluidized-Bed Equipment

Edited by William L. James
President
Fennell Corp.

FLUIDIZED-BED TECHNIQUES are not new. A 19th century American patent describes the roasting of minerals under fluidized-bed conditions. Other established applications include potter's clay and miner's hydraulic slurries. Systems of fluidized solid particles, such as quicksand, occur in nature.

Early attempts to use fluidized beds in heat treatment of metals were limited in the temperatures that could be employed. Electrically heated furnaces capable of maintaining fluidized beds at temperatures up to 500 °C (930 °F) could be produced commercially, but difficulties were encountered when attempts were made to attain higher temperatures. A principal problem was the high rate at which refractory distributors, which distribute the hot fluidizing gases, were consumed.

In early gas-fired fluidized-bed furnace design, gas entered the base of the container after being mixed with air to make it ignitable at the point of entry. With newer designs, the mixtures are introduced separately and so cannot be ignited accidentally. This design eliminates the danger of explosion at the point of entry. The surface of the bed is heated first, and the heating of surface particles causes progressive ignition downward through the container until the entire contents of the bed achieves uniform heat treating temperature.

Newer furnace designs extend fluidized-bed technology into the higher temperature ranges required for most common heat treatments.

Principles of Fluidized-Bed Heat Treating

In fluidization, a bed of dry, finely divided particles, typically aluminum oxide in the heat treating context, is made to behave like a liquid by a moving gas fed upward through the bed. A gas-fluidized bed is considered a dense-phase fluidized bed when it exhibits a clearly defined upper limit or surface. At a sufficiently high fluid-flow rate, however, the terminal velocity of the solids is exceeded, the bed goes into motion, and the upper surface of the bed disappears. This state constitutes a disperse, dilute or lean-phase fluidized bed with pneumatic transport of solids. The general types of fluidized beds are shown in Fig. 1. The majority of beds used for heat treatment are of the aggregative or bubbling type.

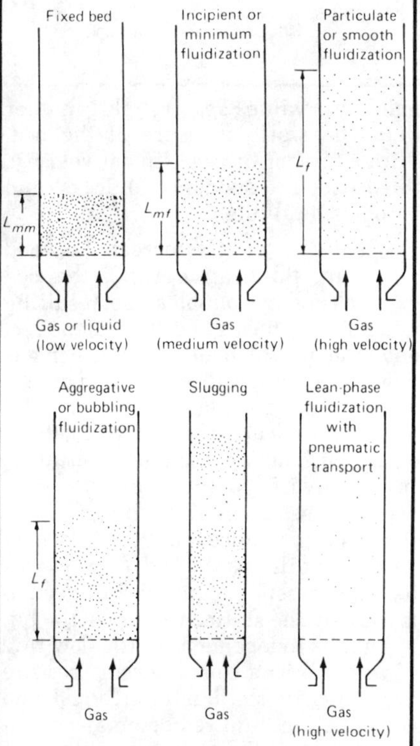

Although the properties of solid and fluid alone determine the quality of fluidization (that is, whether smooth or bubbling fluidization occurs), many factors influence the rate of solid mixing, the size of bubbles and the extent of heterogeneity in the bed. These factors include bed geometry; gas-flow rate; type of gas distributor; and internal-vessel features such as screens, baffles and heat exchangers.

Determination of Fluidization Velocity. In determining the quality of fluidization, a diagram of pressure drop (Δp) versus velocity (μ_o) is useful as a rough indication when visual observation is not possible. A well-fluidized bed will behave as shown in

Fig. 1 Various types of contacting in fluidized beds (Ref 1)

the diagram in Fig. 2, which has two distinct zones. In the first, at relatively low flow rates in a packed bed, the pressure drop is approximately proportional to gas velocity, usually reaching a

Reprinted from Metals Handbook, 9th Edition, Vol. 4, 299-306, © 1981 American Society for Metals

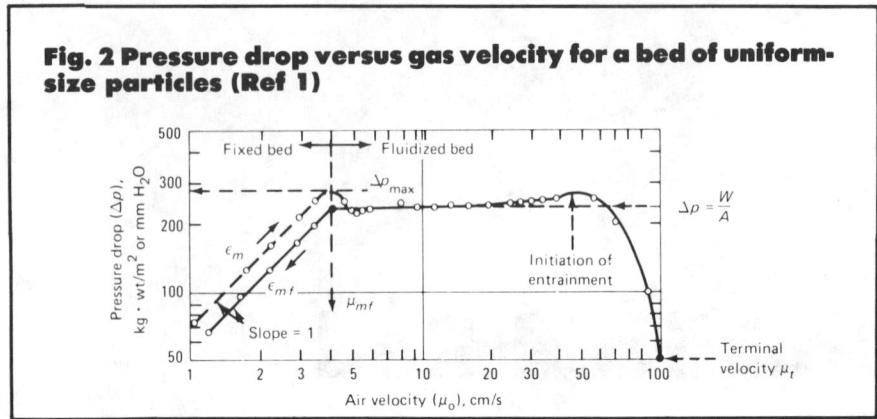

Fig. 2 Pressure drop versus gas velocity for a bed of uniform-size particles (Ref 1)

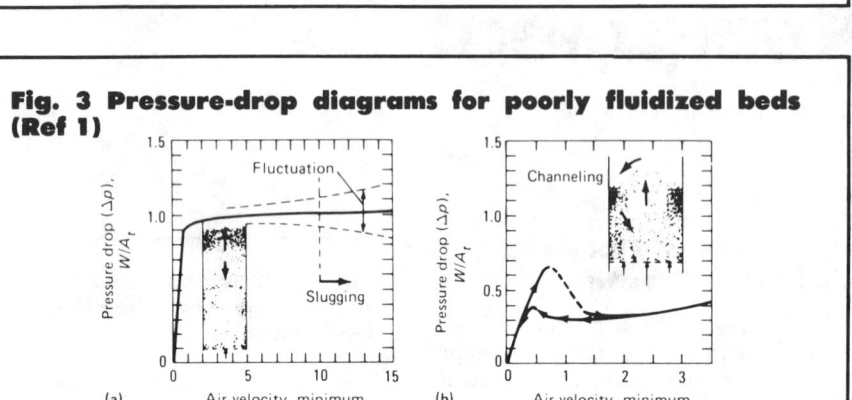

Fig. 3 Pressure-drop diagrams for poorly fluidized beds (Ref 1)

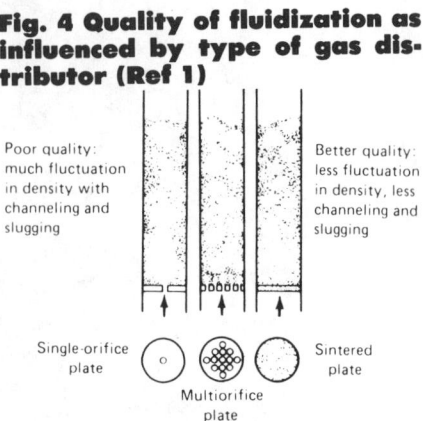

Fig. 4 Quality of fluidization as influenced by type of gas distributor (Ref 1)

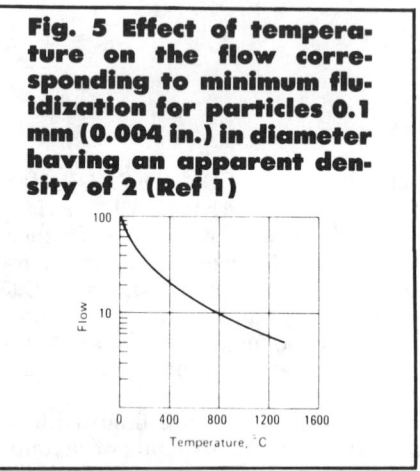

Fig. 5 Effect of temperature on the flow corresponding to minimum fluidization for particles 0.1 mm (0.004 in.) in diameter having an apparent density of 2 (Ref 1)

maximum value (Δp_{max}) slightly higher than the static pressure of the bed. With a further increase in gas velocity, the packed bed suddenly "unlocks" and becomes fluid-like.

When gas velocity increases beyond minimum fluidization (μ_{mf}), the bed expands and gas bubbles rise, resulting in a heterogeneous bed. This is the second zone, in which, despite a rise in gas flow, the pressure drop remains practically unchanged. The dense gas-solid phase is well aerated and can deform easily without appreciable resistance. In its hydrodynamic behavior, the dense phase can be likened to a liquid. If a gas is introduced into the bottom of a tank containing a liquid of low viscosity, the pressure required for injection is roughly the static pressure of the liquid and is independent of the flow rate of the gas. The constancy in pressure drop in both the bubbling liquid and the bubbling fluidized bed may be taken intuitively to be analogous.

The diagrams in Fig. 3 show poorly fluidized beds. The large pressure fluctuations in Fig. 3(a) suggest a slugging bed. In Fig. 3(b), an absence of the characteristic sharp change in slope at minimum fluidization and the abnor-mally low pressure drop suggest incomplete contacting, with particles only partly fluidized.

One of the most important factors influencing the quality of fluidization is the type of distributor. Figure 4 illustrates this schematically.

Temperature Effect on Minimum Fluidization Velocity. One of the most important parameters of a fluidized bed is the minimum fluidization velocity. In simplified terms, minimum fluidization velocity (μ_{mf}) approximates to a function of the square of the particle diameter (d) and a linear function of particle mass (p) as follows:

$$\mu_{mf} \simeq d^2 p$$

In the design of heat treating furnaces, the effect of temperature must be considered. Figure 5 shows how the flow of gas required for fluidization decreases rapidly with increases in temperature.

Defluidization. One of the common misconceptions about fluidized beds is that, because of their principle of operation, they are not well suited for large solid parts with horizontal surfaces that remain stationary in the bed. With parts of this type, a cap of nonfluidized particles collects on the horizontal surfaces, forming a thermal screen. The higher the temperature of operation, however, the greater the energy and agitation of the bed, and the smaller the likelihood that the bed will collapse. Moreover, various methods can be used to overcome this apparent disadvantage, and these are designed into most fluidized beds. These methods are:

- Movement of the part being treated
- Introduction of additional agitation in the zone of fluidization around the part, either by localized injection of fluidizing gas or by careful design of the outline of the basket that holds the parts
- Increased fluidizing velocity.

Selective Heat Treatment. Bed collapse can be turned to advantage for special heat treatments where one area must be hard and tough while the remainder must be soft and more ductile, as in the case of the engineered parts of the shape described above. In this case, after uniform heating, the

Fig. 6 Relative heat-transfer rates (Ref 1)

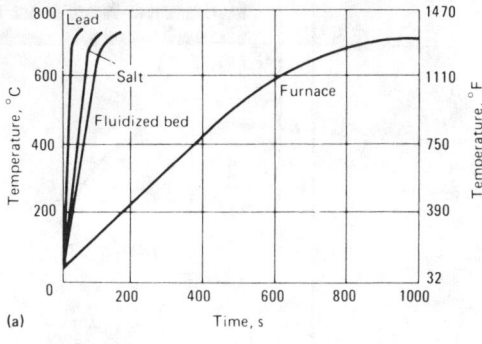

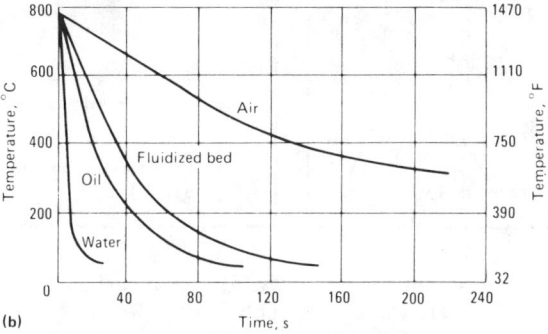

(a) Heating rates for 16-mm (0.6-in.) -diam steel bars in lead, in salt, in a fluidized bed and in a conventional furnace. (b) Quenching rates for 16-mm (0.6-in.) -diam steel bars in air, in oil, in water and in a fluidized bed

Table 1 Comparison of the effects of hardening and isothermal quenching of type D3 tool steel in salt baths and in fluidized beds

Heating or cooling medium	Diameter of test pieces mm	in.	Preheating temperature °C	°F	Total time for final heating and holding at 960 °C (1760 °F), min	Hardness, HRC At surface	At center
Salt bath...........	80	3.2	500	930	44	65.5	65
Fluidized bed(a).....	80	3.2	490	915	51	65	65
Salt bath...........	40	1.6	540	1000	36	64.5	64
Fluidized bed.......	40	1.6	500	930	41	64.5	64

(a) Small parts of the same steel but with a diameter of 8 mm (0.3 in.) were treated at the same time; hardness of these parts was 66 HRC.

product is submerged in a fluidized quenching bed, with the part to be hardened facing downwards. The top horizontal surface becomes covered with a cap of particles that form a thermal screen, which retards the vigorous cooling caused by the fluidized bed.

Heat Transfer in Fluidized Beds

An important characteristic of fluidized beds is high-efficiency heat transfer. The turbulent motion and rapid circulation of the particles in the fluid furnace provide a heat-transfer efficiency comparable with that of conventional salt- or lead-bath equipment.

The heat-transfer coefficient of a fluidized bed is typically between 120 and 1200 W/m²·°C (21 and 210 Btu/ft²·h·°F). The turbulent motion and rapid circulation rate of the particles and the extremely high solid-gas interfacial area account for this feature. The following factors are important in heat transfer:

Particle Diameter. Of all the pa-

Fig. 7 Recovery rates for 25-mm (1-in.) -diam steel parts in a 0.3-m³ (10-ft³) fluidized bed (Ref 1)

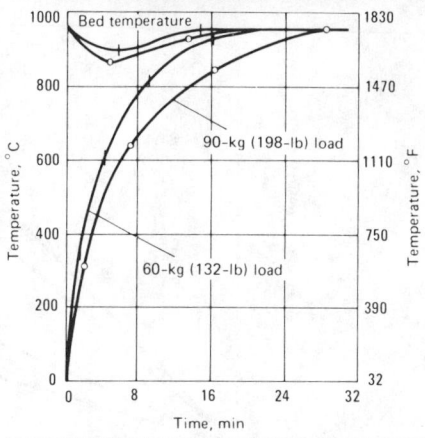

rameters that affect the heat-transfer coefficient in fluidized beds, particle diameter exerts the greatest influence. Particle diameter generally should be as small as possible; however, below a certain size, entrainment and carry-out effects may cause problems.

Bed Material. The governing physical property of any bed material is its density. There appears to be an optimum density for bed materials—about 1280 to 1600 kg/m³ (80 to 100 lb/ft³). High-density materials tend to produce lower heat-transfer coefficients and also require more power for fluidization. Carry-out problems occur with low-density materials. Other properties, such as thermal conductivity and specific heat, are less important.

Fluidization Velocity of Gas. It is essential to use the optimum flow rate that provides maximum heat-transfer rates for a particular particle density and diameter. Generally, this flow rate is considered to be between two and three times the minimum fluidization velocity. Too high a velocity leads to particle entrainment, high consumption of fluidizing gas and poor heat transfer; too low a velocity leads to poor heat transfer and lack of uniformity in processing.

Heating Rates. Relative heating rates of a 16-mm (0.6-in.) steel bar in salt, in lead, in a fluidized bed and in a conventional furnace are illustrated in Fig. 6(a); relative cooling rates for air, oil, water and a fluidized bed are shown in Fig. 6(b). Figure 7 presents heating and recovery rates for a fluidized bed. Results of both hardening and isothermal quenching of type D3 tool steel

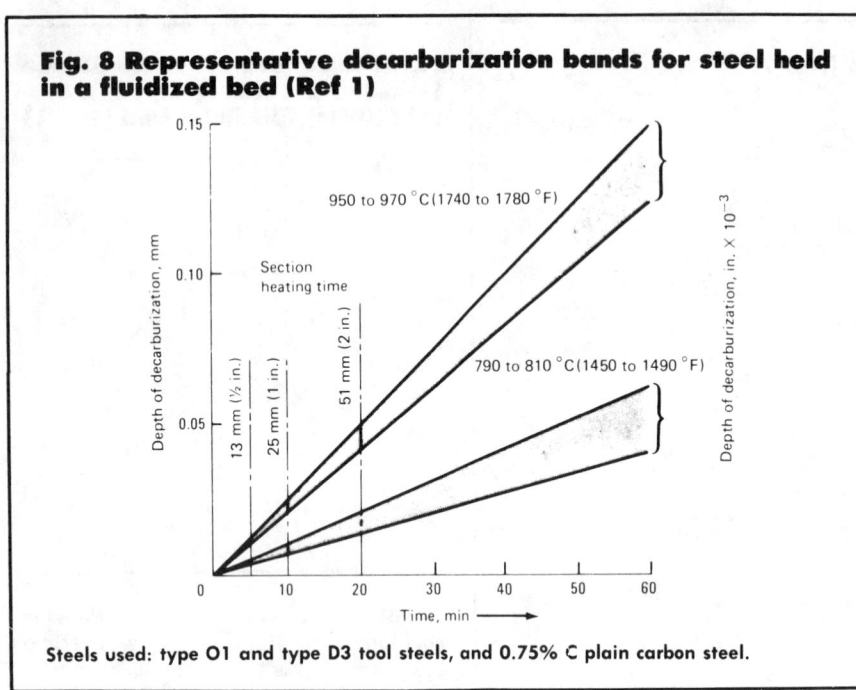

Fig. 8 Representative decarburization bands for steel held in a fluidized bed (Ref 1)

Steels used: type O1 and type D3 tool steels, and 0.75% C plain carbon steel.

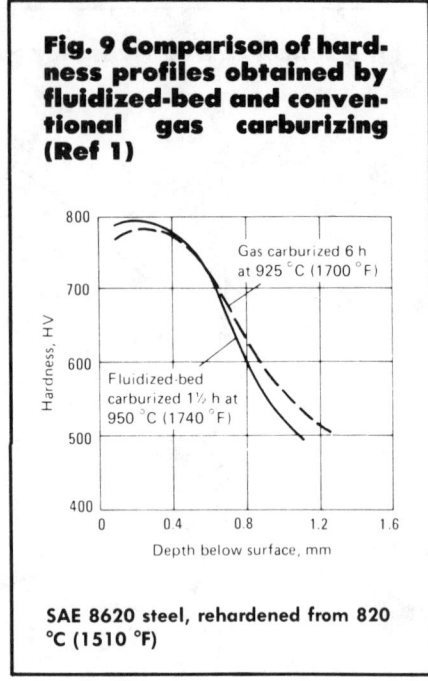

Fig. 9 Comparison of hardness profiles obtained by fluidized-bed and conventional gas carburizing (Ref 1)

SAE 8620 steel, rehardened from 820 °C (1510 °F)

with salt baths and with fluidized beds are given in Table 1. The difference between the two installations in total time for final heating and holding resulted from a difference in preheating conditions.

Control of Atmospheres

A full range of atmospheres can be used within the work zones of fluidized beds. The volume of gas used is clearly dictated by particle size, temperature of operation and optimum fluidization velocity. However, it can be shown that, with careful design and use of low-cost carrier gases such as nitrogen, even low-temperature surface treatments can be both effective and economical. In addition, one of the major advantages of a fluidized bed is that expensive gas need not be consumed while there is no work in the bed. Atmosphere conditioning is rapid: within about 30 to 60 s after an inert gas is introduced into the bed, the purity of the atmosphere is equivalent to that of the gas supply. In internal-combustion fluidized beds, two types of atmospheres can be obtained, as discussed below.

Reducing or Oxidizing Atmosphere. Adjustment of the gas/air mixture to the bed so that it is either gas-rich or oxidizing causes some decarburization or oxidation reactions in the materials being processed (the gas-rich mixture produces somewhat less severe reactions). However, these are time-dependent reactions, and, because of the rapid heating rates of parts being processed and the subsequent short immersion times needed to obtain the correct structure and through hardness, little surface effect other than discoloration and slight scaling is exhibited in section sizes up to 25 mm (1 in.). For larger sizes, the user must be aware of surface reactions that can occur, particularly as the processing temperature increases. Figure 8 illustrates the relative decarburization bands for steels held in a fluidized bed.

Neutral Hardening and Carburizing. In electrically heated or special gas-heated beds, atmospheres for neutral hardening of tool steels or carburizing of low-carbon steels can be used for bed flotation. This will allow oxygen-free heating of tool steels. However, caution must be exercised during transport of workpieces to the quench tank, to prevent decarburization or oxidation. For carburizing, the atmosphere can be collected, rejuvenated and recycled for efficient usage of the carbon-carrying vehicle.

Surface Treatments

Fluidized beds, using atmospheres composed of ammonia, endothermic gas and oxygen, or similar combinations, are capable of performing low-temperature nitrocarburizing treatments equivalent to conventional salt-bath processes or other atmosphere processes. High-speed tools oxynitrided in a fluidized bed are comparable to similar tools treated by the more conventional gaseous process. Carburizing and carbonitriding in a fluidized bed can yield results similar to those achieved in conventional atmosphere furnaces.

Mixtures of propane and air produced the results shown in Fig. 9, which compares the case depths obtained on SAE 8620 steel bearing rings carburized in a fluidized bed and by the conventional atmosphere process. An effective case depth of 1 mm (0.04 in.) was achieved in 1.5 h using the fluidized-bed technique.

Fundamental work on this process is still being performed, but sufficient knowledge exists to compare the mechanisms of conventional gas carburizing and the fluidized-bed process.

Conventional Gas Carburizing. Carburizing occurs through catalytic decomposition of CO according to

$$CO + H_2 \rightarrow C_{Fe} + H_2O$$

Propane enrichment aids this reaction according to

$$C_3H_8 + 3CO_2 \rightarrow 6CO + 4H_2$$

and

$$C_3H_8 + 3H_2O \rightarrow 3CO + 7H_2$$

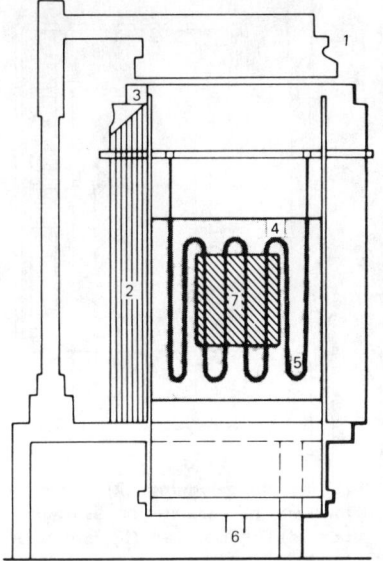

Fig. 10 Fluidized-bed furnace with internal heating by electrical-resistance elements (Ref 1)

(1) Pivoting cover in two parts. (2) Insulating lagging. (3) Refractory material. (4) Fluidized bed. (5) Heating elements. (6) Intake for fluidizing gas. (7) Parts to be treated

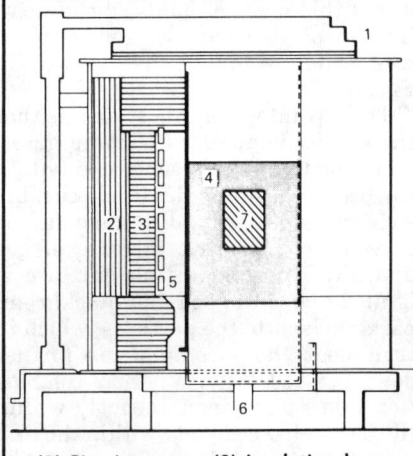

Fig. 11 Fluidized-bed furnace with external heating by electrical-resistance elements (Ref 1)

(1) Pivoting cover. (2) Insulating lagging. (3) Refractory material. (4) Fluidized bed. (5) Resistance elements. (6) Intake for fluidizing gas (air or nitrogen). (7) Parts to be treated

Fluidized-Bed Carburizing. The relatively large volumes of propane consumed during fluidized-bed carburizing, together with high gas velocities, favor carburization by the thermal decomposition of propane to precipitate carbon in accordance with

$$C_3H_8 \rightarrow C \downarrow + 2CH_4$$

The amount of carbon precipitated is proportional to the number of carbon atoms in the hydrocarbon fuel gas; that is, propane forms more carbon than methane. In addition, the purity of propane is important, especially with respect to unsaturated hydrocarbon content, which increases its carbon-forming capability.

The precipitated carbon reacts instantaneously with the oxidizing products of combustion:

$$C_3H_8 + 5O_2 \rightleftharpoons 3CO_2 + 4H_2O$$

to form carbon monoxide and hydrogen:

$$C + H_2O \rightarrow CO + H_2$$

and

$$C + CO_2 \rightarrow 2CO$$

Carburization then proceeds by the catalytic decomposition of CO by H_2 as in conventional carburizing. It is possible that carburization is further complemented by thermal dissociation of the methane formed during carbon precipitation:

$$CH_4 \rightarrow C_{Fe} + 2H_2$$

The carbon potential of the atmosphere varies with the air-to-gas ratio. For each type of hydrocarbon gas, a relationship can be established among air-to-gas ratio, temperature and carbon potential. Control of the reaction and carbon potential of the atmosphere by conventional gas analysis is possible, and fluidized-bed furnaces are equipped with sample ports and probes so that suitable measurements can be taken.

Types of Furnaces for Heat Treating With Fluidized Beds

The type of fluidized bed most widely used for heat treatment is the dense-phase type, although units have been constructed based on the dispersed-phase bed, with particle circulation for heat treatment of long, thin metal parts such as shafts and plates. In a typical dense-phase fluidized bed, the parts to be treated are submerged in a bed of fine solid particles held in suspension, without any particle entrainment, by an upward flow of gas.

Liberation of adequate quantities of heat within fluidized beds is a prime consideration in adapting them for metal processing. Because transfer of heat from the bed to the workpiece is usually much more efficient than transfer of heat from the heat source to the fluidizing medium, the greatest difficulty is encountered in transferring suitable quantities of heat to the fluidizing medium. In addition, the major part of the heat loss from any practical fluidized system is the heat content of the spent fluidizing gas. In instances in which thermal efficiency is unduly influenced by this factor, recirculation of the fluidizing gas or installation of a recuperative system may be justified, and each has been used in practical applications.

Heat input to a fluidized bed can be achieved by several different methods which are described in the following paragraphs.

Internal-Resistance-Heated Fluidized Beds. In this type of unit, the

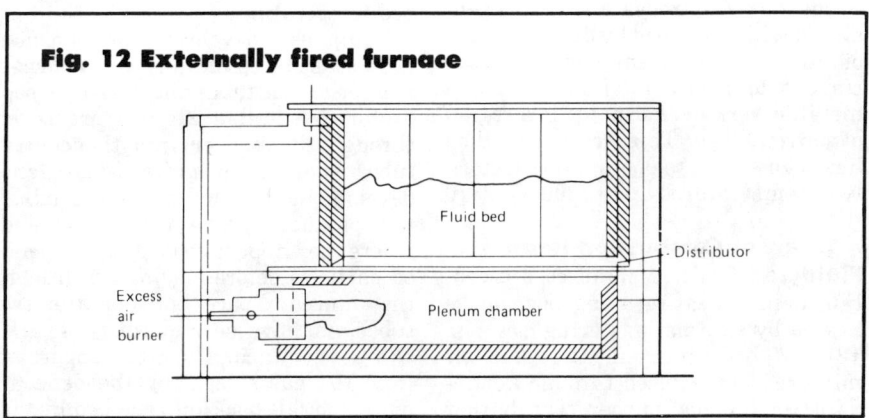

Fig. 12 Externally fired furnace

gas and particles are heated by suitably sheathed internal-resistance-heated elements (see Fig. 10). For high-temperature operation between 500 and 1000 °C (930 and 1830 °F), silicon carbide elements can be used, but they must be sheathed to prevent reactions between the elements and the bed material. At lower temperatures, a mineral-insulated heater with an integral metal sheath can be used, particularly where a heater with greater structural integrity is required. With a system of this type, it is essential to ensure that there are no areas of poor fluidization close to the element. Such areas cause localized overheating and buildup of fused material on the heater, resulting in failure. This problem normally can be avoided, and such heaters provide a simple method of heating fluidized beds up to approximately 500 °C (930 °F). However, heat-up rates (from cold to operating temperature) and heat-recovery rates are relatively slow in this type of bed, particularly for higher temperatures.

External-Resistance-Heated Fluidized Beds. A fluidized bed contained in a heat-resisting pot can be heated by external resistance elements, as shown in Fig. 11. Waste-heat recovery can be used to increase thermal efficiency, and the fluidizing gas can be maintained at any desired composition. Although this appears to be a good method of applying heat to the fluidized bed, there is a severe limitation on the rate of heat input that can be achieved through the wall of the pot. Heat-up rates from cold to operating temperatures of 700 to 800 °C (1290 to 1470 °F) can be as long as 5 to 6 h.

Direct-resistance-heated fluidized beds employ an electrically conducting material such as carbon powder or silicon carbide as the bed material. Power is applied directly to the bed by means of electrodes, and heat is generated within the bed by direct passage of an electric current. However, the current distribution is influenced by a metallic workpiece situated in a region of current flow. This heating method has been shown to operate satisfactorily at temperatures up to 1300 °C (2370 °F).

External-Combustion-Heated Fluidized Beds. A fluidized bed contained in a heat-resisting pot can be heated by external gas firing (see Fig. 12). In this arrangement, a fuel/air mixture is introduced through a standard commercial burner. The burner

Fig. 13 Controlled-atmosphere fluidized-bed furnace heated by submerged combustion (Ref 1)

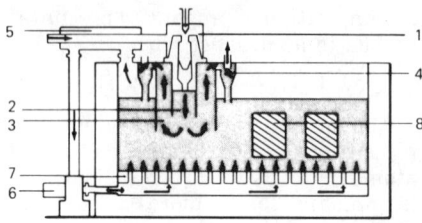

(1) Burner. (2) Combustion tube. (3) Tube through which combustion gases and particles rise. (4) Particle separators. (5) Heat exchanger. (6) Gas recycle compressor for fluidization. (7) Distributor plate. (8) Parts to be treated

can be controlled very accurately down to low temperatures for low-temperature tempering. The products of combustion are passed through a perforated metallic plate to achieve flotation of the bed. By introduction of tempering air after mixing and burning of the fuel, the burner can be operated stoichiometrically for optimum efficiency and prevention of aldehyde formation.

Submerged-Combustion Fluidized Beds. The technique of submerged combustion consists of passing the combustion products directly through the mass to be heated. This method provides an excellent rate of heat transfer and is now well established for a wide range of liquid heating applications, from heating of swimming pools to concentration of acid solutions. Application of this method to the heating of a fluidized bed requires that the burner be used in such a way as to provide strong agitation of the suspended particles, thereby achieving the desired properties of outstandingly good heat transfer and uniformity of bed temperature.

Equipment developed for this purpose consists essentially of a burner, two concentric tubes and a particle separator. A suitable gas mixture is fed through the burner into the central tube, where it is ignited. The flame develops in the tube and the combustion products escape at its lower end, where they impart heat to the suspended particles before moving up through the annular space between the two tubes. As they rise, a quantity of particles is entrained. These are separated from the gas stream by the deflector plate and fall back into the bed by vir-

Fig. 14 Gas-fired fluidized-bed furnace with internal combustion (Ref 1)

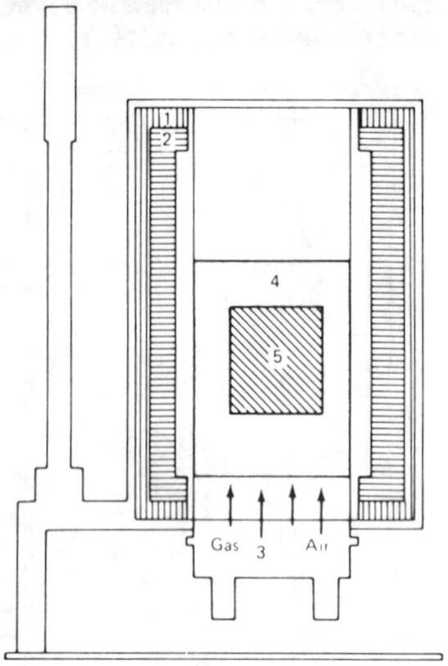

(1) Insulating lagging. (2) Refractory material. (3) Air and gas distribution box. (4) Fluidized bed. (5) Parts to be treated

tue of gravity. Figure 13 illustrates a system that incorporates submerged combustion together with a controlled atmosphere for low-temperature treatment of metals.

Internal-Combustion Gas-Fired Fluidized Beds. A major development in the heating of fluidized beds occurred when an air/gas mixture was used for fluidization and ignited in the bed, generating heat by internal combustion. Prior to this breakthrough, many technical difficulties prevented use of this mode of fluidized-bed heating. A typical furnace design incorporating this technique is shown in Fig. 14.

The advantage of this system is that the bed is fluidized by burning gases, and thus the heat is generated within the bed. In gas-fired fluidized beds, the supporting gas or fluidizing medium is a near-stoichiometric mixture of gas and air. This combustible mixture is ignited above the bed and quickly imparts its heat to the particles, which in turn heat the incoming gas further down the bed. After a period, combustion takes place spontaneously within the bed, being complete within the first 25 mm (1 in.) of the diffuser once the spontaneous combustion temperature for the gas being used is reached. This temperature commonly varies between 600 and 800 °C (1110 and 1470 °F). If the vessel is well insulated, the bed

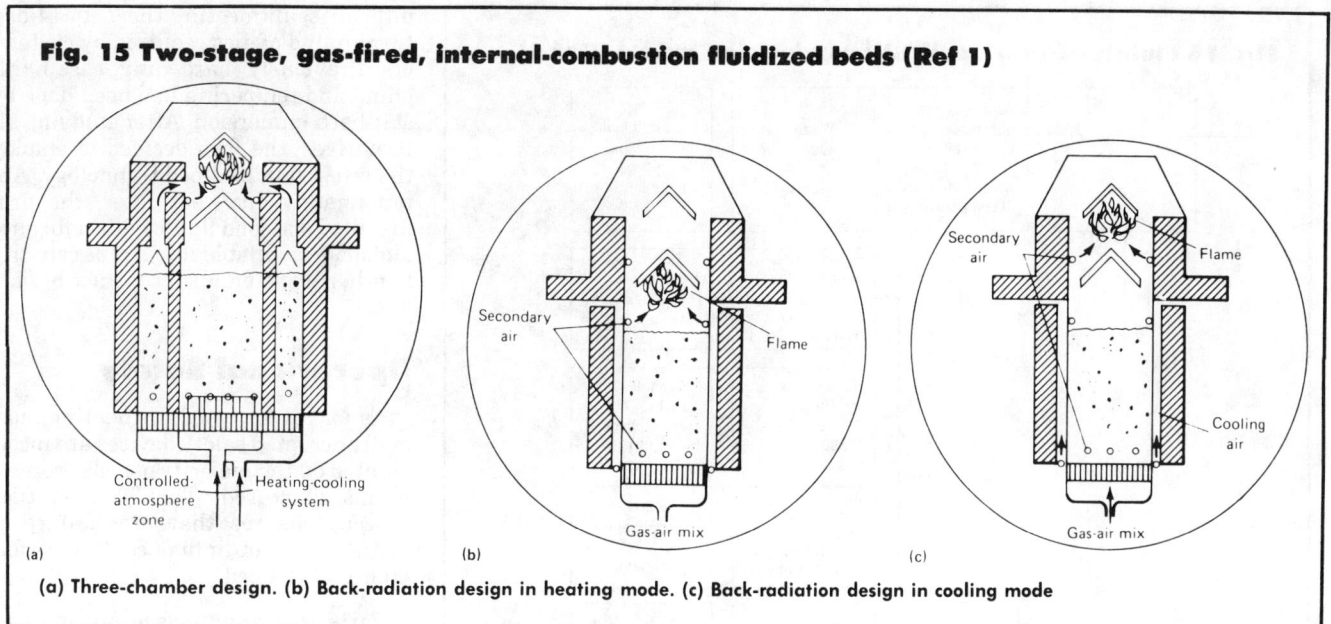

Fig. 15 Two-stage, gas-fired, internal-combustion fluidized beds (Ref 1)

(a) Three-chamber design. (b) Back-radiation design in heating mode. (c) Back-radiation design in cooling mode

temperature can rise to a theoretical combustion temperature, and heat-up rates from cold to 800 °C (1470 °F) are typically between 1 and 1½ h. However, the following problems are inherent to the basic technique and must be considered:

- The bed is fluidized by burning gases. To obtain good temperature control and optimum fluidizing conditions, however, it is desirable that the fuel-input rate and fluidizing velocity be independently variable.
- Combustion is somewhat unstable below the spontaneous combustion temperature.
- Very high temperatures can occur in the immediate vicinity of the distributor/diffuser tile. When the bed is incorrectly fluidized, so that this heat cannot be removed from the top of the distributor, theoretical flame temperatures are achieved with consequent deterioration of the distributor. The thermal stresses of expansion and contraction on the distributor tile at these high temperatures tend, even with the best fixing techniques available, to cause failure of joints, which enhances the problem.

Two-Stage Internal-Combustion Gas-Fired Fluidized Beds. The basic problem of separating control of heat input from control of fluidizing velocity has been overcome in two alternative designs, shown in Fig. 15. In both designs, the initial heat-up from cold to operating temperature is carried out by two-stage internal combustion. A noncombustible mixture of gas and air is introduced beneath the distributor tile. Secondary air is added to make up a stoichiometric or slightly gas-rich mixture immediately above the tile by means of jet holes drilled into heat-resisting tubes. This is done to reduce the possibility of explosion and to avoid high flame temperatures at the surface of the tile. The technique has an adverse effect on good fluidization, but this is unimportant during initial heat-up, in which the prime objective is to raise the bed to operating temperature as quickly as possible.

Once the bed has reached operating temperature, the remaining problem is that of isolating heat-up control from control of the fluidizing velocity. This is achieved in two ways:

- *Three-Chamber Design:* In this design (Fig. 15a), the heat-control outer chambers are separated from the treatment zone by a muffle. The fluidizing velocity and atmosphere are independently controlled in the inner chamber, while the outer two zones are still supplying heat by internal combustion. To achieve adequate heat input, fluidization levels in these outer chambers are above the optimum for heat transfer and surface reactions, but this fact is relatively unimportant.
- *Back-Radiation Design:* When fuel-rich gases are permitted to burn by the injection of secondary air immediately above the control chamber of the fluidized bed, a back-radiation effect causes a rise in bed temperature. This design (shown operating in the heating/controlling and cooling modes in Fig. 15b and c) makes use of this effect and, at the same time, utilizes heat that is normally dissipated when gases are burned outside the furnace. It is therefore more economical in fuel usage. In principle, the gas-rich mixture is supplied to the central chamber, and extra air is added to produce stoichiometric conditions during initial heating of the bed. When cold work is loaded for treatment, the extra air is injected above the bed to produce a radiating flame and recover bed temperature. Should bed temperature exceed set temperature, the extra air is switched to the outside of the furnace wall to provide cooling and finally is mixed with the rich gas/air to produce combustion at the top of the specially constructed hood.

Applications of Fluidized-Bed Furnaces

The potential applications of fluidized-bed technology to heat treating are many. Figure 16 specifies those applications in which fluidized beds can compete with conventional furnaces.

Applications of fluidized-bed furnaces to heat treatment of metals include continuous units for all types of

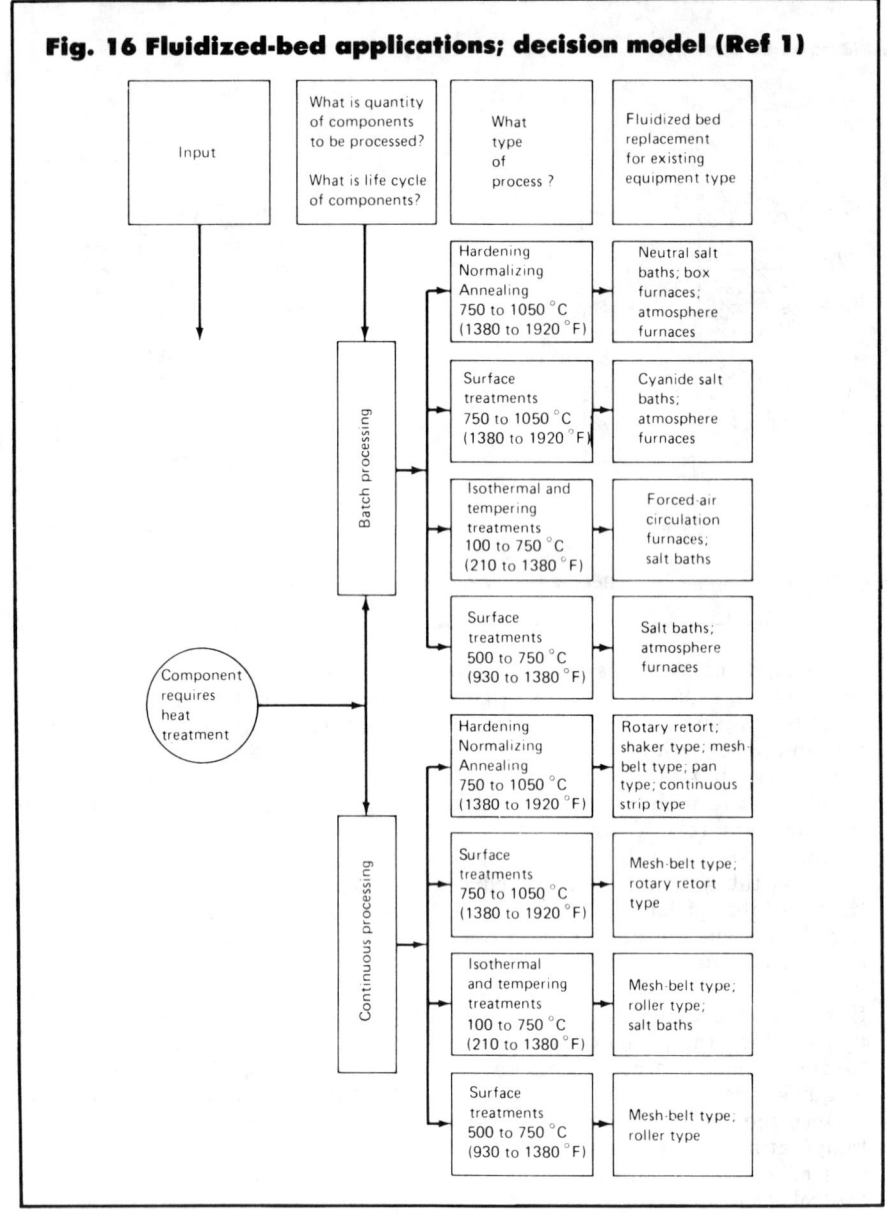

Fig. 16 Fluidized-bed applications; decision model (Ref 1)

wire and strip processing (patenting, austenitizing, annealing, tempering, quenching, etc.); continuous rotary types for fasteners, bearings and other small parts; and all configurations of batch-type units for general heat treating applications. A typical batch-type unit with an output of approximately 150 kg/h (330 lb/h) is available as a standard furnace. Using mechanical handling equipment, it can be auto- mated into a continuous heat treat- ment line. The following example de- scribes how one firm decided to install fluidized-bed furnaces for heat treat- ment.

Example 1. A company specializing in design and production of aluminum extrusion dies had relied on subcon- tract heat treatment facilities for hard- ening of dies. The decision to install in-house facilities came as a result of difficulties in meeting the 7- to 14-day turnaround of dies required by custom- ers. Previously, hardening, case hard- ening and tempering had been done by salt-bath immersion. After studying al- ternatives, the firm decided to employ the latest fluidized-bed technology. Ap- proximately one year later, the firm installed a second fluidized-bed furnace and made available its surplus capacity to other firms on a subcontract basis.

Operational Safety

As for all forms of gas heating, nor- mally accepted safety devices are incor- porated on the majority of beds present- ly manufactured. The "flexible tile" concept ensures that any failure of joints does not influence the perfor- mance of the bed.

Parts carrying surface oil or mois- ture do not create an explosion risk because the contaminants simply va- porize and are removed with the waste gas, as in conventional furnaces.

Cleaning Operations

Fluidized solids are nonabrasive and noncorrosive, and they do not wet im- mersed objects. There is some drag-out loss of the aluminum oxide, however, because some particles accumulate on flat surfaces occurs as workloads are removed from the fluidized bed. These particles can be removed in part by agi- tation, bouncing, or blowing with an air pipe. Particles can be reused by being dried, sieved and returned to the bed. When parts already scaled or preoxi- dized are placed in a fluidized bed, par- ticles tend to adhere to the scale to a greater degree than if the workpieces were clean. These particles can be re- moved by water spraying.

REFERENCE

R. W. Reynoldson, "Controlled Atmo- sphere Fluidised Beds for the Heat Treatment of Metals," *Heat Treatment of Metals,* University of Aston in Bir- mingham, 1976

SECTION XI
Atmospheres and Their Generation

Atmosphere Control

By the ASM Committee on
Atmosphere Control*

THE PURPOSE of atmosphere control is to maintain consistent levels of constituents in a protective atmosphere and/or to determine the changes in those levels that are required in order to produce a desired result under a given set of conditions. Controls are required for various heat treating operations in many different atmospheres. All methods of atmosphere control can effectively be divided into two groups: those involving control of the atmosphere-generating system, and those involving control of the atmosphere once it is inside the furnace. Both are important in maintaining a controlled condition throughout the heating process.

Most operations performed in the heat processing industry can be done under endothermic or exothermic protective atmospheres. Effective control of the generators that produce such atmospheres may be accomplished by means of certain types of controllers— namely, combustibles controllers, infrared CO_2 controllers and oxygen probes. The required accuracy of control is directly proportional to the economics of the individual atmosphere system.

Most common among combustibles controllers is one that operates on the principle of catalytic combustion. A sample drawn from the gas generator or from the furnace is mixed with air (to supply oxygen) and then is passed over a detector. Any combustibles in the gas burn catalytically, raising the temperature of the detector and increasing the electrical resistance. The detectors in the catalytic chamber make up half of a balanced electrical circuit. A corresponding imbalance in the bridge resistance develops, and the resultant electrical output voltage proportional to the concentration of total combustibles in the sample is read out on the panel meter.

An infrared-type analyzer can be utilized for specific analysis of simple or complex mixtures of vapors and liquids and can be used to detect any components that absorb infrared energy. An infrared-type analyzing system could be of a design utilizing two separate or twin infrared radiation cells. One is referred to as a reference source, and gives a known signal in response to the reference gas. The other cell is used as a sample cell, and would vary from the known reference cell depending upon the elements in the gas contained inside that cell. When the gases in both cells are the same, they are in balance. If the gas absorbing infrared radiation in the sample cell is increased, more infrared radiation is absorbed, and consequently the beams become unequal. Movement of a membrane within the detector varies a condenser's electrical capacity, in turn resulting in an electrical signal proportional to the difference between the two beams. The signal is amplified and fed into an indicating meter.

These same principles are utilized for both furnace control and atmosphere-generator control. Infrared analyzers may be used to control systems other than carbon monoxide and carbon dioxide in furnace applications. For example, another form of atmosphere control would be dew-point control (water vapor). Because many heat treating systems are based on carbon dioxide analysis, infrared control is very common. Carbon dioxide control and/or dew-point control can be used for determining carbon potential in carburizing atmospheres. Infrared analyzers can also be used to monitor ammonia in nitriding atmospheres and carbon monoxide in other applications.

Some applications involving heat processes require the use of analyzers to determine the oxygen content of a given processed gas. The most common practice for oxygen analysis in heat treating furnaces is to determine the lack of oxygen. Depending on the level of oxygen desired or permitted in the processed gas, specific ranges of concentration may have to be determined.

The magnetic oxygen analyzer is a common analyzing system used with heat treating systems. Oxygen has an affinity for magnetic fields; most other gases do not. By adding a sample gas to a magnetic field and a detector, the change in resistance can be measured in a cell. The output from this cell is fed to a meter or a recorder and quite often is used as a permissive circuit in an atmosphere-control system.

As mentioned previously, use of a

*Paul L. Huber, *Chairman,* Manager of Sales, Sunbeam Equipment Corp.; Jeffrey W. Boswell, Project Engineer, Sunbeam Equipment Corp.; G. B. Zuber, Sales Manager, Instrument Division, Mine Safety Appliance Co.

dew-point measuring system is another method of controlling the carbon potential of a furnace atmosphere. It can also be utilized to determine the water-vapor content in any given atmosphere. One type of dew-point recorder consists of an element with two parallel electrical conductors helically wound on an insulated tube and connected to electrical power. The insulation of the tube is saturated with a hydroscopic salt and absorption of the moisture by the salt permits the electrical current to flow between the conductors, causing heating to an equilibrium point where moisture absorption ceases. That point where equilibrium exists is fed to an indicator and corresponds to the dew point of the major sample.

The above-mentioned control systems can be modified and incorporated together to form an automatic control system for any given set of conditions, utilizing normal instrumentation and final control devices. The system is ultimately a product of the needs of the user and justifiable economics. As metallurgical reproducibility of any given process becomes more important, the need for atmosphere control to produce chemical stability becomes a necessity. A second imperative not to be overlooked in the application of atmosphere-control devices is the safety aspects of operating atmosphere generators and industrial furnaces.

Analysis and Control of Endothermic and Exothermic Atmospheres

For success in metallurgical atmosphere control, gas analysis instrumentation must be applied where the atmosphere starts—at the gas generator. This is true for endothermic as well as exothermic atmosphere applications. Analyzers and associated control systems applied to a furnace should never be expected to correct deficiencies in the basic atmosphere produced by the gas generator. Instrumentation dedicated to the furnace is designed to provide information and control relative to operation of the furnace and to the product being processed. Likewise, instrumentation dedicated to the generator is designed to provide information and control relative to operation of the generator and to the atmosphere produced by it.

Because the heat treated product emerges from the furnace, there is a tendency to emphasize furnace atmosphere control and overlook the generator. In many cases, automatic atmosphere control of the gas generator will resolve heat treating atmosphere problems, simplify control of the furnace atmosphere, and in some cases reduce furnace monitoring to occasional spot checks.

Endothermic gas is used in heat treating furnaces as a protective atmosphere for hardening, stress relieving, carbon restoration and carburizing. The most recognized guides to endothermic generator operation are CO_2 and dew point. The ultimate guide is CO_2 and dew point. Any deviation from the relationship between the two indicates a generator problem, such as a leak or a carbon-coated catalyst. Figure 1 shows proper carbon dioxide/dew point relationship.

The nominal operating ranges for endothermic-atmosphere generators are -7 to $+16$ °C (20 to 60 °F) dew point and 0.2 to 0.7% CO_2. Generally, maintenance of an endothermic generator can be reduced by operating the generator near the lower end of the CO_2 range or near the higher end of the dew-point range within the limits required by the furnace process. As a rule of thumb, the dew point will rise approximately 6 °C (10 °F) between a sample taken at the generator and a sample taken at the furnace before enriching gases are admitted to produce the desired carbon potential within the furnace chamber.

The following list describes typical endothermic generator problems that can be identified, corrected, or avoided through application of proper analysis and control:

- Carbon on catalyst
- Worn or clogged carburetor or mixer
- Poor temperature control
- Combustion products leaking into retort
- Barometric pressure change
- Humidity change
- Hydrocarbons from quench oil or engine exhaust drawn into generator with air
- Change in natural gas composition
- Sticking regulators
- Dirty flowscope orifices.

When it is determined that carbon is present on the catalyst, a "burnout" procedure is followed to restore catalyst activity. Burnout is accomplished by turning off the natural gas and allowing air to pass through the catalyst. Oxygen combines with the excess carbon to form CO_2. Using the CO_2 analyzer to follow the decrease in CO_2 during burnout allows prompt determination of burnout endpoint. Operators often allow burnout to continue for excessive periods of time. Following the CO_2 decrease shows that burnout usually can be completed in less than one hour. The necessity for burning out a generator is greatly reduced when the

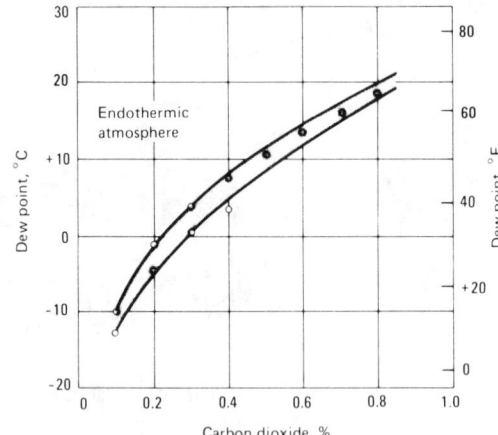

Fig. 1 Variation in the relation between dew point and carbon dioxide in generation of an endothermic atmosphere, as obtained from four plants

The generator in each plant was operated at a different temperature, all within the range from 1005 to 1095 °C (1840 to 2000 °F)

generator is controlled by CO_2 or dew-point analysis, because large upsets commonly induced by manual adjustments are eliminated. Reports have been received indicating that endothermic generators have been controlled at 0.20% CO_2 (dew point of −4 °C or +25 °F) for one year without burnout. This is exceptional and somewhat contingent on generator design.

Exothermic-type atmospheres are used where inert atmospheres are required. Applications are common in the metal treating industry for bright annealing of copper, annealing of motor laminations, processing of aluminum, and annealing of coils of wire and steel sheet. Exothermic atmospheres are also used as an inert blanket in production of glass, in food processing and in rubber curing. Various chemical processes and storage facilities employ inert atmospheres produced by exothermic-gas generators.

Typical analyses of various exothermic atmospheres are listed below, along with recommended analyzers. Total combustibles analyzers are applied most frequently to monitor and/or control exothermic atmospheres. Very rich (20% or higher total combustibles) and very lean (less than 1% total combustibles) exothermic atmospheres are most accurately monitored and/or controlled by an infrared analyzer sensitized to an appropriate range of carbon monoxide.

Rich exothermic atmosphere

- 10 to 25% combustibles (approximately equal parts CO and H_2), 5% CO_2, 3% H_2O, rem N_2
- Recommended analyzers to control and/or monitor rich exothermic atmosphere:
    Infrared analyzer calibrated for 0 to 10 or 20% carbon monoxide
    Total combustibles analyzer (catalytic type) calibrated for 0 to 15%, 20 or 25% combustibles

Lean exothermic atmosphere

- 1 to 10% combustibles (approximately equal parts CO and H_2), 12% CO_2, 3% H_2O, rem N_2
- Recommended analyzers to control and/or monitor lean exothermic atmosphere:
    Total combustibles analyzer (catalytic type) calibrated for 0 to 1, 2, 5 or 10% combustibles
    Infrared analyzer calibrated for 1, 2, 5 or 10% carbon monoxide

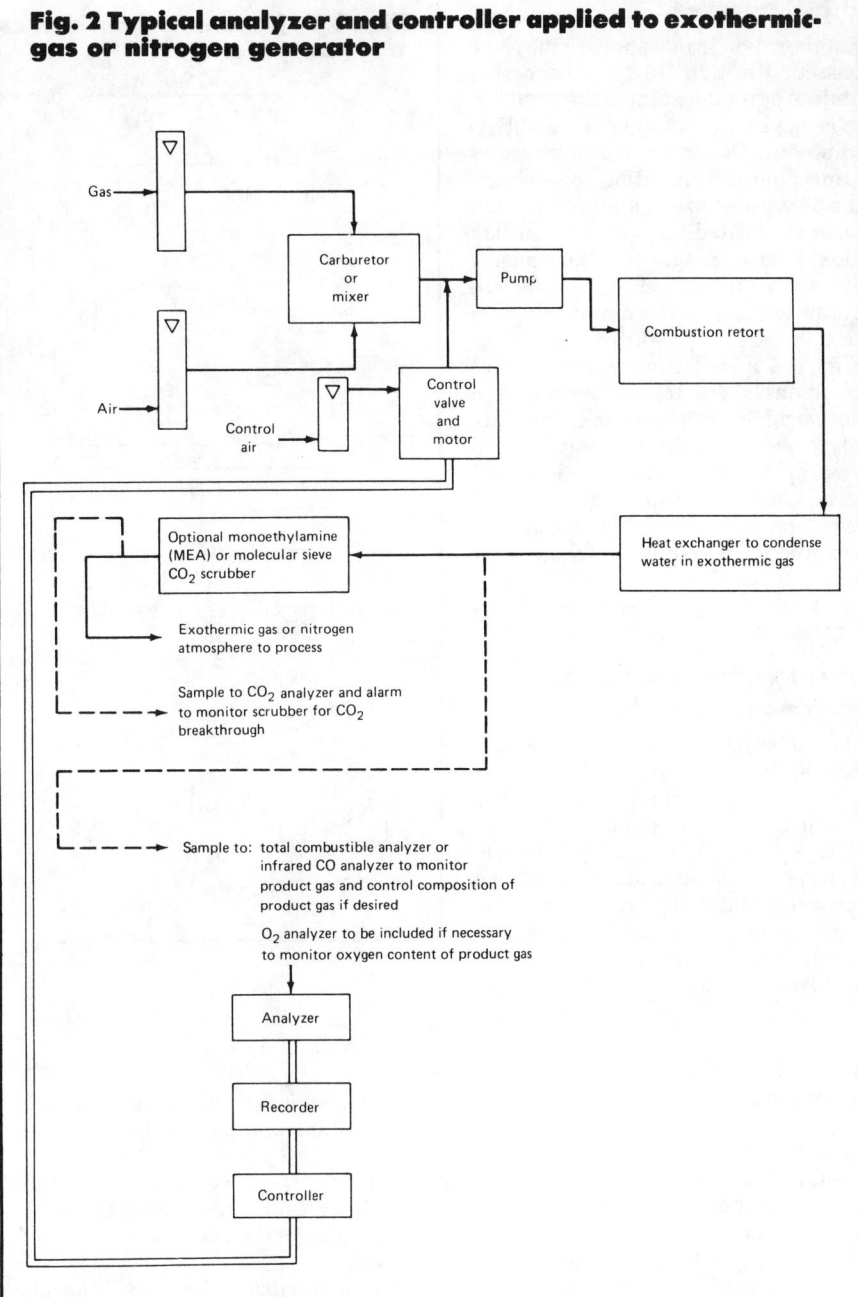

Fig. 2 Typical analyzer and controller applied to exothermic-gas or nitrogen generator

Gas

Carburetor or mixer

Pump

Combustion retort

Air

Control air

Control valve and motor

Heat exchanger to condense water in exothermic gas

Optional monoethylamine (MEA) or molecular sieve CO_2 scrubber

Exothermic gas or nitrogen atmosphere to process

Sample to CO_2 analyzer and alarm to monitor scrubber for CO_2 breakthrough

Sample to: total combustible analyzer or infrared CO analyzer to monitor product gas and control composition of product gas if desired

O_2 analyzer to be included if necessary to monitor oxygen content of product gas

Analyzer

Recorder

Controller

Oxygen analyzer for 0 to 1.0% or 2% oxygen to monitor lean atmospheres to ensure minimum oxygen

Nitrogen (lean exothermic) atmosphere

- 0.2% combustibles, 0.1% oxygen, 12% CO_2, 3% H_2O, rem N_2
- Recommended analyzers to control and/or monitor nitrogen atmospheres

Recommended analyzers are similar to lean exothermic analyzers but with lower ranges specified for greater accuracy; infrared analyzers calibrated for 0 to 0.1% or 0 to 0.5% carbon monoxide are generally preferred over total combustibles analyzers when best accuracy is required

Exothermic atmospheres with CO_2 and H_2O removed

- Exothermic atmospheres may be passed through MEA (monoethyl amine) or a molecular sieve scrubber to remove CO_2 (<1000 ppm) and H_2O to produce N_2 or N_2 + combustibles atmosphere; in addition to applying the above analyzers, an infrared analyzer calibrated for 0 to 0.1% carbon dioxide provides an excellent means of monitoring CO_2 scrubber efficiency as well as prompt indication of carbon dioxide breakthrough.

Figure 2 and 3 summarize application of analyzers to exothermic- and endothermic-gas generators, indicate analyzer sample take-off locations and show proper locations for installation of control valves for automatic atmosphere control. The following adjustment procedure should be followed for proper setup of a control valve and motor when they are first installed on a gas generator for automatic control.

Control Valve and Motor Adjustment Procedure

The effectiveness of an automatic atmosphere analysis control system is directly related to the proper setup of the control valve and motor. Proper-control valve port size and driven-motor speed are important to system success. The following procedure is provided to avoid problems that are easily overlooked during initial setup of control valves and drive motors.

- Disconnect control-valve arm from motor linkage.
- Loosen control-valve locknut.
- Move control-valve arm to open position.
- Adjust control-valve port size to pass amount of air equal to 10% of air normally flowing into generator.
- Tighten control-valve locknut.
- Move control-valve arm toward close position, observe point where air flow starts to decrease, this is now considered "full open" position.
- Move control valve to locate "full close" position.
- Drive motor to "full close" position with manual control.
- Connect linkage.
- Loosen linkage arm at motor shaft.
- Rotate motor arm by hand to determine that full valve travel (close to open) corresponds to full motor travel. Adjust linkage as required to obtain above action.
- Drive motor from "full close" position

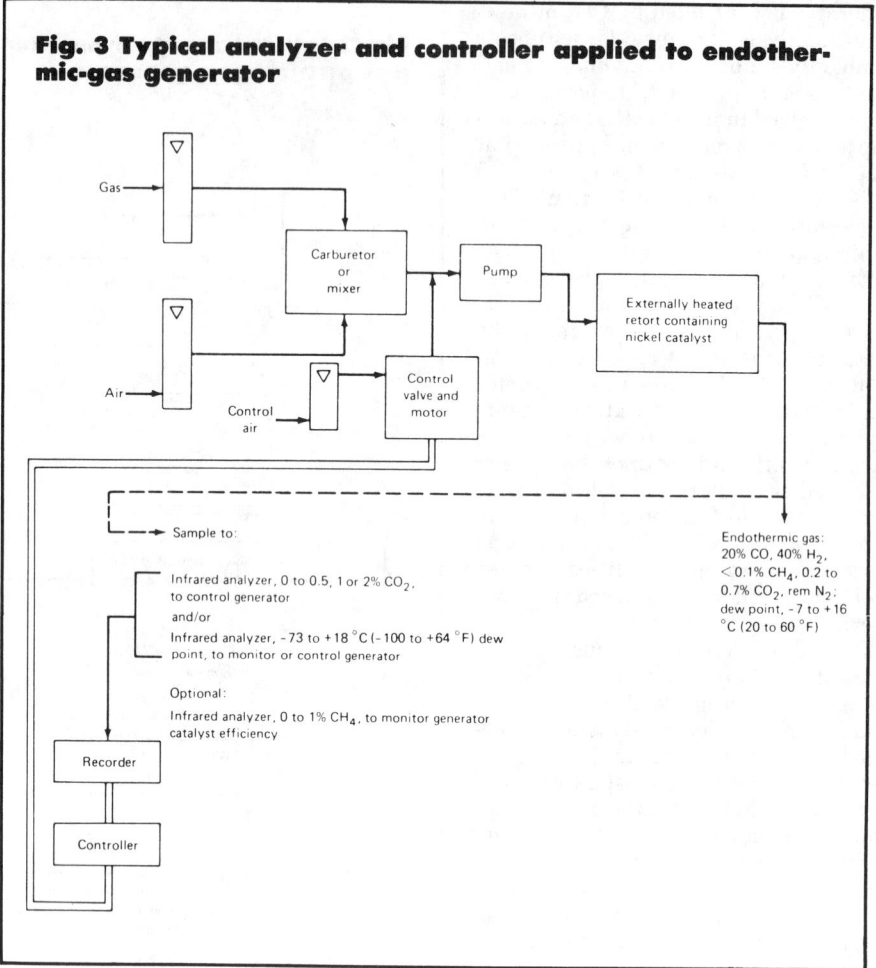

Fig. 3 Typical analyzer and controller applied to endothermic-gas generator

Gas

Carburetor or mixer

Pump

Externally heated retort containing nickel catalyst

Air

Control air

Control valve and motor

Sample to:

Infrared analyzer, 0 to 0.5, 1 or 2% CO_2, to control generator
and/or
Infrared analyzer, -73 to +18 °C (-100 to +64 °F) dew point, to monitor or control generator

Optional:
Infrared analyzer, 0 to 1% CH_4, to monitor generator catalyst efficiency

Recorder

Controller

Endothermic gas: 20% CO, 40% H_2, <0.1% CH_4, 0.2 to 0.7% CO_2, rem N_2; dew point, -7 to +16 °C (20 to 60 °F)

toward open. Valve must start to open immediately without "dead travel".
- Continue to drive motor to "full open" position. Valve must arrive at "full open" as motor reaches end of travel.
- Drive motor from "full open" toward close. Valve must start to move immediately without "dead travel". "Dead travel" is when motor moves and valve does not move or moves in opposite direction; condition is corrected by adjusting linkage.
- Drive motor to mid-position, air flow is one-half of "full open" setting.
- Adjust generator carburetor so CO_2 or dew point is at desired set point.
- Move set point to a higher setting and observe direction in which controller drives valve when system is placed in automatic control mode.

Valve should drive open when analysis point is below set point. If set point is moved below analysis point, valve should drive closed. If valve drives in wrong direction, wires to motor must be reversed. Increasing air to endothermic generator must increase CO_2 and dew point. Increasing air to exothermic generator must decrease CO or total combustibles. During first week or two of operation, control valve on endothermic generator may keep seeking a position at full motor travel and it will be necessary for the operator to manually reset to midposition, adjust carburetor or mixer to obtain desired set point and switch back to automatic. This is a normal sequence caused by gradual cleaning and conditioning of catalyst in the endothermic generator.

Oxygen Probes

The oxygen probe is perhaps the newest technique used today in the heat processing industry for monitoring and controlling different atmospheres such as endothermic gas, exothermic gas, steam with small traces of hydrogen, direct-firing (fuel-rich) atmospheres, and hydrogen-bearing gases with low to medium water contents. Oxygen probes have been in use long enough to provide significant operating experience in heat treatment process control. Sufficient output data have been accumulated and correlated to provide practical usages that offer significant advantages over other forms of monitoring and control such as infrared and dew-point analyzers (Ref 1).

The oxygen probe is based in theory on a hot ceramic electrochemical cell. The probe will respond to oxygen, hydrogen, carbon monoxide, water and carbon dioxide and thus can determine the oxidization potential of a gas. The output of the oxygen probe is a direct measurement of the oxidation potential of the atmosphere at the process temperature of the furnace. Therefore, when the probe temperature is close to the furnace temperature, the response of the probe is a direct indication of whether the atmosphere will oxidize or reduce steel, provided that the composition of the atmosphere with regard to the proportions of carbon gases and hydrogen is known. Under such conditions, the probe will give a reliable indication of the oxidation/reduction situation for all furnace temperatures. Thus, with simple mechanical methods, additive gases or liquids can be introduced into the furnace in order to control the oxidation/reduction potential of the atmosphere.

An oxygen probe is a closed-end tube usually constructed of lime-stabilized zirconia or yttriastabilized material for temperatures up to 1600 °C (2900 °F). When such a probe is subjected to elevated temperatures, the nonporous sheath material acts as a solid electrolyte which permits the passage of oxygen ions when the inner and outer surfaces are subjected to atmospheres of different oxygen partial pressures—for example, a reference gas (such as air, because its O_2 content is constant at 20.9% by volume at sea level) and the process furnace atmosphere, respectively. The electromotive force (emf) thus generated, and measured via the electrodes attached to the sheath, is related directly to, and provides an accurate quantification of, atmosphere characteristics in terms of its oxidizing/reducing or, in some endothermic-atmosphere applications, carburizing/decarburizing tendencies at a known temperature (Ref 2). The electrodes mentioned above are in physical contact with the zirconia on both the inside and outside of the tube, and usually are constructed of platinum because of its superior chemical resistance at elevated temperatures. When the probe is subjected to elevated temperatures, there is a difference between the oxygen partial pressures on the two sides of the probe, and electricity will flow through a circuit connecting the two sides. This flow of electricity is from the higher pressure to the lower. If the oxygen pressure on one side is known, the oxygen pressure on the other side can be determined:

$$E = K \times T \log \frac{[O_2] \text{ known}}{[O_2] \text{ unknown}}$$

where T is absolute temperature, $[O_2]$ is oxygen partial pressure, K is a constant and E is the electromotive force generated. This measurement can be made using a simple panel meter without electronics (Ref 3).

Oxygen probes have been extensively used for control of carburizing furnaces. Accuracy of carbon-potential control through use of oxygen probes is estimated to be approximately ±0.05% carbon in actual practice because of the limitations of temperature control and temperature variations in typical heat treating furnaces. Because of the fast response rate, on/off control systems utilizing solenoid valves for regulating propane, natural gas, or liquid enrichment are adequate for batch furnaces. In situ or ex situ probes control carbon potential by adjusting the set point on the control instrument to the desired carbon level wanted in the furnace. The oxygen probe supplies to the control station an electrical signal related to the carbon potential. High- and low-deviation contacts are adjusted through the control instrument and either contact is made by the electrical signal from the probe. The low-deviation contact controls the solenoid valve that supplies enriching gas or liquid to the furnace. The high-deviation contact can be made to control the solenoid valve to add air or an oxidizer to the furnace. Carbon-concentration re-producibility of ±0.02% is frequently achieved using systems of this type, provided that the cycle, temperature and furnace conditions remain constant.

Continuous carburizing furnaces and straight hardening furnaces usually use proportional control in conjunction with the oxygen-probe control. Experience in control of continuous furnaces has shown that proportional control is normally required in order to cope with disturbances involving movement of work, opening of doors and the like. Compensation for these disturbances are rapid because of the rapid response rate of the oxygen probe. In the proportional-control system, the output voltage is supplied to a two-mode controller, and the controller positions a motorized or pneumatic control valve that regulates the flow of enriching gas to the furnace. Such carbon control is advantageous when short treatment cycles of 20 min or less are used, because compensations of 2 to 3 min or less are required for disturbances in such applications.

Oxygen probes often are used to monitor a furnace atmosphere while control is accomplished manually by adjusting flowmeters on enriching lines to the furnace. This is often used when the cost of control systems cannot be justified. The cost of replacing the consumable probe and the cost of simple on/off control systems must be weighed against human error, process accuracy and reproducibility.

Endothermic generators also can be controlled by use of the oxygen probe. Ex situ probes are generally used because of the difficulty of placing an in situ probe in the retort of the generator. In the case of the ex situ probe, a sample is taken from the output line of the generator. The electrical signal from the probe is wired to the control instrument that activates the solenoid valve located on the air-bypass line to the mixer. Thus, the air/gas ratio is automatically adjusted to give the desired properties of the endothermic gas.

Stainless steels, nickel, brass, titanium, aluminum, Incoloy and other metals and alloys are frequently annealed using exothermic gas. The reducing power of the gas necessary to protect these metals varies considerably; in general, copper alloys require a lean exothermic gas and stainless steels require a rich gas.

Usual practice for control of such a

process is to adjust the gas composition by varying the air/gas ratio prior to combustion in the gas generator. However, in practical experiments and in theoretical calculations, it is shown that it should be possible to control the air/fuel ratio closely by measurement of oxygen potential in either the gas generator or the annealing furnace (Ref 4).

By means of the oxygen probe, for any atmosphere, it should be possible to provide very early warnings of the ingress of air into the furnace from leaking seals or of ingress of water from water-cooled bearings or fans. It is also anticipated that this instrument may replace the conventional oxygen analyzer as a device for providing very early warnings of possible hazardous conditions in the furnace.

REFERENCES

1. The Application of Free-Energy-Temperature Diagrams and High Temperature Electrochemical Cells in the Field of Furnace Atmospheres, by L. H. Fairbank: *Metallurgia*, Vol 79, No. 425, May 1969, p 179–185

2. Recent Developments in the Design and Use of the Oxygen Probe for Furnace Atmosphere Monitoring and Control, by L. H. Fairbank: *Heat Treatment of Metals*, Vol 4, 1977, p 1–12

3. A New Type of Gas Sensor for Combustion Work and Metal Treating Atmospheres, by D. A. Sayles and J. L. Cotter: paper presented at the 20th National ISA Iron and Steel Conference, Pittsburgh, PA, Mar 1970, and published in *Instrumentation for the Iron and Steel Industry*, Vol 20, p 57–66

4. Control of Carburizing Furnace Atmospheres Using O_2 Potential Measurements, by R. G. H. Record: *Metallurgia and Metal Forming*, Vol 39, No. 12, Dec 1972, p 413–416 and Vol 40, No. 1, Jan 1973, p 19–23

CONTROL OF CARBON POTENTIAL IN HEAT TREATMENT*

By
J. N. Dutta

Abstract

Control of carbon potential in precision ferrous heat treatment is indispensable for ensuring least possible day-to-day and lot-to-lot variations in Metallurgical quality. Various methods, e.g. Oxygen potential, Infra Red, Dew Point, Carbon resistance, Orsat analysis, etc. are now available for determination and control of Carbon potential in Generator and Furnace atmospheres and combination of some techniques presumably provides the best control. With efficient carbon potential control, surface decarburisation during hardening and other Heat Treatments would be absolutely eliminated. In case of carburising, possibility of formation of high surface carbon concentrations, carbide network and retained austenite etc. could be ruled out. Automation in carbon potential control further provides consistent carbon concentration and case depth values that ultimately minimise distortion and make unavoidable distortions more predictable.

A comparative study of various carbon potential control methods with brief references to their relative merits, demerits, capabilities etc. has been made and added benefits derived from such controls are primarily discussed in this paper.

Introduction

Control of carbon potential of protective furnace atmospheres is required to perform one or both of two important functions for the purpose of ensuring better quality in steels that are heated in the furnace. These functions are preventions of surface oxidation and decarburisation. When surface oxidation is eliminated, post treatment cleaning methods like pickling and sand blasting may not be necessary. When decarburisation is eliminated, heat treated and quenched ferrous parts do not have a soft skin at the surface and thus have optimum strength and resistance to fatigue and wear, without requiring machining or grinding to get down to the parent metal. Secondly, protective furnace atmospheres like dry endothermic gas can be enriched in the furnace by additions of hydrocarbons like methane or propane to supply carbon and to promote carbon enrichment in low carbon or carburising steels. Steels processed in this manner will case harden in quenching. Control of carbon potential of such carburising atmosphere is essential as surface carbon content has a pronounced effect on the properties of the case hardened parts.

Generally a case of eutectoid or slightly higher surface carbon content of 0.8 to 1% is considered to be quite satisfactory. Higher surface carbon content leads to hypereutectoid case with excess carbides which not only give rise to brittle net work, large amount of retained austenite, low hardness, possibility of grinding cracks etc, but also promote transformation products other than martensite and remove part of the carbide forming elements from austenite and thus decrease effective hardenability during quenching. All of these may lead to ultimately easy service failures of case hardened components.

On the other hand, when the surface carbon content is significantly low, the parts will not develop desired hardness on quenching. If the surface carbon in case is gradually increased, the hardness will also rise until it reaches a maximum. The decrease in hardness with increasing carbon from this point is due to retention of austenite. There is then an optimum carbon content to give maximum surface hardness, as shown in figure 1 more carbon is required as the carburising temperature increases for a nickel-chrome type of carburising steel such as SAE-8620. This type of information is only available for a limited number of steel composition but is of obvious importance if the optimum surface properties are to be obtained after high-temperature carburising. These results may be modified if quenching is carried out at a temperature lower than the carburising temperature.

To obtain a desired surface carbon content, it is necessary to maintain the carbon potential of the atmosphere by a gas atmosphere control system.

* Presented at National Metallurgists Day Celebration on 16-11-77 at IIT, Madras.

** Manager, Metallurgical Services, New Allenberry Works, Calcutta.

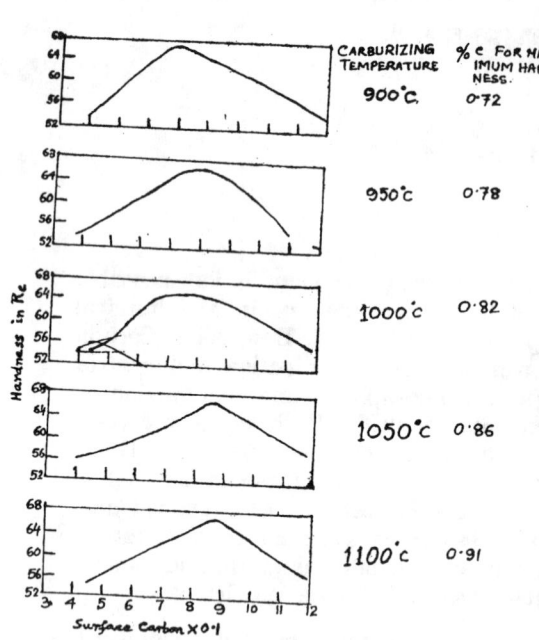

Fig. 1. The Effect of Carbon on the Surface Hardness of 8620 Steel Carburized and Direct Quenched from Temperatures up to 1100°C

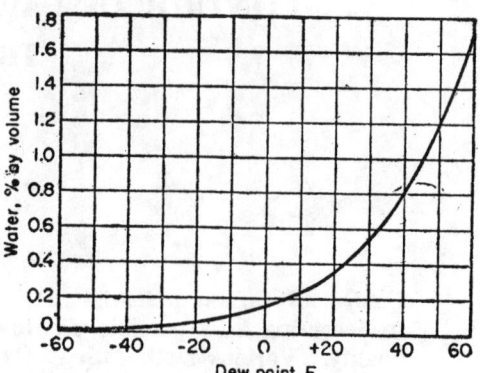

Fig. 2. Dew point versus percentage H₂O by volume

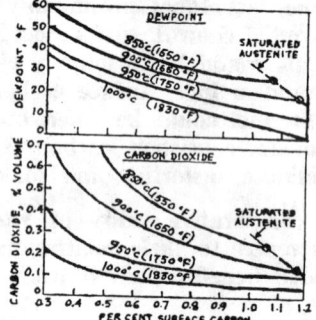

Fig 3. Experimental Equilibrium Between Carbon Steels and Carburizing Atmospheres Generated from Propane.

A carburising atmosphere is so controlled as to make the atmosphere be in equilibrium with a desired carbon potential. It is difficult to measure carbon potential directly and usually some other convenient parameter is measured which can be thereafter related to the carbon potential. The carburising reactions occur as follows:

$$2\ CO \rightleftharpoons C + CO_2 \text{ (in iron)}$$

and

$$CH_4 \rightleftharpoons C + 2H_2 \text{ (in iron)}$$

Therefore,

$$\text{Carburising potential} \quad \alpha = \frac{(CO)^2}{CO_2}$$

and

$$= \frac{CH_4}{(H_2)^2}$$

For a particular carrier gas the concentration of carbon monoxide and hydrogen is roughly constant over a wide range of carbon potential, and hence, increasing the amount of methane or other hydrocarbon or reducing the carbon dioxide content would increase the carburising potential. The carbon dioxide would react with hydrogen as under:

$$CO_2 + H_2 \rightleftharpoons CO + H_2O$$

Therefore, the carbon dioxide content and water content or dew point are interdependent. A high dew point would promote the formation of carbon dioxide and thus reduce the carburising potential.

Methane and other hydro-carbons would react with water in the following manner:—

$$CH_4 + H_2O \rightleftharpoons CO + 3H_2$$

Thus hydrocarbons may be added to reduce the dew point and increase the carburising potential. Hence it is obvious that the carburising potential can be determined by the measurement of carbondioxide or dewpoint. The relationship between these factors and surface carbon content is shown in figures 2 and 3 and Table 1.

It is inevitable that there would be a temperature gradient through a work load during heating and, if carburising is allowed to proceed during the heating period, some lack of case depth uniformity is likely to result. If close control is desired, it may be necessary to commence carburising after the entire load has reached carburising temperature.

Various Methods for Determination & Control of Carbon Potential

Orsat Analysis: This has been a principal tool for control of carbon potential for a long time but now is replaced by speedier and more sensitive methods like dew point, analysis, chromatography, carbon resistant and Infra Red oxygen potential etc.

A sample of atmosphere is bubbled at a fixed sequence through a series of solutions each of which is meant to absorb one of the gas constituents. Thus full analysis of all constituents like CO2, CO, O2, H2, CH4 & C2H4 etc. that are normally present in common Heat Treatment atmospheres is possible within reproducibility of 0.2%.

This permits easy detection of air leaks, water leaks or degeneration of catalyst in the generator etc. Apart from this, the instrument is very cheap, portable and simple to operate without any complicated electrical circuit etc. The major disadvantages are long analysis time (5 minutes for CO2, & O2 and 30 minutes for full analysis) and relatively large error is possible in determining the CO2 content of the furnace atmospheres. A small error in CO2 content measurement would give rise to a relatively large inaccuracy in ratio of CO to CO2 used for determining carbon potential. Orsat analysis can not provide any automation for controls.

Table - 1
Carbon Potential in Endothermic Atmospheres

Carbon Potential, % C	Volume %, H_2O	Dew Point, F	Volume %, CO_2
Equilibrium Temperature (1,700 F) 925°C			
0.17	1.75	60	0.72
0.26	1.21	50	0.49
0.37	0.825	40	0.33
0.52	0.553	30	0.23
0.75	0.367	20	0.15
1.05	0.236	10	0.10
Equilibrium Temperature (1,550 F) 850°C			
0.35	1.75	60	0.94
0.47	1.21	50	0.69
0.63	0.825	40	0.54
0.84	0.553	30	0.40
1.1	0.367	20	0.32

Table - 2
Comparative Performance Characteristics of Carbon Potential Control Systems

Characteristic	System		
	Dew Point	Infrared	Carbon Resistance
Accuracy	±1 F (dew point)	±0.002% CO₂ at set point of 0.05% CO₂	±0.05% C
Response speed	85 sec in range 45 to −8 F	90% final reading in 5 sec	90% final reading in 2 min
Detector type	Lithium chloride (automatic) or fog chamber (manual)	Thermopile connected in series opposition	Fine-gage, iron-carbon wire
Control mode	Manual or automatic	Manual or automatic	Automatic
Length of sample lines, maximum, ft	25	1,000	150
Ambient temperature range, F	Depends on sample gas temperature	40-115	Not applicable
Control points	Single or multipoint	Single or multipoint	Single point
Maintenance costs	High	Low	Medium
Operating costs	High	Medium	Medium
Equipment costs	Low (manual) Medium (automatic)	High to medium	Medium

Gas Chromatography: This is a relatively fast method. Full analysis can be completed within about 5 minutes and an additional 5 to 7 minutes may be necessary for water vapour determination. Reproducibility is + 1% of full scale reading which is 12% for H2, 3% for H2O and 1% for other gases. Reproducibility of water vapour is, however, + 2%. Automatic control utilising gas chromatography is limited to one constituent because of the intermittent nature of the analysis. Major limitations are that its discontinuous nature of the analysis. Major limitations are that its discontinuous nature provides only periodic analysis of a gas stream and it is difficult to interpret results. The concentration of each constituent is recorded as a bar on the recording chart and each should first be identified from the sequence of recording. Additionally as full scale range varies for the different constituents, percent of full scale has to be converted to percent concentration.

Flame Temperature Analysers: With controlled known volume of air gas entry and after thorough mixing it is possible to determine & even automatically control the carbon potential from flame temperature. The burner is designed for a specific size & shape of flame and a chromel-Alumel thermocouple is to be accurately placed in flame. Flame temperature is related to Dew point and can be read out from the graph that has to be established based on experimental observations first over a significant period of operations.

Dew Point: Manual or automatic dew point systems determine carbon potential indirectly. A sample of atmosphere is generally extracted from the furnace and its water vapour content is measured and expressed as the dew point temperature.

Manual readings can be readily taken with compact and portable instruments operating on the fog chamber principle. The gas sample is first compressed and then quickly expanded. If the expansion cools down the gas below its dew point a fog forms in the chamber. The process should be repeated until the fog disappears. At this point, a pressure ratio gauge reading is taken and converted to dew point temperature.

Automatic dew pointers usually employ sensors utilising the hygroscopic properties of lithium chloride.

The salt absorbs moisture from the gas and becomes a conductor and causes a current flow between two wires, raising the temperature of the sensor until equilibrium is obtained. This equilibrium temperature is then indicated on a calibrated recorder.

The major drawback to dew point systems is that an accurate calibration is not available. In addition, sampling lines are very sensitive to dirt and moisture requiring frequent cleaning.

Infrared: It is well known that Gases absorb infrared energy in an amount dependent on their composition. Variation in the percentage of even one

Source: Tool & Alloy Steels, July 1982, 247-253

353

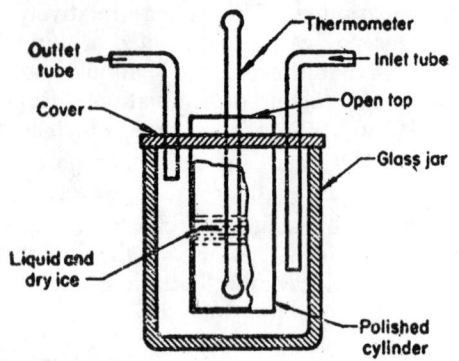

Fig. 4: Elements of dew cup apparatus for measuring dew point

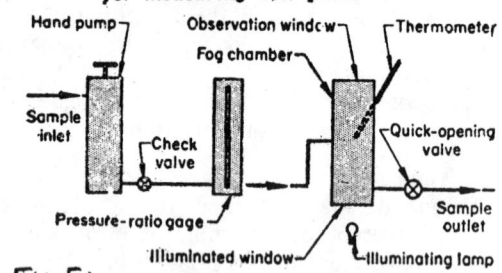

Fig 5: Elements of a fog chamber apparatus for measuring dew point

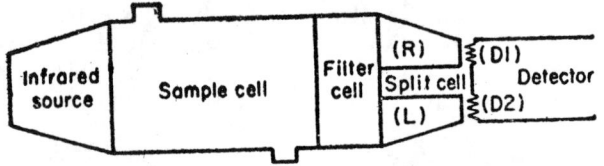

Fig. 6. Elements of a negative-filtering type of infrared analyzer for measuring carbon monoxide, carbon dioxide and methane contents of an atmosphere

element of a gas mixture would cause a corresponding change in the amount of energy absorbed.

Infrared (IR) atmosphere control systems operate on the basis of this principle and are primarily specified for measuring CO, CO2 and CH4 in endothermic atmospheres.

Furnace gas is allowed to flow continuously through the sample cell. When it is sensitised for CO2 measurement, one half of the split cell is filled with CO2 and the other half with a nonabsorbing gas.

When furnace sample contains no CO2, the maximum possible infrared radiation passes through the non-absorbing gas-filled half and none passes through the CO2—filled half.

If the sample contains CO2, the amount of radiation falling on the portion of the detector under the non-absorbing gas filled cell is reduced by an amount proportional to the CO2 concentration. Signals from the thermopile assembly are then transmitted to a recorder calibrated in percentage of CO2.

The filter cell can be filled with an interfering gas to counter the effects of sample components possessing absorption properties similar to those of CO2.

Advantages of IR systems are fast response, low maintenance costs, and suitability for multipoint applications. They are, however, quite expensive.

The major limitations are that it is expensive, complex, maintenance & repair require the skills of a trained electronic specialist. Malfunctioning is not readily detectable. The unit may continue to operate despite the failure of several components thus producing erroneous readings. Automatic control valves must be checked to make that these are operating properly and not in permanently open or closed position.

Unit has to be adjusted quite often to reflect changes in minor variables e.g. a change in energy content of the gas. Hence to ensure the accuracy of the system, trained personnel must recalibrate the unit daily.

Finally if very high dew point is encountered considerable amount of moisture may condense in lines & be carried to the sample cell to damage the cell or to render the system inoperative until the cell has been dismantled and thoroughly cleaned. Damage caused by moisture can be prevented by incorporation of a moisture trap or an electrical warning system ahead of the sample cell.

Carbon Resistance: The only direct method for control of carbon potential is by carbon resistance techniques. The detector is an iron wire placed in the work chamber so that it is exposed to the same temperature as well as the carbon potential as the load.

Samples of furnace atmosphere are continuously made to pass over the detector. Variations in atmosphere carbon potential cause changes in the detector's carbon content. The detector's electrical resistance varies linearly with carbon content.

Effects of temperature on resistance are compensated for by a slidewire in the temperature recorder. Hence the resistance changes measured and recorded are a direct indication of carbon potential.

In practice, a control point setting of 0.90% C would maintain that carbon potential at temperatures of (1,450 to 1,850 F.) 800°C to 1000°C.

Apart from the direct measurement advantage, carbon resistance systems incorporate self-calibration of the sensor and automatic circuit readjustment.

On the minus side, availability of the system is only limited and is not suitable for carbonitriding control applications.

Major limitations are:

(i) The fine iron-alloy wire, is very fragile.

(ii) Instruments are very sensitive to contaminants by either work or work container.

(iii) To obtain accuracies work pieces & work container must be free from all types of dirts, alkalies, sulpher oils and lead compounds etc.

Typical illustrations of various important analysers so for discussed and their comparative performance characteristics are shown in figures 4 to 6 and a comparative statement as regards performance characteristics of three most competing methods—infrared, dew point and carbon resistance is given, in table 2.

Carbon Sensor/Oxygen Potential Method: This is most recent development in this area. Carbon sensor measures furnace atmosphere carbon potential by determining the partial pressure of Oxygen (Po_2) in the furnace atmosphere. This can be performed in the furnace work area without any need of drawing a gas sample. The sensor probe which is inserted through the furnace wall generates an electromotive force (emf) that is a function of the temperature and the oxygen partial.

It would be readily understood from the following reaction that carbon potential in a typical heat treating atmosphere may be controlled by fixation of oxygen, partial pressure.

$CO = C + \frac{1}{2} O_2$, where C is carbon in solution in iron. The equilibrium constant, K for above equation would be given by the relation:

$$K = \frac{ac \; Po_2\frac{1}{2}}{Pco} ,$$

where ac is the activity of carbon and Po_2 and Pco are partial pressures of oxygen and carbon monoxide respectively, ac is related to carbon potential and K is temperature—dependent only. Hence, by controlling partial pressure of oxygen at a particular temperature, carbon potential is fixed, as because Pco is essentially constant in most of the carburising atmospheres.

Carbon sensor operates on the principle that stabilised Zirconia produces en emf when it seperates two gases of different oxygen contents. For utilising this principle, the probe is constructed with one end closed, stabilised zirconia tube, encased in a protective ceramic sheath. Two windows are kept near the tip of the sheath to allow free circulation of the furnace atmosphere on the outside of the stabilised zirconia tube. Inside of zirconia tube is closed to the furnace atmosphere and is purged with 0.030 m³/hr i.e. 1cu. ft/hr. of room air approximately which serves as oxygen reference. To both sides of the stabilised Zirconia tube ends are attached platinum wiremesh electrodes. Platinum leads are laid from the electrodes to terminals in the sensor head that is outside the furnace.

Zirconia tube is actually a solid electrolyte which is capable of conducting electricity through migration of o^{2-} ion. The emf produced by the cell is related to the oxygen pressures at the inner reference electrode P ref, the outer measuring electrode (Po_2) and temperature as given by the following equation.

$$E = 0.0496 \; T \log \frac{Po_2}{Pref} \; \text{volts,}$$

Where T is in degree kelvin and E in milli volts.

For inner electrode Air is used as reference gas and its oxygen partial pressure is assumed to be 0.21 atmosphere. Hence, it may be noted that if the emf is measured and the furnace temperature is known, the oxygen partial pressure of concentration (Po_2) can be determined. As the probe is located directly in the hot zone of the furnace near parts, the oxygen concentration in the furnace atmosphere can be calculated from the voltage output of the cell with a known furnace temperature; or with the help of converter with temperature compensation from a thermocouple, a recorder can be driven which directly gives the oxygen concentration.

In case of carburising, it may be advantageous to relate probe output to oxygen potential, for which following relation has proved to be quite useful:

$E = 10.840 \; O_2 \; Pot. + 40.0,$

where E is in milli volts and O_2 potential is in Kcal mole $^{-1}$.

Gas equilibrium relationships would reveal that oxygen potential is related to CO/CO_2 ratio as under:

$O_2 \; Pot = 0.04150 \; T — 135.00 — 0.009150 \; T \log (CO/CO_2),$

Where T is in degree kelvin. Hence carbon potential may also be controlled by controlling oxygen potential.

Relative effectiveness for oxygen probe has been evaluated as compared to infrared and dew point systems by incorporation of all these systems together, close to each other, in the same installation and it was felt that the oxygen probe method offers same degree of accuracy as infrared equipment. Positively, however, response time of oxygen probe is much faster, which is extremely helpful for trouble shooting furnace operating problems and provides the room for more simplified control systems in contrast with gas sampling systems having many inherent weak points liable to cause error. For example, if gas sample was not cooled fast enough while leaving furnace, gas composition could change showing erroneous potential.

These equipments would be extremely flow-rate dependent and fluctuations in sample gas flow would vary results. During starting of furnace or when in cold weather running with low carbon potential, condensation in sample line may render the equipments useless. There are needs for frequent checking and changing of line filters and there might be many spots where air leaks could develop. As opposed to these difficulties in other methods, Oxygen probe method could be considered as a major break through in this area cf controlling carbon potential.

Source: Tool & Alloy Steels, July 1982, 247-253

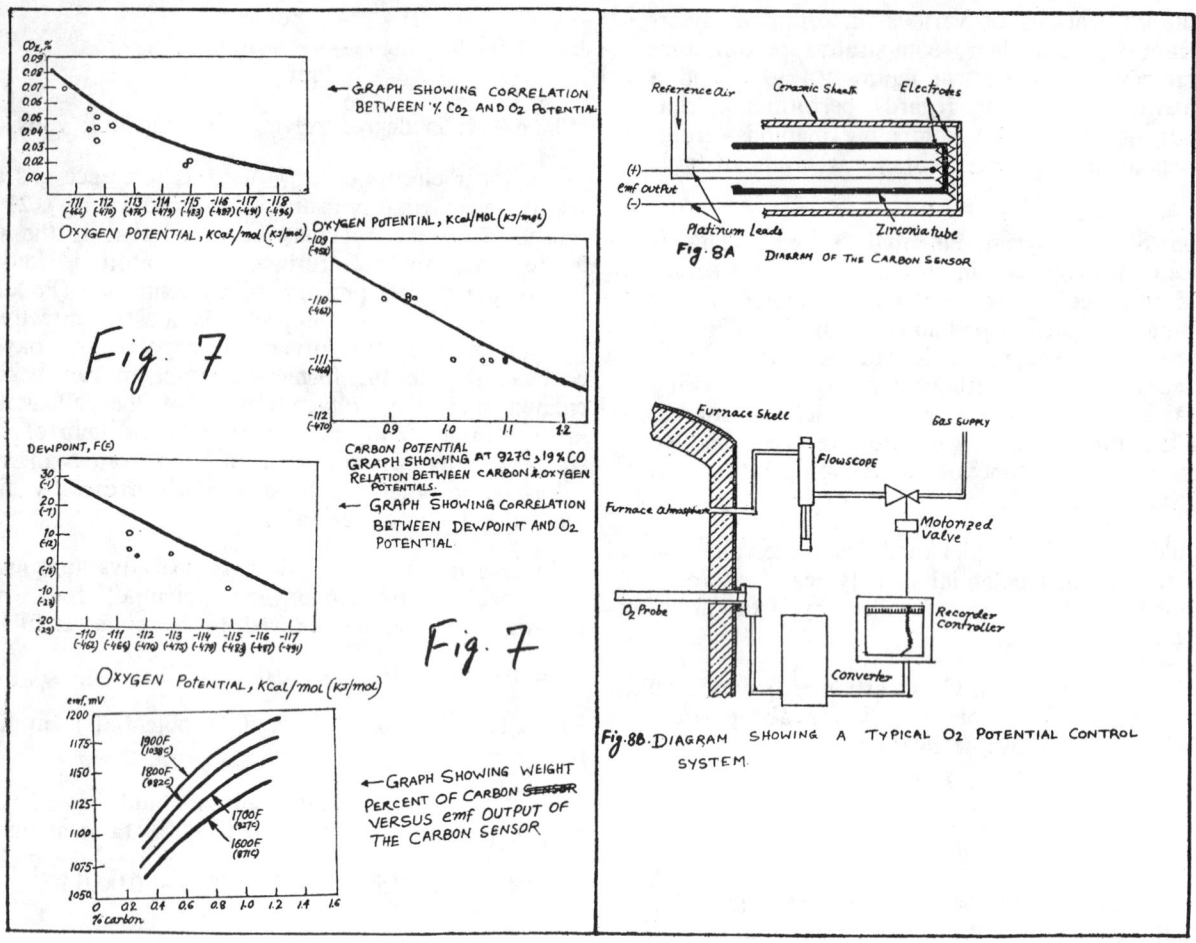

Fig. 7

Fig. 7

←— GRAPH SHOWING CORRELATION BETWEEN % CO₂ AND O₂ POTENTIAL

CARBON POTENTIAL GRAPH SHOWING AT 927C, 19% CO RELATION BETWEEN CARBON & OXYGEN POTENTIALS.

←— GRAPH SHOWING CORRELATION BETWEEN DEWPOINT AND O₂ POTENTIAL

←— GRAPH SHOWING WEIGHT PERCENT OF CARBON SENSOR VERSUS emf OUTPUT OF THE CARBON SENSOR

Fig. 8A   DIAGRAM OF THE CARBON SENSOR

Fig. 8B. DIAGRAM SHOWING A TYPICAL O₂ POTENTIAL CONTROL SYSTEM.

Additionally, because of faster response time, leaks become apparent very readily, enabling maintenance of furnaces to good working condition. In applications involving a high degree of sample contaminations, such as from sooting, the oxygen probes would require much less attention and maintenance than the infra red method. Again effect of small variations of furnace temperature, say upto 10°C, on maintaining carbon potential is experienced to be insignificant, Manufacturer's stated accuracy of the probe is + 0.05% C, although it was found in practice to average better than + 0.02% C.

The only disadvantage associated with the carbon sensor is that the probe itself is a consumable item. The platinum electrode that is exposed to the furnace atmosphere becomes fragile and subject to easy breakage. While breakages for quite a few have been reported to be with in only a few months' time, the life of some has exceeded even 2 years while operating at 925°C in carburising atmosphere. Average probe life is stated to be approximately one year. Initial high cost for oxygen probes would be reasonably justified by reduced maintenance cost while compared with maintenance of infra red unit. Above all, it is considered that oxygen probe method is quite competitive with infrared control and appears to have an edge over other methods in many applications.

Interrelations among oxygen potential emf output, CO₂%, Dew point and Carbon potential are shown in four different graphs pereaining to Figure no. 7 and diagrams showing carbon consor as well as a typical oxygen potential control system are depicted in Figure no. 8.

Selection

From the foregoing it is seen that not a single method alone can claim a smooth, trouble—free, consistent, easy and economic determination as well as control of the process in a production Heat Treatment Shop, although admittedly controls through automatic infrared, Dew point analysers and recently carbon sensors utilising oxygen potential methods are much sophisticated to-day. These are, however, difficult to be maintained properly and expertise is required for efficient operation, frequent calibrations and rectifications of instruments.. It appears that much remains to be done towards instrumentation for achieving a smooth, economic and continuous trouble-free service necessary for a production shop like heat treatment. The most appropriate control method for a specific application is rarely a cut and dried choice. The relative merits and demerits of various processes must be carefully evaluated in relation to the overall requirements of the heat treating system ivolved. For most gas carburising operations it is now widely accepted that automatic infra-red analysers and/or carbon sensors supplemented by occasional dew point and Orsat analysis would provide extremely efficient carbon potential for all practical purposes.

Automatic equipment for controlling atmosphere mixture ratio & carbon potential is generally more reliable than manual equipment, which varies in effectiveness with the skill of the operator.

Generator output & composition of furnace atmospheres are controlled by recording analysers & these should have proportional type rather than an on-off or two position control, because of the time lag in the sample lines & the slow response of some analyser. For controlling carburising atmospheres, the preferred method is to hold the flow of carrier gas constant & to control the flow of the enriching gas. The usual procedure for generators is to set the carburetor (or mixer) to produce a rich full-air ratio & then to control the flow of air in a by pass line around the carburetor.

When the carbon potential of an atmosphere is to be controlled and one constituent of the atmosphere is measured, the set point must be determined on the basis of the chemical analysis of the carbon content of either shim stock or turnings from test bars. Equilibrium data can be used to determine the approx. set point, but equilibrium conditions do not normally exist in furnace atmospheres.

Therefore, the set point must be adjusted until the work meets specifications.

On batch-type furnaces with a programmed time and temperature, the atmosphere set point also must be programmed. Shims removed at the end of temperature period would serve as a guide in determining the correct set point. Both control & monitoring facilities may be built within the same unit.

Considering various aspects, it may not be presumably wrong to say that the recent oxygen potential method would be most advantagous in many applications and infrared and Dew point methods could be placed next in order of merit.

Added Benefits

With an efficient automatic carbon potential control of atmosphere backed up by scheduled quality control procedures like (a) daily calibration of infra-red and/or carbon sensor equipments, (b) Double checks on furnace and generator atmospheres with dew point equipments, (c) Monthly calibration of temperature controllers by an outside instrument service, (d) Examining integrity of sample gas lines, leakages etc. daily, and (e) periodic running of test pieces and carbon concentration bars with furnace loads and the examination of the same by metallographic, microhardness traverses & surface carbon analysis at 0.005" intervals would provide a number of benefits some of which are listed below:

(1) Consistent carbon and case depth values which help in minimising distortion and making unavoidable distortion more predictable.

(2) Rapid and accurate recovery of furnace equilibrium specially when charging loads with different case depths. This minimises time lost considerably between the loads.

(3) Close carbon concentration control in higher-nickel containing carburising steels that helps control retained austenite and eliminates the need of time-consuming sub-zero treatments.

(4) Close control also permits to carburise deep case jobs at relatively higher carburising temperatures without sacrificing carbon concentration or casedepth precisions, and this at the time reduces processing time and contributes signnificantly to increased productivity.

(5) Improved quality is not the only benefit provided by better atmosphere control. As specification can be met more easily, it is no longer necessary to aim at the high side of the case depth ranges to make sure that the values fall within it. This can contribute also directly to shorter cycle times and increased production.

(6) Other cost benefits would include enhanced life for alloy furnace parts, trays, fixtures and charge carriers and longer refractory life in carburising furnaces.

Acknowledgement

Author is grateful to Mr. J. V. Raghavan, Works Manager of New Allenberry Works, Calcutta for his guidance, encouragement and according permission for preparation, presentation and publication of this paper.

References

1) "ASM Metals Handbook"—Vol. II—8th Edition, American Society for metals.
2) "Heat Treatment of Gears"—J. N. Dutta, H. T. Seminar Booklet, IIM, March 1978, Pages 8-12.
3) "Gear Failures for Heat Treatment and Metallographic reasons"—J. N. Dutta, TISCO Tech. journal, Oct. 1977 issue, pages 143-150
4) "Instrument practice", March 1970—R.G.H. Record, pages 161-167.
5) "Metallurgia and Metal forming", December 1972—R.G.H. Record—pages 413-416.
6) "Gas Carburising"—1964 edition—American Society for Metals.
7) "Control of Carbon potential in Heat Treating Atmospheres using a Carbon Sensor" R. N. Bluementhal and A. T. Melville, Industrial Heating, Vol. XLIII, No. 5, 1976.
8) "Composition of Atmospheres Inert to Heated Carbon Steel"—R. W. Gurry, Trans. AIME, Journal of Metals, Vol. 188, 1950.
9) "Metal Treat" Jan-Feb, 1957—H. N. Ipsen.
10) "Metal Treat"—Drop forging—Vol. 30, 1963—M. A. H. Howes.

Strategies for Efficient Operation of Furnaces Using Nitrogen-Base Atmospheres

C. A. STICKELS

To minimize the cost of operating furnaces using protective atmospheres formed from nitrogen plus a reducing gas, both the flow rate of atmosphere gas and the amount of added reducing gas must be controlled. Optimum parameters for steady state operation are determined by (1) the type of atmosphere desired—nonoxidizing (but decarburizing), or neutral, (2) the necessity for maintaining a positive furnace pressure, and (3) the rate of air leakage into the furnace. In general, low flow rates of atmosphere gas and small additions of reducing gas are best for steady state operations.

For purging furnaces during start-up, on the other hand, it is more efficient to operate with high fractions of reducing gas. Mathematical models describing furnace purging with pure nitrogen, nitrogen-hydrogen, and nitrogen-methanol are derived. From these models it can be shown that (1) for nitrogen-hydrogen blends, there is usually a particular hydrogen fraction, less than one, which gives a minimum purging cost, (2) for nitrogen-methanol blends, the purging cost usually decreases continuously as the methanol fraction increases, and (3) purging cost is virtually independent of gas flow rate as long as the rate of air leakage into the furnace is low. Control of atmospheres for subcritical annealing and for neutral hardening of steel are discussed and cost-effective strategies for purging and for steady state furnace operations are recommended.

INTRODUCTION

The increasing cost of natural gas and the increasing availability of cryogenic nitrogen have led to the substitution of nitrogen-base atmospheres for endothermic and exothermic gas in such operations as subcritical annealing, sintering, and neutral hardening. For these operations, nitrogen is blended with a reducing gas to offset the effects of small amounts of air entering the furnace. The reducing gas is considerably more expensive than nitrogen, so low fractions, *e.g.*, five to ten percent, are usually used.

Operating furnaces with nitrogen-base atmospheres is different in many respects from operating with endothermic or

exothermic gas. Many of the differences can be to the operator's advantage—for example, he has much more flexibility in controlling atmosphere composition and flow rate. Very little atmosphere gas need be wasted; gas generators, on the other hand, must often be run at a higher output than can be utilized. Another difference is noticed during start-up. When the furnace is filled with air initially, a much longer purging time is often required with nitrogen-base atmospheres to produce a furnace atmosphere suitable for processing parts, than when exo- or endo-gases are used. This is partly due to the fact that flow rates tend to be lowered for nitrogen-base atmospheres, but it is primarily due to the lower fraction of reducing gases in the nitrogen-base atmosphere. (Endothermic gas contains about 60 pct hydrogen and CO; rich exothermic gas about 12 pct hydrogen plus CO.)

C. A. STICKELS is with the Scientific Research Laboratory, Ford Motor Company, P. O. Box 2053, Dearborn, MI 48121.

Reprinted from Journal of Heat Treating, December 1982, 359-371, © 1982 American Society for Metals

In this paper some mathematical models for furnace purging will be developed which can be used to predict the effects of atmosphere flow rate, amount of reducing additive, air leakage, and furnace temperature on purging time. Using these models, strategies for minimizing operating costs with nitrogen-base atmospheres during purging and during steady state operations will be suggested. Three types of atmospheres will be considered: (1) nitrogen alone, (2) nitrogen-hydrogen blends, and (3) nitrogen-methanol blends.

PURGING MODELS

Pure Nitrogen

Purging with a nonreactive gas involves only mixing; the "perfect mixing" model, used below, is often described in textbooks.[1] Figure 1 is a sketch of a furnace with volume V, operating at temperature T, and filled with air (21 pct oxygen) at time $t = 0$. It is assumed that nitrogen enters the furnace at a flow rate f (measured at ambient temperature and pressure) and that nitrogen and oxygen leave the furnace at the same flow rate. The dependence of furnace oxygen content C on time is found by solving the differential equation which arises from a mass balance on oxygen in the system.

$$\begin{array}{c}\text{Rate of Oxygen}\\\text{Accumulation}\end{array} = \text{Oxygen In} - \text{Oxygen Out}$$

$$V\frac{dC}{dt} = f(0) - f(C), \quad \text{or} \quad \frac{dC}{dt} + \frac{f}{V}C = 0 \quad [1]$$

The oxygen content of the gas leaving the furnace is assumed to be the same as the average oxygen content in the furnace (perfect mixing).

Using the initial condition

$$C = 0.21 \text{ when } t = 0, \quad [2]$$

the solution to Eq. [1] is

$$C = 0.21 \exp\left(-\frac{f}{V}t\right) \quad [3]$$

This equation is most often used in the form

$$t = -\frac{V}{f(T)} \ln\left(\frac{C}{0.21}\right) \quad [4]$$

where $f(T)$ is the flow rate of nitrogen measured at the furnace temperature. It is sufficient to assume ideal gas behavior, so that

$$f(T) = \frac{T}{300}f \quad [5]$$

where T is the furnace temperature in degrees Kelvin and ambient temperature is taken to be 300 K.

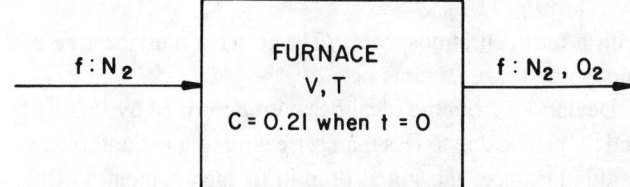

Fig. 1 — Furnace purging using nitrogen. The furnace is filled with air initially (21 pct oxygen concentration), and the oxygen content falls exponentially with time as purging proceeds.

The quantity $V/f(T)$ in Eq. [4] is t_m, the mean residence time of the gas in the furnace. This follows from the definition

$$t_m = \frac{\int_0^\infty Ct\,dt}{\int_0^\infty C\,dt}$$

If there are slow reactions within the furnace, the reactions more nearly approach equilibrium as t_m increases. The mean residence time is also a measure of the response time of the furnace to changes in inlet gas composition. Provided that there are no reactions within the furnace which affect the volume of gas, the change to the furnace gas composition caused by a change in inlet gas composition is 95 pct complete in three mean residence times.

The form of Eq. [4] shows that all the oxygen can never be eliminated from a furnace by purging with nitrogen. In fact, since most furnaces have some air leaks, very low oxygen contents can never be attained. A zirconia oxygen sensor is a convenient way of monitoring nitrogen-base atmospheres for oxygen content. By recording the response of the oxygen sensor as a function of time during purging, it is possible to estimate the magnitude of the air leaks.

Let f' be the flow rate of air into the furnace due to leaks. Then a mass balance on oxygen leads to the equation

$$V\frac{dC}{dT} = 0.21f' - (f + f')C \quad [6]$$

and the change in oxygen content with time is described by

$$C = 0.21\left(1 - \frac{f'}{f + f'}\right)\exp\left(-\frac{f + f'}{V}t\right) + \frac{0.21f'}{f + f'} \quad [7]$$

After very long times the concentration of oxygen approaches C' as a lower limit, where

$$C' = \frac{0.21f'}{f + f'}$$

So, if it is found that the lowest oxygen content attainable after long purging times is C', the magnitude of the air leak is

$$f' = f\left(\frac{C'}{0.21 - C'}\right) \quad [8]$$

This method of determining air leak rate can be used with any furnace, whether or not the furnace will be operated with a nitrogen atmosphere. (The furnace must be free of soot so that no reactions occur.)

Deviations from the ideal behavior described by Eqs. [4] and [7] can be due to less than perfect mixing in the furnace. In this instance, the initial drop in oxygen content will be faster than expected, but it will take longer than expected to reach very low oxygen contents.

Sealed quench heat treating furnaces are equipped with unheated vestibules which also must be purged. Generally, the door between the furnace chamber and the vestibule is not sealed, so there is no pressure difference maintained between the furnace and vestibule. As a result, atmosphere gas, driven by the temperature difference between the furnace and vestibule, circulates between the furnace and vestibule. By measuring gas composition as a function of time during purging, it is possible to estimate the rate of circulation between furnace and vestibule.[2] However, for estimating purging times, it should be sufficient simply to add the temperature corrected volume of the vestibule to the furnace volume and use Eq. [4]. The temperature corrected vestibule volume is the physical volume of the vestibule multiplied by the ratio of the absolute furnace and vestibule temperatures.

Nitrogen Plus Hydrogen

Nitrogen-base furnace atmospheres are most frequently used with a small addition of a reducing gas, such as hydrogen, to offset the effects of small amounts of air entering the furnace. Nitrogen-hydrogen atmospheres can be used to prevent oxidation, but they will always, in principle, be more or less decarburizing, since carbon-bearing compounds are not part of the atmosphere supplied. As a practical matter, however, whether or not significant carbon loss will occur depends on such factors as the atmosphere gas flow rate, the total surface area of the parts in the furnace, the treatment time, and the treatment temperature.

The process of purging air from a furnace using nitrogen-hydrogen mixtures can be divided into two stages. In the first stage (Figure 2), all of the hydrogen entering the furnace reacts with the available oxygen to form water vapor. The first stage of purging ends when all of the oxygen is consumed and the furnace atmosphere consists of just nitrogen and water vapor. In the second stage of purging (Figure 3), the hydrogen content of the furnace atmosphere progressively increases and the water vapor content decreases. Purging is assumed to be complete when the hydrogen/water ratio is high enough to prevent oxidation of steel.

First Stage

For each volume of hydrogen entering the furnace, one volume of water vapor is produced and one-half volume of

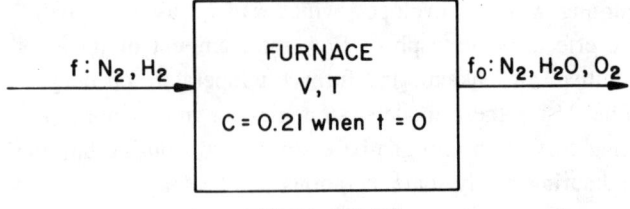

Fig. 2—Furnace purging with nitrogen-hydrogen. During the first stage of purging, all the hydrogen entering the furnace reacts with the oxygen present.

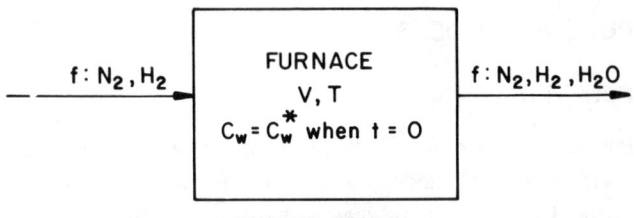

Fig. 3—Furnace purging with nitrogen-hydrogen. During the second stage of purging, the water vapor content of the furnace atmosphere decreases exponentially and the hydrogen content increases exponentially with time. Purging is complete when the H_2/H_2O ratio is high enough to prevent oxidation of steel.

oxygen is lost. As a result, the flow rate leaving the furnace, f_o, is less than the flow rate into the furnace, f. If x is the fraction of hydrogen in the inlet gas, then

$$f_o = \left(1 - \frac{x}{2}\right)f \qquad [9]$$

The time t_1 needed to eliminate all oxygen is found by taking an oxygen balance:

$$\text{Rate of Oxygen Accumulation} = \text{Oxygen In} - \text{Oxygen Out}$$

$$- \text{Oxygen Lost by Reaction}$$

$$V\frac{dC}{dt} = 0 - f_o C - \frac{x}{2}f \qquad [10]$$

where C is the furnace oxygen content and V is the furnace volume. The differential equation to be solved is found by combining [9] and [10].

$$\frac{dC}{dt} + \frac{f}{V}\left(1 - \frac{x}{2}\right)C = \frac{xf}{2V} \qquad [11]$$

Using the initial condition, $C = 0.21$ when $t = 0$, the solution is

$$C = \left(0.21 + \frac{x}{2 - x}\right)\exp\left[-\frac{f}{V}\left(1 - \frac{x}{2}\right)t\right] - \frac{x}{2 - x} \qquad [12]$$

The value of t_1 is found by setting $C = 0$ in this equation and solving for t.

$$t_1 = -\frac{V}{f}\frac{2}{(2 - x)}\ln\left(\frac{x}{0.21(2 - x) + x}\right) \qquad [13]$$

To find the water vapor content at time t_1 it is necessary to make a material balance on water:

$$\text{Rate of Water Accumulation} = \text{Water In} + \text{Water Generated by Reaction} - \text{Water Out}$$

$$V\frac{dC_w}{dt} = 0 + xf - f_o C_w \qquad [14]$$

where C_w is the water vapor content in the furnace. Inserting Eq. [9] and rearranging gives

$$\frac{dC_w}{dt} + \frac{f}{V}\left(1 - \frac{x}{2}\right)C_w = \frac{f}{V}x \qquad [15]$$

Using the initial condition, $C_w = 0$ when $t = 0$, the solution to Eq. [15] is

$$C_w = \frac{2x}{2-x}\left\{1 - \exp\left[-\frac{f}{V}\left(1 - \frac{x}{2}\right)t\right]\right\} \qquad [16]$$

To find the water vapor content at the end of the first stage, C_w^*, Eqs. [13] and [16] are combined, giving

$$C_w^* = \frac{0.42x}{0.21(2-x) + x} \qquad [17]$$

Second Stage

Let R be the minimum value of the ratio H_2/H_2O needed to prevent oxidation of steel. The value of R is shown in Figure 4 as a function of temperature. To find the second

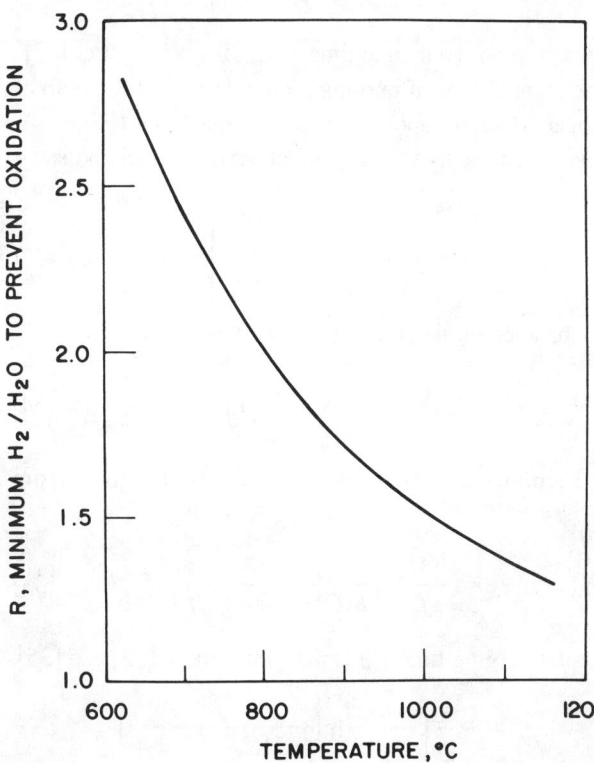

Fig. 4—H_2/H_2O ratio as a function of temperature for the formation of wustite, $Fe_{0.95}O$. Free energies of formation of wustite and H_2O are from Ref. 3.

stage purging time, t_2, expressions must be found for the furnace hydrogen content, C_h, and the water vapor content, C_w, as a function of time. The time, t_2, is found by solving the equation

$$\frac{C_h(t)}{C_w(t)} = R \quad \text{for} \quad t$$

No reactions occur within the furnace during the second stage, so $f_o = f$. A material balance on the hydrogen content yields the equation

$$\frac{dC_h}{dt} + \frac{f}{V}C_h = \frac{xf}{V} \qquad [18]$$

Using the initial condition, $C_h = 0$ when $t = 0$, the solution to [18] is

$$C_h = x\left[1 - \exp\left(-\frac{f}{V}t\right)\right] \qquad [19]$$

A material balance on water vapor content yields the equation

$$\frac{dC_w}{dt} + \frac{f}{V}C_w = 0 \qquad [20]$$

Applying the initial condition, $C_w = C_w^*$ when $t = 0$, gives the solution

$$C_w = C_w^* \exp\left(-\frac{f}{V}t\right) \qquad [21]$$

where C_w^* is given by Eq. [17].

The time t_2 is found by solving

$$R = \frac{C_h}{C_w} = \frac{x\{1 - \exp[-(f/V)t_2]\}}{C_w^* \exp[-(f/V)t_2]}, \quad \text{or}$$

$$t_2 = -\frac{V}{f}\ln\left(\frac{x}{x + C_w^*R}\right) \qquad [22]$$

The total purging time, t_p, is the sum of Eqs. [13] and [22]

$$t_p = -\frac{V}{f}\left[\frac{2}{2-x}\ln\left(\frac{x}{0.21(2-x)+x}\right) + \ln\left(\frac{x}{x + C_w^*R}\right)\right] \qquad [23]$$

By inserting various values for x in [23], one finds that the total purging time decreases as x, the fraction of hydrogen, increases.

The state of the furnace atmosphere can be monitored during purging by measuring the oxygen potential, using a zirconia oxygen sensor, or by measuring dew point. Figure 5 shows the oxygen sensor voltage corresponding to the critical gas ratio R as a function of temperature.

An expression for the water vapor content corresponding to the critical gas ratio R, C_w^+, can be found by inserting the expression for t_2, Eq. [22], into Eq. [21], yielding

$$C_w^+ = \frac{C_w^* x}{x + C_w^*R} \qquad [24]$$

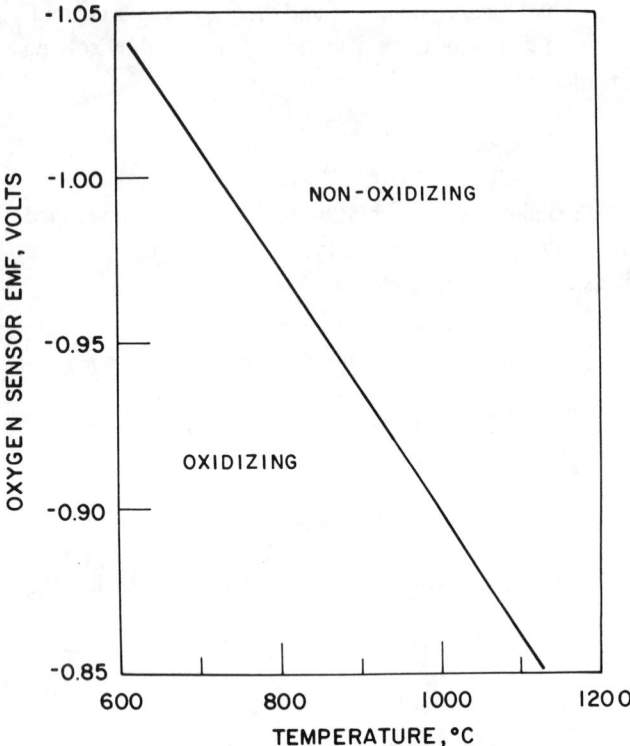

Fig. 5—Oxygen sensor EMF as a function of temperature corresponding to the formation of wustite.

Combining Eqs. [24] and [17] gives

$$C_w^+ = \frac{x}{1 + R + 1.881x} \qquad [25]$$

Note that C_w^+ depends on x as well as R—as x increases, so does C_w^+. An expression for the dew point corresponding to C_w^+ is obtained by fitting a curve to measurements of water vapor pressure as a function of temperature.[4]

$$\text{Dew Point, }°C = \frac{-5422.2}{\ln(C_w^+) - 14.732} - 273.16 \quad [26]$$

Air Leakage

Let f' be the rate of air leakage into the furnace. For simplicity of calculation, the flow of air is combined with the nitrogen-hydrogen inlet gas, and reaction between oxygen in the air and hydrogen in the inlet gas is assumed, so the problem is set up in terms of a fictitious inlet gas having a flow rate f_e given by*

---

*One volume of oxygen from the air leak reacts with two volumes of hydrogen in the inlet gas, yielding two volumes of water vapor. Thus, the sum of the two flows after reaction is less than the sum of the individual flows.

$$f_e = f + 0.79f' \qquad [27]$$

a water vapor content, w, given by

$$w = \frac{0.42f'}{f + 0.79f'} \qquad [28]$$

and a hydrogen content, y, given by

$$y = \frac{xf - 0.42f'}{f + 0.79f'} \qquad [29]$$

where x, as before, is the actual fraction of hydrogen in the nitrogen-hydrogen inlet gas. Figure 6 shows this schematically.

As before, purging can be regarded as occurring in two stages. In the first stage, the hydrogen entering the furnace reacts completely with the oxygen present, so that the outlet gas consists of nitrogen, oxygen, and water vapor. The outlet flow rate is $(1 - y/2)f_e$. During the second stage, the outlet gas consists of nitrogen, hydrogen, and water vapor with a flow rate f_e.

Proceeding as before, an oxygen balance for the first stage yields the equation

$$\frac{dC}{dt} + \frac{f_e}{V}\left(1 - \frac{y}{2}\right)C = \frac{-f_e y}{2V} \qquad [30]$$

where C is the oxygen content in the furnace. The solution to this equation is

$$C = \left(0.21 + \frac{y}{2 - y}\right)\exp\left[-\frac{f_e}{V}\left(1 - \frac{y}{2}\right)t\right] - \frac{y}{2 - y}$$

$$[31]$$

and the time at which all the oxygen is consumed, t_1, is

$$t_1 = -\frac{2V}{f_e(2 - y)}\ln\left(\frac{y}{0.21(2 - y) + y}\right) \qquad [32]$$

The water vapor content at time t_1, C_w, is given by Eq. [17].

The second stage of purging is complete when the ratio of hydrogen to water vapor in the furnace reaches R (Figure 4). A balance on the hydrogen content during the second stage gives

$$C_h = y\left[1 - \exp\left(-\frac{f_e}{V}t\right)\right] \qquad [33]$$

and a balance on the water vapor content gives

$$C_w = (C_w^* - w)\exp\left(-\frac{f_e}{V}t\right) + w \qquad [34]$$

The second stage purging time, t_2, is obtained from $C_h/C_w = R$.

$$t_2 = -\frac{V}{f_e}\ln\left(\frac{y - wR}{R(C_w^* - w) + y}\right) \qquad [35]$$

The total purging time, t_p, is the sum of Eqs. [32] and [35].

$$t_p = -\frac{V}{f_e}\left[\frac{2}{2 - y}\ln\left(\frac{y}{0.21(2 - y) + y}\right)\right.$$

$$\left. + \ln\left(\frac{y - wR}{R(C_w^* - w) + y}\right)\right] \qquad [36]$$

362

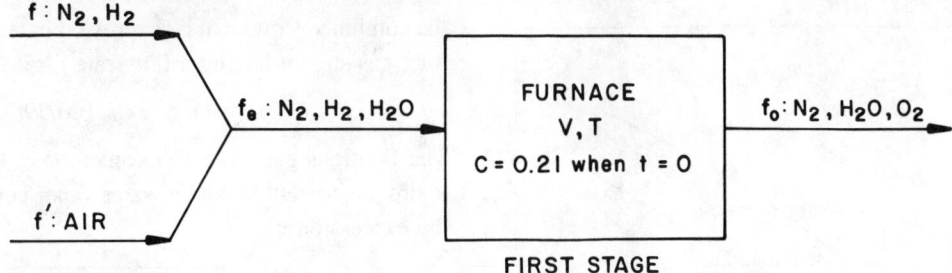

Fig. 6 — Furnace purging with nitrogen-hydrogen with air leakage. For purposes of computation, the inlet flows of nitrogen-hydrogen and air are combined to create a fictitious inlet gas containing nitrogen, hydrogen, and water vapor.

Operating Costs

For steady state operations, there is an economic incentive to minimize the inlet flow rate f and to minimize x, the fraction of hydrogen, because hydrogen is more expensive than nitrogen. However, to maintain steady state non-oxidizing conditions, the ratio y/w must always be greater than or equal to R. Using Eqs. [28] and [29], this leads to the inequality

$$xf \geq f'(1 + R)0.42 \qquad [37]$$

which must be satisfied to prevent oxidation.

If we assume that hydrogen is three times as expensive as nitrogen on a molar basis, then the atmosphere cost per hour for steady state operation is proportional to

$$f(1 + 2x) \qquad [38]$$

While operating costs decrease as f decreases, for each furnace there will be a minimum practical flow rate determined by the necessity for keeping the furnace pressure positive. As long as the furnace pressure, measured at the lowest point in the furnace, is positive, air enters the furnace primarily during door openings, as parts are charged and discharged. The rate of air leakage should be nearly independent of atmosphere flow rate in this case. Therefore, for lowest steady state operating cost, f is kept at the minimum needed to develop a positive furnace pressure, and x is kept just above the lowest value which satisfies inequality [37].

The atmosphere cost for purging is proportional to

$$f(1 + 2x)t_p \qquad [39]$$

where t_p is the total purging time. The cost of purging can display a minimum as a function of x, because the term $(1 + 2x)$ increases linearly with x, while t_p decreases nonlinearly with x. In the absence of air leaks, purging cost is independent of inlet gas flow rate, because t_p varies as $1/f$. With air leaks, the purging cost decreases slightly with increasing flow rate.

For most efficient furnace operation, the following strategy is suggested.

1. Choose the lowest flow rate, f, which will maintain a positive furnace pressure. Choose the lowest fraction of hydrogen, x, for steady state operations which will satisfy [37].

2. For purging, increase the hydrogen flow rate until the fraction of hydrogen is near its optimum value (if one exists), keeping the nitrogen flow rate constant.

3. The furnace atmosphere must be monitored to determine the end of purging. For furnace temperatures above about 650 °C, a zirconia oxygen sensor is the best choice. Below that temperature, dew point measurements can be made.

Nitrogen Plus Methanol

Nitrogen-methanol atmospheres can be used to prevent decarburization of steel, as well as to prevent oxidation, at temperatures at which steel is austenitic. Nitrogen-hydrogen atmospheres are preferred for subcritical annealing of steel, regardless of carbon content, because sooting from methanol can occur at temperatures below about 600 °C.

It will be assumed here that methanol, upon entering the heat treat furnace, decomposes by the reaction

$$CH_3OH \rightarrow CO + 2H_2$$

and that the CO and H_2 produced react with the oxygen present during the first stage of purging to form CO_2 and H_2O. At the end of the first stage of purging, the furnace atmosphere consists of nitrogen, CO_2, and water vapor. During the second stage of purging, no further reaction of the gases is assumed.* Thus, the CO_2 and H_2O contents

---

*Or, more precisely, it is assumed that the equilibrium constant for the reaction

$$CO_2 + H_2 \leftrightarrow CO + H_2O$$

is equal to one. This is true at about 810 °C, but not at higher or lower temperatures. For estimating purging times for furnace temperatures from 700 to 950 °C, this assumption should not cause significant error.

---

progressively decrease, and the H_2 and CO contents increase. When the H_2/H_2O ratio reaches a value R (Figure 4), the atmosphere will protect steel from oxidation. When the quantity $(CO)^2/CO_2$ reaches a value S, given in Figure 7, the atmosphere will protect against decarburization. In principle, to maintain a neutral atmosphere, neither carburizing nor decarburizing, the furnace atmosphere has to be controlled to maintain the quantity $(CO)^2/CO_2$ at the value of S corresponding to the carbon content of the steel and the furnace temperature. In practice, however, with low methanol fractions, low flow rates of atmosphere gas, and a high

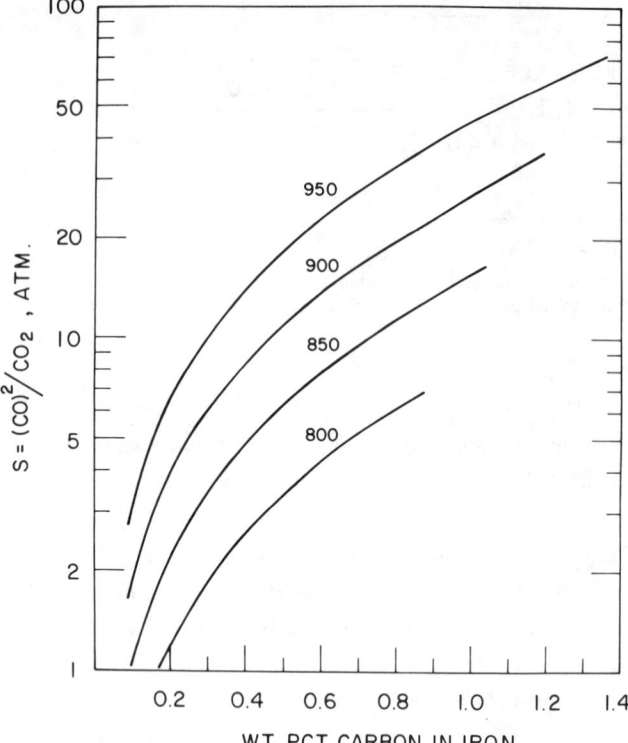

Fig. 7 — The quantity $(CO)^2/CO_2$ corresponding to equilibrium with various concentrations of carbon in austenitic iron at several temperatures. Free energies of formation are from Ref. 3; the relation between carbon activity and wt pct carbon in iron is due to Chipman.[5]

total surface area in the parts charged, the atmosphere will be buffered by carbon exchange with the parts. That is, a very small gain or loss of carbon from the parts will suffice to bring the atmosphere into equilibrium with the parts. So for neutral hardening of steel in the temperature range from 775 to 900 °C, there should be little danger of significant carburization or decarburization. Even for operations like sintering, which employ higher temperatures and longer treatment times, it should be possible to maintain the carbon level of the parts provided that air leakage is kept to a minimum so that methanol fractions can also be low.

The purging model for nitrogen-methanol is similar to the one used for nitrogen-hydrogen. We will consider only the more general case in which air leakage into the furnace also occurs. Let f be the total flow of nitrogen plus methanol and let f' be the rate of air leakage, both measured at ambient temperature and pressure.* Let x be the mole fraction of

*Methanol is actually metered as a liquid, not a gas. Using 0.79 g/cc for the density of liquid methanol, a liter of liquid methanol is equivalent to 0.55 cubic meters of methanol gas.

methanol in the inlet gas. To simplify the calculation, we combine the two flows, f and f', and assume decomposition of methanol and reaction with the oxygen present in the air. Using an equilibrium constant of one for the reaction

$$CO_2 + H_2 \leftrightarrow CO + H_2O$$

the combined flows can be expressed in terms of a fictitious inlet gas stream having a flow rate f_e given by

$$f_e = (1 + 2x)f + 0.79f' \qquad [40]$$

The fictitious gas has a CO content ϕ, a CO_2 content ω, a hydrogen content ψ, and a water vapor content ρ, given by the expressions

$$\phi = \frac{xf - 0.14f'}{f_e}$$

$$\omega = \frac{0.14f'}{f_e}$$

$$\psi = 2\phi$$

$$\rho = 2\omega \qquad [41]$$

The furnace is assumed to be filled with air initially. In the first stage of purging (Figure 8), hydrogen and CO in the inlet gas combine with oxygen present in the furnace to form water vapor and CO_2. These reactions result in a loss of volume, so the flow rate leaving the furnace during the first stage, f_o, is

$$f_o = f_e\left(1 - \frac{3\phi}{2}\right) \qquad [42]$$

Taking an oxygen balance on the furnace leads to the equation

$$\frac{dC}{dt} + \frac{f_e}{V}\left(1 - \frac{3\phi}{2}\right)C = -\frac{f_e}{V}\frac{3\phi}{2} \qquad [43]$$

where C is the oxygen content of the furnace, V is the furnace volume, and t is time. Applying the initial condition $C = 0.21$ when $t = 0$ produces the solution

$$C = \left(0.21 + \frac{3\phi}{2 - 3\phi}\right)\exp\left[-\frac{f_e}{V}\left(1 - \frac{3\phi}{2}\right)t\right] - \frac{3\phi}{2 - 3\phi} \qquad [44]$$

The end of the first stage of purging occurs when $C = 0$. The time, t_1, for the first stage of purging is found by applying this condition to Eq. [44].

$$t_1 = -\frac{V}{f_e}\frac{2}{(2 - 3\phi)}\ln\left(\frac{3\phi}{0.21(2 - 3\phi) + 3\phi}\right) \qquad [45]$$

The increase in water vapor content, C_w, and CO_2 content, C_d, during the first stage are found by taking material balances on H_2O and CO_2, then applying the initial conditions $C_w = C_d = 0$ at $t = 0$.

$$C_w = \frac{2(\rho + 2\phi)}{2 - 3\phi}\left\{1 - \exp\left[-\frac{f_e}{V}\left(\frac{2 - 3\phi}{2}\right)t\right]\right\} \qquad [46]$$

$$C_d = \frac{2(\omega + \phi)}{2 - 3\phi}\left\{1 - \exp\left[-\frac{f_e}{V}\left(\frac{2 - 3\phi}{2}\right)t\right]\right\} \qquad [47]$$

Combining [46] and [47] with [45] yields expressions for the water vapor content, C_w^*, and CO_2 content, C_d^*, of the furnace at the end of the first stage of purging.

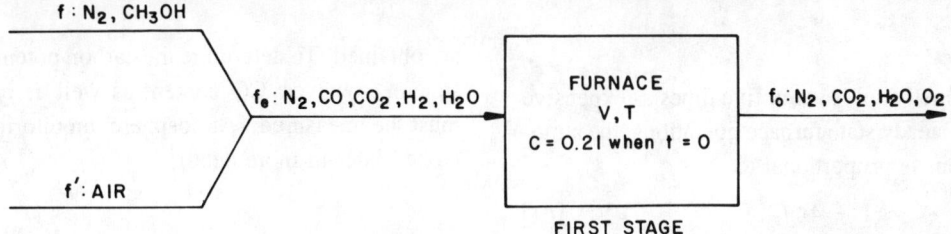

Fig. 8—Furnace purging with nitrogen-methanol with air leakage. For purposes of computation, the inlet flows of nitrogen-methanol and air are combined to form a fictitious inlet gas containing nitrogen, CO, CO₂, H₂, and H₂O.

$$C_w^* = \frac{0.42(\rho + 2\phi)}{0.21(2 - 3\phi) + 3\phi} \quad [48]$$

$$C_d^* = \frac{0.42(\omega + \phi)}{0.21(2 - 3\phi) + 3\phi} \quad [49]$$

In the second stage of purging, the H_2/H_2O and CO/CO_2 ratios progressively increase. To find the time needed for H_2/H_2O to reach the value R, defining nonoxidizing conditions, it is necessary to derive expressions for the hydrogen content, C_h, and the water vapor content, C_w, as a function of time. These expressions are

$$C_h = 2\phi\left[1 - \exp\left(-\frac{f_e}{V}t\right)\right] \quad [50]$$

$$C_w = (C_w^* - \rho)\exp\left(-\frac{f_e}{V}t\right) + \rho \quad [51]$$

The time needed to reach R, t_{2r}, is found by solving the expression $C_h/C_w = R$ for t. The result is

$$t_{2r} = -\frac{V}{f_e}\ln\left(\frac{2\phi - R\rho}{R(C_w^* - \rho) + 2\phi}\right) \quad [52]$$

The total purging time for H_2/H_2O to reach the value R, t_{pr}, is the sum of Eqs. [45] and [52]:

$$t_{pr} = -\frac{V}{f_e}\left[\frac{2}{2 - 3\phi}\ln\left(\frac{3\phi}{0.21(2 - 3\phi) + 3\phi}\right)\right.$$
$$\left. + \ln\left(\frac{2\phi - R\rho}{R(C_w^* - \rho) + 2\phi}\right)\right] \quad [53]$$

To find the purging time needed for the quantity $(CO)^2/CO_2$ to reach the value S (Figure 7), for a neutral atmosphere, expressions for the CO content, C_m, and CO_2 content, C_d, as a function of time during the second stage must be found. These expressions are:

$$C_m = \phi\left[1 - \exp\left(-\frac{f_e}{V}t\right)\right] \quad [54]$$

$$C_d = (C_d^* - \omega)\exp\left(-\frac{f_e}{V}t\right) + \omega \quad [55]$$

The time for C_m^2/C_d to reach S, t_{2s}, is found by writing

$$S = \frac{C_m^2}{C_d} = \frac{\phi^2(1 - E)^2}{(C_d^* - \omega)E + \omega} \quad \text{where } E = \exp\left(-\frac{f_e}{V}t\right) \quad [56]$$

This equation can be written in the standard quadratic form

$$AE^2 + BE + C + 0 \quad [57]$$

where $A = 1$

$$B = -\left(2 + \frac{S(C_d^* - \omega)}{\phi^2}\right)$$

$$C = 1 - \frac{\omega S}{\phi^2}$$

For purging to be possible, $\phi^2/\omega > S$, therefore C in Eq. [57] is a number between 0 and 1. Since $C_d^* > \omega$, B in Eq. [57] will have an absolute value greater than 2. Therefore, the quantity $4AC/B^2 < 1$. We can write the radical $\sqrt{B^2 - 4AC}$ in the form of a power series

$$\sqrt{B^2 - 4AC} = B\sqrt{1 - \frac{4AC}{B^2}}$$

$$= B\left(1 - \frac{2AC}{B^2} - \frac{1}{2}\left(\frac{AC}{B^2}\right)^2 - \cdots\right)$$

So the solution to Eq. [57] is

$$E = \frac{-B \pm \sqrt{B^2 - 4AC}}{2A}$$

$$= -\frac{B}{2A} \pm \frac{B}{2A}\left(1 - \frac{2AC}{B^2} - \frac{1}{2}\left(\frac{AC}{B^2}\right)^2 - \cdots\right)$$

Choosing the + sign in front of the second term, the solution is

$$E = -\frac{C}{B}\left(1 + \frac{1}{4}\frac{AC}{B^2} + \cdots\right) \quad [58]$$

The second term in the expression in parentheses above can never have a value greater than ¹⁄₁₆, therefore the second and all succeeding terms are neglected. Finally, from [58] and the definition of E we obtain

$$t_{2s} = -\frac{V}{f_e}\ln\left(\frac{\phi^2 - \omega S}{2\phi^2 + S(C_d^* - \omega)}\right) \quad [59]$$

and the total purging time, t_{ps}, the sum of Eqs. [45] and [59] is

$$t_{ps} = -\frac{V}{f_e}\left[\frac{2}{2 - 3\phi}\ln\left(\frac{3\phi}{0.21(2 - 3\phi) + 3\phi}\right)\right.$$
$$\left. + \ln\left(\frac{\phi^2 - \omega S}{2\phi^2 + S(C_d^* - \omega)}\right)\right] \quad [60]$$

Operating Costs

On a molar basis, methanol is about five times as expensive as nitrogen. So for steady state furnace operations, the atmosphere cost per hour is proportional to

$$(1 + 4x)f \qquad [61]$$

where x is the mole fraction of methanol and f is the flow rate. Therefore, to minimize cost, both x and f should be as low as possible. The lowest value for f will be that needed to keep a positive furnace pressure. The lowest value for x is that just needed to offset air leakage. If nonoxidizing (but decarburizing) conditions are desired, then the condition $\psi/\rho \geq R$ leads to the expression

$$xf \geq (R + 1)0.14f' \qquad [62]$$

which can be used to determine the most economical value of x. If a neutral atmosphere is desired, then $\phi^2/\omega = S$. From the Eqs. [41] a quadratic expression in x is obtained.

$$x^2 - (1 + S)0.28\frac{f'}{f}x$$
$$+ 0.14\frac{f'}{f}\left(0.14\frac{f'}{f} - S\left(1 + 0.79\frac{f'}{f}\right)\right) = 0 \qquad [63]$$

When $f'/f \ll 1$, terms $(f'/f)^2$ can be neglected, giving

$$x^2 - (1 + S)0.28\frac{f'}{f}x - 0.14\frac{f'}{f}S = 0 \qquad [64]$$

from which the appropriate value for x can be found when f and f' are known.

Purging costs are proportional to

$$(1 + 4x)ft_p \qquad [65]$$

where t_p is the total purging time given by either Eq. [53] or Eq. [60]. Purging costs decrease continually as x increases; the lowest purging cost is obtained using pure methanol.*

---

*This is true for the assumption that methanol is five times as expensive as nitrogen on a molar basis. If the cost differential is greater, an optimum value for x, less than one, may exist.

Therefore, the most economical mode of operation is similar to that suggested for nitrogen-hydrogen atmospheres:

1. Find the minimum flow rate, f, needed to keep a positive furnace pressure.

2. For steady state operations, choose x using [62] or [64] above, depending upon whether the atmosphere must just be protective against oxidation or neutral for steel with a specified carbon content.

3. During purging, increase the fraction of methanol to 35 pct, keeping the nitrogen flow constant. Higher methanol contents, as long as they can be used without sooting, will reduce the purging cost still further.

4. The furnace atmosphere can be monitored with a zirconia oxygen sensor to determine when nonoxidizing conditions are obtained. To determine the carbon potential of the atmosphere, however, CO content as well as oxygen potential must be measured. Atmosphere monitoring will be discussed later in more detail.

RESULTS

The use of these purging models is illustrated by the following examples.

Example I

An annealing furnace with a volume of 3 m³ is operating with nitrogen-hydrogen at 702 °C (1295 °F). From Figure 4, the value of R is 2.41. Assume air leak rates of 0.06, 0.03, and 0.00 m³ per hour. Compute the operating variables for steady state operations and for purging which provide the most economical furnace operation.

Steady State

From furnace trials it is found that a flow rate of 3 m³ per hour of atmosphere gas is sufficient to keep the furnace pressure positive. The lowest possible value for x, from [37] and the relative cost are given in Table I for flow rates of 3 and 4.5 m³ per hour. So if the air leak rate is 0.06 m³ per hour, a hydrogen content of 3 pct will maintain nonoxidizing conditions for a flow rate of 3 m³ per hour. The relative cost is 3.17. Note that the operating cost is nearly proportional to flow rate.

Purging

From Eq. [36] the total purging time can be computed for various hydrogen fractions and the relative cost can be found using [39]. Figure 9 is a plot of the results for a flow rate of 3 m³ per hour. Note that the purging time is about the same for various rates of air leakage as long as the fraction of hydrogen is high. The minimum purging cost is for a hydrogen fraction of about 0.55. However, the minimum is shallow; there is not much difference in cost for values of x in the range $0.35 < x < 1.0$. The location of the minimum is not measurably affected by the air leak rate; therefore, in

Table I. Minimum Hydrogen Fraction (Relative Cost)

Air Leak Rate	0.06	0.03	0.00
Inlet Flow Rate			
3	0.029	0.014	0.0
	(3.17)	(3.08)	(3.0)
4.5	0.019	0.010	0.0
	(4.67)	(4.59)	(4.5)

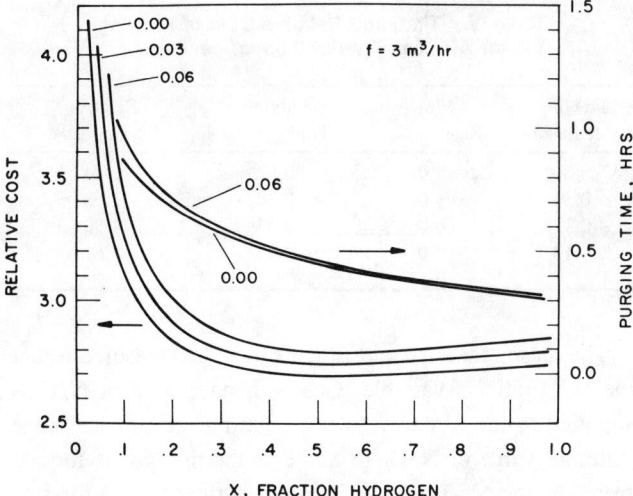

Fig. 9—Purging cost and purging time as a function of hydrogen fraction at a flow rate of 3 m³/hr (Example I). Curves for air leak rates of 0, 0.03, and 0.06 m³/hr are presented.

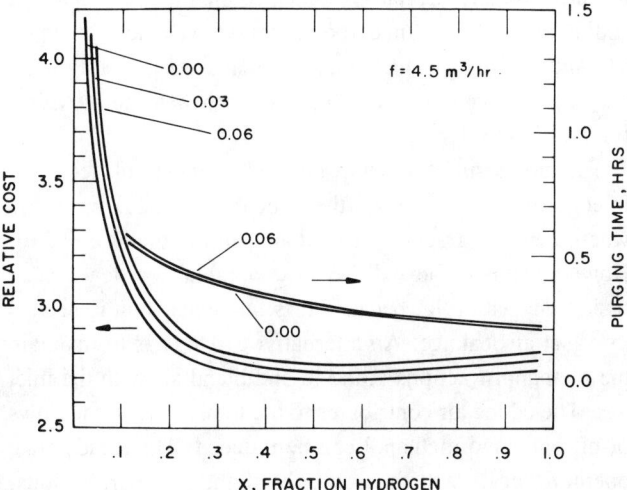

Fig. 10—Purging cost and purging time as a function of hydrogen fraction at a flow rate of 4.5 m³/hr (Example I). Curves for air leak rates of 0, 0.03, and 0.06 m³/hr are shown.

this case, an efficient value of x can be found without knowing the precise rate of air leakage.

Let us take $x = 0.35$ for purging. Following the strategy previously outlined, the nitrogen flow for purging will be maintained at the value used for steady state operations (2.91 m³ per hour) and the hydrogen flow will be increased until the fraction of hydrogen is 35 pct of the total flow. This increases the total flow to 4.5 m³ per hour. The purging time (0.37 hour) and relative cost (2.79) at this higher flow rate are shown in Figure 10. The form of the curves has not changed, and the purging cost is slightly lower than at a flow rate of 3 m³ per hour.

This purging strategy is much more economical than simply increasing the flow rate of atmosphere gas at the same composition that is used for steady state operations. For an air leak of 0.06 m³ per hour, Table II compares purging

costs when (1) the hydrogen fraction is raised, but the flow rate is held constant, and (2) the hydrogen fraction is held constant, while the flow rate is raised.

Example II

A sealed quench furnace with a volume of 3 m³ is operating with nitrogen-methanol at 850 °C (1560 °F) for neutral hardening of 0.4 pct C steel parts. From Figure 7, the value of S is 4.80. Assume air leak rates of 0.06, 0.03, and 0.00 m³ per hour. Compute the operating variables for steady state operations and for purging to provide the most economical mode of operation.

Steady State

From furnace trials it is found that a flow rate of 3 m³ per hour of atmosphere gas is sufficient to keep the furnace pressure positive. The lowest value of x, from [64], for various air leak rates is shown in Table III. So if the air leak rate is 0.06 m³ per hour, a methanol mole fraction of 0.13 will maintain neutral conditions for a 0.4 pct C steel at 850 °C.

Purging

From Eq. [60] the total purging time can be computed for various methanol fractions and, using [64], a relative cost can be found. The top two sets of curves in Figure 11 show how the purging cost and purging time vary with methanol fraction when the goal is to produce an atmosphere neutral to 0.4 pct C steel. The lower two sets of curves in Figure 11 show the purging time and cost for achieving a nonoxidizing atmosphere ($R = 1.85$).

The purging cost decreases as the fraction of methanol increases. However, often there is an upper limit to the

Table II. Comparison of Purging Times and Relative Costs

Fraction Hydrogen	Flow Rate, m³/hr	Purging Time, Hours	Relative Purging Cost
0.35	3.0	0.56	2.83
0.35	4.5	0.37	2.79
0.03	6.0	0.70	4.44
0.03	9.0	0.44	4.19

Table III. Methanol Fraction and Relative Cost for 3 m³ per Hour Inlet Flow

Air leak rate, m³/hr	0.06	0.03	0.00
Methanol fraction	0.13	0.09	0.00
Relative cost per hour	4.60	4.09	3.00

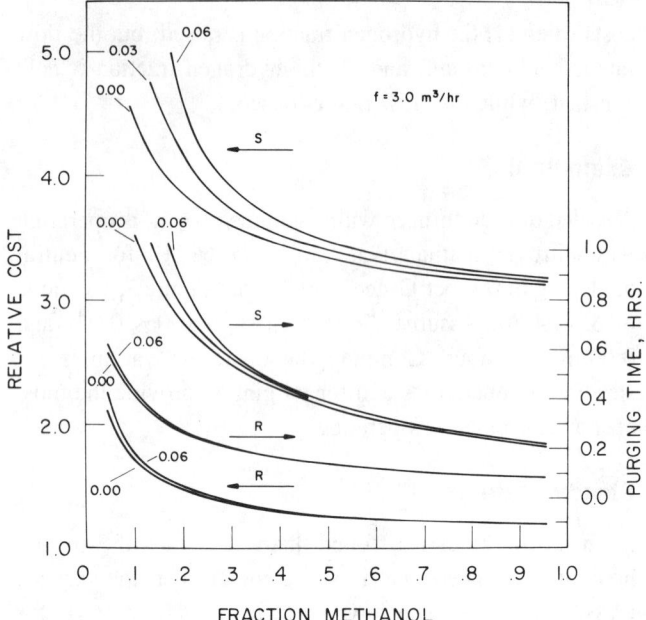

Fig. 11 — Relative cost and purging time as a function of methanol fraction for a nitrogen-methanol flow rate of 3 m^3/hr (Example II). Curves for 0, 0.03, and 0.06 m^3/hr air leakage are shown. The top 2 sets of curves are cost and time to reach a 0.4 pct carbon potential. The lower 2 sets of curves are cost and time to achieve nonoxidizing conditions.

Table IV. Time and Relative Cost of Purging for Air Leak Rate of 0.06 m^3 per Hour

Fraction Methanol	Flow Rate, m^3/hr	Purging Time, Hours	Relative Purging Cost
0.35	3.0	0.52	3.74
0.35	4.0	0.38	3.68
0.13	6.0	0.54	4.90
0.13	9.0	0.34	4.61

amount of methanol that can be added, because the decomposition of methanol to CO and H_2 is endothermic. Too much methanol added at one location in the furnace will cause local cooling, promoting decomposition to soot. A methanol fraction of 0.35 should be satisfactory — blends of nitrogen with 35 pct methanol are used commercially to form furnace atmospheres similar in composition to endothermic gas derived from methane.[6]

If the nitrogen flow is held constant, increasing the methanol fraction from 0.13 (steady state) to 0.35 for purging increases the total inlet flow to 4.0 m^3 per hour. This causes the purging time to be less than that shown in Figure 11, but has very little effect on the cost of purging (Table IV).

Table IV also shows that increasing the methanol fraction during purging is more economical than just increasing the flow rate with the methanol fraction unchanged.

DISCUSSION

Monitoring Nitrogen-Methanol Atmospheres

When nitrogen-methanol is used to produce atmospheres neutral to steel of specified carbon content, two constituents of the atmosphere must be monitored to find the carbon potential. For example, measurements of either CO and CO_2 content or CO content and oxygen potential would allow computation of a carbon potential. This is in contrast to the more familiar case of endothermic gas atmospheres in which monitoring of only one constituent suffices to determine carbon potential.

The reason that two constituents must be measured is that the N/O ratio is variable. Gas with one value of N/O is supplied to the furnace, but air entering the furnace has a different value of N/O. (Since endothermic gas is formed from air, the N/O ratio in endo gas is the same as for air.)

Commercial devices employing microprocessors are available to monitor two gases and compute a carbon potential. So, for steady state operations, the methanol flow can be automatically regulated to balance the rate of air leakage and maintain a constant carbon potential. (Devices based on CO and oxygen potential measurements are probably best, because CO_2 contents will be very low when the methanol fraction is low.)

The monitoring and control problems are simplified if the inlet gas has N/O = 3.78, the same ratio as air. This occurs when the mole fraction of methanol in the blend is 0.346. However, this blend will be uneconomical for steady state operations as a rule, because it is far richer than is needed to offset air leakage. An alternative strategy is to maintain the methanol fractions at 0.346 and blend air with the inlet gas. The added air contributes to the total flow, so the flows of nitrogen and methanol can be reduced. The steady state operating costs will then be only slightly higher, as illustrated in the following example.

Example III

In Example II it is found that for steady state operations, with an inlet gas flow rate of 3 m^3 per hour and an air leak rate of 0.06 m^3 per hour, the fraction of methanol needed to maintain $S = 4.8$ was 0.13. In this case, the reacted gas flow (Eq. [40]) is 3.83 m^3 per hour and the relative cost per hour (Table III) is 4.6.

For comparison, let x be 0.346 and let enough air be added to make $(CO)^2/CO_2$ equal to 4.8. Rearranging Eq. [63] we obtain

$$\left(\frac{f'}{f}\right)^2 - \left(\frac{2x(1 + S) + S}{0.14 - 0.79S}\right)\frac{f'}{f} + \frac{x^2}{0.14(0.14 - 0.79S)} = 0 \quad [66]$$

from which, for $S = 4.8$ and $x = 0.346$ we find that

$$\frac{f'}{f} = 0.0934 \quad [67]$$

So when $f = 3$, f', the sum of the air leakage and the air intentionally added, is 0.28 m^3 per hour. From Eq. [40], the reacted gas flow rate is 5.3 m^3 per hour. Since it is the reacted gas flow rate which is important in maintaining furnace pressure, the comparison should be made at constant reacted gas flow rate. Using [40] and [67] we find that to have a reacted gas flow rate of 3.83 m^3 per hour (the same as Example II) with $x = 0.346$, we need $f = 2.17$ and $f' = 0.20$ m^3 per hour. For these flows, the relative cost per hour, from [61] is 5.17, about 12 pct greater than when nitrogen — 13 pct methanol is used.

To summarize, the atmosphere system nitrogen-methanol-air, with the nitrogen and methanol balanced to give N/O = 3.78, has the advantage that its carbon potential can be determined by a single measurement, *e.g.*, oxygen potential.* In steady state operations, nitrogen and

---

*If the goal is to produce a nonoxidizing atmosphere, there is no advantage to maintaining a fixed nitrogen/oxygen ratio because the oxygen potential can be measured directly. For this situation one can show that for temperatures below 860 °C ($R = 1.82$) steady state operations are least costly when the smallest possible methanol fraction is used (assuming methanol is five times as expensive as nitrogen). That is, there is no advantage to deliberate additions of air. Above 860 °C, however, the lowest cost steady state atmosphere is one formed from air and methanol. Adding nitrogen increases the cost!

---

methanol flows could be held constant while the air flow was automatically regulated to maintain a constant carbon potential. For purging, only nitrogen and methanol would be used. The disadvantages to this system are:

1. Steady state operating costs are almost always higher than if nitrogen with a low fraction of methanol is used. The cost penalty is greatest for "tight" furnaces with small air leaks.

2. Close atmosphere control is necessary because buffering of the atmosphere by carbon exchange with the parts is more likely to cause significant carburization or decarburization the higher the methanol fraction. The situation in this respect should be similar to neutral hardening in endo gas.

Use of By-Product Nitrogen

Cryogenic nitrogen supplied for heat treating operations usually contains less than 10 ppm oxygen. However, high purity nitrogen is not actually needed for furnace heat treating, because entry of air into the furnace can never be completely eliminated. Some reducing gas must always be added to offset the effect of the air.

Impure liquid nitrogen, containing one percent oxygen or more, is often available as a by-product of plants making liquid oxygen. The cost of this product is low; it is used mainly for cooling at present. In Example III above it is shown that the cost penalty attached to use of a high methanol fraction with air additions is not large for neutral hardening. Therefore, if impure nitrogen was available at a much lower cost than pure nitrogen, it could be used instead with

a significant cost savings. As long as the oxygen content of the impure nitrogen was reasonably constant, *e.g.*, 1 to 3 pct oxygen, a methanol fraction could be chosen which would keep the N/O ratio close enough to 3.78 so that the atmosphere carbon potential could be monitored with a single measurement.

Limitations of the Model

The mathematical models used here to find optimum operating conditions for furnaces employing nitrogen-base atmospheres are intended to be general enough so that they can be applied to any type of furnace. However, in making computations of relative cost, it has been implicitly assumed that the rate of air entry into the furnace is independent of atmosphere flow rate, as long as the furnace pressure is positive. For some types of furnaces, *e.g.*, continuous belt furnaces, open at each end, this assumption is probably incorrect. (It may be uneconomical to operate some of these furnaces at a high enough flow rate to develop a positive pressure.) If the rate of air leakage as a function of inlet gas flow rate for a particular furnace can be quantified by experimental measurements, then this data can be incorporated into the computations to find optimum flow rates and fractions of reducing gas.

Another limitation on the generality of some of these results is the relative costs assumed for the various gases. Hydrogen and methanol will always be more expensive than nitrogen on a molar basis, but relative costs can vary significantly. In applying these results to any specific situation, actual gas costs should be used to modify Eqs. [38], [39], [61], and [65].

SUMMARY

Economical operation of furnaces with nitrogen-base atmospheres requires an understanding of the importance of the flow rate of atmosphere gas, the fraction of reducing additive, and the magnitude of the air leakage into the furnace on atmosphere composition. For efficient operation, atmosphere monitoring and control are just as important as with endothermic and exothermic atmospheres.

Mathematical models for purging of furnaces with nitrogen, nitrogen-hydrogen, and nitrogen-methanol blends have been derived. Strategies for efficient (and convenient) operation of furnaces using these atmospheres are suggested based on the models.

For nitrogen-hydrogen atmospheres, the cost of steady state operation is least when the flow rate of atmosphere gas is just sufficient to maintain a positive furnace pressure and the fraction of hydrogen is just sufficient to balance the entry of air, maintaining an oxygen potential below that which will oxidize steel. The cost of purging is not greatly affected by the flow rate of atmosphere gas, but it is strongly dependent upon the fraction of hydrogen. In many cases there will be an optimum value of the hydrogen fraction, less than one, for least cost purging.

For nitrogen-methanol atmospheres, two strategies for efficient operation are suggested. The first strategy is similar to that proposed for nitrogen-hydrogen. Steady state operation is at the lowest flow rate and lowest methanol fraction possible, maintaining a positive furnace pressure and a specified carbon potential. For purging, the highest methanol fraction that can be used without sooting is the most economical. Carbon potential control requires monitoring two constituents of the atmosphere, *e.g.*, CO and oxygen potential. The second strategy uses a nitrogen-methanol blend balanced to produce the same N/O ratio as exists in air. Air is deliberately added to the blend to reduce the carbon potential of the atmosphere to the level desired. This approach is somewhat more costly than the first strategy suggested, but it has the advantage that measurement of a single constituent of the atmosphere — CO_2, oxygen potential or dew point — suffices to determine the carbon potential.

REFERENCES

1. R. Schuhmann, Jr.: *Metallurgical Engineering, Vol. I, Engineering Principles,* Addison-Wesley Press, Inc., Cambridge, MA, 1952.
2. C. A. Stickels and C. M. Mack: *Metall. Trans. B*, 1980, vol. 11B, pp. 481-84.
3. D. R. Stull and H. Prophet: *JANAF Thermochemical Tables, 2nd edition,* U. S. Department of Commerce (NSRDS–NBS 37), Washington, DC, June 1971.
4. R. C. Weast, ed.: *CRC Handbook of Chemistry and Physics, 62nd edition,* 1981, CRC Press, Inc., Boca Raton, FL, pp. D-168.
5. J. Chipman: *Metall. Trans.*, 1972, vol. 3, pp. 55-64.
6. S. Jansen and R. H. Kohler: *Heat Treatment '79*, 1980, The Metals Society, London, pp. 82-89.

SECTION XII
Quenchants

Selecting
and
Handling
Quenching Fluids

By WILLIAM H. NAYLOR

Four types of oils — mineral, low viscosity, hot quenching,
and marquenching (high viscosity) — provide a wide range
of cooling rates in baths from 120 to 350 F.
Tests to determine cooling rates, viscosity, flash and fire points,
acid number, and other properties reveal service characteristics
of new and used oils.

QUENCHING CONSISTS of immersing a hot component in a liquid to lower its temperature rapidly. To date, little has been published about cooling mediums and their roles, probably because field testing is the only valid method of evaluating them.

To describe the way an oil acts when quenching and to outline the information needed to select a proper oil, we must know how many factors interrelate. First, consider the steel. Its isothermal transformation curve shows how it develops various structures at different stages of cooling. Pearlite and bainite begin to form at high temperatures and require specific times for complete development. Martensite, the structure most heat treaters aim for, starts to develop at lower temperatures and increases in content as the part approaches room temperature.

The quenching cycle consists of three stages.

Mr. Naylor is technical representative, metalworking oils section, Sun Oil Co., Philadelphia.

During the vapor-blanket phase, an envelope of oil vapor (generated by the heated piece as it is immersed) slows heat transfer to the surrounding oil. As part temperature drops, the barrier collapses as rapidly boiling oil contacts the metal, raising the rate of heat transfer sharply. The last phase begins when the temperature of the piece drops below the boiling range of the oil. Then oil constantly contacts the metal surface, cooling the part by conduction and convection. When the part reaches bath temperature, quenching stops. Figure 1 shows that an oil must cool rapidly through the first two phases to prevent pearlite and bainite from forming and then cool slowly to minimize distortion and cracking.

Effects of Temperature

Changes in bath temperature do not usually alter the cooling curve during the vapor or critical cooling phase; the maximum effect occurs in slow cool-

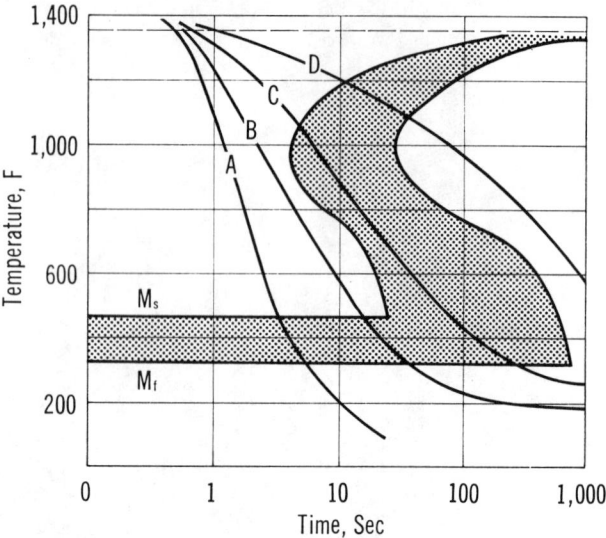

Fig. 1—To harden austenitized steel, the oil must cool it past the Ac₃ (dashed line) to below the nose of the transformation curve (the shaded area is one for AISI 1080) fast enough to prevent austenite from transforming. Then, slow cooling below the M_s temperature precludes cracking. Cooling curve A, brine; B, fast quenching oil (Sunquench 1070); C, straight quenching oil; D, air. Specimen: stainless steel, ½ in. diameter by 4 in. long (same for Fig. 2, 4, and 5).

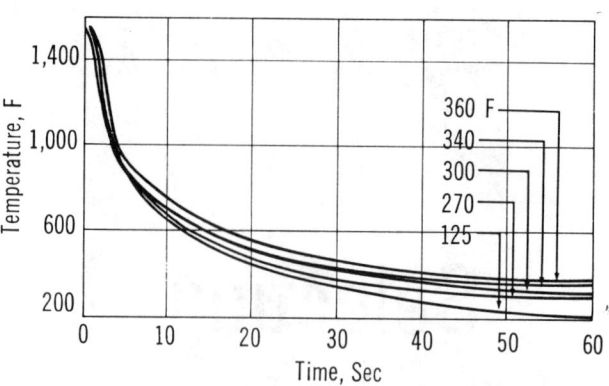

Fig. 2 — Raising the bath temperature generally alters cooling rate only during the third phase. In hotter oil, parts cool more slowly and are less apt to crack. Oil, Sunquench 1021 (agitated); viscosity, 250 SUS; flash point, 440 F.

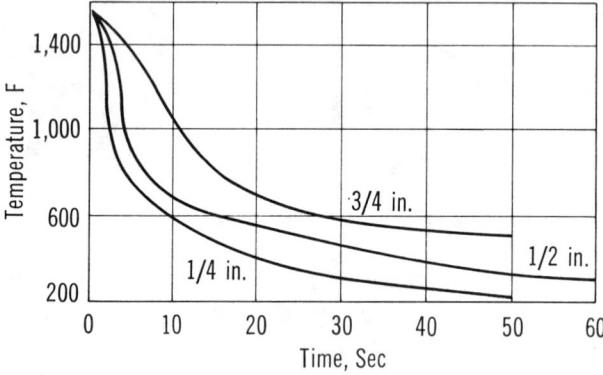

Fig. 3 — The larger the part being quenched, the slower it cools. Note that all phases of the cooling curve shift to the right. Oil, Sunquench 1070 (unagitated); viscosity, 90 SUS; 125 F. Specimens: 4-in. stainless steel bars of indicated diameters.

Fig. 4 — Agitation raises the cooling rate, especially during the vapor and boiling phases. Below the boiling range, oil always contacts the piece, and agitation has little effect. Oil, Sunquench 1070; 125 F.

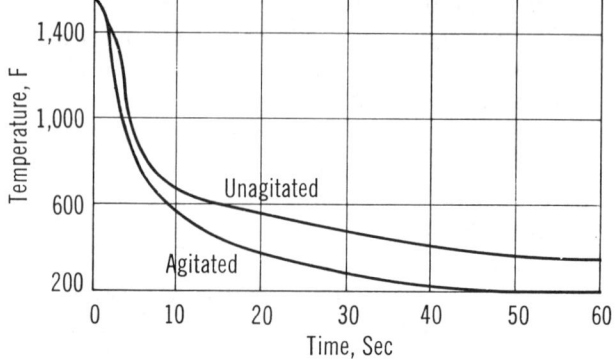

ing. The higher the bath temperature, the faster the oil and specimen temperatures reach equilibrium. Then, cooling becomes slower, as indicated by Fig. 2. This effect helps to minimize distortion and cracking.

Because the cooling curve of a straight mineral oil of a given viscosity varies in relation to the mass of the part being quenched (large parts cool more slowly than small parts), all phases of the cooling curve are affected (Fig. 3). Resulting steel hardness will be determined by the extent of transformation.

The rate of cooling throughout a part is determined mainly by the cooling rate at its surface. Agitation forces oil to flow against surfaces, causing the surrounding vapor barrier to collapse. With this collapse, the rate of cooling, both external and internal, rises (Fig. 4). More agitation increases the speed of quenching, the major effect taking place during the vapor and boiling phases. Once the piece temperature drops below the boiling range, the oil constantly contacts the work, and the effectiveness of agitation decreases rapidly.

Characteristics of Quenching Oils

Mineral oil with a viscosity range of 70 to 300 SUS (Saybolt Universal Seconds) at 100 F is most widely used. The most common type is rated at 100 SUS at 100 F. Bath temperature is 120 to 150 F

for high-carbon and high-alloy steels having low critical rates. Typical grades: AISI 4140, 52100, 6150. Desired hardnesses are obtained, yet stresses are minimized due to slow cooling. Baths are held at 150 to 200 F when slower cooling rates are desired in the third phase of quenching.

A fast quenching oil normally has low viscosity (70 to 125 SUS at 100 F). Inhibited mineral oils exhibit excellent thermal and oxidation stability while promoting maximum heat transfer in the first two stages of cooling. Heated to between 120 and 150 F, these oils are used for low-carbon and lean-alloy steels (such as AISI 1010 and 1020) which have high critical cooling rates. Uniform quenching at maximum speed is needed to produce desired hardnesses and to minimize stresses and distortion. Bath temperatures of 150 to 200 F are sometimes employed; these higher temperatures tend to negate the full effect of the speed additive, while assuring uniform quenching and stability of oil.

Hot quenching oils normally include medium to high viscosity (250 to 3,000 SUS at 100 F) mineral oils, straight or inhibited to some degree. (Inhibitors, generally proprietary compounds, combine with oxygen to retard deterioration.) Used between 200 and 300 F, these stable oils generally minimize distortion and cracking in parts with intricate contours. Greater tensile and fatigue strengths can be attained in thin-wall parts of alloy steels such as AISI 8620.

Marquenching oils usually have high viscosity, 2,000 SUS at 100 F and above. Being inhibited, they offer excellent oxidation and thermal stability. Promoting uniformity of cooling at bath temperatures above 300 F, such oils are used for steels having low critical cooling rates and where an absolute minimum of cracking and distortion is desired. Large pieces are normally quenched in this manner to insure similar core and surface hardnesses.

Properties Vary Widely

Physical properties of oil include viscosity, flash and fire points, stability, and demulsibility. Taking them in turn, viscosity relates to flow rate. Generally, low-viscosity oils (70 to 125 SUS at 100 F) are used when the heat treater wants to retard loss of oil due to dragout. Heat treaters employ oils with high viscosity in hot-oil systems because higher flash points provide a margin of protection. (An oil with low viscosity, 70 SUS, can have a flash point of 345 F and a fire point of 395 F; with high viscosity, 2,000 SUS, a flash point of 550 F and a fire point of 600 F.)

The flash temperature at which an oil flashes into flame should be about 150 F above the bath temperature to insure adequate safety during the operation. Also, the oil must be stable, both chemi-

cally and thermally, so that it does not break down or flame when contacting the hot piece. Further, this stability is required over a wide temperature range to assure durability for long times, a uniform rate of cooling and minimum formation of deposits. Demulsibility, the ability of oil to separate from water, is required so that a contaminated system can be corrected with minimal loss of production.

Quenching additives basically comprise wetting agents, oxidation resistors, detergents, and dispersant materials. Wetting agents, in general, increase quenching speed by causing the vapor barrier to collapse more rapidly during the first and second phases (Fig. 5). They do this by helping to bring the liquid into quick contact with the steel, inciting faster boiling and more rapid heat transfer. For practical purposes, any material which tends to speed up cooling can be considered a wetting agent. Additives which promote oxidation resistance are growing in importance because heat treaters are using higher bath temperatures — and the oxidation rate doubles for every 18 F (10 C) rise in temperature. Another factor is the heat treater's desire to extend intervals between oil changes, which are needed when undesirable oxidation products drop out of the bath as sludge. Oxidation inhibitors greatly retard the oxidation rate, making the oil more serviceable for longer times. Detergents and dispersant additives control sludge that results from resins, scale, and external contaminants. When they are used, a filtration system is needed to remove contamination from the oil.

Quenching in Hot Oil

Hot quenching (in 200 to 300 F oil) minimizes dimensional changes and reduces stress and cracking by controlling the cooling rate through the first and second phases of quenching. Marquenching calls for an oil bath at 300 to 350 F. After being quenched at a controlled rate to equilibrium, the part will be above the M_S temperature. When it is removed and cooled in air, it transforms to martensite slowly and at a uniform rate to preclude distortion or cracking.

When steels with critical cooling rates must be quenched rapidly, oil temperatures are in the 200 to 300 F range. Under this condition, additive oils, as well as lower-viscosity types, may be used. As before, the part is heated to form austenite and immersed in the bath. Since the bath temperature is below the M_S, some transformation takes place before the part is taken out to cool in air. Though small stresses form, the plasticity of the untransformed austenite relieves them. Again, after temp-

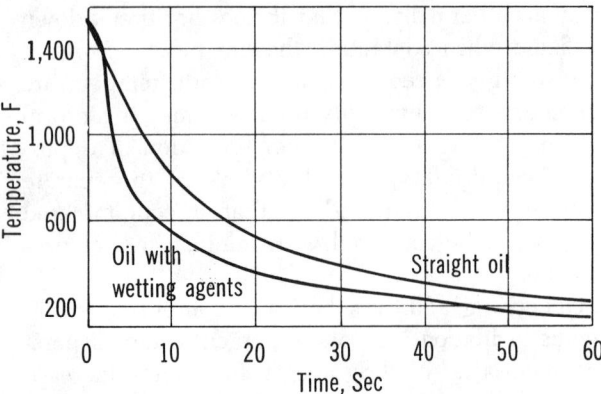

Fig. 5 — Wetting agents added to straight oils raise the quenching rate, particularly in the first and second phases. Oils, Sunquench 7 (straight), Sunquench 1070 (fast); 125 F; agitated.

eratures equalize in the bath, air cooling completes transformation to martensite.

Testing Oils in Service

Oils in quenching baths can be checked by magnetic quenchometer and hot-wire tests. The first method is described in "Measuring Quenching Rates With the Electronic Quenchometer," by H. J. Gilliland, *Metal Progress,* October 1960, p. 111. A nickel ball is heated to 1,625 F, surrounded by a magnetic field, and quenched in the oil. When the ball cools to 670 F (the Curie point of nickel), it regains its magnetism. The time required for cooling through this range indicates the speed of quenching.

The hot-wire test requires a wire (of specific material with a standard resistance) suspended between two terminals. The wire is lowered into the oil, and power is stepped up 5 amp every 5 sec until the wire burns out. The oil's speed of quenching varies directly with the amperage that the wire will carry. (An oil that cools rapidly will convey more heat from the wire, allowing it to conduct a greater current.)

Testing Used Oil

A quenching system must be controlled closely so that results will remain constant and predictable. In particular, a bath of used oil has to be tested regularly for viscosity, flash and fire point, acid number, deposit formation, cooling rate, and appearance to determine its condition.

After oils are used for a while, their cooling characteristics change. In time, oils of the straight (non-additive) type usually develop faster cooling rates, and final hardness of quenched parts rises to some degree. This often occurs because soluble resins

and acids form. They tend to increase wettability.

A used oil accelerates the temperature drop of the piece being quenched by causing the vapor barrier to collapse more readily. The third phase is reached faster than it is with fresh oil. Additions of fresh oil tend to swing the cooling curve back to its original pattern, acting to stabilize the quenching rate and giving a more uniform rate of cooling.

Additive-type oils react differently in that the wetting agent is depleted. However, depletion does not always reduce the speed of quenching. Instead, soluble resins and acids form due to higher temperatures, tending to offset the loss of additive. Adding fresh oil returns the system to its original state, stabilizing the cooling rate.

A higher viscosity rating normally signifies oxidation, polymerization, and thermal cracking of the oil. Drops in flash and fire points indicate possible contamination by volatile materials that would create fire hazards. A rise in the total acid number reveals oxidation. And the oil should be checked for viscosity to determine if it can still be used.

The degree of deposit formation is important because large amounts of sludge clog pipes and drains. Products of oxidation and contamination should be measured and analyzed to determine whether contaminants are entering from the outside. Decreases in cooling rate act to lower the hardness of quenched pieces. (Conversely, a drop in the hardness of the quenched piece indicates that the oil has lost some cooling capability.)

Finally, the quenched piece is examined for appearance. Surface discoloration due to varnish and lacquer deposits indicates the oil has changed. Parts which are dirty or loaded with scale usually show that more adequate filtration is needed.

Contaminants and Their Effects

As a contaminant of quenching oil, water is a major problem because it causes irregular cooling rates. Quenched parts may have higher hardnesses in spots where the quenching speed has been accelerated. Some steels quenched in such a system may crack and distort due to high stresses. Furthermore, excessive water can create a fire hazard by causing foaming.

Water can get into oil baths through tank condensation, leaks from coolers, and seepage. Its effect is greatest in the slow-cooling phase of the cycle. In circulating systems, water is more of a problem because it can be entrapped in a suspended form. Water in still baths settles out. An easy removal method: heat the system to over 250 F and hold at that temperature for about 4 hr. ⊕

Applying Synthetic Quenchants to High-Strength-Alloy Heat Treatments

Thomas R. Croucher

THE QUENCHING portion of heat treatment has relied mainly upon immersion quenching into either water or oil. Numerous other quenching media such as liquid nitrogen, water sprays, salt brine, polyvinyl alcohol, and woods metal, have been used in research and limited production. However, the ability of these techniques to achieve variability in quenching rates or improved heat transfer properties has been limited due to the single, basic, characteristic quenching value (H) for each. Water, although an inexpensive, universal, rapid quenchant, is too rapid for many applications and causes severe distortion, excessive stresses, and cracking. By varying water temperature, flexibility in quenching characteristics can be achieved, but precise temperature control is cumbersome in production, difficult to control, and only moderately successful. Oil quenching, although reducing the quench rate and quench cracking tendencies, is not flexible and is generally too slow to achieve adequate properties in many aplications. Oil quenching can also cause severe fire, smoke, and pollution problems.

To achieve flexible, yet more consistent results, the synthetic quenchants were developed. The development was undertaken with three primary objectives in mind:

1. Greater control over quenching speeds, i.e., to compare with a fast water quench or a slow oil quench.
2. Better heat transfer characteristics.
3. Reduced fire and pollution characteristics.

The most useful of available synthetics are polyalkylene-glycol-base materials. They have been successful in achieving the aforementioned objectives. These products are water-soluble, and by simple variation of water dilution, a wide range of quenching speeds can be achieved. As a result of the unique characteristic of inverse solubility, a thin film of concentrated glycol wets the hot part as it is quenched, and a more uniform rate of heat transfer is achieved, resulting in reduced quench distortion. Because of a high flash point, the fire and smoke problems connected with many oil-quenching operations are virtually eliminated. Polyalkalene glycols also have the additional advantages of:

- No quenchant heating is necessary, as in some systems. Slower cooling, if required, can be achieved by increasing the concentration.

- Parts, baskets, and fixtures can be cleaned by simple water rinsing. No detergents or oil-soluble materials are necessary.
- The material is nontoxic.
- Quenching systems may be used with or without agitation, depending on the application.

Development of Background Data

Synthetic quenching efforts at Progressive Metallurgical Industries (PMI) have been mainly in research and production aluminum quenching. Some development work has been conducted in steel, and a titanium program is being formulated. Our main effort has been the evaluation of Ucon Quenchant A, although some work has been conducted on other products. Based on our work thus far, we believe that synthetic quenchants are one of the most useful tools ever made for the heat treater. However, they are just a tool, although a highly technical tool. With proper understanding and utilization, this tool can achieve heretofore unachievable cost savings and to fully heat treat previously unheat-treatable parts.

At PMI, 18 months were spent in the laboratory developing technical data and experience before production was established. A total systems concept based on an engineering aproach was necessary to achieve the optimum results attainable—that is, full mechanical properties with reduced distortion and residual stresses. The engineering systems approach used is based on the following:

- Quench sensitivities of the alloys.
- Expansion characteristics.
- Geometry of the part.
- Quenching characteristics of the quenchant.
- Immersion profile.*
- Knowledge of the fabrication sequence prior to heat treatment.

In applying this concept, Fig. 1 through 5 illustrate some of the data we developed and utilize in production. Figure 1 shows a series of cooling curves from quenching ½-in. plate samples in various concentrations of Ucon Quenchant A. Figure 2 shows cooling rates from water at various temperatures. These two figures illustrate the variation in quench rates achieved by varying the temperature of water or the concentration of the polyalkylene-glycol

---

Mr. Croucher is President and General Manager, Progressive Metallurgical Industries, Gardena, Calif.

* Encompasses racking attitude, racking spacing, angle of entry, and rate of entry into the quench medium.

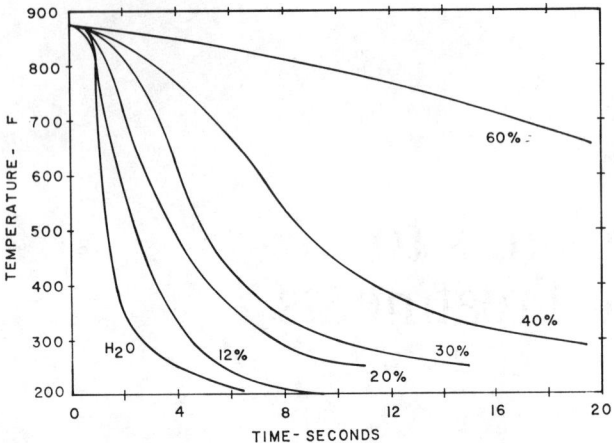

Fig. 1. Quench rate curves for various concentrations of Ucon quenchant A measured at the center of a ½-in. 7075 plate.

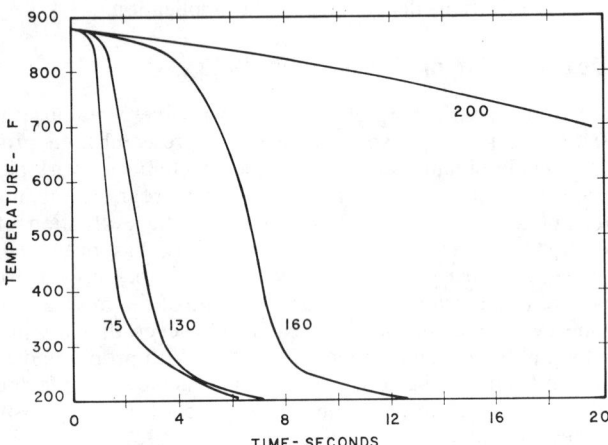

Fig. 2. Quench rate curves for various water temperatures measured at the center of a ½-in 7075 plate.

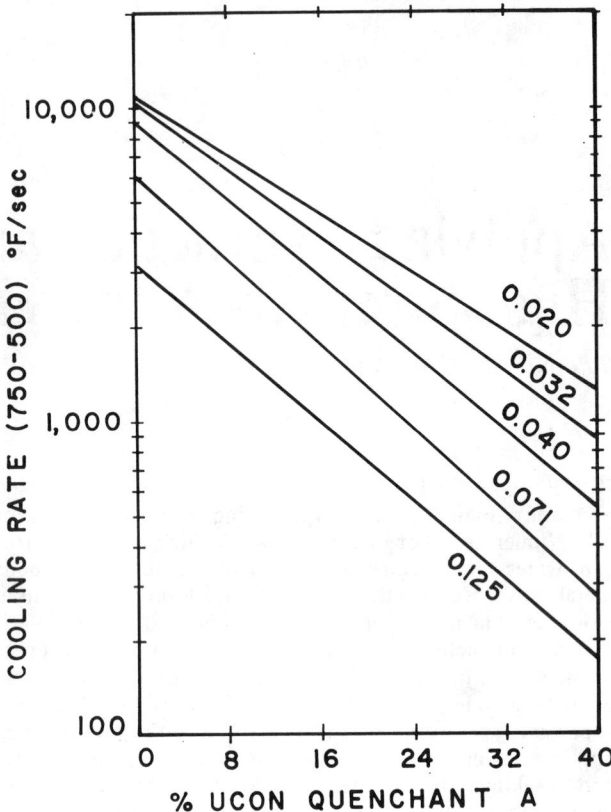

Fig. 3. Variations in quench rate for sheetmetal at various concentrations of Ucon Quenchant A.

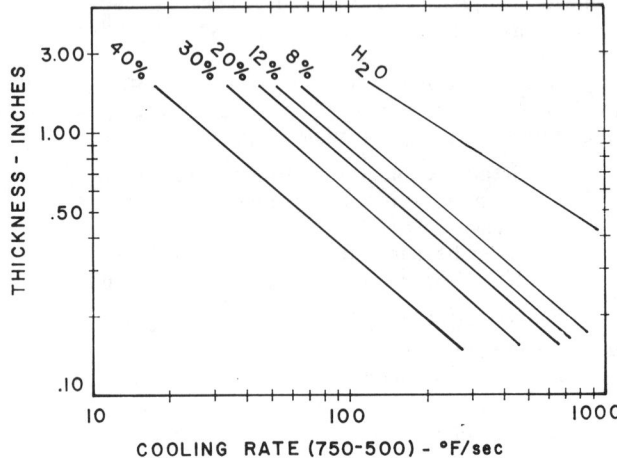

Fig. 4. Variation in quench rate for metal plate at various concentrations of Ucon quenchant A.

mixture. Figures 3 and 4 show the effect of thickness and glycol concentration on cooling rates for various sheet and plate gages. Figure 5 shows the effect that various geometries can have in obtaining a given cooling rate. These figures represent just a small percentage of the technical data developed and necessary to utilize synthetic quenchants in production.

Aluminum Parts

PMI has found that all forms of aluminum products can be quenched in synthetic quenchants. Some of the parts quenched and the results obtained are illustrated in the following discussion.

PMI has heat-treated approximately 15 different configurations of aircraft window frame forgings, Fig. 6, and have maintained consistently a 0.008-0.015 movement throughout the operation. To achieve this degree of tolerance, however, it was necessary to utilize approximately ten different techniques on the fifteen configurations. Varia-

tions included, among others, glycol concentration, racking spacing, racking attitude, angle of entry, and rate of entry into the quench tank. Figure 7 shows a precision die forging which normally experienced a movement of approximately 0.125 in. On a first prototype production part, the quench distortion was reduced to approximately 0.020 in. Upon modification of the quenching procedure, the total distortion was further reduced to 0.008 in.

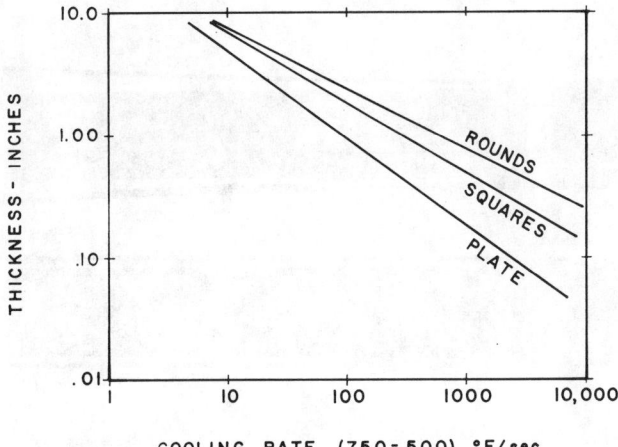

Fig. 5. Effect of geometrical shape on cooling rate when quenching with 70 F water.

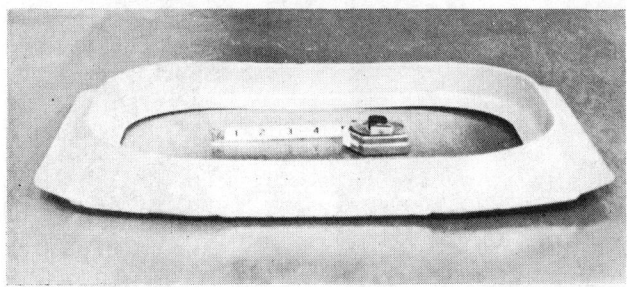

Fig. 6. Aircraft window-frame forging made from 7075-T6. Normal movement during glycol quenching is 0.010-0.015 in.

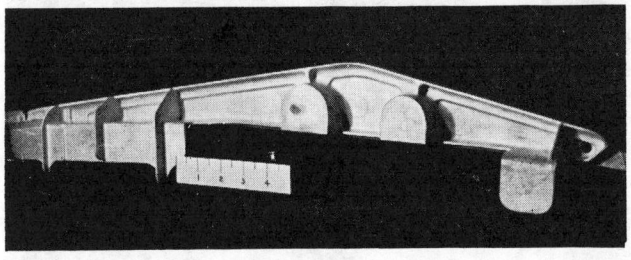

Fig. 7. Test die forging of 7075-T6. Distortion was reduced from 0.125 to 0.008 in.

Fig. 8. Formed pod of 7075-T6 sheet. Sheet is 0.125-in. thick by 92-in. long.

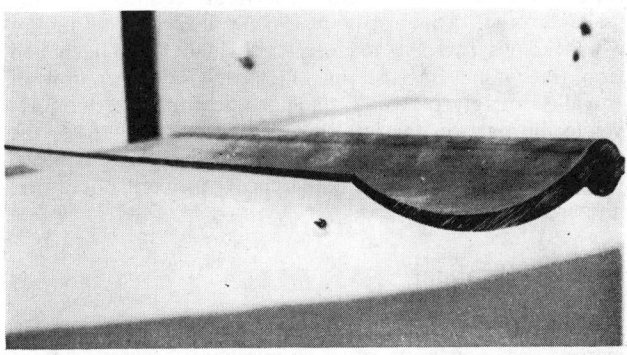

Fig. 9. Prototype extrusion of 7075-T6, 12-in. long.

Figure 8 shows a 92-in. long pod formed from 0.125-in., 7075 plate which. When quenched by standard techniques, it distorted so badly that it could not be salvaged. In quenching this part in Ucon Quenchant A, an acceptable part was achieved.

Extruded parts have been successfully produced. A 7075 extrusion maintained a 0.003-in. contour tolerance in a prototype operation, Fig. 9. A 7075 machined hinge extrusion had a 0.015 flatness tolerance and a 0.003 hole tolerance maintained during production quenching of 275 parts.

Possibly the greatest advantage of synthetic quenching is the close tolerance that can be achieved while heat treating machined parts. Extreme care must be exercised when attempting to heat treat them to ensure that the residual stresses are kept to a minimum and that no stress relief movement occurs. Figure 10 shows two machined bars. The top bar illustrates typical distortion which may occur in a normal machining/heat-treat sequence using a water quench. End movement of this part throughout the operation was 0.394 in., of which approximately 0.200 in. was due to the relief of machining stresses during solution heat treatment. A slight change in machining to lower residual stresses and proper utilization of synthetic quenching through selection of quenchant and control of the immersion profile, reduced total movement of the part on the bottom to 0.003 in. The valve body in Fig. 11 demonstrates the effectiveness of this technique in production operations. This part is an A-356 casting which required a complete reheat treatment after finish machining and threading of holes. The tolerances were held so that after quenching and aging to premium properties, the castings were reassembled and completed a full life-cycle test. A precision,

Source: Metals Engineering Quarterly, May 1971, 226-231

premium-quality casting of an electronic chassis has been heat-treated in production to premium properties, maintaining low distortion. By proper selection of quenchant concentrations, synthetic quenchants can be used not only for standard heat-treat cycles of aluminum castings, but can also be used in achieving premium-quality properties as required by MIL-A-21880.

Controlling distortion in welded-and-brazed assemblies has been extremely difficult. Through proper utilization of concentrations as high as 60%, tight tolerances can be maintained. A complicated welded/brazed chamber was successfully heat-treated maintaining a rigid tolerance callout of ±0.005 in., Fig. 12. A dip-brazed chassis housing, Fig. 13, is illustrative of the point that closer tolerances and a more economical end product can be achieved by heat treating after machining is complete.

The parts shown here illustrate the various forms, aluminum alloys, and configurations to which synthetic quenching can be applied. In all cases, check and straighten operations cannot be completely eliminated, but they can be significantly reduced.

The quenchants are applicable to all forms of aluminum products and most section sizes. Every imaginable form, product, and thickness from 0.020 to 3.75 in. has been successfully quenched, with completely acceptable results. However, their use must be coupled with an understanding of their application, technical data to select products and concentrations, and precision heat-treat techniques in the furnace room. This last point can be illustrated by fig. 14 which shows two sheetmetal parts which were quenched from the same furnace into the same concentration of polyalkylene glycol using standard heat-treat techniques. While the part on the left showed acceptable flatness, the part on the right was severely distorted. The difference was attributed to subtle differences in controlling the immersion profile.

Steel Applications

Applications of synthetic quenchants to steel are numerous, as a variety of cooling characteristics can be achieved. For most applications, direct substitution can be made for oil quenching. Figure 15 shows some cooling curves obtained by quenching 1065 steel into various quench media after austenitizing at 1400 F. Specific applications at PMI have been limited to three laboratory-size cases:

1. Quenching of dual-faced armor plate samples which resulted in full hardness of both faces and no quench cracking.
2. Quenching of 17-4 pinions to achieve full hardness following machining operations.
3. Quenching of 1065 street sweeper-broom segments to achieve hardness and ductility.

The armor plate was a sandwich structure of two carbon grades of a Cr-Ni-Mo steel. The lower-strength, high-toughness face was made from a 0.20 carbon alloy; while the higher-strength, hard face was made from a 0.60 carbon alloy. Because of the different hardenabilities and martensitic transformation ranges, difficulty was reported in normal heat treating. Either severe quench cracking occurred or low properties were developed on the low-carbon side. By proper utilization of synthetic quenching, full

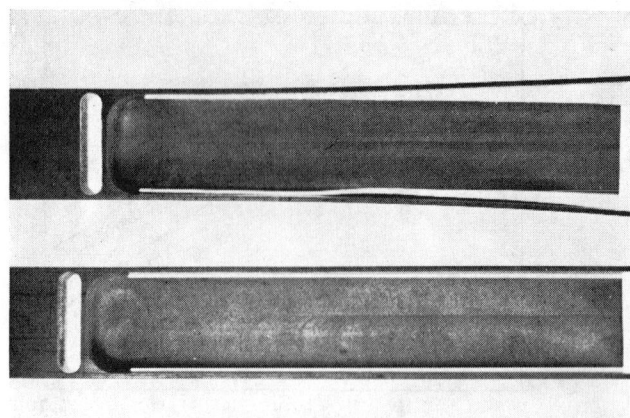

Fig. 10. Machined bars of 7075-T6 that have been quenched in water and 20% UQA.

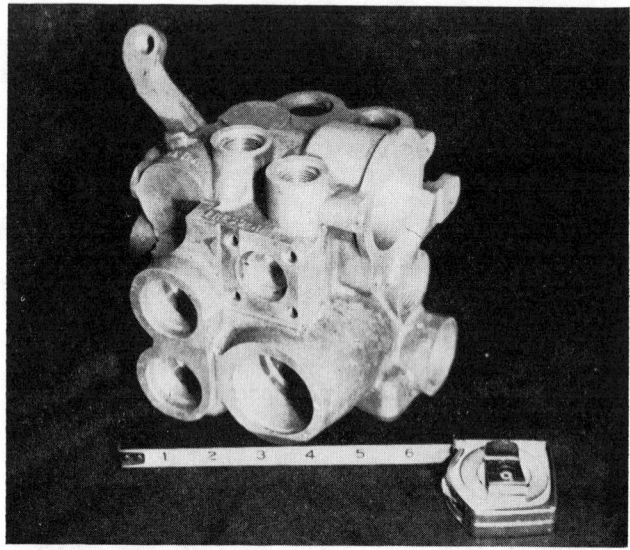

Fig. 11. Machined valve body of A-356, heat-treated in fully machined condition.

Fig. 12. Welded/brazed altitude chamber of 6061 which held a ±0.015-in. heat-treat tolerance.

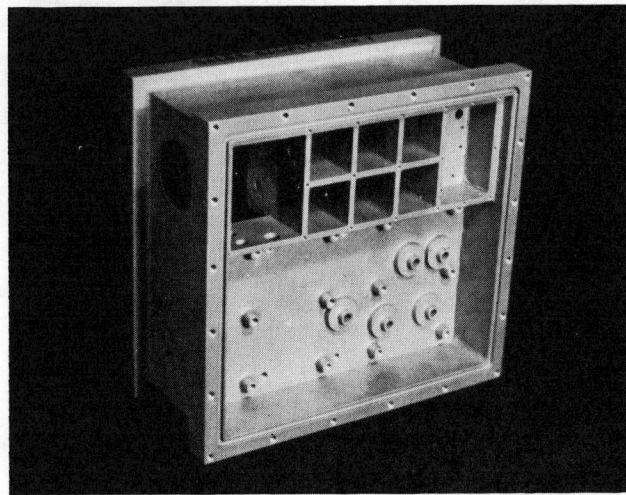

Fig. 13. Dip-brazed chassis housing of 6061 heat-treated after machining.

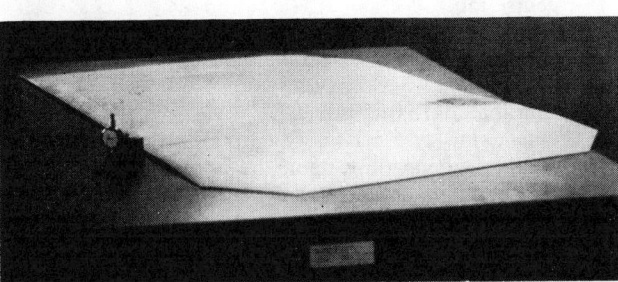

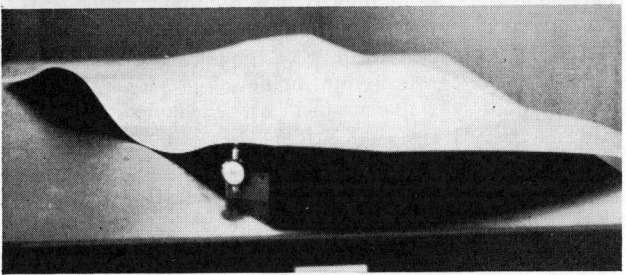

Fig. 14. Two glycol-quenched parts illustrating the need for technical understanding and precise techniques.

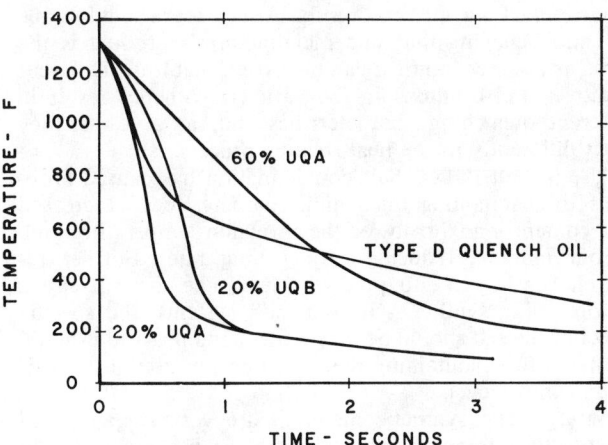

Fig. 15. Cooling characteristics of 1065 steel wire quenched in Ucon Quenchant A, Ucon Quenchant B, and a standard quenching oil.

hardness was achieved in both faces with no quench cracking in evidence.

Because of the relative quench insensitivity of the 17-4 alloy, high concentrations can be used with good results for the pinions.

The 1065 street sweeper-broom segments were quenched after developing cooling rate data which was shown in Fig. 15. Good results were obtained by time-quenching into a 60% concentration and direct-aging in a lead bath, achieving a hardness of Rc 40-44. However, a noticeable difference, particularly in the ductility of the wire, was observed between the developed technique using synthetic quenchants versus a normal oil quench. This was true despite the fact that full hardness of Rc 60-62 was developed on initial quench regardless of the quenchant used. This was initially explained by the difference in the shape of the cooling curve for the different quenchants used.

Further effort is planned to positively determine the cause of the ductility difference.

Titanium Applications

Synthetic quenching is applicable to any quench-sensitive material. Applications to titanium alloy is particularly appealing, especially alloys 6-4 and 6-6-2 which are normally water-quenched.

The problem with utilizing synthetic quenchants is that little reliable critical-temperature-range or quench-sensitivity data is available from which to preselect proper quenching concentrations and temperatures for different alloys. Thus, basic studies of cooling characteristics and continuous-cooling transformation phenomena are necessary before extensive use of synthetics can be used in titanium heat treatment. PMI plans to initiate some of this work in the near future.

Limitations of Synthetic Quenchants

Like most beneficial tools, synthetic quenchants have their disadvantages, and these should be recognized and weighed before deciding on their use. Some of these are as follows:

Quenchant Cost. Cost can be high, particularly when compared to water-quenching. A thorough study of the potential savings must be made to justify conversion of an existing system. In analyzing the cost of various synthetic quenchants, the final decision should be made on the basis of cost per quenching power or cost per distortion reduction, not cost per gallon.

Lack of Technical Data. Because of the complexity of the quenchant system when coupled with materials of different quench sensitivities, data must be available to the user. Some data is available in the literature, and some is available from the manufacturers; but in most cases this is not adequate to establish a complete system or to ensure proper controls in a system. Tremendous results can be achieved, but by utilization of an improper concentration, improper control of temperature, or incorrect immersion profile, inadequate properties, intergranular corrosion,

Source: Metals Engineering Quarterly, May 1971, 226-231

quench cracking, etc., can result. Also, a user should not be led into believing that one manufacturer's product is the same as another's until it can be proven that it is. Different additives and dilutions in the various synthetics result in different quenching characteristics and, as a result, they react differently in the heat-treating process.

Use in Salt Baths. Salt contamination has caused problems in maintaining quenching characteristics. Increased salt content tends to lower the inversion temperature and, in our opinion, reduce the quenching rate. Further research is necessary to precisely determine the salt effect. Before using synthetics in salt bath systems, the specific effect of the salt should be determined, and provision should be made for maintaining the salt concentration at a maximum control level.

Safety. The synthetic materials are very slippery, and safety precautions must be taken in the heat-treat area to avoid accidents.

Flaming Problems. Although no ignition problem exists during quenching, the material will burn if reloaded into a furnace. As a result, use of hollow-tube baskets and spacers should be prohibited, and baskets should be adequately rinsed before re-entry into the furnace.

Rinsing. Adequate rinsing facilities for parts must be provided after quenching.

Contamination. To ensure proper life of the quenching bath, a filtration system should remove errant particles. This is particularly true if used in aluminum forging applications where forging lubrications are left on the part during heat treatment. Excessive build-up of contaminants in the bath can vary quenching characteristics and can even cause corrosion problems.

Temperature Control. A cooling tower or other means of cooling is necessary to cool the tank. Tank heating is usually not necessary and is, therefore, an excessive expense. If slower quenching rates are required, these can be achieved by higher concentrations of the synthetic quenchant.

Complexity. A complete system usually requires more than one concentration. We recommend three to four. This entails duplicate tanks or furnaces, or a means of changing a tank concentration rapidly.

Direct Substitution. The material is not a direct substitute for oil quenchant. As yet, exact cooling characteristics cannot be duplicated. By varying the temperature and agitation parameters of the quenching operation, an adequate substitute can be achieved in most cases, but there are certain cases where a glycol quench cannot achieve the same results as an oil quench.

Summary

Use of synthetic quenchants allows for the first time a truly engineered heat-treat system. By utilizing the ability to preselect a desired cooling rate based on quench sensitivity and geometry, desired engineering properties can be achieved with the following advantages:

- Significant reduction in distortion.
- Reduction in residual stresses.
- Reduced need for spray-quenching.
- Elimination of need for cryogenic quenching systems.
- Elimination of preheated quench tanks. Slower cooling can be achieved by increasing the concentration.
- Simple water rinse is required to clean parts, baskets, and fixtures. No detergents or water-soluble materials are necessary.
- The material is nontoxic.
- Quenching systems may be used with or without agitation, depending upon the application.
- Reduced fire, smoke, and other pollution compared to oil-quenching.
- Reduced total cost of end item.

One point must be re-emphasized. Synthetic quenchants are simply a tool, not a panacea. Their usage must be understood, and coupled with good engineering and shop practice. However, results not achieved previously in the heat-treat industry can be attained, with the added benefit of significant cost savings. As further data is developed and products are introduced, synthetic quenchants will find many new applications, thus making their future in the heat-treat industry very bright.

Reclaiming quench oil–and $$– may be easier than you think

A relatively simple oil recovery system pioneered by a husband-and-wife engineering team has already saved one heat treater more than $20,000 in one year.

In the summer of 1966, Gene and Betty Brill were enjoying the sunshine in their backyard in a Cleveland suburb when they noticed a garden hose in their wading pool was accumulating some spilled tanning lotion off the water's surface.

Few people would have given this a second thought. But the Brills, both engineering graduates of the Case Institute of Technology, were inspired by what to them was a chance demonstration of a principle that could be applied to industry for the reclamation of oils. They developed the appropriate hardware, formed a company called Oil Skimmers Inc. in Cleveland, and began marketing their products to a wide variety of industries including heat treating, in which the skimmers can be used to collect quench oils.

About a year and a half ago, AMAC Enterprises Inc., a Cleveland-based company whose business is 40 percent heat treating (the rest is plating) incorporated the Oil Skimmers technique into a system to reclaim its K65 quench oil. Until then, up to 120 gallons a day of this valuable oil was going down the drain after being separated from the wash water used to clean the parts after quenching. The water was reclaimed, but the used oil was discarded and had to be replaced with new oil that cost $2.60 a gallon, resulting in an annual bill of $60,000. By reclaiming the old oil from the wash water, and then having it re-refined, the com-

First of two Oil Skimmers 1500-gallon holding tanks for wash water installed at AMAC is at rear; tank for picking up remaining oil is in foreground.

pany paid only $1.10 a gallon, a saving of $1.50. Within a year and a half, AMAC has saved $32,000 in new oil purchased.

The initial investment for the complete system was $25,000, $18,000 of which was material costs, the rest, labor.

"On an investment of this type, we generally like to see a payback in about a year and a half," says Tom Chimples, vice president and son of the founder, George Chimples, who launched the plating business in 1951

Conveyor belt at AMAC carries fasteners out of the quench tank; quench oil is washed off in a 10.5-pH cleaning solution heated to 190°F before fasteners enter tempering furnace.

and expanded into heat treating seven years later. "This system was paid off in nine months. From March 1982, when we brought the system on stream, to March 1983, we purchased 21,397 gallons of oil at the re-refined price of $1.10 a gallon. With the $1.50 savings, we only had to purchase 16,633 gallons to recoup our original $25,000 investment.

"The company that re-refines our oil (Drubo Oil Co., Cleveland) gets it back up to the same properties as the new oil and will provide a certificate to back up this claim," he says. "It doesn't have quite as long a life span, but thus far, we haven't had to make any new oil purchases."

The AMAC installation

The main components of the AMAC reclamation system are two 65-pound Oil Skimmers model 5-H units (see photo), which are mounted on two 1500-gallon holding tanks. These units use a floating tube system. The closed-loop tube, which is made of a specially formulated plastic, attracts oil but not water—mush as the Brills' garden hose accumulated the suntan lotion. The 5-H's automatic skimmer continuously draws the oil-covered tube through scrapers and returns the clean tube to the water surface to gather more oil.

All of AMAC's washers flow first into one of the holding tanks. About 80% of the oil is recovered here. The water then flows from the bottom of the first tank to the second, where the other 5-H skimmer picks up the remaining sheen.

The recovered oil is pumped to a 1000-gallon-capacity collection tank. Once a week, a truck from Drubo drives up to the tank and takes the oil away for re-refining. Later the oil is brought back to AMAC for reuse. About 800 gallons a week are recovered in this manner.

To ensure that virtually no water is carried along with the recovered oil to the collection tank, the system includes a small intermediate tank from which 20% of the contents are purged from the bottom and pumped back into the first holding tank.

(Almost) no soap

In addition to the money saved from oil reclamation, AMAC's system is cutting the cost of water, gas and soap.

The water bill is cut because more wash water can be reclaimed with the new installation than was possible with the previous procedure. After separation from the oil, the water is simply pumped from the second holding tank back to the washer for reuse. As a result, less fresh city water has to be pumped in.

The cost of gas is lowered because the wash water that goes through the oil recovery system cools down only to 120°F, and thus requires less heating than new water to reach the 190°F necessary for washing.

The soap bill was reduced because the wash solution now requires only two to three quarts of alkaline-base cleaner every week. By comparison, the greater inflow of fresh water prior to the skimmers' installation necessitated the addition of two quarts daily.

AMAC says the use of Oil Skimmers equipment fits in well with management's current strategy of adding equipment that will cut operating costs without sacrificing quality. AMAC has found that in addition to saving oil, water, natural gas and soap, the system itself is simple and easy to maintain. In addition, Oil Skimmers says, its units are engineered to have an operational life of at least 15 to 20 years.

AMAC's strategy seems to have paid off. If business continues at the pace set in the first half of 1983, which was 32% over the first half of last year, the current year will end approximately 35% ahead of 1982, according to Chimples.

Presently Oil Skimmers units can be found in steel and aluminum mills; large manufacturing operations and other heavy industrial producers; food, chemical and oil processors; and the rail, air and trucking service industries. These systems remove not only oils but also grease and floating sludge from settling ponds, drainage ditches, scale pits, sumps, central coolant systems, tanks and vats. **HT**

SECTION XIII
Energy Conservation, Safety, and Ecology

HEAT RECOVERY IN THE STEEL PROCESSING INDUSTRY

By

JAY T. WARE
C-E Air Preheater
Wellsville, New York, U.S.A.

Introduction

With increasing fuel costs and the restriction on fuel supplies throughout the world, interest in conserving heat energy has risen dramatically. Today, almost every energy consuming facility or manufacturing operation has either an individual or committe assigned to study ways to conserve energy and reduce fuel costs. In many cases the recovery of waste heat from process exhaust streams represents one of the more significant ways in which energy can be conserved.

This paper discusses two types of recuperative type heat exchangers as manufactured by the C-E Air Preheater Company and their application in the steel and aluminium process industries.

A brief summary of the benefits of utilizing waste heat recovery equipment to preheat combustion air in these processes is as follows:

1. A reduction in furnace fuel rate of up to 35 percent can be achieved with 1000°F preheated air. The curves included for both oil and gas as fuel illustrate savings as a function of furnace outlet temperature and preheat for both Hazen and Cor-Pak recuperative type heat exchangers (Figure 1).

2. Furnace production rate can be increased due to improved combustion kinetics and increased heat transfer to the product.

3. It is widely accepted that preheating combustion air enhances the burning of byproduct fuels and heavier grade fuel oils. As fuel quality deteriorates, air preheating becomes more important since it improves burner flame pattern and leaves less unburned carbon on product and furnace walls.

General Review of C-E Air Preheater's Recuperative Type Heat Exchangers

Hazen Metallic Recuperator: The Hazen metallic recuperator has been applied on furnaces with 434 million BTU per hour total capacity; 8.5 million BTU per hour is the practical minimum heat rate for application of this recuperator to a furnace.

The Hazen metallic recuperator is a gas-to-gas heat exchanger that has been widely used for 37 years in the U.S. steel industry for high-temperature heat recovery applications. In 1943, the first Hazen unit was installed in the United States on a slab mill furnace at an 80-inch billet and slab reheat steel mill at Gary, Indiana. Since that time we have installed more than 200 units on soaking pits, along with another 100 units installed on other types of furnaces. This recuperative air heater design has been used satisfactorily in the steel and aluminium industries.

The basic Hazen recuperator element is a tube-within-a-tube. Banks of tubes are arranged vertically or horizontally in the waste gas stream and appropriate air ducts are integrated into the design (Figure 2). The tubes are attached at only one end so that each tube is free to expand or contract with temperature changes No expansion joints or bellows are necessary. Air to be preheated enters the cold air chamber and flows down through the plenum between the walls of the inner and outer tubes. A formed end at the bottom of the outer tubes directs the air back up through the inner tube (Figure 3). The preheated air moves then through a chamber and to connecting ductwork. Hot flue gas flows over the outside of the outer tube. Multiple pass flow arrangements can be provided to include cross flow, simple counter-or-parallel-flow, and cross/counter-or-cross/parallel-flow. Additional variations can be obtained by air mixing and designing for multiple passes. Tube air flow area is optimized to achieve desirable operating metal temperatures and to maintain high convection coefficients in the annular space.

Depending upon the operating conditions, the Hazen recuperator may vary in length from a unit approximately three feet to one that is 26 feet in the direction perpendicular to gas flow. Each tube set is suspended from a common air plenum box. The boxes are designed so that an upper section can be unbolted and removed from the lower section to allow access for mantenance (Figure 4).

The overall efficiency of the Hazen recuperator may range from a maximum value of 75 percent to one that is as low as 43 percent. The pressure drops may range from 2.5 inches W.G. to 20 inches on the air side. The gas side pressure drop range may be from .15 inches W.G. to 2 inches maximum. Fuel savings may range from 15 to 35 percent depending upon the equipment selection and the operating conditions imposed on the recuperator.

Presented at the International Seminar on Industrial Furnaces, Ranchi, November 1980.

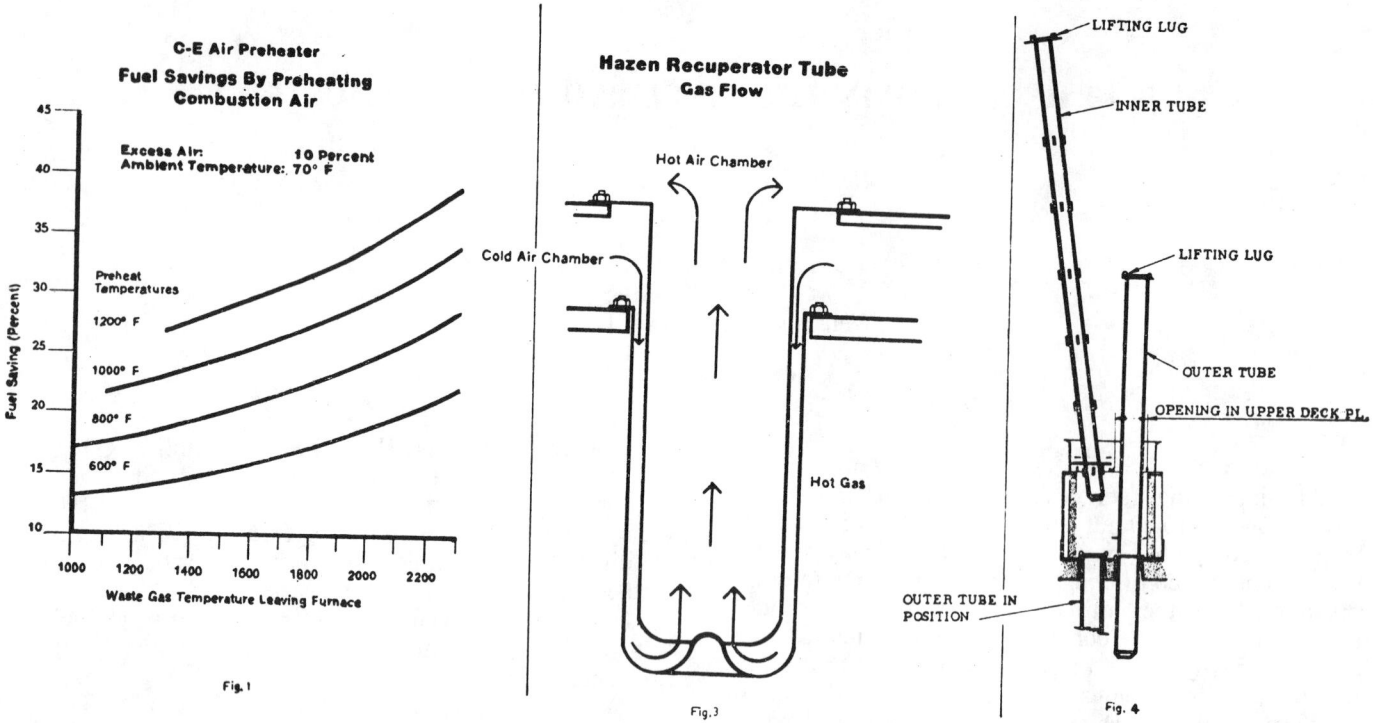

Fig. 1

C-E Air Preheater
Fuel Savings By Preheating Combustion Air

Excess Air: 10 Percent
Ambient Temperature: 70° F

Preheat Temperatures
1200° F
1000° F
800° F
600° F

Fuel Saving (Percent)

Waste Gas Temperature Leaving Furnace

Fig. 3

Hazen Recuperator Tube
Gas Flow

Hot Air Chamber
Cold Air Chamber
Hot Gas

Fig. 4

LIFTING LUG
INNER TUBE
LIFTING LUG
OUTER TUBE
OPENING IN UPPER DECK PL.
OUTER TUBE IN POSITION

Material selection is based primarily on obtaining the best materials to comply with acceptable corrosion rates and limiting creep stresses based on the specifics of each job. Directly comparable process or analogous process experience is always factored in. This experience is based on 37 years of service with more than 300 applications, including units where tube substitutions or replacements have been made. Basic to proper material selection is the availablity of accurate fuel analysis data and data on the presence of gaseous and solid process constituents.

Tube materials are varied along the length of the flue gas passage in accordance with the gradient condition of the gases as to temperature, entrainment, and ash or chemical attack, if present. More than 20 different materials are available to satisfy a wide range of design requirements.

The Hazen metallic recuperator is a radiation/convection heat exchanger with maximum gas entering temperature limited to 2400°F. The heat transfer components are 40 per cent radiation and 60 percent convection at 1400°F and shift to 58 percent radiation and 42 percent convection at 2200°F. During a typical furnace start-up, convection is most significant during the first 15 minutes of the cycle until gas inlet temperatures reach about 1800°F. Radiation then becomes the pre-dominant factor for the remainder of the heating cycle.

Hot-zone tube metal temperature is monitored by a thermocouple embedded in an inner tube wall, protected from the sever conditions encountered by the outer tube wall. The thermocouple is used to actuate a dilution or bleed system for temperature control (Figures 5 and 5a).

Hazen recuperator tubes are flanged and bolted to the header boxes so that they can be replaced iindividually in a short time. Tubes can be rotated for a nearly double service life if surface impingement reduces wall thickness. If fuel or operating conditions change, substitute tubes can be easily added for optimum operation in the new atmosphere.

Cor-Pak Recuperator: In 1960, C-E Air Preheater introduced the Cor-Pak heat exchanger to the industrial market as a lightweight and compact high-temperature (up to 1500°F) heat exchanger (Figure 6). The Cor-Pak metallic recuperator has been applied on furnaces with a total apacity of 231 million BTU per hour. The minimum firing rate is 10 million BTU per hour. More than 200 units are in service with about 75 percent of the applications on thermal oxidizers for hydrocarbon emission control and indirectly fired air heaters for various processes.

The Cor-Pak recuperative tubular heat exchanger has one-inch by five-inch corrugated rectangular tubes offered in standard lengths of five and eight feet. The tubes are spaced one-half inch apart across the flow and two inches apart normal to that and are end-welded into membrane tube sheets. These tube sheets are two tubes high and the width varies to accommodate specific designs. Spacer bars at the tube midpoint maintain spacing and prevent vibration. Normally, hot flue gas flows over the tubes and air flow is through the tubes (Figure 6). The tube arrangement can be either horizontal or vertical. Tube sheet sealing and casing insulation methods vary depending on the operating temperature and allowable leakage. Internal insulation is used for high temperatures and a low-leakage packed joint or seal-welded expansion loop seal can be provided. Tube material is varied along the gas path based on the metal temperature in each zone.

**Hazen Tube Metal
Temperature Control**

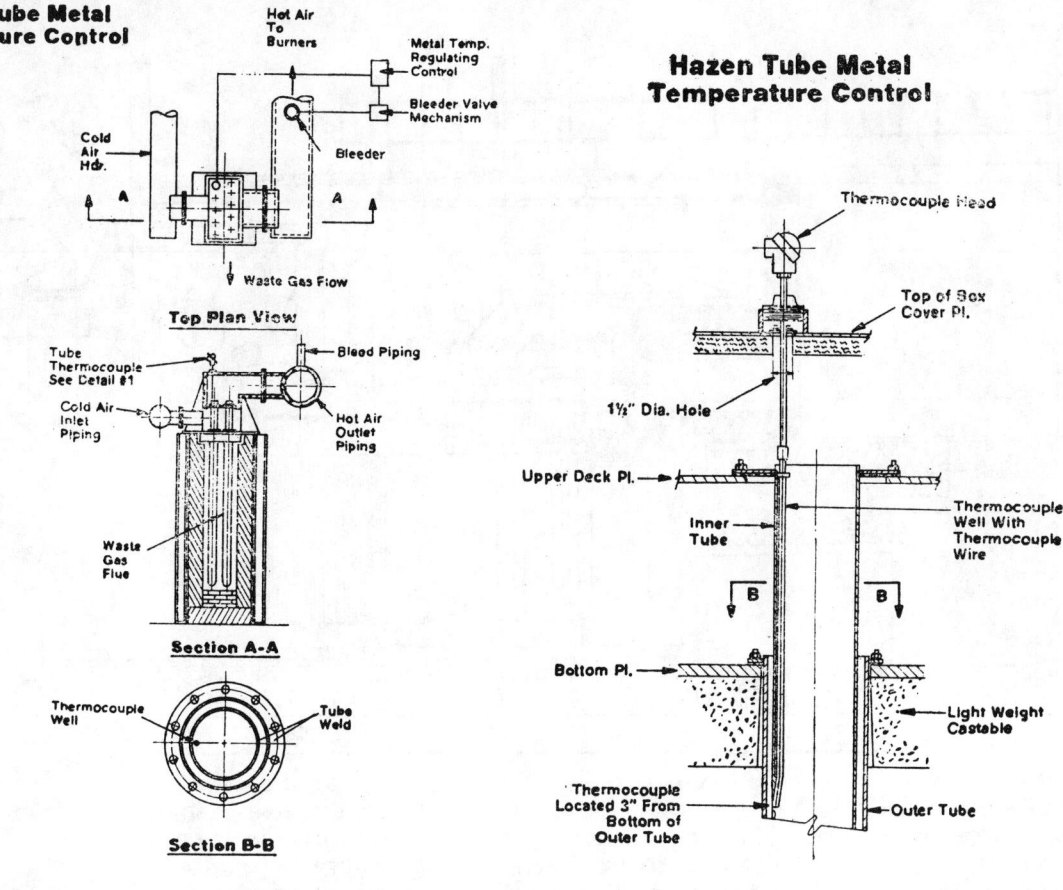

**Hazen Tube Metal
Temperature Control**

Fig. 5

Fig. 5a

The overall efficiency of the Cor-Pak heat exchanger ranges from 45 to 65 percent depending upon the operating conditions of the furnace. The fuel saving that may result from using this type heat exchanger on an industrial furnace is approximately 25 to 30 percent.

Typical Furnaces that use Heat Exchangers

Today's market is constantly changing because of the ever increasing cost of fuel. As the price of fuel goes up, a greater interest in the use of heat recovery equipment is being generated in a wide variety of industries. The newer furnaces now being looked at include box, car-bottom, and rotary furnaces.

These are in addition to the furnaces that have been the main target for recuperator installations for years. These include walking beam, slab heating and reheating furnaces, as well as soaking pits and heat treating furnaces in general.

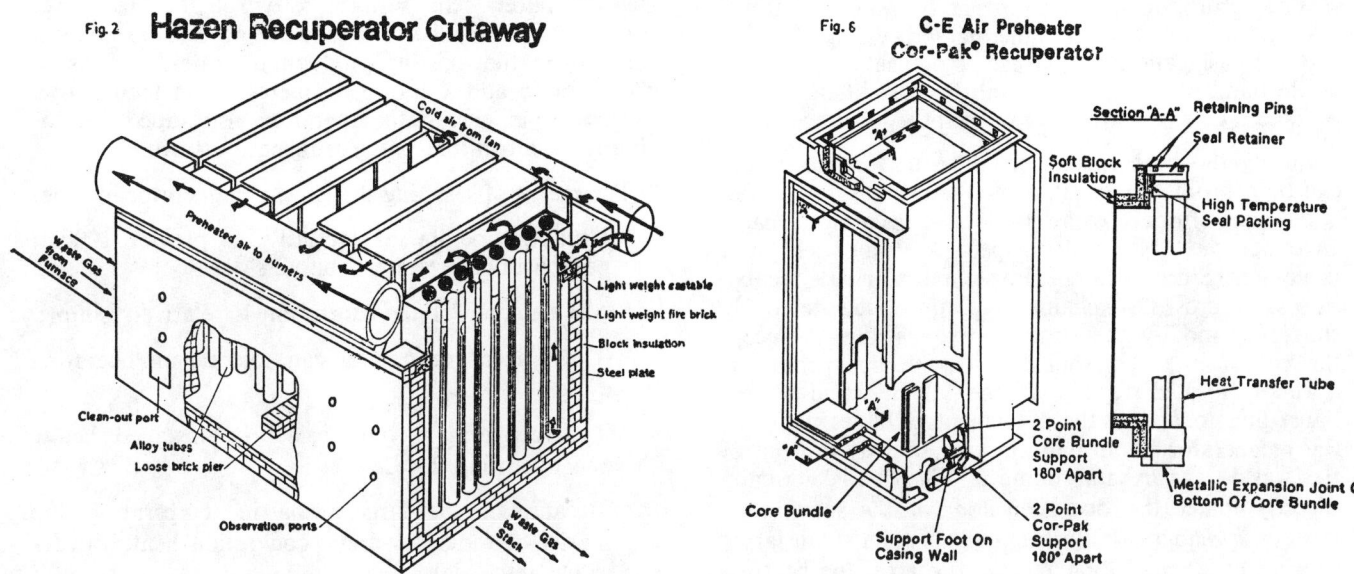

Fig. 2 **Hazen Recuperator Cutaway**

Fig. 6 **C-E Air Preheater
Cor-Pak® Recuperator**

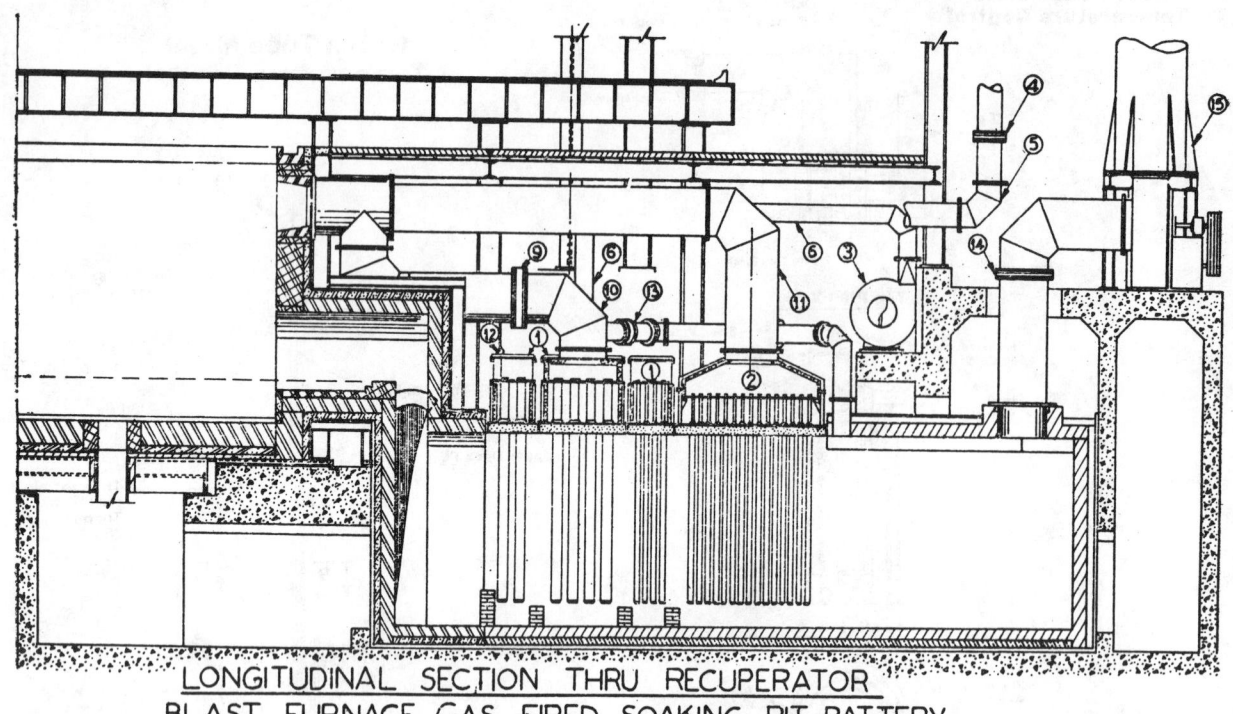

LONGITUDINAL SECTION THRU RECUPERATOR
BLAST FURNACE GAS FIRED SOAKING PIT BATTERY

Fig. 7

① - AIR RECUPERATOR
② - GAS RECUPERATOR
③ - COMBUSTION AIR FAN
④ - SHUT-OFF VALVE
⑤ - COLD B.F. GAS PIPING

⑥ - COLD AIR PIPING
⑦ - BUTTERFLY VALVE
⑧ - WARM AIR HDR.
⑨ - FLOW CONTROL VALVE
⑩ - HOT · AIR PIPING

⑪ - HOT B.F. GAS PIPING
⑫ - BLEEDER CONTROL THERMO.
⑬ - BLDR. CONT. VALVE & PIPING
⑭ - DAMPER
⑮ - STACK AND FAN

The industries that have traditionally recognized the advantages of heat recovery equipment are primary metals industries, ie., steel or aluminium. Potential applications for the Hazen and Cor-Pak heat exchangers include:

1. Glass .... batch and continuous process furnaces
2. Fibreglass .... continuous process furnace
3. Iron ore pelletizing.... continuous process furnace
4. Grey iron cupolas..... furnace burner preheating
5. Materials processing ··· thermal processing
6. Coal gasification .... reactor preheat
7. Portland cement .... alkali bypass system
8. Zinc .... reverberatory furnace

Fitting the heat exchanger into the furnace system can be a problem regardless of whether the installation is on a new furnace or a retrofit to an existing furnace. In either case, plant floor space is always held at a premium. Since both the Hazen and Cor-Pak recuperators are cross-flow channel type units, it is relatively simple to modify the furnace exhaust flue to accept the equipment. On soaking pits, the recuperator is normally located directly in front of the waste gas exhaust flue to accept the equipment. On soaking pits, the recuperator is normally located directly in front of the waste gas exhaust damper. The flue is normally directly under the operation floor of the soaking pit battery. Figugre 7 illustrates this point. On larger flow furnaces, including reheat furnaces, the heat ex-

changer is located directly obove the hearth line and furnace roof. This arrangement eliminates the involvement of additional plant floor space and takes advantage of the buoyancy of the hot gases as they exhaust to the stack. Figure 8 illustrates this point.

Operating History

Hazen Metallic Recuperator On An Alumnium Remelt Furnace:—The furnace to which the Hazen recuperator is applied had a capacity of 165,000 pounds. To expand the existing production rates, a decision was made to add a Hazen recuperator. In their efforts to increase efficiency, the operators also added fuel-to-air ratio control and pressure furnace control.

The merits of selecting the Hazen recuperator include:

* Unit can operate satisfactorily to provide 1000°F combustion air for the burners.

* Maintenance of the equipment is relatively simple.

* The Hazen recuperator can tolerate an operating cycle.

* The recuperator could easily be installed below the operating floor.

* Because of the large swing in temperature, the merits of radiation and convection heat transfer come into play.

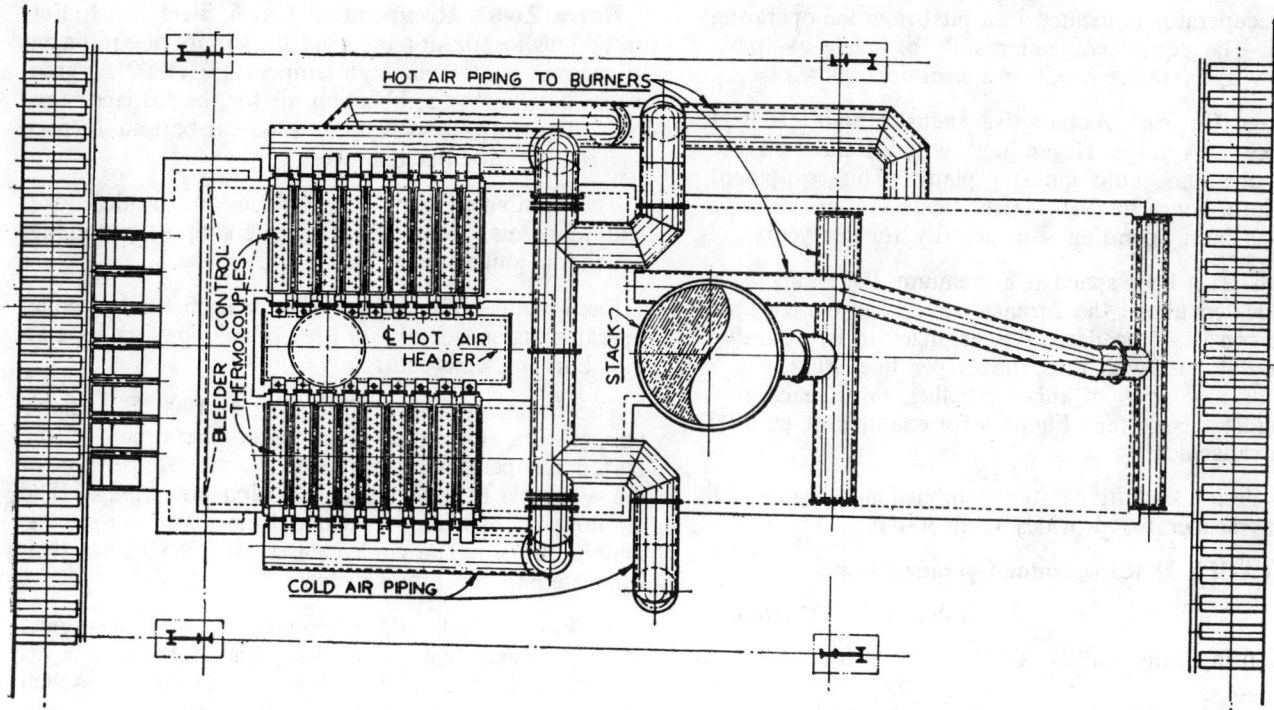

Fig. 9   5 ZONE - PUSHER TYPE - SLAB HEATING FURNACE
PLAN VIEW

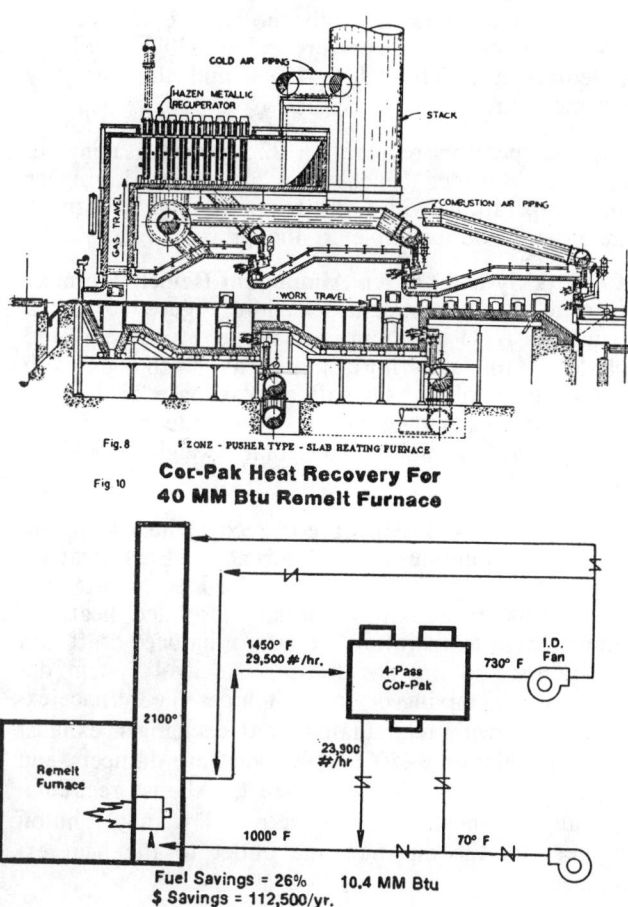

Fig. 8   5 ZONE - PUSHER TYPE - SLAB HEATING FURNACE

Fig 10   **Cor-Pak Heat Recovery For 40 MM Btu Remelt Furnace**

Fuel Savings = 26%   10.4 MM Btu
$ Savings = 112,500/yr.

(10.4 x $3.00 x 3,000 hrs.) - $93,600
(2.1 x $3.00 x 3,000 hrs.) - $18,900
5:1 Btu Turndown

TABLE I Recuperator Design Conditions

Thermal input to the burner	40 million BTU per hour
Equivalent thermal input (total energy)	48 million BTU per hour
Gas flow	7500 scfm
Air flow	7000 scfm
Air temperature, in	70°F
Air temperature, out	1060°F
Gas temperature, in	2200°F
Gas temperature, out	1508°F
Air resistance	7.5″ W.G.
Gas resistance	0.11″ W.G.
Fuel, natural gas	1000 BTU per cubic foot

When chlorine gas is used to flux the alloys produced, the heat exchanger is bypassed. A pneumatic damper directs the waste gas flow around the recuperator, directly to the stack.

Two methods of high-temperature protection are now in use on this recuperator. The first is a dilution air system that adds enough dilution air to the flue gas stream to lower the overall temperature of the gases flowing over the heat exchanger tubes. The second is a means of protecting the tubes from overheating by measuring the inner tube metal temperature. If the metal temperature exceeds the safe limits, a motorized damper in the hot air piping, operated by a controller will open causing hot air to bleed off. The added cooler air stream checks the excursion and protects the recuperator tubes.

The space required to house he recuperator is approximately 18 feet high by 20 feet wide by 24 feet long.

Source: Tool & Alloy Steels, May 1982, 145-151

391

The recuperator is located in a pit below the operating floor. The equipment can readily be made available for service by the removal of a steel deck plate.

Hazen On An Automotive Industry Slab Reheat Furnace: A large Hazen unit was installed on the furnaces at an auto industry plant. This equipment went into service in 1980. A duplicate unit at the same site has been operating satisfactorily for six years.

With plant floor space at a premium, the recuperator was located above the furnace roof. The two recuperators consist of header boxes arranged in two parallel horizontal flues with three boxes per flue. There is a total of nine rows of tubes per flue, with each row having 24 tubes. (See Figure 9 for example of parallel flue arrangement.)

At design conditions, the combustion air that the Hazen recuperator provides is at 800°F.

TABLE II Recuperator Operating Conditions

	Design	Maximum
Total firing rate, million BTU per hour	735	959
Hot air firing rate, million BTU per hour	672	868
Gas flow, scfm	139,702	182,278
Air flow, scfm	116,304	150,224
Temperature, air in, °F	70	70
Temperature, air out, °F	800	730
Temperature, gas in, °F	2,100	2,100
Temperature, gas out, °F	1,644	1,695
Gas resistance, "W.G.	0.52	0.86
Air resistance, "W.G.	10	16
Fuel: natural gas		

The recuperator unit occupies a space 21 feet high by 80 feet wide by 12 feet long.

The waste gases flow up a common uptake and enter parallel flues. A separate natural draft stack was provided for each flue.

It was necessary to provide a proportionately controlled bleeder, connected to the hot air duct between the recuperator discharge and the burners, for equipment protection. The bleeder has a one-point modulating control which is actuated by the thermocouple located in the inner wall of one of the tubes contacted by the hottest gases. One thermocouple is provided for each flue. The protection equipment insures that a sufficient quantity of air will flow through the heat transfer elements at all times to prevent overheating.

The unit's performance is based on flues arranged so that the waste gas is distributed of uniform velocity and temperature over the tubes. To insure this further, loose brick piers are located between each bank of tubes and extend across the full width of the flue.

Hazen metallic recuperators (11 units) have been installed to date on aluminum remelt furnaces with seven different U.S. companies.

Hazen Zoned Recuperators On A Steel Batch Furnace: This installation involved the supply of equipment to recover heat from high-temperature (1800°F) gases by preheating the combustion air for the furnace burners. The furnace is a three-zone car-bottom furnace and is used in a batch process.

The equipment supplied includes the header boxes and tube sets, along with the tube support panel into which the outer tubes nest.

Each recuperator system (one for each furnace zone) operates independently of the others, having its own fan, burners, and controls.

The furnace exhausts 1768 scfm at 1800°F. At this temperature, the gases flow over the tubes of the Hazen recuperator, transferring some 32 percent of the available heat to the combustion air supply. The air flow of 1610 scfm is heated from 70°F to approximately 800°F. The cooled gases, at 1285°F, are then exhausted to the atmosphere.

A series of tube thermocouples is used to protect the equipment from excessively high temperature (as referenced in Figures 5 and 5a). The control system allows a supply of cold air to flow through the recuperator and this flow must be maintained at all times. This means that when the burner is off during the cycle, the heat exchanger must be protected by passing enough cold air flow through the heat exchanger to keep the air outlet temperature below 830°F until the gas temperature falls below 450°F and the fan may be turned off.

The recuperators represent a 22.5 percent savings in fuel usage. Added benefits are also noted in higher flame temperatures with a greater percentage of furnace heat input made available to the process work.

Cor-Pak System On An Aluminium Remelt Furnace: A typical Cor-Pak system selection (Figure 10) for a 40 million BTU per hour batch-type furnace would include a four-pass horizontal heat exchanger with five-foot long tubes. Overall approximate dimensions for the heat exchanger are 9.5 feet wide by 9.5 feet long by 5.5 feet high. The total weight would be about 4.75 tons.

Heat exchanger systems are to recover heat from the off gases of aluminium remelt furnaces. Each heat recuperation system consists of a Cor-Pak high-temperature (1500°F design), internally insulated heat exchanger, high-temperature (900°F) induced-draft fan with motor and hot gas dilution control system designed with a mixing device to temper the furnace exhaust gases with recirculated heat exchanger exhaust gases to maintain 1450°F. Included are dampers and operators for the I.D. fan exhaust, exhaust recirculation and combustion air bypass and manual shutoff dampers for the air inlet and outlet to the heat exchanger.

The function of the aluminium remelt furnace is to recover usable metallic aluminium from residual aluminum scrap. The process involves preparation, including fluxing and melting of the scrap. The melting process results in the emission of high-temperature exhaust. The customer requested that we review this high-temperature exhaust for potential heat recovery to reduce fuel usage in the furnace. The decision was that we could recover some of this energy by preheating the combustion air to the furnace. Review of the burner operation indicated that a maximum of 1000°F preheated air could be accepted; therefore, the Cor-Pak system was designed to provide this air preheat temperature. As shown on Figure 10, portion of the 2100°F gas is automatically controlled to 1450°F through dilution of the furnace exhaust gas with Cor-Pak heat exchanger exhaust gas recirculation. The combustion air temperature is also maintained automatically at 1000°F with an air bypass system.

Cor-Pak recuperator systems (nine systems) have been installed to date on aluminum remelt furnaces with two different U.S. companies.

The Cor-Pak was designed with the flexibility to rotate the heat exchanger 180° for installation on adjacent furnaces. Another feature of the Cor-Pak design has a bolted cover which allows removal of the tube bundle for possible cleaning or periodic tube inspection.

Cor-Pak Recuperator System On A Heat Treating Furnace: This application involves the supply of equipment to recover heat from high-temperature (2000°F) gases by preheating combustion air for furnace burners. The furnace is octagon-shaped and is used for annealing large steel rings.

The equipment supplied includes an exhaust-to-exchanger transition, a cooling section with two small alternate start fans, a 1500°F design Cor-Pak heat exchanger, a fan transition duct, induct-draft high-temperature fan with exhaust damper, hot start motor and drive, and the controls necessary to monitor and control the cooling air input and furnace hood pressure as induced by the exhaust fan.

The system operation takes 30,000 pounds per hour of gases exiting the furnace at 1950°F and cools them to 1500°F by forced mxing of the gases with 11,300 pounds per hour of ambient air. At this temperature. the gases are drawn over the tubes of the heat exchanger transferring approximately 35 percent of the available heat to the combustion air supply. This flow of 27,250 pounds per hour is heated from 70°F to approximately 850°F. The furnace gases, cooled now to 1000°F, are exhausted to the atmosphere. Control circuitry maintains a 1500°F gas temperature to the exchanger and operates the system exhaust damper to maintain a very slight positive pressure in the furnace outlet.

The system represents a 24 percent savings in fuel usage. Added benefits of using preheated combustion air include an increased rate of heat transfer due to higher percentage of the furnace heat input being made available for transfer to the metal.

Hazen and Cor-Pak are registered trade marks.

Fuel consumption in reheat furnaces can be reduced by improving insulation coverage, obtaining positive control of the furnace, adding a ceramic fiber veneer to the furnace walls, eliminating air infiltration, using preheated combustion air, adjusting the furnace thermal profile and charging billets while hot. Fuel-cost savings obtainable with these methods are described.

Improving energy utilization in steel reheat furnaces

L. Joseph Grafe, Sales Engineer, Bloom Engineering Co., Inc., Homewood, Ill.

INDUSTRIAL energy costs will be significantly higher in the years ahead and will continue to represent a substantial part of the total cost of steel mill end products. Consequently, prompt attention should be given to controlling and reducing fuel consumption.

This article examines ways of reducing fuel consumption in a large energy consumer, the steel reheat furnace, using a 3-zone pusher and a top-fired walking beam furnace as examples. The basic methods of saving energy that will be identified for these furnaces apply to other types of furnaces as well.

Base furnace conditions

It is assumed that certain steps have already been taken: the furnaces have a fuel-air ratio system which can hold 10% excess firing through the operating range; a reasonable furnace-pressure control system; a temperature control system for each zone of control; and, in the case of the pusher furnace, an annual outage at which time the water-cooled support pipes are fully insulated. In addition, it is assumed that efforts have been made to maximize hearth coverage.

The methods of reducing energy usage that will be reviewed are: insulation practice; positive control of the furnace; application of a ceramic fiber veneer to the walls; reduction of air infiltration by the use of doors that make a positive seal; use of preheated combustion air; adjusting the thermal profile of the furnace; and hot charging of the steel. The following assumptions are also made:

- Natural gas cost is $4/million Btu.
- Furnaces operate at 68% of the maximum capacity on a monthly average.
- Steel heated is 4½ in. sq x 30 ft long. The walking beam furnace allows a spacing/thickness ratio of 1.5 to 1.
- Pusher furnace has 1200 sq ft of water-cooled support pipe surface.
- Furnaces are 60 ft long x 33 ft wide.
- Maximum design production rate is 125 tons/hr for the pusher furnace and 100 tons/hr for the walking beam furnace.
- Fuel and production rates are based on charged tons.

A representation of the pusher furnace under the foregoing conditions, with 70% of the pipe insulation and typical temperature control, is shown in Fig. 1. The fuel rate for the operation shown is 2.37 million Btu/ton or $9.48/ton. All the data shown in the bar chart are in terms of gross input, ie, the net value for each component of the fuel rate has been divided by the furnace efficiency.

A representation of the walking beam furnace based on a

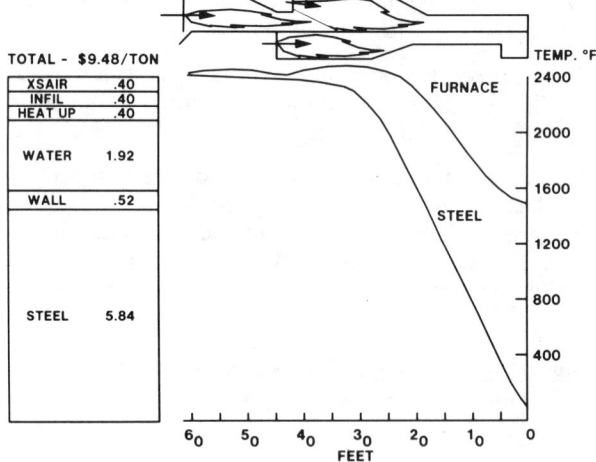

Fig. 1 — Pusher furnace. Base case: poor control; 70% insulation; 2.37 million Btu/ton.

typical operation is shown in Fig. 2. The monthly fuel rate for this operation is 1.97 million Btu/ton or $7.88/ton.

Energy conservation: Series I

Insulation — One simple but effective way to reduce the fuel rate in the pusher furnace is to increase the average amount of coverage of water-cooled pipe, either by making

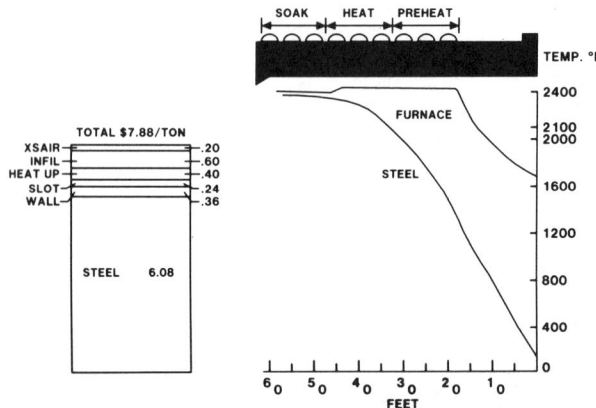

Fig. 2 — Walking beam furnace. Base case: poor control, 1.97 million Btu/ton.

Courtesy Association of Iron and Steel Engineers, extracted from Iron and Steel Engineer magazine, January 1985, 43-47

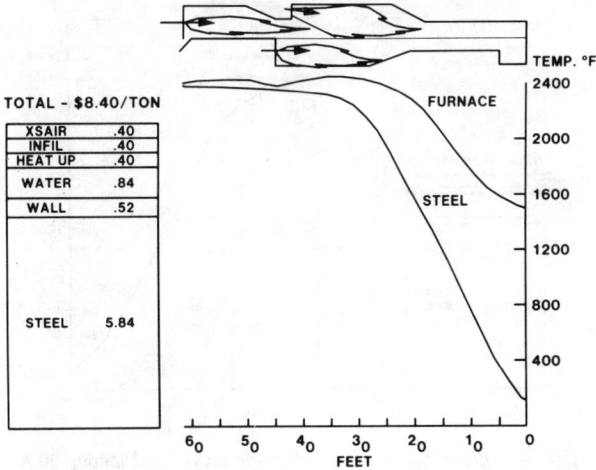

TOTAL - $8.40/TON

XSAIR	.40
INFIL	.40
HEAT UP	.40
WATER	.84
WALL	.52
STEEL	5.84

Fig. 3 — Pusher furnace. Condition: improved insulation, 90%; poor control; 2.1 million Btu/ton.

special shapes of pipe insulation to increase the initial coverage or by maintaining the insulation on a regular basis. The latter method is by far the most effective method. The amount of insulation coverage on a furnace that is insulated only once a year is 70% on the average, and in many cases, much less.

By increasing maintenance to a quarterly schedule and using special shapes, the average coverage can be increased to 90%. As shown in Fig. 3, regular maintenance of insulation results in a fuel cost savings of $1.08/ton, without modification to the furnace. The cost of the additional insulation and its installation can be written off in a month. This illustration was based on 1¼-in. thick material. The use of 2-in. thick pipe insulation or of low thermal conductivity material in the charge end of the furnace can further enhance the fuel savings. It should be noted, however, that the single most important factor in obtaining benefits from insulation is to maintain a high percentage of coverage.

Positive furnace control — A second improvement is the use of positive temperature control to heat the steel as late as possible. Poor furnace control leads to heating the steel early and wasting energy by creating high flue gas temperatures; because furnace operators do not want to get caught with cold steel, the furnace setpoint is maintained during delays, which is costly. However, in most cases, the operators do not have the tools to control the furnace correctly.

One way to help operators achieve more positive furnace control is by the use of a management discipline system that receives many inputs from the furnace operation, analyzes these inputs and adjusts the zone setpoints to optimize furnace efficiency. A management discipline system can be a set of predetermined recipes programmed into a microprocessor or a set of rules that the furnace operator follows. Relatively inexpensive systems are available for applying sound control logic to these furnaces without installing a complete computer control system.

Another method by which operators can achieve better furnace control is by controlling the furnace directly from the steel surface temperature. In a pusher furnace, a special optical pyrometer can be used in conjunction with a relatively simple logic system, which maintains the steel at the proper temperatures at the pyrometer sighting point and allows the thermal profile of the furnace to vary as required. Although the spaces between billets make it difficult to adapt this method to a walking beam furnace, attempts have been made to approximate this type of control on some walking

beam furnaces by locating the hot junction of the control thermocouples as close to the steel surface as possible.

The first method described is the most effective, and there is a strong tendency toward the use of process controllers; however, the second method can achieve good results on pusher furnaces. Direct steel temperature measurement requires only the use of a special optical pyrometer to replace zone thermocouples and a simple logic system.

The ultimate furnace control system uses a microprocessor dedicated to heating strategy coupled with a direct steel temperature measurement system to complete the control function. Generally, on a multi-zoned furnace, the cost of this type of system is approximately half the cost of a complete computer control system. The benefits derived from achieving positive furnace control are shown in Fig. 4 for the pusher furnace and in Fig. 5 for the walking beam furnace. In these examples, fuel cost savings of $1.20 and $1.00/ton, respectively, were obtained.

Ceramic veneer — A third area of possible savings is in the application of a ceramic fiber veneer to the hot face of the furnace walls. Ceramic fibers reduce the heat loss by conduction through the walls and decrease furnace heat-up time (by as much as 50%) because of their low thermal conductivity and low heat storage properties. The effect of veneering is shown in Fig. 6 for the pusher furnace (a savings of $0.40/ton) and in Fig. 7 for the walking beam furnace (a savings of $0.36/ton).

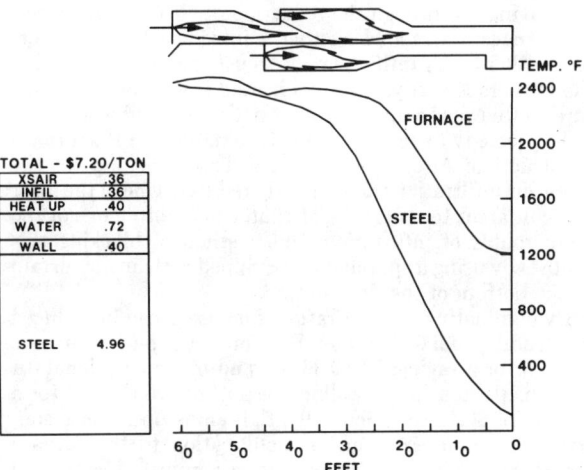

TOTAL - $7.20/TON

XSAIR	.36
INFIL	.36
HEAT UP	.40
WATER	.72
WALL	.40
STEEL	4.96

Fig. 4 — Pusher furnace. Condition: improved insulation, 90%; good control; 1.80 million Btu/ton.

Fig. 5 — Walking beam furnace. Condition: good control; 1.72 million Btu/ton.

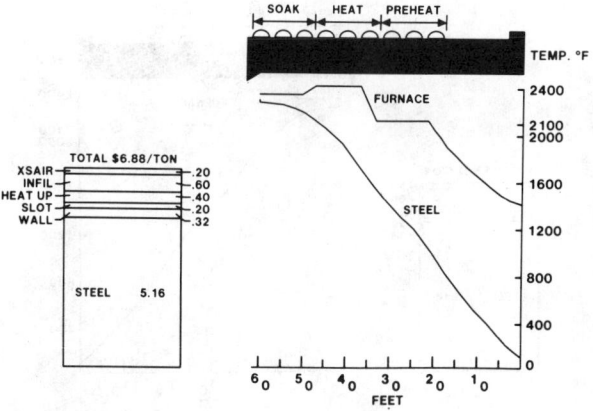

TOTAL $6.88/TON

XSAIR	.20
INFIL	.60
HEAT UP	.40
SLOT	.20
WALL	.32
STEEL	5.16

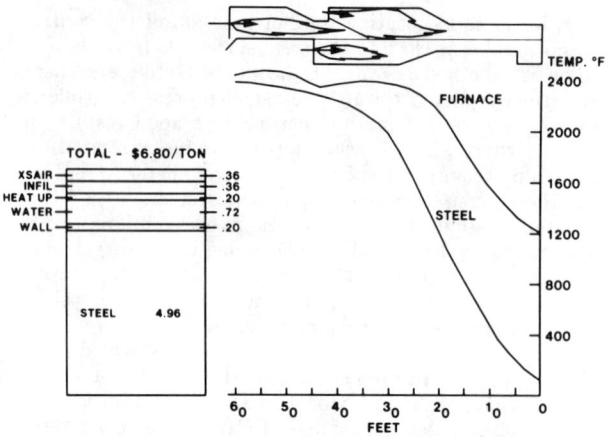

Fig. 6 — Pusher furnace. Condition: improved insulation, 90%; good control; veneer on walls; 1.70 million Btu/ton.

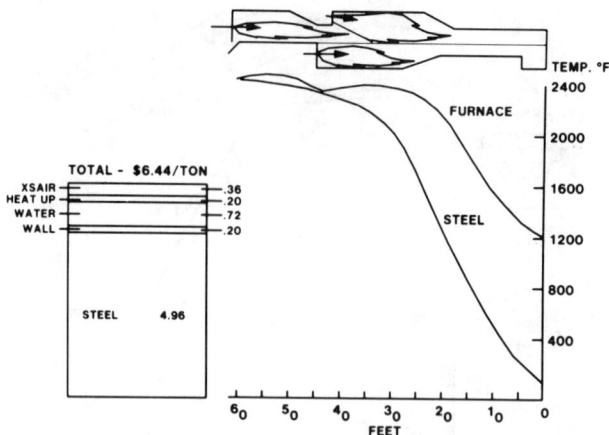

Fig. 8 — Pusher furnace. Condition: improved insulation, 90%; good control; veneer on walls; door seal; 1.61 million Btu/ton.

Fuel savings are also enhanced by veneering when used in conjunction with a positive furnace temperature system, ie, the low heat storage of the ceramic fibers allows the furnace profile to fluctuate readily as required by the control system. While this advantage is hard to quantify, it can be substantial.

Reduction of air infiltration — The fourth potential for fuel savings is through the reduction of air infiltration into the furnace. Extra cold air that is introduced into the furnace either by air infiltration or by using excess air in firing the burners is costly; the air is heated from room temperature to flue gas temperature but does no useful work.

The remedy to air infiltration is to tighten up the furnace construction. Also, by eliminating direct openings to the room, air infiltration can be greatly reduced. One of the most difficult areas to seal is the discharge door, often one of the main routes of infiltration. One method of blocking this route is by using a special door equipped with an air curtain at the bottom of the drop-out door.

By eliminating air infiltration, fuel usage can be reduced by an additional 0.09 million Btu/ton in the pusher furnace (Fig. 8) for a savings of $0.36/ton and by an additional 0.1 million Btu/ton in the walking beam furnace (Fig. 9) for a savings of $0.40/ton. Generally, tightening up furnace construction requires no major modification to the furnace structure and is relatively easy to accomplish. Installing a positive sealing discharge door can be a cost-effective way to reduce energy consumption.

By using special shapes of pipe insulation, increasing the frequency of maintenance of the insulation to achieve higher average coverage, using positive temperature control of the furnace, adding a ceramic veneer and eliminating air infiltration, the fuel rate can be reduced from 2.37 to 1.61 million Btu/ton for the pusher furnace and from 1.97 to 1.53 million Btu/ton for the walking beam furnace, representing savings of $3.04 and $1.76/ton, respectively.

Energy conservation measures: Series II

Generally, Series I conservation measures require no major modification of the furnace structure, are relatively easy to accomplish and can be cost effective in reducing energy consumption.

In addition to the Series I measures, there are three additional methods of reducing energy consumption: preheated combustion air; adjusting the furnace temperature profile to optimize fuel utilization; and hot charging of steel. For this discussion, it is assumed that the four Series I modifications have been made. Each method in Series II will be discussed as if only one of the three would be implemented on the example furnaces.

Preheated combustion air — Recuperation, or the use of preheated combustion air, has been used for a considerable period of time to save energy. In recent years, recuperator manufacturers have attempted to obtain higher effectiveness so as to achieve higher air preheat temperatures. The

Fig. 7 — Walking beam furnace. Condition: good control; veneer on walls; 1.63 million Btu/ton.

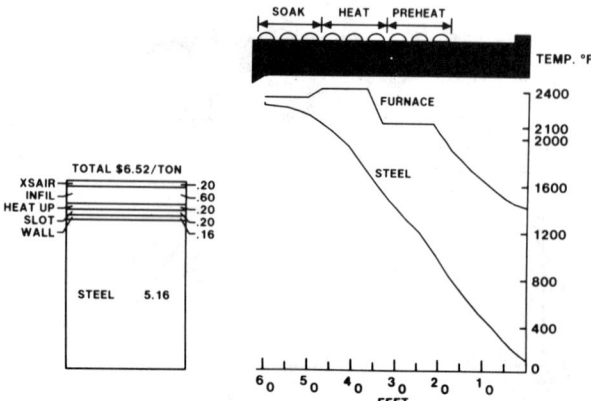

Fig. 9 — Walking beam furnace. Condition: good control; veneer on walls; door seal; 1.53 million Btu/ton.

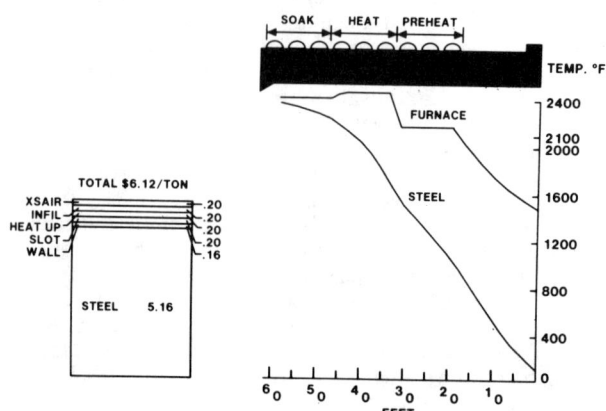

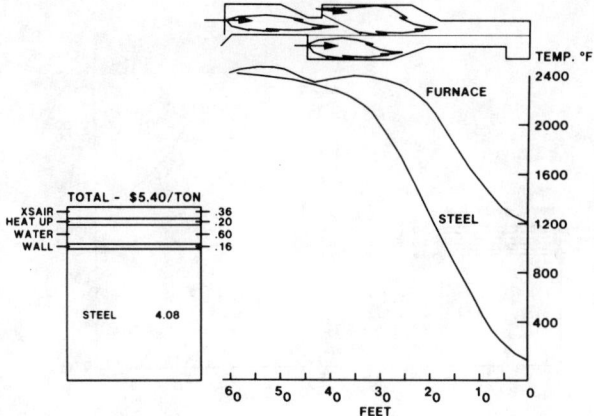

Fig. 10 — Pusher furnace. Condition: improved insulation, 90%; good control; veneer on walls; door seal; air preheat, 700°F; 1.35 million Btu/ton.

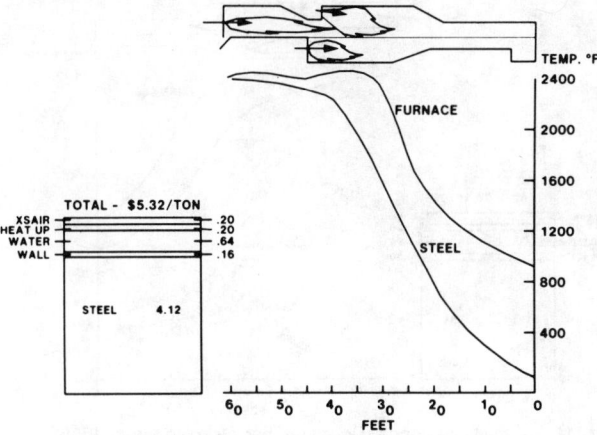

Fig. 12 — Pusher furnace. Condition: adjustable thermal profile burner with Series I modifications, 1.33 million Btu/ton.

higher the air preheat, the greater is the fuel savings. However, as air preheat temperatures exceed 800°F and approach 1200°F, the cost of installing such a system also increases because of higher cost for the high effectiveness recuperator, for the interior lined pipe or stainless steel ductwork that must be used and for burners to utilize the high air preheats.

For the most cost-effective installation, recuperators with an effectiveness in the 40 to 50% range were used in the two example furnaces, with 700°F air preheat delivered to the burners. Under these conditions, the fuel savings obtained were $1.04/ton for the pusher furnace and $0.92/ton for the walking beam furnace (Fig. 10 and 11). Fuel savings due to air preheat depend on the specific air preheat temperature and the overall thermal effectiveness of the furnace. Therefore, each furnace must be evaluated independently to determine the exact fuel savings.

Furnace temperature profile — The ability to adjust the thermal profile of the furnace can also yield substantial savings. The method of achieving this feature is different on the longitudinal-fired pusher furnace than on the roof-fired walking beam furnace. The concept, however, is identical; basically, the required heat is released as far from the flue as possible so that, in effect, the carry-over section of the furnace is lengthened. On the pusher furnace, this means installing burners capable of adjusting the flame length with the firing rate of the burner. A conventional long-flame burner generally fires with approximately the same flame length from high fire to 50% firing rate. The flame then

shortens with a decrease in firing rate. Also, the conventional burner usually loses good mixing abilities at approximately 25% of ratio firing rate, so it is biased excess air. The adjustable thermal release burner can be made to operate so that the flame length can be varied to meet the production requirements. Also, because this type of burner has better turndown capabilities than the conventional burners, the excess air at turndown can be reduced.

In the walking beam furnace, the adjustment of the thermal profile is obtained by shutting off rows of burners starting with those nearest the flue. This type of cascade control on a roof-fired furnace usually requires additional input and decision-making capabilities not found on the conventional furnace. However, during times of reduced production, the ability to adjust the thermal profile of the furnace can be rewarding; as shown in Fig. 12 and 13, fuel cost savings were $1.12/ton for the pusher furnace and $0.60/ton for the walking beam furnace, in the same range as the savings with the recuperator. Note that with cascade control on a walking beam furnace, the burners that remain on are firing at higher firing rates.

Hot charging of steel — In plants that have continuous casters or other means of obtaining hot steel billets, an inexpensive, but effective way of reducing fuel usage is to charge the billets into the reheat furnace as soon as possible; the thermal energy stored in the billets can greatly reduce the fuel consumed in the reheat furnace to reheat the steel to rolling temperatures. For example, when 1000°F billets were

Fig. 11 — Walking beam furnace. Condition: good control; veneer on walls; door seal; air preheat, 700°F; 1.30 million Btu/ton.

Fig. 13 — Walking beam furnace. Condition: cascade control with Series I modifications, 1.38 million Btu/ton.

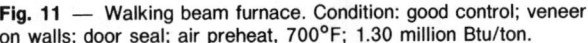
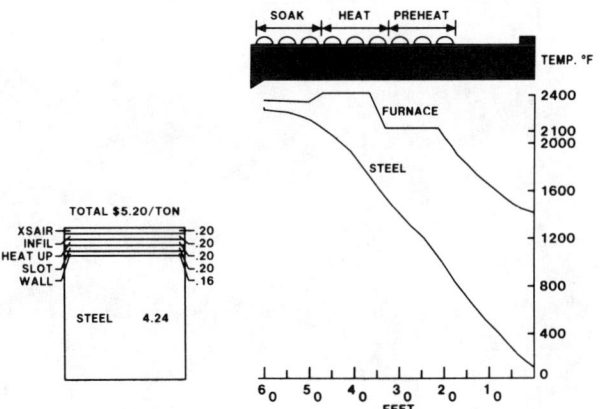

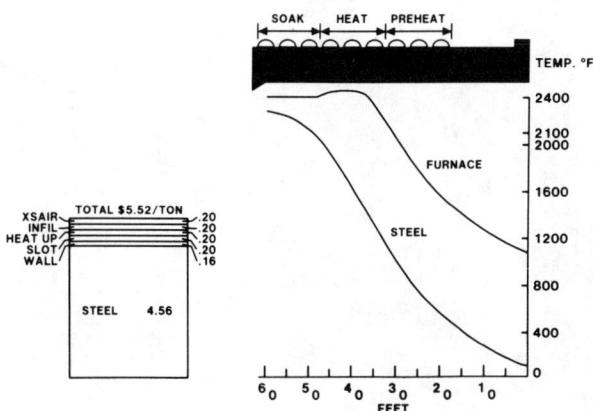

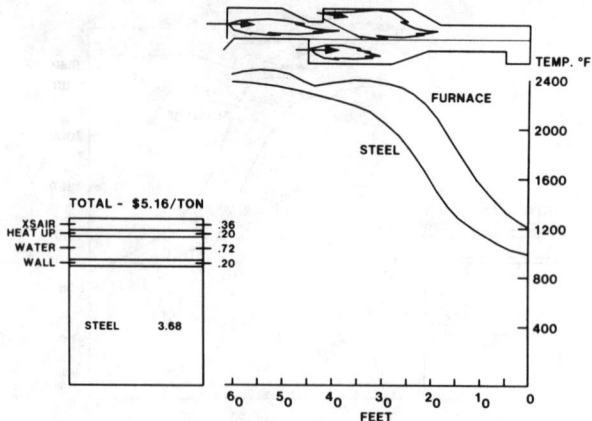

Fig. 14 — Pusher furnace. Condition: hot charging steel, 1000°F, with Series I modifications, 1.29 million Btu/ton.

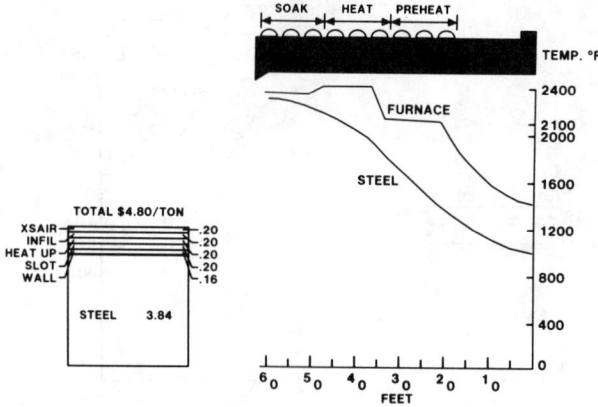

Fig. 15 — Walking beam furnace. Condition: hot charging steel, 1000°F, with Series I modifications, 1.20 million Btu/ton.

charged into the furnace, the fuel cost savings were $1.28/ton for the pusher furnace and $1.32 for the walking beam furnace (Fig. 14 and 15).

Other methods — Other posssible ways of saving or recuperating energy in reheat furnaces include oxygen enrichment of the combustion air, in cases where low-cost oxygen is available, and cogeneration of energy using the hot flue products. These methods have not been dealt with in this article, however, because its purpose is to present only methods that can be implemented relatively easily and that directly affect furnace performance.

Summary

Energy can be saved in reheat furnaces by improving insulation coverage, obtaining positive control of the furnace, adding a ceramic fiber veneer, eliminating air infiltration, using preheated combustion air, adjusting the thermal profile of the furnace and hot charging the billets. Savings in fuel costs that can be realized through these procedures are summarized in Table I.

TABLE I Comparison of fuel costs

	Pusher furnace, $/ton	Walking beam furnace, $/ton
Base rate without modifications	9.48	7.88
Series I modification		
90% insulation	8.40	—
Good control	7.20	6.88
Veneer	6.80	6.52
Door seals	6.44	6.12
Series I modifications with		
Air preheat, 700°F or	5.40	5.20
Adjustable thermal profile, or	5.32	5.52
Hot charge, 1000°F	5.16	4.80

Modernization of Rotary Hearth Forge Furnace and Fuel Savings Documented

By STEVEN R. FRIED
Combustion and Instrumentation
Engineer
Bethlehem Plant
Bethlehem Steel Corp.
Bethlehem, Pa. 18016

No. 47 rotary hearth furnace in the drop forge shop of Bethlehem Steel Corporation's Bethlehem plant heats stock for closed die forging by the 10,000 lb steam driven No. 47 hammer. The steel billets range to 300+ lb and must be heated at a rate of 10,000 lb per hr to 2350°F.

The furnace, Fig. 1, was originally installed during World War II as designed by Hagan Furnace, and was rebuilt in 1951. It comprised a 17 ft, water sealed rotary hearth with a steel shell anchoring a firebrick wall and domed roof.

Eight low pressure atomizing, automatic proportioning burners fired bunker "C" fuel oil. Combustion air was provided by a Spencer turbine blower, which also supplied atomizing air to the burners. (In the early 1960's the eighth burner was removed, because flame would shoot out of the open furnace door.) Furnace control was originally designed to automatically regulate the proportioning burners by mechanically linking the individual valves of each zone's burners, and controlling the temperature by operating a power positioner connected to the linkage. Through the years, this system was abandoned and the furnace burners were adjusted manually. Around 1970, the fuel oil system was purged of heavy oil, and the use of no. 2 fuel oil was instituted.

The original furnace was also equipped with furnace pressure control which deteriorated through the years and was abandoned.

The maximum fuel firing rate of the original furnace was approximately 23.6 million Btu per hour.

Stages of Evaluation

Evaluation of this furnace covers the unimproved furnace and three stages of rebuild as follows:

(1) Base period of unimproved furnace.

(2) Rebuild of furnace with flat roof, ram refractory, and coneless hearth.

(3) Addition of ceramic fiber refractory veneer.

(4) Installation of recuperative burners and complete combustion control system.

Furnace performance data were taken after each stage of rebuild so that the benefits of each change in the furnace structure could be examined. In addition, the theoretical performance of the recuperators was examined to investigate the benefits of the recuperation separately from the other furnace changes so that the minimum acceptable lifetime of the actual recuperation modules could be determined.

Stage 1 — Base Period Performance

The base period chosen to document the performance of the furnace is the entire year of 1979 because the furnace structure then was reasonably sound and good data were available.

The raw data collected during the base period is presented on a logarithmic plot of Btu/lb vs. production in Fig. 2. Since the data points appear to suggest a linear plot on the log scales, a statistical (SAS) regression was run on the data to produce a smooth line that represents the operation of the furnace.

Since the average production run during the final data period was 19,354 lb, the furnace performance for each stage was evaluated at that point. At this production rate, the fuel use based on the statistical analysis was 8,644 Btu per pound. For comparison purposes, the smoothed data is shown in Fig. 3 with the base period represented as Rate 1.

Stage 2 — Rebuild: Flat Roof, Rammed Refractory

In spring of 1981, while a capital expenditure study was under way for the rebuild and modernization of the furnace, matters were precipitated by the collapse of the hearth and a large portion of the furnace refractory. It was decided to completely rebuild the hearth, walls, and roof of the furnace utilizing lightweight materials — basically ceramic fiber and rammed plastic refractories. A flat roof, designed by Brent Rohlfing of the plant combustion department, was installed on the furnace by running 12 in. structural beams across the furnace buckstays and hanging 3 in. beams from the major structural beams. Stainless refractory hangers were attached to the 3 in. beams to provide support for the rammed plastic.

The new refractory make-up of the roof was: 2 in. of fiber material (Kaowool) backup; 9 in. of rammed plastic (Plibrico RAM 55S).

The new refractory composition of the furnace walls was: 1 in. of fiber lining (Kaowool); 2 in. of International

Fig. 1 Rotary furnace in forge shop of Bethlehem Steel Corp. has been modernized with changes in refractories, burners and the addition of recuperative combustion system to achieve significant energy savings.

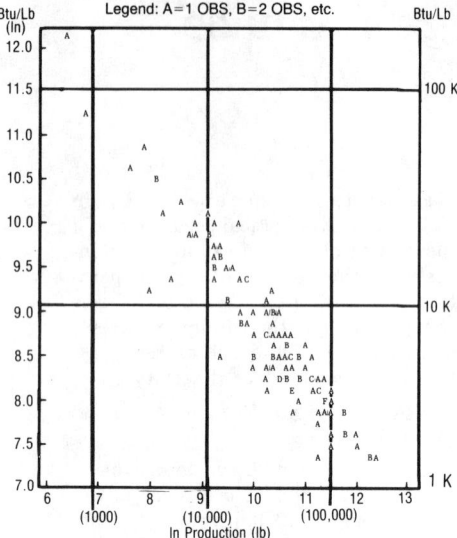

Fig. 2 Plot of base period data—Log Plot of No. 47 Rotary Furnace Data Before Modernization.

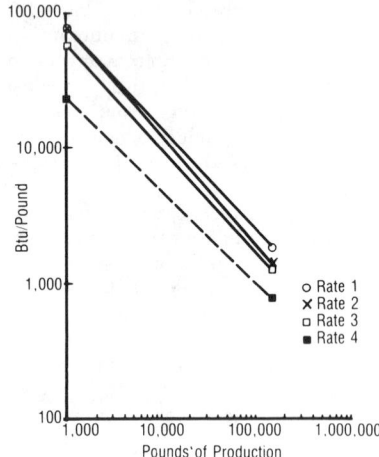

Fig. 3 Plot of furnace performance at each stage. (Btu per lb vs. lb production for No. 47 rotary furnace).

Vermiculite IV20 backup; 10.5 in. of rammed plastic (RAM 55S).

While the previous roof design contained numerous wall flues, the new design incorporated a 30x42-in. flue in the center of the roof, simplifying the refractory installation considerably. The rammed refractory was installed with the intent of gluing ceramic fiber modules to its surface, after the furnace had been fired for a few weeks to cure the lining.

The old hearth had supported a large cone of dense firebrick that covered the middle 10 ft of hearth diameter. This cone was structurally unsound, and would crumble shortly after being rebuilt. Elimination of the cone presented two distinct advantages: (1) the reduction of nuisance shutdowns needed to remove crumbling refractory; (2) the removal of a large quantity of thermal mass from the furnace interior which had to be heat-

ed every time the furnace was lit.

The 36 in. perphery of the hearth was composed of 4.5 in. of K20 insulating brick, topped with 14 in. of rammed plastic (RAM 55S). The remaining center portion of the hearth was built up as follows: 10.5 in. of K-20 brick; 3 in. of high duty clay brick; 4.5 in. of RAM 55S.

Results of Flat Roof Rebuild

During the few weeks that the furnace was fired to cure the rammed plastic and catch up on production, data were collected on the furnace operation. One of the first characteristics noted after the change was a reduction in the amount of time needed for the initial furnace heat-up: Light-up times were quickly reduced from 4 hr to 3 hr before the start of the forging shift. Removal of the cone eliminated the down time that was associated with the cone's crumbling. Also, it was noted that the removal of the cone did not seem to significantly interfere with temperature gradients across the hearth, i.e., hot billets that were ready for forging were not affected by the charging of cold steel into the furnace.

A linear regression was run against data plotted in similar fashion as shown in Fig. 2, and the resulting smooth line was added to the graph in Fig. 3 as Rate 2.

The fuel rate for a production run of 19,354 lb was 7,223 Btu per pound. This represents a reduction of 16.4% from the base period criteria.

Stage 3 — Rebuild: Ceramic Fiber Veneer

After firing the furnace for four weeks to cure the rammed plastic and clear some production, a 3 in. veneer of ceramic fiber (Durablanket CH) was installed. The veneer was applied to the entire furnace refractory face, and also to the portions of the hearth that do not hold steel.

After the addition of the veneer, the furnace came up to heat faster, with the result that the light-up time was reduced to two hours before the start of the shift. Data were collected on this furnace configuration for five months until the furnace was taken down for the final stage of rebuild. Once again, a statistical regression was run on the new data plotted similarly to that in Fig. 2, to obtain the resulting smooth line identified in Fig. 3 as Rate 3.

After installation of the veneer, a fuel rate of 6006 Btu per lb of production resulted for the production rate of 19,354 lb. This represents a reduction of 16.8% from the same furnace without the veneer, and a 30.5% reduction in fuel use from the base observations.

Stage 4 — Final Rebuild

After approximately five months of operation, the furnace was again taken down for rebuild with the best available technology, consisting of the following:

Combustion System

(1) Use of a combustion air fan (Spencer), 32 osi, 3,000 scfm.

(2) Installation of 8 self recuperating burners (Selas DNS-R-2500), capable of firing either natural gas or light oil. The recuperator section, provided by GTE, consists of layers of small rectangular ceramic channels. Alternating layers are stacked at 90 degrees to provide cross flow heat exchange. The entire recuperator "cube" is held together under compression.

Control System

(1) Control of all combustion processes for both zones is handled by a controller (Westinghouse model 1500 GPC). Each zone's temperature control activates an air primary fuel/air control configuration to maintain the temperature setpoint. DP cells on the fuel and air lines to each zone provide sensing of the flows for mass balance air/fuel ratio control, and for recording and integrating fuel flows.

(2) Burner light-up and flame supervision control, furnished by Selas, utilizes UV flame sensors (Protection Controls). The relay panel (Selas) interfaces with the microprocessor for many of the timing cycles, limit contacts, and other functions.

(3) Furnace pressure control (North American EPIC II) maintains a pressure setpoint by controlling the operation of eductors which draw furnace combustion products through the recuperators. When the furnace firing rate falls below a certain threshold, the microprocessor can operate air dampers to "bottle up" the recuperator exhaust.

Observations on Results of Final Rebuild

The current burners operate extremely efficiently, with almost all combustion processes occurring directly within the burner well itself. Measurements of the burner refractory temperature at high fire exceeded 2800°F, creating conditions for excellent radiant heat transfer.

With the zones each set up to fire at 10% excess air and with the combustion process confined to the burner cup itself, there was a marked improvement in regard to the scale formation on the billets being heated. Previously, a very heavy, adhesive scale had formed. Under current conditions, scale formation has been limited to a very fine, light scale that has proven to be very easy to strike off, resulting in a better finished appearance and less scale rejection of finished pieces.

Once the new system was installed and operating, the initial light-up of the furnace with full automatic control brought the furnace up to heat in 23 minutes. Noise problems attributed to resonance between the burners at high firing rates forced the maximum firing rate of the burners to be cut back. Zone 1 burners were cut back to approximately 1.3 mm Btuh per burner, while the burners of zone 2 were limited to about 1.5 mm Btu. The furnace now comes up to heat in approximately 45 minutes after a cold start, and light-up time has been reduced to one hour before the start of the forging shift.

In Fig. 3 the smooth line resulting after running a regression analysis against the raw data from a plot similar to that shown in Fig. 2 is presented as Rate 4. The fuel rate at the average production of 19,354 lb, which has been used as the production rate for comparison in all phases of the rebuild, is 2,895 Btu per lb. This represents a reduction of 51.8% from the observations on the furnace for the five months before the rebuild, and a reduction of 66.5% from the base period. An additional 17% savings was realized from the fuel conversion.

Evaluation of Recuperation

Even under the best circumstances, recuperators will fail and be in need of replacement. For that reason, it is

Table I Estimated Minimum Acceptable MTBF for Various Production Rates

Rate (lb)	Savings	MTBF (Shifts of Operation)
19354	37.7%	31
45000	37.5%	24
70000	37.2%	21

Table II Synopsis of Furnace Changes in Various Stages and Resulting Savings

Evaluation Stage	BTU/LB (19354 lbs)	% Fuel Reduced from Base
1. Base Period	8644	———
2. Flat Roof, RAM Refractory	7223	16.4%
3. Ceramic Fiber Veneer added	6006	30.5%
4. Sealed burners with Recuperation and microprocessor control	2895	66.5%

Table III Savings Generated by Major Furnace Changes

Furnace Change	Fuel Savings	Cumulative Savings
Refractory change to lightweight RAM with fiber blanket veneer.	30.5%	30.5%
Sealed burners and good automatic controls.	22.6%	46.2%
Recuperative preheat of combustion air to 1200 F.	37.7%	66.5%

important to investigate the cost benefits of the recuperation aside from the overall improvement in furnace operation. The basis for such investigation and calculation is presented in the original documentation.

Calculations based on the 19,354 production rate denote an estimated fuel savings attributed to the recuperation as being 37.7%.

Dividing the estimated cost of replacing a single recuperator module by the savings generated per production run (shift of operation) gives a ballpark estimate of the minimum acceptable mean time between failures (MTBF) rate of individual recuperators necessary to make continued replacement of the units cost effective (see Table I).

Conclusion

Significant savings have been realized by modernizing one of the forging furnaces at Bethlehem plant's drop forge. The savings gained from each phase of this study are clearly indicated by the curves in Fig. 3 and data in Table II. The amount of fuel savings that can be expected from various furnace modifications is presented in Table III.

The performance that has been documented within this article clearly shows how well a modern, "state of the art" furnace can operate. The data indicate significant savings in fuel costs, even though the figures are computed at historically low rates of production. The condition of the base period furnace is typical of the condition that many furnaces are presently in, and savings similar to those documented herein are to be expected with the modernization of any of these furnaces.

Information Retrieval

For your guidance and filing convenience the following subjects are included in this article:
Key: forge reheat furnaces, heat treat furnaces, ceramic fiber insulation, furnace lining installation, recuperators, electric arc furnaces

Retrofitting a Box Forge Furnace with Ceramic Recuperators Considerably Reduces Fuel Use

By KENT H. KOHNKEN
Ceramic Product Manager
GTE Sylvania
Towanda, Pa. 18848

A box type furnace for reheating steel billets for forging at McInnes Steel in Corry, Pa. (Fig. 1) was retrofitted with ceramic recuperators (GTE) in January 1981 under a jointly funded demonstration program involving DOE, GTE Sylvania and McInnes Steel.

DOE/GTE jointly provided funds for the refractory burners, recuperators, furnace pressure control and furnace mass flow control. Mr. Don Whipple of North American Manufacturing, Buffalo, N.Y., provided design and startup assistance. Mr. Dave Dearborn, McInnes plant engineer, supplied other related equipment and installation labor. Mr. Frank Holden of Battelle Columbus Laboratories, under another DOE contract, analyzed the operating data of the furnace before recuperation, after recuperation and compared the recuperated furnace with that of a companion unrecuperated furnace operating at the same time and conditions.

Furnace Design and Operation

The furnace retrofitted was one of four identical box type forge billet heating furnaces. It is used for heating various size billets to 2250 to 2450°F and consists of a refractory lined enclosure surrounded by a steel frame and a front door which is raised to permit loading and unloading of billets. Energy is supplied to the process by 4 burners set high on one side (see Fig. 1) and firing slightly down on the load of billets being heated.

The exhaust is flued from the floor opposite the burners up through the wall to 4 ceramic recuperators located directly

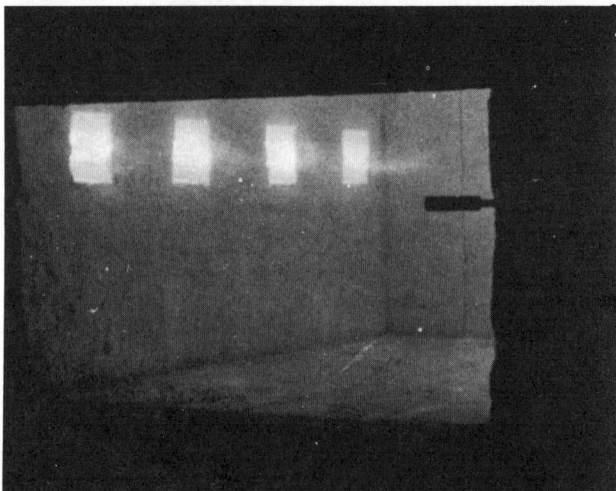

Fig. 1 Box furnace for reheating steel billets for forging has four gas-fire burners. The furnace had just been retrofitted with four ceramic recuperators which combined with insulated piping and special refractory lined high temperature tempered air burners reduced the energy requirements by 58%.

above the flues. The exhaust passes through the recuperators and out of area through an exhaust fan. The cold clean air passes through the ceramic recuperator, is heated to 1100 to 1300°F and is piped through insulated piping across the top of the furnace and down to special refractory lined high temperature tempered air burners. Furnace pressure control is utilized to govern regulation of the exhaust fan and to turn the combustion system on low fire when the door opens. The retrofitted furnace also utilizes a system (North American Marc) which automatically keeps the combustion stoichiometry within satisfactory limits.

The furnace operates on 1 or 2 shifts per day, 5-7 days/week, during which steel billets are loaded into the 1100-1200°F furnace and heated to forging temperature. Billets are then unloaded as required and the furnace idled both when full at forging temperature and when empty waiting for the next load of forgings.

The furnace is 16 years old and utilizes 4 hot air burners—each 1.5 million Btu (North American Series 6825) and 4 ceramic recuperators (GTE Model R1500 TPX). Fig. 2 is a schematic diagram of the combustion system.

Recuperator Design and Operation

The ceramic recuperator consists of a ceramic module (Fig. 3) which is inserted into a 3000°F castable lined housing (Fig. 4) which is ready for insertion into the exhaust and air piping. The recuperator is composed of many sets of exhaust passages (.200″ high X .680″ long) separated by a .050″ ceramic wall from many sets of air passages (.125″ high X .680″ long). The ceramic removes the heat energy from the hot dirty exhaust and transfers this energy to heat the cold incoming combustion air. This room temperature combustion air is then heated to 1100-1300°F and is transmitted to special refractory lined hot air/gas burners.

The recuperators are placed on exhaust flues and the hot air piped through insulated piping to the burners. It also is possible to directly attach the recuperator to the burners, as shown in Fig. 5. High fire data for each recuperator is: combustion air flow—8500 scfh; exhaust air flow—9350 scfh (products of combustion); recuperator exhaust inlet temperature—2150°F; recuperator air preheat temperature—1240°F; preheat air temperature at burner—1180°F.

Expenditures

The total cost of the installation was $94,450 which included all equipment, refractory and labor costs, as itemized in Table I. It was estimated by Battelle Columbus Labora-

Table I Summary List of Components for Total Retrofit of Forge Furnace

Ceramic Recuperators	$ 20,000
Refractory Burners	4,200
Furnace Pressure Control	6,000
Furnace Mass Flow Control	4,250
Combustion Hardware	2,600
Refractory	17,500
Flame Monitoring	7,000
Fuel Monitoring	4,900
Electrical & Misc.	10,600
Engineering	2,000
Labor	15,400
	$94,450

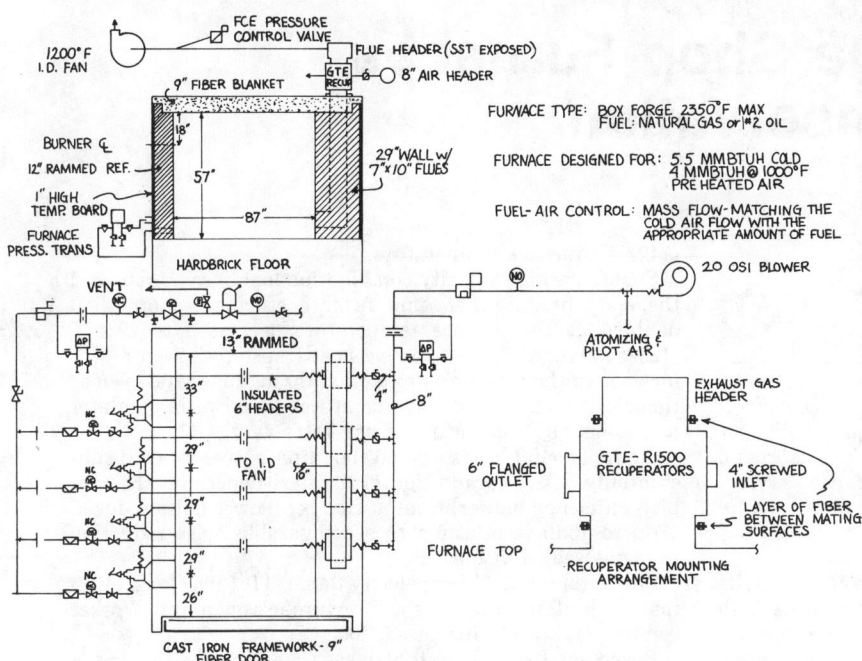

Fig. 3 This 12 x 12 x 12 cu in. cube is the heart of the ceramic recuperator. It features a crossflow corderite (magnesium alumina silicate) matrix composed of alternate layers of ceramic passages oriented at right angles to each other. Hot exhaust gases are directed through the matrix and the stored heat energy is transferred to the alternate layer of passages through which flows the cold clean incoming air used for combustion. The corderite is capable of handling high temperatures (2500°F) and has excellent resistance to thermal shock.

Fig. 2 Schematic diagram of combustion system showing insulation and controls.

Fig. 4 Fully assembled recuperator on the left illustrates how a spring-loaded cover firmly seats the corderite matrix cube in compression against a 3000°F castable lined housing. The second unit, which has the cover removed, shows how the housing design allows for the cleaning and replacement of the matrix. Both the cover and housing are fabricated from welded cold rolled steel.

Fig. 5 Ceramic recuperator is attached directly to a refractory lined burner. The ceramic removes the heat energy from the hot dirty exhaust and transfers this energy to heat the cold incoming combustion air.

tories that $48,500 of the total amount is attributed to the recuperator retrofit system.

Energy Savings

The box forge furnace (05-23) was placed in operation on January 12, 1981. Table II shows the Battelle Columbus findings after 6 months of furnace operation compared to data acquired for several months prior to rebuild and also compared to data for a similar existing furnace (05-17) located near the rebuilt recuperated furnace. At a daily average production level of 15,000 lb the recuperated furnace experienced a 58% fuel saving when compared to its operation before recuperation. The recuperated furnace experienced a 50% savings when compared to its "side by side" companion furnace (without recuperation).

The recuperated furnace was located close to a large forge hammer and vibration did not affect operations or perform-

Table II Energy Used for Furnaces Operating with and without Recuperation

Furnace #	Btu/Lb*
05-23 Before Recuperation	7400
05-23 After Recuperation	3100
Savings	58%**
05-17 Without Recuperation	6200
Savings	50%

*Values include fuel used for low fire periods during furnace idling and for startup after shutting down on weekends, for a period of 6 months.
**Savings of 40% have been determined by Battelle to be directly attributable to the ceramic recuperators.

ance of the recuperators. There were also no maintenance costs associated with the ceramic recuperator during the six months of operation. ■

Retrofitting Forge Shop Furnaces for Energy Conservation

By JOHN WEST
Marketing Manager
Hauck Manufacturing Co.
Lebanon, Pa. 17042

Editor's Note: *Case histories and energy-saving considerations involving furnaces for heating to forging and heat treating temperatures are covered in this article. It is from a paper presented at the Forging Industry Association's Heating Equipment and Energy Conservation Symposium in Chicago.*

A reduction in forge furnace temperature typically results from increased convective combustion equipment. Thus, high velocity burners can maintain or even increase furnace production rates while furnace temperatures decrease some 350 to 400°F (177 to 204°C). This may affect refractory material selection. Furthermore, lower furnace temperatures resulting from convective heat transfer may allow selection of different instruments and thermocouples.

Case Histories

The need to modernize their furnaces was recognized by Berwick Forge and Fabricating (BFF), a division of Whittaker Corp. in central Pennsylvania. This company operates a relatively large forging, heat treating and fabricating operation and processes many parts for railroad cars and locomotives, as well as builds complete boxcars, coal cars, and gondolas.

Box forge furnace #510 measures 6 ft (1.8m) wide by 17 ft (5.2m) long, with a solid wall in between; basically, two furnaces separated by a common wall. Two doors per half operate pneumatically by foot pedals. A burner on each end fires towards the middle. Table I describes the original furnace while Table II lists the changes after the retrofit.

Previously, oil for the open fired burners was well-atomized, but the volume of combustion air was not precisely controlled.

Individual furnace operators could bypass the burner's proportional capability and modulate oil only. Either excess oil or excess air could result.

Also, furnace combustion noise passed back through the mounting bracket opening into the shop. And heat transferred to the work by radiation only, requiring a 2600°F

(1427°C) furnace temperature.

Sealed-in, high velocity combination fuel burners replaced the open fired burners and furnace efficiency more than doubled as the furnace temperature decreased to 2240°F (1232°C). Increased convective heat transfer allowed a 25% increase in the furnace's production throughput rate, even at the 360°F (182°C) lower temperature. An oil ratio regulator maintains air/fuel ratio as firing rates vary.

Normal sound levels around the furnace were lowered substantially with the addition of a filter-silencer on the new high efficiency blower. Combustion air blower noise reduced 13db to 15db in addition to a comparable noise reduction from the sealed-in burner.

Furthermore, the high velocity flame, 3 ft (.9m) long, eliminated the flames coming out of furnace openings. Worker comfort and safety increased, too.

Based on both laboratory measurements, Fig. 1, and a number of similar field installations reported in the original paper, the 360°F (182°C) temperature reduction in the forge furnace can be attributed to the high velocity burner. This increased the furnace's fuel availability and reduced the wall losses compared to a long luminous flame burner. The production increase at a 360°F (182°C) lower furnace

Table I Original Furnace 510 Details

Item	Description
Burners	Individually proportional, oil only, with air opening in mounting bracket. 5:1 Turndown. Operator could manually adjust to excess air or excess fuel.
Blower	16 Ounce, medium efficiency, with no filter or silencer.
Refractories	Hard brick.
Production Rate	8 RCS Commutator Caps per hour (9" x 7½" x 7½")
Furnace Temperature	2600°F
Furnace Efficiency	4½% including morning warm-up, approximately 7% to 8% during production.
Sound Levels	Relatively high as combustion noise from furnace, and blower noise, passes unattended to shop floor.
Worker Productivity & Safety	Flames came out of furnace. Hot steel furnace shell dangerous.

Table II Retrofitted Furnace 510 Details

Item	Description
Burners	Sealed-in nozzle mix high velocity (25,000 FPM), combination fuel Enerjet burner. Fourteen inch tile is the approximate thickness of the refractory wall. 8½:1 Turndown. Cross-connected ratio regulator holds burner at 10% excess air as operator manually adjusts a butterfly valve to modulate fuel input. The Enerjet can burn natural gas if it becomes available.
Blower	20 Ounce, high efficiency blower with filter-silencer.
Refractories	Plastic with back-up block insulation.
Production Rate	10 RCS Commutator Caps per hour (25% Increase) (9" x 7½" x 7½")
Furnace Temperature	2240°F (360°F Reduction)
Furnace Efficiency	9% Including morning warm-up, 15% during production
Sound Levels	Should be reduced approximately 10 to 15 db, since sealed-in burner with thick tile muffles combustion noise, as does filter-silencer on the blower.
Worker Productivity & Safety	Three foot, high velocity flame remains inside furnace. Lower furnace temperature and changed refractory wall material combine to substantially lower furnace shell temperature.

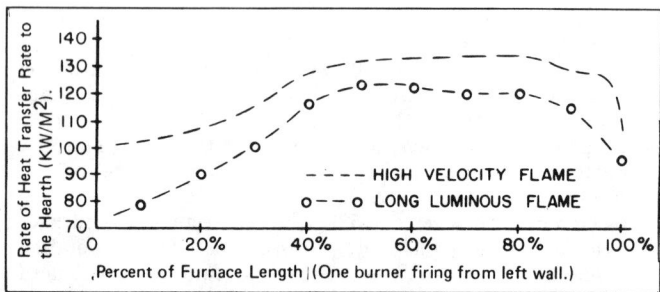

Fig. 1 Superior heat transfer with high velocity burner as compared to a long luminous flame burner. (International Flame Research Foundation's "Report on the M-3 Trials," document number F36/a/6, Ijmuiden, Oct. 1975.)

Reprinted with permission from Industrial Heating, July 1982, 26, 31, 32, © 1982 National Industrial Publishing Co.

temperature is attributable to the increased convective heat transfer from the high velocity burners. The motion of the furnace gases "scrubbing" more surface area boosts furnace efficiency. Most of the 25% increase in productivity as well as 15% to 20% of the fuel savings can be attributed to the high velocity burner.

Considerations in Retrofitting with HVB

In selecting a high velocity burner its refractory tile should be about as thick as the furnace wall (see Fig. 2). Since the tile is under positive pressure, it needs support around the outside. By having the burner's jet discharging right at the inside furnace wall, maximum convection results.

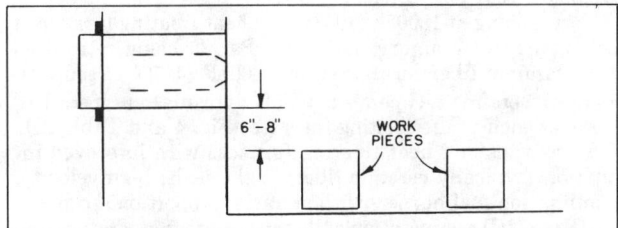

Fig. 2 A schematic cross-section of a high velocity burner tile mounted on furnace end wall.

With the lower furnace temperature, less heat is stored in the walls during morning warmup. If warmup is a manual process, personnel must start up the high velocity burners later than the others because heatup time of the former is less.

Temperature uniformity in the furnace can be improved with the high velocity burner. The work pieces at each end heat up quicker.

When selecting refractory materials, note that the furnaces operate at lower temperature. This results in longer furnace life between rebuilds.

If automatic temperature controls are used, the lower furnace temperature may allow use of less expensive and more reliable thermocouples. A flame safety/supervision system may include UV scanner on the burner, blocking valves, pressure switches and panel. These devices allow startup under NFPA guidelines, as well as worry-free operation.

Another consideration involves combination fuel, high velocity burners. Natural gas, when available, currently offers lower cost combustion but at the expense of flame luminosity. A combination burner, though, allows the burning of both natural gas and oil simultaneously. A small percentage of oil added to a natural gas flame significantly increases luminosity and flame radiation.

Recuperation today is more practical than ever before. With the furnace temperature now below 2300°F (1260°C) there are available heat exchangers which require no dilution air. Also, with the lower furnace temperatures, 800°F (427°C) preheated air on high velocity burners often becomes optimal. Mild steel piping is suitable and less expensive compared to the stainless steel required with higher temperatures.

Recuperation has been successfully demonstrated, for example, at Ajax X-Ray, Sayre, Pa. The change from nozzle mix burner to high velocity burner reduced fuel usage 14%. When an 800°F (427°C) recuperator was added, fuel consumption reduced another 31% for total savings of 45% on a 2000°F (1093°C) furnace. The ceramic recuperator (GTE Sylvania) bolts directly to the existing flue. Flexible pipe connections allow for expansion, and contraction during daily on-and-off cycles. After one year of operation, the burner assembly, recuperator, and other components are in excellent condition. Recent inspection of a similar installation in Europe—going strong after 3 years of operation—shows no evidence of wear or needed component replacement.

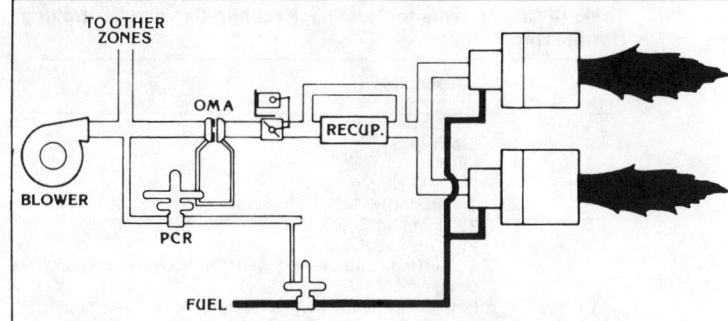

Fig. 3 Schematic of a simple system for providing preheated combustion air.

Electronics has a history of short life in certain forge plants and therefore must be carefully considered. A simple, reliable control system, used with hot air, high velocity combustion systems is shown in Fig. 3. The mass flow system senses a pressure drop on the cold air side of the recuperator, by a pilot control regulator. Its valve varies a signal pressure to a ratio regulator. This economical system holds proper ratio for a long time.

Conclusion

By refurbishing existing forge furnaces with improved insulation/tight seals; high velocity, combination fuel burners; 800°F (427°C) recuperation and appropriate temperature & pressure controls, significantly higher furnace efficiencies —in the 20%+ range—can be achieved economically.

Heat Treating Furnaces

Heat treating furnaces also can be improved with high velocity burners. One firm, X-tek Corp., Cincinnati, Ohio,

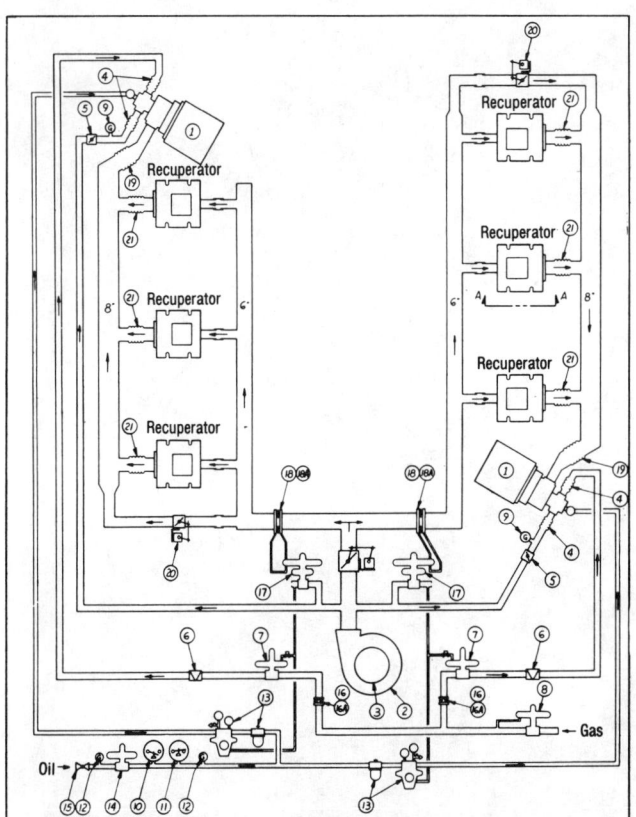

Fig. 4 Diagram of heating, recuperating and controlling system for normalizing furnace.

Table III Components for Heating, Recuperating and Controlling Normalizing Furnace

Item No.	Qty.	Description
1	2	Burner Unit, NMC H 240
2	1	Blower Assy.
3	1	Filter
4	4	2½" Flex Nipple
5	2	2½" Butterfly Vlv.
6	2	2½" Limiting Orifice Vlv.
7	2	2½" Gas Regulator
8	1	2" Gas Regulator, #166-5 w/1" Orifice & 2psig. Spring
9	2	Gauge, 0-55" WC
10	1	Low Oil Press. SW., DAF-81-3 Rg. 23K
11	1	High Oil Press. SW., DAF-81-2 Rg. 23M
12	2	Gauge, 0-100 Psi
13	2	½" Regulator Assy.
14	1	¾" Oil Regulator
15	1	¾" Globe Needle Vlv.
16	2	3" Orifice Meter
16A	2	Plate, Orifice, 5285 CFM
17	2	¾" Pilot Control Regulator
18	2	6" Orifice Meter
18A	2	Plate, Orifice, 370 CFM
19	2	6" STSTL Flex. Nipple
20	2	4" Butterfly Vlv. w/Control Mtr.
21	6	4" Ststl. Flex. Nipple

has 14 car bottom heat treating furnaces. One furnace has been instrumented for test and evaluation.

X-tek typically carburizes steel at 1800°F (982°C). Several years ago when the furnace had hard brick walls and nozzle mix burners, fuel consumption averaged 1400 scfh of natural gas. As a first energy-saving step, the brick was replaced with ceramic fiber insulation. Fuel consumption decreased to 1100 scfh (average). Then, the six long flame nozzle mix burners, which ran on-ratio at high fire and on excess air at low fire, were replaced with high velocity burners. They were installed to run on-ratio at all firing conditions. The high velocity circulation in the furnace allowed process uniformity even at low fire. Fuel consumption now averages 600 scfh, a 45% reduction by changing burners and employing suitable control techniques. Overall, fuel consumption was reduced 60%, representing about $20,000 a year per furnace.

Normalizing at 1900°F (1038°C), a heat treating furnace at Lebanon Steel Foundry, Lebanon, Pa., is being retrofitted with ceramic fiber insulation and 800°F (427°C) recuperation. Six ceramic recuperators (GTE Sylvania) are installed, one over each of the existing flues (see Fig. 4 and Table III).

By refurbishing heat treating furnaces with improved insulation (typically ceramic fiber)/tight seals; high velocity, combination fuel burners with on ratio, proportional control; 800°F (427°C) recuperation and appropriate temperature and pressure controls substantial fuel reductions can result, with low capital expenditures. □

Refractory Fiber Modules Saves Fuel in a Walking Beam Billet Heating Furnace

By R.C. BRAUN, Field Engineer and
W.H. PARKER, Field Engineer
Manville Products Corporation
Denver, Colorado

Editor's Note: *The advantages of a refractory fiber furnace lining system as opposed to hard refractories in specific applications have been well documented. However, the shrinkages inherent in all refractory fiber products, and a lack of suitable anchoring hardware, have limited the use of fiber in some high temperature applications encountered in the forging industry. This article which is from a paper presented at Forging Industry Association's 1981 Heating Equipment & Energy Conservation Symposium, describes a modular refractory fiber system which proved itself in a forging industry application where direct mechanical abuse is not involved.*

When Portec Inc., Oakbrook, Ill., decided to install a new, high efficiency furnace in their railway products division plant, many practical considerations were involved. The use of refractory fiber modules (Z-Blok®) as part of the furnace lining was the result of a carefully considered choice among several alternatives. Since this was a new application for refractory fiber modules, it was also considered a somewhat daring choice by the furnace designers. The new furnace was needed to replace an older unit used to heat steel stock prior to a forging operation.

The old, dense refractory lined pusher type furnace had been in use for a number of years and it was becoming increasingly expensive to operate and maintain from the standpoints of labor and fuel. Rising fuel prices and fuel allocations made fuel conservation imperative. The new furnace was expected to meet certain criteria: low heat loss through the refractories, a low heat storage refractory lining, a walking beam method of transporting the stock through the furnace, a low operating labor requirement, increased throughput, low fuel consumption, and ease of maintenance.

The Salem Furnace Co., Carnegie, Pa., was awarded the contract for the new walking beam furnace. Initial planning called for a nine inch refractory fiber lining and roof mounted burners which fan flames out parallel to the roof. This type burner heats the roof which then radiates the heat down to the work. After layout of overall design and completion of the structural and piping engineering, refractory specialist and designer, Harry Frederick of Salem, decided to use accordion pleated refractory fiber modules (Z-Blok) for the roof and upper sidewall sections.

The furnace internal dimensions are 52' long x 10' wide x 3-to-5' high (see Fig. 1). Billets are transported through the length of the furnace, Fig. 2, by a walking beam mechanism in about two hours. The two outer hearth sections remain stationary, while the center section moves in a rectangular pattern: up, forward, down and back (see Fig. 3). The upward movement picks up the steel billets; they move forward, about 12", resting on the movable beam; then are deposited on the stationary sections on the down movement.

The hot zone discharge end operates at 2350°F. It has a cus-

tom designed furnace lining composed of 9", 2400°F modules (Z-Blok made from Cerablanket® refractory fiber blanket) in the charge end or cooler end of the furnace, and 9", 2600°F modules (Z-Blok made from Cerachrome® refractory fiber blanket) in the hot zone. The modules are installed in the flat suspended arch and upper sidewalls. The wear band on the sidewalls and the hearth are monolithic abrasive-resistant refractories. The new furnace requires

Fig. 1 Interior view of gas-fired walking beam furnace showing hearth with its movable center section and the side walls and roof which are lined with ceramic fiber modules in parquet fashion.

Fig. 2 Side view of walking beam furnace showing location of some of the burners.

three men to operate versus seven men for the old unit. A demonstrated 66% fuel savings over the old unit translates to $225,000 per year in fuel savings. Throughput is 33% greater than for the old unit.

Considerations for Lining System

Fuel cost and availability dictated an efficient unit. Therefore, low heat loss through the wall was important. The furnace is used for production on only one shift per day, it is idled for two shifts, and cooled on weekends. This thermal cycling operation makes heat storage a major consideration. The speed of heatup is also important.

The lining had to be resistant to the high temperature and high velocity gas flow from the roof mounted burners. A positive and solid attachment system was needed to secure the lining to the shell of the furnace. The constant thermal cycling of the operation puts a strain on refractory attachment systems because of differential thermal expansion. Mechanical vibrations, common in forge shops, add to this strain on attachments. Ease and speed of installation was expected and also ease and speed of repair, if needed.

Since the basic structural, piping and burner design had already been engineered and laid out, the lining system had to be compatible with this existing design.

Module Design

The modules are made from needled refractory fiber blankets. Twelve inch wide blanket is accordion pleated and alloy steel support beams are inserted in the pleats (see Fig. 4). The tabs on the beam protrude through the back of the block. A stainless channel is mounted on the exposed metal tabs. The blanket is compressed, covered with fiber board protectors and banded to hold the module under compression.

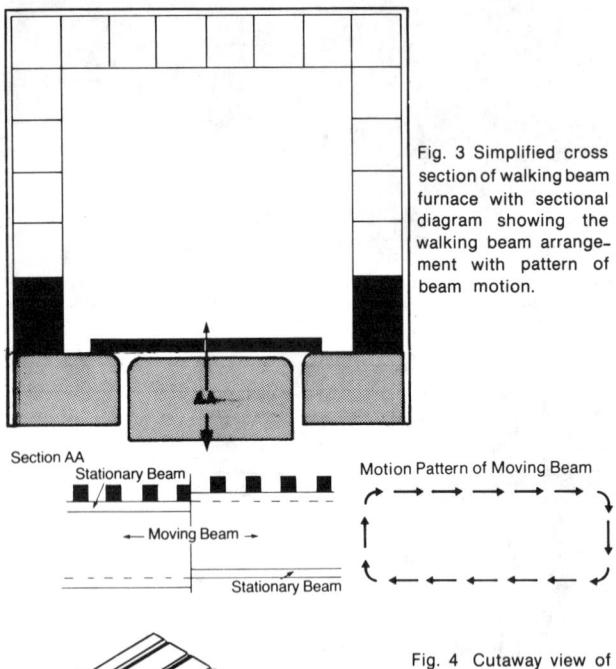

Fig. 3 Simplified cross section of walking beam furnace with sectional diagram showing the walking beam arrangement with pattern of beam motion.

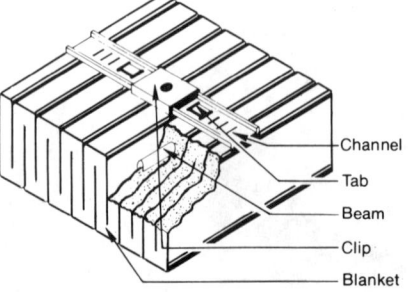

Fig. 4 Cutaway view of ceramic fiber module showing design for ready installation with attachments away from the hot face.

Channel
Tab
Beam
Clip
Blanket

Ease of Installation

Installation of modules is quick and simple if proper planning and layout is done in advance. Chalk lines are applied to the furnace shell indicating exact center positions for the modules. Clips are attached to the shell by any desired means. Possibilities include a nut and bolt, weld-on threaded stud and nut, nail gun, pop rivets, weld-on cotter pins, and knock off studs. Each block is installed by sliding the channel on the clip. The blocks are applied in a parquet pattern. The bands and fiber boards are removed and the surface pounded. This assures that each block expands against those next to it. This installation design and technique assures a virtually gap free lining.

Fiber Vs. Brick

Refractory modules are inappropriate for load bearing situations, areas which are subjected to direct mechanical abuse, or severe chemical attack situations. However, within these limitations, a refractory fiber module is a versatile and reliable alternative to hard refractories.

A Z-Blok lining is 75% lighter than insulating firebrick and 95% lighter than high duty fireclay brick. Energy used to heat the furnace lining to operating conditions is significantly reduced, and a greater portion of the heat reaches the workload. The furnace heats and cools more quickly. Thermal conductivity of refractory fiber is much lower than that of dense refractories. Re-radiation from the surface means better response to temperature control, and better thermal uniformity. Lighter steelwork is required for the furnace construction.

Resilient fiber is not damaged by vibration — a major consideration for the forging industry —, and there is no degradation due to thermal shock. Z-Blok required no curing or dry out after installation, no allowance is needed for expansion and contraction of shell or insulation, and plant personnel can perform maintenance to the lining, if required. Refractory fiber is less fatiguing for workers to install overhead, and lighter doors and frames are possible. The modules also provide quieter furnace operation. Furnaces can be shop-built and shipped to the plant completely lined, or with the clips in place ready for the modules to be field installed.

Advantages of Lining System

Hardware degradation and thermal shocking of ceramic anchoring materials is avoided because all hardware of the Manville Z-Blok system is within 1″ of the furnace shell, unaffected by furnace operating conditions. The anchor layout is simple, only requiring one anchor/sq ft. The full lining thickness is applied at one time, resulting in fast installation. Modules are easily handled and maintenance is simple and fast. Z-Blok modules offer a margin of safety in that they can withstand short periods of overheating without damage and are easily replaceable should damage occur.

These modules are installed under uniform 14% to 25% compression, adjusted to suit the application, and each module expands tightly against adjoining ones. The fiber on the hot face shrinks but the fiber behind the hot face is protected from the high temperatures.

The pleats continue to push against adjacent blocks and maintain a tight lining. The Z-Blok modules are in compression across the full width of the block not just at a butt joint. At temperatures above 1800°F this compression feature is imperative to maintain a tight joint between modules.

Other features of this lining system which contribute to proper installation and long life are: proper uniform positioning of the modules; clip attachment can be inspected before blocks are installed to assure that the attachment is strong; numerous choices of attachment techniques are possible; packing between blocks, when required, is simply accomplished without chance of tearing material from lath and damaging the attaching hardware; cardboards protect modules during shipping and installation and reduce worker contact with fiber; material is easily field fabricated to special sizes and shapes; customized blocks can be made in the plant rather than on the job; product is readily available and serviced.

Lining the Furnace

The decision to use the refractory fiber module system was made after the basic furnace design and detail work such as burner size and spacing was done. This required extensive deviation from the parqueted 12″ x 12″ modules usually recommended and illustrates the flexibility of the system. The burner blocks were 16″ x 16″ with spacing of 42, 44 and 40 inches between the blocks. In order to maintain the parquet arrangement which best assures a tight gap free lining, a design of custom sizes and shapes was planned to accommodate the burner block installation (see Fig. 5). Harry Frederick of Salem Furnace Co., designed the system using 9″ x 15″, 9″ x 12″, 15″ x 9″, and 12″ x 9″ modules with standard 12″ x 12″ modules to obtain the parquet pattern. This provided a lining so tight that additional stuffing around the burner block has been required only once in 20 months.

Fig. 5 Ceramic fiber modules and blanket insulation packed around the burner block.

New Furnace Vs. Old One

The new furnace is somewhat larger than the old one—52′ long compared to 45′ long. The width is the same. The discharge end height of the new furnace is 5′ versus 4′ for the old one. Both charge ends are 3 ft.

Sidewalls of the old units were 24″ of dense refractories. The new furnace has 9″ of refractory fiber modules in the upper sidewalls with castables in the wear areas. The old furnace has a roof of 9″ of dense firebrick in the cooler part of the furnace and 18″ of dense brick in the hot zone. The new furnace has 9″ of 2400°F refractory fiber modules at the charge end and 9″ of 2600°F refractory fiber modules at the hot discharge end.

Dense refractories were used in the hearth areas of both furnaces.

Operating conditions were changed slightly. The old unit was operated at 2350°F in the discharge end for eight hours per day for five days per week. At night it was turned off and would cool down to about 1000°F and on weekends, much lower. Upon start-up, the dense refractories would have to be heated back up to operating temperature, a substantial waste of heat. The new furnace operates on the same schedule but is held at 1500°F overnight. The charging end temperatures are lower as the burners are all at the discharge end and the cold billets enter at the charge end.

The thermal performance of the new furnace is dramatically greater than the old one. The calculated heat losses due to the refractories of the old unit were 12.4 billion Btu/year of which 2.0 was lost through the walls and 10.4 lost by heat storage in the refractories. The new unit experiences 6.8 billion Btu/year of losses of which 2.4 is heat flow and 4.4 billion Btu/year is in heat storage. A study of the heat losses in the new unit is instructive: 23% of the heat flow loss is through the refractory fiber modules and 77% is through the castables. Only 3% of the heat storage loss is in the modules whereas 97% is in the castables.

In this facility, the major improvement in thermal performance is due to the low heat storage of refractory fiber—a particularly significant feature because of the cyclic operation of the furnace. Heat storage is the heat that must go into the refractories in heating the furnace to operating temperatures. The refractory fiber has the effect of cutting the total heat storage by more than half.

The total calculated heat losses through the wall plus the heat losses required to heat the refractory show a 45% savings on the new furnace versus the old furnace. This compares rather favorably with the 66% overall fuel savings which includes not only these heat loss differences but also fuel reduction from improved burner design, improved overall furnace design, the effect of the walking beam rather than the pusher system, and several other considerations. ∎

A Process for Disposal of Salt Bath Wastes

By QUENTIN D. MEHRKAM

Contaminated quench and rinse waters are economically and efficiently treated with the Cyanil system. Free cyanide concentrations can be reduced to less than 1 ppm, satisfying even the toughest regulation.

SODIUM AND POTASSIUM cyanide — common ingredients in salt baths for surface hardening treatments — serve as sources of carbon and/or nitrogen in the diffused case. Cyanide concentrations range from 10 to 70% of the molten bath by weight. Because cyanide dragout contaminates waters used in the quenching, washing, and rinsing operations, environmental regulations state that these cyanides must be destroyed.

An economical and efficient system for destroying free cyanides and cyanates has been developed by Cyanil Co., Kitchener, Ont. The electrochemical oxidation process was honored with the Chemical Institute of Canada's "1975 Environment Improvement Award." It features:

● The lowest operating cost, for a nominal capital investment, among alternative processes.
● No requirement for chemicals, their storage or monitoring.
● Safe, easy-to-control operation.
● No need for dilution or chemical treatment of effluent before processing because the process is most efficient at high cyanide concentrations.

Ajax Electric Co. was recently granted an exclusive U.S. license to manufacture and sell the Cyanil system.

Case History — Various types of bearings, rollers, and other parts are salt bath carburized and hardened at Federal-Mogul Corp., Lancaster, Pa. FM operates three semiautomatic bearing carburizing and hardening lines. Bearings are racked, and then immersed at 1700 to 1750 F (925 to 955 C) in the carburizing salt bath. Subsequent steps include treatment in a stabilizing cyanide at 1450 to 1500 F (790 to 815 C), brine quench, power wash, and a continuous-oven stress relief.

The process generates 10 000 gal

Mr. Mehrkam is vice president, Research & Development, Ajax Electric Co., Philadelphia, Pa.

(37 850 l) of cyanide-contaminated water per month with cyanide concentrations of 2000 to 4000 ppm.

FM formerly paid about $1000 a month for an outside chemical service to remove and treat the effluent. The fee was increasing by about 20% annually. Solid wastes (cyanide-type salts) were trucked to a sanitary landfill for a charge. Now FM uses a Cyanil system.

Table (p. 106) compares costs for contract disposal, the Cyanil unit, and two chemical alternatives. Data are based on the operating and capital costs of destroying 6200

lb (2800 kg) of cyanide per year.

As mentioned above, projected cost increases for the removal service were significant. Note too the substantial operating costs for chemical destruction and alkaline chlorination. The higher initial capital cost of the electrochemical process was partially offset by an operating cost saving in the first year. After the first year, the Cyanil cyanide destruction process is competitive and offers the lowest over-all cost.

Operation — A major advantage of the Cyanil process is that only electricity is needed; approximately 4 to 8 kWh/lb of cyanide destroyed. Costwise, this equates to $0.10 to $0.15/lb. No chemical pretreatment or post-treatment of waste is required.

A typical unit is pictured in Fig. 1, top, and shown schematically in Fig. 1, bottom.

Cyanide-containing waters from the quench, wash, and rinse operations are periodically pumped to the storage tank. When the tank's full — FM's holds 5000 gal (18 900

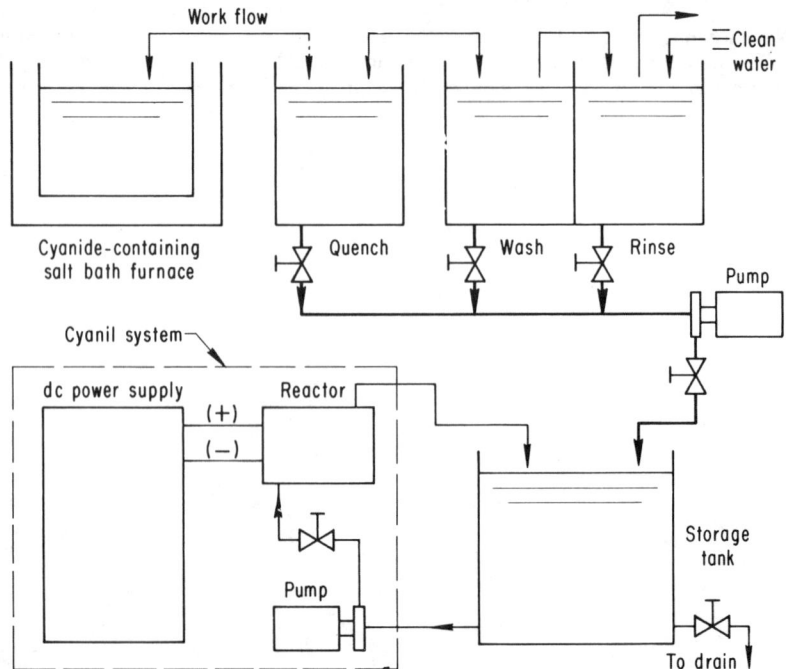

Fig. 1 — A typical Cyanil unit (right) incorporates a 20 kVA solid-state power supply. The primary 460 V, three-phase, 60 Hz power is converted by silicon rectifiers to full wave-rectified dc. The cabinet on the right contains two reactor cells. The pump, located between the cabinets, circulates cyanide-contaminated water between the reactors and the storage tank (see schematic below).

Reprinted from Metal Progress, September 1975, 103-104, 106, © 1975 American Society for Metals

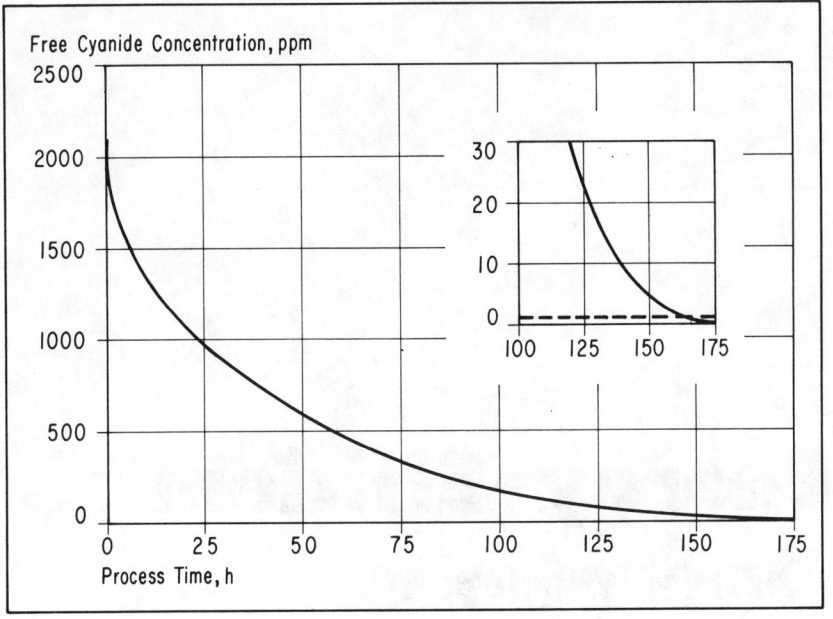

Fig. 2 — The Cyanil process is most efficient at high free-cyanide concentrations. However, as this example shows, CN⁻ levels as low as 1 ppm can be reached given enough processing time.

Alternative Cyanide Destruction Process Costs

Process	Annual Operating Cost	Capital Cost
Electrochemical oxidation	$ 2 332	$21 000*
Chemical destruction	$10 337	$10 800
Alkaline chlorination	$10 944	$14 000
Removal service	1974	$ 9 800
	1975†	$11 745
	1976†	$16 324

*Includes one spare reactor.
†Estimated.

1) — the in-feed is stopped. Then, the system operates in a closed-loop mode between the storage tank and destruction unit until all the cyanide has been destroyed.

Although not shown, solid cyanide salt wastes are also dissolved in water and added to the process for destruction.

Reactions causing the electrochemical oxidation of cyanide in water are given below:
Anode reactions:

$$CN^- + 2OH^- = CNO + H_2O + 2e^-$$
$$2CNO^- + 4OH^- = 2CO_2 + N_2 + 2H_2O + 6e^-$$
$$4OH^- = 2H_2O + O_2 + 2e^-$$
$$CNO^- + H_2O = NH_4^+ + CO_2 \text{ (by hydrolysis)}$$

Cathode reaction:

$$2H^+ + 2e^- = H_2$$

The unit can oxidize free cyanide to less than 1 ppm in heat treating waste water. This capability satisfies even the most stringent regulation.

The destruction rate is inversely proportional to the logarithm of the concentration. If reaction time becomes a problem, the system can be augmented with chemical (hypochlorite) treatment when cyanide concentration has been reduced to less than 200 ppm. This combination provides maximum utilization of time and energy.

A typical destruction cycle is plotted in Fig. 2. Free cyanide concentration in the 4800 gal (18 200 1) batch was reduced from 2100 ppm to less than 1 ppm in 168 h.

Process byproducts are nitrogen, carbon dioxide, and a trace of ammonia; all are vented from the top of the storage tank to the atmosphere. There is no noticeable odor.

Process Control — Because the electrolytic process is self-regulating, it's impossible to overtreat effluent. After the initial voltage, amperage, and flow settings are made, only occasional inspection and minor voltage adjustments are needed to keep the system running efficiently.

The solution is analyzed at the beginning of the process and several times during a run. The plot of cyanide concentration versus time (Fig. 2) obeys the relationship:

$$C(t) = C(I) e^{-kt}$$

where $C(t)$ = cyanide concentration (ppm) at time, t; $C(I)$ = initial cyanide concentration (ppm); k = hydraulic rate constant; t = elapsed time (h).

Once an average hydraulic rate constant, k, is determined, the formula can be used to predict the time needed to reduce cyanide concentration to a given level for a given initial cyanide concentration and reactor capacity. Figure 2 shows that the process is most efficient at high cyanide concentrations.

Maintenance — Reactor service life is generally rated at 5000 h. The used reactor is simply replaced with a rebuilt unit at a fraction of a new reactor's cost. If multiple reactors are hooked up in parallel, a used reactor can be replaced without shutting the system down.

The circulation pump is the only other system component requiring periodic inspection and maintenance.

Other potential applications for the Cyanil electrochemical process are the oxidation of nitrites to nitrates, and eventually electroplating waste treatment, still being developed. 🜨

Source: Metal Progress, September 1975, 103-104, 106

Energy-Efficient Operations

By Ross Shingledecker
Metallurgical Director
Manufacturing Service
Ladish Co.

ENERGY COSTS have increased dramatically since 1973, and supplies have become less certain, both of which give heat treaters ample reason to seek out energy-efficient processes.

Considerable efforts have been made by various agencies to advise all energy consumers, including overseas users, of ways of reducing energy usage by up to 15% in comparison with 1973 levels. The readers of this article are referred to these agencies and their publications for assistance. Savings can be achieved in a variety of ways, including the introduction of more advanced technological approaches whereby expenditure of the necessary capital can yield corresponding energy savings.

Theoretical Evaluation of Energy Requirements

Perhaps the most fundamental definition of energy relates various types of energy to the amount of work being done or to the amount of potential work that energy in a particular form is capable of doing. Heat is a form of energy that, when applied to a body, increases the energy of that body. Because molecules are in constant motion, they possess kinetic energy. Because neither increases in potential energy nor increases in kinetic energy can be measured for that body as a

whole, it is concluded that heat energy must have been imparted to the molecules of the body. Moreover, almost all solids and liquids expand when heated. This means that work (energy) must have been exerted on the molecules to separate them in opposition to the forces of cohesion. Heat is, therefore, a form of energy that exists as kinetic energy, especially in gases. In solids and liquids, heat also includes potential energy due to expansion.

Kinetic energy can be demonstrated by heating one end of a steel rod. The entire rod warms as heat is conducted along the rod. According to the kinetic theory of heat, the molecules of the metal in the heated end of the rod are set into rapid motion. These molecules strike neighboring molecules and impart kinetic energy to them, and this process continues throughout the length of the rod. The amount of kinetic energy is greatest at the heated end and least at the opposite end.

The potential theory of heat is demonstrated by the expansion and contraction of solids with temperature changes. In many instances, this property of metals is put to work in products and in production processes. The application of heat energy to a body can produce other dramatic effects, although these effects may not be as visible as expansion or contraction. In

some instances, the temperature of an object does not increase when the object is heated, although the over-all internal energy of the body is increased. Basic examples of this phenomenon are the formation of steam when water is heated continuously at 100 °C (212 °F) or the formation of liquid when heat is applied to ice at 0 °C (32 °F). Thus, when heat is applied to an object at a specific temperature, a change of state can occur without an increase in temperature—a change from solid to liquid or from liquid to vapor, for example.

Heat treatment of metals deals with a material in its solid state. Thus, energy requirements to produce changes of state, although substantial in quantity, are of little consequence in a normal heat treating process. Further, except in the broadest concepts of heat treatment, the energy requirements related to expansion or contraction of the solid are not major concerns. In heat treating, the desired changes in properties or conditions result from changes in structure. Energy is used to change residual stress within a structure. More frequently, the changes desired in heat treating involve changes in the nature, form, size or distribution of the structural constituents. Changes in the nature of the constituents result from the effect of temperature on phase equilibrium or from changes in the

Reprinted from Metals Handbook, 9th Edition, Vol. 4, 337-342, © 1981 American Society for Metals

composition of material, as in carburizing or nitriding. The forms of constituents are affected by the conditions under which they separate from the solid solution and by their tendencies to assume certain shapes at temperatures that permit diffusion and rearrangement, as in spheroidizing. Changes in the arrangement or distribution of constituents can be brought about by solution and reprecipitation.

The energy requirements of any particular heat treatment are finite, predictable and usually unalterable. The requirements are proportional to the temperature level at which that treatment is carried out. Further, the quantity of heat that a substance can contain or hold for each degree of temperature rise is known as its heat capacity. The ratio of this heat capacity to that of water is known as the specific heat of a substance. By use of this value, the level of energy (temperature) and the quantity of energy (heat content or thermal capacity) in a process can be measured. By use of these energy requirements in conjunction with the time to achieve the energy level (temperature) desired, the time the material is held at that level awaiting change of structure, and the prescribed time to achieve cooling of the substance, the total energy requirements for a particular heat treating process can be calculated.

The potential energy due to any given phase change or other structural change is calculable. Although such a change may require extended time to be accomplished, the kinetic energy due to the specific heat capacity of the material and the temperature level of the process is calculable and is also not time-related. The function of time relates to the over-all energy requirement because of thermal inefficiencies in heat treating processes. The inherent inefficiencies of a process continue throughout the time required to complete changes during the heat treatment. Generally, the inefficiency of a particular heat treatment, and the amount of energy consumed, is proportional to the treatment temperature.

An equipment designer or heat treater cannot alter the energy requirement of a specific heat treatment, but a great deal of latitude exists in selection of heat treatment processes. In this choice lies the opportunity for energy conservation and cost control.

Accounting Practice

Because it is an intensive user of energy, the heat treating industry remains a virtual bonanza of energy savings awaiting energy-conscious heat treatment operators. Part of the reason for inaction often relates to accounting systems. In many shops, cost systems do not account for the costs of process energy as a separate entity. These costs often are obscured within burden rates or machine center costs. If listed as a utility cost, the cost of process energy frequently is not separated from the costs of building heat and electricity. The energy cost for a particular heat treating operation in the past typically has been a rather insignificant part of the total product cost. In some situations, nearer the end-use product form, even the total cost of heat treatment became insignificant when viewed against the total cost of the product. Even with high energy prices, energy savings are measured in terms of cost effectiveness relative to other means of reducing costs rather than being seen as ends in themselves. Many corporations and large, privately held companies have reorganized their approaches to energy utilization, however. Through the energy manager approach, they have isolated and analyzed energy costs in great detail. Energy requirements are analyzed for each operation and for each phase of an operation.

Energy-Saving Strategies

Although many methods of reducing energy may seem elementary, experience has shown that simple or obvious solutions are often overlooked—particularly where quality and productivity are paramount goals, as in the heat treating industry. The following areas should be examined for possible improvements or changes:

- Product, process or specification restraints
- Modification of process
- Modification of equipment
- Alternate equipment

Product or Process. When faced with the task of improving any operation, an obvious question should be asked: "Why perform the operation at all?" Examples of the results of this type of analysis can be found in almost every sector of the heat treating industry. At one time, for example, all manufacturers of large gears specified that all forgings be double austenitized— that is, normalized, reaustenitized, quenched and tempered. When the mechanical properties of normalized, reaustenitized, quenched and tempered large rolled gear blanks or gear tires were compared with those of normalized, quenched, and tempered gear tires, however, absolutely no difference in properties was found. Now, large marine gear tires are normalized, quenched and tempered. The net savings in energy resulting from elimination of a heat treating operation have been sizable. In many such instances, product and process specifications were written when energy was comparatively inexpensive, and it was common practice to specify that certain operations be done twice to ensure that they were done once correctly.

Design criteria sometimes can be altered to reduce the energy required for heat treatment. For example, a case depth of about 0.75 mm (0.030 in.) can be attained in a 0.1 to 0.15% carbon steel in 3 h in a gas-carburizing atmosphere, whereas a case depth of 0.25 mm (0.010 in.) can be attained in the same steel in less than 1 h of exposure in the same atmosphere. Costly energy can be conserved through use of shallower case depths, even considering that the designer may have to specify a base material with a slightly higher carbon content and consequently higher cost. Further savings could accrue if designers would standardize carburized case depths: deep, medium and shallow, for example. Commercial heat treaters often process batch loads with less than optimum loading simply because other parts awaiting carburization require slightly different case depths. Any sort of overspecification usually results in increased use of energy and increased cost. Carburizing is just one example. Marked savings also can be achieved by changing combinations of both material and process. For instance, ball-bearing screws have been changed from carburized 4615 steel to 1045 steel with induction hardened threads.

Sometimes, energy can be saved by making changes in cleaning specifications. Considerable development work has been done recently on room-temperature cleaners, which usually are aqueous solutions for removing soil from the product without the aid of heat. Not all types of work can be cleaned acceptably with room-temperature cleaners, but all such cleaners achieve significant savings in energy.

Conservation Checklist. All areas of a heat-treatment-related specification should be examined for energy-saving opportunities. The following can be used as a checklist to help locate these opportunities:

- Are the service requirements of the part clearly known?
- In design of the part, has every effort been made to either simplify or eliminate one or more heat-processing operations?
- Is selection of material based on actual property requirements rather than on traditional selection practice?
- Have the possibilities of using alternative materials been investigated, for reasons of lowering cost or increasing availability?
- Has selective heat treating been considered?
- Has the possibility of using pretreated material been considered?
- Has the required case depth been reviewed?
- Is unnecessary normalizing being done?
- Are multiple tempering operations of questionable value being performed?
- Can tempering be eliminated without sacrificing quality?
- Is the heat treating equipment being upgraded in line with other equipment?
- Is a sound maintenance program being enforced in the heat-processing shop?
- Has quenching practice been thoroughly reviewed, and has consideration been given to use of alternative quenching media?
- If double quenching is being used, has the necessity for it been thoroughly reviewed?
- Is direct quenching used when possible?

Modification of Heat Treating Practices

A major equipment manufacturer modified his link-manufacturing operation to allow the link steel to transform fully in air at 540 °C (1000 °F). This modification permitted additional continuous operations to be performed at that temperature, thereby utilizing the heat in the link. The project was subsequently abandoned for mechani-

cal reasons, but the change was metallurgically sound, and the energy savings were significant. Use of residual heat from a previous operation should be explored as a conservation strategy. Material can be fully transformed after a forging or casting operation and still retain substantial amounts of energy. It could be scheduled for immediate entry into a heat treating furnace as soon as the transformation has been fully completed. New process lines often are built with these features in mind, but practices in existing plants can also be changed to achieve the same energy savings. Large savings in energy can be accomplished by optimizing loading of batch furnaces. Too often, batch furnaces are cycled with only a portion of the full design load. In most batch-type furnaces without recuperative devices, 65% or more of the maximum energy input for the cycle is required for the furnace, regardless of the size of the load. Sometimes loads can be combined even with large differences in material thicknesses. The total furnace time is set for the largest cross section, but the reduction in energy consumption can be considerable, when compared with the energy required for processing two separate loads. A careful review of heat treating records may reveal that combining loads or improving load density could save three or more additional cycles per month.

Lowering of Cycle Temperatures. A heat treating operation that is carried out at the lowest temperature normally will require the least energy, and the temperature of a cycle should be reduced whenever possible. A traditional practice in metallurgy has been to normalize at a temperature 110 °C (200 °F) above the upper critical temperature and to austenitize at 55 °C (100 °F) above the upper critical, but this may not be necessary. Instead of normalizing at 110 °C above the upper critical, a temperature 55 °C above it may be sufficient for a particular job. For hardening, instead of 55 °C above the upper critical, a temperature 28 °C (50 °F) above it may be appropriate. The furnace operator must carry out instructions as to which temperature to use, but reduced furnace temperatures can be specified.

Energy can also be saved by minimizing heating times or holding times (soak times). Many specifications require holding times for austenitizing of one hour or more per inch of cross sec-

tion. This is a conservative requirement, compared with other hardening processes (such as flame hardening and laser hardening, for example). From a metallurgical viewpoint, all that is required is that the metal be heated to a temperature above the hardening temperature. Unfortunately, design specifications are written to include excessive holding times. The metallurgist wants to achieve a certain microstructure or a certain combination of properties in a material. In repetitive situations, it would be worthwhile to discover the minimum time required to produce that microstructure or those properties.

Newer instruments for control of furnace temperature permit presetting of the desired heat-up rate. Not all material can or should be heated to temperature at the same rate or within the same time span. When the heat-up rate is specified, maximum energy savings can be realized by bringing the workpieces to temperature as quickly as possible within the constraints of the capability of the furnace and the necessity of avoiding deleterious metallurgical effects.

Trays and Fixtures. In any heat treating operation, the furnace charge often is considered to be limited to the parts being heat treated. The furnace charge, however, consists of all material that must be brought to temperature, including trays and fixtures. In job-shop operations, it often would be worthwhile to have several sets of trays and fixtures. Lighter parts could be treated in lighter-weight trays or baskets, and heavier parts could be handled in more substantial containers. This problem has been observed in batch treating of titanium parts where the net weight of the high-alloy furnace trays exceeded the net weight of the parts being treated. In another shop, shaft-type parts were being treated on a tray in a batch furnace with a grooved hearth when the workpieces could have been treated adequately without a tray. The most energy-efficient load is the one whose net-charge weight most nearly approaches the gross-charge weight.

Furnace Utilization. Most industrial furnaces are designed to be used at a specific production rate. At that rate, they will achieve the expected design thermal efficiency, but any departure from that specific design rate will result in a decrease in thermal efficiency. Once a furnace reaches a steady-state

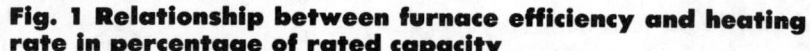

Fig. 1 Relationship between furnace efficiency and heating rate in percentage of rated capacity

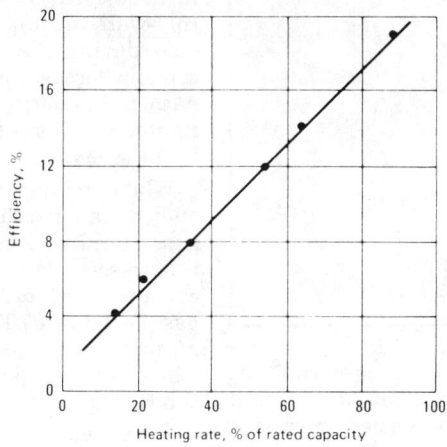

operating condition, the furnace losses will remain constant provided that the operating conditions are unchanged, regardless of whether the furnace is being operated empty or with a capacity load. This constant loss reduces the relative efficiency of the furnace when it is operated at less than its rated capacity. The curve in Fig. 1 is based on data obtained from metering of the gas used in a pusher-type heat treating furnace operating at 950 °C (1750 °F). This curve shows that furnace efficiency dropped when the furnace was operated below its design capacity.

In certain instances, heat treaters conserve energy by stacking, nesting or otherwise consolidating workpieces so as to use a furnace at its design capacity without any deleterious effect on the product. In other instances, energy can be saved by selecting the proper furnace for the job at hand. For instance, a batch furnace should never be used when a fully loaded continuous furnace can be used. Savings result not only from full utilization of a furnace but also from heating in an inherently more efficient furnace. Batch furnaces are always less efficient than continuous furnaces, in which some of the exhaust-gas heat is imparted to incoming cold workpieces.

Idling of Furnaces. Another practice that often can be modified for energy savings is idling of furnaces when not in use. For example, a batch-type furnace idling at 815 °C (1500 °F) requires 9.3 m³/h (327 ft³/h) of natural gas. Reducing the temperature to 760, 705 or 650 °C (1400, 1300 or 1200 °F)

reduces gas consumption to 6.9, 6.1 or 5.2 m³/h (245, 215 or 184 ft³/h), respectively. Thus, a 165 °C (300 °F) reduction in idling temperature reduces gas consumption and can produce a savings of 43.9% each weekend, or an annual savings of 8.6%. In this example, the furnace could produce 515 load cycles in 50 five-day weeks of operation, requiring 220 m³ (7776 ft³) of natural gas per load. If the same furnace were operated seven days a week, the same 515 loads could be processed in 30 weeks. Furthermore, the gas required per load would be reduced to 198 m³ (6995 ft³), an 11.2% savings. In addition, there would be 20 weeks of free time for repairs or for other work.

In another example, in an industrial plant with a captive heat treating facility, operators stored all flanges forged each month in front of the heat treating furnace. In the first week of the next month, the previous month's production was heat treated on a 24-h-a-day, 7-day-a-week basis and then sent to the machine shop. This energy-efficient operation was possible because this manager could stage his in-process inventory at whatever point in the process that provided the lowest over-all manufacturing cost. Although other factors, such as premium pay and urgent delivery schedules, may preclude 24-h-a-day, 7-day-a-week operation, plant managers often have introduced such a practice and have reaped significant energy savings.

If intermittent operation is necessary, the relative costs of idling a furnace and of reheating it after shutdown

can be determined with the following empirical formulas:

$$Q/Q_1 = B$$
and
$$B/K = t$$

where Q is the total amount of energy consumed in bringing the furnace from ambient temperature to operating temperature; Q_1 is an average of the combined fuel consumed per hour while idling at a given temperature and the fuel used to return from idling temperature to operating temperature; B is break-even time; K is a constant (1.7 is suggested for tube-fired furnaces, and 1.5 for direct-fired furnaces); and t is the maximum idling time in hours, that will still result in energy savings.

Thus, if it took 1223 m³/h (43 200 ft³/h) of natural gas to heat a direct-fired furnace from ambient temperature to operating temperature, and 25.2 m³/h (890 ft³/h) to idle it and to return it from idling temperature to operating temperature, the break-even point would be 1223/25.2 (43 200/890), or 48.5 h. Because the furnace is direct-fired, the optimum idling time would be 48.5/1.5, or 32.33 h. The break-even point in terms of energy savings is 48.5 h; but, considering damage to the furnace, alloy life and all other factors, the optimum idling time is 32.33 h.

Equipment Considerations

Improved Use of Equipment. In industry, if high quality and good production are being maintained, there is a natural reluctance to tamper with the equipment. But this reluctance can postpone correction of unnecessary losses of energy. Production equipment should be examined in a search for opportunities to conserve energy. With fuel-fired furnaces, the importance of proper fuel-to-air ratios cannot be overemphasized. In products of combustion, the presence of 1% CO or H_2 corresponds to a 5% loss in heat release. The problem of energy remaining in the unburned fuel can be corrected through simple adjustments of the ratio-control equipment. The effects of improper ratios (both excess fuel and excess air) are shown in Fig. 2.

If an insufficient amount of air is supplied, not all of the energy will be extracted from the fuel, as shown in Fig. 2. If too much air is supplied, additional fuel will be required to heat the

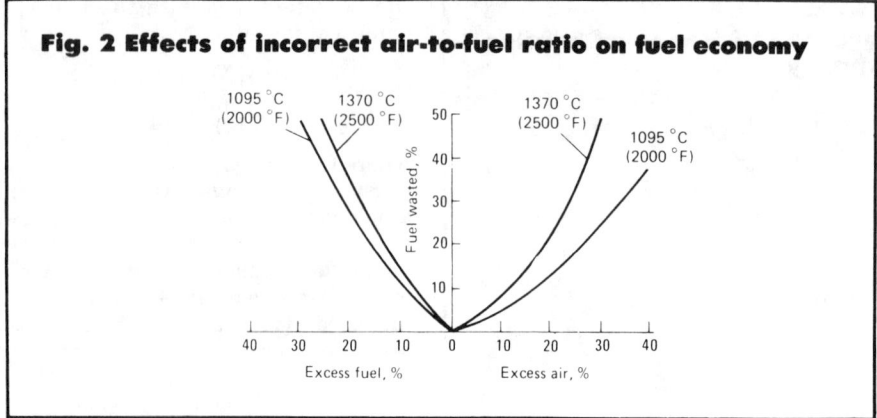

Fig. 2 Effects of incorrect air-to-fuel ratio on fuel economy

excess air. If either condition exists, a substantial loss of energy results.

Too often, furnace doors are opened too wide or left open too long, or both, when furnace charges are moved in and out. A high rate of heat loss, primarily by radiation, occurs when a furnace door is opened, and this rate of heat loss can be calculated; the following formula is often used:

$$ q = 0.173 \ Ae \left[\left(\frac{T_o}{100} \right)^4 - \left(\frac{T_a}{100} \right)^4 \right] $$

where q is rate of heat loss, A is effective area of door opening, e is emissivity, T_o is absolute temperature, in furnace, and T_a is absolute temperature of air.

Examination of this formula reveals that the amount of heat lost is directly proportional to the effective area of the door opening. The advantages of limiting door openings to that area actually required for proper access to the furnace are obvious.

Convective heat losses also are proportional to the effective area of the door opening. This loss can be accentuated when both doors are opened at the same time on a continuous furnace, because a tunnel effect is created. Convective losses also result around closed doors and through other openings in fuel-fired furnaces. To conserve energy, all unintentional openings should be sealed. The number of intentional openings should be reduced to the absolute minimum, and these openings should be sealed when not in use. Even when not permitting convective losses, openings in furnaces can permit infiltration of cold air, which must be heated to exit-gas temperature. Proper control of furnace pressure will help prevent these convective losses through necessary openings.

Proper application of refractories in reworked furnaces often can reduce or eliminate constant heat losses that may occur in water-cooled members. Water-cooled skid pipes can be replaced with alloy load supports if maximum temperature of operation permits, or with skid blocks made of high-strength, high-temperature refractories. Walking beam rails sometimes can be topped with refractory shoes rather than with noninsulating alloy shoes.

Marked energy savings sometimes can be achieved by improving heat treating quality controls. Production scrap results in a twofold energy loss; not only the energy required to treat the replacement part but also the energy required to produce and process the part to the point of heat treatment. Retreatments due to poor quenching, missed draw temperatures and any other irregularity are very expensive in terms of both time and energy.

Equipment Replacement or Modification. As the cost of energy increases, the number of economically feasible opportunities to replace or alter heat treating equipment also increases. The following items are not intended to be all-inclusive, but they will point the reader in the direction of greater savings and more efficient operation.

Heat treating equipment that is sturdy and substantially built will undoubtedly provide longer life with less downtime. Such equipment may be slightly higher in cost than equipment of less substantial design, but the difference in cost often can be justified through avoidance of energy costs that continue when equipment is out of service for minor, repetitive repairs. The most efficient furnace is one requiring the least fuel per operating hour, not necessarily the one with the best hour-

ly design efficiency. Fuel cost per unit weight of product produced should be the critical measure instead of cost per furnace-cycle hour. For instance, a cold-wall vacuum furnace uses energy only during the actual heat treating cycle, whereas a conventional box furnace uses energy from the time it is lighted until it is shut down at the end of the week.

When the opportunity arises, much energy can be saved by altering the gases used in atmosphere heat treating. In some cases, an inert carrier gas can be added to the normal working gas. Perhaps a 100% substitution with an inert gas is possible during periods when working gas is not required, but the furnace must be kept "bright".

Other energy-saving opportunities may be realized by improving the insulation systems in heat treating furnaces. The energy lost through a furnace wall is a function of area, operating temperature and composition of the insulation. The first two factors are fixed, but heat flow through and heat storage in the insulating system can be reduced by addition of insulation. Heat storage, which is the amount of heat contained in the wall, can be greatly reduced by using newly developed insulating materials, principally ceramic fibers and mineral wools. At the same time, the thickness of a wall for a given heat flow (loss) also can be changed significantly. One major heat treating firm not only reduced energy requirements by using ceramic-fiber insulating materials, but also was able to heat treat larger rolls because of the decrease in required wall thickness, which in turn provided greater furnace work-zone width.

Addition of alloy fans to existing furnaces sometimes can change a stagnant atmosphere into a high-velocity stream. This increase in ambient velocity breaks up boundary layers of furnace gases that surround the workpieces and shortens the heating time. This reduced heating time is a result of the change from heating only by radiation heat transfer to a combined radiation and convection transfer. The energy savings accrue through reduced furnace time required per cycle.

Thermodynamic thermal efficiency also can be improved by use of waste combustion gases. A particular batch furnace, heated to 980 °C (1800 °F) with natural gas, with the flue gas exiting at or near that temperature, operates at only about 35% efficiency. Through use of a recuperator, the incoming combustion air can be heated by the waste products of combustion. If the combustion air is heated to 425 °C (800 °F), a 23% savings in fuel usage can be achieved. This method is perhaps the least efficient way of reusing heat from the products of combustion; sometimes, however, it is the only practical way.

Because the coefficient of heat transfer between combustion products and a solid is twice as high as between combustion products and incoming combustion air, it is better to use waste gases to preheat the workpieces, as is done in properly designed continuous direct-fired furnaces. Other commonly used recuperative methods also are employed successfully on continuous furnaces, but some of these methods employ a lower heat-transfer coefficient. Methods may vary, but increases in the cost of energy are likely to move the heat treating industry more and more toward use of waste gases.

Furnace Safety

By the ASM Committee
on Furnace Safety*

HEAT TREATING FURNACES require safety procedures common to all industrial installations, but in addition, they have requirements specific to the use of high-temperature energy sources and potentially explosive gases and liquids used as aids to chemical processing.

Because these heat treating processes require careful control for optimum technical results as well as for safety, proper training of operating personnel is a primary consideration. Proper equipment design is also critical.

The information presented here is not intended to be interpreted as a safety standard but is offered only as a set of guidelines. Safety standards for furnaces are maintained by the National Fire Protection Association, by the U.S. Occupational Safety and Health Administration and by insurance underwriters.

All equipment should be installed and operated with awareness of the potentials for fire and explosion and the hazards to operators and equipment. Equipment designs should ensure reliable, safe operation over the expected maximum life of the equipment.

Fuel-Fired Furnaces

Fuel-fired furnaces for heat treating have several major control requirements that depend on (a) whether the process must be direct or indirect fired; (b) whether heat treating is to be done under a particular pressure or vacuum, or in a controlled atmosphere; and (c) whether the product uses some special type of precoat or laminant. In all situations, there are fundamental control variables, and instrumentation is available to identify and control change and drift, thus achieving the desired results.

The main control elements are the three requirements for proper combustion: a source of heat, an oxidizing agent and time.

Fuel-fired furnaces can be automated to the extent that manual intervention is not required for normal operation. Many processes and operations require manual control, however; thus, furnace controls range from almost completely manual devices to highly sophisticated computer-controlled devices.

The major control variables and types of instrumentation used in each instance are described below, in sequence, from start-up through final cycle control.

Electrical Power for Fuel-Fired Furnaces

The safe use of electrical energy employed in heat treating control processes requires adherence to National Electrical Codes and to local requirements of states and communities. Good practice dictates that a circuit breaker be positioned within view of the operator. Control cabinets must be designed to ensure that operators cannot inadvertently become a path to ground. Wiring type should be based on the environment of use, and wiring for all motor and control circuits should be contained in appropriate conduits. Numbers of wires within specific conduits should be governed by the fact that elevated temperatures may be encountered. All enclosures for electrical apparatus should be designed to protect the contents from the environment. The furnace itself should be grounded for proper control.

The source of electrical power to the furnace installation should be equipped with fuses. Each motor also should be equipped individually with fuses and protected with thermal-overload elements based on operating temperatures. Motor selection should be based on such conditions of use as temperature, weather, dust, dirt, atmosphere and humidity. Manufacturers should be consulted on motor design and selection. Control-panel power should be fused and provided with externally operated switches to allow safe entry by authorized personnel.

Electronic-signal wiring from the flame-safety circuits should have its own conduit, free from the "noise" and induction present in normal power and

*Raymond Ostrowski, *Chairman*, Sales Manager, Protection Controls, Inc.; Fred J. Bartkowski, Vice President, Marshall W. Nelson & Associates, Inc.; Roger G. Blocks, President, Chem-Al, Inc.; Don G. Ensweiler, President, Heat Process Associates, Inc.; Ross Shingledecker, Metallurgical Director, Manufacturing Services, Ladish Co.

Reprinted from Metals Handbook, 9th Edition, Vol. 4, 378-385, © 1981 American Society for Metals

control circuits. Thermocouple wiring also should be contained in its own conduit to avoid creating induced errors from such random sources as power lines, signal wiring, motors and ballasts.

Control Circuits for Fuel-Fired Furnaces

Combustion-Air Blower Control. Combustion-air blowers must be interlocked with the combustion-limit circuits to shut down the process in the event of failure. The flow of combustion air must always be proven before and during a processing cycle with two independent sources of information. The motor should be protected from short circuits with fuses and from overheating or amperage draw with thermal breakers (heaters). The motor starter should be wired so that it will disconnect when any phase is interrupted or when the motor malfunctions. It should not be assumed that the blower is providing combustion air just because the blower motor is operating; combustion-air flow must be proven. At one time, an end switch, or rotary switch, on the motor was a common indirect method of gaining this information. A better method is to use a pressure switch in the air line for direct sensing. A sail or flag switch, although not quite as good because of the mechanical movements required, also can be used to sense air flow directly.

Gas-Pressure Control. Fuel must arrive at the burner in the correct quantity and at the correct time for safe combustion. Fuel pressure thus must be proven within an allowable range. Gas-pressure switches for both high and low gas limits are installed in the main gas lines. Visual pressure gages also are helpful to operators in setting burners and in verifying that the fuel is being supplied within the range desired and that pressure limit switches are not malfunctioning. Mercury-wetted relay pressure switches are recommended for their ease of setup and maintenance and for their reliability.

Pressure Regulators. Pressure of gaseous fuel is most commonly regulated by pressure-regulating diaphragms. Good, safe design normally requires one regulator for pilot fuel and one regulator for main-burner fuel. The pilot gas, if taken from the main fuel line, should be drawn from a point between the gas supply and the regulator for the main fuel. Thus, the pilot and main burner can be set up optimally, safely and independently. The regulators should be vented to a safe location outside the plant to ensure safety if a regulator diaphragm is damaged in service. Good practice and manufacturers' recommendations show that diaphragm life can be substantial if regulators are shielded from thermal radiation and are used below their maximum design limits. Positive lock-up regulators are recommended to prevent downstream pressure buildup during shutdown periods.

Valves. Blocking valves normally are closed valves that are energized only by the combustion-control circuits. The pilot-gas blocking valve is placed downstream of the pressure regulator and a hand-operated gas cock. A pipe union should be inserted just ahead of the electrically operated blocking valve to allow safe removal if repairs are needed. The blocking valve is opened to the pilot assembly only after the furnace is purged.

Purging of Fuel-Fired Furnaces

The furnace must be purged of any possible combustible materials. This is best accomplished by opening the furnace doors, which should be equipped with a limit switch to ensure that they are opened adequately. Once the doors are open, the combustion blower or exhaust fans can be timed to allow for a minimum of four changes of air in the combustion chamber. This purge cycle is required for safe start-up and is standard practice for all well-managed operations.

Pilot Control

Pilot assemblies can be of either the atmospheric or the blast type. The atmospheric type is similar to an atmospheric burner, in which the air is inspirated from the atmosphere by the gas stream.

In the blast type, air and gas are brought to a mixer under pressure. The gas is then reduced to atmospheric pressure and pulled into the mixer by the pressurized air stream. This is the most positive means of pilot-gas control.

Ignition

For ignition trials, a high-voltage transformer is used in conjunction with a spark plug designed for the pilot or burner assembly. The control circuit causes the pilot valve to open and a spark to be produced. The spark continues for a short period (normally 15 s) and establishes a flame, which can be detected. If the flame is not established, because the flame or signal is inadequate, the cycle returns to the purging stage.

The voltages normally employed are approximately 5000 to 6000 V, and the high-voltage transformer is normally mounted on the furnace and grounded to it. The spark in turn is grounded to the pilot assembly, then to the furnace; hence, a well-grounded furnace is an important safety requirement.

Flame Detection

A thermocouple junction placed in intimate contact with the pilot flame is perhaps the most common means of flame detection, but thermocouples are useful only on very small pilot assemblies or burners, and they are not useful after a burner becomes hot. The flame may no longer be present, but a hot burner block or refractory may retain heat and slow the rate of thermocouple cooling. Thus, thermocouple junctions are not recommended except for quench-tank heaters of the constant-pilot, open-grid burner design or for small atmospheric burners.

Flame electrodes which are small anodes of heat-resisting alloy placed in intimate contact with the normal pilot flame, work on the principle that flame causes ionization within the burner atmosphere and thus allows a circuit to be formed to ground. The flow of a minute amount of current, at low voltage, is sufficient to sense and communicate the presence of a flame.

Flame electrodes are common on all industrial heat treating furnaces where the flame is kept on-ratio or slightly oxidizing. The flame electrode tends to become carbon coated in a reducing flame, a condition that can cause nuisance shutdowns.

Ultraviolet scanners are the third common device for sensing flame. They are normally dependable if the lens viewing the flame is kept clean. The UV scanners must not be used in any application where ultraviolet light is present from a source other than the burner in question. The UV scanner is a useful and practical device for any clean-flame, clean-furnace operation, if it is located and aimed properly. A flow of clean, filtered cooling air across the scanner face aids in keeping it clean and cool, extending scanner life appreciably.

Depending on the burner used, the application, and property-insurance requirements, it may be necessary to monitor both the main burner and the pilot flame independently.

Common and serious errors in flame detection are made by operators who circumvent flame-safety equipment rather than correcting the usually minor problems that cause nuisance shutdowns. Flame-safety equipment that uses totally enclosed relays is recommended in favor of types with accessible relays that may be kept open, for example, with a piece of paper. This point, however trivial it may seem, has been profoundly recognized by those firms who have lost operators, furnaces and product as a result of poorly designed flame-safety equipment that can be circumvented easily. Any employee found tampering with this equipment should receive disciplinary action, and all employees should be trained in use of flame-safety equipment.

Burner Operation

The main fuel supply for fuel-fired heat treating furnaces normally is natural gas, propane-air, propane, butane or one of the fuel oils. Although this discussion centers on natural gas, the same principles apply to the other gases and oils.

The main gas valve may be of the manual-reset type, requiring an operator, or may be fully automatic. The manual type usually is preferred when the furnace is run intermittently or when operators must perform some other function, such as opening doors. When the operator opens the valve, he is in effect making a conscious decision that conditions are ready for the main burner heat. The valve may be made automatic when the furnace is designed and interlocked to preclude an unsafe condition.

For furnaces with capacities greater than 422 MJ/h (400 000 Btu/h), it is recommended that a second blocking valve be inserted into the main gas line and that a normally open vent valve be installed between the main blocking valves. The vent valve should be vented out of doors away from any openings such as windows, air intakes and exhaust louvers in the building walls or roof. This "double block and bleed" arrangement prevents faulty valves or dirty valve seats from causing leakage

of fuel into the furnace between operating cycles. It is not totally foolproof: the open vent may malfunction and allow valuable fuel to leak into the atmosphere during operations, or the internal mechanisms may malfunction because their design is unsuited to operating factors such as environmental conditions. Good preventive maintenance is required. One survey showed that 1 out of every 10 blocking valves was faulty after operating for no more than 10 years.

Burner Control. The gas-air ratio ordinarily is controlled to about 10 parts air to 1 part natural gas for good combustion efficiency. There are several devices involved in control of this ratio. Typically, the amount of blower air is varied by a butterfly valve to satisfy the demands of a temperature-control device. A pulse or static pressure line is connected from the combustion air line, downstream of the butterfly valve, to a proportionator valve located in the gas line. The gas is then regulated by the ratio-control valve in proportion to the air flow, and the air-to-gas ratio remains constant throughout the firing range. The devices used to regulate the ratio fall into two broad categories: the diaphragm type, or proportionator, which uses the pulse line to keep air and gas at a specified ratio; and the mechanical-linkage type. Both are effective and common, but the diaphragm type is the more positive, because there are no linkages that can slip and require adjustment. Also, if air lines should become dirty, resulting in a lessening of air pressure, the gas pressure will follow, maintaining the correct ratio.

Ratio control alone is not sufficient to ensure safe start-up of the main burners. It is recommended that the burners be set to a low firing rate when the main burner is started. This may be done either automatically or manually. Once the main burners have been started, the furnace doors may be shut and the furnace brought up to temperature.

Temperature Control

Temperature-control devices fall into two categories: primary controls and process limiting devices. Safe operation, especially when furnace practices require long cycles and little operator attention, dictates that limits be placed on the process to ensure adequate alarm and perhaps to shut down the operation to prevent destruction of the

product, the furnace or the plant itself. Whether an analog device, a strip chart recorder, digital readout or a printout is used is a matter of operator preference and depends on the nature of the product.

The typical temperature sensor is either a thermocouple or a resistive temperature device (RTD). The thermocouple is most common. Several types of thermocouple junctions are available, with the choice depending on such factors as temperature range and furnace atmosphere. They are comparatively inexpensive and can be easily protected from atmospheres with protective "wells", which are immersion tubes that project into the furnace zone to be controlled. RTDs, although more accurate than thermocouples by factors ranging from 10 to 1 up to 50 to 1, are expensive and less rugged. For most purposes, thermocouples are satisfactory. Some firms are using heat-flow sensing to remotely ascertain interior temperatures and to provide an element of redundancy for protection of furnaces and their contents. Good temperature-sensing devices will detect failure of a thermocouple or RTD, cause the process firing rate to be reduced to its minimum rate, and perhaps provide an alarm.

Furnace temperature can be regulated by one of two very common procedures. Simple high and low firing rates are used when temperature can be allowed to vary within a fairly large range. More common, in heat treating, is the use of proportional control, wherein the temperature is held nearly constant through the use of a bridge circuit. This circuit balances the signal between the controller and the butterfly valve and holds the latter at the proper opening to maintain the desired temperature. The latter scheme, although more costly, is required for close control.

Waste-Heat Recovery

Recuperative devices used for conserving energy present special problems in safety and control instrumentation. Typically, these recuperators use the products of combustion for preheating of the combustion air. Shell and tube heat exchangers are normally used in this type of arrangement. Because preheated air becomes less dense, the air temperature must be sensed, and control of the gas-to-air ratio must be based on this temperature-density function. Experience with these devices has shown that such factors as poor design, poor gasketing, leaks in heat-exchange surfaces, overheating of burner blocks, and failure to allow for expansion and contraction have caused numerous operational problems that constitute safety hazards. Tracking of air-gas ratios can be affected by leakage of mechanical seals, and products of combustion can enter the combustion air. Thus, it is recommended that oxygen analyzers be used periodically to check the combustion air immediately ahead of the burner block. This analysis will reduce the likelihood of erroneous and perhaps hazardous conditions in the furnace, will give clues to potential design changes needed, and will give warning of part deterioration. Further, it is recommended that heat treating operators monitor the room atmosphere for carbon monoxide on a periodic basis. Although exhaust may be provided for products of combustion and sufficient air exchanges may be occurring to satisfy state regulations, there may be a temperature inversion that can cause leaking of products of combustion inside the building, resulting in a potentially dangerous situation.

Supervisory Gas-Cock System

A supervisory gas-cock system is used to ensure a safe "lightoff" procedure on a manually ignited, multiburner furnace that does not have flame-safety equipment and a programmed sequence of piloting the main burners.

The system consists of specially designed gas valves that have inlet and outlet passages for a checking pressure medium such as air or gas. Air or gas—usually air from a combustion blower—can pass only through the valve when the valve is fully closed. When the individual burner valves and the main gas line valve are closed, the air flow enters a pressure switch that closes and completes an electrical or pneumatic circuit. This allows the main gas valve, usually of the manual-reset type, to be opened. The burners are then individually manually ignited.

Supervisory gas-cock systems are used on radiant-tube furnaces and other furnaces where flame-safety systems are difficult to apply. Fewer of these systems are being used on new furnaces, because most burners now are adapted to flame safety and automatic ignition.

Electric Furnaces

Electric-furnace installations are made up of various electrical and mechanical components, many of which are water cooled and equipped with protective devices.

Furnace manufacturers generally issue instructions concerning safe practices, and these instructions should never be ignored. Potential hazards can be avoided by ensuring that operating personnel are trained thoroughly and that installations conform to safety practices and local codes.

Original equipment usually contains devices for preventing overloads and short circuits. In addition, ground detectors and surge detectors protect motor-generator units from faulty coil or transformer installations at heating stations and from breakdown of insulation in the generator windings.

Protective devices commonly used with induction-heating radio-frequency generators are as follows:

- Door interlocks
- Grounding devices to ground high-voltage circuit when furnace doors are open
- Warning lights
- Warning signs
- Circuit breaker for entire unit
- Overload relays
- Water-flow switches
- Water-temperature switches
- Time-delay relays (tube warmup)
- Grid overload relays
- Control-circuit overload relays
- Arc gaps on blocking and tank capacitors
- Surge protection
- Electronic crowbar.

Operators should become familiar with these safety devices and should inspect them periodically to ensure that they are in good working condition.

Electrical Power. Although motor generators account for the largest total power output of installed induction-heating equipment, vacuum-tube oscillators probably occur in the greatest numbers of units. Many small vacuum-tube oscillators are required to account for as many kilowatts as one 1250-kW, 3-kilocycle motor-generator set. Although many vacuum-tube oscillators for induction heating are made in small sizes, 25-kW and 50-kW outputs are also common ratings. Some have been constructed for special applications with ratings as high as 500 kW. Many

small composite (custom-built, or home-made) vacuum tube units are in use also.

Power Interlocks. Most systems produced by reputable manufacturers are designed to be completely "fail-safe". These systems also have interlocks that automatically shut down the power supply if a fault develops during operation. The system cannot be restarted until the fault is corrected. Interlocking systems also are used to increase production of induction-heating machines.

Induction-heating machines are expensive; therefore, steps are taken to keep them busy as much of the time as possible. If a hardening process, for example, requires 5 s of heating, followed by 5 s of quenching (before the part is moved from the inductor), and if another 5 s are required on the average for the operator to load and unload the part, then the generator itself is only in use one-third of the time. Production can be increased appreciably by having the same basic equipment supply two or more individual work coils (Fig. 1). This can be done by arranging the control circuit so that, if one of the work coils is demanding heat, none of the others can be started. If an operator pushes the start button at one station while another is heating, a relay withholds the actual start of heating until the first station has completed its high-frequency power demand cycle.

It is also customary in the case of interlocked multistation operation of motor-generator equipment to preset the alternator field current for the various stations. Each station has its own field-adjusting autotransformer or potentiometer, which is automatically switched into the circuit when that station has the power. This is feasible because, with only one station on at a time, it is not necessary to use the same voltage at each.

Fixtures. As introduction of automated systems increases in induction heating, the need for safety controls increases beyond the greater need for such devices as power interlocks.

For example, with highly automated induction-heating machines, a part completely foreign to the part to be treated may enter the automatic feed hopper or bin. If the fixture tries to feed this part into the coil, mechanical jamming and damage may result. If the part does get pushed into the coil itself, assorted problems may result, especial-

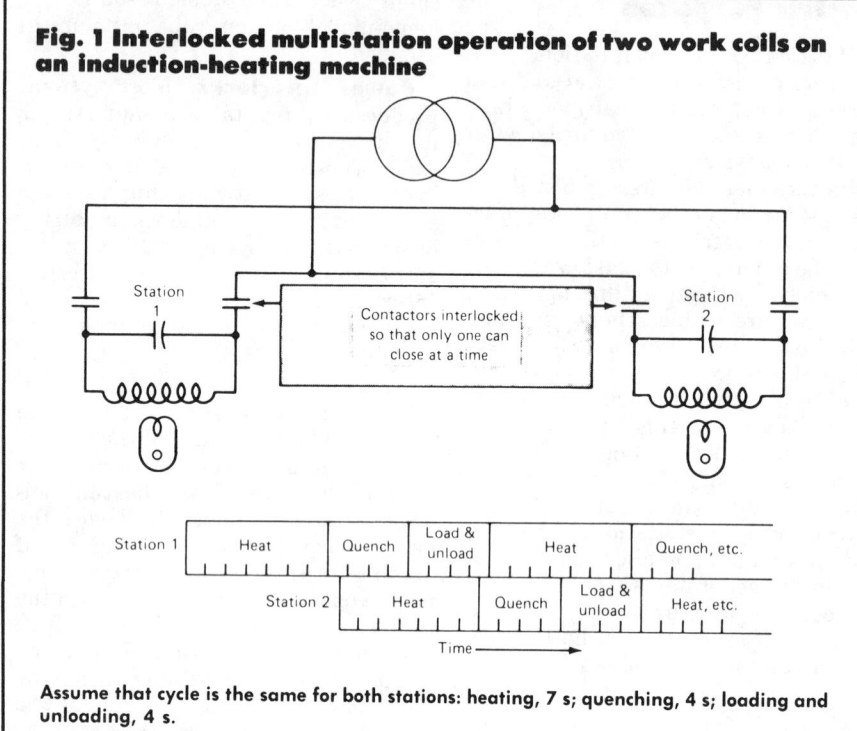

Fig. 1 Interlocked multistation operation of two work coils on an induction-heating machine

Station 1

Station 2

Contactors interlocked so that only one can close at a time

| Station 1 | Heat | Quench | Load & unload | Heat | Quench, etc. |
| Station 2 | | Heat | Quench | Load & unload | Heat, etc. |

Time ──────▶

Assume that cycle is the same for both stations: heating, 7 s; quenching, 4 s; loading and unloading, 4 s.

ly if the unwanted part is larger than the inside diameter of the coil.

The greater the degree of automation, the greater the necessity for safety devices that further complicate the machine. Some automatic induction-heating machines incorporate templates through which the workpiece must pass before being fed to the more delicate parts of the mechanism. If the part touches the template, an electric circuit stops the machine until the error is manually corrected.

Spurious Radiation. Industrial heating equipment using radio-frequency energy is, in many instances, governed by rules of the Federal Communications Commission (FCC). FCC rules apply to induction-heating equipment operating at 10 kHz or above. Any operation whatsoever in frequency bands reserved for international distress equipment is strictly prohibited. One of these bands is from 490 to 510 kHz.

The best way for a user of induction-heating equipment to become familiar with the FCC rules is to acquire a copy of Part 18 of the rules and study them carefully.

Vacuum tube oscillators and some motor-generator equipment fall into the FCC-designated categories. The rules state that operation must either be within certain narrow frequency bands (in which any amount of energy may be radiated) or be restricted in field strengths. All such equipment must be certified by a competent engineer, such certification being based upon actual measurements of field strength made around the equipment. In some circumstances, prototype models of industrial high-frequency heating equipment may be tested at the manufacturer's plant and a certificate issued to cover other equipment of the same design. Even though a specific piece of equipment may have been properly certified, and even though its spurious radiation may fall below the prescribed limits, FCC rules state that, if it interferes with communications equipment, further corrective action must be taken. The mere existence of a certificate, therefore, does not necessarily absolve the user of further responsibility.

In terms of output, induction-heating machines sometimes rival the largest communications transmitters. The frequencies, and harmonics thereof, used by many induction-heating oscillators fall within the range utilized by their more delicate counterparts in radio and television services. If only a small portion of the power output of high-frequency heating machines were to be broadcast as unwanted (spurious) radiation, the results would be catastrophic.

Historically, dielectric heating machines have caused more interference than have induction-heating machines; they operate at higher frequencies and are more difficult to shield. However, induction-heating machines, especially vacuum tube oscillators, have also caused trouble. It is necessary to observe certain precautions in their design and operation to ensure that they do not create interference. Reputable manufacturers of induction-heating equipment take precautions to protect users of their equipment from this type of trouble. They house their equipment in heavy steel cabinets, and provide instructions which, if followed by the user, will ensure conformity with FCC rules. However, it must be emphasized that the final legal responsibility for a piece of equipment lies with the user.

Maintenance. Electrical heat treating equipment is expensive, and standby equipment generally is not maintained. Thus, preventive maintenance is critical, and ready availability of replacement parts is highly desirable.

Dust, dirt, moisture and high ambient temperatures are the primary causes of electrical equipment failures; these conditions are commonly present in industrial locations where induction-heating units are installed. In any maintenance program, warnings should be highly visible and clearly stated. The following is a typical warning:

"If the interlocks are disabled and the main circuit breaker is on with the door open, potentially lethal voltages are exposed. There is always 460 V ac present behind the control circuit breaker and on the line side of the main circuit breaker: care should be exercised at all times when the door is open. Power should be removed by opening the feed breaker or disconnect switch external to supply before working within cabinet. Solid-state circuit breaker board and isolator board are connected directly to 300 V ac. Turn off all breakers and allow one minute for capacitors to discharge before working on these boards."

Special Heat Treating Processes

Certain special heat treating processes using such systems as lasers, electron-beam heating, plasma carburizing and ion nitriding have their own unique safety requirements in addition to standard safeguards associated with high-temperature processing.

Safety of personnel is paramount, but safety of equipment often ensures personnel safety. Thus, proper care and use of equipment cannot be over stressed and frequently becomes almost synonymous with safety.

In this section, ion nitriding and plasma carburizing are used as examples of special processes and the safety precautions related to them. For all heat treating systems, however, special safety problems can be solved through sound training programs for operators and through effective and regular maintenance.

Ion Nitriding. In this system, also known as "glow-discharge" nitriding and as "ionitriding", parts are connected to the cathode for processing, and the retort is the anode. After the retort is evacuated of atmospheric gases, nitrogen and hydrogen are bled slowly into it. The glow discharge is produced when the parts are heated by electric current to approximately 500 °C (930 °F), although specified temperatures can be as high as 565 °C (1050 °F).

The retort becomes heated by radiation from the parts; additional heat is not required. The glow discharge ionizes the nitrogen, and the electrical potential accelerates the movement of the ionized nitrogen toward the parts.

Although ion nitriding is faster and produces a more ductile and fatigue-resistant case with less white layer, the extra handling and precautions it requires is an important factor in over-all cost.

Plasma Carburizing. The normal range of electrical power used for plasma-arc processing is 25 to 50 kW. Most systems are of "fail safe" design and are interlocked to shut down the power supply automatically if a fault develops during operation.

One of the most serious hazards associated with plasma-arc operation is radiation caused by electromagnetic high-electron excitation. Such radiation ranges from radio frequencies to the far ultraviolet, and it includes infrared and visible radiant-energy light rays.

The radiation produced by the plasma is capable of producing severe eye and skin burns. The plasma should never be observed with the naked eye.

Furnace Protection. The primary safety feature of a surface ion nitriding furnace is the arc-control system. Successful ion processing requires application of 300 to 1000 V dc to the workpiece. In these voltage ranges, the potential for an arc to form between the cathode and the anode is quite high during the initial part of the cycle. These arcs can be quite small; under certain circumstances, however, major arcs can occur that may be powerful enough to rupture vessel walls.

One method of controlling arcing and protecting equipment is to place a large resistance in series between the furnace vessel and the power supply. This prevents overloading of the power supply. An electrical device used to construct a type of "arc-shutdown" circuit is a Saturable Core Reactor (SCR), which operates with a low-resistance load at steady voltage.

When an arc begins to form there is an initial rapid increase in current prior to formation of the arc. Upon arc formation, the voltage drops drastically (dv/dt); the SCR senses the voltage change (dv/dt increase) and increases the resistance, thus protecting the power supply. Additionally, the SCR is normally connected to another electrical device, such as a bridge rectifier. Once the change in voltage signifying arc formation is detected, the rectifier damps the current supply. This damping is normally sufficient to shut down the arc by allowing redistribution of the energy on the portion of the part that was arcing.

This type of circuit reacts after the arc has formed, however, and the potential for ruined workpieces or holes in anodes or vessel walls remains. This is a particular problem if the unit requires operators to manually shut down the power supply after observing dead shorts, formed by misloaded parts that create short circuits.

Arc Suppression. Some equipment used in ion nitriding and plasma carburizing does not rely on an arc-shutdown circuit as the primary safety factor. Rather, true arc-suppression circuits are employed. Such a circuit senses the change in current just prior to the formation of an arc and shuts the power off completely, preventing the arc from forming. The power is then proportionally ramped back on, allowing redistribution of energy.

In one system, a counter in the controlling microprocessor tallies the number of times the power is turned off and on.

If the potential for arcing is too large, as determined by the logic preprogrammed into the microprocessor as part of the executive command package, the system shuts down, prints a fault message on both the cathode-ray tube and data logger, sounds an alarm and siren, and lights a warning light. All this occurs automatically without an arc forming and protects both the part and the equipment. The microprocessor detects a dead short by sensing and reporting the rapid frequency of shutdown/start-up cycles and prints a different "short" warning message, also with alarms, siren and lights. This sequencing is also totally automatic and requires no operator interface. The equipment has a backup system that operates on the dv/dt principle. Additionally, all leads through the vessel wall that could carry current to the power supply are triple-protected. As an example, the thermocouple has primary protection through an outer ceramic insulator. This is followed by a second powder ceramic insulator. The third protection is a high-voltage isolating amplifier between the thermocouple and the processor.

Thermocouples. The problems associated with passing a thermocouple, or any other lead, through a vessel wall in ion-processing equipment are as follows:

- The lead can become metallized, creating a pathway for catastrophic arcing. This problem becomes increasingly severe as a function of vessel use.
- The material of construction used for seals through the vessel wall is quite critical. The conditions of ion processing can affect the sealing materials, allowing increased leakage as a function of vessel use.
- Temperature uniformity throughout the entire workload in the vessel is quite critical. Proper design of thermocouple insulators that are calibrated to give true temperature readings, as well as proper design of fixturing, is necessary to achieve uniform temperatures.

Fixture Design. Design of the fixture is critical to the successful application of ion processing. Poorly designed fixtures can allow overheated or underheated parts. If equipment without an arc-suppression circuit is used, parts can be ruined because of poor fixture design. The fixture often allows simple masking and, through proper design, can minimize or eliminate the hollow-cathode effect. This effect is signified by either (a) failure of the glow to uniformly penetrate the interior surface of a hole or cylinder, or (b) overheating and possible melting of the part, caused by overlapping glows.

Atmosphere Furnaces

Atmosphere furnaces must be considered in any discussion of furnace safety because of the potential explosion hazard produced by introduction of special flammable atmospheres.

Although many of these furnaces are supplied with "inert" gas purging and standby emergency purging, training of operators in manual "burn-out" procedures is extremely important in the event of failure of automatic controls. These emergency instructions may vary with equipment design, and thus the importance of consulting and understanding emergency procedures, as outlined by the furnace manufacturer's instructions, should not be minimized.

Protective Controls. Protective devices should be installed and interlocked and should include the following:

- A safety shutoff valve on the atmosphere supply line to the furnace
- An atmosphere gas-supply monitoring device that permits the operator to visually determine the adequacy of atmosphere gas flow at all times
- A sufficient number of temperature-monitoring devices to determine temperature in all zones of the furnace; these should be interlocked to prevent opening of the atmosphere-gas-supply safety shutoff valve until all zones are at or above 760 °C (1400 °F).
- An automatic safety shutoff valve for flame curtain burner supply gas; this should be interlocked to prevent opening of the valve when furnace temperature is below 760 °C (1400 °F).
- Audible and/or visual alarms to alert the furnace operator of abnormal conditions
- Manual door-opening facilities to permit operator control in the event of power failure.

Operator Training. The most essential safety consideration is the selection of alert and competent operators. Their knowledge and training are vital to continued safe operation. New operators should be instructed thoroughly and required to demonstrate an adequate understanding of the equipment and its operations.

Regular operators should receive scheduled retraining to maintain a high level of proficiency and effectiveness, and all operators should have ready access to operating instructions at all times. An outline of these instructions should be posted near the furnace.

Operating instructions generally are provided by the equipment manufacturer, and these instructions include schematic piping and wiring diagrams. All such instructions should include procedures for light-up, shutdown, emergencies and maintenance.

Operator training should include instructions in:

- Combustion of air-gas mixtures
- Explosion hazards
- Sources of ignition and ignition temperature
- Atmosphere gas analysis
- Handling of flammable atmosphere gases
- Handling of toxic atmosphere gases
- Functions of control and safety devices
- Purpose and basic principles of atmosphere-gas generators.

This listing is intended only to serve as a guideline; specific requirements are covered in the following standard issued for furnaces by the National Fire Protection Association: "NFPA 86C, Industrial Furnaces, Special Atmospheres, 1977".

Process Cooling

Heat treating of metals includes controlled cooling or quenching of the heated metal; metals are cooled from the specific treatment temperature in a variety of media which include air, oils, salts, water and synthetic fluids.

As a general rule, furnace equipment does not include instrumentation for control of the safety aspects of the quench media. Normal practice in layout of plant and facilities will provide for isolation of air-cooling areas, for the usual pedestrian protection at pits and for the necessary building protection should an uncontrolled conflagration

occur due to the quenching operation. However, certain equipment can be specified to ensure safe and controllable operation in specific cooling processes.

The greatest concern exists for fires associated with oil quenching. All of the ingredients for a dangerous fire—fuel, oxygen and a source of ignition—exist at the surface of an oil tank.

The most common type of fire occurs when movement of a hot workpiece is obstructed as it enters the quench oil. The result is sustained ignition and vaporization that continues as the liquid is locally heated above its flash point. Prompt immersion removes the source of local vaporization, and local flashing is extinguished by normal agitation of the oil.

A second type of fire occurs when the main body of oil is heated above the flash point because of malfunction of heating or cooling equipment, or when the quench load is greater than that for which the system was designed. When an ignition source is supplied, the resultant fire soon reaches full intensity and is very difficult to extinguish.

A third and less likely type of fire occurs because of material-handling accidents that involve spills on or near heated furnaces or cooling equipment.

Equipment is available for detection of fires and release of control media. Automatic water-spray systems are usually recommended in buildings of fire-resistant or noncombustible construction, and areas adjacent to quench-oil tanks can be protected with automatic sprinklers. Quench oil and water should not be mixed, however. High-value oils can be protected with automatic carbon dioxide or dry chemical systems. In general, these automatic systems are specified for centralized large-capacity quench-oil systems.

Special equipment is available for totally enclosed systems that operate in special atmosphere furnaces. Because water and quench oil do not mix, however, this incompatibility can be a source of trouble in these units. Water in oil rapidly turns to steam when locally heated beyond 100 °C (212 °F). This steam can cause violent boilover, increased pressure within the enclosed system, and forcible ejection of burning oil from small openings. Commercial safety equipment capable of detecting even small quantities of water in oil can be arranged to alert the furnace operator and to interrupt the quench process.

Quench-tank heating systems should be equipped with all of the safety devices normally used in conjunction with the particular heating method chosen. Overtemperature safety systems are essential. In addition, system interlocks between heating media actuators and both agitators and pumps will prevent local overheating of the bath.

Mechanical Equipment

Material handling cannot be separated from other safety considerations associated with heat treating operations. Many mechanical operations must be performed before, after, and sometimes during the actual temperature-induced transformation that usually occurs while the work is in the heat treating furnace. Doors must be opened and closed, conveyors started, rolls advanced and mechanical handling equipment activated. All of these mechanical operations constitute hazards of varying degrees of severity that must be evaluated. Some are serious enough to warrant the introduction of safety equipment to preclude serious injury of personnel and damage of expensive furnace equipment.

Furnace doors are often interlocked with other components of a facility through use of limit switches that prevent inappropriate opening or closing. For instance, in an atmosphere furnace, inner doors cannot be opened until proper vestibule ambient conditions have been restored after parts are removed. Interlocks on doors may include complex designs to prevent inadvertent opening during a power failure, on a hydrogen atmosphere furnace. There are simpler designs that prevent closing of a door on a simple normal-izing furnace before the extractor is removed. Some door interlocks are connected to pressure- or temperature-sensing devices. Such devices are used with vacuum furnaces in which inadvertent exposure of the molybdenum-graphite heating elements to air at high temperature would be disastrous. Moreover, without door interlocks, roller hearth or walking-beam furnaces could advance workpieces into an unopened discharge-zone door.

Moving parts of furnaces represent potential hazards that can be neutralized rather easily with simple time-delay relays connected to audible alarms, which in turn are connected to the start buttons of electric motors. For instance, the conveyors on some large continuous furnaces cannot be advanced until an alarm has been sounded and a timer has allowed sufficient time for workmen to stand clear.

SECTION XIV
Testing and
Quality Control

Microprocessor-based control + ion nitriding = high quality

The precise parameter manipulation that ensures
consistent metallurgical results is a major
reason one automaker opted for this system.

by RICHARD CREAL

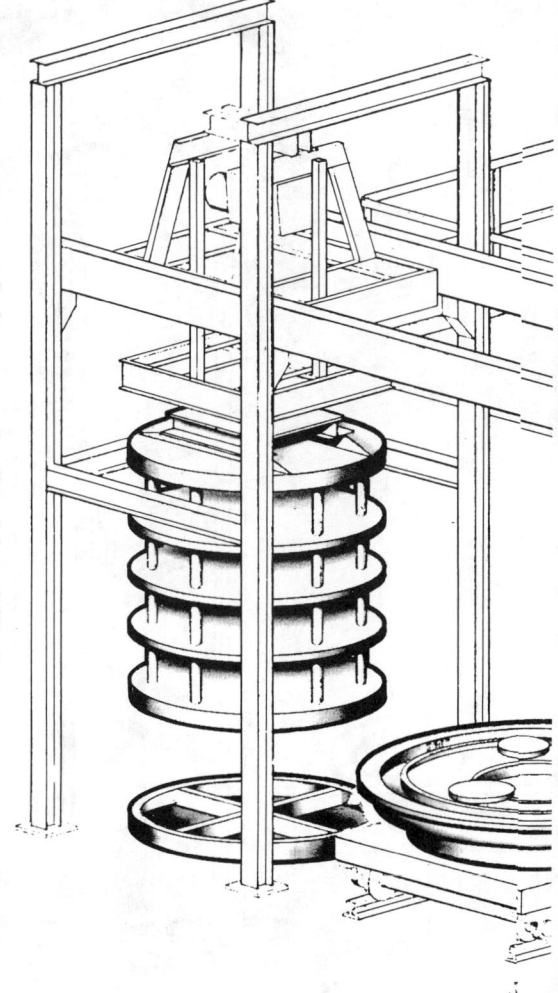

Microprocessor-based controls have greatly improved quality assurance in heat treating, and one process in particular that has benefited from this is ion nitriding.

Ion nitriding's well-documented versatility—it can case-harden all cast iron and steel materials, with or without a compound layer—is due to the wide variations that can be made in a large number of process parameters: time, temperature, warm-up rate, vacuum pressure, electrical current intensity, and gas composition, to name the more important ones.

Getting consistent, specified results in terms of surface hardness, wear resistance, fatigue life, corrosion resistance and/or reduced distortion means that the process variables must be maintained at their correct levels at each point in every cycle, and in proper coordination with each other.

In view of the precision that's required, manual controls leave a lot of room for human error; microprocessor-based controls practically eliminate it.

A good example of the application of microprocessors to ion nitriding can be found at Wellman Thermal Systems Corp., Shelbyville, Ind., a manufacturer of heat treating systems and other industrial equipment. Wellman recently invited the trade press to view an ion nitriding unit the firm was preparing to ship to a major automotive manufacturer.

Among the features on this unit is a completely automatic microprocessor-based controller. Wellman designed and built the controller in-house (rather than buying it from an outside firm) specifically for ion nitriding applications. The controller, which incorporates a standard Intel 8080 microprocessor, is the product of a development program Wellman began in 1979. With certain programming variations (Wellman writes all the software), the controller can be adapted to any number of ion nitriding systems.

One of the reasons the customer decided to buy the ion nitriding system, aside from the lack of pollution problems posed by gas and salt bath nitriding, is the repeatability of results the controller ensures. Prior to bringing the nitriding function in-house to speed up production, the customer had been sending some of its gears to commercial shops that used a low-temperature carbonitriding process. The metallurgical results, according to Wellman metallurgist Robert L. Chaney, showed some inconsistencies; "the results were acceptable, but not great," he says.

The new ion nitriding system, with its microprocessor-based controller, promises to improve significantly on this situation.

The system

The system itself is a high-production, large elevator type that can process workloads up to 6000 pounds a cycle. Its work dimensions are 68 inches in diameter by 72 inches high, while the overall system dimensions are approximately 20 feet wide by 20 feet long by 20 feet high. The system includes an ingenious material handling system (8 feet wide by 24 feet long by 18 feet high) that minimizes operator movement.

The chamber is of hot-wall (single-wall) construction and has auxiliary heating and forced gas quench. These features, Wellman says, provide shorter cycle times and lower cycle costs than conventional cold-wall ion nitriding systems. The design also includes a closed-loop, recirculated cooling water system with an external heat exchanger.

Specs and parameters

The customer's specifications call for wear resistance on the gears in the form of a nitrided case with an Epsilon compound layer ($Fe_{2-3}N$) to a minimum depth of .0004 inch.

To assure this same result within specified limits on about 600 gears in every load and throughout the entire affected surface of each gear, it is necessary for the large number of variables in the ion nitriding system to be precisely controlled, just as they would need to be in any ion nitriding application. The microprocessor-based unit provides this kind of control accurately and constantly.

In this particular case, the gears must be nitrided at 1070°F with a ±10°F uniformity throughout the chamber. The controller interfaces with a thermocouple that has been precisely placed within the chamber to assure a reading that will not exceed the uniformity limits. The controller is also hooked up to the power supply. Once the controller receives the reading from the thermocouple, it can make constant adjustments in the power input to maintain the ion bombardment at a certain intensity, and thus keep the temperature at the desired level.

The controller also interfaces with the gas mixture controller to ensure that the required proportions of nitrogen, hydrogen and hydro-carbon are surrounding the workpiece. This is important to get the desired composition in the compound layer.

Another important variable is the pressure within the chamber, which must be held at different levels during a cycle. (Ion nitriding is performed in a vacuum, usually at 1 to 5 torr.) A sensor tells the controller what the pressure in the chamber is at any given moment, and the controller, in accordance with its programmed instructions sends the appropriate signals to the pressure-control valve.

The microprocessor unit also controls the instantaneous temperature rate. If the temperature falls behind the expected ramp, the controller will not try to "catch up," thus eliminating the possibility of damaging parts with an excessive warm-up rate.

Unique to ion nitriding is the problem of arc prevention. During the

Sketch shows Wellman ion nitride system as it will be installed at an automaking plant. Microprocessor controls and power supply are on bottom level.

process, the bombardment of nitrogen ions creates a visible glow discharge around the workpiece. Ideally, the glow discharge should emanate evenly from the entire surface. However, an unchecked overintensity of the electrical current that causes

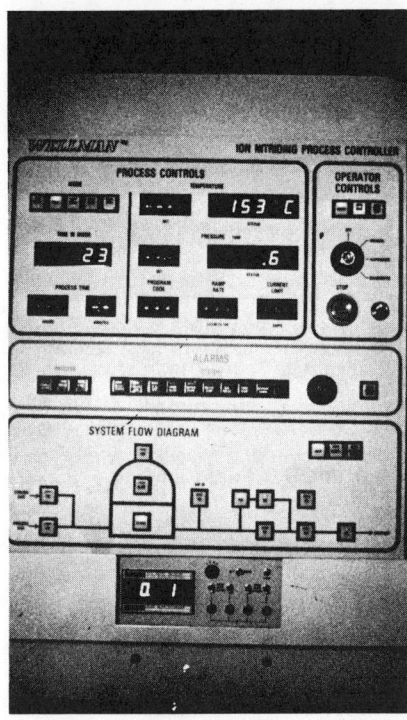

Wellman decided to develop its own microprocessor-based control unit to assure maximum capabilities for ion nitriding. The controls can be dedicated to one process or used to control a variety of processes.

the ion bombardment could result in the formation of an arc which could damage the workpiece or the chamber.

"You can imagine what a 100-amp arc will do inside a furnace," says Chaney. "It's going to start shooting holes in everything it can get near. The microprocessor eliminates that possibility."

In ion nitriding's earlier days, an operator had to stand at the chamber window and watch for arcs throughout the entire cycle. The microprocessor controller, however, has eliminated the need for that. It senses the

magnitude and quantity of overintensities and automatically adjusts the process program to impede arc formation. If an arc begins to develop anyway, the controller will cut off the power from the power supply for a few milliseconds. And if all these arc suppression measures exceed a certain limit, the controller will sound an alarm and terminate the cycle.

Operator attention

Alarms are also provided for any parameter abnormality that might require operator attention. If the operator is summoned by an alarm, he can refer to the controller's digital display of temperature, pressure, operating mode, time-in-mode, and vacuum system status. If he needs to know what went on before, he has access to the controller's strip chart recorder. And if he needs to enter different process parameters, he can do so with the controller's digital thumbwheel switches, a process that requires little training and provides scant opportunity for error. All these features help assure the load will be processed correctly.

In-house development

To provide the controller with maximum capabilities for ion nitriding, Wellman decided to design its own.

"The controllers on the market had many excellent features," says Chaney. "But they didn't have exactly what we wanted for ion nitriding. Ours is almost a totally dedicated unit."

Chaney says the Wellman controller has a better response time, memory, and programmability for ion nitriding applications than the general heat treating controllers on the market.

"Variables such as response times are extremely important in a situation like arc formation," says Chaney. "That's why we opted to develop our own controller rather than buy one off the shelf."

Different combinations

The ion nitriding system going to the auto manufacturer is a dedicated, high-production unit. It is designed

to nitride just one kind of gear and always impart the same properties. The microprocessor-based controller can keep the process parameters on target and produce the correct results. But it can also be applied to an ion nitriding system, earmarked for a wide variety of parts, materials, configurations, and specifications, such as would be found in a commercial heat treating shop.

As previously noted, ion nitriding is very versatile. By varying the parameters, the ion nitrider can produce a nitrided case with different depths and types of layers in just about any ferrous material. By altering the temperature and gas composition, for example, the process will produce a case that may consist of a diffusion layer only; a Gamma prime (Fe_4N) layer on top of the diffusion layer; or an Epsilon ($Fe_{2.3}N$) layer on top of a diffusion layer. Or, by varying the pressure in the chamber, the controller can assure that parts with awkward configurations and hard-to-reach surfaces will come out properly nitrided.

Whatever combinations of parameters are called for, the microprocessor-based controller will be able to set, monitor, and hold them. An operator can simply punch in the job number, and the controller will come up with the right "recipe," which will then produce the same level of consistency seen on a dedicated system.

"We have it set up right now where you can put in a hundred different programs," says Chaney.

The implications are obvious for commercial heat treaters who have always needed versatility and today need a high level of quality assurance as well. In addition, ion nitriding units are extremely energy efficient, a point of great interest to commercial shops nowadays.

In fact, Chaney expects the best market for ion nitriding to be in is the commercial heat treating industry. "The heat treating shops are setting the pace for ion nitriding, and I think they'll continue to do so," says Chaney. "All our new stuff has been going to heat treaters. That's why this controller was built—with the commercial shops in mind." **HT**

Cryogenic Treatment Improves Properties of Drills and P/M Parts

By RICK FREY
M. G. Industries
Valley Forge, Pa.

Editor's Note: *Controlled transformation of retained austenite by cryogenic treatment, using liquid nitrogen, improves the properties of both wrought and P/M parts as described in this article. It is from a paper presented by the author during the 1983 Annual Powder Metallurgy Conference & Exhibition in New Orleans, last May.*

The concept of improving the properties of steel by exposure to low temperature is very old. It is believed that old Swiss watchmakers improved the properties of their watch components by exposing them to the harsh winter temperature in the Alps. Today the technique of cold treatment has become less of an art and more of a science; however, there is still very little documentation concerning all the benefits available.

A cold treatment following quench hardening generally results in an increased rate of the austenitic-martensitic transformation. There have also been experimental test results showing a reduction of grain size due to exposure to very low temperatures, but the transformation of retained austenite is believed to be the major advantage. Controlled transformation of retained austenite to martensite by cold treatment results in an increase in hardness and wear resistance. Also, since retained austenite is a very unstable structure, uncontrolled transformation which occurs during normal service conditions can be extremely detrimental. Uncontrolled transformation causes internal stresses because of the volume change and a very brittle, untempered martensite.

Influence of Carbon on Ms and Mf

The martensitic transformation in steel which occurs during rapid cooling from high temperatures, starts at a definite temperature, Ms, and continues until the martensitic finish temperature, Mf, is reached. Fig. 1 shows the relationship of Ms and Mf with carbon content in plain iron-carbon steels. According to Fig. 1, steels containing more than .6 wt % of carbon must be cooled below ambient temperatures to reach the martensitic finish temperature and would therefore require special handling to complete transformation. In fact, high carbon alloys can have Mf temperatures lower than –350°F, and will have substantial amounts of retained austenite if not cooled below ambient temperatures.

Benefits of Liquid Nitrogen Cooling

There are three basic methods of achieving metallurgical refrigeration in common use today: dry ice, mechanical refrigeration, and liquid nitrogen cooling. The most advantageous method for most powder metal parts producers would be the liquid nitrogen method for several reasons:

(1) Nitrogen although used for sintering as a gas is usually supplied as a liquid for economy of transportation. The inherent cooling energy is usually wasted to the atmosphere outside the plant, but can easily be captured and controlled thereby providing "free" energy.

(2) Cryogenic temperatures below –300°F (–184°C) are easily obtained. This is essential for optimum treatment of high carbon alloys.

(3) Liquid nitrogen systems have very low maintenance costs.

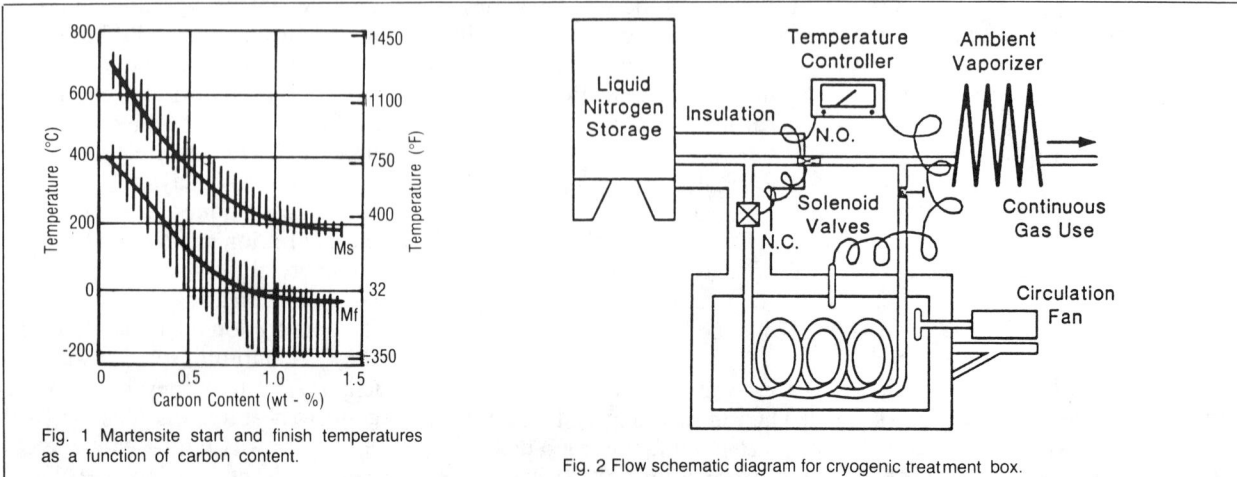

Fig. 1 Martensite start and finish temperatures as a function of carbon content.

Fig. 2 Flow schematic diagram for cryogenic treatment box.

Reprinted with permission from Industrial Heating, September 1983, 21-23, © 1983 National Industrial Publishing Co.

(4) Liquid nitrogen systems have relatively low capital costs.

Fig. 2 shows a typical flow schematic for a cryogenic treatment box. The cooling potential is obtained from a bypass off a continuous nitrogen gas use. The schematic includes two solenoid valves tied in with a thermocouple and temperature controller allowing easy control of soak temperatures. Long stem valves and an appropriate thermometer can be used for manual operation to provide economy. Either way, the system is relatively simple and does not require a large capital outlay to implement.

By controlling the flow of liquid nitrogen into the cold box, the temperature and cooling rate can be controlled. In Fig. 3 a temperature profile for a 20-hr soak at –320°F (–196°C) is shown. This was easily accomplished with manually controlled valves. Since the boiling point of liquid nitrogen is –320°F (–196°C), it is very easy to maintain true cryogenic temperatures.

Improved Drill Life

The first benefit of cryogenic treatment examined was the treatment of tooling used for secondary machining of p.m. parts. (Carbon City Products, St. Marys, Pa., was instrumental in helping to obtain this information.)

Comparative data was gathered on a specific drilling operation in which a copper infiltrated powder metal part receives a 21/64 inch diameter hole through a 3/8 inch (.95 cm) section of the part. Conventional high speed steel and cobalt drills were cryogenically treated and compared with untreated drills of the same lot or package. The cryogenic treatment temperature cycle resembled that of Fig. 3. Quantitative drilling tests were then conducted.

A comparison of the average number of holes drilled with the cryogenically treated bits undoubtedly favors the treated drills. (See Fig. 4). The cryogenically treated high speed drill bits averaged 2.83 times as many holes as the untreated high speed steel, while the cryogenically treated cobalt drills averaged 3.42 times as many holes as the untreated cobalt drill bits. The cryogenic treated drills in this comparison were not tempered after treatment. Tempering after the cryogenic treatment slightly reduced wear resistance for the 21/63 inch diameter drills. However, for very small drills, 1/8 inch (.3 cm) diameter, tempering was necessary to prevent breaking. It seems that if the drills were large enough to survive breaking, the newly formed martensite was best left untempered.

The potential economic advantage resulting from the production increases is apparent. The cryogenically treated drills reduce production interruptions

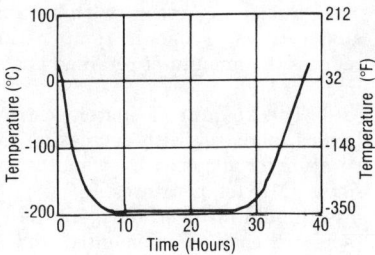

Fig. 3 Typical cooling cycle for cryogenic treatment.

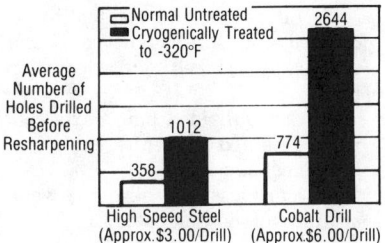

Fig. 4 Effect of cryogenic treatment on drill life.

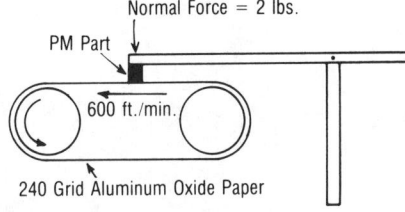

Fig. 5 Abrasive wear testing apparatus.

required to change dull drill bits. More important, the average cost per hole is drastically reduced. The treated drills will bore more holes before resharpening. It is also reported that less than one half of the material is removed from the cutting edge when resharpening the cryogenically treated materials. Therefore, the cold treated drills should accept twice as many resharpenings.

A qualitative analysis was then performed on the drills in order to determine the actual internal changes. X-ray detection of retained austenite was performed at the metallurgical laboratory of Penn State University. By cryogenic treatment, the retained austenite in the high speed steel was reduced from approximately 12% to 6.3%; while the cobalt drills showed a change from approximately 11.3% to 6.8% retained austenite. The cobalt drills exhibited slightly less conversion of retained austenite, possibly because they were already at a more stable state than the high speed steel before cryogenic treatment, also because the cobalt inhibits austenitic transformation.

Improved Properties of P/M Parts

Cryogenically treated powder metal parts were also examined. Two types

of parts, a spur and a gear, containing approximately .9% combined carbon were heat treated by two different methods. Half were neutral hardened and the other half were carbonitrided. Several parts from both batches were then cryogenically treated. They were cooled slowly, held at –320°F (–196°C) for approximately 8 hr then slowly warmed to ambient temperatures. All parts were then annealed in the same cycle to assure an even comparison. Again, X-ray analysis was used to determine the amount of retained austenite. The cryogenically treated parts contained less retained austenite than the untreated parts. (See Table I). The carbonitrided p.m. parts contained much more retained austenite than the hardened parts. This is to be expected because of the nature of the heat treatment and because the X-ray analysis was performed near the surface of the parts. Therefore, the carbonitrided parts exhibited more transformation of retained austenite due to cryogenic treatment than did the hardened parts. The resulting change in hardness is also greater for the carbonitrided parts. (See Table II). The hardened parts gained approximately 1.5 on the Rc scale while the carbonitrided parts average about a 4.5 increase in hardness due to cryogenic treatment.

Wear tests were then conducted on the p.m. parts to determine any increase in wear resistance. Fig. 5 illustrates the method of testing. The parts were subjected to a 240 grid aluminum oxide sanding belt traveling approximately 600 ft/min (304.8 cm/sec) across the parts. A normal force of approximately 2 lb (0.9 kg) was exerted on the part. Testing times ranged from 1 to 3 minutes. The weight loss as a function of time was then recorded to simulate resistance to abrasive wear. Table III shows the results of the wear tests. As expected, the carbonitrided parts exhibit more resistance to abrasive wear because of the cryogenic treatment than the hardened parts do. This occurs simply be-

Table I Percent Retained Austenite in P/M Parts

	Hardened	Carbonitrided
Normal Untreated PM Parts	4.9%	10.2%
Cryogenically Treated Soaked @ –320°F Approximately 8 Hrs.	3%	5.1%

Table II Rockwell Hardness of P/M Parts, Untreated vs. Cryogenically Treated

	Hardened	Carbonitrided
Normal Untreated PM Parts	39.5	36.8
Cryogenically Treated PM Parts	41	41.3
Increased Hardness from Cryogenic Treatment	1.5	4.5

Source: Industrial Heating, September 1983, 21-23

433

Table III Results of Abrasive Wear Tests

| | Average Weight Loss per Minute | |
	Hardened (gms/min)	Carbonitrided (gms/min)
Normal Untreated PM Parts	.338	.320
Cryogenically Treated PM Parts	.282	.203

cause the carbonitrided parts experienced more transformation of retained austenite. In either case, however, the cryogenic treatment showed an improvement in the properties of the p.m. parts.

Conclusions

(1) The amount of retained austenite increases with the amount of carbon and alloying elements in most steels, including p.m. parts.

(2) Subjecting steels with retained austenite to cryogenic temperatures reduces the amount of retained austenite.

(3) The amount of austenite transformed increases with decreasing temperature for alloys with low martensitic finish (Mf) temperatures.

(4) Liquid nitrogen refrigeration systems are easily implemented and are very economical for powder metal parts producers using nitrogen based atmospheres.

(5) Life expectancy of tooling is increased and subsequent costs can be drastically reduced by cryogenic treatment.

(6) Carbonitrided p.m. parts contain more retained austenite than neutral hardened p.m. parts.

(7) Hardness and abrasive wear resistance of heat treated, high carbon powder metal parts can be improved through cryogenic treatment.

Information Retrieval

For your guidance and filing convenience the following subjects are included in this article:

Key: cryogenic treatment, liquid nitrogen, mechanical refrigeration, sintering, solenoid valves, temperature controller, thermocouple, high speed drill bits, X-ray, cobalt drills, powder metal parts, carbonitrided P/M parts

Methods of Measuring Case Depth

CASE HARDENING may be defined as a process by which a ferrous material is hardened in such a manner that the surface layer, known as the case, becomes substantially harder than the remaining material, known as the core. Case hardening processes include carburizing, nitriding, carbonitriding, cyaniding, and induction and flame hardening. In every instance, case hardening affects chemical composition or mechanical properties, or both.

An accurate and repeatable method of measuring case depth is essential for quality control of the case hardening process and for evaluation of workpieces for conformance with specifications, such as might be done during a failure analysis. This article (sources: Ref 1 and 2) describes various methods for measuring the depth to which a change has been made in either composition or mechanical properties. Each method has its own area of application established through proven practice, and no single method is recommended for all purposes. Relationships among case depths determined by the different methods can vary extensively. Some of the factors that affect these relationships are case characteristics, parent-steel composition and quenching conditions.

Because measurements made by the various methods are not necessarily taken at the same location in a case, confusion and misunderstanding can result if the method of measurement is not specified. Specific descriptions,

such as "total case depth", "effective case depth to 50 HRC" and "case depth to 0.40% carbon" will help to avoid misunderstandings.

Typical hardness surveys taken on cross sections at the pitch line, root fillet and root land of a tooth in a carburized and hardened gear made of 8620H steel are shown in Fig. 1. These data illustrate the importance of well-defined specifications by showing that

there are variations in effective case depth even among three areas of the same gear tooth.

The methods employed for measuring case depth are chemical, mechanical or visual, and the specimens or parts may be subjected to testing in either the soft or the hardened condition. The measured case depth then may be reported as either effective or total case depth for hardened speci-

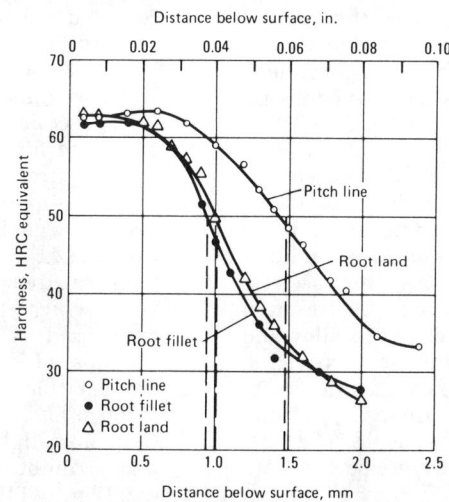

Fig. 1 Variation in hardness with distance below surface for a carburized and hardened 8620H steel gear

Effective case depths to 50 HRC: 0.94 mm (0.037 in.) at root fillet; 1.02 mm (0.040 in.) at root land; 1.46 mm (0.057 in.) at pitch line

mens and as either total case depth or equivalent effective case depth for unhardened specimens.

Effective case depth is the perpendicular distance from the surface of a hardened case to the deepest point at which a specified level of hardness is maintained. The hardness criterion, except when otherwise specified, is 50 HRC.

Total case depth may be defined as the perpendicular distance from the surface of a hardened or unhardened case to the point at which differences in chemical or physical properties of the case and core can no longer be distinguished. Total case depth sometimes is considered to be the distance from the surface to the deepest point at which the carbon content is 0.04% higher than the carbon content of the core.

Chemical Method

The chemical method of measuring case depth generally is used only for carburized cases but may be used for cyanided or carbonitrided cases as well. This method consists of determining the carbon content (and, when applicable, the nitrogen content) by chemical analysis at incremental depths below the surface. The chemical method is considered to be the most accurate method of measuring total case depth. One of two common methods is used to analyze for carbon content: combustion analysis or spectrographic analysis. Combustion carbon analysis currently is the most widely employed.

Procedure for Carburized Cases. If test specimens are used they should be of the same grade of steel as that of the parts being carburized. Specimens may be actual parts, rings or bars, and the carburized surface should be flat or otherwise suitable for accurate machining to obtain chips for subsequent carbon analysis. To ensure maximum uniformity of the carburizing process among various types of furnaces, large heat treatment facilities often use test specimens. These specimens are often standardized with respect to alloy and configuration to establish carburizing schedules for various case depths and to ensure maximum uniformity among various furnaces. Case depths of actual parts then can be correlated to the standard test specimen.

Test specimens should be carburized with actual parts or in a manner representative of the procedure to be used for actual parts. In cooling of test specimens after carburizing, care should be exercised to avoid distortion and decarburization. When parts and test specimens are quenched after being carburized, they should be tempered at approximately 600 to 650 °C (1100 to 1200 °F). Time at temperature should be minimized to avoid excessive carbon diffusion. The parts and specimens should be straightened to 0.038-mm (0.0015-in.) max total indicator reading (TIR) before machining is attempted.

Test specimens must have clean surfaces and should be machined dry, taking the necessary precautions to avoid burning, in predetermined increments of depth. The increments of depth usually chosen vary from 0.05 to 0.25 mm (0.002 to 0.010 in.) depending on desired accuracy and expected case depth.

The furnace load containing the test specimens should approximate actual production conditions in terms of load density, configuration and surface area to be carburized. These three variables affect atmosphere flow, temperature uniformity and carbon demand. Differences in these conditions between production loads and the load that contains the specimens can lead to errors in correlation of case depths. A typical procedure for obtaining specimens for carbon analysis is as follows:

1 Prepare a bar of suitable material in the configuration shown in Fig. 2. Identify the bar in some manner, such as by stamping a number on the end.
2 Carburize and then quench or cool the bar as required. If bar is slowly cooled, steps 3 through 7 can be omitted.
3 Wash bar with soap and water. Rinse with methanol and dry.
4 Cut a section from the 25-mm-diam end for examination of microstructure.
5 Record as-quenched surface hardness of large-diameter end.
6 Temper bar for the time and at the temperature specified for the part with which the test bar was carburized. Record as-tempered hardness of large-diameter end.
7 Temper for 1½ h at 650 °C (1200 °F).
8 Grit blast lightly, clean centers, and straighten bar to 0.03-mm (0.001-in.) TIR taken in three places.
9 Wash bar with soap and water. Rinse with methanol and dry.
10 For case depths less than 5.0 mm

Fig. 2 Nominal configuration of standard test bar used for chemical method of case-depth measurement

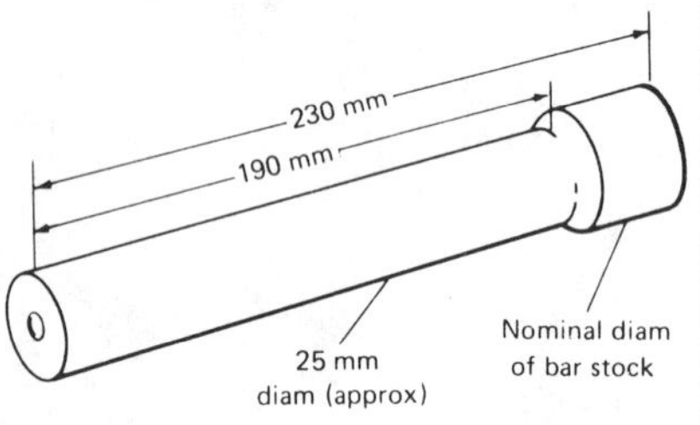

230 mm
190 mm
25 mm diam (approx)
Nominal diam of bar stock

The 25-mm-diam end is finished with 80-grit sandpaper.

(0.200 in.), machine approximately 3.8 mm (0.15 in.) from the 25-mm-diam end to a depth of 5.0 mm, to ensure that the case on the end does not contaminate the specimens for carbon analysis.
11 Machine bar. Before each machining operation, record diameter of bar as measured with a micrometer. Maximum allowable taper of machined area is 0.03 mm (0.001 in.) on the radius. Machine a maximum of 0.05 mm (0.002 in.) from the radius to clean the surface. Save chips for analysis. Next, machine increments of 0.13 mm (0.005 in.) from the radius to a depth of 0.25 mm (0.010 in.) below the maximum expected case depth. Take three more increments of 0.25 mm (0.010 in.) from the radius. Save chips from each increment for separate analysis. Take precautions to ensure that chips from each cut are not burnt or contaminated by dirt, paper, oil or chips from preceding cuts.
12 Calculate and plot the carbon-gradient curve. A sample data sheet and a carbon-gradient curve are presented in Table 1 and Fig. 3, respectively.

Spectrographic Analysis. Carbon content may be determined accurately by spectrographic analysis. This method makes use of a vacuum spectrometer, which permits measurement of spectral lines in the ultraviolet region where air would ordinarily absorb much of the emitted radiation.

Many critical items must be assessed for surface carbon content to ensure uniform properties after heat treat-

Table 1 Sample data sheet for computing case-depth values for a carbon-gradient plot

Data are for 8620H steel, carburized at 925 °C (1700 °F) in a 19-tray continuous pusher furnace with infrared control of carbon dioxide content in zones 2, 3 and 4. See text for explanation of procedure, and see Fig. 3 for plot of carbon gradient.

Cut No.	D_l	D_r	A_l	A_r	C_l	C_r	X	M	P	Carbon, %
0	25.35	25.36	...	...	...	...	...	...	...	...
1	25.20	25.23	0.15	0.13	0.15	0.13	0.07	0.03	0.03	0.987
2	24.98	24.99	0.22	0.24	0.37	0.37	0.18	0.06	0.13	0.953
3	24.76	24.76	0.22	0.23	0.59	0.60	0.30	0.06	0.24	0.918
4	24.49	24.47	0.27	0.29	0.86	0.89	0.44	0.07	0.37	0.871
5	24.22	24.22	0.27	0.25	1.13	1.14	0.57	0.06	0.50	0.818
6	23.94	23.91	0.28	0.31	1.41	1.45	0.71	0.07	0.64	0.787
7	23.69	23.65	0.25	0.26	1.66	1.71	0.84	0.06	0.77	0.717
8	23.41	23.38	0.28	0.27	1.94	1.98	0.98	0.07	0.91	0.675
9	23.10	23.10	0.31	0.28	2.25	2.26	1.13	0.07	1.05	0.627
10	22.80	22.78	0.30	0.32	2.55	2.58	1.28	0.08	1.21	0.583
11	22.49	22.48	0.31	0.30	2.86	2.88	1.43	0.08	1.36	0.540
12	22.19	22.17	0.30	0.31	3.16	3.19	1.59	0.08	1.51	0.483
13	21.87	21.87	0.32	0.30	3.48	3.49	1.74	0.08	1.67	0.444
14	21.59	21.56	0.28	0.31	3.76	3.80	1.89	0.07	1.81	0.401
15	21.25	21.27	0.34	0.29	4.10	4.09	2.05	0.08	1.97	0.365
16	20.80	20.75	0.45	0.52	4.55	4.61	2.29	0.12	2.17	0.328
17	20.27	20.24	0.53	0.51	5.08	5.12	2.55	0.13	2.42	0.283
18	19.72	19.68	0.55	0.56	5.63	5.68	2.83	0.14	2.69	0.245

Fig. 3 Carbon gradient for carburized test bar of 8620H steel

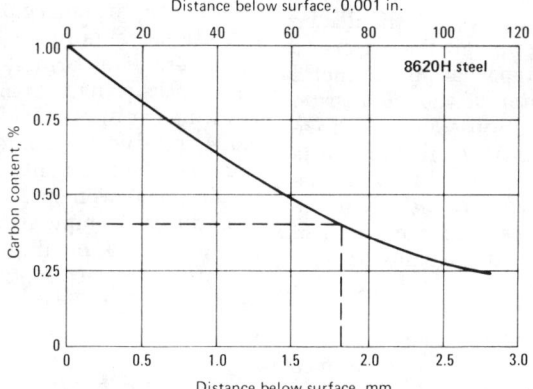

A test bar similar to the one shown in Fig. 2 was carburized at 925 °C (1700 °F) in a 19-tray continuous pusher furnace with infrared control of carbon dioxide content in zones 2, 3 and 4. Effective case depth to 0.40% carbon is 1.82 mm (0.0715 in.), and is indicated by dashed lines. See text for explanation of procedure for calculating plot points, and see Table 1 for sample data sheet containing data for this figure.

ment. The spectrographic carbon method normally uses flat test specimens that can be taper ground, step ground or reground incrementally after each carbon determination. A very small amount of material is ground from the surface (to remove oxides). Successive cuts are made and analyses are performed after each cut. Each test takes less than 2 min.

Special care must be taken for accurate measurement of the depth corresponding to each carbon determination. Case depth determined on flat or round test specimens will often be different from case depth determined directly on workpieces, because of the difference in shape.

Whereas carbon determination by the combustion method provides an average carbon content for the amount of material removed by machining, the spectrograph determines the local carbon content of the specimen to a depth of 0.03 mm (0.001 in.). A comparison of carbon values obtained from five specimens by spectrographic methods is presented in Table 2.

Mechanical Method

In the mechanical method of measuring case depth, hardness traverses are taken on the case and core of a specimen that has been prepared by one of three procedures. It is considered the most accurate method of measuring effective case depth (depth to 50 HRC). This method also is preferred for measuring total depth of thin cases (0.25 mm or less).

For measurement of effective case depth, read to point of specified hardness, which is 50 HRC (or approved equivalent) except for selectively hardened cases, for which the following values are recommended:

Carbon content, %	Case hardness, HRC
0.28 to 0.32	35
0.33 to 0.42	40
0.43 to 0.52	45
0.53 and over	50

Hardness testers that produce small, shallow impressions should be used for all of the following procedures, so that the hardness values obtained will be representative of the surface or area being tested. Testers that produce Vickers or Knoop microhardness numbers with loads of at least 0.5 kg are recommended, although testers using heavier loads (such as Rockwell superficial) can be used in some instances.

Considerable care should be exercised during preparation of specimens for case-depth determination by the mechanical method to prevent grinding or cutting burn. The use of an etchant for burn detection is recommended as a general precaution, because of the serious error that can be introduced by the presence of metal whose metallurgical condition has been altered during specimen preparation.

Cross-Section Procedure. Cut specimens perpendicular to the hardened surface at a critical location, being careful to avoid any cutting or grinding practice that would affect the original hardness.

Both shim stock and workpieces were heat treated in a continuous-belt furnace with an endothermic-base atmosphere (class 301; dew point, −9 to −1 °C, or +15 to +30 °F).

	Amount of carbon present, %		
	Shim shock		Workpiece surface (spectro-graphic analysis)
Speci-men No.	Spectro-graphic analysis	Combustion analysis	
1.	0.36	0.36	0.38
2.	0.24	0.27	0.25
3.	0.22	0.24	0.225
4.	0.35	0.35	0.34
5.	0.30	0.30	0.305

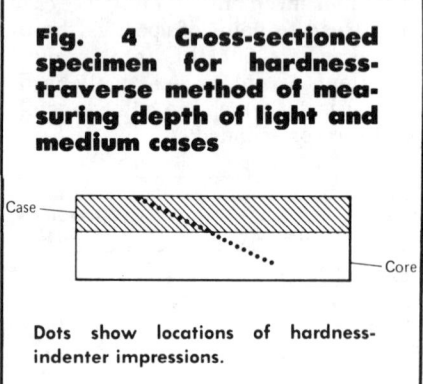

Fig. 4 Cross-sectioned specimen for hardness-traverse method of measuring depth of light and medium cases

Dots show locations of hardness-indenter impressions.

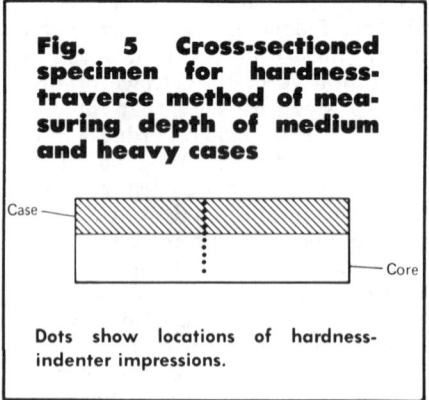

Fig. 5 Cross-sectioned specimen for hardness-traverse method of measuring depth of medium and heavy cases

Dots show locations of hardness-indenter impressions.

Grind and polish the specimen. The surface of the area to be traversed should be polished finely enough so that hardness impressions are unaffected (the lighter the indenter load, the finer the polish necessary).

The procedure illustrated in Fig. 4 is recommended for measurement of light and medium cases. The alternative procedure illustrated in Fig. 5 is recom-

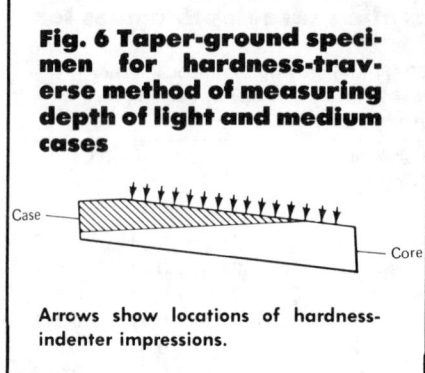

Fig. 6 Taper-ground specimen for hardness-traverse method of measuring depth of light and medium cases

Arrows show locations of hardness-indenter impressions.

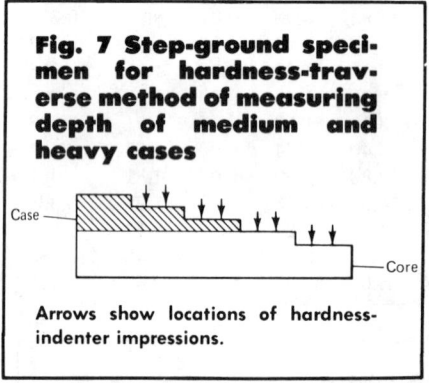

Fig. 7 Step-ground specimen for hardness-traverse method of measuring depth of medium and heavy cases

Arrows show locations of hardness-indenter impressions.

mended for measurement of medium and heavier cases.

The hardness traverse should be started far enough below the surface to ensure proper support from the metal between the center of the impression and the surface. Subsequent impressions are spaced far enough apart so as not to distort hardness values. The distance from the surface of the case to the center of the impression is measured on a calibrated optical instrument, micrometer stage, or other suitable measuring device.

Taper-Grind Procedure. This procedure, illustrated in Fig. 6, sometimes is used for measurement of light and medium cases.

A shallow taper is ground through the case, and hardness measurements are made along the surface thus prepared. The angle is chosen so that readings spaced equal distances apart will represent the hardnesses at the desired increments below the surface of the case.

Step-Grind Procedure. This procedure, illustrated in Fig. 7, is recommended for measurement of medium and heavy cases. It is essentially the same as the taper-grind procedure, with the exception that hardness readings are made on steps that are known distances below the surface.

A variation on this procedure is the step-grind method in which two predetermined depths are ground. If the hardness is greater than 50 HRC on the shallow step and less than 50 HRC on the deep step, the effective case depth to 50 HRC lies somewhere between the two steps. This variation frequently is used to ensure that the effective case depth is within specified limits.

Visual Methods

These methods employ any visual procedure, with or without the aid of magnification, for reading the depth of case produced by any of the various processes. Specimens may be prepared by combinations of fracturing, cutting (with water cooling to prevent burning), grinding and polishing. Etching with a suitable reagent normally is required to produce a contrast between the case and the core. Nital (concentrated nitric acid in alcohol) of various strengths frequently is used as the reagent for producing this contrast.

Visual methods have been classified into two general categories: macroscopic and microscopic. In macroscopic procedures, specimens normally are ground no finer than through No. 000 metallographic emery paper, and magnifications usually do not exceed 20 diameters. The Brinell glass, a hand-held optical instrument with retical markings at intervals of 0.1 mm (about 0.004 in.) and 20-diameter magnification, is a convenient tool for macroscopic measurement. In microscopic procedures, complete metallographic polishing and etching generally are required, and case depths normally are read at a magnification of 100 diameters.

Macroscopic Visual Procedures

Macroscopic methods for measuring case depth are recommended for routine process control, primarily because of the short time required for determinations and because of the minimum of specialized equipment and trained personnel that are needed. Although these methods normally are applied to hardened specimens, they have the additional advantage of being applicable to measurement of unhardened cases as well. However, the accuracy of such measurements can be improved by correlation with results of other methods. A variety of etchants may be employed with equal success, but the following procedures are typical and widely used:

- *Fracture:* Prepare part or specimen by fracturing. Examine at a magnification not exceeding 20 diameters, with no further preparation.
- *Fracture and etch:* Water quench part or specimen directly from the carburizing temperature. Fracture, then etch in 20% nitric acid in water for a time established to develop maximum contrast. Rinse in water, and read while wet.
- *Fracture or cut, and rough grind:* Prepare specimen by either fracturing or cutting (with water cooling), and then rough grinding. Etch in 10% nital for a time established to provide a sharp line of demarcation between case and core. Examine at a magnification not exceeding 20 diameters (Brinell glass), and read all of the darkened area for approximate total case depth.
- *Fracture or cut, and polish or grind:* Prepare specimen by fracturing or cutting (with water cooling). Polish, or grind through No. 000 or finer metallographic emery paper, or both. Etch in 5% nital for approximately 1 min. Rinse in two clean alcohol or water rinses. Examine at a magnification not exceeding 20 diameters (Brinell glass), and read all of the darkened zone. After correlation, effective case depth can be determined by reading from external surface of specimen to a selected line of the darkened zone. An alternative etching procedure is to etch in 25% nital for 30 s, wash in concentrated picral, rinse in alcohol, blow dry, and read as described above.
- M_s *method:* This method of case-depth measurement utilizes the fact that the martensite-start temperature (M_s point) varies with carbon content. Quenching and holding the steel for a short time at the M_s point corresponding to a given carbon content will temper the martensite formed at all lower carbon levels. Subsequent water quenching transforms austenite at all higher carbon levels to untempered martensite. Then polishing and etching of the test piece will reveal a sharp line of demarcation between tempered and untempered martensite; this line is normally read at 20-diameters magnification (Brinell glass) to a precision of ± 0.05 mm (± 0.002 in.).

The case depth is not sensitive to small temperature changes in the quenching bath. Final selection of quenching temperature usually is done statistically to produce an equal plus-and-minus distribution of error about known carbon curves.

The main factors that affect the accuracy of this method are pearlite formation during quenching to the M_s point, and time at M_s temperature. The specimen size should be sufficiently small to ensure that the severity of quench will transform all austenite of lower carbon levels to martensite without any formation of pearlite. (Specimen size may be critical for low-hardenability steels.) The time at M_s temperature should be sufficiently short so as not to allow formation of bainite, which interferes with the sharpness of the line of demarcation upon etching and can obliterate it completely.

For additional information on the M_s technique, see Ref 3.

Microscopic Visual Procedures

Microscopic methods generally are used for laboratory measurement of case depth and require complete metallographic polishing and etching suitable for the material and the process. The most common magnification used for examination is 100 diameters.

Carburized Cases. Microscopic methods may be used for laboratory determinations of total and effective case depths of material in the hardened condition. When the specimen is annealed properly, the total case depth can be determined quite precisely. For certain applications involving alloy steels of moderate to high hardenability that contain 0.4 to 0.8% carbon, the M_s method of determining case depth to specific carbon level has been found effective.

Procedure for hardened condition

1 Fracture or cut specimen (water cool when cutting) at right angles to the surface.
2 Prepare specimen for microscopic examination and etch in 2 to 5% nital.
3 For effective case depth, read from surface to metallographic structures that have been shown to be equivalent to 50 HRC. Often, the structure that is equivalent to 50 HRC consists of about 85% martensite and 15% intermediate quench products.

4 For total case depth, read to the line of demarcation between the case and the core. In alloy steels that have been quenched from a high temperature, the line of demarcation is not sharp. Read all of the darkened zone that indicates a difference in carbon content from that of the uniform core structure.

Procedure for annealed condition (for specimens previously hardened or not cooled under controlled conditions)

1 The specimen to be annealed may be protected by copper plating or any other suitable means of preventing carbon loss.
2 Anneal specimens in a protective atmosphere, or pack in a small thin-wall container with a suitable material such as charcoal, spent chips or pitch coke.
3 Heat specimens to a temperature about 25 to 50 °C (about 50 to 100 °F) above the upper critical temperature (Ac_3) for the core. (Generally, an annealing temperature of 870 to 925 °C, or 1600 to 1700 °F, is satisfactory.) Hold specimen at temperature only long enough to transform completely to austenite; otherwise, excessive diffusion of carbon may lead to inordinately high estimates of actual total case depth.
4 Cool from the annealing temperature at the following rates:
Carbon steels: A normally satisfactory cooling rate for most plain carbon steels such as 1010, 1015 and 1018 is 150 °C/h (270 °F/h) from the annealing temperature to 425 °C (800 °F). For high-manganese steels (1500 series), boron steels, and steels with high residual alloy contents, cooling may have to be slower. Cool as desired below 425 °C.
Alloy steels: For most alloy steels, best results are obtained from isothermal transformation. For some steels, however, a low cooling rate such as 75 °C/h (135 °F/h) from the annealing temperature to 425 °C (800 °F) is satisfactory. If martensite is retained in the structure, better contrast after etching may be obtained by tempering the specimen at 540 to 590 °C (1000 to 1100 °F). Cool as desired after tempering.
5 Section, prepare and etch specimen as described under "Procedure for hardened condition".

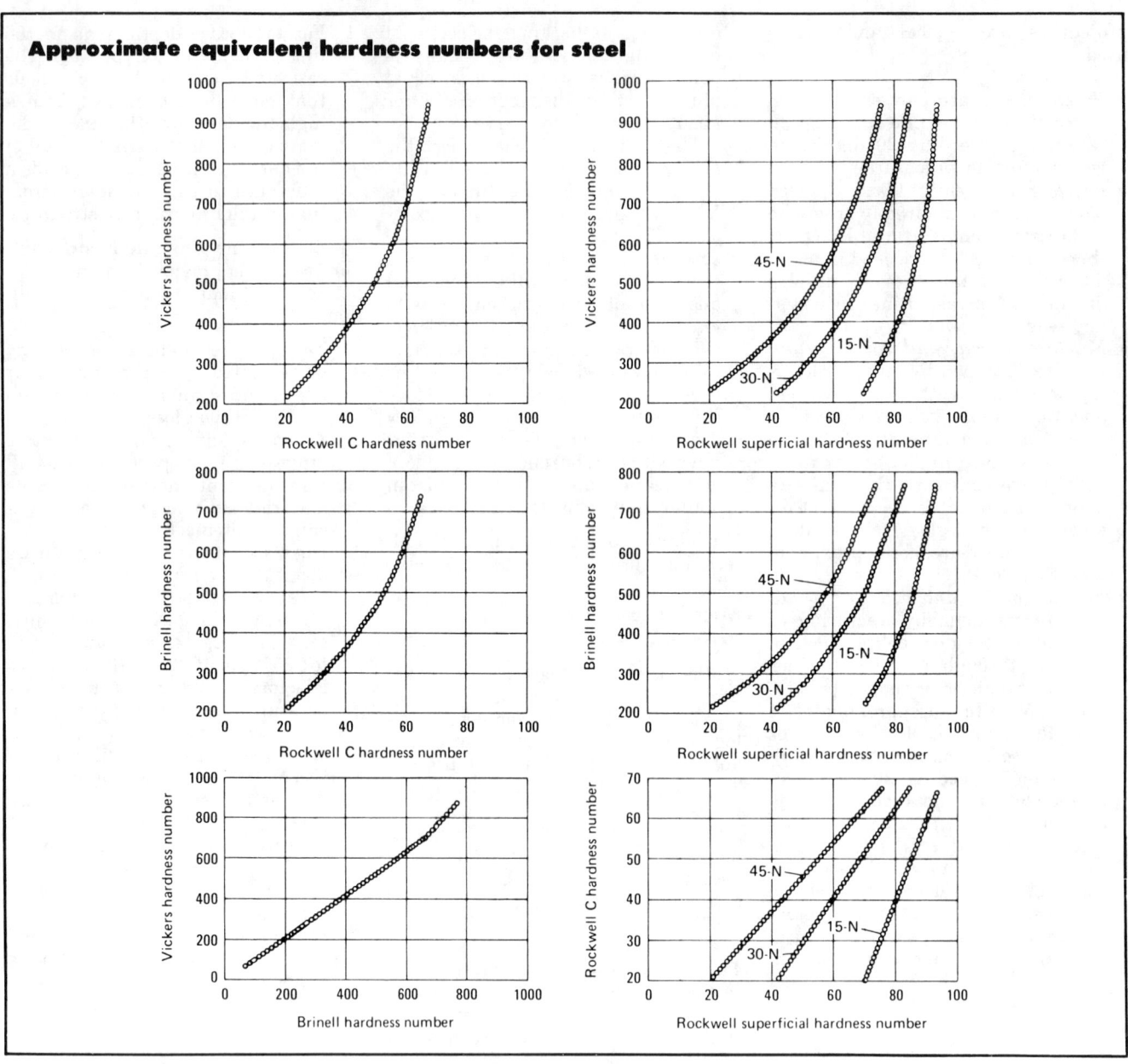

Approximate equivalent hardness numbers for steel

6 For total case depth, read the depth at which no further change in microstructure occurs.

Production carburizing schedules often have cooling rates similar to those described above under "Procedure for annealed condition". Specimens treated in this manner may be prepared and examined without being reheated after carburizing, and the results can be accurately correlated to a standard specimen.

Carbonitrided cases are measured for total case depth in the hardened condition. High quenching temperatures, high alloy content of the steel and high carbon content of the core

decrease the accuracy of readings obtained by this method.

Procedure

Section, prepare, etch and read as described above for carburized cases, under "Procedure for hardened condition".

Cyanided cases are thin, and only the microscopic method is recommended for accurate case-depth measurement. The usual cyanided case contains a light-etching layer followed by a totally martensitic constituent, which in turn is followed by martensite accompanied by increasingly extensive networks of other constituents, de-

pending on the type of steel. Cyanided cases are read in the hardened condition only, and results are reported as total case depth.

Procedure

1 Section, prepare and etch specimen as described above for carburized cases, under "Procedure for hardened condition".
2 Read to the line of demarcation between the case and the core. (When a sharp line of demarcation does not exist, a hardness traverse such as that described under "Mechanical Method" is recommended.)

Nitrided Cases. For measuring the depths of nitrided cases, the microscopic method is used chiefly in those situations where the available sample cannot readily be prepared for the more desirable hardness-traverse method.

Procedure

1 Section and prepare specimen as described above for carburized cases, under "Procedure for hardened condition".
2 Etch in 10% nital.
3 Read all of the darkened zone for total case depth.

Selectively Hardened Cases. Because no compositional change occurs in selective hardening (induction hardening, for example), readings must be taken on material in the hardened or the hardened-and-tempered condition only. A procedure for reading effective case depth may be established by correlating microstructures with a hardness-traverse method. A minimum hardness of 50 HRC is commonly used, but some other value may be selected or required—for example, in lower-carbon steels that do not reach 50 HRC when fully hardened (see the in-text table correlating carbon content with effective-case-depth hardness, under "Mechanical Method"). The microstructure at the selected location will vary depending on steel composition, prior treatment, and hardness level chosen.

Procedure

1 Section, prepare and etch specimen as described above for carburized cases, under "Procedure for hardened condition".
2 For total case depth, read the entire zone containing structures hardened by the process.
3 For effective case depth, read to selected microstructure correlated with specified hardness.

REFERENCES

1. Methods of Measuring Case Depth: SAE Recommended Practice J423a, Society of Automotive Engineers, Warrendale, PA, 1963 (reaffirmed 1970)
2. Measurement of Case Depth: Chapter 4 in *Carburizing and Carbonitriding,* prepared under the direction of the ASM Committee on Gas Carburizing, American Society for Metals, Metals Park, OH, 1977
3. The Application of M_s Points to Case Depth Measurement, by E. S. Rowland and S. R. Lyle: *Transaction of ASM,* Vol 37, 1946, p 27

Glossary of Selected Terms Related to Heat Treating

A

A$_{cm}$, A$_1$, A$_3$, A$_4$. Same as Ae$_{cm}$, Ae$_1$, Ae$_3$ and Ae$_4$.

Ac$_{cm}$, Ac$_1$, A$_3$, Ac$_4$. Defined under *transformation temperature*.

acicular ferrite. A highly substructured nonequiaxed ferrite that forms upon continuous cooling by a mixed diffusion and shear mode of transformation that begins at a temperature slightly higher than the temperature transformation range for upper bainite. It is distinguished from bainite in that it has a limited amount of carbon available; thus, there is only a small amount of carbide present.

Ae$_{cm}$, Ae$_1$, Ae$_3$, A$_4$. Defined under *transformation temperature*.

age hardening. Hardening by aging, usually after rapid cooling or cold working. See *aging*.

age softening. Spontaneous decrease of strength and hardness that takes place at room temperature in certain strain hardened alloys, especially those of aluminum.

aging. A change in the properties of certain metals and alloys that occurs at ambient or moderately elevated temperatures after hot working or a heat treatment (quench aging in ferrous alloys, natural or artificial aging in ferrous and nonferrous alloys) or after a cold working operation (strain aging). The change in properties is often, but not always, due to a phase change (precipitation), but never involves a change in chemical composition of the metal or alloy. See also *age hardening, artificial aging, interrupted aging, natural aging, overaging, precipitation hardening, precipitation heat treatment, progressive aging, quench aging, step aging*.

air-hardening steel. A steel containing sufficient carbon and other alloying elements to harden fully during cooling in air or other gaseous mediums from a temperature above its transformation range. The term should be restricted to steels that are capable of being hardened by cooling in air in fairly large sections, about 2 in. (50 mm) or more in diameter. Same as self-hardening steel.

alpha ferrite. See *ferrite*.

alpha iron. The body-centered cubic form of pure iron, stable below 910 °C (1670 °F).

Alumel. A nickel-base alloy containing about 2.5 Mn, 2 Al and 1 Si used chiefly as a component of pyrometric thermocouples.

annealing. A generic term denoting a treatment, consisting of heating to and holding at a suitable temperature followed by cooling at a suitable rate, used primarily to soften metallic materials, but also to simultaneously produce desired changes in other properties or in microstructure. The purpose of such changes may be, but is not confined to: improvement of machinability, facilitation of cold work, improvement of mechanical or electrical properties, and/or increase in stability of dimensions. When the term is used without qualification, full annealing is implied. When applied only for the relief of stress, the process is properly called stress relieving or stress-relief annealing.

In ferrous alloys, annealing usually is done above the upper critical temperature, but the time-temperature cycles vary widely in both maximum temperature attained and in cooling rate employed, depending on composition, material condition, and results desired. When applicable, the following commercial process names should be used: black annealing, blue annealing, box annealing, bright annealing, cycle annealing, flame annealing, full annealing, graphitizing, in-process annealing, isothermal annealing, malleablizing, orientation annealing, process annealing, quench annealing, spheroidizing, subcritical annealing.

In nonferrous alloys, annealing cycles are designed to: (*a*) remove part or all of the effects of cold working (recrystallization may or may not be involved); (*b*) cause substantially complete coalescence of precipitates from solid solution in relatively coarse form; or (*c*) both, depending on composition and material condition. Specific process names in commercial use are final annealing, full annealing, intermediate annealing, partial annealing, recrystallization annealing, stress-relief annealing, anneal to temper.

annealing carbon. Fine, apparently amorphous carbon particles formed in white cast iron and certain steels during prolonged annealing. Also called temper carbon.

annealing twin. A twin form in a crystal during recrystallization.

anneal to temper. A final partial anneal that softens a cold worked nonferrous alloy to a specified level of hardness or tensile strength.

Ar$_{cm}$, Ar$_1$, Ar$_3$, Ar$_4$, Ar′, Ar″. Defined under *transformation temperature*.

artificial aging. Aging above room temperature. See *aging*. Compare with *natural aging*.

athermal transformation. A reaction that proceeds without benefit of thermal fluctuations; that is, thermal activation is not required. Such reactions are diffusionless and can take place with great speed when the driving force is sufficiently high. For example, many martensitic transformations occur athermally on cooling, even at relatively low temperatures, because of the progressively increasing driving force. In contrast, a reaction that occurs at constant temperature is an *isothermal transformation*; thermal activation is necessary in this case and the reaction proceeds as a function of time.

ausforming. Hot deformation of metastable austenite within controlled ranges of temperature and time that avoids formation of nonmartensitic transformation products.

austempering. A heat treatment for ferrous alloys in which a part is quenched from the austenitizing temperature at a rate fast enough to avoid formation of

Reprinted from Heat Treater's Guide, 447-457, © 1982 American Society for Metals

ferrite or pearlite and then held at a temperature just above M_s until transformation to bainite is complete.

austenite. A solid solution of one or more elements in face-centered cubic iron. Unless otherwise designated (such as nickel austenite), the solute is generally assumed to be carbon.

austenitic grain size. The size attained by the grains of steel when heated to the austenitic region; may be revealed by appropriate etching of cross sections after cooling to room temperature.

austenitizing. Forming austenite by heating a ferrous alloy into the transformation range (partial austenitizing) or above the transformation range (complete austenitizing). When used without qualification, the term implies complete austenitizing.

B

bainite. A metastable aggregate of ferrite and cementite resulting from the transformation of austenite at temperatures below the pearlite range but above M_s. Its appearance is feathery if formed in the upper part of the bainite transformation range; acicular, resembling tempered martensite, if formed in the lower part.

baking. (1) Heating to a low temperature to remove gases. (2) Curing or hardening surface coatings such as paints by exposure to heat. (3) Heating to drive off moisture, as in the baking of sand cores after molding.

bark. The decarburized layer just beneath the scale that results from heating steel in an oxidizing atmosphere.

black annealing. Box annealing or pot annealing ferrous alloy sheet, strip, wire. See *box annealing*.

black oxide. A black finish on a metal produced by immersing it in hot oxidizing salts or salt solutions.

blank carburizing. Simulating the carburizing operation without introducing carbon. This is usually accomplished by using an inert material in place of the carburizing agent, or by applying a suitable protective coating to the ferrous alloy.

blank nitriding. Simulating the nitriding operation without introducing nitrogen. This is usually accomplished by using an inert material in place of the nitriding agent or by applying a suitable protective coating to the ferrous alloy.

block brazing. An obsolete brazing process in which the joint was heated using hot blocks.

blue annealing. Heating hot rolled ferrous sheet in an open furnace to a temperature within the transformation range and then cooling in air, in order to soften the metal. The formation of a bluish oxide on the surface is incidental.

blue brittleness. Brittleness exhibited by some steels after being heated to some temperature within the range of about 200 to 370 °C (400 to 700 °F), particularly if the steel is worked at the elevated temperature. Killed steels are virtually free of this kind of brittleness.

bluing. Subjecting the scale-free surface of a ferrous alloy to the action of air, steam, or other agents at a suitable temperature, thus forming a thin blue film of oxide and improving the appearance and resistance to corrosion.

Note: This term is ordinarily applied to sheet, strip, or finished parts. It is used also to denote the heating of springs after fabrication to improve their properties.

box annealing. Annealing a metal or alloy in a sealed container under conditions that minimize oxidation. In box annealing a ferrous alloy, the charge is usually heated slowly to a temperature below the transformation range, but sometimes above or within it, and is then cooled slowly; this process is also called close annealing or pot annealing. See *black annealing*.

brazing. A group of welding processes that join solid materials together by heating them to a suitable temperature and by using a filler metal having a liquidus above 450 °C (840 °F) and below the solidus of the base materials. The filler metal is distributed between the closely fitted surfaces of the joint by capillary attraction.

brazing alloy. See preferred term *brazing filler metal*.

brazing filler metal. A nonferrous filler metal used in *brazing* and braze welding.

brazing sheet. Brazing filler metal in sheet form or flat-rolled metal clad with brazing filler metal on one or both sides.

breaks. Creases or ridges usually in "untempered" or in aged material where the yield point has been exceeded. Depending on the origin of the break, it may be termed a cross break, a coil break, an edge break, or a sticker break.

bright annealing. Annealing in a protective medium to prevent discoloration of the bright surface.

burning. (1) Permanently damaging a metal or alloy by heating to cause either incipient melting or intergranular oxidation. See *overheating*. (2) In grinding, getting the work hot enough to cause discoloration or to change the microstructure by tempering or hardening.

C

calorizing. Imparting resistance to oxidation to an iron or steel surface by heating in aluminum powder at 800 to 1000 °C (1472 to 1832 °F).

carbonitriding. A case hardening process in which a suitable ferrous material is heated above the lower transformation temperature in a gaseous atmosphere of such composition as to cause simultaneous absorption of carbon and nitrogen by the surface and, by diffusion, create a concentration gradient. The process is completed by cooling at a rate that produces the desired properties in the workpiece.

carbonization. Conversion of an organic substance into elemental carbon. (Should not be confused with car*bur*ization.)

carbon potential. A measure of the ability of an environment containing active carbon to alter or maintain, under prescribed conditions, the carbon level of the steel. Note: In any particular environment, the carbon level attained will depend on such factors as temperature, time, and steel composition.

carbon restoration. Replacing the carbon lost in the surface layer from previous processing by carburizing this layer to substantially the original carbon level. Sometimes called recarburizing.

carburizing. Absorption and diffusion of carbon into solid ferrous alloys by heating, to a temperature usually above Ac_3, in contact with a suitable carbonaceous material. A form of *case hardening* that produces a carbon gradient extending inward from the surface, enabling the surface layer to be hardened either by quenching directly from the carburizing temperature or by cooling to room temperature, then reaustenitizing and quenching.

carburizing flame. A gas flame that will introduce carbon into some heated metals, as during a gas welding operation. A carburizing flame is a *reducing flame*, but a reducing flame is not necessarily a carburizing flame.

case. That portion of a ferrous alloy, extending inward from the surface, whose composition has been altered so that it can be *case hardened*. Typically considered to be the portion of the alloy (*a*) whose composition has been measurably altered from the original composition, (*b*) that appears dark on an etched cross section, or (*c*) that has a hardness, after hardening, equal to or greater than a specified value. Contrast with *core*.

case hardening. A generic term covering several processes applicable to steel that change the chemical composition of the surface layer by absorption of carbon, nitrogen, or a mixture of the two and, by diffusion, create a concentration gradient. The processes commonly used are carburizing and quench hardening; cyaniding; nitriding; and carbonitriding. The use of the applicable specific process name is preferred.

cementation. The introduction of one or more elements into the outer portion of a metal object by means of diffusion at high temperature.

cementite. A compound of iron and carbon, known chemically as iron carbide and having the approximate chemical formula Fe_3C. It is characterized by an orthorhombic crystal structure. When it occurs as a phase in steel, the chemical composition will be altered by the presence of manganese and other carbide-forming elements.

checks. Numerous, very fine cracks in a coating or at the surface of a metal part. Checks may appear during processing or during service and are most often associated with thermal treatment or thermal cycling. Also called check marks, checking, *heat checks*.

Chromel. (1) A 90Ni-10Cr alloy used in thermocouples. (2) A series of nickel-chromium alloys, some with iron, used for heat-resistant applications.

close annealing. Same as *box annealing*.

coarsening. An increase in the grain size, usually, but not necessarily, by *grain growth*.

coherency. The continuity of lattice of precipitate and parent phase (solvent) maintained by mutual strain and not separated by a phase boundary.

coherent precipitate. A crystalline precipitate that forms from solid solution with an orientation that maintains continuity between the crystal lattice of the precipitate and the lattice of the matrix, usually accompanied by some strain in both lattices. Because the lattices fit at the interface between precipitate and matrix, there is no discernible phase boundary.

cold treatment. Exposing to suitable subzero temperatures for the purpose of obtaining desired conditions or properties such as dimensional or microstructural stability. When the treatment involves the transformation of retained austenite, it is usually followed by tempering.

columnar structure. A coarse structure of parallel elongated grains formed by unidirectional growth, most often observed in castings, but sometimes in structures resulting from diffusional growth accompanied by a solid-state transformation.

combined carbon. The part of the total carbon in steel or cast iron that is present as other than *free carbon*.

conditioning heat treatment. A preliminary heat treatment used to prepare a material for desired reaction to a subsequent heat treatment. For the term to be meaningful, the exact heat treatment must be specified.

congruent transformation. An isothermal or isobaric phase change in which both of the phases concerned have the same composition throughout the process.

constantan. A group of copper-nickel alloys containing 45 to 60% copper with minor amounts of iron and manganese and characterized by relatively constant electrical resistivity irrespective of temperature; used in resistors and thermocouples.

constitution diagram. A graphical representation of the temperature and composition limits of phase fields in an alloy system as they actually exist under the specific conditions of heating or cooling (synonymous with phase diagram). A constitution diagram may be an equilibrium diagram, an approximation to an equilibrium diagram, or a representation of metastable conditions or phases. Compare with *equilibrium diagram*.

continuous precipitation. Precipitation from a supersaturated solid solution in which the precipitate particles grow by long-range diffusion without recrystallization of the matrix. Continuous precipitates grow from nuclei distributed more or less uniformly throughout the matrix. They usually are randomly oriented, but may form a Widmanstatten structure. Also called general precipitation. Compare with *discontinuous precipitation, localized precipitation*.

controlled cooling. Cooling from an elevated temperature in a predetermined manner, to avoid hardening, cracking, or internal damage, or to produce desired microstructure or mechanical properties.

cooling curve. A curve showing the relation between time and temperature during the cooling of a material.

cooling stresses. Residual stresses resulting from nonuniform distribution of temperature during cooling.

core. In a ferrous alloy prepared for *case hardening*, that portion of the alloy that is not part of the *case*. Typically considered to be the portion that (*a*) appears light on an etched cross section, (*b*) has an essentially unaltered chemical composition, or (*c*) has a hardness, after hardening, less than a specified value.

critical cooling rate. The rate of continuous cooling required to prevent undesirable transformation. For steel, it is the minimum rate at which austenite must be continuously cooled to suppress transformations above the M_s temperature.

critical point. (1) The temperature or pressure at which a change in crystal structure, phase or physical properties occurs. Same as *transformation temperature*. (2) In an equilibrium diagram, that specific value of composition, temperature and pressure, or combinations thereof, at which the phases of a heterogeneous system are in equilibrium.

critical strain. The strain just sufficient to cause *recrystallization*; because the strain is small, usually only a few percent, recrystallization takes place from only a few nuclei, which produces a recrystallized structure consisting of very large grains.

critical temperature. (1) Synonymous with *critical point* if the pressure is constant. (2) The temperature above which the vapor phase cannot be condensed to liquid by an increase in pressure.

critical temperature ranges. Synonymous with *transformation ranges*, which is the preferred term.

Curie temperature. The temperature of magnetic transformation below which a metal or alloy is ferromagnetic and above which it is paramagnetic.

cyaniding. A case hardening process in which a ferrous material is heated above the lower transformation range in a molten salt containing cyanide to cause simultaneous absorption of carbon and nitrogen at the surface and, by diffusion, create a concentration gradient. Quench hardening completes the process.

cycle annealing. An annealing process employing a predetermined and closely controlled time-temperature cycle to produce specific properties or microstructures.

D

dead soft. A *temper* of nonferrous alloys and some ferrous alloys corresponding to the condition of minimum hardness and tensile strength produced by *full annealing*.

decalescence. A phenomenon, associated with the transformation of alpha iron to gamma iron on the heating (superheating) of iron or steel, revealed by the darkening of the metal surface owing to the sudden decrease in temperature caused by the fast absorption of the latent heat of transformation. Contrast with *recalescence*.

decarburization. Loss of carbon from the surface layer of a carbon-containing alloy due to reaction with one or more chemical substances in a medium that contacts the surface.

degrees of freedom. The number of independent variables (such as temperature, pressure or concentration within the phases present) that may be altered at will without causing a phase change in an alloy system at equilibrium; or the number of such variables that must be fixed arbitrarily to define the system completely.

delta ferrite. See *ferrite*.

differential heating. Heating that intentionally produces a temperature gradient within an object such that, after cooling, a desired stress distribution or variation in properties is present within the object.

diffusion. (1) Spreading of a constituent in a gas, liquid, or solid, tending to make the composition of all parts uniform. (2) The spontaneous movement of atoms or molecules to new sites within a material.

diffusion coefficient. A factor of proportionality representing the amount of substance diffusing across a unit area through a unit concentration gradient in unit time.

direct quenching. (1) Quenching carburized parts directly from the carburizing operation. (2) Also used for quenching pearlitic malleable parts directly from the malleablizing operation.

discontinuous precipitation. Precipitation from a supersaturated solid solution in which the precipitate particles grow by short-range diffusion, accompanied by recrystallization of the matrix in the region of precipitation. Discontinuous precipitates grow into the matrix from nuclei near grain boundaries, forming cells of alternate lamellae of precipitate and depleted (and recrystallized) matrix. Often referred to as cellular or nodular precipitation. Compare with *continuous precipitation, localized precipitation*.

double aging. Employment of two different aging treatments to control the type of precipitate formed from a supersaturated matrix in order to obtain the desired properties. The first aging treatment, sometimes referred to as intermediate or stabilizing, is usually carried out at higher temperature than the second.

double tempering. A treatment in which a quench-hardened ferrous metal is subjected to two complete tempering cycles, usually at substantially the same temperature, for the purpose of ensuring completion of the tempering reaction and promoting stability of the resulting microstructure.

drawing. A misnomer for *tempering*.

dry cyaniding. (obsolete) Same as *carbonitriding*.

E

embrittlement. Reduction in the normal ductility of a metal due to a physical or chemical change. Examples include *blue brittleness*, *hydrogen embrittlement*, and *temper brittleness*.

enantiotropy. The relation of crystal forms of the same substance in which one form is stable above a certain temperature and the other form stable below that temperature. Ferrite and austenite are enantiotropic in ferrous alloys, for example.

end-quench hardenability test. A laboratory procedure for determining the hardenability of a steel or other ferrous alloy; widely referred to as the *Jominy test*. Hardenability is determined by heating a standard specimen above the upper critical temperature, placing the hot specimen in a fixture so that a stream of cold water impinges on one end, and, after cooling to room temperature is completed, measuring the hardness near the surface of the specimen at regularly spaced intervals along its length. The data are normally plotted as hardness versus distance from the quenched end.

equilibrium diagram. A graphical representation of the temperature, pressure and composition limits of phase fields in an alloy system as they exist under conditions of complete equilibrium. In metal systems, pressure is usually considered constant.

eutectic. (1) An isothermal reversible reaction in which a liquid solution is converted into two or more intimately mixed solids on cooling, the number of solids formed being the same as the number of components in the system. (2) An alloy having the composition indicated by the eutectic point on an equilibrium diagram. (3) An alloy structure of intermixed solid constituents formed by a eutectic reaction.

eutectic carbide. Carbide formed during freezing as one of the mutually insoluble phases participating in the eutectic reaction of ferrous alloys.

eutectic melting. Melting of localized microscopic areas whose composition corresponds to that of the eutectic in the system.

eutectoid. (1) An isothermal reversible reaction in which a solid solution is converted into two or more intimately mixed solids on cooling, the number of solids formed being the same as the number of components in the system. (2) An alloy having the composition indicated by the eutectoid point on an equilibrium diagram. (3) An alloy structure of intermixed solid constituents formed by a eutectoid reaction.

extra hard. A *temper* of nonferrous alloys and some ferrous alloys characterized by tensile strength and hardness about one-third of the way from *full hard* to *extra spring* temper.

extra spring. A *temper* of nonferrous alloys and some ferrous alloys corresponding approximately to a cold worked state above *full hard* beyond which further cold work will not measurably increase the strength and hardness.

F

ferrite. A solid solution of one or more elements in body-centered cubic iron. Unless otherwise designated (for instance, as chromium ferrite), the solute is generally assumed to be carbon. On some equilibrium diagrams, there are two ferrite regions separated by an austenite area. The lower area is alpha ferrite; the upper, delta ferrite. If there is no designation, alpha ferrite is assumed.

ferritizing anneal. A treatment given as-cast gray or ductile (nodular) iron to produce an essentially ferritic matrix. For the term to be meaningful, the final microstructure desired or the time-temperature cycle used must be specified.

file hardness. Hardness as determined by the use of a file of standardized hardness on the assumption that a material that cannot be cut with the file is as hard as, or harder than, the file. Files covering a range of hardnesses may be employed.

final annealing. An imprecise term used to denote the last anneal given to a nonferrous alloy prior to shipment.

finishing temperature. The temperature at which hot working is completed.

flame annealing. Annealing in which the heat is applied directly by a flame.

flame hardening. A process for hardening the surfaces of hardenable ferrous alloys in which an intense flame is used to heat the surface layers above the upper transformation temperature, whereupon the workpiece is immediately quenched.

flame straightening. Correcting distortion in metal structures by localized heating with a gas flame.

fog quenching. Quenching in a fine vapor or mist.

free carbon. The part of the total carbon in steel or cast iron that is present in elemental form as graphite or temper carbon. Contrast with *combined carbon*.

free ferrite. Ferrite that is formed directly from the decomposition of hypoeutectoid austenite during cooling, without the simultaneous formation of cementite. Also proeutectoid ferrite.

freezing range. That temperature range between liquidus and solidus temperatures in which molten and solid constituents coexist.

full annealing. An imprecise term that denotes an annealing cycle to produce minimum strength and hardness. For the term to be meaningful, the composition and starting condition of the material and the time-temperature cycle used must be stated.

full hard. A *temper* of nonferrous alloys and some ferrous alloys corresponding approximately to a cold worked state beyond which the material can no longer be formed by bending. In specifications, a full hard temper is commonly defined in terms of minimum hardness or minimum tensile strength (or, alternatively, a range of hardness or strength) corresponding to a specific percentage of cold reduction following a full anneal. For aluminum, a full hard temper is equivalent to a reduction of 75% from *dead soft*; for austenitic stainless steels, a reduction of about 50 to 55%.

furnace brazing. A mass-production *brazing* process in which the filler metal is preplaced on the joint, then the entire assembly is heated to brazing temperature in a furnace. Usually, a protective furnace atmosphere is required, and wetting of the joint surfaces is accomplished without using a brazing flux.

fusion. A change of state from solid to liquid; melting.

G

gamma iron. The face-centered cubic form of pure iron, stable from 910 to 1400 °C (1670 to 2550 °F).

gas cyaniding. A misnomer for *carbonitriding*.

grain growth. An increase in the average size of the grains in polycrystalline metal, usually as a result of heating at elevated temperature.

grain size. For metals, a measure of the areas or volumes of grains in a polycrystalline material, usually expressed as an average when the individual sizes are fairly uniform. In metals containing two or more phases, the grain size refers to that of the matrix unless otherwise specified. Grain sizes are reported in terms of number of grains per unit area or volume, average diameter, or as a grain-size number derived from area measurements.

graphitization. Formation of graphite in iron or steel. Where graphite is formed during solidification, the phenomenon is called primary graphitization; where formed later by heat treatment, secondary graphitization.

graphitizing. Annealing a ferrous alloy in such a way that some or all of the carbon is precipitated as graphite.

growth. In cast iron, a permanent increase in dimensions resulting from repeated or prolonged heating at temperatures above 480 °C (900 °F) due either to graphitizing of carbides or to oxidation.

H

half hard. A *temper* of nonferrous alloys and some ferrous alloys characterized by tensile strength about mid-way between that of *dead soft* and *full hard* tempers.

hardenability. The relative ability of a ferrous alloy to form martensite when quenched from a temperature above the upper critical temperature. Hardenability is commonly measured as the distance below a quenched surface where the metal exhibits a specific hardness (50 HRC, for example) or a specific percentage of martensite in the microstructure.

hardening. Increasing hardness by suitable treatment, usually involving heating and cooling. When applicable, the following more specific terms should be used: *age hardening*, *case hardening*, *flame hardening*, *induction hardening*, *precipitation hardening* and *quench hardening*.

hard temper. Same as *full hard* temper.

heat-resisting alloy. An alloy developed for very high temperature service where relatively high stresses (tensile, thermal, vibratory, or shock) are encountered and where oxidation resistance is frequently required.

heat tinting. Coloration of a metal surface through oxidation by heating to reveal details of the microstructure.

heat treatable alloy. An alloy that can be hardened by heat treatment.

heat treating film. A thin coating or film, usually an oxide, formed on the surface of metals during heat treatment.

heat treatment. Heating and cooling a solid metal or alloy in such a way as to obtain desired conditions or properties. Heating for the sole purpose of hot working is excluded from the meaning of this definition.

homogeneous carburizing. Use of a carburizing process to convert a low-carbon ferrous alloy to one of uniform and higher carbon content throughout the section.

homogenizing. Holding at high temperature to eliminate or decrease chemical segregation by diffusion.

hot quenching. An imprecise term used to cover a variety of quenching procedures in which a quenching medium is maintained at a prescribed temperature above 70 °C (160 °F).

hydrogen brazing. A term sometimes used to denote brazing in a hydrogen-containing atmosphere, usually in a furnace; use of the appropriate process name is preferred.

hydrogen embrittlement. A condition of low ductility in metals resulting from the absorption of hydrogen.

I

induction brazing. *Brazing* in which the required heat is generated by subjecting the workpiece to electromagnetic induction.

induction hardening. A surface-hardening process in which only the surface layer of a suitable ferrous workpiece is heated by electromagnetic induction to above the upper critical temperature and immediately quenched.

induction heating. Heating by combined electrical resistance and hysteresis losses induced by subjecting a metal to the varying magnetic field surrounding a coil carrying alternating current.

intermediate annealing. Annealing wrought metals at one or more stages during manufacture and before final treatment.

interrupted aging. Aging at two or more temperatures, by steps, and cooling to room temperature after each step. See *aging*, and compare with *progressive aging* and *step aging*.

interrupted quenching. A quenching procedure in which the workpiece is removed from the first quench at a temperature substantially higher than that of the quenchant and is then subjected to a second quenching system having a different cooling rate than the first.

isothermal annealing. Austenitizing a ferrous alloy and then cooling to and holding at a temperature at which austenite transforms to a relatively soft ferrite carbide aggregate.

isothermal transformation. A change in phase that takes place at a constant temperature. The time required for transformation to be completed, and in some instances the time delay before transformation begins, depends on the amount of supercooling below (or superheating above) the equilibrium temperature for the same transformation.

J

Jominy test. See *end-quench hardenability test*.

L

ledeburite. The eutectic of the iron-carbon system, the constituents being austenite and cementite. The austenite decomposes into ferrite and cementite on cooling below the Ar_1.

liquid phase sintering. *Sintering* a powder metallurgy compact under conditions that maintain a liquid metallic phase within the compact during all or part of the sintering schedule. The liquid phase may be derived from a component of the green compact or may be infiltrated into the compact from an outside source.

localized precipitation. Precipitation from a supersaturated solid solution similar to *continuous precipitation*, except that the precipitate particles form at preferred locations, such as along slip planes, grain boundaries, or incoherent twin boundaries.

M

malleablizing. Annealing white cast iron in such a way that some or all of the combined carbon is transformed to graphite or, in some instances, part of the carbon is removed completely.

maraging. A precipitation-hardening treatment applied to a special group of iron-base alloys to precipitate one or more intermetallic compounds in a matrix of essentially carbon-free martensite. Note: The first developed series of maraging steels contained, in addition to iron, more than 10% nickel and one or more supplemental hardening elements. In this series, aging is done at 480 °C (900 °F).

marquenching. See *martempering*.

martempering. (1) A hardening procedure in which an austenitized ferrous workpiece is quenched into an appropriate medium whose temperature is maintained substantially at the M_s of the workpiece, held in the medium until its temperature is uniform throughout—but not long enough to permit bainite to form—and then cooled in air. The treatment is frequently followed by tempering. (2) When the process is applied to carburized material, the controlling M_s temperature is that of the case. This variation of the process is frequently called marquenching.

martensite. A generic term for microstructures formed by diffusionless phase transformation in which the parent and product phases have a specific crystallographic relationship. Martensite is characterized by an acicular pattern in the microstructure in both ferrous and nonferrous alloys. In alloys where the solute atoms occupy interstitial positions in the martensitic lattice (such as carbon in iron), the structure is hard and highly strained; but where the solute atoms occupy substitutional positions (such as nickel in iron), the martensite is soft and ductile. The amount of high temperature phase that transforms to martensite on cooling depends to a large extent on the lowest temperature attained, there being a rather distinct beginning temperature (M_s) and a temperature at which the transformation is essentially complete (M_f).

martensite range. The temperature interval between M_s and M_f.

martensitic transformation. A reaction that takes place in some metals on cooling, with the formation of an acicular structure called *martensite*.

McQuaid-Ehn test. A test to reveal grain size after heating into the austenitic temperature range. Eight standard McQuaid-Ehn grain sizes rate the structure, No. 8 being finest, No. 1 coarsest.

metallurgy. The science and technology of metals and alloys. Process metallurgy is concerned with the extraction of metals from their ores and with the refining of metals; physical metallurgy, with the physical and mechanical properties of metals as affected by composition, processing, and environmental conditions; and mechanical metallurgy, with the response of metals to applied forces.

M_f temperature. For any alloy system, the temperature at which martensite formation on cooling is essentially finished. See *transformation temperature* for the definition applicable to ferrous alloys.

microhardness. The hardness of a material as determined by forcing an indenter such as a Vickers or Knoop indenter into the surface of a material under very light load; usually, the indentations are so small that they must be measured with a microscope. Capable of determining hardnesses of different microconstituents within a structure, or of measuring steep hardness gradients such as those encountered in case hardening.

mill scale. The heavy oxide layer formed during hot fabrication or heat treatment of metals.

monotropism. The ability of a solid to exist in two or more forms (crystal structures), but in which one form is the stable modification at all temperatures and pressures. Ferrite and martensite are a monotropic pair below Ac_1 in steels, for example. May also be spelled monotrophism.

M_s temperature. For any alloy system, the temperature at which martensite starts to form on cooling. See *transformation temperature* for the definition applicable to ferrous alloys.

N

natural aging. Spontaneous aging of a supersaturated solid solution at room temperature. See *aging*, and compare with *artificial aging*.

neutral flame. A gas flame in which there is no excess of either fuel or oxygen in the inner flame. Oxygen from ambient air is used to complete the combustion of CO_2 and H_2 produced in the inner flame.

nitriding. Introducing nitrogen into the surface layer of a solid ferrous alloy by holding at a suitable temperature (below Ac_1 for ferritic steels) in contact with a nitrogenous material, usually ammonia or molten cyanide of appropriate composition. Quenching is not required to produce a hard case.

nitrocarburizing. Any of several processes in which both nitrogen and carbon are absorbed into the surface layers of a ferrous material at temperatures below the lower critical temperature and, by diffusion, create a concentration gradient. Nitrocarburizing is done mainly to provide an antiscuffing surface layer and to improve fatigue resistance. Compare with *carbonitriding*.

normalizing. Heating a ferrous alloy to a suitable temperature above the transformation range and then cooling in air to a temperature substantially below the transformation range.

O

optical pyrometer. An instrument for measuring the temperature of heated material by comparing the intensity of light emitted with a known intensity of an incandescent lamp filament.

overaging. Aging under conditions of time and temperature greater than those required to obtain maximum change in a certain property, so that the property is altered in the direction of the initial value. See *aging*.

overheating. Heating a metal or alloy to such a high temperature that its properties are impaired. When the original properties cannot be restored by further heat treating, by mechanical working, or by a combination of working and heat treating, the overheating is known as *burning*.

oxidizing flame. A gas flame produced with excess oxygen in the inner flame.

P

packing material. Any material in which powder metallurgy compacts are embedded during the presintering or sintering operations.

partial annealing. An imprecise term used to denote a treatment given cold worked material to reduce the strength to a controlled level or to effect stress relief. To be meaningful, the type of material, the degree of cold work, and the time-temperature schedule must be stated.

patenting. In wiremaking, a heat treatment applied to medium-carbon or high-carbon steel before the drawing of wire or between drafts. This process consists of heating to a temperature above the transformation range and then cooling to a temperature below Ae₁ in air or in a bath of molten lead or salt.

pearlite. A metastable lamellar aggregate of ferrite and cementite resulting from the transformation of austenite at temperatures above the bainite range.

postheating. Heating weldments immediately after welding, for tempering, for stress relieving, or for providing a controlled rate of cooling to prevent formation of a hard or brittle structure.

pot annealing. Same as *box annealing*.

precipitation hardening. Hardening caused by the precipitation of a constituent from a supersaturated solid solution. See also *age hardening* and *aging*.

precipitation heat treatment. *Artificial aging* in which a constituent precipitates from a supersaturated solid solution.

preheating. Heating before some further thermal or mechanical treatment. For tool steel, heating to an intermediate temperature immediately before final austenitizing. For some nonferrous alloys, heating to a high temperature for a long time, to homogenize the structure before working. In welding and related processes, heating to an intermediate temperature for a short time immediately before welding, brazing, soldering, cutting, or thermal spraying.

presintering. The heating of a powder metallurgy compact to a temperature lower than the normal temperature for final sintering, usually to increase the ease of handling or forming the compact or to remove a lubricant or binder before sintering.

process annealing. An imprecise term denoting various treatments used to improve workability. For the term to be meaningful, the condition of the material and the time-temperature cycle used must be stated.

progressive aging. Aging by increasing the temperature in steps or continuously during the aging cycle. See *aging* and compare with interrupted aging and step aging.

pseudocarburizing. See *blank carburizing*.

pseudonitriding. See *blank nitriding*.

pusher furnace. A type of continuous furnace in which parts to be heated are periodically charged into the furnace in containers, which are pushed along the hearth against a line of previously charged containers thus advancing the containers toward the discharge end of the furnace, where they are removed.

Q

quarter hard. A *temper* of nonferrous alloys and some ferrous alloys characterized by tensile strength about midway between that of *dead soft* and *half hard* tempers.

quench-age embrittlement. Embrittlement of low-carbon steel evidenced by a loss of ductility on aging at room temperature following rapid cooling from a temperature below the lower critical temperature.

quench aging. Aging induced by rapid cooling after *solution heat treatment*.

quench annealing. Annealing an austenitic ferrous alloy by *solution heat treatment* followed by rapid quenching.

quench cracking. Fracture of a metal during quenching from elevated temperature. Most frequently observed in hardened carbon steel, alloy steel, or tool steel parts of high hardness and low toughness. Cracks often emanate from fillets, holes, corners, or other stress raisers and result from high stresses due to the volume changes accompanying transformation to martensite.

quench hardening. (1) Hardening suitable alpha-beta alloys (most often certain copper or titanium alloys) by solution treating and quenching to develop a martensitic-like structure. (2) In ferrous alloys, hardening by austenitizing and then cooling at a rate such that a substantial amount of austenite transforms to martensite.

quenching. Rapid cooling. When applicable, the following more specific terms should be used: *direct quenching, fog quenching, hot quenching, interrupted quenching, selective quenching, spray quenching,* and *time quenching.*

R

recalescence. A phenomenon, associated with the transformation of gamma iron to alpha iron on the cooling (supercooling) of iron or steel, revealed by the brightening (reglowing) of the metal surface owing to the sudden increase in temperature caused by the fast liberation of the latent heat of transformation. Contrast with *decalescence*.

recarburize. (1) To increase the carbon content of molten cast iron or steel by adding carbonaceous material, high-carbon pig iron, or a high-carbon alloy. (2) To carburize a metal part to return surface carbon lost in processing; also known as carbon restoration.

recovery. Reduction or removal of work-hardening effects, without motion of large-angle grain boundaries.

recrystallization. (1) The formation of a new, strain-free grain structure from that existing in cold worked metal, usually accomplished by heating. (2) The change from one crystal structure to another, as occurs on heating or cooling through a critical temperature.

recrystallization annealing. Annealing cold worked metal to produce a new grain structure without phase change.

recrystallization temperature. The approximate minimum temperature at which complete recrystallization of a cold worked metal occurs within a specified time.

recuperator. Equipment for transferring heat from gaseous products of combustion to incoming air or fuel. The incoming material passes through pipes surrounded by a chamber through which the outgoing gases pass.

reducing flame. A gas flame produced with excess fuel in the inner flame.

refractory. (1) A material of very high melting point with properties that make it suitable for such uses as furnace linings and kiln construction. (2) The quality of resisting heat.

refractory alloy. (1) A heat-resistant alloy. (2) An alloy having an extremely high melting point. See *refractory metal*. (3) An alloy difficult to work at elevated temperatures.

refractory metal. A metal having an extremely high melting point; for example, tungsten, molybdenum, tantalum, niobium (columbium), chromium, vanadium, and rhenium. In the broad sense, it refers to metals having melting points above the range of iron, cobalt, and nickel.

regenerator. Same as *recuperator* except the gaseous products of combustion heat brick checkerwork in a chamber connected to the exhaust side of the furnace while the incoming air and fuel are being heated by the brick checkerwork in a second chamber, connected to the entrance side. At intervals, the gas flow is reversed so that incoming air and fuel contact hot checkerwork while that in the second chamber is being reheated by exhaust gases.

resist. (1) A material applied to a part of the surface of an article to prevent reaction of metal from that area during chemical or electrochemical processes. (2) A material applied to prevent the flow of brazing filler metal into unwanted area.

resistance brazing. Brazing by resistance heating, the joint being part of the electrical circuit.

reverberatory furnace. A furnace with a shallow hearth, usually nonregenerative, having a roof that deflects the flame and radiates heat toward the hearth or the surface of the charge.

Rockwell hardness test. An indentation hardness test based on the depth of penetration of a specified penetrator into the specimen under certain arbitrarily fixed conditions.

rotary furnace. A circular furnace constructed so that the hearth and workpieces rotate around the axis of the furnace during heating.

S

selective heating. Intentionally heating only certain portions of a workpiece.

selective quenching. Quenching only certain portions of an object.

self-hardening steel. See preferred term, *air-hardening steel*.

shrink forming. Forming metal wherein the inner fibers of a cross section undergo a reduction in a localized area by the application of heat, cold upset, or mechanically induced pressures.

siliconizing. Diffusing silicon into solid metal, usually steel, at an elevated temperature.

sinter. To heat a mass of fine particles for a prolonged time below the melting point, usually to cause agglomeration.

sintering. The bonding of adjacent surfaces in a mass of particles by molecular or atomic attraction on heating at high temperatures below the melting temperature of any constituent in the material. Sintering strengthens a powder mass and normally produces densification and, in powdered metals, recrystallization. See also *liquid phase sintering*.

slack quenching. The incomplete hardening of steel due to quenching from the austenitizing temperature at a rate slower than the critical cooling rate for the particular steel, resulting in the formation of one or more transformation products in addition to martensite.

slot furnace. A common batch furnace where stock is charged and removed through a slot or opening.

snap temper. A precautionary interim stress-relieving treatment applied to high-hardenability steels immediately after quenching to prevent cracking because of delay in tempering them at the prescribed higher temperature.

soaking. Prolonged holding at a selected temperature to effect homogenization of structure or composition.

soft temper. Same as *dead soft* temper.

solution heat treatment. Heating an alloy to a suitable temperature, holding at that temperature long enough to cause one or more constituents to enter into solid solution, and then cooling rapidly enough to hold these constituents in solution.

sorbite. (obsolete) A fine mixture of ferrite and cementite produced either by regulating the rate of cooling of steel or by tempering steel after hardening. The first type is very fine pearlite difficult to resolve under the microscope; the second type is tempered martensite.

spheroidite. An aggregate of iron or alloy carbides of essentially spherical shape dispersed throughout a matrix of ferrite.

spheroidizing. Heating and cooling to produce a spheroidal or globular form of carbide in steel. Spheroidizing methods frequently used are:

1 Prolonged holding at a temperature just below Ae_1
2 Heating and cooling alternately between temperatures that are just above and just below Ae_1
3 Heating to a temperature above Ae_1 or Ae_3 and then cooling very slowly in the furnace or holding at a temperature just below Ae_1
4 Cooling at a suitable rate from the minimum temperature at which all carbide is dissolved, to prevent the reformation of a carbide network, and then reheating in accordance with method 1 or 2 above. (Applicable to hypereutectoid steel containing a carbide network.)

spinodal structure. A fine homogeneous mixture of two phases that form by the growth of composition waves in a solid solution during suitable heat treatment. The phases of a spinodal structure differ in composition from each other and from the parent phase but have the same crystal structure as the parent phase.

spray quenching. Quenching in a spray of liquid.

spring temper. A *temper* of nonferrous alloys and some ferrous alloys characterized by tensile strength and hardness about two-thirds of the way from *full hard* to *extra spring* temper.

stabilizing treatment. (1) Before finishing to final dimensions, repeatedly heating a ferrous or nonferrous part to or slightly above its normal operating temperature and then cooling to room temperature to ensure dimensional stability in service. (2) Transforming retained austenite in quenched hardenable steels, usually by *cold treatment*. (3) Heating a solution-treated stabilized grade of austenitic stainless steel to 870 to 900 °C (1600 to 1650 °F) to precipitate all carbon as TiC, NbC, or TaC so that *sensitization* is avoided on subsequent exposure to elevated temperature.

Stead's brittleness. A condition of brittleness that causes transcrystalline fracture in the coarse grain structure that results from prolonged annealing of thin sheets of low-carbon steel previously rolled at a temperature below about 705 °C (1300 °F). The fracture usually occurs at about 45° to the direction of rolling.

step aging. Aging at two or more temperatures, by steps, without cooling to room temperature after each step. See *aging,* and compare with *interrupted aging* and *progressive aging*.

stoking. (obsolete) Presintering, or sintering, in such a way that powder metallurgy compacts are advanced through the furnace at a fixed rate by manual or mechanical means; also called continuous sintering.

stop-off. See *resist*.

stopping off. (1) Applying a *resist*. (2) Depositing a metal (copper, for example) in localized areas to prevent carburization, decarburization, or nitriding in those areas.

strain-age embrittlement. A loss in ductility accompanied by an increase in hardness and strength that occurs when low-carbon steel (especially rimmed or capped steel) is aged following plastic deformation. The degree of embrittlement is a function of aging time and temperature, occurring in a matter of minutes at about 200 °C (400 °F) but requiring a few hours to a year at room temperature.

stress relieving. Heating to a suitable temperature, holding long enough to reduce residual stresses, and then cooling slowly enough to minimize the development of new residual stresses.

subcritical annealing. A process anneal performed on ferrous alloys at a temperature below Ac_1.

superalloy. See *heat-resisting alloy*.

supercooling. Cooling below the temperature at which an equilibrium phase transformation can take place, without actually obtaining the transformation.

superheating. Heating above the temperature at which an equilibrium phase transformation should occur without actually obtaining the transformation.

surface hardening. A generic term covering several processes applicable to a suitable ferrous alloy that produces, by quench hardening only, a surface layer that is harder or more wear resistant than the core. There is no significant alteration of the chemical composition of the surface layer. The processes commonly used are induction hardening, flame hardening, and shell hardening. Use of the applicable specific process name is preferred.

T

temper. (1) In heat treatment, reheating hardened steel or hardened cast iron to some temperature below the eutectoid temperature for the purpose of decreasing hardness and increasing toughness. The process also is sometimes applied to normalized steel. (2) In tool steels, temper is sometimes used, but inadvisedly, to denote the carbon content. (3) In nonferrous alloys and in some ferrous alloys (steels that cannot be hardened by heat treatment), the hardness and strength produced by mechanical or thermal treatment, or both, and characterized by a certain structure, mechanical properties, or reduction in area during cold working.

temper brittleness. Brittleness that results when certain steels are held within, or are cooled slowly through, a certain range of temperature below the transformation range. The brittleness is manifested as an upward shift in ductile-to-brittle transition temperature, but only rarely produces a low value of reduction of area in a smooth-bar tension test of the embrittled material.

temper carbon. Same as *annealing carbon*.

temper color. A thin, tightly adhering oxide skin (only a few molecules thick) that forms when steel is tempered at a low temperature, or for a short time, in air or a mildly oxidizing atmosphere. The color, which ranges from straw to blue depending on the thickness of the oxide skin, varies with both tempering time and temperature.

thermocouple. A device for measuring temperatures, consisting of lengths of two dissimilar metals or alloys that are electrically joined at one end and connected to a voltage-measuring instrument at the other end. When one junction is hotter than the other, a thermal electromotive force is produced that is roughly proportional to the difference in temperature between the hot and cold junctions.

thermomechanical working. A general term covering a variety of processes combining controlled thermal and deformation treatments to obtain synergistic effects such as improvement in strength without loss of toughness. Same as thermal-mechanical treatment.

three-quarters hard. A *temper* of nonferrous alloys and some ferrous alloys characterized by tensile strength and hardness about midway between those of *half hard* and *full hard* tempers.

time quenching. Interrupted quenching in which the time in the quenching medium is controlled.

total carbon. The sum of the free and combined carbon (including carbon in solution) in a ferrous alloy.

transformation-induced plasticity. A phenomenon, occurring chiefly in certain highly alloyed steels that have been heat treated to produce metastable austenite or metastable austenite plus martensite, whereby, on subsequent deformation, part of the austenite undergoes strain-induced transformation to martensite. Steels capable of transforming in this manner, commonly referred to as TRIP steels, are highly plastic after heat treatment, but exhibit a very high rate of strain hardening and thus have high tensile and yield strengths after plastic deformation at temperatures between about 20 and 500 °C (70 and 930 °F). Cooling to −195 °C (−320 °F) may or may not be required to complete the transformation to martensite. Tempering usually is done following transformation.

transformation ranges. Those ranges of temperature within which a phase forms during heating and transforms during cooling. The two ranges are distinct, sometimes overlapping but never coinciding. The limiting temperatures of the ranges depend on the composition of the alloy and on the rate of change of temperature, particularly during cooling. See *transformation temperature.*

transformation temperature. The temperature at which a change in phase occurs. The term is sometimes used to denote the limiting temperature of a transformation range. The following symbols are used for iron and steels.

Ac$_{cm}$. In hypereutectoid steel, the temperature at which the solution of cementite in austenite is completed during heating.

Ac$_1$. The temperature at which austenite begins to form during heating.

Ac$_3$. The temperature at which transformation of ferrite to austenite is completed during heating.

Ac$_4$. The temperature at which austenite transforms to delta ferrite during heating.

Ae$_{cm}$, Ae$_1$, Ae$_3$, Ae$_4$. The temperatures of phase changes at equilibrium.

Ar$_{cm}$. In hypereutectoid steel, the temperature at which precipitation of cementite starts during cooling.

Ar$_1$. The temperature at which transformation of austenite to ferrite or to ferrite plus cementite is completed during cooling.

Ar$_3$. The temperature at which austenite begins to transform to ferrite during cooling.

Ar$_4$. The temperature at which delta ferrite transforms to austenite during cooling.

Ar′. The temperature at which transformation of austenite to pearlite starts during cooling.

M$_f$. The temperature at which transformation of austenite to martensite finishes during cooling.

M$_s$, (or Ar″). The temperature at which transformation of austenite to martensite starts during cooling.

Note: All these changes except the formation of martensite occur at lower temperatures during cooling than during heating, and depend on the rate of change of temperature.

TRIP steel. A commercial steel product exhibiting *transformation-induced plasticity.*

troostite. (obsolete) A previously unresolvable rapidly etching fine aggregate of carbide and ferrite produced either by tempering martensite at low temperature or by quenching a steel at a rate slower than the critical cooling rate. Preferred terminology for the first product is tempered martensite; for the latter, fine pearlite.

U

undercooling. Same as *supercooling.*

Useful Diagrams and Tables

Iron-Carbon Equilibrium Diagram

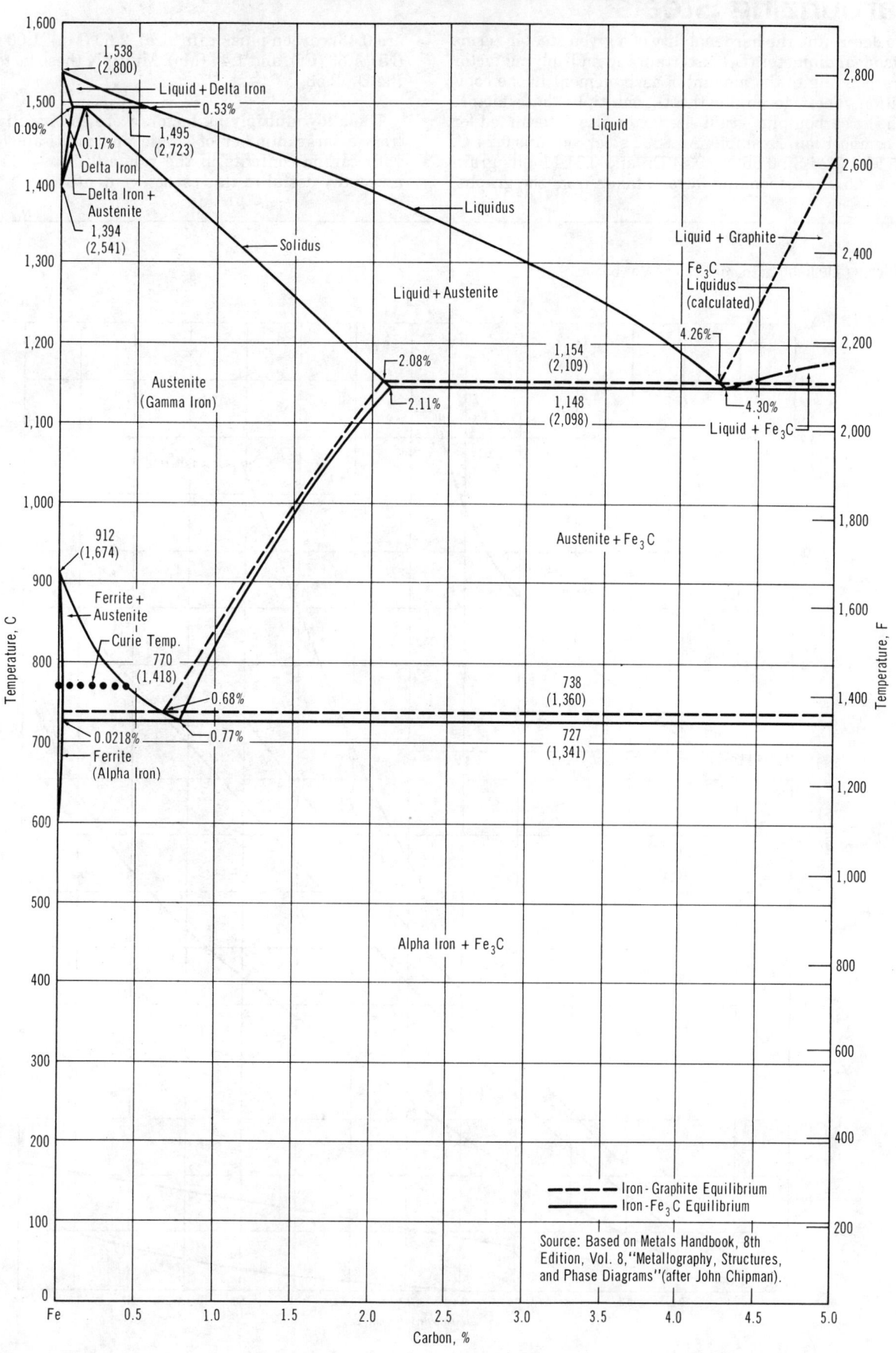

Multiplying Factors for Predicting Hardenability of Carburizing Steels

To determine the hardenability of a given steel in terms of its ideal diameter (D_1), ascertain the multiplying factor corresponding to the amount of each element in the composition. Then, to obtain the D_1, multiply these factors with the carbon-plus-grain size factor also determined for the composition. Example: An 8822 steel contains 0.24 C, 0.95 Mn, 0.28 Si, 0.56 Ni, 0.44 Cr, and 0.31 Mo; its grain size is 7.5. Corresponding factors, taken from the graphs, are 0.43 (carbon plus grain size), 2.60 (Mn), 1.00 (Si), 1.18 (Ni), 1.33 (Cr), and 1.44 (Mo). Multiply these factors to get the D_1, 2.53.

The alloy multiplying factors were empirically derived from a large number of single and multi-alloyed steels with carbon contents in the range 0.15 to 0.30 and are especially useful in this carbon content range.

Alloying Elements. (Source: Climax Molybdenum)

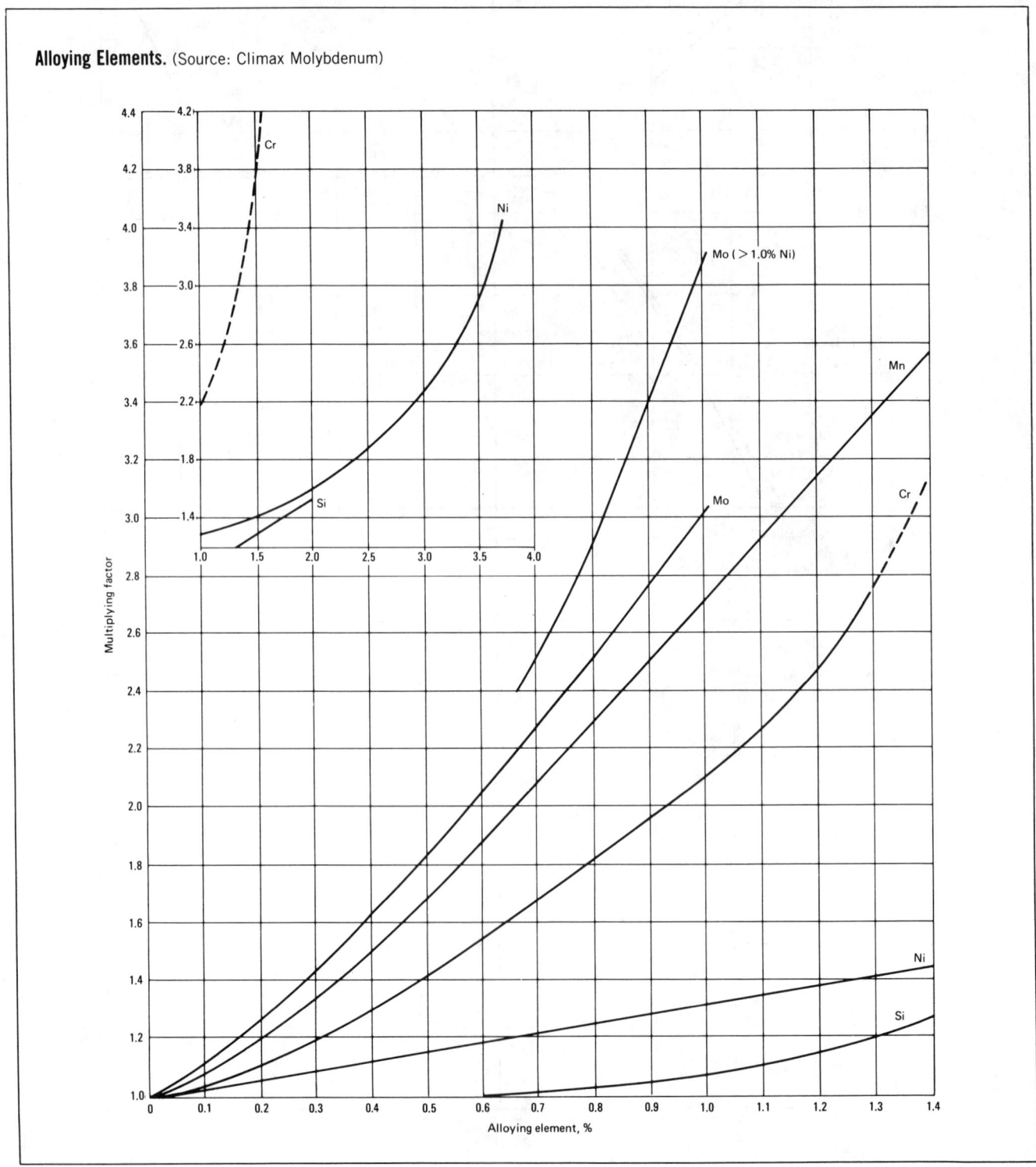

Reprinted from Heat Treater's Guide, 460-465, © 1982 American Society for Metals

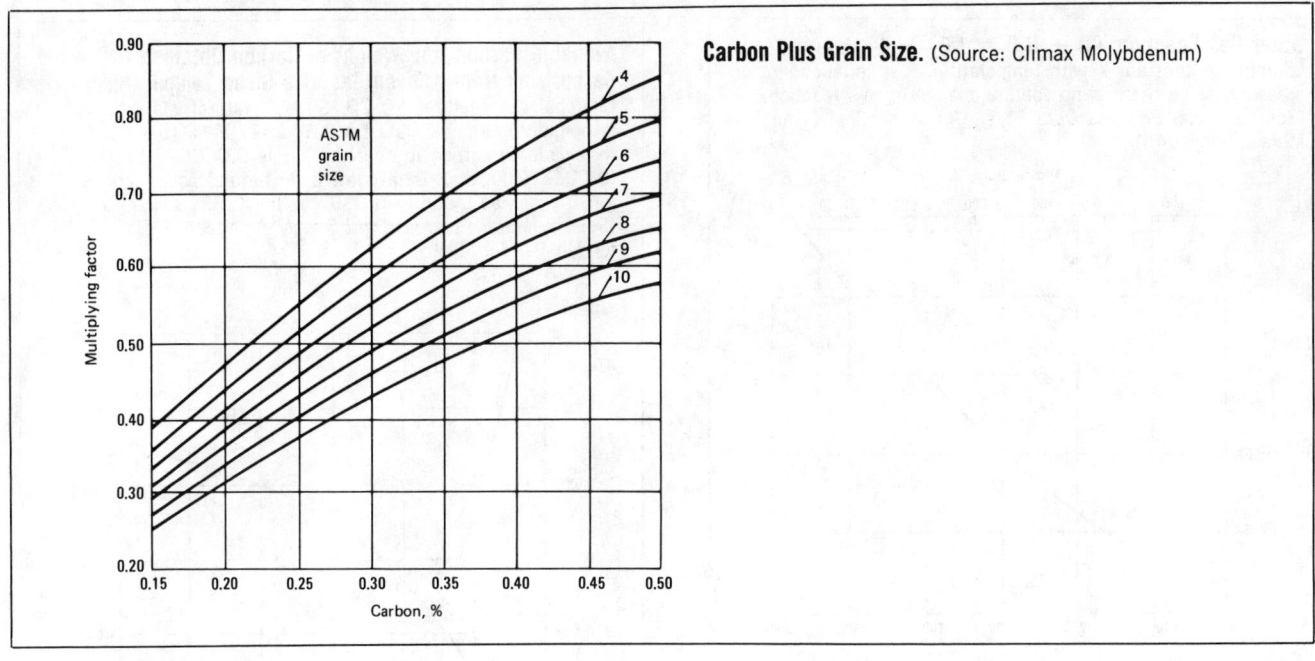

Carbon Plus Grain Size. (Source: Climax Molybdenum)

Equilibria for Gas Carburizing

Relation Between Dew Point and Moisture Content of Gases. Hydrogen can be purified by a room-temperature catalytic reaction that combines oxygen with hydrogen, forming water. Then, all water vapor is removed by drying to a dew point of −60 °F (−51 °C). (Source: *Metals Handbook*, 8th ed., Vol 2, American Society for Metals, 1964)

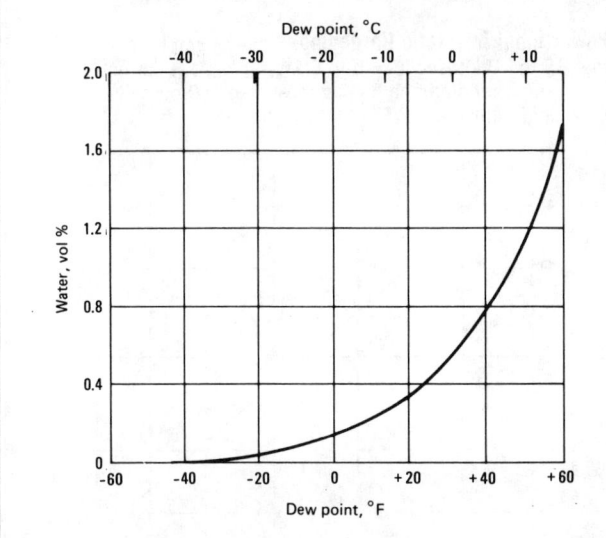

Iron Oxides from CO_2 or H_2O. Data point 1: an atmosphere consisting of 75 H_2 and 25 H_2O will reduce scale on iron (FeO or Fe_3O_4) at 1400 °F (760 °C). Data point 2: same atmosphere will scale metal at 900 °F (480 °C). (Source: *Metal Progress Data Sheet*, American Society for Metals, Jan 1945)

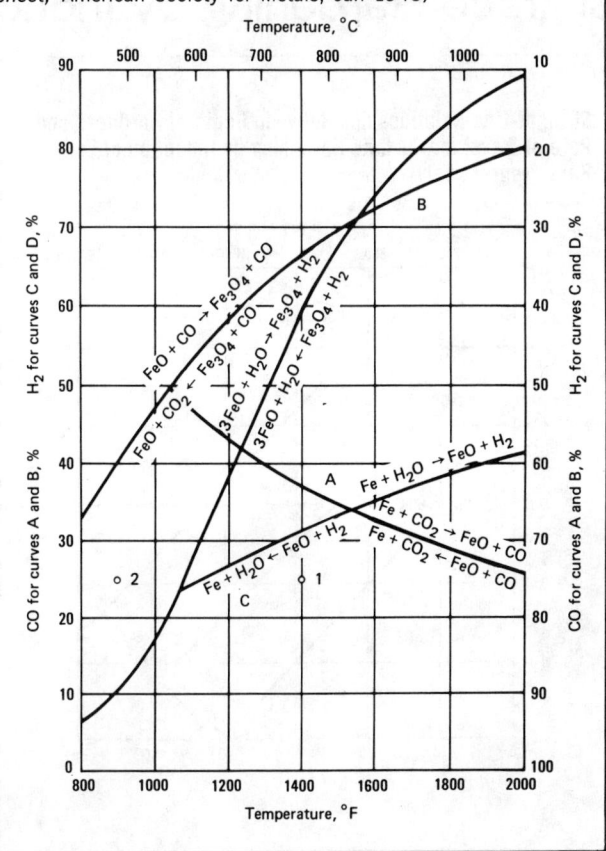

Water Gas Reaction, $CO + H_2O \rightleftarrows CO_2 + H_2$. Variation of equilibrium constant K with temperature. K is independent of pressure, since there is no volume change in this reaction. (Source: *Metal Progress Data Sheet,* American Society for Metals, Jan 1945)

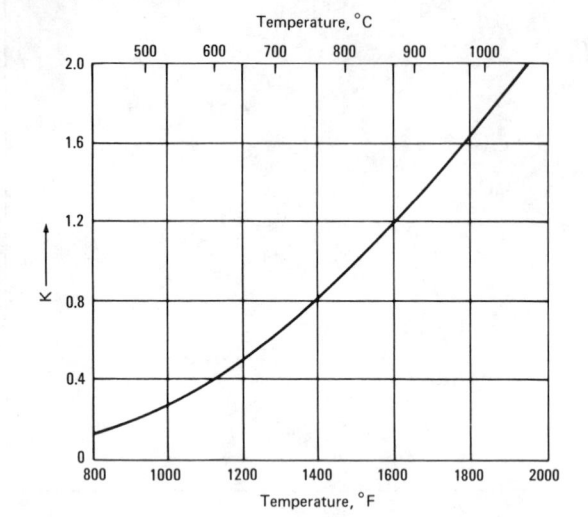

Available Carbon (the Weight of Carbon Obtained for Carburizing from a Given Gas at a Given Temperature). Charcoal gas analyzes 20 CO, 80 N_2. Natural gas is principally methane. Data point 1: at 1700 °F (925 °C), the available carbon in charcoal gas is 0.0000272 lb/ft^3 (0.004357 kg/m^3). Data point 2: in natural gas, there is 1200 times as much or 0.0337 lb/ft^3 (0.5398 kg/m^3). (Source: *Metal Progress Data Sheet,* American Society for Metals, Jan 1945)

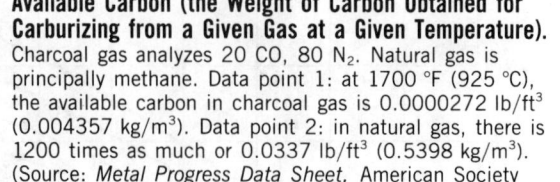
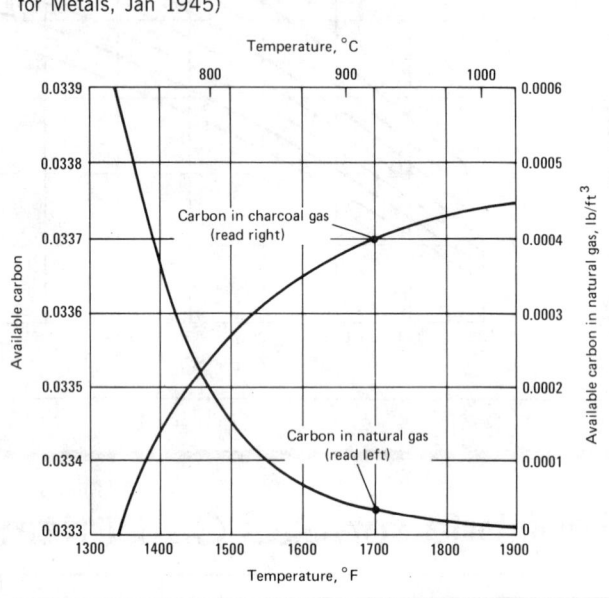

Surface Hardening by Induction

Straight-Line Relationships Between Depth of Hardness and Rate of Travel for Surface Hardening by Induction of Long Bars Progressively. (Source: Park-Ohio Industries)

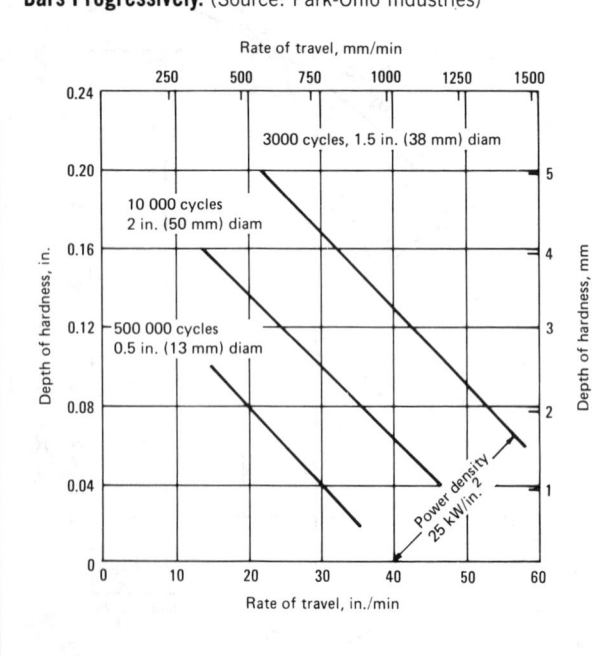

Power Input for Static Hardening. Slope of graph indicates that 35 to 40 kW-sec/in.^{2} (5 to 6 kW-sec/cm^2) is correct power input for static hardening most steels. (Source: Park-Ohio Industries)

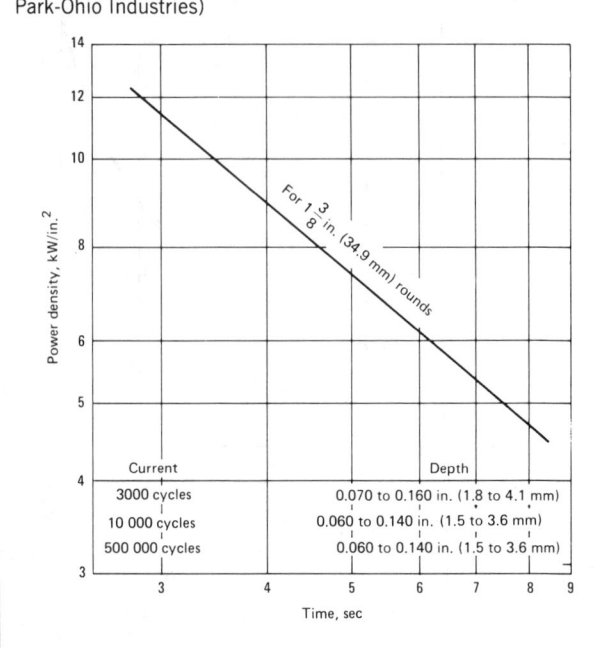

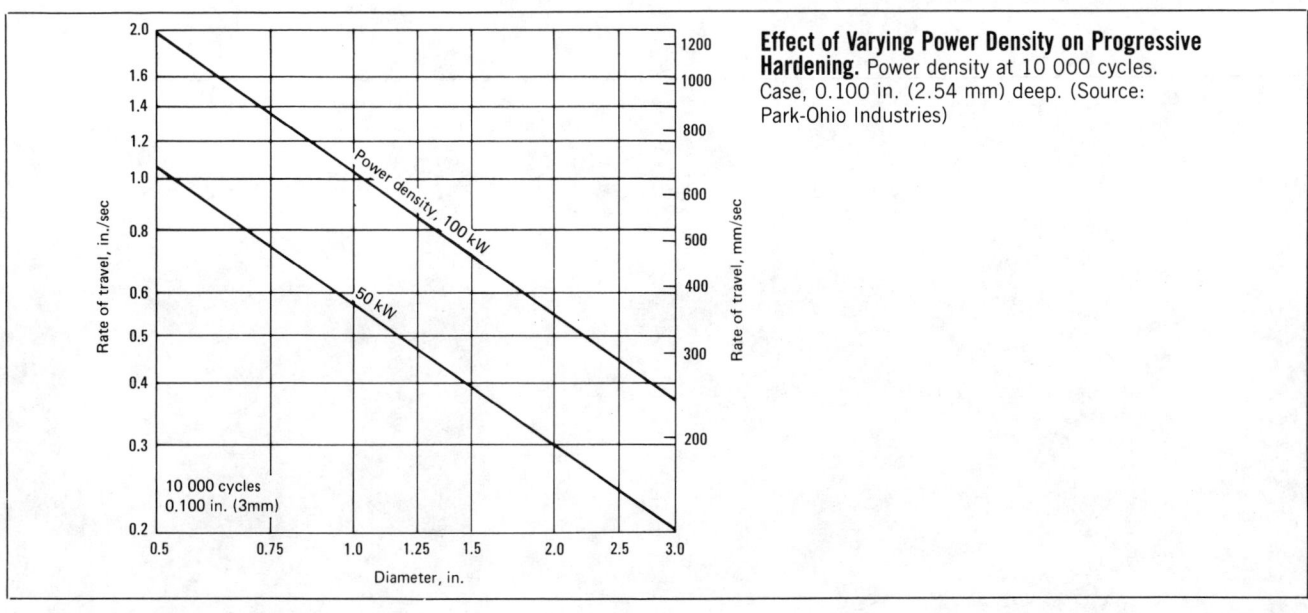

Minimum Power Density Versus Stock Diameter for Static Hardening and Versus Rate of Travel for Progressive Hardening. (Source: Park-Ohio Industries)

Rate of travel, in./sec

Power density, kW/in.²

Min diam for sorbite

Straight line relation to 100 kW/in.²

Conditions for
3000 cycles, 0.160 to 0.200 in. (3 to 5 mm) depth
10 000 cycles, 0.100 to 0.150 in. (3 to 3.8 mm) depth

Power density versus rate of travel
Read up

3000 and 10 000 cycles, 0.050 to 0.080 in. (1.3 to 2.03 mm)

500 000 cycles, 0.040 to 0.050 in. depth (1.0 to 1.3 mm)

3000 and 10 000 cycles
0.100 to 0.200 in. depth (3 to 5 mm)

Stock diameter, in.

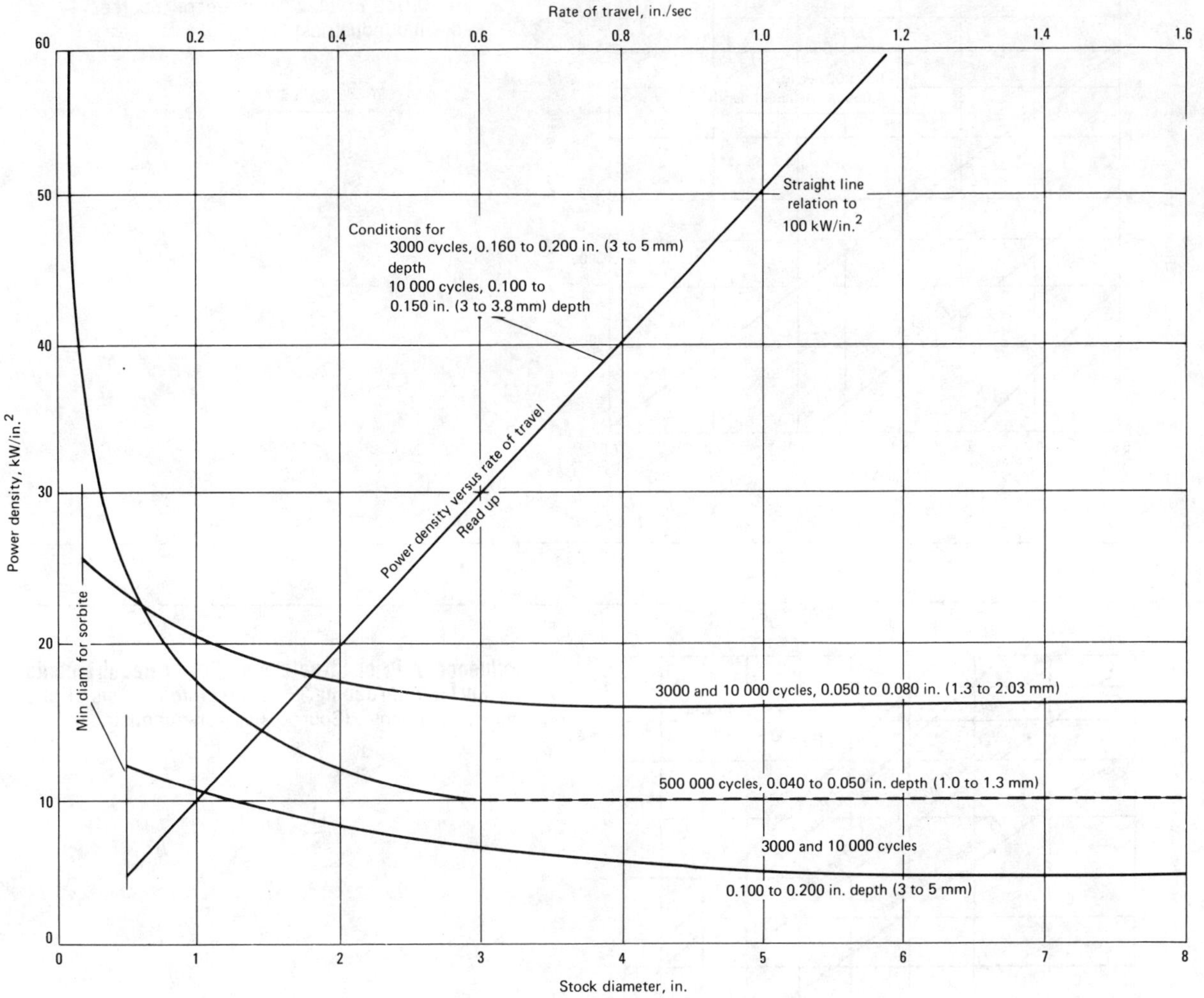

Effect of Varying Power Density on Progressive Hardening. Power density at 10 000 cycles. Case, 0.100 in. (2.54 mm) deep. (Source: Park-Ohio Industries)

Rate of travel, in./sec

Rate of travel, mm/sec

Power density, 100 kW

50 kW

10 000 cycles
0.100 in. (3mm)

Diameter, in.

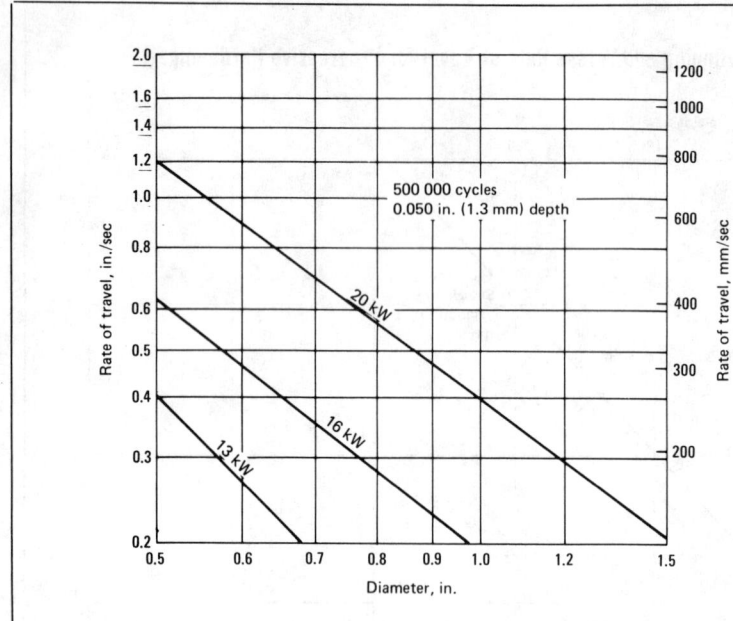

Effect of Varying Power Density on Progressive Hardening. Power density at 500 000 cycles. Case, 0.050 in. (1.27 mm) deep. (Source: Park-Ohio Industries)

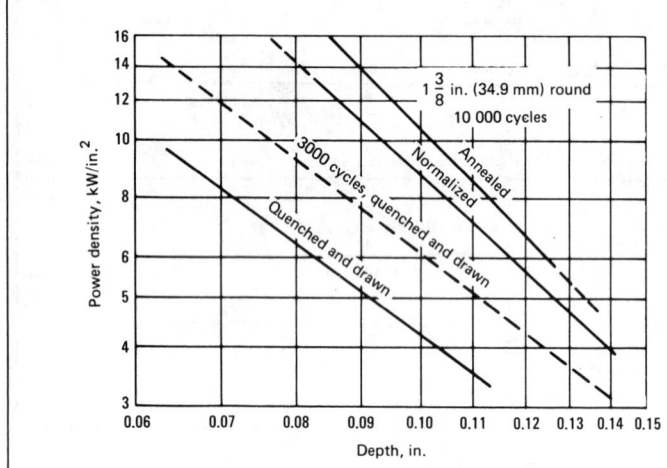

Influence of Prior Structure on Power Requirements for Surface Hardening. Prior structure consists of fine microconstituents. (Source: Park-Ohio Industries)

Exothermic and Endothermic Furnace Atmospheres

Exothermic and endothermic atmospheres are produced in a generating unit which consists primarily of a refractory-lined, gastight combustion chamber fitted with one or more burners to which a mixture of gas and air is delivered from controlled ratio pumping equipment. The generator is equipped with a cooler through which the products of combustion are discharged. The cooler removes a portion of the water produced from the reaction.

The generator produces an atmosphere that is a mixture of many elements and compounds, including hydrogen, carbon monoxide, nitrogen, carbon dioxide, and water vapor. Minor amounts of sulfur dioxide, hydrogen sulfide, and other gases may also be present. Quantities of various gases in an atmosphere vary widely with the fuel-air ratio. Curves are for gas analyzed at room temperature, except where noted.

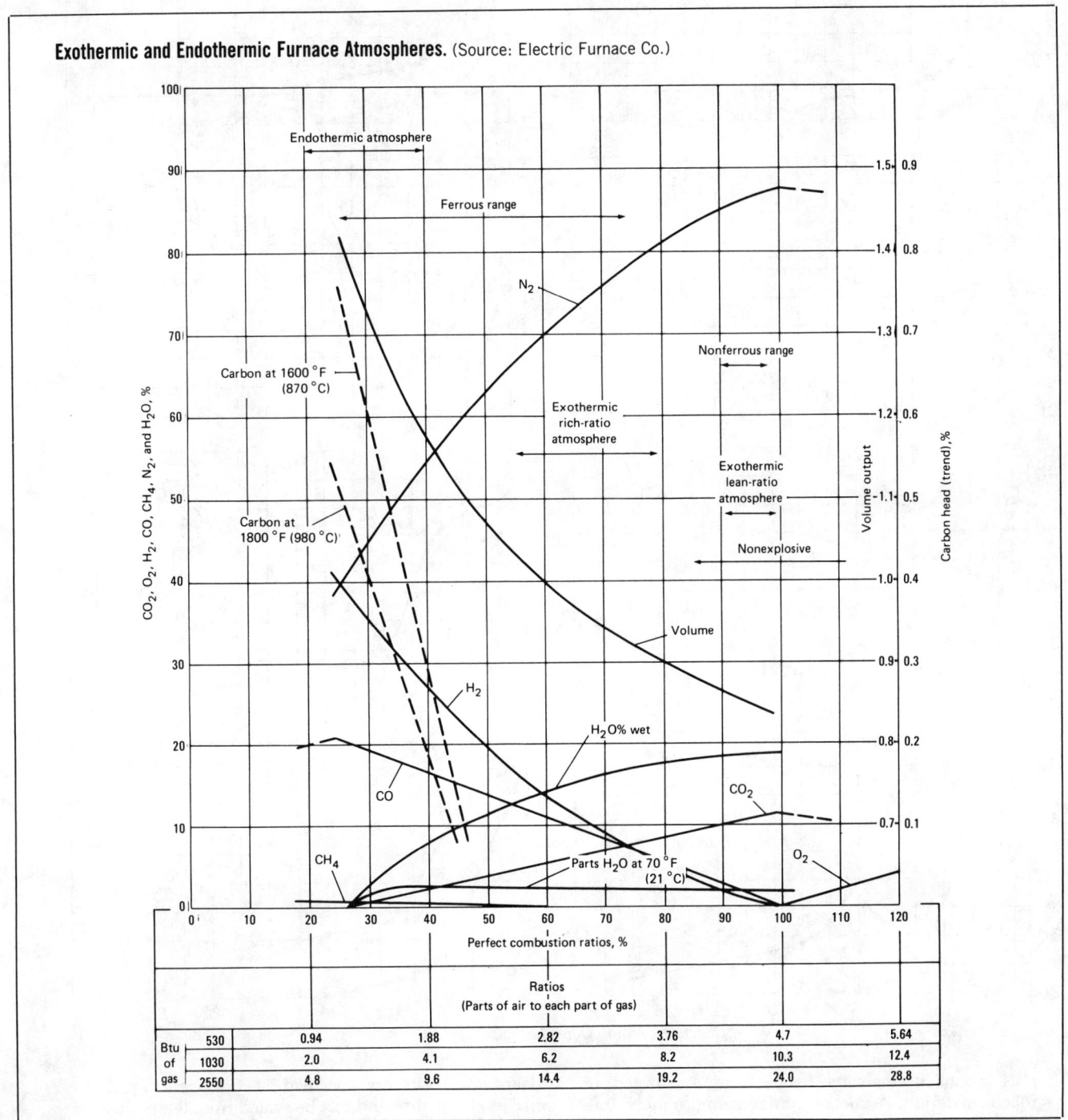

Exothermic and Endothermic Furnace Atmospheres. (Source: Electric Furnace Co.)

Btu of gas						
530	0.94	1.88	2.82	3.76	4.7	5.64
1030	2.0	4.1	6.2	8.2	10.3	12.4
2550	4.8	9.6	14.4	19.2	24.0	28.8

Ratios
(Parts of air to each part of gas)

Hardenability Bands for Steels 1038-H to 1541-H

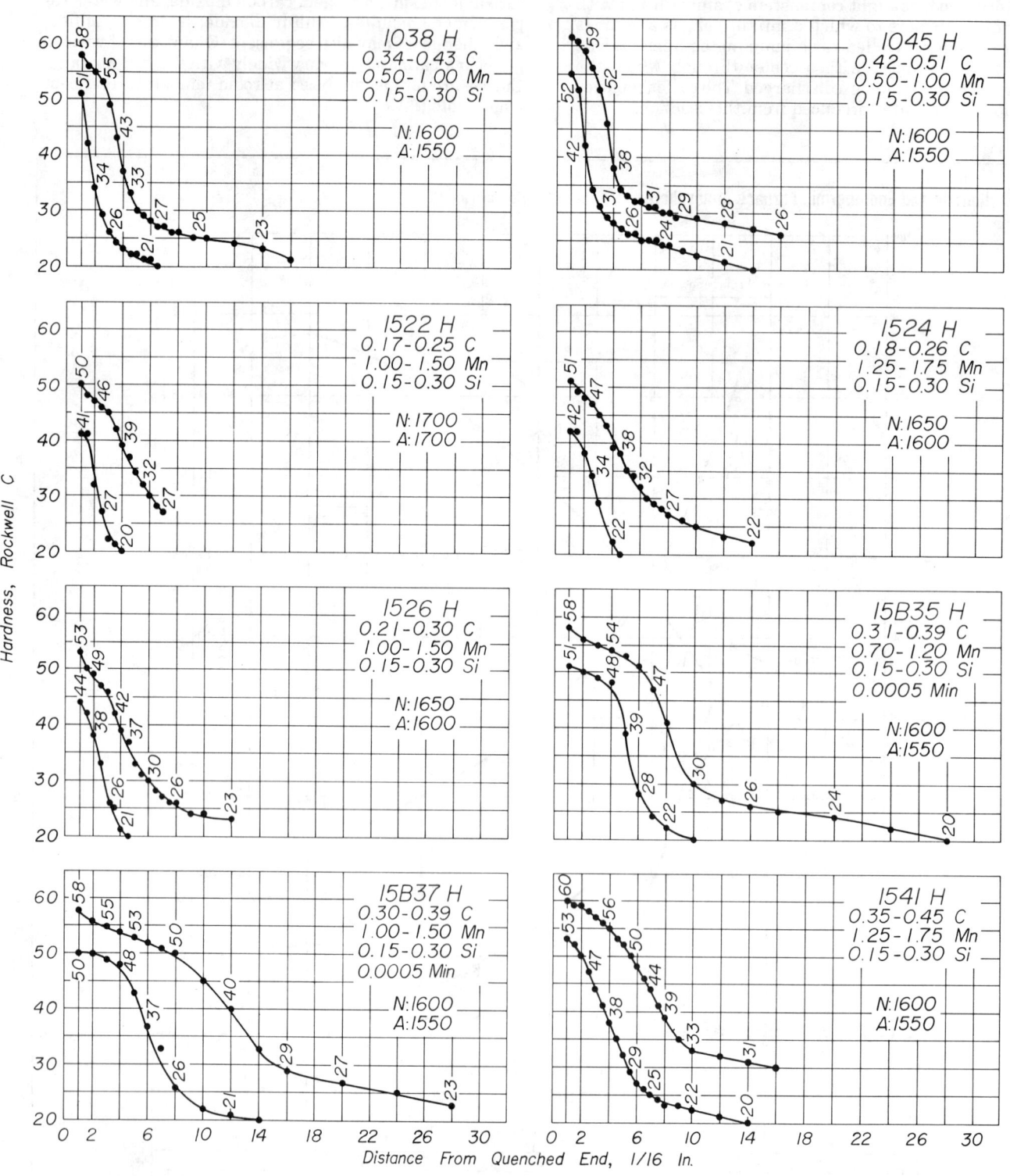

N means normalizing temperature for forged or rolled material; A means austenitizing temperature (both as recommended by S.A.E.)

Hardness limits are specified in Rockwell C-scale units (no fractions) and can be scaled from the plotted points where not labeled at even sixteenths.

Reprinted from Metal Progress Databook, mid-June 1974, 151, 153, 157, 159, 161, 163, © 1974 American Society for Metals

Hardenability Bands for Steels 1330-H to 4047-H

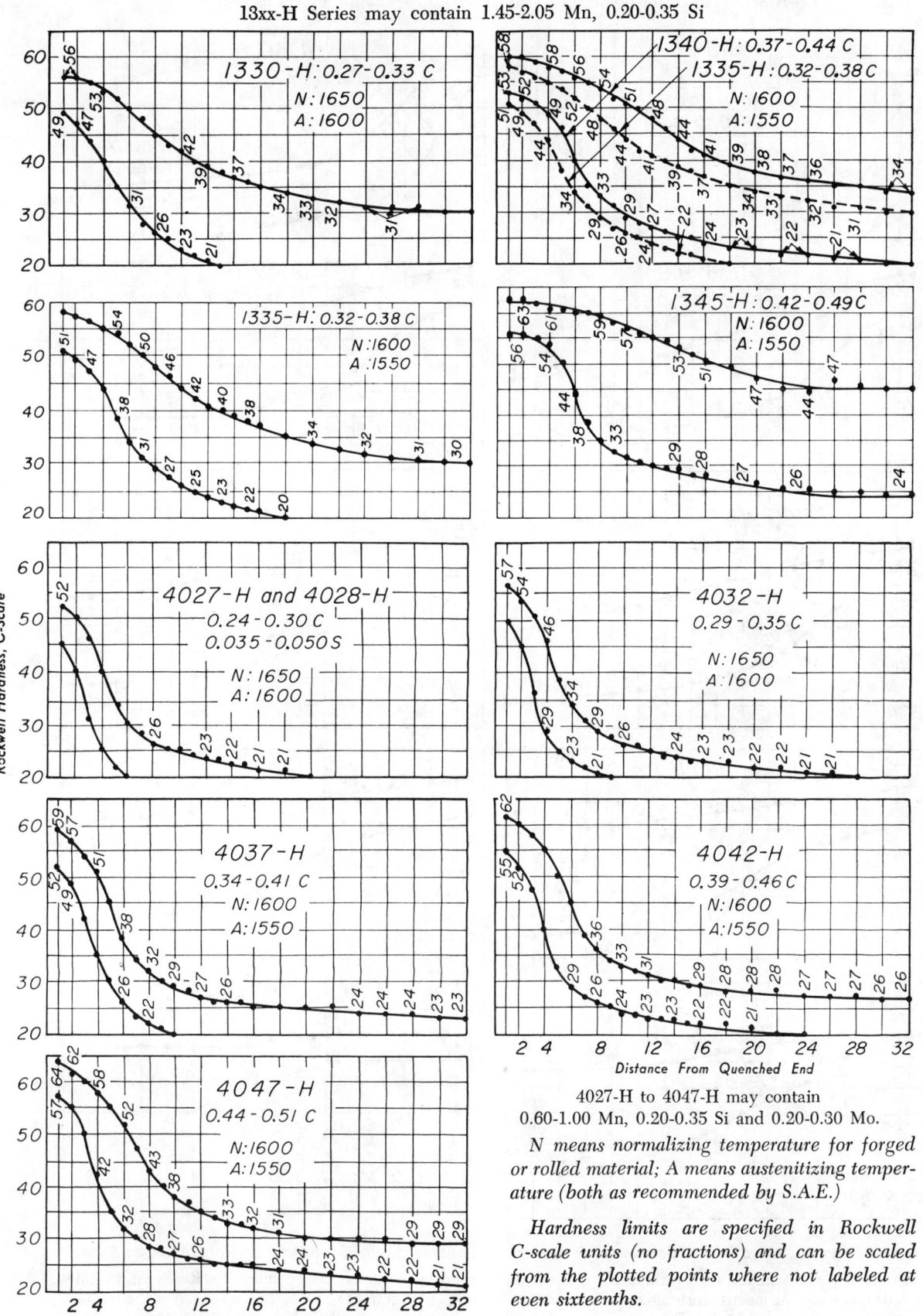

13xx-H Series may contain 1.45-2.05 Mn, 0.20-0.35 Si

1330-H: 0.27-0.33 C
N: 1650
A: 1600

1340-H: 0.37-0.44 C
1335-H: 0.32-0.38 C
N: 1600
A: 1550

1335-H: 0.32-0.38 C
N: 1600
A: 1550

1345-H: 0.42-0.49 C
N: 1600
A: 1550

4027-H and 4028-H
0.24-0.30 C
0.035-0.050 S
N: 1650
A: 1600

4032-H
0.29-0.35 C
N: 1650
A: 1600

4037-H
0.34-0.41 C
N: 1600
A: 1550

4042-H
0.39-0.46 C
N: 1600
A: 1550

4047-H
0.44-0.51 C
N: 1600
A: 1550

Rockwell Hardness, C-Scale

Distance From Quenched End

4027-H to 4047-H may contain
0.60-1.00 Mn, 0.20-0.35 Si and 0.20-0.30 Mo.

N means normalizing temperature for forged or rolled material; A means austenitizing temperature (both as recommended by S.A.E.)

Hardness limits are specified in Rockwell C-scale units (no fractions) and can be scaled from the plotted points where not labeled at even sixteenths.

Source: Metal Progress Databook, mid-June 1974, 151, 153, 157, 159, 161, 163

Hardenability Bands for Steels 4118-H to 4161-H

41xx-H Series has 0.20-0.35 Si, 0.75-1.20 Cr, and 0.15-0.25 Mo except as shown.

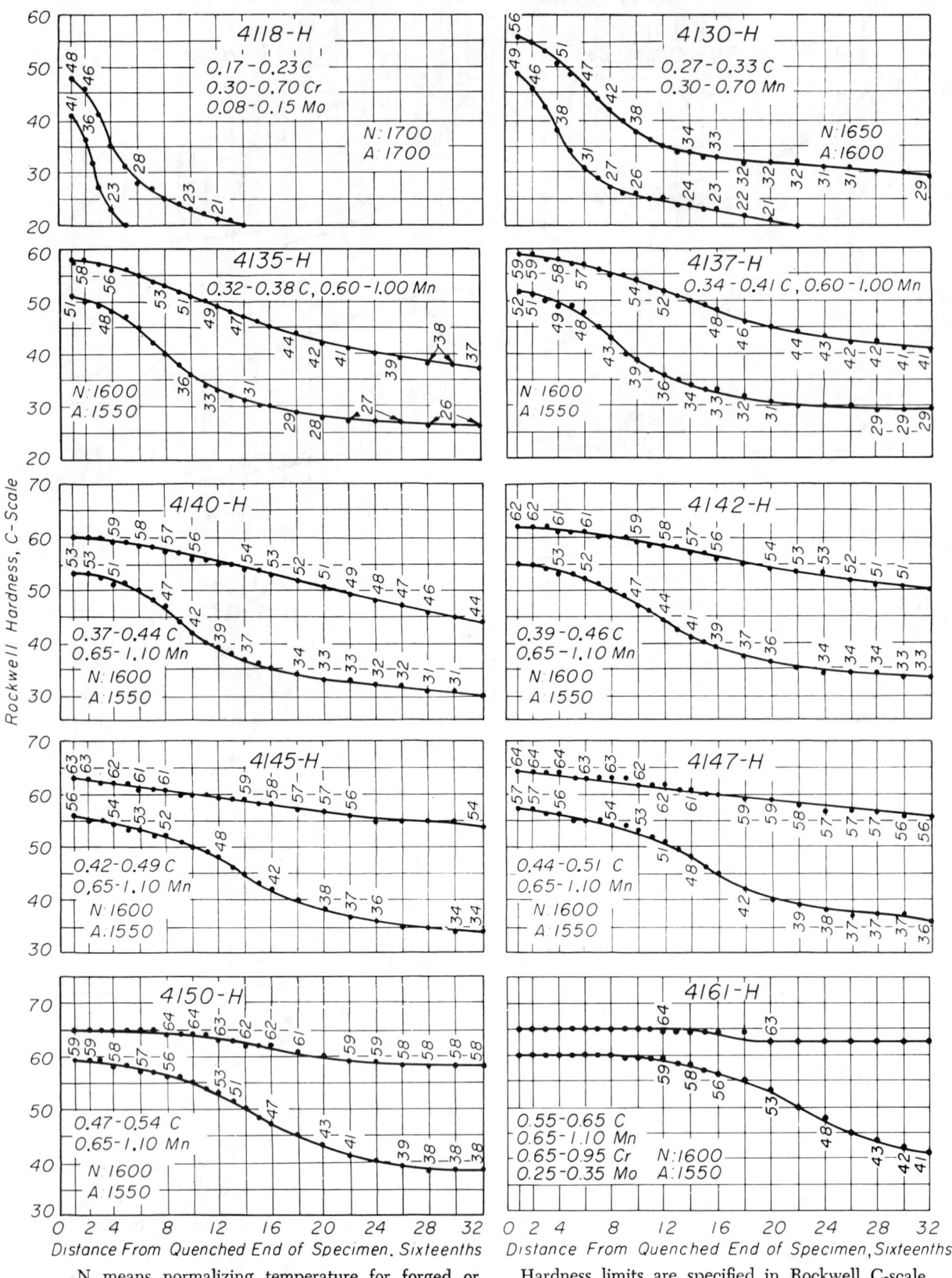

4118-H
0.17-0.23 C
0.30-0.70 Cr
0.08-0.15 Mo
N:1700
A:1700

4130-H
0.27-0.33 C
0.30-0.70 Mn
N:1650
A:1600

4135-H
0.32-0.38 C, 0.60-1.00 Mn
N:1600
A:1550

4137-H
0.34-0.41 C, 0.60-1.00 Mn
N:1600
A:1550

4140-H
0.37-0.44 C
0.65-1.10 Mn
N:1600
A:1550

4142-H
0.39-0.46 C
0.65-1.10 Mn
N:1600
A:1550

4145-H
0.42-0.49 C
0.65-1.10 Mn
N:1600
A:1550

4147-H
0.44-0.51 C
0.65-1.10 Mn
N:1600
A:1550

4150-H
0.47-0.54 C
0.65-1.10 Mn
N:1600
A:1550

4161-H
0.55-0.65 C
0.65-1.10 Mn
0.65-0.95 Cr N:1600
0.25-0.35 Mo A:1550

Rockwell Hardness, C-Scale

Distance From Quenched End of Specimen, Sixteenths

N means normalizing temperature for forged or rolled material; A means austenitizing temperature (both as recommended by S.A.E.).

Hardness limits are specified in Rockwell C-scale units (no fractions) and can be scaled from the plotted points where not labeled at even sixteenths.

464

Hardenability Bands for Steels 4320-H to 4718-H

43xx-H Series may contain 0.20-0.35 Si, 1.55-2.00 Ni, 0.20-0.30 Mo

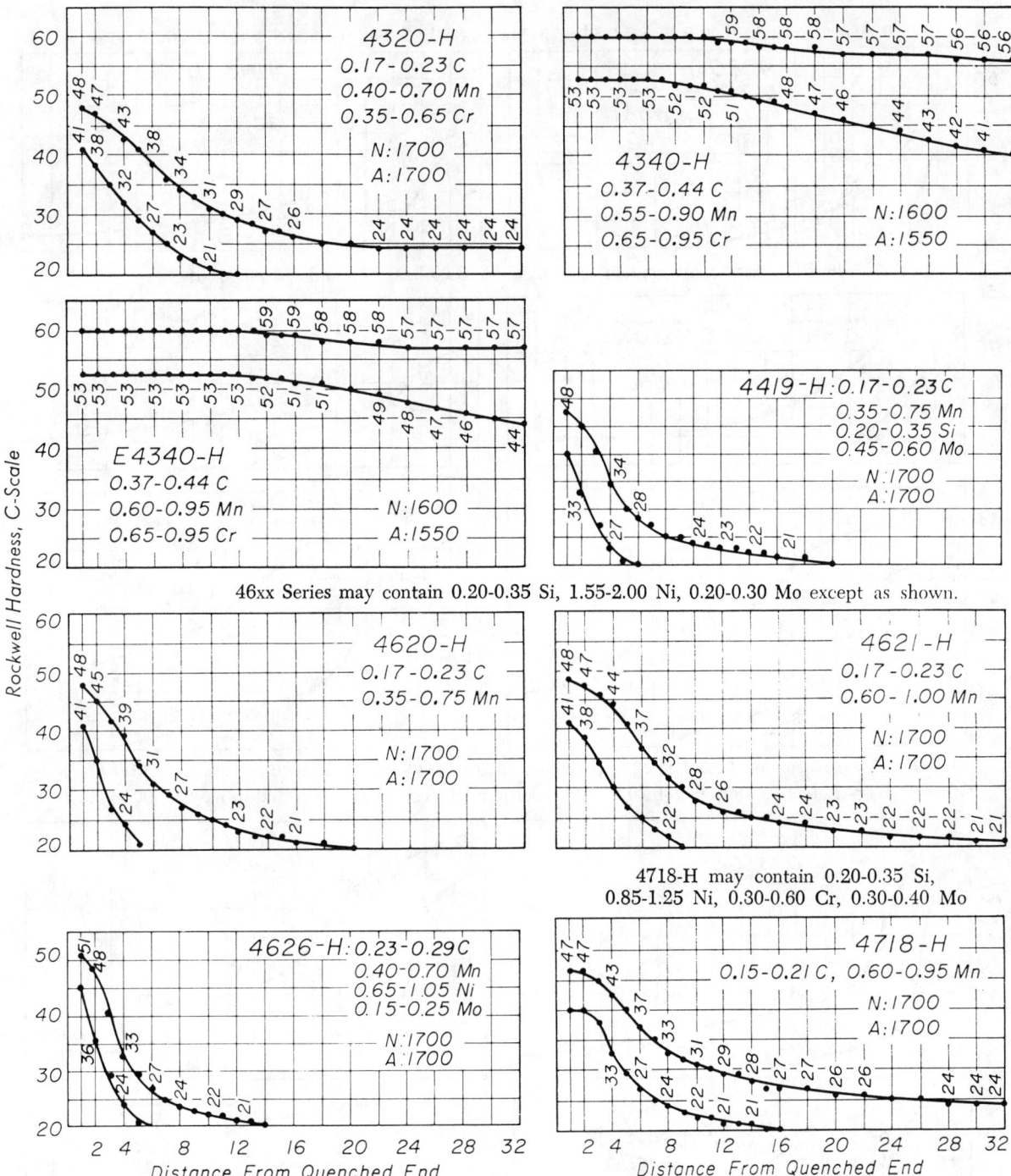

N means normalizing temperature for forged or rolled material; A means austenitizing temperature (both as recommended by S.A.E.). Hardness limits are specified in Rockwell C-scale units (no fractions) and can be scaled from the plotted points where not labeled at even sixteenths.

Diameters of Rounds With Same as Quenched Hardness									Location in Round	Quench
3.8									Surface	Mild Water Quench
1.1	2.0	2.9	3.8	4.8	5.8	6.7			3/4 Radius From Center	
0.7	1.2	1.6	2.0	2.4	2.8	3.2	3.6	3.9	Center	
0.8	1.8	2.5	3.0	3.4	3.8				Surface	Mild Oil Quench
0.5	1.0	1.6	2.0	2.4	2.8	3.2	3.6	4.0	3/4 Radius From Center	
0.2	0.6	1.0	1.4	1.7	2.0	2.4	2.8	3.1	Center	

0   2   4   6   8   10   12   14   16   18   Distance From Quenched End

Source: Metal Progress Databook, mid-June 1974, 151, 153, 157, 159, 161, 163

Hardenability Bands for Steels 4720-H to 50B60-H

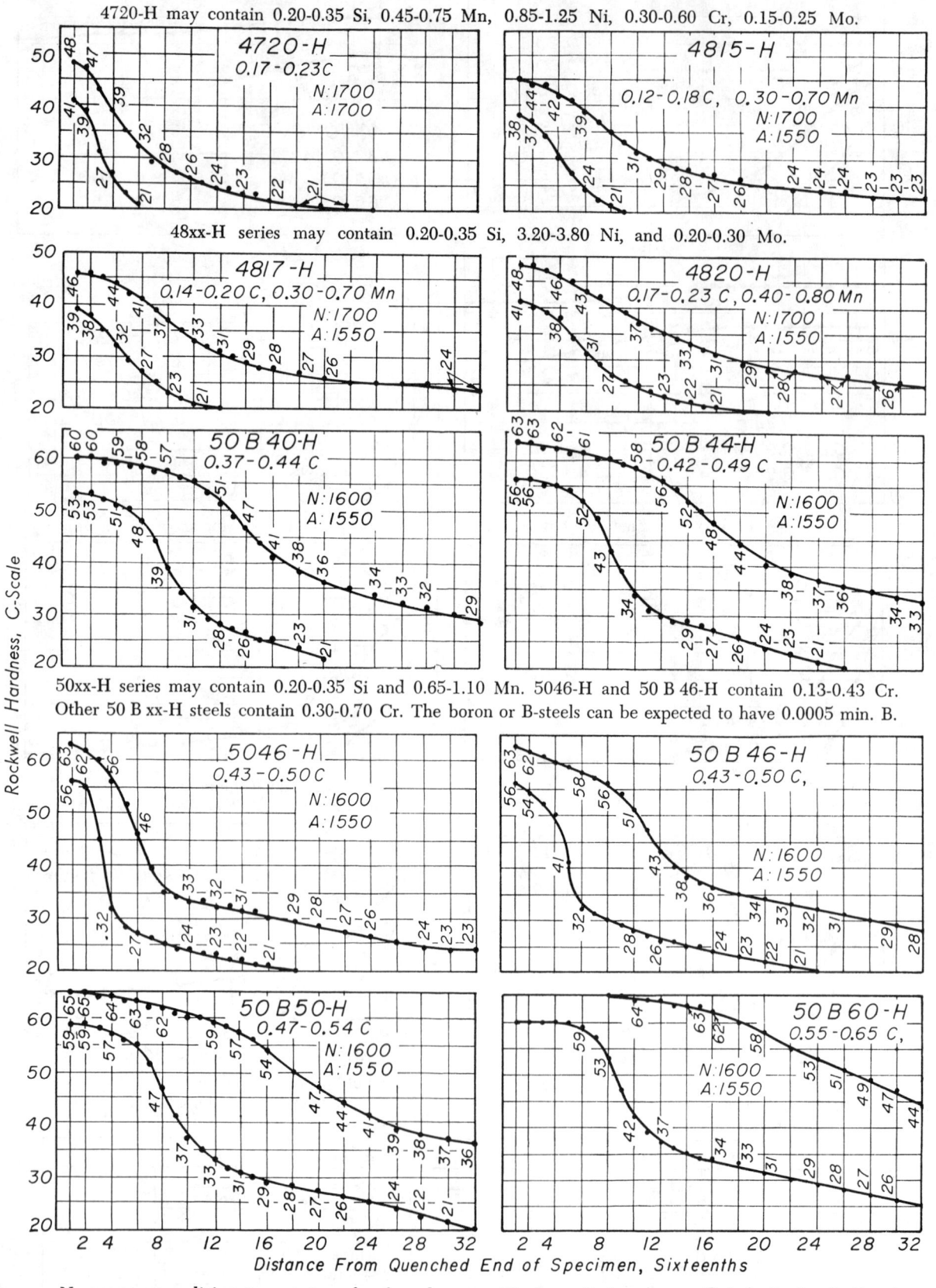

4720-H may contain 0.20-0.35 Si, 0.45-0.75 Mn, 0.85-1.25 Ni, 0.30-0.60 Cr, 0.15-0.25 Mo.

4720-H 0.17-0.23C N:1700 A:1700

4815-H 0.12-0.18 C, 0.30-0.70 Mn N:1700 A:1550

48xx-H series may contain 0.20-0.35 Si, 3.20-3.80 Ni, and 0.20-0.30 Mo.

4817-H 0.14-0.20 C, 0.30-0.70 Mn N:1700 A:1550

4820-H 0.17-0.23 C, 0.40-0.80 Mn N:1700 A:1550

50B40-H 0.37-0.44 C N:1600 A:1550

50B44-H 0.42-0.49 C N:1600 A:1550

50xx-H series may contain 0.20-0.35 Si and 0.65-1.10 Mn. 5046-H and 50B46-H contain 0.13-0.43 Cr. Other 50Bxx-H steels contain 0.30-0.70 Cr. The boron or B-steels can be expected to have 0.0005 min. B.

5046-H 0.43-0.50C N:1600 A:1550

50B46-H 0.43-0.50 C, N:1600 A:1550

50B50-H 0.47-0.54 C N:1600 A:1550

50B60-H 0.55-0.65 C, N:1600 A:1550

Rockwell Hardness, C-Scale

Distance From Quenched End of Specimen, Sixteenths

N means normalizing temperature for forged or rolled material; A means austenitizing temperature (both as recommended by S.A.E.).

Hardness limits are specified in Rockwell C-scale units (no fractions) and can be scaled from the plotted points where not labeled at even sixteenths.

Hardenability Bands for Steels 5120-H to 51B60-H

51xx-H series may contain 0.20-0.35 Si; C, Mn and Cr as shown

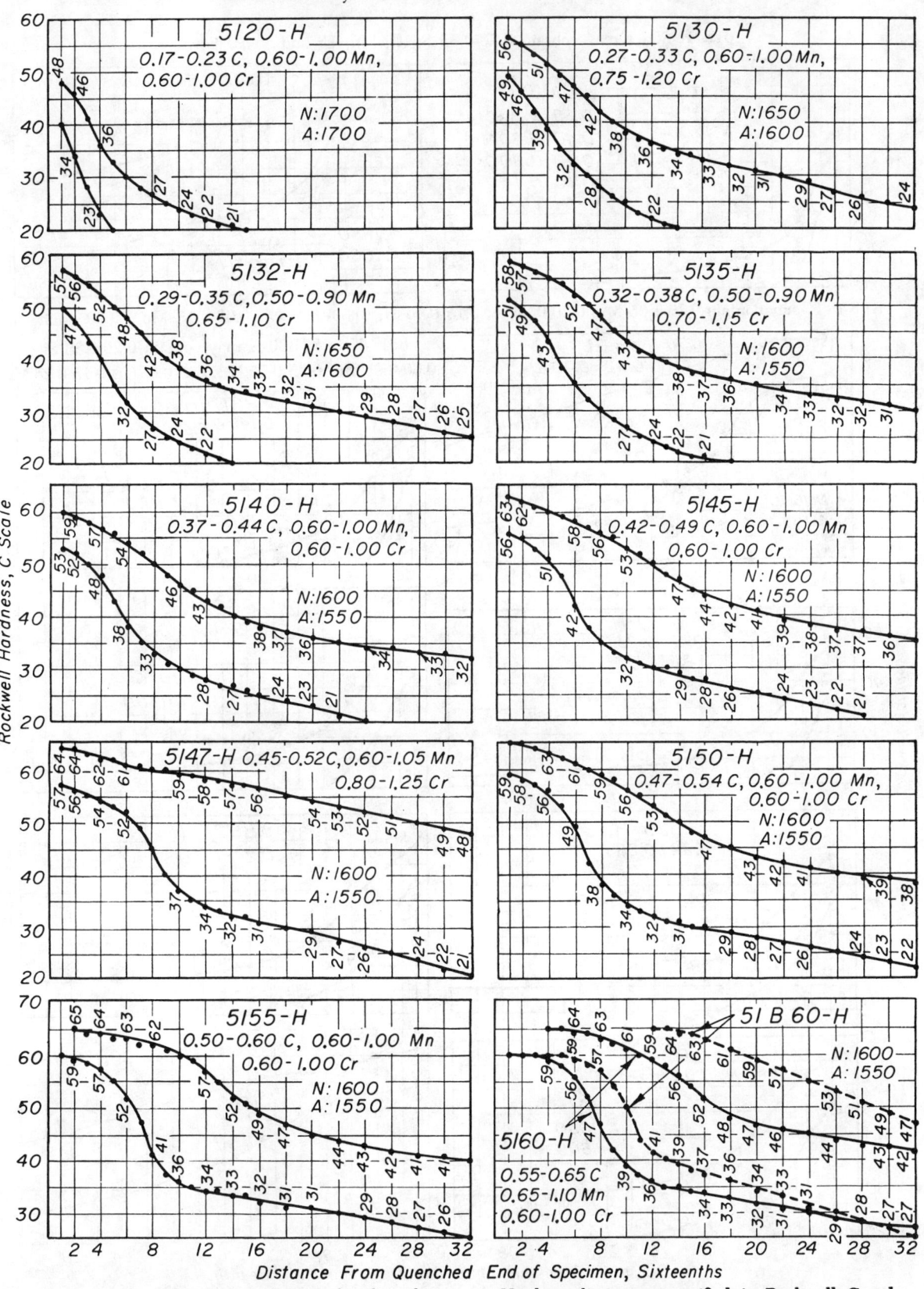

N means normalizing temperature for forged or rolled material; A means austenitizing temperature (both as recommended by S.A.E.).

Hardness limits are specified in Rockwell C-scale units (no fractions) and can be scaled from the plotted points where not labeled at even sixteenths.

Hardenability Bands for Steels 6118-H to 86B30-H

61xx-H Series may contain 0.20-0.35 Si; C, Mn, Cr and V as shown

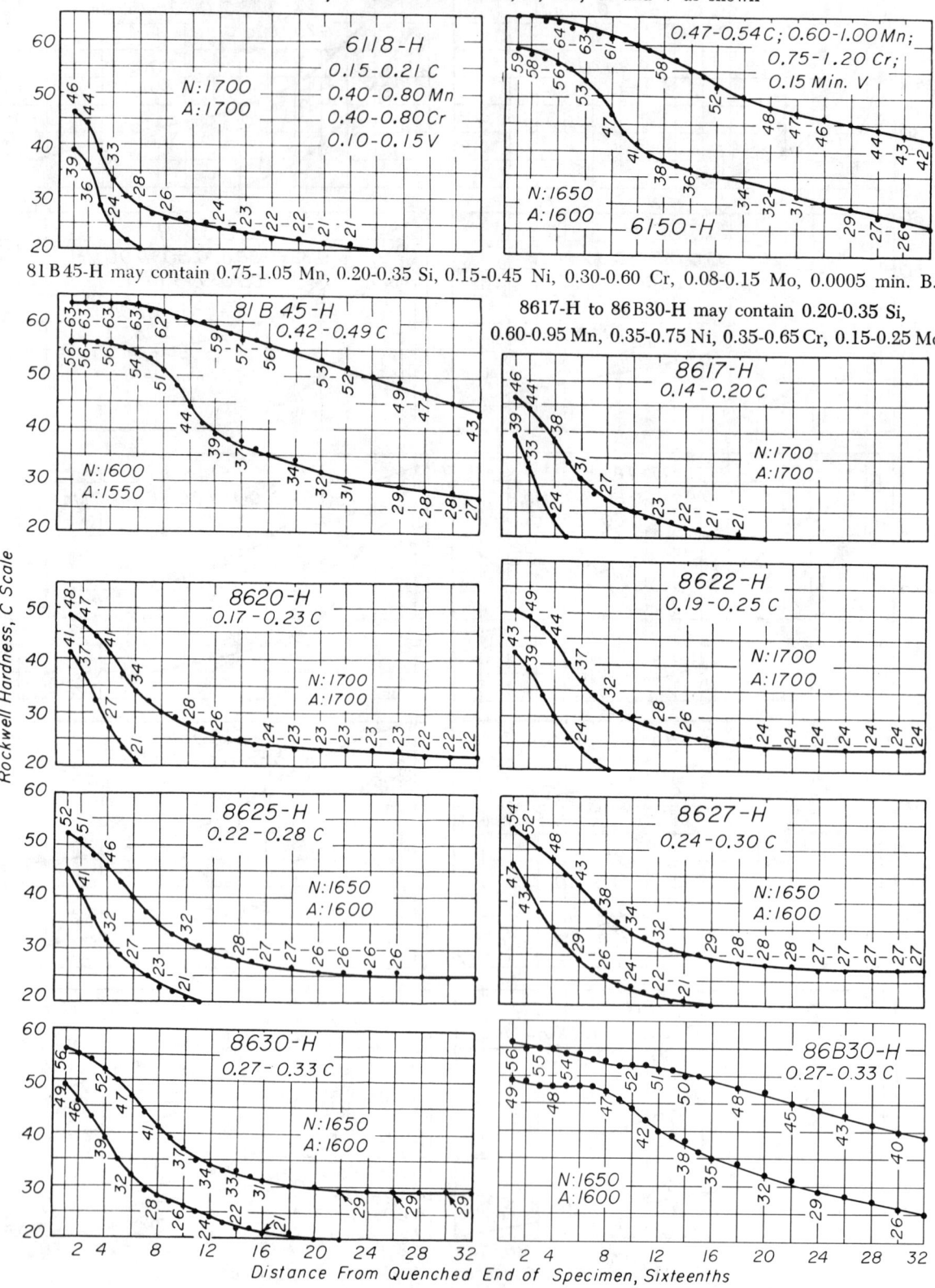

81B45-H may contain 0.75-1.05 Mn, 0.20-0.35 Si, 0.15-0.45 Ni, 0.30-0.60 Cr, 0.08-0.15 Mo, 0.0005 min. B.

8617-H to 86B30-H may contain 0.20-0.35 Si, 0.60-0.95 Mn, 0.35-0.75 Ni, 0.35-0.65 Cr, 0.15-0.25 Mo

N means normalizing temperature for forged or rolled material; A means austenitizing temperature (both as recommended by S.A.E.)

Hardness limits are specified in Rockwell C-scale units (no fractions) and can be scaled from the plotted points where not labeled at even sixteenths.

Hardenability Bands for Steels 8637-H to 8742-H

8635-H to 8660-H may contain 0.70-1.05 Mn, 0.20-0.35 Si, 0.35-0.75 Ni, 0.15-0.25 Mo
86 B 45-H can be expected to have 0.0005% min. boron.

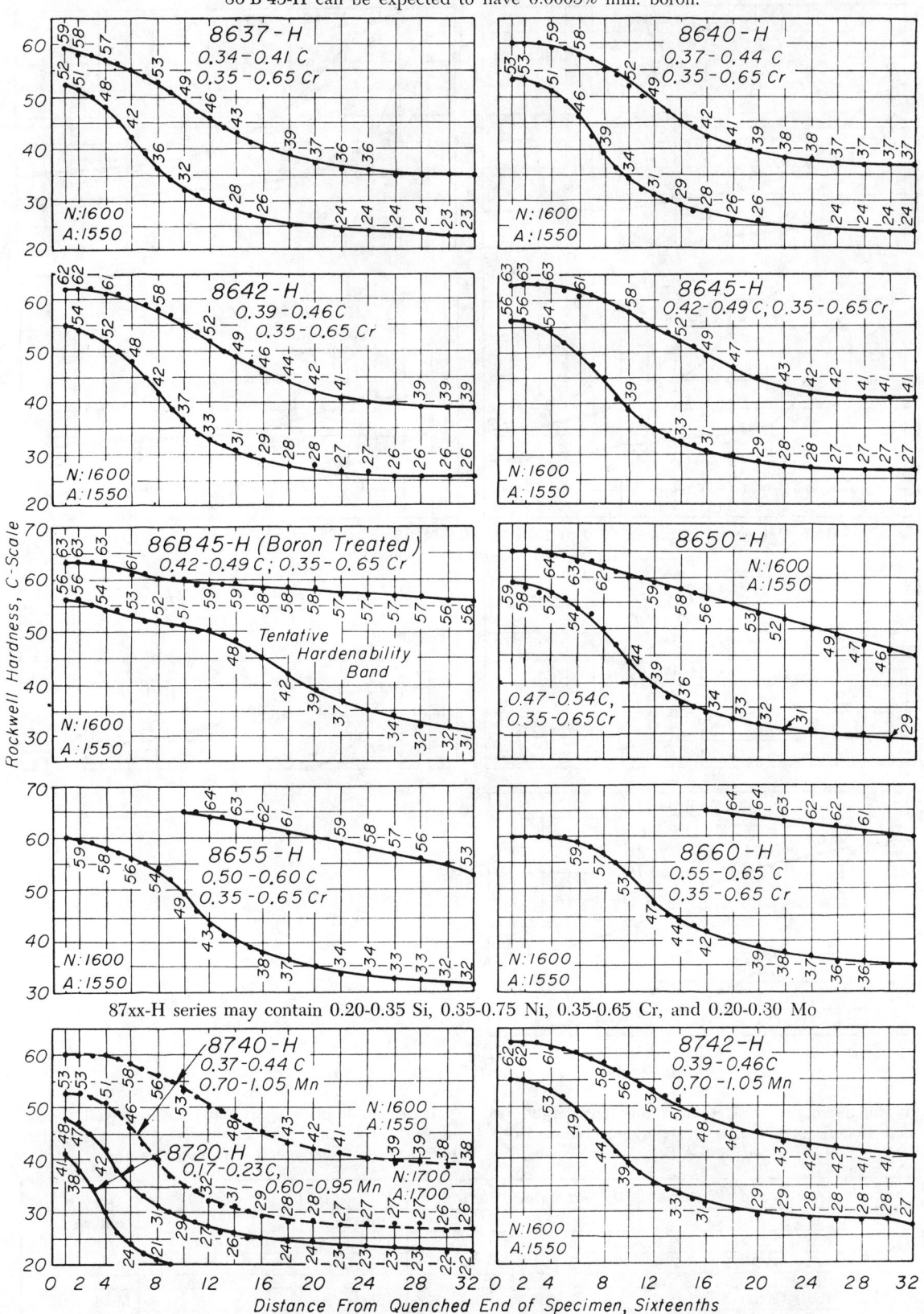

87xx-H series may contain 0.20-0.35 Si, 0.35-0.75 Ni, 0.35-0.65 Cr, and 0.20-0.30 Mo

N means normalizing temperature for forged or rolled material; A means austenitizing temperature (both as recommended by S.A.E.).

Hardness limits are specified in Rockwell C-scale units (no fractions) and can be scaled from the plotted points where not labeled at even sixteenths.

Source: Metal Progress Databook, mid-June 1974, 151, 153, 157, 159, 161, 163

Hardenability Bands for Steels 8822-H to 94B30-H

926x-H Series varies as to chromium. It may contain 0.55-0.65 C, 0.65-1.10 Mn, 1.70-2.20 Si.

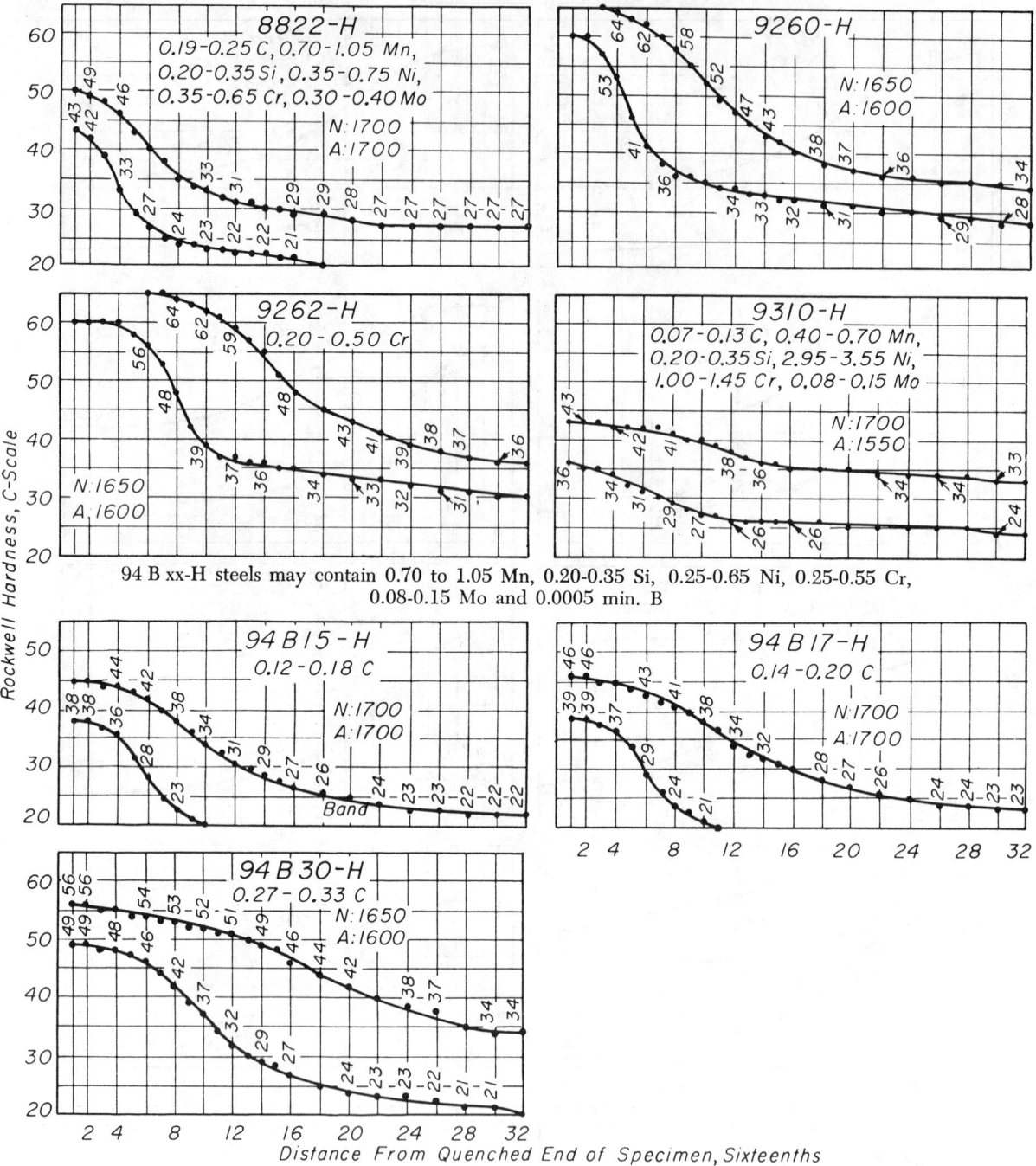

94 B xx-H steels may contain 0.70 to 1.05 Mn, 0.20-0.35 Si, 0.25-0.65 Ni, 0.25-0.55 Cr, 0.08-0.15 Mo and 0.0005 min. B

Distance From Quenched End of Specimen, Sixteenths

N means normalizing temperature for forged or rolled material; A means austenitizing temperature (both as recommended by S.A.E.).

Hardness limits are specified in Rockwell C-scale units (no fractions) and can be scaled from the plotted points where not labeled at even sixteenths.

Diameters of Rounds With Same as Quenched Hardness									Location in Round	Quench
3.8									Surface	Mild Water Quench
1.1	2.0	2.9	3.8	4.8	5.8	6.7			3/4 Radius From Center	
0.7	1.2	1.6	2.0	2.4	2.8	3.2	3.6	3.9	Center	
0.8	1.8	2.5	3.0	3.4	3.8				Surface	Mild Oil Quench
0.5	1.0	1.6	2.0	2.4	2.8	3.2	3.6	4.0	3/4 Radius From Center	
0.2	0.6	1.0	1.4	1.7	2.0	2.4	2.8	3.1	Center	

0   2   4   6   8   10   12   14   16   18   *Distance From Quenched End*

Hardenability Curves for Selected EX Steels

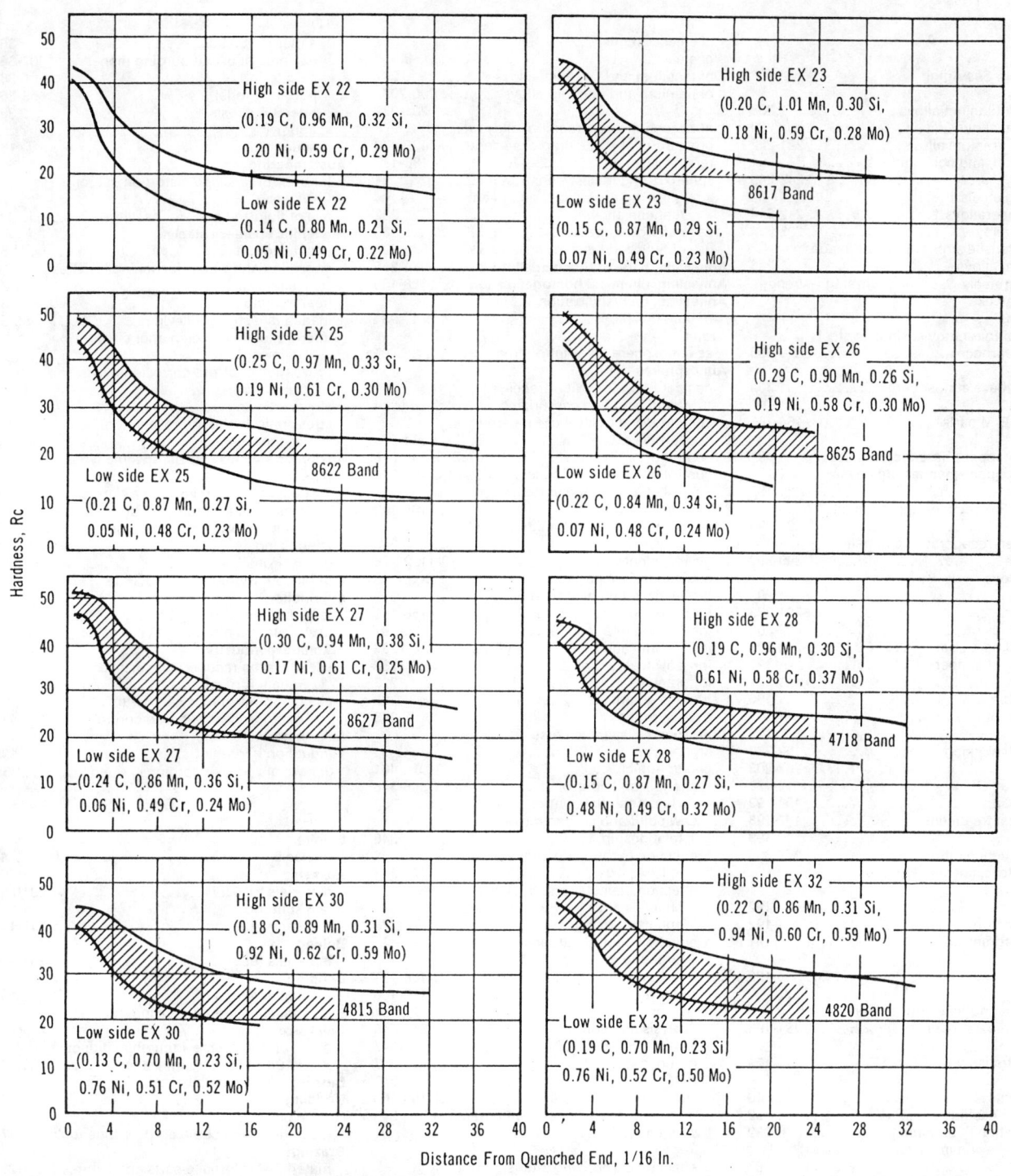

Distance From Quenched End, 1/16 In.

Hardness, Rc

High side EX 22 (0.19 C, 0.96 Mn, 0.32 Si, 0.20 Ni, 0.59 Cr, 0.29 Mo)

Low side EX 22 (0.14 C, 0.80 Mn, 0.21 Si, 0.05 Ni, 0.49 Cr, 0.22 Mo)

High side EX 23 (0.20 C, 1.01 Mn, 0.30 Si, 0.18 Ni, 0.59 Cr, 0.28 Mo)

Low side EX 23 (0.15 C, 0.87 Mn, 0.29 Si, 0.07 Ni, 0.49 Cr, 0.23 Mo)

8617 Band

High side EX 25 (0.25 C, 0.97 Mn, 0.33 Si, 0.19 Ni, 0.61 Cr, 0.30 Mo)

Low side EX 25 (0.21 C, 0.87 Mn, 0.27 Si, 0.05 Ni, 0.48 Cr, 0.23 Mo)

8622 Band

High side EX 26 (0.29 C, 0.90 Mn, 0.26 Si, 0.19 Ni, 0.58 Cr, 0.30 Mo)

Low side EX 26 (0.22 C, 0.84 Mn, 0.34 Si, 0.07 Ni, 0.48 Cr, 0.24 Mo)

8625 Band

High side EX 27 (0.30 C, 0.94 Mn, 0.38 Si, 0.17 Ni, 0.61 Cr, 0.25 Mo)

Low side EX 27 (0.24 C, 0.86 Mn, 0.36 Si, 0.06 Ni, 0.49 Cr, 0.24 Mo)

8627 Band

High side EX 28 (0.19 C, 0.96 Mn, 0.30 Si, 0.61 Ni, 0.58 Cr, 0.37 Mo)

Low side EX 28 (0.15 C, 0.87 Mn, 0.27 Si, 0.48 Ni, 0.49 Cr, 0.32 Mo)

4718 Band

High side EX 30 (0.18 C, 0.89 Mn, 0.31 Si, 0.92 Ni, 0.62 Cr, 0.59 Mo)

Low side EX 30 (0.13 C, 0.70 Mn, 0.23 Si, 0.76 Ni, 0.51 Cr, 0.52 Mo)

4815 Band

High side EX 32 (0.22 C, 0.86 Mn, 0.31 Si, 0.94 Ni, 0.60 Cr, 0.59 Mo)

Low side EX 32 (0.19 C, 0.70 Mn, 0.23 Si, 0.76 Ni, 0.52 Cr, 0.50 Mo)

4820 Band

Bands for corresponding SAE grades are also shown.
Source: Climax Molybdenum Co.

Source: Metal Progress Databook, mid-June 1974, 151, 153, 157, 159, 161, 163

INDEX

481